C0-ATA-035

Fundamentals of Mycology

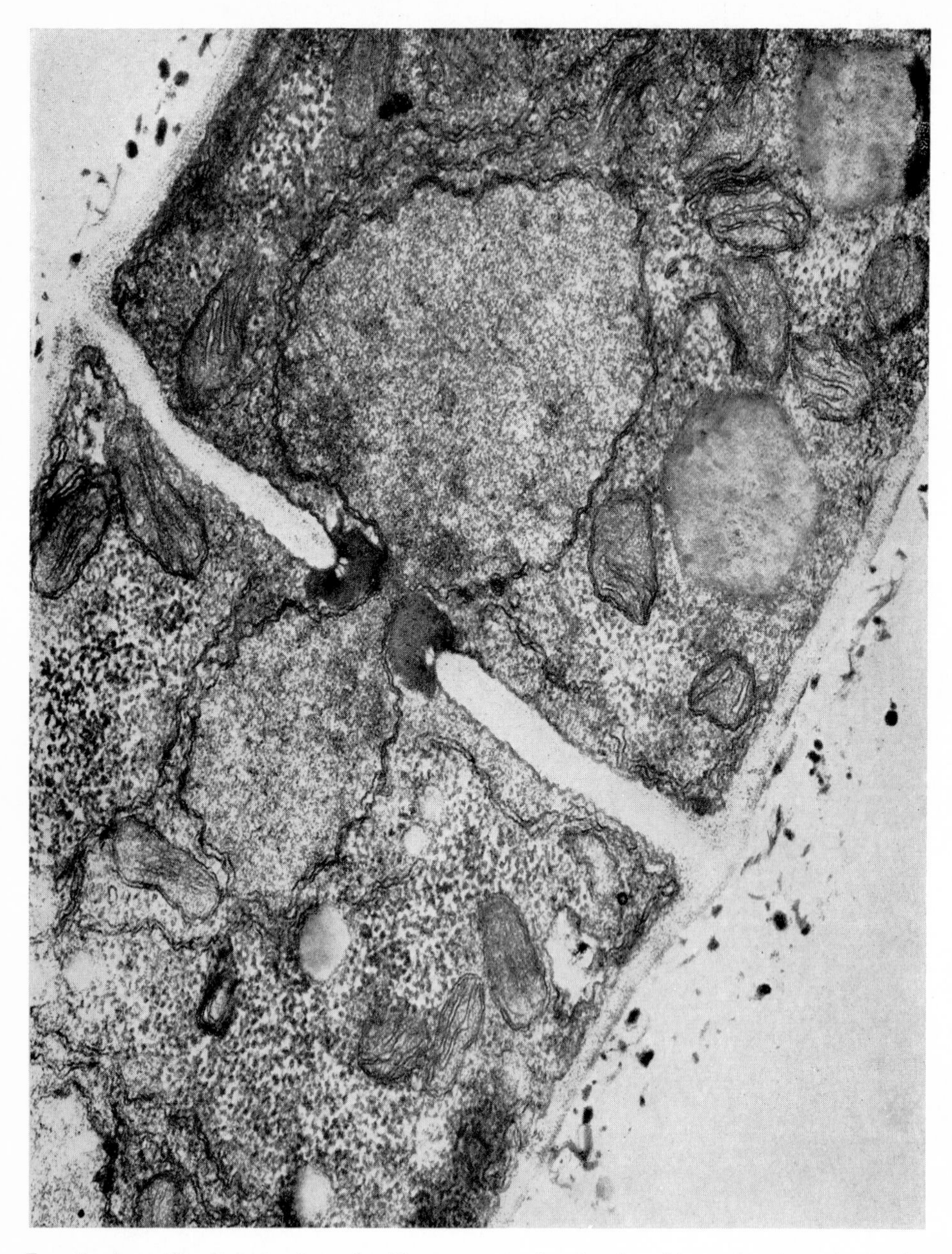

Frontispiece. The fine structure of a *Neurospora* hypha. An ultra-thin section through a hypha of *Neurospora crassa* grown in submerged culture. Note wall with adherent loose material on outside; simple pore in the septum lined with a ring of electron-dense material. In the cytoplasm, note, cytoplasmic membrane; regions of endoplasmic reticulum; detached, particulate ribosomes; mitochondria with long, well developed cristae; fat globules; an hexagonal crystalline inclusion. The nucleus is passing through the septal pore, note its double membrane and considerable constriction as it squeezes through. (× *c.* 42,000; original by M. Bourne)

Fundamentals of Mycology

J. H. Burnett M.A., D.Phil.

Regius Professor of Botany, University of Glasgow

Edward Arnold (Publishers) Ltd. London

First published 1968
Reprinted 1970

Books edition SBN: 7131 2203 x
Paper edition SBN: 7131 2221 8

Printed in Great Britain by
William Clowes and Sons, Limited, London and Beccles

Preface

The existence of living organisms constructed wholly of fine tubes is novelty enough and when coupled with the remarkable metabolic properties and other attributes which these organisms possess ensures that the fungi are of potential interest to a wide range of biologists and other scientists. Since that great milestone in the history of mycology, ANTON DE BARY's *Morphologie und Biologie der Pilze, Fletchen und Myxomyceten* of 1866, this potential interest has developed in innumerable ways. For example, in the last 20–30 years alone, aspects of mycology such as antibiotics, hallucinogenic drugs, *Neurospora* genetics, aerobiology, the rhizosphere and the biochemistry of yeast have emerged and developed so greatly that they have made major contributions to biology and its concepts.

Excellent books are available which deal with these and other specialist topics in great detail but, as far as I know, there is no modern text which attempts, in a limited compass, to enumerate the general features common to all (or most) fungi and to expound the broad trends in structure, function and behaviour which can be discerned in the group. This book attempts to do these things. I believe the attempt to be worth while for three reasons. Firstly, to provide the increasing numbers of scientists who are not primarily mycologists but who use fungi for experimentation or exploitation with a perspective of the fungi as a whole. Secondly, to draw the attention of professional mycologists to the many unsolved general problems in the hope of deflecting some from the attractive minutiae and specialist attractions of particular topics. It is as true of fungi as of physics that a science advances as rapidly as its data can be reduced to laws and systems of general validity! Finally, although a science may tend to fragment in the short term, I believe that this should be avoided in the long term and that periodical attempts should be made to construct a general picture. This book is such an attempt.

Because my aim has been to bring together all things fungal and in their proper proportions, I have omitted details of the extraordinary diversity of structure manifested by fungi. This is deliberate. I have tried to give instead some account of the principles of construction, for I hold that detailed knowledge is better culled by personal observation in field or laboratory than at second-hand. Such observations cannot fail to reveal areas of wonder and beauty which for long have been the delight of mycophiles and which for professional mycologists are an essential complement to the matter of this book.

12th December, 1967 J.H.B.

NOTE ON NOMENCLATURE

The progress of fungal taxonomy is such that the correct names for many common fungi are not yet satisfactorily settled. In this book, the names used are normally those employed by the authors of the works cited. As a result, a fungus is occasionally referred to by two different names but this can be checked by reference to the appendix on pages 494–503 and this convention enables readers to refer to the authors' work more readily.

Acknowledgements

I am grateful to Professors C. F. Robinow and N. T. Flentje and Dr. G. W. Gooday for unpublished information concerning nuclear division, hyphal interactions, and sexual hormones in *Mucor*, respectively.

I am equally grateful and fortunate to be able to reproduce unpublished photographs as plates supplied by:

Professor N. T. Flentje (Plate 4: 2 and 3), Dr. M. Girbardt (Plate 9: 1), Dr. P. S. Knox-Davies (Plate 5: 1–8), Professor G. Peyton and C. Bowen (Plate 2: 1), Professor C. F. Robinow (Plate 4: 1) and Professor M. Westergaard and D. von Wettstein (Plate 9: 2, 3).

Photographs of published plates were supplied and the editors of the journals concerned have granted permission to reproduce as follows:

Professor P. R. Day, Dr. M. Giesey and *American Journal of Botany* (Plate 3: 1, 2); Professor R. T. Moore and Dr. A. J. McAlear and *Nova Hedwigia* (Plate 2: 3); Professor K. Mühlethaler and Dr. H. Moor and *Journal of Cell Biology* (Plate 1: 2, 3; Plate 3: 3, 4); Professors R. D. Preston and J. M. Aronson and *Journal of Cell Biology* (Plate 2: 1); Professor C. F. Robinow, *Canadian Journal of Microbiology* and *Journal of Cell Biology* (Plate 5: 9–14; Plate 6: 1–9); the late Professor P. A. Roelofsen and Springer-Verlag (Plate 1: 1, 4); the executor of the late Professor J. R. Singleton and *American Journal of Botany* (Plates 7 and 8: 1–19).

Figures and tabular material have been reproduced with, or without modification through the courtesy of the publishers of the books, or editors of the journals listed below, to whom I express my gratitude:

Academic Press: *Chemical Activities of Fungi*: Foster, J. W. (1953): Figure 2.3c. Tables 7.3a; 11.1; 11.3; 11.4.

The Fungi: I and II, ed. Ainsworth, G. C. and Sussman, A. S. (1965, 1966): Figures 8.4; 11.4; 13.2; 18.4. Tables 8.2; 8.4; 8.12; 10.1.

International Review of Cytology: Figures 8.5; 8.15; 8.16.

Plant Pathology, ed. Horsfall, J. G. and Dimond, A. E. (1960): Figure 16.3.

Advances in Genetics: Table 17.7.

Biochemistry of Industrial Micro-organisms, ed. Rainbow, C. and Rose, A. H. (1963): Table 11.2.

American Association for the Advancement of Science: *Sex in Micro-organisms*: ed. Weinrich, D. H. *et al.* (1954): Figures 1.1; 15.2.

Edward Arnold: *An Introduction to the Biology of Micro-organisms*, Hawker, L. E. *et al.* (1960): Figures 10.1; 10.2; 10.4.

Blackwell Scientific Publications: *Fungal Genetics*, 2nd edn., Fincham, J. R. S. and Day, P. R. (1963), Figures 1.3; 1.9; 15.8.

N. Boubée et Cie., Paris: *Les Champignons de l'Europe*, Heim, R. (1957): Figures 5.11; 6.1; 18.2. Table 18.1.

Burgess Publishing Co.: *Illustrated Genera of Imperfect Fungi*: Barnett, H. L. (1960): Figure 5.4.

Butterworths: *The Fungus Spore*, ed. Madelin, M. F. (1966): Figure 5.2.

Cambridge University Press: *SEB Symposia* (1948): Figure 11.11.

Symbiotic Associations, ed. Nutman, P. S. and Mosse, B. (1963): Figure 12.13.

Faber & Faber: *The Friendly Fungi*, Duddington, C. L. (1957), Figure 4.14.

Gebrüder Borntraeger: *Anatomie der Asco- und Basidiomyceten*, Lohwag, H. (1941): Figures 5.15; 12.3.

Die Pflanzenfamilien, ed. Engler, A. and Prantl, K. (1935): Figure 5.12.

Leonard Hill: *The Biology of Mycorrhiza*. Harley, J. L. (1959): Table 8.5a, b.

The Microbiology of the Atmosphere, Gregory, P. H. (1961): Figures 6.10; 6.14. Table 6.2.

Holden-Day Inc.: *Methodology in Basic Genetics*, ed. Burdette, W. J. (1963): Figure 14.7.

Longmans, Green & Co.: *Researches on Fungi*, I–VI, Buller, A. H. R. (1909–1934): Figures 2.1b; 4.3; 4.9; 4.10; 5.9; 6.5e–r; 12.2.

McGraw-Hill & Co.: *Comparative Morphology of the Fungi*: Gäumann, E. A., trans. Dodge, C. W. (1928): Figures 4.16; 5.8; 5.14; 6.4; 18.3.

The Lower Fungi, Fitzpatrick, H. M. (1930): Figure 4.12.

The Biosynthesis of Natural Products, Bu'lock, J. D. (1965): Figure 11.10.

Masson et Cie: *Précis de Mycologie*, Langeron, M. and Vanbreuseghem, R. (1952): Figure 5.3.

John Murray: *Ecology of Soil-borne Plant Pathogens*, ed. Baker, K. F. and Snyder, W. C. (1965): Table 12.2.

Oliver & Boyd Ltd.: *The Absorption of Solutes by Plant Cells*, Jennings, D. H. (1963): Figures 8.7; 8.10. Table 8.6.

Oxford University Press: *Comparative Morphology and Biology of the Fungi Mycetozoa and Bacteria*, de Bary, A. (1887): Figures 2.3e, 2; 4.13b.

Spore Discharge in Land Plants, Ingold, C. T. (1939): Figure 6.9.

Dispersal in Fungi, Ingold, C. T. (1953): Figures 6.6; 6.7.

Spore Liberation, Ingold, C. T. (1965): Figures 6.5a–d; 6.8.

Nucleo-cytoplasmic relations in Micro-organisms. Ephrussi, B. (1953): Figure 17.2.

A Monograph of Cantharelloid Fungi, Corner, E. J. H. (1967): Figure 5.13b.

Pergamon Press: *Advances in Enzymic Hydrolysis of Cellulose and Related Materials*, ed. Reese, G. T. (1963): Figures 12.7; 12.8.

Ronald Press: *Genetics of Sexuality in Higher Fungi*, Raper, J. R. (1966): Figures 15.11; 15.12.

Plant Pathology, Stakman, E. L. and Harrow, J. G. (1957): Figures 17.5; 17.6. Tables 17.5; 17.6.
Royal Society of London, *Proceedings*: Figure 14.6. Tables 2.3; 14.1; 14.3. *Transactions*: Tables 7.2; 7.3b; 11.5; 14.1.
Royal Society of Edinburgh, *Proceedings*: Table 17.8.
Royal Society of Victoria, *The Evolution of Living Organisms* (1959): Table 15.1.
Scottish Plant Breeding Station, Annual Report: Table 17.9.
Springer-Verlag: *Incompatibility in Fungi*, ed. Esser, K. and Raper, J. R. (1965): Figures 16.2; 17.1; 17.3; 17.4. Tables 14.2; 15.3; 16.3.
J. Wiley & Sons Ltd.: *Introductory Mycology*, Alexopoulos, C. J. (1962): Figures 1.2; 1.4; 1.5; 1.7; 1.10; 1.11; 15.9.

American Journal of Botany: Figures 2.2; 3.5; 4.1; 4.13e; 15.7. Tables 2.1; 3.3; 3.4; 4.4; 5.2; 9.7; 13.2; 15.2.
Annals of Botany: Figures 4.15; 5.19; 5.22; 6.3; 13.1; 13.3. Tables 4.1; 4.5; 4.6; 4.7; 5.3; 5.4; 6.1; 6.3; 8.3; 9.8.
Archiv für Mikrobiologie: Figures 9.3; 15.5. Tables 6.5; 9.4.
Bacteriological Reviews: Figure 3.6.
Beihefte zu den zeitschriften des Schweizerischen Forstvereins: Figure 3.8.
Bericht der Deutschen botanischen Gesellschaft: Figure 3.2.
Biochimica et Biophysica Acta: Figures 3.9; 3.11; 8.14. Tables 8.8; 8.9.
Botanical Gazette: Figures 4.13c; 9.2; 15.3. Table 16.2.
Botanische Abhandlungen (Goebel): Figure 15.4a.
Canadian Journal of Botany: Figures 6.2a–f; 6.16. Table 17.4.
Chromosoma: Figures 13.8; 15.13.
Cold Spring Harbor Symposia (1947): Figure 16.1.
Compte rendu hebdomadaire des séances de l'Académie des Sciences, Paris: Figures 4.5; 4.6; 14.2.
Compte rendu des travaux du Laboratoire de Carlsberg: Figure 13.7.
El Aliso: Figures 2.1c; 5.1; 18.1.
Evolution: Table 17.13.
Experimental Cell Research: Figures 3.3; 11.6. Table 11.7.
Flora: Figure 2.3a.
Genetics: Figure 11.5.
Heredity: Table 17.2.
Jahrbüche für wissenschaftliche Botanik: Figure 3.1.
Journal of Agricultural Research: Table 17.14.
Journal of Bacteriology: Figures 8.1; 8.11; 8.12; 8.13. Tables 6.4; 8.10; 8.11.
Journal of Cellular and Comparative Biology: Figures 3.8; 8.3; 12.4. Table 12.1.
Journal of Experimental Biology: Figure 12.5.
Journal of Experimental Botany: Figure 2.4a, b. Table 8.14.
Journal of General Microbiology: Figures 2.4c, d; 14.5. Tables 2.2; 2.3; 4.2.
Journal of General Physiology: Table 6.4.
Journal of General and Applied Microbiology: Figures 4.11. Table 6.5.

Journal of Heredity: Figure 17.9.
Journal of the Linnean Society: Figure 3.7.
Lloyddia: Figure 15.1.
Mycologia: Figures 2.3b; 3.3; 5.5; 5.16; 5.20; 6.15; 14.4; 17.7; 17.8. Tables 4.3; 8.1; 17.11; 17.12.
Nature: Figures 2.1a; 4.4; 14.3. Table 9.2.
New Phytologist: Figures 3.4; 8.6; 8.8; 9.4; 12.9; 15.14. Tables 8.7; 9.1; 9.6; 12.3; 12.4; 12.5; 12.6; 17.10.
Physiologia Plantarum: Figure 8.9. Table 9.5.
Phytomorphology: Figure 5.13a, c.
Phytopathology: Figure 14.1.
Plant Physiology: Figure 12.1. Tables 5.6; 6.4; 7.1.
Planta: Figures 2.3a, 1; 4.7; 13.6; 15.6.
Quarterly Review of Biology: Figure 3.8.
Revue générale de botanique: Figure 2.3d; 15.4b, c.
Transactions of the British Mycological Society: Figures 4.13a; 5.7; 5.10; 5.17; 5.18; 6.2g–i; 6.11; 6.12; 6.13; 12.6; 12.10; 12.11; 12.12. Table 5.1.
University of California Publications in Botany: Figure 4.13d.
Wentia: Figure 5.21. Tables 2.3; 5.5.
Zeitschrift für Botanik: Figure 9.5. Table 9.10.
Zeitschrift für Vererbungslehre (formerly *Z. für induktive Abstammungs und Vererbungslehre*): Table 17.1.

Finally, I acknowledge with gratitude the assistance of my wife who drew or re-copied all the illustrations. I thank also Mrs. F. Hunston and Miss Janis Hall for their cheerful and accurate typing and frequent re-typing of much of the manuscript.

Contents

SECTION II. Function

SECTION III. Recombination

SECTION I Structure and Growth

Chapter 1
An Introduction to the Fungi

THE GENERAL FEATURES OF FUNGI

It is not easy to define the limits of the fungi and, indeed, not all mycologists would accept the same limits. However, all fungi are heterotrophic organisms and the vast majority are constructed of more or less microscopic, cylindrical filaments, or hyphae, with well defined cell walls. Heterotrophy and hence a saprophytic or parasitic mode of life sets the fungi clearly apart from the autotrophic green plants. Likewise, stress upon the hyphal form of construction separates the fungi from most bacteria, actinomycetes and protozoa. A number of organisms, collectively termed 'slime moulds', characterized by the possession of a multinucleate, motile protoplasmic phase or plasmodium, are included by some authors (e.g. ALEXOPOULOS, 1962) within, or are thought to be related in part to, the fungi. In this book, however, forms with a free-living plasmodial or pseudo-plasmodial stage will be excluded, although more from the point of view of convenience than from taxonomic conviction!

Hyphae, characteristically, grow at their tips and can usually produce lateral branches in their older parts. In many fungi, the tips of the hyphae may fuse so that ultimately the whole mass of hyphae, or mycelium, forms a three-dimensional network which ramifies through the substrate. The mycelium may remain microscopic or develop easily visible organized structures such as strand- or cord-like rhizomorphs, compact resting bodies (sclerotia) or, eventually, the more familiar spore-bearing structures commonly called mushrooms, toadstools, brackets, puff-balls, stink-horns, truffles and so on. All these structures, like the mycelium itself, may be evanescent or perennial and the hyphae of which they are composed may show greater or lesser degrees of structural specialization. All growing hyphae, at least, are filled with cytoplasm, frequently vacuolated and either aseptate and hence forming a single multinucleate cell (coenocyte), or divided into segments by transverse walls (septa) which may, however, be perforated. The nuclei have well defined nuclear membranes and this feature reinforces the differences between fungi and bacteria, or actinomycetes. Despite their apparent similarity to the nuclei of higher organisms, somatic divisions of fungal nuclei frequently differ in a number of respects from mitosis, as usually described, and meiosis too may show some unusual features. For this reason the terms somatic, and meiotic, divisions will be used in this book. The dimensions of the hyphae, the thickness and composition of their walls, the absence or presence and degree of development of their septa, the extent of their branching and anastomoses, and their rate of growth are all characters which can differ in different fungi and vary between different regions of the same mycelium.

Reproduction

Almost without exception fungi reproduce by spores in appropriate conditions. The modes of origin, shapes, sizes, modes of liberation and kinds of spores are legion; their common characteristic is that they are all produced asexually. Three main types of spore occur:

Conidia are the characteristic asexual reproductive units of the fungi. They are usually produced terminally or laterally on hyphae or special structures, conidiophores, and are borne externally as single, separable cells.

Sporangiospores are produced within a specialized structure, the sporangium, often derived from a sporangiophore, and may be motile (zoospores) or non-motile. In some cases, meiosis occurs in a sporangium prior to spore formation; such spores may be termed meiospores, in contrast to spores produced solely by somatic divisions which are termed mitospores.

Chlamydospores are formed by the rounding off of the contents of one or more hyphal segments. They are thick-walled, are not easily detached from the mycelium and may remain viable yet dormant for considerable periods, up to several years.

The kind of spore produced differs according to the location on the mycelium, the environmental conditions and the stage in the life cycle. Several fungi are capable of producing more than one kind of spore, a feature which confused the early mycologists and which is still capable of causing confusion today.

Sexual reproduction, i.e. nuclear fusion and meiosis, usually in that sequence, is widespread in fungi. Fertilization can be brought about in a variety of ways. In aquatic forms, the fusion of free-living gametes or of a free-living gamete with a gametangium is common. In other fungi there may be fusion of gametangia, which either differ clearly in form and function (as oogonium and antheridium), or in size only, or not at all. In the vast majority of fungi, however, fusion of undifferentiated hyphae enables potentially conjugant nuclei to migrate and penetrate the hyphae, frequently reciprocally, until fusion ultimately occurs in a cell of a specialized and often pre-formed structure, e.g. a protoperithecium. Meiosis is, thereafter, confined to a specialized cell type and is frequently followed by the production of spores from these structures, or their equivalents in parthenogenetic forms. Two such cell types, of great importance both for classification and biology of the fungi, are the ascus and the basidium. Both are normally terminal, or sub-terminal, somewhat ovoid cells. The four nuclei derived by meiosis in an ascus usually undergo a further somatic division and the eight products become incorporated in endogenously formed ascospores which are subsequently discharged or released. A similar sequence of fusion followed by meiosis occurs in a basidium but here the basidiospores generally arise immediately after meiosis at the tips of four tapering processes, the sterigmata, which are produced at the free, rounded end of the cell. Basidiospores are said to be discharged violently and inspection usually confirms this claim.

Life cycles

The occurrence, sequence of processes and types of products involved in asexual and sexual reproduction as well as the occurrence of hyphal fusions have profound effects upon the life cycles and hence the whole biology of the fungi. Five basic life cycles, reduced from the seven described by RAPER (1954), can be recognized (Fig. 1.1). They are:

1. *Asexual*, in which sexual reproduction is apparently lacking entirely. Since this definition is based upon the absence of a phase, it is somewhat artificial but it has great convenience and accurately describes the situation in many fungi, viz, Fungi Imperfecti.
2. *Haploid*, in which meiosis immediately follows nuclear fusion and the meiotic products are then dispersed. The diploid phase is, therefore, of minimal duration. This cycle is shown by many Phycomycetes and some Ascomycetes.
3. *Haploid-dikaryotic*, similar to the preceding save that paired, potentially conjugant kinds of nuclei persist in close physical association in the same hyphal segment (hence dikaryon) and divide synchronously, for a greater or lesser period. At one extreme the association may be for a few cell generations only, e.g. in many Ascomycetes, binucleate ascogenous hyphae are developed just prior to ascus development, and such a dikaryon cannot apparently exist independently of the haploid phase (RAPER's 'Haploid cycle with restricted dicaryon'). At the other extreme is the condition where the meiospores fuse to reform a dikaryon so that the fungus is dikaryotic throughout its life cycle, save for the moment of fertilization and during the subsequent meiosis. This can occur in yeasts (Saccharomycetales) but is more frequent in smut fungi (Ustilaginales). An intermediate and highly characteristic type of life cycle is shown by the

majority of Basidiomycetes. Here the mycelium derived from germination of a meiospore may persist in the haploid condition, as a monokaryon, but once a dikaryon is formed, through hyphal fusion for example, it shows potentially unrestricted and independent growth so that it may well comprise the most long-lived phase of the life cycle (RAPER's 'Haploid-dicaryotic cycle'). For example, the dikaryotic phase of various fairy-ring fungi has probably persisted for several centuries (RAMSBOTTOM, 1953).

4. *Haploid-diploid,* in which these phases alternate regularly, an unusual condition in fungi restricted almost entirely to a few species of aquatic Oomycetes.
5. *Diploid,* in which the haploid phase is restricted to the gametes or gametangial phase. It seems likely that the majority of Oomycetes may well conform to this pattern (SANSOME, 1963).

CLASSIFICATION

The development of the mycelium, the kinds of spores and sporangia, the nature of the life cycle and the presence, or absence, of sexual reproduction are features employed in defining the major groups of the fungi. The classification of fungi is not in a settled state and, at present, it is probably best to adopt some particular system and use it for operational purposes until there is greater stability. Suitable systems are described in a number of general texts and reference works (BESSEY, 1950; AINSWORTH, 1961; ALEXOPOULOS, 1962; GÄUMANN, 1964). It will be convenient at this stage to give a simple outline classification on which to base the accounts of life cycles which conclude this chapter. A more detailed classification, of which this is an abbreviated form, is given in the Appendix where all the fungi referred to in this book are assigned to their appropriate places.

The outline scheme is set out below and excludes organisms with a free-living plasmodium or plasmodium-like structure in their life cycle:

FUNGI

Phycomycetes

Mycelium absent, or rudimentary, or, when present, usually aseptate, at least when young. Sporangiospores characteristically produced, more rarely conidia. Sexual reproduction by fusion of gametes and/or gametangia.

(a) Zoospores produced.

(i) Zoospores with 1 anterior flagellum:
HYPHOCHYTRIDIOMYCETES—Water moulds.

(ii) Zoospores with 1 posterior flagellum:
CHYTRIDIOMYCETES—Water moulds.

(iii) Zoospores biflagellate $\pm$ equal in length:
OOMYCETES—Water moulds, blights, downy mildews.

(iv) Zoospores biflagellate unequal in length:
PLASMODIOPHOROMYCETES—Club-root organisms.

(b) Only non-motile spores produced:
ZYGOMYCETES—Pin moulds, bread moulds.

Ascomycetes

Mycelium usually septate, frequently producing conidia, rarely ± unicellular and reproducing by budding. Ascospores characteristically produced within asci. Sexual fusion by gametangia or through somatic hyphae.

(a) Asci born singly or, at least, not in special structures (ascocarps).
HEMIASCOMYCETES—Yeasts, leaf curls.
(b) Asci developed from ascogenous hyphae in ascocarps.
(i) Ascocarp lacking a pore (ostiole)—a cleistothecium:
PLECTOMYCETES—Black moulds, blue moulds, powdery mildews.
(ii) Ascocarp characteristically pyriform or flask-shaped—a perithecium:
PYRENOMYCETES—'Flask' fungi.
(iii) Ascocarp characteristically cup- or saucer-shaped—an apothecium:
DISCOMYCETES—Cup fungi, morels, truffles.

Basidiomycetes

Mycelium usually septate, simple or complex, with or without clamp connexions, basidiospores characteristically delimited outside the basidium. Sexual reproduction generally by somatic hyphal fusion.

(a) Basidia septate or deeply divided, often elongated; basidiospores sometimes capable of germinating by repetition.
HETEROBASIDIOMYCETES—Smuts, rusts, jelly fungi.
(b) Basidia simple, unicellular, usually ± clavate or urniform; basidiospores on germination usually give a mycelium directly.
HOMOBASIDIOMYCETES.
(i) Hymenium present, exposed before spores mature:
HYMENOMYCETES—Toadstools, brackets, crusts.
(ii) Hymenium present or absent, enclosed until spores released from basidia, or permanently so:
GASTEROMYCETES—Puff-balls, earth stars, stink-horns, birds'-nest fungi.

Fungi Imperfecti

Mycelium usually septate bearing conidia on isolated conidiophores or aggregated to form an acervulus or in a flask-shaped pycnidium. Sexual reproduction totally absent.

LIFE CYCLES OF SELECTED FUNGI

The individual life cycles now to be described are selected both because they are characteristic and because they are of fungi which will be used to illustrate particular matters later in this book.

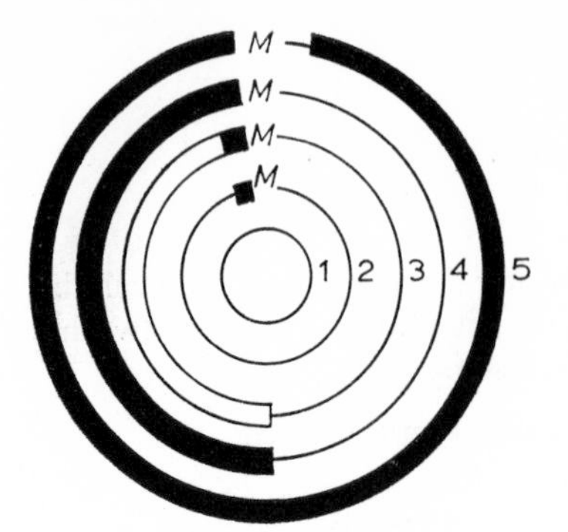

Fig. 1.1. A diagram to illustrate the five basic life cycles in fungi. Each circle represents a life cycle and should be followed clockwise; *M* represents meiosis, ——— a haploid phase, ═══ a dikaryotic phase and ▬▬▬ a diploid phase. The life cycles shown are, 1, asexual; 2, haploid; 3, haploid-dikaryotic; 4, haploid-diploid; 5, diploid. (Based on Raper, 1954)

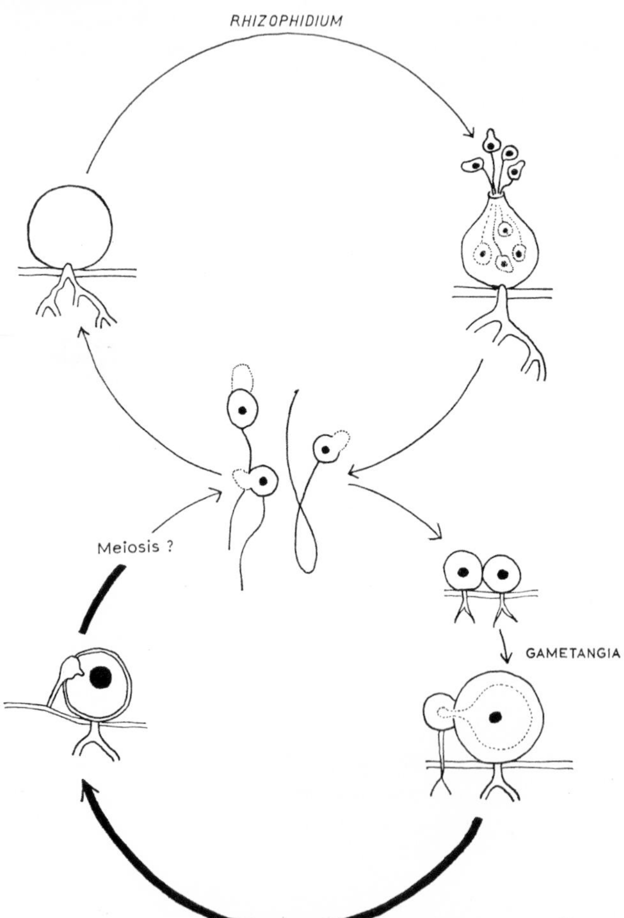

Fig. 1.2. Life cycle of *Rhizophydium* (Phycomycetes–Chytridiomycetes). This microscopic fungus parasitizes algae and possesses a haploid life cycle. The thallus only shows typical hyphal growth in its rhizoids. It is wholly converted into either a sporangium or gametangium. Dispersal is by zoospores each with a posterior flagellum. (Based on Alexopoulos, 1962)

The wealth of diversity described in so many texts, or better, observed by even the most casual scrutiny of living fungi, is deliberately avoided here where the emphasis is on common features and general principles.

In the twelve examples given, a common lay-out is employed. The life cycle should be read in a clock-wise direction and the main circles describe the principal order of development. Variants are shown by smaller circles or arcs within the main circles. A common feature throughout is the use of a single thin line to indicate a haploid phase, a double thin line to indicate a dikaryophase and a thick single line to indicate a diploid phase.

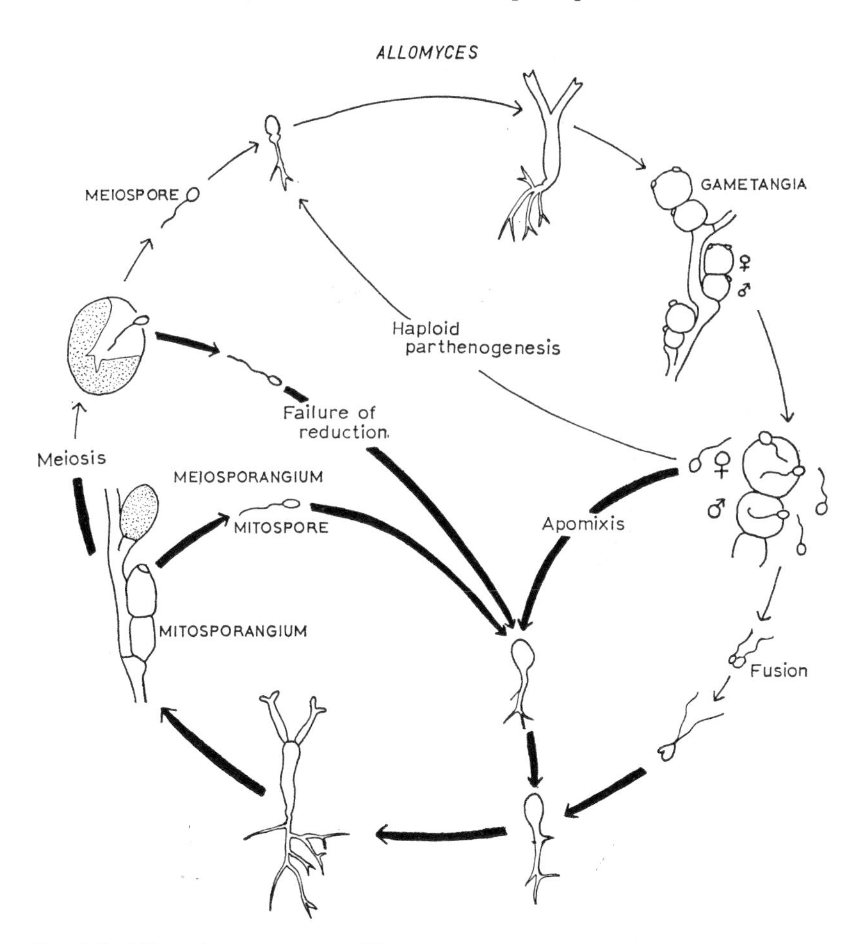

Fig. 1.3. Life cycle of *Allomyces* (Phycomycetes–Chytridiomycetes). A water mould, this fungus can just be detected by the unaided eye. It has a haploid–diploid life cycle but the normal alternation of nuclear phases may be interrupted as a result of haploid parthenogenesis, apomixis or the failure of meiosis in the zygote. Here typical hyphal growth occurs and gametes and zoospores are posteriorly flagellate. (Based on Fincham and Day, 1962)

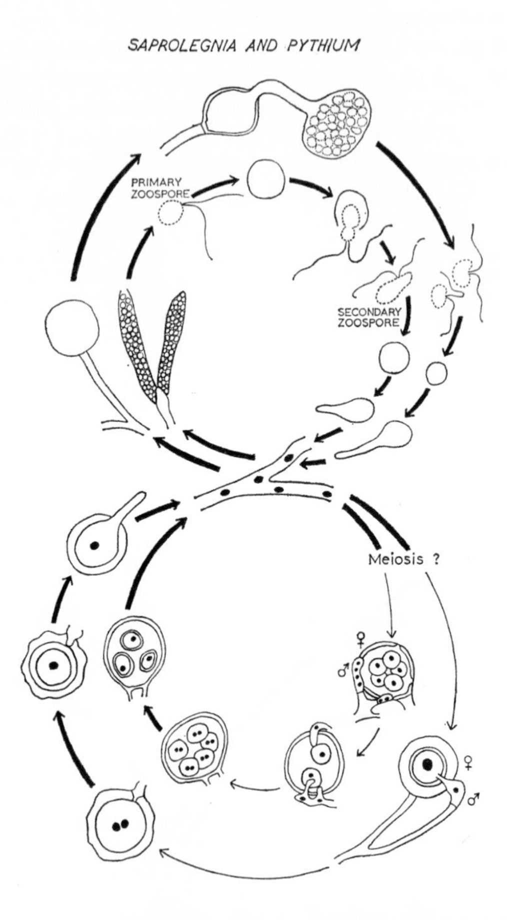

Fig. 1.4. Life cycles of *Saprolegnia* (inner circles) and *Pythium* (outer circles) (Phycomycetes-Oomycetes). The former is a common water mould known to many as a cause of fish disease, the latter includes the common 'damping-off' fungi of seedlings. In fact, many saprophytic species of both genera may be isolated from water or soil. In this interpretation the life cycle is shown as diploid and it is assumed that meiosis occurs just prior to gametangium formation. (Meiosis may occur at the germination of the zygote, when the life cycle would then be haploid.) In these fungi the hyphal habit is well established although they are coenocytes. Specialized, frequently terminal, sporangia give rise to biflagellate zoospores which may encyst and reappear with lateral flagella (diplanetism) as in *Saprolegnia*. In relatives of *Pythium* the whole sporangium may become detached and function as a single unit of dispersal and propagation. Well differentiated gametangia are characteristic of these fungi and their allies. The zygotes are often difficult to germinate. (Based on Alexopoulos, 1962)

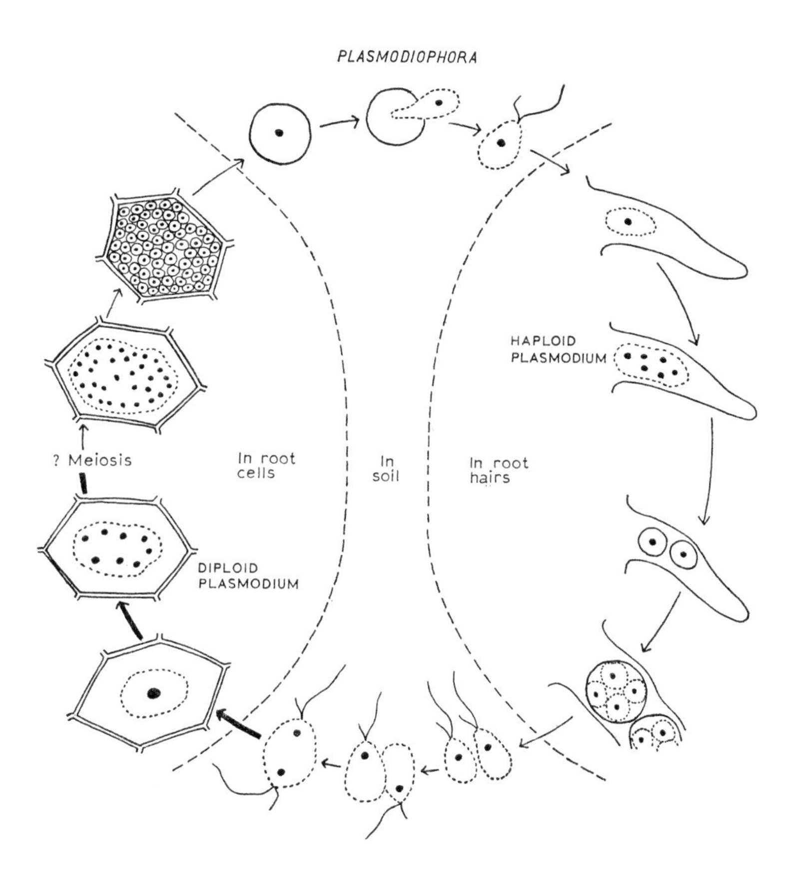

Fig. 1.5. Life cycle of *Plasmodiophora* (Phycomycetes-Plasmodiophoromycetes). The causative organism of 'club-root' disease of crucifers. The life cycle of the fungus is not yet completely understood; however, it is probably haploid. A feature of especial interest is that it exists only as a parasitic, multinucleate plasmodium—lacking anything approaching a mycelial organization. Its units of dispersal are unequally biflagellated zoospores. (Based on Alexopoulos, 1962)

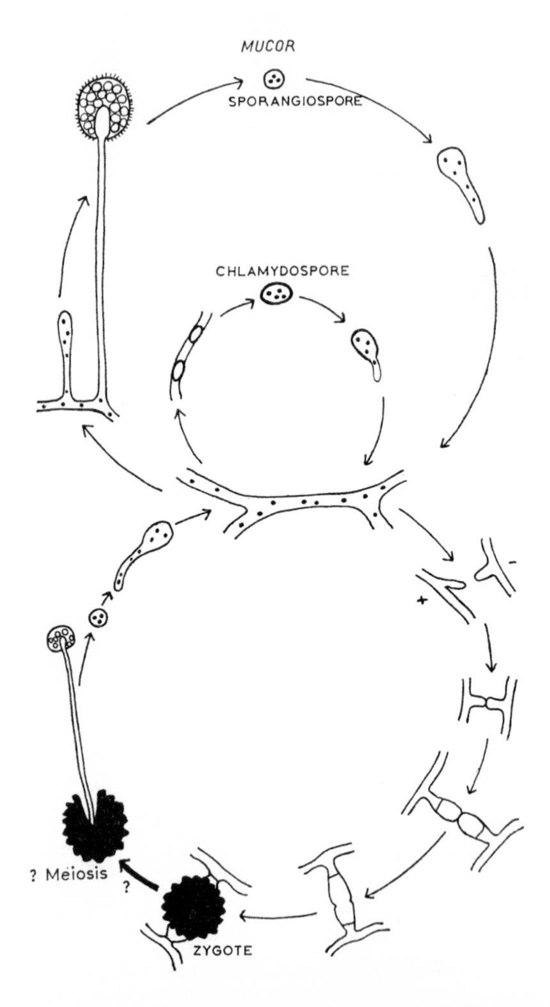

Fig. 1.6. Life cycle of *Mucor* (Phycomycetes-Zygomycetes). A representative of the 'pin moulds', common soil saprophytes but often noticed more readily on damp organic matter such as bread, jam, leather, etc. The life cycle is probably haploid. The hyphal habit is well developed but spores are non-motile, they may be sticky or dry and be dispersed by insects or wind. Gametangial differentiation is minimal and only rarely do the gametangia differ in size so that compatible strains are designated + and −. The zygotes are difficult to germinate and unequivocal evidence that meiosis occurs within them is lacking.

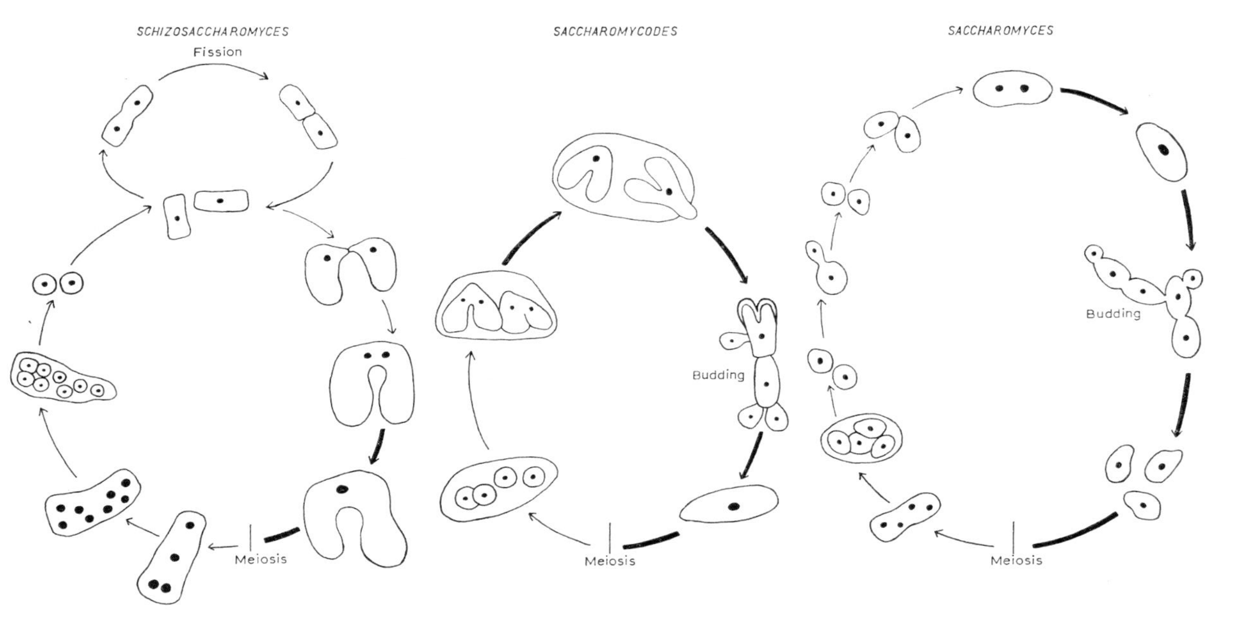

Fig. 1.7. Life cycle of selected yeasts, *Schizosaccharomyces*, *Saccharomycodes* and *Saccharomyces* (Ascomycetes-Hemiascomycetes). These fungi all possess the characteristic ascus which sets the Ascomycetes apart from other fungi. *Schizosaccharomyces* grows by fission, its life cycle is haploid. *Saccharomycodes* and *Saccharomyces* grow by budding and their life cycles are diploid and haploid-diploid respectively, although the life cycle of the latter is not necessarily regular. Morphological sexual differentiation is virtually absent and compatible types are again designated by symbols only, e.g. $+/-$, A/a or a/α. (Based on Alexopoulos, 1962)

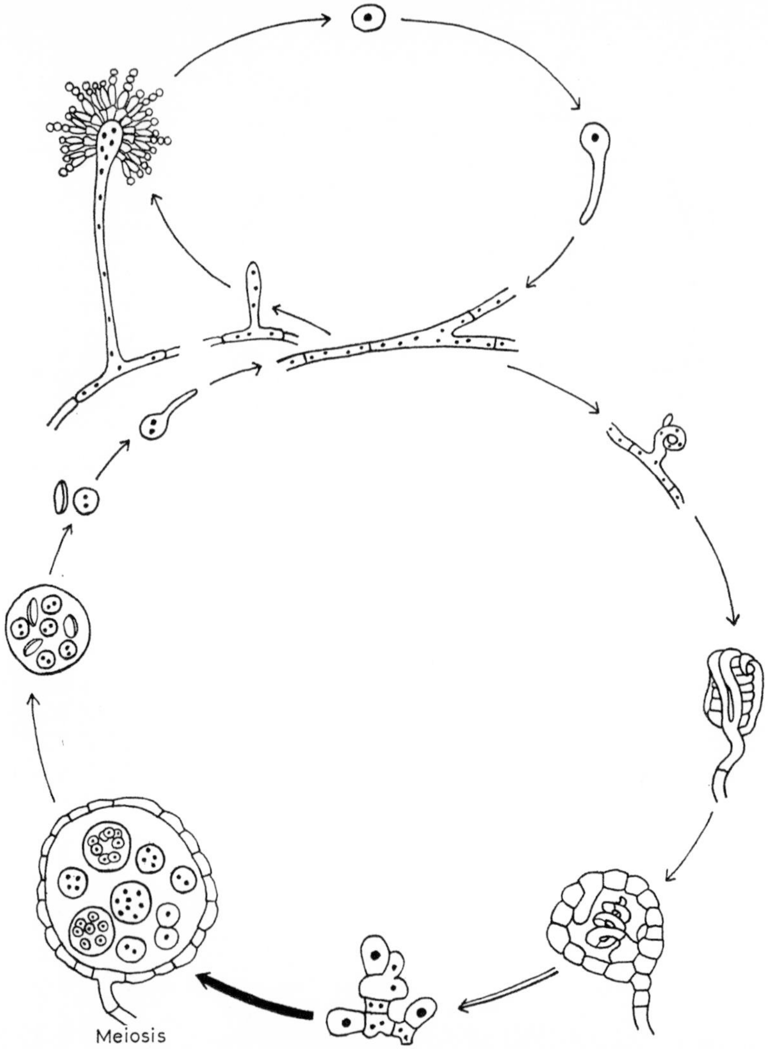

Meiosis

Fig. 1.8. Life cycle of *Aspergillus* (Ascomycetes-Plectomycetes). This genus includes several common green and black moulds. The life cycle shown is haploid-dikaryotic but some species, e.g. *A. niger*, entirely lack a sexual phase (the bottom circle) and possess, therefore, an asexual life cycle. These fungi are septate with simple pores in their cross-walls. Asexual reproduction is by the production of enormous numbers of conidia which are responsible for the widespread dispersal of these moulds. Sexual differentiation is minimal and the twisted whorl of branches merely serves to locate the position of the future cleistothecium in which the asci develop. Ascospores are liberated by the irregular rupture of both the asci and the cleistothecium. The gametes are represented only by nuclei which may come together by nuclear migration through the hyphae. Nuclei which will eventually fuse divide conjugately in the binucleate ascogenous hyphae, so initiating a brief dikaryophase. (Based on Fincham and Day, 1962)

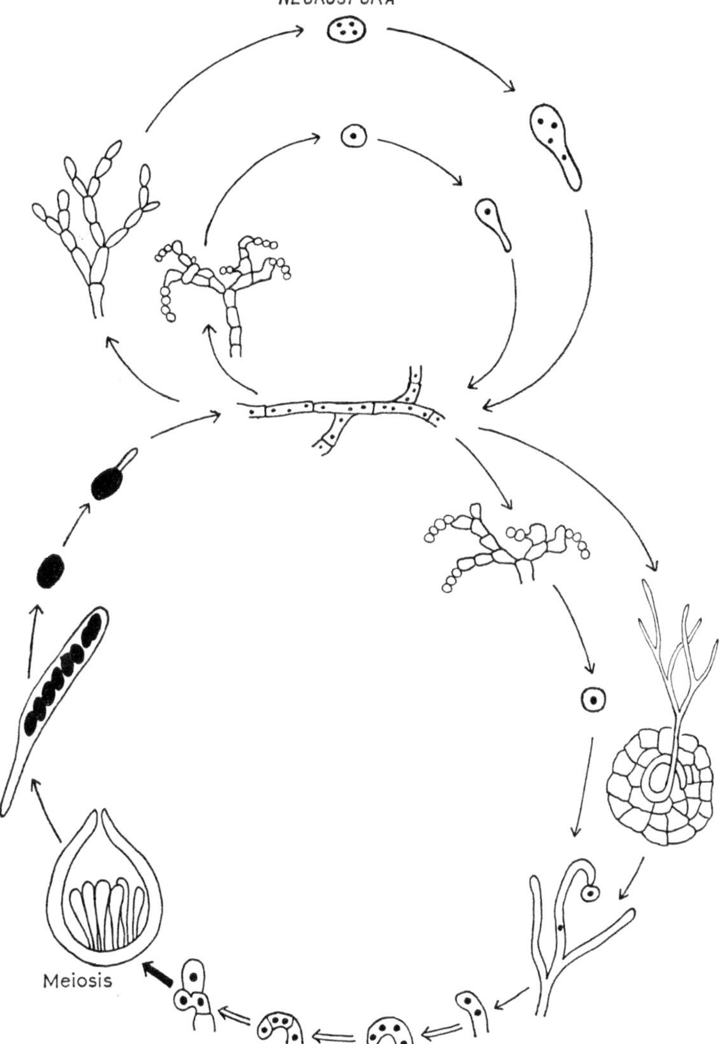

Fig. 1.9. Life cycle of *Neurospora* (Ascomycetes-Pyrenomycetes). The red bread mould now well known because of the numerous genetic studies made upon it; it used to be a common 'mould-weed' of bakeries. The life cycle is haploid-dikaryotic, the dikaryophase usually being of short duration in the ascogenous hyphae. In this genus conidia of two kinds, micro- and macroconidia, occur. Sexual reproduction can be initiated by bringing nuclei together through hyphal fusion, or the fusion of conidia of either type with the delicate hyphae of the trichogynes produced by protoperithecia. These latter structures determine the sites of perithecium formation if a successful fusion occurs. Colonies from single conidia are generally self-incompatible and two morphologically indistinguishable compatible strains are necessary for successful sexual reproduction. Dikaryotic ascogenous hyphae arise within the young perithecium and give rise to asci. The ascospores are discharged violently from the ascus through an ostiole in the 'neck' of the perithecium which is usually phototropically sensitive. Ascospores will not normally germinate without a high temperature, or specific chemical, shock treatment to break their dormancy. Discomycetes are similar in their life cycle but here the perithecium is replaced by an apothecium. (Based on Fincham and Day, 1962)

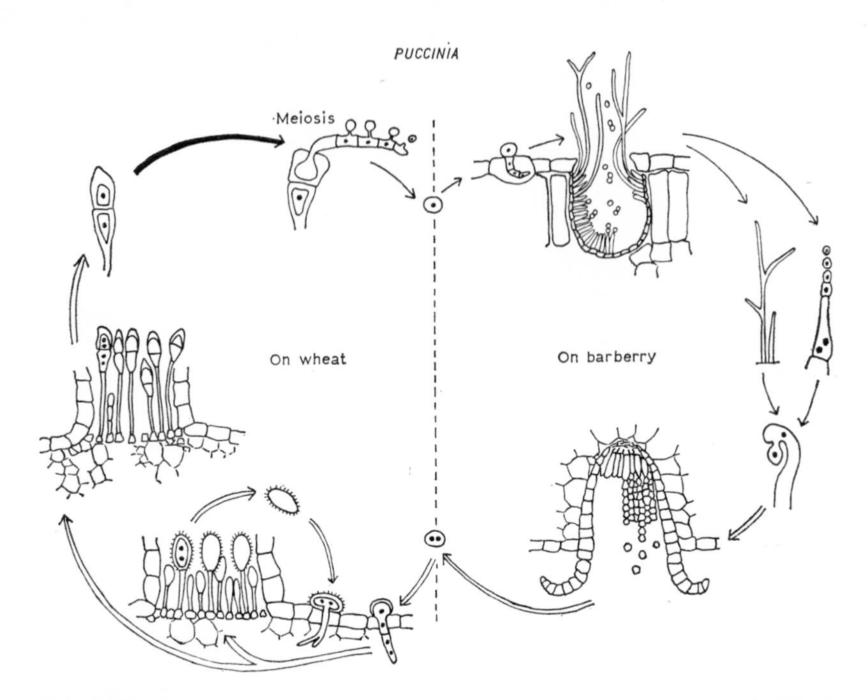

Fig. 1.10. Life cycle of *Puccinia* (Basidiomycetes-Heterobasidiomycetes). The life cycle shown is that of *P. graminis*, an important obligate pathogen of wheat and other cereals causing 'rust' disease. The life cycle is haploid-dikaryotic, the dikaryophase now being considerably prolonged; it is also complicated through involving two hosts and a variety of different spore types. The dikaryotic teleutospores are liberated from lesions on the leaves of wheat in the late summer and undergo nuclear fusion and meiosis. A basidium (the characteristic of Basidiomycetes) is produced and the basidiospores discharged violently. They infect the leaves of barberry and give rise to pycnidia on the upper surface and protoaecidia on the lower surface. In the following spring, the pycnidia develop masses of pycnidiospores and a drop of nectar-like liquid is exuded; slender, 'flexuous' hyphae grow out through this drop. The basidiospores are of two compatible types and if pycnidiospores of one mating type come into contact with flexuous hyphae of the other mating type they fuse. This apposition may come about through the spread and mixing of adjacent nectar droplets on the barberry leaves or through the passive transfer of pycnidiospores on the bodies of insects visiting the nectar drops. The nuclei, after the cytoplasmic fusion, migrate to the proto-aecidia and there initiate the dikaryophase. Each aecidium now develops and the aecidiospores, which infect wheat leaves when liberated, are binucleate possessing one nucleus of each mating type. Infection of wheat leaves is followed by the development of uredosori containing binucleate uredospores which serve to disperse the pathogen and infect other wheat plants in the early summer. Later they are replaced by teleutosori in which teleutospores are formed and the complex cycle recommences. In other species and genera the life cycle, although still essentially haploid-dikaryotic, may become simplified through the elimination of a second host and some of the spore types. (Based on Alexopoulos, 1962)

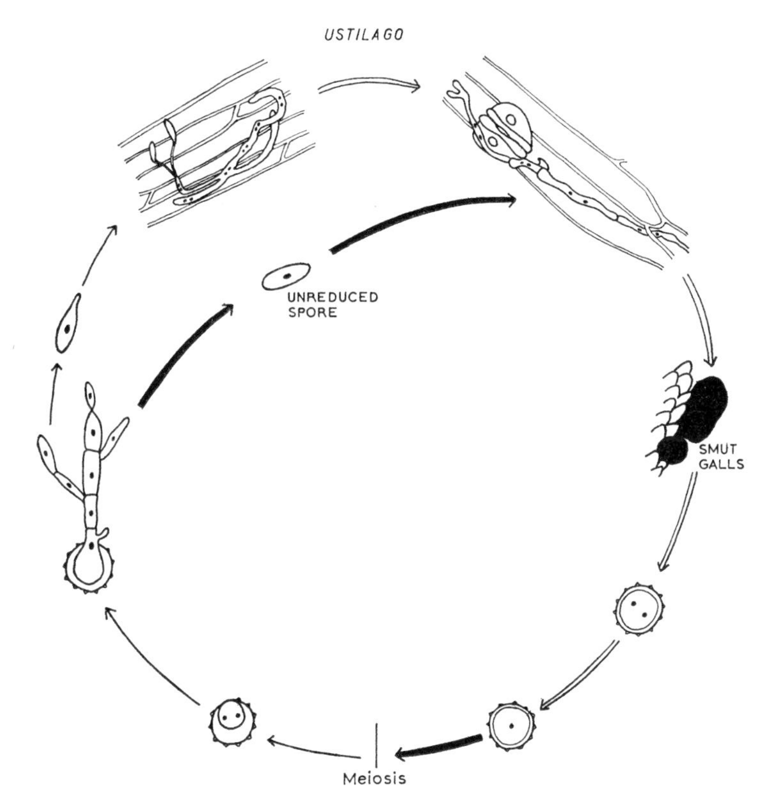

Fig. 1.11. The life cycle of *Ustilago* (Basidiomycetes-Heterobasidiomycetes). The important cause of 'smut' diseases of cereals in which galls develop in and around the fruits and ovaries of the infected plant. The life cycle is essentially haploid-dikaryotic but the production of occasional unreduced diploid spores provides a variant. Such spores are capable of re-infecting cereals without further fusion, normally a prerequisite of infection, and give rise to what are termed 'solopathogenic lines'. Characteristically the basidiospores proliferate in a manner comparable to the multiplication of yeasts so that in normal circumstances haploid spores of different compatibilities are dispersed intermixed. *Ustilago maydis* forms characteristic black, charcoal-like galls in place of the fruits on the cob of maize. (Based on Alexopoulos, 1962)

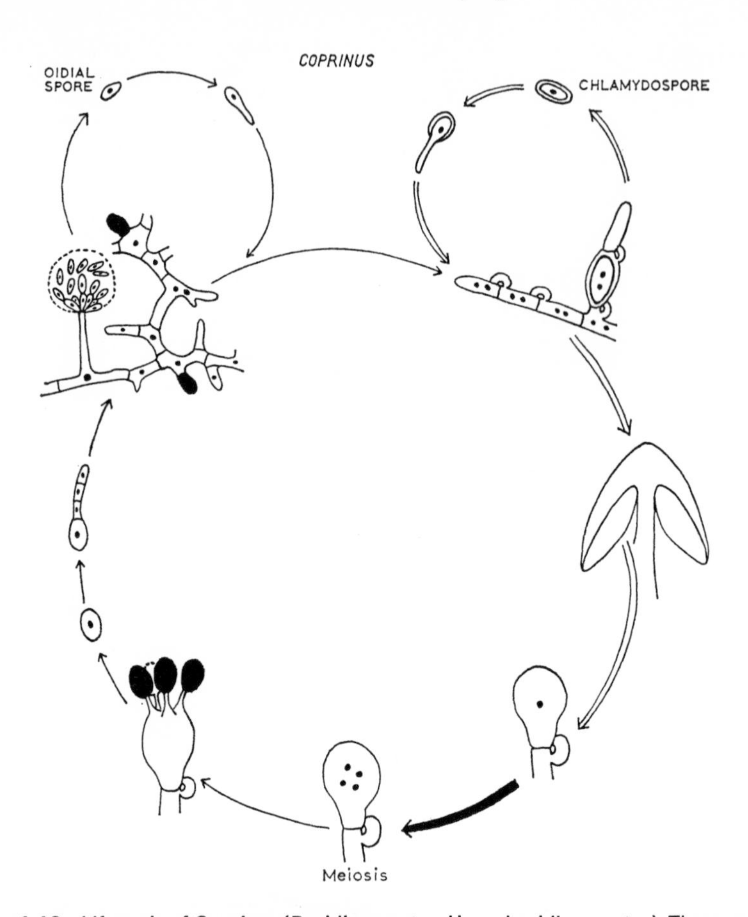

Fig. 1.12. Life cycle of *Coprinus* (Basidiomycetes-Homobasidiomycetes). The common ink-cap toadstool and characteristic, in its life cycle, of all the larger mushrooms, toadstools and bracket fungi. The life cycle is haploid-dikaryotic. Since the vegetative mycelium is characteristically binucleate and perennial, the dikaryophase is of great duration and is often recognizable by the production of clamp connexions at the cross-walls and by the formation of complex pores in the walls. Asexual reproduction may be by oidiospores or chlamydospores, but the former are not widespread in the Homobasidiomycetes and rare, or lacking, in the dikaryophase. This and related fungi are readily recognized by their macroscopic spore-bearing structures with stalk, cap and gills, etc. In terminal, hyphal cells on the gills the two nuclei fuse and each cell becomes a basidium. After meiosis four spores are generally formed and are discharged violently. Compatibility is determined in a complex way in these fungi but hyphal fusions between mycelia derived from two spores of compatible mating types restores the dikaryophase. Oidia, if developed at all, usually occur on the monokaryon prior to hyphal fusion. In related fungi most basic features are similar but in some clamp connexions are lacking, in others the cells are not regularly binucleate and in others, notably the puff balls, earth-stars and stink-horns, the basidiospores may not be discharged violently.

Chapter 2
Structure and Fine Structure of Fungal Cells

HYPHAE AND NON-MOTILE UNICELLS

Hyphae vary enormously in their overall length and diameter, the latter ranging from less than 0·5 μm to 1 mm, as in the developing sporangiophores of *Phycomyces blakesleeanus*. In this fungus the sporangiophores can maintain themselves erect to a height of 30 cm and, ultimately, may grow twice as long although they cannot then stand alone. It is clear that hyphae can reach much greater lengths in other fungi, for example, rhizomorphs and fungal strands comprising masses of parallel-running hyphae have been found extending 10 m or more (FINDLAY, 1951; GRAINGER, 1962).

Under some conditions at least, therefore, the hyphal cell walls of fungi must be remarkably strong and rigid structures. Yet, although the walls of growing hyphae are detectable because of their refractive properties, they are not very thick, often 0·2 μm or less. They often become thicker after their formation, even immediately behind the growing apex, as in the sporangiophores of *Albugo candida*. Hyphae with especially thick walls frequently occur in the sporophores of toadstools or bracket fungi, e.g. *Fomes* spp., where they form skeletal or binding hyphae of great mechanical strength. Hyphal differentiation will be discussed later (Chapters 4 and 5).

Wall structure

The structural causes of the strength of fungal cells are not immediately apparent. Walls of growing hyphae are normally optically homogenous although distinct layers are often visible in thick-walled resting cells such as chlamydospores, or zygotes of Phycomycetes. However, it is possible to demonstrate lamellate or fibrillar structures in certain spore-bearing hyphae using special techniques and these observations are substantiated by studies of their ultrastructure with the electron microscope.

CARNOY (1870) treated sporangiophores of *Phycomyces* with hot KOH and demonstrated the dissolution of an external cuticle which had prevented the entry of various reagents. Elegant photographic evidence for the presence of this cuticle was provided sixty years later (Plate I, Fig. 4) and, in the following year, it was shown, in electron micrographs, to be a tenuous, structureless membrane (ROELOFSEN, 1950, 1951). Carnoy followed his alkali treatment by staining with chlor-zinc-iodine and showed that the sporangiophore wall was three-layered except at the tip where only the outermost layer persisted. Subsequent treatment with strong H_2SO_4 rendered some 10–12 lamellae visible in the outer and middle layers and the same treatments applied to species of *Pilobolus*, *Rhizopus* and *Mucor* revealed somewhat fewer lamellae. Later, studies of the birefringence of torn fragments of sporangiophores of *Phycomyces* have supported the notion that the wall is three-layered (OORT and ROELOFSEN, 1932). Somewhat earlier, FREY (1927), using similar optical techniques, had provided evidence that the walls of the conidiophores of *Aspergillus* contained fibrous elements and THOM and RAPER (1945) have provided direct evidence for this by breaking and crushing the walls which then show a frayed, fibrillar appearance.

The development of electron microscopy since 1950 has provided clear evidence for the occurrence of microfibrillar, structural elements in the walls of fungal cells. Ultra-thin sections of hyphae of *Rhizopus* spp. revealed a striated appearance, the striations parallel to the surface, perhaps the basis of Carnoy's 'lamellae'. No difference could be detected between the outer, presumed older region and the inner, presumed younger region of the wall (HAWKER and ABBOT, 1963b). Hyphae of *Mucor rouxii* also showed a layer of clearly defined microfibrils parallel to the surface but in the induced, yeast-like phase the walls became ten times thicker, less compact and apparently two-layered (BARTNICKI-GARCIA and NICKERSON, 1962). There is suggestive evidence from other fungi that walls may have two or more layers. In true

Plate I. Wall structure in fungi

Fig. 1. Electronmicrograph of shadowed hyphal tip of *Phycomyces blakesleeanus* grown in shake-culture. Note reticular pattern of microfibrils of chitin (or perhaps chitosan). (×24,500)

Fig. 2. Surface of protoplasmic membrane of *Saccharomyces cerevisiae* prepared for electron-microscopy by freeze-etching. Note the elongated folds in the membrane and the characteristic hexagonal arrays of particles from which microfibrils seem to originate. (×70,000)

Fig. 3. More highly magnified illustration of hexagonally arrayed particles showing microfibrils apparently connecting particles with wall at top of picture. (×110,000)

Fig. 4. Drops of water which have collected outside the sporangiophore wall but inside the lipid cuticle of *Phycomyces blakesleeanus*. (×250)

[Fig. 1 supplied by Houwink, from Roelofsen, 1959; Figs. 2 and 3 by Mühlethaler, from Moor and Mühlethaler, 1963 and unpublished; Fig. 4 by Roelofsen, from Roelofsen, 1959]

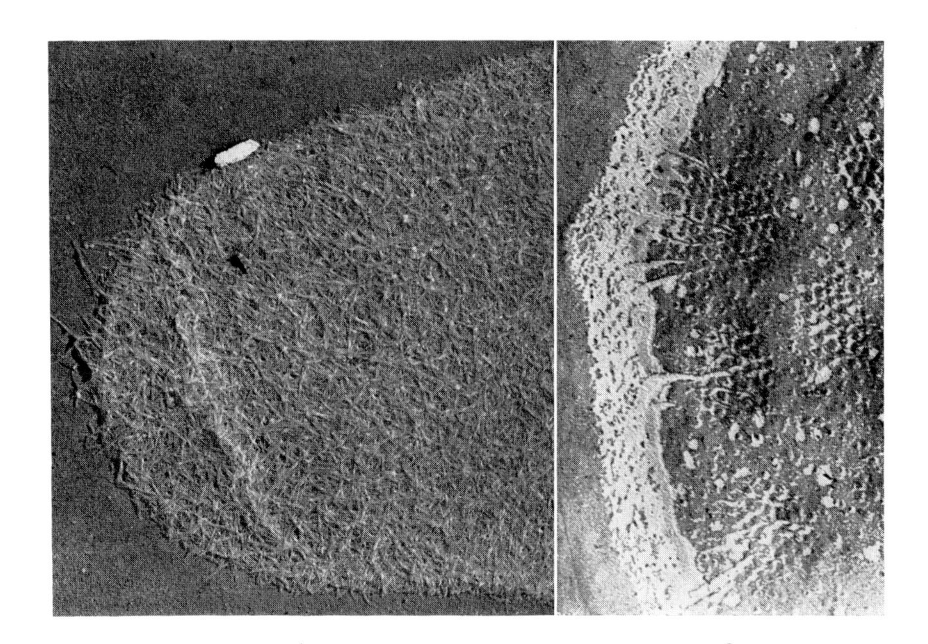

1 3

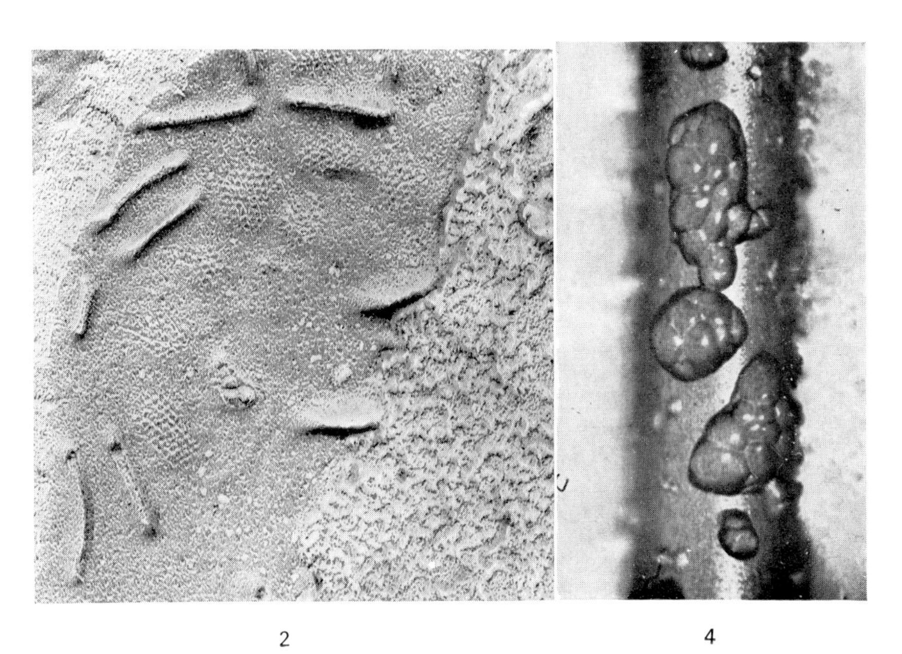

2 4

1
2
3

yeast (*Saccharomyces cerevisiae*) early pictures showed an outer, electron-dense layer and an inner, less dense layer containing microfibrils (AGAR and DOUGLAS, 1955) and the occurrence of an outer granular, non-fibrillar layer overlying an inner fibrillar layer has been demonstrated in the yeast-like *Candida tropicalis*, the aquatic fungus *Allomyces macrogynus* and the dermatophytes *Microsporun* and *Trichophyton* (HOUWINK and KREGER, 1953; ARONSON and PRESTON, 1960a; TAMAKI, 1959). How clearly defined these layers are is not clear. Other workers with *Saccharomyces* have been unable to demonstrate two layers and it has been claimed that such an appearance is a fixation artefact (VITOLS *et al.*, 1961). A preparation of the walls of *Neurospora crassa*, superficially similar to those of many other fungi, has been interpreted by SALTON (1960) as showing microfibrils embedded in a homogeneous matrix. No detailed information is available for Basidiomycetes but electronmicrographs appear to show a thin outer and a broad inner dense layer on either side of a clear central region (e.g. *Polystictus versicolor*: GIRBARDT, 1958; *Coprinus lagopus*: GIESEY and DAY, 1965). In the Basidiomycete *Rhizoctonia solani* the hyphal tips are said to be one layer thick but further back this thickens inwards to form a series of lamellae. Outside this lamellate layer is an amorphous region some 100 nm thick (BRACKER and BUTLER, 1963). Photographic evidence is not yet available for all these statements. Thus, apart from Basidiomycetes, there is clear evidence that fungal cell walls include microfibrillar components associated with non-fibrillar material. Studies of the orientation of the microfibrillar material will be discussed later, in connexion with cell growth (Chapter 3). It is appropriate now to deal with the chemical nature of the materials which have been observed.

The chemical heterogeneity of the cell walls of hyphae has been recognized for several decades, since DE BARY (1866, 1887) distinguished 'true cellulose' in *Perenospora* and *Saprolegnia* (Oomycetes) from 'fungal cellulose' in other forms. Progress has been handicapped by two restrictions, the availability of appropriate techniques and the purity of cell wall material. Steady progress has been made in respect of the former problem but the second is of an

Plate II. Wall and cytoplasmic structures in fungi (continued in Plate III, p. 50)

Fig. 1. Electronmicrograph of part of shadowed wall of hypha of *Allomyces macrogynus* to show differences in alignment of microfibrils of inner and outer lamellae. Inner microfibrils are aligned parallel to the main axis of the hyphae, the outer ones variously orientated but, in general, more transverse. ($\times$8,550)

Fig. 2. Lomasome in hypha of *Perenospora manshurica* in the leaf cell of *Glycine max*. The hyphal wall runs across the picture and is separated from the cytoplasmic membrane of the host by a layer of unknown origin. Part of a mitochondrion and chloroplast can be seen in the host cell. The clear areas in the fungus are probably fat droplets. ($\times$25,700)

Fig. 3. Conformation of endoplasmic reticulum in developing ascospore of *Neobulgaria pura* resembling that of the dictyosome (golgi body) of other organisms. Although endoplasmic reticulum is very apparent in ultra-thin sections of many fungi and assumes various conformations, it is unusual to find this particular pattern. ($\times$48,000)

[Fig. 1 supplied by R. D. Preston, from Aronson and Preston, 1960b; Fig. 2 by G. Peyton and C. Bowen, unpublished; Fig. 3 by R. T. Moore, from Moore and McAlear, 1963a.]

altogether different and more fundamental kind. It presupposes not only that walls can be obtained free from contaminating cellular material but that walls really are structures independent of the underlying cytoplasm. There is some justification for both views. Ultra-thin sections examined in the electron microscope usually reveal a wall separable and distinct from an immediately underlying distinct cell membrane. Such an appearance could be an artefact due to a split developing along a plane of weakness during processing. These difficulties are emphasized by a recent study of the yeast cell by means of the frozen-etched technique. MOOR and MÜHLETHALER (1963) have been able to demonstrate characteristic elongated invaginations, 0·3 μm long, 0·02–0·03 μm wide and 0·05 μm deep, on the inner surface of the wall and occurring as frequently as 15 per mm^2 in old cells (Plate I, Fig. 2). Moreover, in those parts of the cell membrane not associated with the invaginations, groups of hexagonally arranged particles occur from which emerge outwardly-directed microfibrils each some 50 10^{-10} m in diameter. These latter regions covered some 25% of the membrane surface (Plate I, Figs. 2 and 3) so that, in an old cell, in all nearly 40% of the wall was in the most intimate contact with either invaginations or fibrils from the cytoplasm. No comparable observations are available for any other fungus. Despite the reservations which such observations must invoke it is generally believed that modern methods of mechanical disintegration followed by thorough washing in water and, possibly, mild chemical or enzymatic treatments provides both clean and intact wall material (for methods, see SALTON, 1964). Some measure of the effectiveness of these treatments compared with the older methods is given for *Allomyces* in Table 2.1, while Table 2.2 sets out the most sophisticated analysis available to date. Table 2.3, however, is of more general interest for it presents a conspectus of the principal materials which have been found using modern analytical methods (including x-ray diffraction for crystalline substances) in a range of fungi.

Table 2.1. The composition of walls of *Allomyces macrogynus* isolated by alkaline digestion and sonic oscillation (Aronson and Machlis, 1959)

Component	Per cent of dry walls: Alkaline digestion	Per cent of dry walls: Oscillation
Nitrogen	4·7	5·5
Acetyl	15·5	—
Moles N/Moles acetyl	0·9	—
Chitin	68	58
Glucan	8	16
Protein	—	10
Ash	10	8
Totals	86 [1]	92 [1]

[1] The totals are the sum of the last four entries.

Table 2.2. The composition of the walls of *Mucor rouxii* grown in the filamentous phase and the yeast-like phase. Wall material broken mechanically, washed, lyophilized and subsequently analysed (Bartnicki-Garcia and Nickerson, 1962)

Component (%)	Phase		Component (%)	Phase	
	Hyphal	Yeast		Hyphal	Yeast
Chitin	9·4	8·4	Protein	6·3	10·3
Chitosan	32·7	27·9	Purine and pyrimidine	0·6	1·3
Undet. 2-amino sugars	2·4	3·1	Lipid	2·0	0·8
Fucose	3·8	3·2	Bound lipid	5·8	4·9
Mannose	1·6	8·9	PO_4	23·3	22·1 [1] (PO₄, Ca, Mg)
Galactose	1·6	1·1	Ca	1·0	
Other carbohydrates	1·7	0·9	Mg	1·0	
Totals 93·2			92·9		

[1] 18·5, 15·6 Total ash.

Three classes of compound, cellulose, chitin and insoluble (yeast) glucan can occur as microfibrillar elements and are presumably the basis of the structural rigidity of fungal cell walls. The microfibrils of cellulose and chitin are of indefinite length and similar diameter, 150–250 10^{-10} m; those of insoluble glucan are said to be thinner, down to 60 10^{-10} m, and perhaps shorter (HOUWINK and KREGER, 1953). The others, chitosan, soluble glucan, mannans, proteins, lipids and minerals are apparently non-crystalline or only weakly crystalline. For example, synthetic chitosan prepared by de-acetylating chitin is apparently crystalline but native *Phycomyces* chitosan is said to be amorphous (KREGER, 1954). There is suggestive evidence for polysaccharide-protein complexes in the walls of yeasts (NICKERSON *et al.*, 1961) and possibly for a chitin-glucan linkage in the Basidiomycete *Schizophyllum* (WESSELS, 1965). The composition of fungal cell walls appears to show some variation between different taxonomic groups, although generalizations made with the available data are likely to undergo alteration as more information becomes available.

In the Phycomycetes cellulose is said to occur in the biflagellate Oomycetes (FREY, 1950; PARKER *et al.*, 1963) and the anteriorly uniflagellate Hyphochytridiomycete *Rhizidiomyces* (NABEL, 1939; FULLER and BARSHAD, 1960) but not, as was claimed earlier (VON WETTSTEIN, 1921), in the Chytridiomycetes (FREY, 1950; ARONSON and PRESTON, 1960b). Even when cellulose supposedly occurs it may well not be the predominant material. In *Saprolegnia*, for example, unidentified glucans or mannans may be involved and in *Rhizidiomyces* chitin also occurs. Glucans may well be the predominant material in the walls of Oomycetes for it is clear from recent rather precise analyses of the walls of species of *Phytophthora* and *Pythium* that cellulose is a minority component or even lacking altogether (Table 2.3; BARTNICKI-GARCIA, 1966; MITCHELL and

Table 2.3. The principal components of the cell walls of hyphae of various fungi. Only determinations by modern or reliable methods are included (Data from Bartnicki-Garcia, 1966; Bartnicki-Garcia and Nickerson, 1962; Kreger, 1954; Northcote and Horne, 1952; Parker, Preston and Fogg, 1963; Phaff. 1963; Wessels, 1965)

Fungus	Components (%)								
	Chitin	Chitosan	Cellulose	Insol. glucan	Sol. glucan	Mannan	Protein	Lipid	Ash
Saprolegnia spp.	?		10–15	?+	?+	?+			
Allomyces	58		–	16			10		8
Mucor	9·4	32·7	–			?+	6·3	7·8	25·3
Phytophthora	0·3		~20?	~50	~18	0·6	3·5	2·5	+
Phycomyces	27	10	–					55	
Saccharomyces	1		–	28·8	?tr	31	13	8·5	0·3
Candida	+		–		+	+		1·0	
Nadsonia	+		–	?+	60	–			
Sporobolomyces	10		–	–		–			
Schizosaccharomyces	–		–	–	30	–			
Endomycopsis } *Eremascus* }	20–25		–	+		–			
Neurospora	10·7		–	89·3					
Schizophyllum	5		–	81 (ratio $\frac{\text{sol.}}{\text{insol.}}$ 55.8:1)		–	2·3		0·5

SABAR, 1966). It has been suggested that, 'the notion of having true cellulose as a chemical entity separate from other wall glucans may be artificial, for it is possible that the hyphal walls of *Phytophthora* consist of a single glucan species with amorphous regions and a cellulose-like core'. There is suggestive evidence that the two components comprise an homogenous 1:4β linked cellulose-like polymer and an amorphous component with 1:3β and 1:6β linkages (BARTNICKI-GARCIA and LIPPMAN, 1966). Chitin, or perhaps, chitosan (as in *Mucor rouxii*) predominates in the Chytridiomycetes and Zygomycetes. In the Ascomycetes and Basidiomycetes it looks as though insoluble glucans may be the predominant structural material although chitin may be present in quite appreciable amounts, e.g. 10–20% in *Neurospora* but, even here, glucans usually account for 80–90% of the polysaccharide material. Amongst the yeasts and their allies there is a suggestion of a correlation with habit. Filamentous or pseudo-filamentous yeasts, e.g. *Endomycopsis* and *Eremascus*, possess more chitin in their walls than unicellular forms and, indeed, in some of these chitin is virtually (*Saccharomyces*), or completely (*Schizosaccharomyces*) absent. It is remarkable that in *Schizosaccharomyces* none of the known microfibrillar components appears to be present yet the cells maintain their shape and rigidity. That this is possibly a meaningful correlation is born out by the fact that the chitin and chitosan contents of the walls of the hyphal and yeast-like phases of *Mucor rouxii* are 42·1% and 36·3% respectively. Further consideration of changes in the composition of cell walls as they age or develop will be deferred until the next chapter. There is little evidence concerning the spatial distribution of these materials in cell walls but the treatment of cell walls with purified enzymes seems to be a promising approach. A good deal of work of this nature has been done with impure enzyme mixtures, e.g. digestive juice of snails. Isolation of the products of lysis have provided information about the relative amounts of the different glucans present (e.g. TANAKA, 1963). POTGEITER and ALEXANDER (1965), however, treated *Neurospora* hyphae with purified 1:3β glucanase and chitinase. About 50% of the bound glucose was liberated by the former but very little *N*-acetyl glucosamine by the latter unless the glucanase was present; then the glucosamine yield was more than doubled. They suggested that this indicates that a chitin core is covered by an outer layer of glucan, a suggestion in keeping with the electronmicrographs described earlier (p. 19) by Salton. The wall remained very largely intact after these treatments and they concluded, therefore, that other glucans or non-chitinous, acetyl hexosamine polymers were present.

It is clear that, at the present time, knowledge of the cell walls of fungi is inadequate and uneven but that great possibilities are inherent in the combined use of the techniques of chemical and enzymatic analysis and electron microscopy. A further aspect of wall structure is in an equally uncertain state, the nature of cross-walls or septa. These will now be considered.

Septal structure

The majority of Phycomycetes are aseptate except when reproductive structures are delimited or when a wall is developed to seal off a damaged region of hypha. In septate Phycomycetes and all other fungi, transverse septa are of general occurrence but in the majority of cases they are found to possess a central pore. More rarely the septa possess more than one pore, or none at all, e.g. certain specialized hyphae of sporophores of large Basidiomycetes may lack septa altogether (CORNER, 1953).

The widespread occurrence of perforated septa, through which cytoplasm, cytoplasmic organelles or even nuclei can regularly pass, indicates that the distinction between coenocytes and septate fungi is not so profound as has sometimes been imagined. The septa can, in fact, be regarded primarily as strengthening features of the hyphae. The regular occurrence of annuli of rigid materials provides a maximum increase in mechanical strength with a minimum of interference with the movement of the cell contents. The recent discovery of micropores, possibly in groups, in the apparently intact transverse septa of *Geotrichum candidum* (WILSENACH and KESSEL, 1965a) stresses the importance of intercellular contacts. The size of the micropores, which are at their narrowest in the centre of the septum and flare-out on the sides next to the cells, is $300\ 10^{-10}$ m at their narrowest, to $600\ 10^{-10}$ m at their widest (Fig. 2.1a). This presumably precludes the possibility of the passage of cell organelles through them, although not the occurrence of protoplasmic strands. All septa, in septate Phycomycetes, Ascomycetes, Deuteromycetes and Basidiomycetes appear to originate in the same way: as a rim of material which develops centripetally either to cut off a segment completely or, more usually, to stop development before this happens, thus leaving the characteristic central pore (Fig. 2.1b) (WAHRLICH, 1893; BULLER, 1933). The same process occurs when a hypha is sealed off in an aseptate Phycomycete but the sporangiophores of the Zygomycetes are apparently cut off by the deposition of columella-wall material in a cleavage plane between two masses of protoplasm. Septum formation by centripetal growth is rapid. BULLER (1933) gives 20–25 minutes for the complete sealing of the normally aseptate *Rhizopus*, 10 minutes for the Basidiomycete *Rhizoctonia* and 6 minutes for the Ascomycete *Ciboria*.

Electron microscopy has now revealed what Buller almost saw with superb light microscopy (see p. 146 in BULLER, 1933) and what can be detected with phase microscopy in favourable cases, namely that the septal pores of some Basidiomycetes can be more complex than those of other fungi (Fig. 2.1d). These complex pores were discovered by GIRBARDT (1958) in the bracket fungus *Polystictus versicolor* and have subsequently been seen in several others but so far have not been detected in rust and smut fungi. Since the published account of the formation of the septal pore of *Rhizoctonia solani* includes an account of the development of the septum and, since this was the fungus so carefully examined by Buller, its development will be described here (BRACKER and BUTLER, 1963, 1964). The lateral wall is composed of an outer amorphous layer some 100 nm thick, overlying a lamellate layer. An electron-dense annulus with a central clear layer arises from the innermost lamella of the wall and develops centripetally into the hypha. The annulus walls thicken from the cell wall side centripetally by becoming multi-lamellate. Around the rim of the central pore, which is left by the stoppage of the growth of the

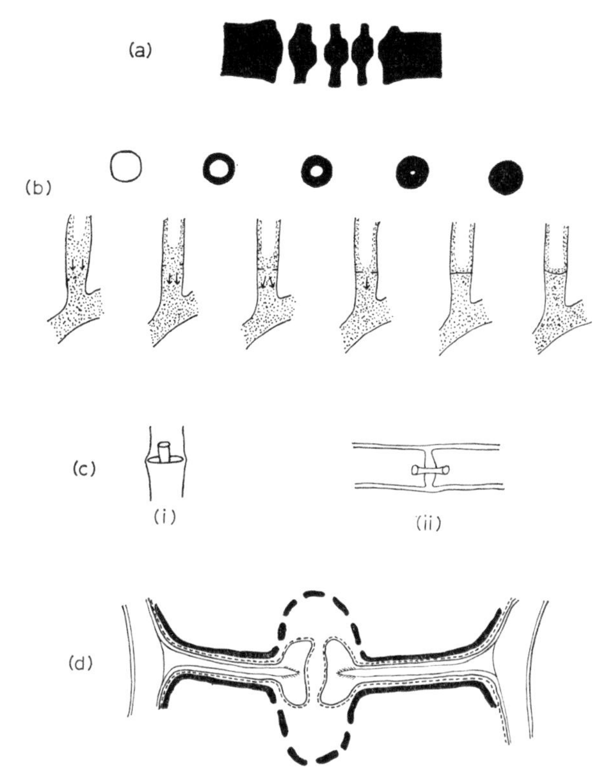

Fig. 2.1. Septa and pores of fungal hyphae. (a) Drawings based on an electron-micrograph to show the micropores in *Geotrichum* (×37,000. (b) Simple pore as found in the septa of most Phycomycetes and Ascomycetes, e.g. *Rhizopus*; section and face view. Note that, in effect the septum is an annular thickening developed from the walls of the main hypha. Arrows indicate the flow of cytoplasm. (×400, after Buller) (c) The complex pores of *Piptocephalis* (i) and *Dispira* (ii). There are found in the cross-walls of the branched reproductive structures and less commonly in the vegetative hyphae. Their function is unknown. (×c.700, from Benjamin) (d) Reconstruction of a complex pore in the transverse septum of a Basidiomycete as seen by electron micrography. Note the lamellate septum with its dumb-bell like swellings surrounding the actual pore. Septum and annular swelling are invested by the protoplasmic membrane (– – – –). A septal pore-cap (the parenthosome) is developed at either side of the pore and is believed to be derived from the endoplasmic reticulum (**– – –**). (×*c*. 20,000, based on various authors)

annulus before closure, amorphous material develops as a swelling. This swollen region thickens and the material also extends over the annular septum, the lamellate tips of which spread out into the material as a series of concentric plates. The septum and its swollen pore rim are invested closely by the cytoplasmic membrane of the cell which remains intact throughout. The endoplasmic reticulum on either side of the pore gives rise to a characteristic, cup-shaped, perforated pore cap as soon as the septum has formed. It is claimed that this complex septal pore is not a barrier to the passage of nuclei, mitochondria or cytoplasm, as observed by phase-contrast or normal optics. Some doubt has been cast upon this by recent observations of GIESEY and DAY (1965) in the toadstool *Coprinus lagopus*. Here complex septal pores are developed in both monokaryons and dikaryons. However, when strains between which nuclear migration is known to occur were allowed to anastomose and then examined, simple pores were found (Plate III, Figs. 1 and 2). In addition, some ultra-thin sections were interpreted as stages in the breakdown of the complex pore, e.g. the disappearance of pore-rim material and the pore cap and possibly some softening and distortion of the septum itself. The diameters of complex pores ranged from 0·09–0·18 μm, those of simple pores from 0·4–1·2 μm. Such observations suggest that the passage of cell organelles may be associated with the development of simple pores but, if this is the case, the function of the elaborate septal pore structure is unclear. Even more of a functional puzzle are the septal pores of certain Zygomycetes such as *Piptocephalis* and *Dispira* (Fig. 2.1c). In species of the former genus the rim of the central pore develops a tubular growth which projects into the cell on the upward side of the hypha; in the latter genus the septa have median disciform cavities which contain solid, colourless more or less biconvex plugs bearing small knob-like projections at their ends (BENJAMIN, 1959). Further investigations are clearly needed to resolve these problems; meanwhile, the development of a septum remains a remarkable example of localized biochemical activity within a hypha.

Cell contents, organelles and inclusions

CYTOPLASMIC MEMBRANE AND ENDOPLASMIC RETICULUM

Reference has already been made to the fact that the cell contents are enclosed within a cytoplasmic membrane. This has been demonstrated by the effects of cell damage, by the liberation of cell contents as free-living 'protoplasts' and, directly, by electron microscopy. In older electron micrographs the membrane appeared as a single line but the characteristic tripartite structure of two electron-dense layers on either side of a less dense, or clear, central region, has now been shown in a wide range of fungi where better fixation and resolution have been achieved. In some regions the membrane is separated from the wall. Sometimes the membrane appears to form an infolded, convoluted pocket, in others it appears to be a pouch enclosing deposits of granular or vesicular material. Such regions were termed lomasomes by MOORE and MCALEAR (1961) although they were first discovered by GIRBARDT (1958). The involuted type is especially well shown in Oomycete haustoria, e.g. *Perenospora manshurica* (Plate II, Fig. 2). Their function is unknown, but, as Girbardt has suggested, they could provide a greatly increased surface area in a localized region of the cytoplasmic membrane. In haustoria such a condition could be associated with the localized transfer of materials from host to parasite (or vice versa). However, WILSENACH and KESSEL (1965b) believe that lomasomes are implicated in the production of the ascospore wall in *Penicillium vermiculatum* and that they arise by the accumulation of small vesicles in association with the endoplasmic reticulum. This is plausible and a similar notion has been developed by MARCHANT *et al.* (1967) in connexion with wall formation in fungal hyphae generally. Such lomasomes are of the pouch-like type which encloses vesicles or granular material and the term should perhaps be restricted to such structures. Much more information is necessary before it will be possible to classify the kinds of lomasomes and their functions must still be regarded as speculative.

Within the cytoplasmic membrane, endoplasmic reticulum has been demonstrated although, possibly as an artefact of fixation, it seems to be looser and more irregular than in the cells of green plants or animals. Adjacent cells often vary enormously and apparently erratically in their appearance. In several fungi the endoplasmic reticulum is composed of sheets or microtubular structures, e.g. *Neurospora, Schizophyllum,* and these may be interspersed by many granules which fall within the characteristic dimensions of ribosomes. As yet there has been no clear demonstration of 'rough' and 'smooth' endoplasmic reticulum and it is, indeed, unusual for the granules to be attached to the membranes even in cells of *Neurospora* engaged in active protein synthesis (unpublished, see Frontispiece). In many fungi, especially in young zygotes or apical growing regions, the endoplasmic reticulum is highly vesicular. This is particularly well shown in Stage IV sporangiophores of *Phycomyces.* Here, PEAT and BANBURY (1967) have described three types: (a) hollow, membrane-bounded spheres of average diameter 300 µm, with faintly granular contents. They may be arranged in chains running radially to the walls; (b) large, multi-vesicular bodies up to 600 nm in diameter often near the cytoplasmic membrane and resembling vesicular lomasomes; (c) a distension of a strand of endoplasmic reticulum to produce a highly pleomorphic vesicle containing numerous, small, irregular electron-dense granules. It is believed by these workers that such vesicles are intimately concerned with the processes of hyphal growth (MARCHANT *et al.*, 1967).

The association of strands or elongated vesicles of endoplasmic reticulum with the nuclear membrane has been seen in a number of fungi but has not yet received detailed study.

A further very remarkable feature is the occurrence of aggregations of complex concentric lamellae associated with an invagination of the cytoplasmic membrane or free in the cytoplasm. First described in *Lenzites saepiaria* (HYDE and WALKINSHAW, 1966), similar aggregations have been seen in *Ascobolus* and *Sporobolomyces roseus* (MOORE and MCALEAR, 1963a; PRUSSO and WELLS, 1967). They also occur extensively in *Neurospora* and *Schizophyllum* (unpublished). MORRÉ and MOLLENHAUER (1964) have described concentric rings of lamellae in connexion with the golgi apparatus of higher green plants but their appearance is more regular than the complex fungal aggregations. Concentric lamellar structures are known in a wide range of animal cells and their function or functions are obscure, although in slime moulds they appear to be associated with food vacuoles (HOHL, 1965). There is no evidence from fungi that these lamellae are associated either with the golgi apparatus or vacuoles but the membranes do resemble those of the endoplasmic reticulum.

DICTYOSOMES

The golgi apparatus, as revealed by silver impregnation or osmic-fixation methods, was believed to be homologous with the vacuolar system in many fungi, especially by French workers (see GUILLIERMOND, 1941). The application of the term has been restricted, since the advent of electron microscopy, to a 'stack' of plate-like, membrane-bound sacs (cisternae) with the edges of which are associated small, more or less spherical vesicles believed to be derived from the cisternae and often termed the dictyosome. This type of structure is rare in the fungi so far studied. In the Oomycetes, dictyosomes resembling those in green plants seem not to be uncommon. They have been described and illustrated in hyphae of *Pythium debaryanum* (HAWKER, 1963; HAWKER and ABBOT, 1963a) and haustoria of *Phytophthora* spp., *Perenospora manshurica* and *Albugo candida* (CHAPMAN and VUJEČIC, 1965; EHRLICH and EHRLICH, 1966; PEYTON and BOWEN, 1963; BERLIN and BOWEN, 1964a). In contrast, dictyosomes have not yet been identified in haustoria of *Erysiphe graminis*, *Uromyces caladii*, *Puccinia graminis* or *Cronartium ribicola* (see EHRLICH and EHRLICH, 1966). In *Puccinia podophylli* and *Neobulgaria pura* MOORE (1963; MOORE and MCALEAR, 1963a) has described as a dictyosome a stack of cisternae-like structures lacking the characteristic peripheral vesicles (Plate II, Fig. 3). LU (1966) has described a series of elongated-flattened vesicles radiating from a common centre in the basidium of *Coprinus lagopus* and claims that it is not dissimilar from the golgi apparatus found in some higher green plants. The three flattened sacs surrounded by many bubble-like structures found in frozen-etched *Saccharomyces* cells suggest a more normal golgi apparatus (MOOR and MÜHLETHALER, 1963). At present, it seems clear that the Oomycetes, which differ from other fungi in a number of respects such as the composition of their walls and their probable diploid condition, possess structures similar to those of the golgi apparatus as generally understood but that there is less certainty in other fungi where a different organization may occur. There is much less ambiguity about the structure and organization of the other classes of cell organelle, the mitochondria and the vacuoles. These have been studied in some detail by light microscopy for some decades, especially by Guilliermond and his school (see GUILLIERMOND, 1941).

MITOCHONDRIA

These are remarkable amongst the filamentous fungi both for their size and for their ability to change shape. They may be small spherical structures at, or below, the level of optical resolution, which may elongate to 30 μm, form unequally thickened structures resembling a row of beads on a thread, or even branch quite considerably and may then apparently break up into a mass of discrete, smaller bodies. In living, growing hyphae they exhibit the most active motion; this has been brilliantly recorded in a film of *Polystictus* by GIRBARDT (1955, 1957), made with phase contrast. Studies with the electron microscope have not revealed any obvious fundamental structural differences between fungal mitochondria and those of other organisms (MOORE and MCALEAR, 1963b and Frontispiece), although HAWKER (1965) is of the opinion that they possess fewer, flatter and possibly more irregular cisternae than those of green plants. There is some evidence from biochemical properties and vital staining that not all mitochondria possess the same properties but whether this indicates that the differences are constant or that mitochondria may exist in different physiological states is not clear (EPHRUSSI and SLONIMSKI, 1955; AVERS *et al*, 1965; AVERS, 1967).

VACUOLES

These are not apparent in the apices of hyphae but can readily be seen to arise further back, to enlarge and to show a tendency to coalesce. In *Saprolegnia*, however, GUILLIERMOND (1920) using the vital stain, neutral red, has illustrated the vacuolar system as a mass of anastomosing canaliculi which extend very nearly to the hyphal apex. A similar development was shown for germinating zoospores (CASSAIGNE, 1931). In most fungi the vacuoles remain as distinct structures often apparently located near to the cell walls, their number increasing with the age or degree of degeneration of the cells. There seems little doubt that they are bounded by a distinct membrane, the tonoplast. Evidence for this is provided by the movement of vacuoles through septal pores of smaller diameter than the vacuoles, which are thus capable of constriction without rupture (BULLER, 1933), and direct evidence exists for a single unit-membrane in electronmicrographs of yeast and other fungi, e.g. *Rhizopus* (HAWKER and ABBOTT, 1963a). Vacuoles may contain pigments or inclusions of various kinds both crystalline and amorphous. Little is known of the chemical nature of these substances but glycogen is commonly held to be stored in the vacuoles of yeast. In this organism particles and aggregates of particles, each 100 10^{-10} m in diameter, have been described as occurring on the cytoplasmic side of supposed vacuolar membranes. It has been suggested that these particles are ribosomes but this is not certain and the observation is an isolated one (MOOR and MÜHLETHALER, 1963, Frontispiece and Plate III, Fig. 4).

NUCLEUS

Fungal nuclei are usually small; the majority range from 2–3 μm in diameter but in some they are much larger, e.g. 25 μm in the Zygomycete *Basidiobolus ranarum* (ROBINOW, 1963, and see Plate VI, Figs. 1 and 9); 20 μm in the bracket fungus *Fomes fomentarius* (GIRBARDT, 1960). In living cells, as seen by phase contrast, they comprise a central dense area surrounded by a clear area, around which some observers believe they can detect a definite nuclear membrane. The nuclei do not, in general, stain readily apart from the central body which stains heavily with iron haematoxylin. The central body is usually Feulgen-negative and appears as an amorphous, or granular, mass in electron-micrographs. It is usually described as the nucleolus but whether it is really comparable to that structure in higher green plants is uncertain. This body was not detectable with Unna's RNA stain in nuclei of *Neurospora* (ZALOKAR, 1959b) even after treatment with ribonuclease, although it stained with iron haematoxylin. Its basophily may be due to protein since a strong reaction was given by Barnett and Seligman's test for protein-bound SH groups. By contrast, the deep affinity for basophilic dyes, e.g. Toluidine blue, was greatly reduced by treatment with ribonuclease (TURIAN and CANTINO, 1960) in *Blastocladiella* and BAKERSPIGEL (e.g. 1960) has claimed that it contains RNA. The central body behaves differently in different fungi during nuclear division and, at present, it might be better to continue to describe this organelle as the 'central body' except where its behaviour parallels the behaviour of the nucleoli of higher green plants. Nuclear division will be described in Chapter 13. Electron microscopy has provided clear evidence for a typical three-layered nuclear membrane, the central region being least electron-dense (Frontispiece and Plate IX, Fig. 2). The membrane is perforated by pores and there may be as many as 200 per nucleus in old yeast cells, occupying 6·8% of the surface area of the nucleus (Plate III, Fig. 3). Electron microscopy has helped in the location of the actual nucleus in yeast, a matter of long-standing controversy (see NAGEL, 1946). It is now clear that there is a large, more or less central vacuole with which is closely associated a much smaller, globular nucleus (MOOR and MÜHLETHALER, 1963). There does not seem to be a structure comparable to the 'nucleolus' or 'central body' of other fungi although one part of the nucleus does stain more heavily with iron haemotoxylin than the rest (NAGEL, 1946). In life, fungal nuclei can vary greatly in size and shape and can move, possibly at great rates, through hyphae. Their ability to pass through simple septal pores, although becoming constricted considerably in the process, provides circumstantial evidence for the presence of a nuclear membrane. The number of nuclei per segment varies considerably and frequently there is a difference between the apical segment and those lying behind it. There may be as many as 20 to 30 crowding a cell, occupying some 20 to 25% of the total volume, as in *Phycomyces* or *Penicillium* (Plate IV, Fig. 1), or there may be one or two nuclei per cell, as in the monokaryotic and dikaryotic hyphae of Basidiomycetes, and occupying but 0·5% of the total volume of the segment. It is not clear what determines the nuclear/cytoplasmic ratio in fungi; it is possible that this ratio is constant for any one species of fungus.

CENTROSOME AND CENTRIOLES

Centrioles are not common in hyphae. When they have been observed it has usually been in a sporangium, especially the ascus during meiosis or ascospore formation (see SHARP, 1934, for earlier references). Recently Knox-Davies has described successive somatic divisions in the developing conidia of the Deuteromycete *Macrophomina phaseoli* (Plate V, Figs. 1–8). The spore-mother cells are relatively large and uninucleate; the nucleus is large and centrioles can readily be seen after division has commenced. Successive divisions of resulting nuclei are synchronous, the nuclei becoming progressively smaller and the details more difficult to see until in many cases the centrioles were

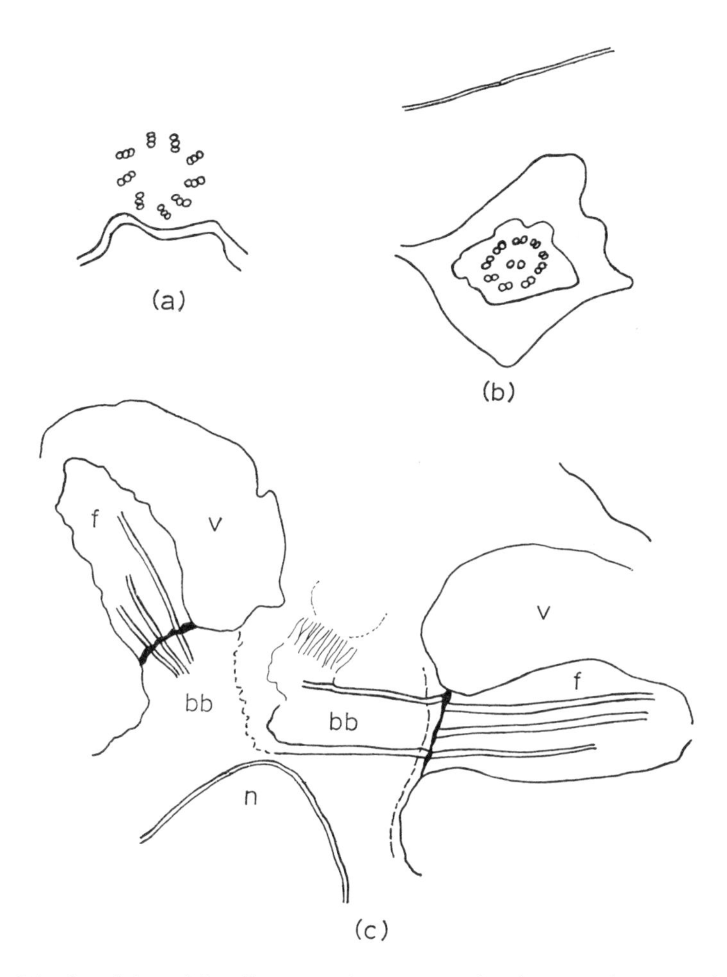

Fig. 2.2. Centriole and flagellar connexions, as seen by electron microscopy, in peripheral cytoplasm of sporangia of *Albugo* which are beginning to differentiate zoospores. (a) The centriole shows the characteristic circular group of 9 tripartite structures each tilted in a clockwise direction. (b) A transverse section through a flagellum traversing a vacuole shows the outer ring of 9 fibres surrounding the central pair. (c) An approximately longitudinal section through the base of the flagellum shows the two flagella (f) extending laterally into the vacuoles (v). At the connexion with the cytoplasm are basal plates and the flagella are joined at the basal bodies (bb), in close proximity to the nucleus (n). Spindle fibres in the cytoplasm can be seen above the basal bodies. (×54,000, from Berlin and Bowen)

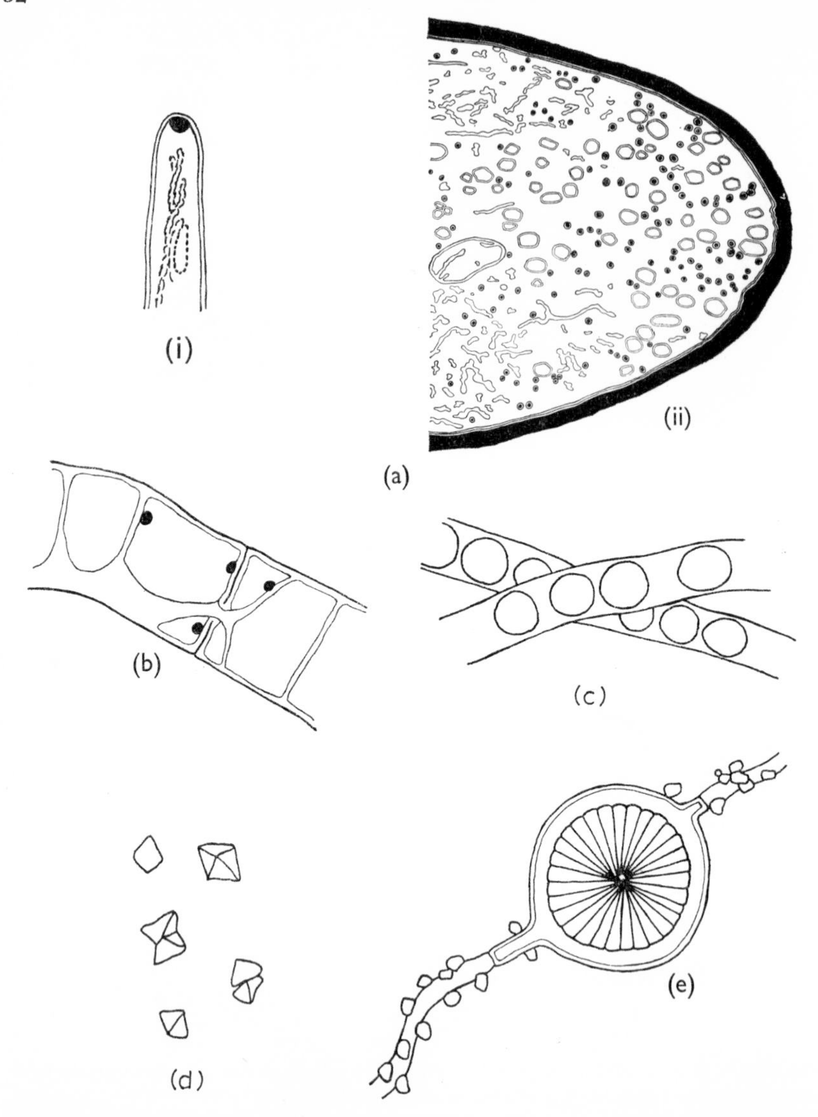

Fig. 2.3. Structures occurring in fungal hyphae. (a) The apical granule of *Polystictus* (ai) and its probable equivalent in an electron micrograph (aii). Note in the latter its resolution into large electron-transparent and small electron-opaque vesicules. (i, × 1000; ii, ×25,000, based on Girbardt). (b) 'Woronin bodies', small, highly refringent bodies seen in the cytoplasm or adhering to the walls in several Ascomycetes, e.g. *Ascobolus* (×1500, after Dodge). (c) Lipid globules in hyphae of *Fusarium* in submerged culture. Note how globules fill the whole diameter of the hyphae (×1450, after Damm). (d) and (e) Crystals from the mature zygotes of *Mucor mucedo* (d) and on and within hyphae from the surface of a mycelial strand of *Mutinus* (e). The nature of the crystals is not often known but is generally claimed to be either proteinaceous or of calcium oxalate (d, ×1000, from Ling Yong: e, ×600 from de Bary)

barely discernible. In the vegetative hyphae they could not be resolved and here the nuclei also were smaller. It is suggested that centrioles are, nevertheless, present. It is surprising, if this is the general situation, that they have not been seen more frequently in electronmicrographs. In the one case, *Albugo candida* (BERLIN and BOWEN, 1964b), where they have been detected as discs 200 nm in diameter, they were adjacent to the nuclei in the sporangium. They showed a characteristic group of nine obliquely arranged triplets connected to a central tube by radially directed elements (Fig. 2.2). They were associated with the development of the flagella (see Motile cells, pp. 34–37). Such structures have only been seen in other fungi in association with flagella formation. It may well be that, as with the 'nucleolus', more than one kind of centriole can exist in the fungi.

GRANULES AND INCLUSIONS

A structure visible by light microscopy at the hyphal apex of many fungi appears as a small, deeply staining granule of variable shape. This was first seen by Brunswik in species of the toadstool *Coprinus* (BRUNSWIK, 1924), who described it as a 'Spitzenkorper' which I shall here term the 'apical granule'. It has been detected in a wide range of Hymenomycetes and Gasteromycetes by Girbardt, who has made a detailed study of it in *Polystictus versicolor* (GIRBARDT, 1955, 1957, 1958; Fig. 2.3a) and he has also seen it in *Penicillium* and *Aspergillus*. I have examined a wide range of Phycomycetes and Ascomycetes without becoming convinced that an apical granule is always present. Girbardt has shown that the apical granule in *Polystictus* lies just within the tip of actively growing hyphae only. Damage to the hyphae by various reagents or a localized check to growth, as brought about by sudden exposure of the tip for 40 sec to high intensity light, results in its movement back from the apex within 25 sec of exposure. In another 25 sec it is no longer visible by phase contrast and seems to have disappeared: a short time later it re-develops. In *Aspergillus* MCCLURE *et al.* (1968). first resolved the microscopic image of the granule, which is a refraction artefact, into an apical mass of microvesicles. In *Polystictus* there are *c.* 1000 electron-transparent and c. 3000 smaller, electron-opaque vesicles, all with double membranes (GIRBARDT, unpublished).

Another type of granule is that described by Buller as a 'Woronin body', after their discoverer. They are highly refractive oval bodies scattered through the cytoplasm of elongating terminal cells or attached to the cross-walls. Their movements are independent of those of streaming cytoplasm. Their nature is obscure and, to date, they have only been described clearly from Ascomycetes, e.g. *Ascobolus* (DODGE, 1912), *Pyronema* (BULLER, 1933) (Fig. 2.3b).

In addition to these numerous cell organelles a variety of inclusions may be found in hyphae (Fig. 2.3c and d). Lipid globules often containing pigments, such as carotenoids, are often conspicuous, especially in certain metabolic conditions. Granules of proteinaceous material, volutin, a metaphosphate polymer ($NaPO_3$), and the reserve carbohydrate glycogen are also present, although the last may occur in vacuoles (cf. p. 29). Hyphae may contain, or more often, are encrusted by, crystals which are generally said to be of calcium oxalate but in most cases there is very little convincing evidence for this assertion. Knowledge of the nature, or function, of such crystal formations in nature or in culture is almost non-existent.

MOTILE CELLS

The motile cells of fungi are either zoospores or gametes of uniflagellate or biflagellate Phycomycetes. A good deal is known about their flagellation but much less about their cell structure.

COUCH (1938, 1941) and VLK (1939) clearly demonstrated the presence of two kinds of flagella in fungi. One narrows more or less abruptly towards the tip, which forms a thin extension of variable length: this is the 'whiplash' flagellum or Peitchgeissel. The second does not obviously narrow and has fine lateral hairs almost to the tip: this is the 'tinsel' flagellum or Flimmergeissel (Fig. 2.4a). It is now known that flagella occur which differ from both of these, resembling the whiplash but lacking the narrow tip; these may be described as blunt.

Flagellation is related to taxonomy in Phycomycetes (COUCH, 1941). The Chytridiomycetes have a single posterior flagellum of the whiplash type; the Hyphochytridiomycetes a single anterior flagellum of the tinsel type. Amongst the biflagellates, the Oomycetes have a forwardly directed tinsel flagellum and a backwardly directed whiplash type; in the Plasmodiophoromycetes the long, forwardly directed flagellum is a whiplash type, the short backwardly directed flagellum a blunt one.

Very little additional information is available from optical studies. Cell shape is usually maintained and may be spherical, pyriform or reniform. From this constancy it may be inferred that the cell is bounded by some form of membrane or pellicle but it must be highly elastic since amoeboid movements can readily be developed. The cells are nucleate, some Oomycetes having a characteristic 'nuclear cap' of basophilic material. Other inclusions are fat droplets, simple or lobed particles and pigments.

A few motile cells have been examined by electron microscopy and have provided more detailed information. Generalizations, save in respect of flagella, can hardly be made at present so that the account here will be confined to specific examples. MANTON and her associates (1951, 1952) have studied the tinsel and whiplash flagella of *Saprolegnia* in some detail. Diplanetism is shown by zoospores of this fungus. Spherical zoospores are liberated from the sporangium, they swim for a short time (1–2 hr), retract their flagella and encyst. This cyst later releases reniform zoospores, somewhat smaller than the first-phase zoospores, with two lateral flagella in a shallow groove on the concave side. They swim for longer than first stage zoospores, encyst and each germinates to give a mycelium. In both types the forwardly directed flagellum (in life) is a tinsel flagellum and is shorter than the backwardly directed whiplash flagellum; in the second phase this size

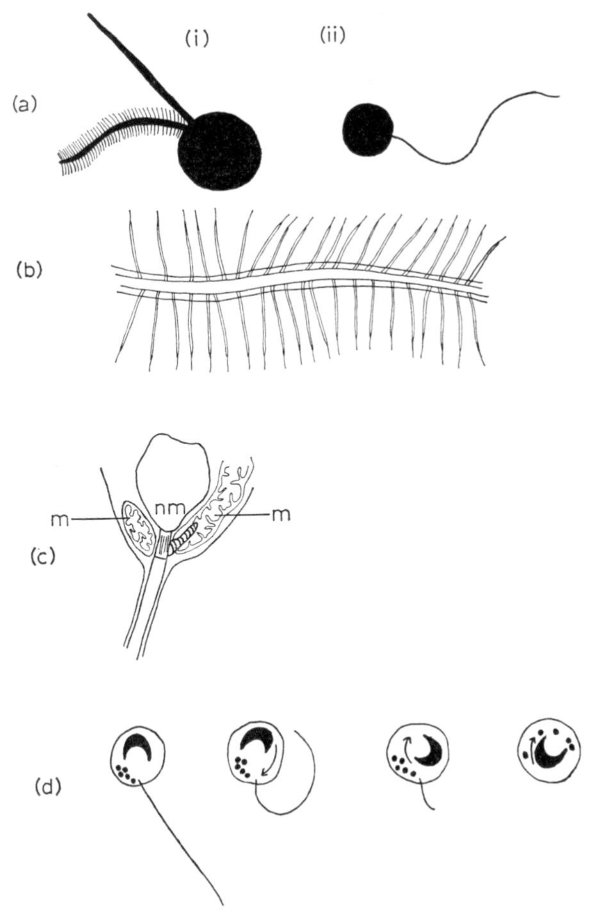

Fig. 2.4. The structure of motile fungal cells and organs. (a) Outlines of zoospores of *Saprolegnia* (ai) and *Olpidium* (aii) to illustrate the (simple) whiplash and the (fringed) tinsel flagellar types (ai, ×500 : aii, ×1500, after Manton *et al.*) (b) Details, from an electronmicrograph, of part of the tinsel flagellum of *Saprolegnia*. Note that the lateral hair-like structures are embedded in the material of the main axis of the flagellum and that they terminate, at their free ends, in sharply demarcated fine tips. (×5500, after Manton *et al.*) (c) The basal attachment of the whiplash flagellum of *Blastocladiella* as shown by electron micrography. The flagellar sheath is confluent with the spore membrane, the two central fibres terminate at, or very near the nuclear membrane (nm). The base of the flagellum also gives rise to a lateral banded structure which is closely associated with the single, enormous mitochondrion (m) of the spore. (×9000' after Cantino *et al.*) (d) Retraction of the flagellum of *Blastocladiella*. Note the initially stiff flagellum which then sweeps round in a wide arc. The spore contents then rotate through some 270° (as evidenced by the movement of the conspicuous, crescentic nuclear cap and the refractile bodies on the mitochondria) and the flagellum appears to be 'reeled in'. (×1000, from Cantino *et al.*)

difference is considerable, the hind flagellum frequently being twice the length of the fore flagellum.

The hind flagellum has a 'core' composed of eleven strands, nine peripheral, two central, and apparently of equal length. These are embedded in an electron-translucent sheath which in life may be flattened. Two rows of exceedingly fine lateral hairs, each about 0·5 μm long, have been detected and may form an internal framework to the sheath. The nature of the narrowed tip of the whiplash flagellum is not yet resolved. Manton believes that it is composed of the eleven strands but that their nature is such that they are more easily destroyed than in the main body of the flagellum. On the other hand, KOLE (1957) and KOCH (1956) working with *Synchytrium* and *Rhizophydium* stated that they believe the tip to comprise the central strands alone which would then exceed the peripheral strands in length.

The fore, tinsel flagellum of *Saprolegnia* (Fig. 2.4b) has a similar core and a massive sheath structure, the core again comprising the characteristic arrangement of eleven strands. The sheath material is apparently glutinous under certain conditions and can be seen to run into blobs of material, often at the tip of the flagellum where they have been described as 'paddles', e.g. especially in *Phytophthora* (KOLE and HORSTRA, 1959). These are detectable in electronmicrographs of *Saprolegnia*. Two rows of lateral hairs 2–3 μm long are embedded in the sheath, they have very fine hair points. The mode of attachment of these hairs is unknown. Essentially the same structures have been recorded for both types of flagella in *Phytophthora* and for the whiplash type in various Chytridiomycetes (MANTON *et al.*, 1952; KOCH, 1956; KOLE, 1957). No details are available for posteriorly uniflagellate or unequally biflagellate forms. However, both MANTON *et al.* (1951) and KOLE and GIELINK (1962) have provided additional information on the cyst stage, of *Saprolegnia* and *Plasmodiophora*, the club-root organism. In both cases the cyst has a delicate wall with projections on its outer surface. In *Saprolegnia* these are numerous, delicate double-headed hooks on long slender stalks, in *Plasmodiophora* short spiny projections. The internal attachment of flagella is not well understood and will be described together with other internal features of the spore body.

The only cells for which information is available concerning their fine structures are the gametes of *Allomyces* (BLONDEL and TURIAN, 1960) and the zoospores of *Blastocladiella* (CANTINO *et al.*, 1963). It is clear that the details of nucleus, nuclear membrane and mitochondria are similar to those in other fungi but certain distinctive features also occur, e.g. both possess a nuclear cap. This has been shown by cytochemical and analytical methods to be rich in RNA and, indeed, in *Blastocladiella* there is little doubt that it is composed of ribosomes (TURIAN, 1956, 1958; LOVETT, 1963). In this same zoospore, pores *c.* 0·1 μm in diameter occur in the nuclear membrane where it is contact with the nuclear cap. The most striking difference, however, lies in the size and distribution of the mitochondria. These are small and numerous in *Allomyces* but in *Bastocladiella* there is but a single, giant, excentrically orientated mitochondrion, which encloses the base of the flagellum and with whose surface are associated lipid globules. This extraordinary structure, unique in the plant kingdom, had previously been seen by optical microscopy and described as a 'side-body' (cf. SPARROW, 1960). In both cell types there are, in addition, a number of strongly osmiophilic bodies but they differ in shape. In *Allomyces* they are lobed, in *Blastocladiella* spherical and probably associated, in life, with carotenoids. In neither cell has a well defined endoplasmic reticulum been described.

There is evidently some variation in the anchoring of flagella in motile cells. In *Blastocladiella* the flagellar sheath appears to be continuous with the cell membrane. The eleven flagellar strands aggregate into a basal body which ends at the nucleus as a series of closely applied processes ('roots'). A banded 'rootlet' arises from the side of the basal body and extends some way into a cavity in the longer lobe of the giant mitochondrion (Fig. 2.4c). A similar banded appearance was noticed at the base of the strands in *Achlya* (MANTON *et al.*, 1952) but not in *Allomyces*. Blepharoplasts, or basal bodies, are said to occur in several motile fungal cells (cf. SPARROW, 1960) but the origin of this structure has only been briefly described for the Oomycete, *Albugo candida* (BERLIN and BOWEN, 1964b). In young sporangia there are four peripheral nuclei with which are associated paired centrioles (see p. 31). Those centrioles on the side next to the sporangium wall elongate to give basal bodies some 0·6 μm in length and from each of these a flagellum develops. This development is similar to that in many animal cells (SLEIGH, 1962). Equally remarkable is the retraction of the flagellum by *Blastocladiella*: it is not known whether this is typical of the way in which all motile cells withdraw (or lose) their flagella (Fig. 2.4d). The process takes 30–40 sec. The zoospore settles and becomes irregularly spherical; its flagellum aligns itself on a wide arc and remains stationary. Within the cell the nuclear cap (and presumably the nucleus) rotates through about 270° in the opposite direction to that taken up by the flagellum. During this process the flagellum shortens but has never been found within the body of the zoospore. Shortly after, the nuclear cap and fat globules adjacent to the mitochondrion, which has remained stationary throughout, disperse. The cell becomes perfectly spherical and germinates to give a mycelium in 10 to 30 minutes. By analogy with *Albugo* it might well be rewarding to search for centriole-like structures both after flagellar retraction and in the sporangium prior to spore formation.

This brief account can only serve to indicate the great dearth of information concerning the structure of these fascinating motile cells.

Chapter 3
Apical Growth

TIP GROWTH

In 1892 REINHARDT claimed, from his own observations and those of others, that growth, i.e. the intussusception of new material, was confined to the extreme apical region. By means of simple transformation diagrams (e.g. Fig. 3.1), he showed how the addition of material at the tip enabled a hypha to extend without the addition of further material at the sides. Table 3.1 gives the ratio of new wall to old wall material for the regions marked in Fig. 3.1, after a period of extension growth. It can be seen that the ratio is largest for those sectors at the tip (11·1 : 1) and declines to 1·0 at the flanks, i.e. in these regions further back no extension growth has occurred. Additional secondary thickening of the walls in these regions is not, of course, excluded.

There have been a number of demonstrations of Reinhardt's assertion. One of the earliest was made by BURGEFF (1915) who used mordanted Indian ink to mark equidistant points along the developing sporangiophore of *Phycomyces* (Fig. 3.2). After a few hours those at the tip had elongated but those further back had not changed. This is a rather special organ since here the growth zone is just below the developing sporangiophore and elongation is associated with rotation (so-called 'spiral growth'). Consideration will be given later to this phenomenon when the sporangiophore of *Phycomyces* is considered as part of a model system for growth (see p. 56).

HENDERSON SMITH (1923) employed a readily reproducible technique with more normally growing hyphae. He measured the distance between the hyphal tip and the first septum and the distances between successive segments over a period of time. Only the former increased; the latter remained constant within the limits of experimental error. In non-segmented fungi he used branch hyphae or other structures as reference points and his observations covered all classes of fungi, e.g. *Rhizopus*, *Phytophthora* (Phycomycetes);

Pyronema, Aspergillus (Ascomycetes); *Botrytis, Fusarium* (Deuteromycetes); *Rhizoctonia* (Basidiomycetes). Subsequent observers have confirmed these observations and there is no doubt that growth is apical although none of these observations locates the precise area of intussusception.

Attempts have been made to stain the walls and to locate new material by locating unstained regions, provided that new material does not incorporate existing material. MIDDLEBROOK and PRESTON (1952a), for example, employed a modification of Noll's stain in which young sporangiophores of *Phycomyces*

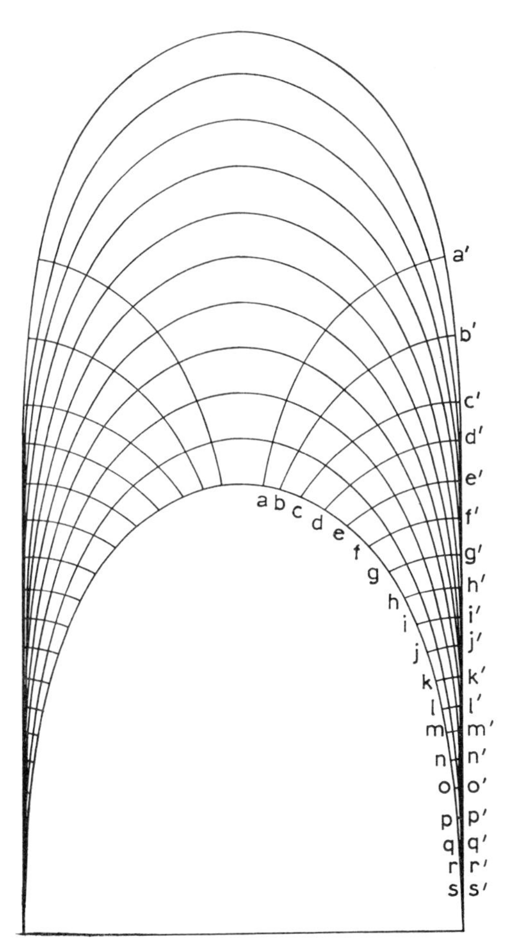

Fig. 3.1. A co-ordinate diagram to illustrate how the growth in length of a fungal hypha can be due to intussusception of material principally at the tip region. The lower curve represents the initial starting points and those above the hyphal apex at successive time intervals. The ratios of the lengths (a′–b′/a–b, b–c, etc.) are shown in Table 3.1. (After Reinhardt)

Table 3.1. Data based on Fig. 3.1 to illustrate how most of the growth may be confined to the hyphal apex of a growing fungus. The apex was divided into equal segments (a–b, b–c, etc.) each 3·9 arbitrary units long and their lengths after a period of growth, illustrated in the figure, determined (i.e. a′–b′, b′–c′, etc.). The final lengths and ratio of final to original lengths are given in the table.

Segment	Length	Ratio Final/Original	Segment	Length	Ratio Final/Original
a′–b′	43·2	11·1	j′–k′	3·9	1·0
b′–c′	11·2	2·6	k′–l′	3·9	1·0
c′–d′	7·9	2·0	l′–m′	3·9	1·0
d′–e′	5·9	1·5	m′–n′	3·9	1·0
e′–f′	5·5	1·4	n′–o′	3·9	1·0
f′–g′	5·0	1·3	o′–p′	3·9	1·0
g′–h′	4·5	1·2	p′–q′	3·9	1·0
h′–i′	4·1	1·1	q′–r′	3·9	1·0
i′–j′	4·0	1·0	r′–s′	3·9	1·0

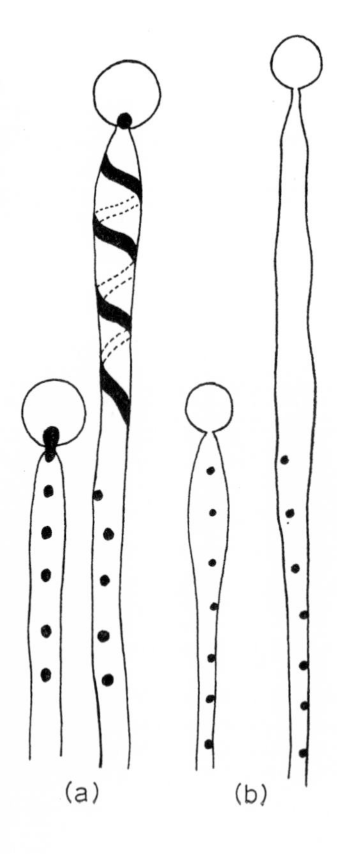

Fig. 3.2. Growth of sporangiophores of *Phycomyces*. Drops of especially mordanted Indian ink were used to put equidistant marks on developing sporangiophores. The two pairs (a) and (b) represent the appearance at the start and about 12 hr later. It will be noted that only the highest marks show any displacement and in pair (a) the uppermost mark has evidently been made on the growing zone itself (× 75, After Burgeff)

were immersed in 0·5% $FeCl_3$ and then treated with potassium ferricyanide. They stained a deep Prussian blue but the new tips were colourless. A similar but more sophisticated technique is now available which may give more precise information; this is the use of fluorescent antibodies. Both yeasts and filamentous fungi have been briefly examined (MAY, 1962; GOOS and SUMMERS, 1964). In this technique cells are exposed to a specific antiserum and then 'stained' with a fluorescent antibody which becomes attached to the antiserum bound on the wall; when viewed by ultra-violet light the antibody-stained material fluoresces brightly indicating precisely where it is bound, via the antiserum, to the wall. If a period of growth is allowed to occur after exposure to antiserum, those parts of the wall to which it is bound may undergo three kinds of development: (i) become incorporated into new material; (ii) be separated by intussusception of new regions of wall, or (iii) take no part in wall formation. When exposed to fluorescent antibody these three developmental possibilities can be distinguished. Reincorporation of old wall material into new will result in a generalized, fainter fluorescence all over the wall; intussusception will result in alternate bright and non-fluorescent areas; and completely new growth will be dark, the old wall still brightly fluorescent. This last alternative was found to be the case in *Schizosaccharomyces pombe* and *Fusarium* (Fig. 3.3). In the former, non-fluorescent caps

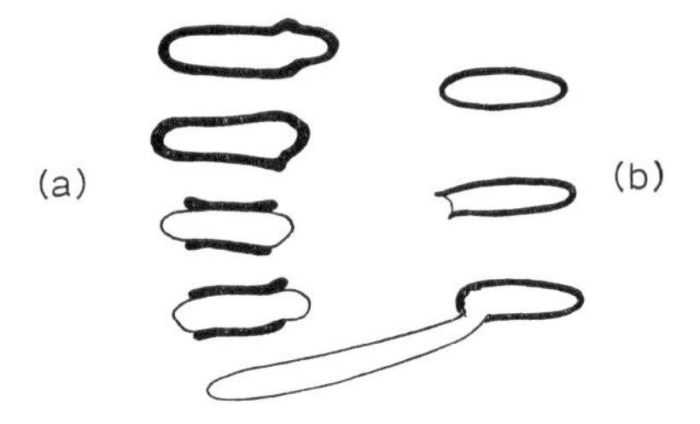

Fig. 3.3. Tracings of cells of (a) *Schizosaccharomyces* (×1100) and (b) *Fusarium* (×2500) to illustrate that new cell wall material (thin line) does not incorporate old material labelled with fluorescent antibody (heavy line) (After (a) May, (b) Goos and Summers)

developed at both ends, confirming earlier measurements which suggested that these yeasts grew at one or both ends (MITCHISON, 1957). Conidia of *Fusarium* exposed to antiserum were germinated; the conidial walls fluoresced brilliantly but the germ tubes totally lacked fluorescence.

These observations are concerned with wall growth but this reflects in some measure the growth of the cytoplasm within since this is at all times closely applied to the inner wall of the apex and is not usually vacuolated in the extreme tip (cf. p. 44). Thus the rates at which hyphae grow must reflect formation of new materials, both wall and cytoplasmic.

Table 3.2. Rates of growth recorded for various fungi.

	Growth rate mm/hr	Growth rate μm/min
Phycomycetes		
Mucor spp.	0·6–0·8	10–13
Phycomyces sporangia	3·9	65
Basidiobolus ranarum	0·26	4·4
Ascomycetes		
Gelasinospora tetrasperma	4·5	75
Humaria sp.	1·5–2·7	25–45
Neurospora crassa	1·5–6·0	25–100
Peziza sp.	1·1–2·0	18–33
Imperfecti		
Penicillium sp.	0·02–0·02	0·17–0·33
Sclerotinia sp.	0·3	5
Botrytis cinerea	0·06–0·09	1–1·5
Basidiomycetes		
Coprinus lagopus	0·10–0·25	1·7–4·2
Trametes vittata	0·28	4·7
Psathyrella disseminata	0·28	4·7
Schizophyllum commune	0·13–0·22	2·2–3·7

Rate of growth

These rates are very variable ranging from 0·1 mm/hr (1·67 μm/min) to 6 mm/hr (100 μm/min) (Table 3.2). The rate may be affected by a variety of external factors but HENDERSON SMITH (1924) drew attention to an important internal factor. This is, that the growth rate of the tip is affected in some way by the parts of the hyphae which lie behind it. Although the plot of the log of the length of individual hyphae, main or lateral, showed an asymptotic curve, the total growth of hyphae from germinating spores of *Botrytis cinerea*, when measured, gave a linear plot for the first 10 hr and

$$\text{Log } \Sigma \text{ Total lengths of Hyphae} \propto \text{Time (Fig. 3.4)}$$

He speculated on the causes of this relationship and suggested that apical growth was regulated by the rates of uptake of nutrients and of their translocation to the apices, the latter being more important.

ZALOKAR (1959b) has developed these notions more precisely with *Neurospora crassa*. Under favourable conditions hyphae of this fungus extend at rates of 100 μm/min and, in the same conditions, the protein content of the mycelium as a whole doubles every 2 hr, for the first 12 hr of growth. He supposed that a region 100 μm long extending back from the apex was responsible for growth. If so, it should double its material every minute but, the rate of protein synthesis obtained suggested that this could not occur for 120 min. Thus the apical 100 μm is not solely responsible for growth;

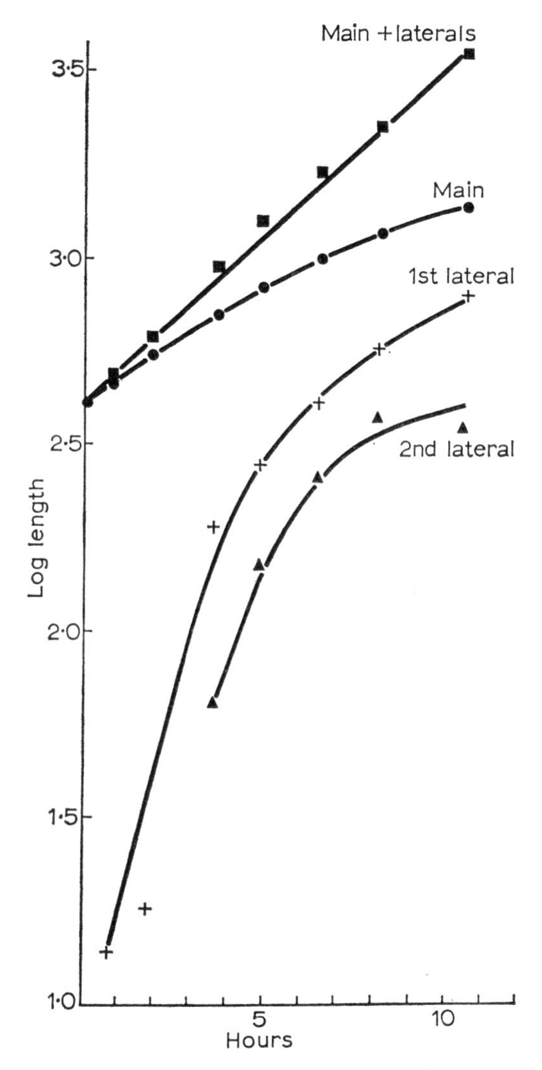

Fig. 3.4. Growth of hyphae of *Botrytis* (data of Henderson Smith)

material must be synthesized over a distance of 12 mm (100 μm × 120) and transferred to the growing region. This is a minimal distance, for laterals are formed within 12 mm of the tip and they too must be supplied with materials. This suggestion provides a rationale for Henderson Smith's earlier measurements. However, other explanations may be provided to account for the observed rate of growth.

(i) It is a common observation in rapidly growing fungi that the older parts are vacuolated so that the rate of synthesis of new cytoplasm may be quite low and the living material moves forward, as it were, within a tubular walled system of its own construction. The regulating factors for growth in such a situation would be the rates of wall formation and protoplasmic movement. A striking and extreme example of this growth pattern is available in *Basidiobolus ranarum* but no rates of synthesis have yet been determined. In young hyphae from germinating spores of this Zygomycete the apical cell grows in length at about 4·4 μm/min and a septum begins to form some way behind. The cytoplasm streams forward and enters the apical cell (300–400 μm long) before the septum is closed, leaving the spore or penultimate segment empty. Thus the cytoplasm-containing apical segment 'moves forward' leaving behind a series of empty tubular cells. When more cytoplasm has been synthesized than can enter an apical cell, it remains in the penultimate cell and develops a lateral, which then shows the same behaviour as the main hypha. A study of the processes regulating growth of this fungus should be rewarding.

(ii) A second alternative is that the synthetic activity at the apex is very much higher than the overall rate measured for the whole mycelium. This would reduce the dependence of the tip on translocation provided that uptake processes were not limiting. ZALOKAR (1959b) has examined the synthetic activity of *Neurospora* hyphae in a number of ways, viz, the distribution of cell organelles, the activity of various enzymes and other substances and their location by histochemical tests, and the rates and sites of incorporation of precursors into RNA and cytoplasmic protein. Some of his results are shown in Fig. 3.5.

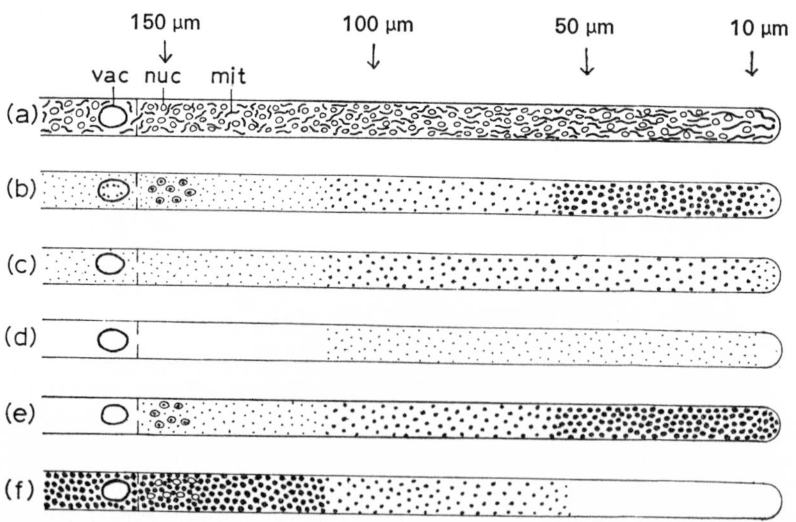

Fig. 3.5. Diagrams to show the localization of various organelles and materials (as shown by histochemical tests) in the apical 100 μm of actively growing hyphae of *Neurospora*. Stippling shows strength of reaction. (a) Organelle distribution. (b) RNA. (c) Aromatic amino acids. (d) Arginine-containing protein. (e) Protein-bound SH groups. (f) Glycogen. (After Zalokar)

The apical growing region is about 100 μm long and is separated by a sub-apical region from an older, heavily vacuolated region. In the apical region there is usually a 10 μm tip region clear of cell organelles (unless an apical granule is present, cf. p. 33). The cells are multinucleate but all the nuclei lie behind this tip region. It is, indeed, remarkable that in all fungi, the nucleus, or nuclei, of the apical segment lie a good distance behind the region of most active tip growth. It is not clear how nuclear control is exerted on these processes. The apical region differs from the sub-apical in being rich in RNA, in arginine-, tyrosine- and histidine-containing proteins and in protein-bound SH groups but it is low in glycogen. In the apical 10 μm tip, however, there was some reduction in the intensity of staining of proteins but not of SH-groups or RNA. Zalokar attempted to measure rates of synthesis of RNA and cytoplasmic protein by exposing mycelia to either ^{14}C-labelled uridine or ^{14}C-labelled proline, for 16 min. He then prepared autoradiographs and counted the grains in the emulsion, due to β-particle emission, over 100 μm lengths of individual hyphae, well separated from others, for the first 20 mm from the tip (Table 3.3).

The ratios of the mean values for protein and RNA for the apical to terminal 100 μm sections are 2·5:1 and 1:4·8 respectively. Various sources of

Table 3.3. The incorporation of ^{14}C-proline and ^{14}C-uridine as protein and RNA respectively by hyphae of *Neurospora*. The number of β-tracks was counted in successive 100 μm lengths of autoradiographs of the hyphae. In some regions counting was not possible because of the proximity of other hyphae emitting β-particles. (From Zalokar, 1959b)

Distance from tip (μm)	[^{14}C]-proline Replicates			Mean	[^{14}C]-uridine Replicates			Mean
100	32	25	27	28·0	9	7	10	8·7
200	31	24	18	24·3	10	14	11	11·7
300	25	23	17	21·7	10	12	9	10·3
400	20	14	—	17·0	11	17	14	14·0
500	20	13	—	16·3	15	19	17	17·0
600	20	12	75	35·7	25	27	20	24·0
700	17	13	22	17·3	19	—	24	21·5
800	15	13	15	14·3	17	32	20	39
900	12	14	19	15·0	25	24	—	24·5
1000	14	11	18	14·3	40	—	—	40·0
1100	15	12	12	13·0	23	—	—	23·0
1200	14	9	11	11·3	30	37	36	34·3
1300	12	8	12	10·7	33	39	—	36·0
1400	—	12	6	9·0	28	—	—	28·0
1500	15	—	11	13·0	43	40	30	37·7
1600	—	—	16	16·0	38	—	—	38·0
1700	—	—	—	—	—	—	—	—
1800	11	—	—	11·0	32	—	—	32·0
1900	14	11	—	12·5	—	—	—	—
2000	14	8	—	16·0	42	—	—	42·0

experimental error led Zalokar to suggest that the data did not indicate any notable difference in rates of incorporation of precursors into RNA or protein over this first 2 cm length. Nevertheless, there is a suggestion of a higher protein value at the apex than that farther back and of the opposite for RNA. This could mean either that more protein was synthesized at the tip or that, after synthesis, it was rapidly translocated to the tip. In 16 min there is ample time for translocation to occur since protein begins to be synthesized within the first 10 sec of exposure to exogenous amino acid (ZALOKAR, 1960b). If translocation is associated with protoplasmic streaming then rates of 500 μm/min are not unusual. Such an interpretation is speculative and would imply that the RNA is not so readily translocated. It may be concluded that Zalokar's results can be interpreted as suggesting that apical growth is sustained both by translocation and, possibly, by higher synthetic rates at the apex.

This, of course, is not necessarily the situation in all fungi. For example, a limiting factor in some fungi which show little or no translocation (SCHÜTTE, 1956) could be either the rates of uptake of materials or the process of wall formation at the apex.

Wall synthesis

Zalokar's data provides little information concerning the synthesis of walls. He suggested that the absence of glycogen in the apical 100 μm was due to its conversion into wall material. This is possible although there is no direct evidence of such synthesis. Indeed, little is known concerning the synthesis of wall components. Until the structure of glucans and mannans is better understood little progress can be expected in the study of their synthesis. In principle, the production of glucose or mannose polymers with a preponderance of 1:3, 1:6 and 1:2-β or α linkages respectively, which are claimed to be mainly involved (e.g. PHAFF, 1963), is not too difficult to envisage. The presence of abundant SH-bound protein groups in *Neurospora* right up to the very tip itself is of interest since Nickerson and his associates (e.g. NICKERSON *et al.*,

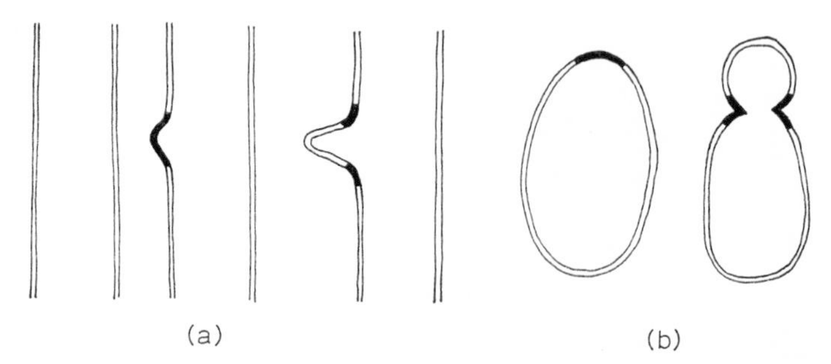

Fig. 3.6. Diagrams to illustrate location of SH-groups in walls of (a) *Eremothecium ashbyii* and (b) *Candida albicans* based on radioautographs using tritiated phenyl mercuric chloride (from Nickerson, after Robson and Stockley)

1961) have provided evidence in yeast that the weakening or strengthening of the wall is related to the activity of a protein disulphide reductase. This controls the SH ⇌ S.S- linkages in polysaccharide-protein complexes which are important wall components in both *Candida* and *Saccharomyces* (Fig. 3.6). Nothing is known of such complexes in other fungi but the presence of glucans, mannans and proteins in the walls of several fungi is suggestive (cf. Table 2.3, p. 22). The formation of chitin from glucose is also quite conceivable and the following reactions have been detected in *Neurospora* and a number of other fungi:

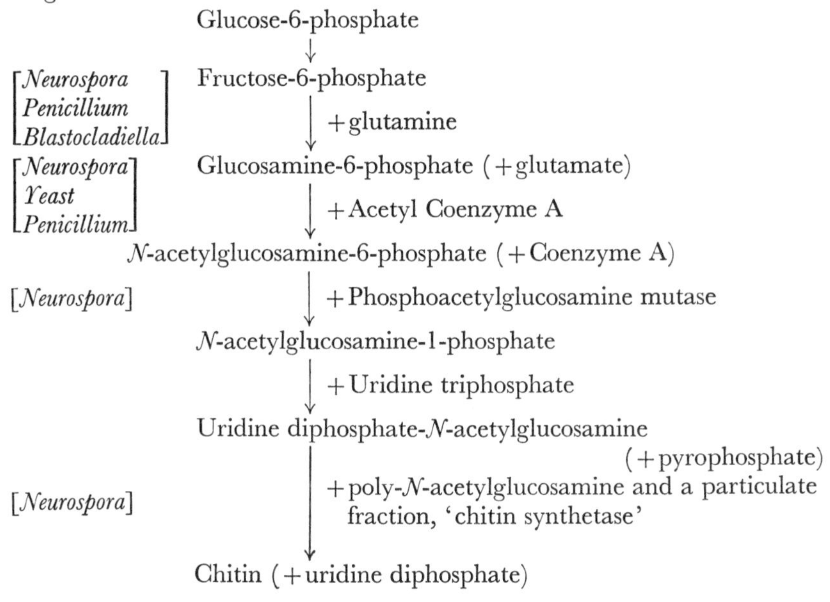

(BLUMENTHAL *et al.*, 1955; BROWN, 1955; DAVIDSON *et al.*, 1957; GLASER and BROWN, 1957a, b; LELOIR and CARDINI, 1953; LOVETT and CANTINO, 1960a; REISSIG, 1956). However, the synthesis of chitin by fungi is far from fully understood at present.

There is also evidence that chitin is implicated in the secondary thickening process of walls. Aronson and Machlis compared the chitin content of walls of *Allomyces* from cultures 50 and 120 hr old. The former contained *c.* 3%, the latter 5% chitin. Unfortunately, there is no information concerning changes in the glucan content. A difference in the balance of glucan and chitin components has been demonstrated in *Neurospora* when showing 'colonial growth'. Here the rate of forward extension is greatly reduced and, in most cases there is an increase in the number of laterals and transverse walls. This growth habit was induced either by using L-sorbose as a carbohydrate source (TATUM *et al.*, 1949), by treatment with snail-juice, or was due to induced mutation. The glucosamine/glucose ratios of cell wall hydrolysates of the walls are shown in Table 3.4 (DE TERRA and TATUM, 1961, 1963).

Table 3.4. Glucose and glucosamine released by 10 mg of cell wall material, cleaned of cytoplasm, by wild-type and mutant strains of *Neurospora crassa* and also a wild-type strain treated with L-sorbose. (From De Terra and Tatum, 1961, 1963)

Growth Type	Glucose mean (mg)	Glucosamine mean (mg)	Total (mg)	Ratio Glucosamine/Glucose	
Wild type	5·64	0·71	6·35	0·13	Mean 0·12
	5·68	0·64	6·32	0·11	
	5·82	0·69	6·51	0·12	
Semi-colonial	5·66	0·67	6·33	0·12	Mean 0·12
	5·40	0·67	6·00	0·12	
Wild type + L-sorbose	3·26	1·05	4·30	0·32	
Colonial—sorbose-like mutants	5·01	1·13	6·14	0·23	Mean 0·27
	5·19	1·24	6·43	0·24	
	4·97	1·77	6·74	0·36	
	5·12	1·23	6·35	0·24	

In wild type *Neurospora* the ratio of glucosamine/glucose is about 0·12 and this is true of the semi-colonial mutants. These mutants have a rate of extension growth intermediate between wild-type mycelia and the colonial mutants but their hyphae resemble those of the latter in being more highly branched than wild-type hyphae. By contrast, the ratio is more than doubled in the colonial mutants and in L-sorbose treated wild-type colonies whose morphology the mutants resemble closely. Since the total content of glucosamine+glucose is more or less the same in the mutants as in the wild-type strains the altered ratios must reflect a change in the balance of the components, presumably their polymeric forms, chitin and probably glucan. Although the overall carbohydrate content is reduced in sorbose treated wild-type mycelia the same trend is shown. SHATKIN (1959) pointed out that hyphae of such mycelia were more brittle than normal hyphae. This and the reduced rate of extension growth suggests that the chitin is the more rigidifying, less extensible wall component and glucan the more plastic, extensible component. No obvious correlation with more dense branching is apparent for this occurs in both semi-colonial and colonial mutants. Comparisons between colonial mutants and sorbose-treated wild-type mycelia are not perhaps wholly valid. In the former, regulation of the balance of carbohydrate moieties of the cell wall is presumably internal. L-sorbose, however, acts in a highly localized manner on walls with which it is in contact. Neither aerial hyphae nor hyphae into which sorbose has been injected develop the abnormal pattern of growth (DE TERRA and TATUM, 1961). Nevertheless, it is clear that hyphal morphology does show some correlation with the chemical composition of the wall.

THE HYPHAL TIP

Both CASTLE (1942) and MIDDLEBROOK and PRESTON (1952a) have shown that the deposition of a particle, as light as a *Lycopodium* spore, on the extreme tip of a young developing sporangiophore-tip of *Phycomyces* is followed by the immediate cessation of growth and a slight swelling. Growth is usually resumed afterwards by a narrower terminal extension or one or more laterals also of smaller diameter. Minute drops of water can induce similar effects provided they are deposited on the growing region (THIMANN and GRUEN, 1960). Similar observations have been made by ROBERTSON (1958, 1959, 1965) on a variety of fungi as a result of flooding the apices, on plate colonies, with water or osmotically active solutions. Once again growth ceases immediately and after about 40 sec a variety of changes may occur: swelling and no further growth; or growth of a terminal hypha of smaller diameter; or the protrusion of one or more laterals from the sub-apical region (Fig. 3.7). The observations

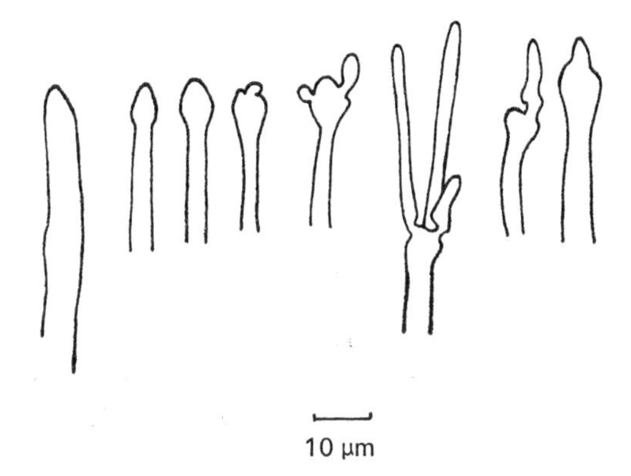

Fig. 3.7. Changes induced in the apices of hyphae of *Fusarium* as a result of flooding them with distilled water or osmotically active agents (From Robertson)

of GIRBARDT (1957) on *Polystictus versicolor* apices exposed to an intense light 'shock' will also be recalled (see p. 33). The apical granule appears to disintegrate, growth is checked and laterals often develop, although some way back from the apices.

There is thus clear evidence of the highly sensitive nature of the extreme hyphal tip. Robertson has speculated on the significance of the changes induced by exposure of apices to water and osmotica. He has pointed out that in normal development very rapid, comparable changes can take place at hyphal tips when they branch or swell to produce sporangiophores or conidia. He has suggested that both induced and normal changes arise as a result of a change in the balance of two processes, 'extension-growth' and 'wall-setting', at the hyphal apex. The former process is responsible for increasing the area of terminal, plastic wall material, the latter process brings about rigidification and stabilization of wall material. He supposes that the two processes are, at least to some extent, independent. Thus induced cessation of extension growth need not prevent wall-setting, which would progress nearer to the apical region and thus leave a smaller, terminal area for renewed growth. Conversely, if the rate of extension-growth increases, the tip could increase in size, being more plastic, and so a swollen tip could arise. He has briefly reported experiments with *Neurospora* whose results can most readily be understood if it is supposed that its wall is two-layered. Evidence bearing on this view has already been adduced (see p. 23). Robertson and Rizvi exposed hyphal tips of *Neurospora* to snail stomach enzyme. If turgid, the tips disintegrated in a manner reminiscent of that induced by a large osmotic shock and there was a suggestion that the disintegration commenced in the immediately sub-apical region. However, if tips were treated with enzymes and then returned to a suitable osmoticum, a spherical protrusion developed on the 'shoulder' of the hyphal apex. The region of origin of this protrusion is identical with that from which laterals develop after flooding hyphal tips with water or from where disintegration probably commences when tips burst. This sub-apical region is, therefore, a region of weakness. It could be the boundary between the plastic and rigidified region of the apex, although this is not claimed by Robertson. Quite different results are obtained if *Neurospora* is grown in an enzyme-containing medium from the outset. The wall is then 'of no measurable thickness' but growth and the behaviour of the cytoplasm, e.g. streaming, can be quite normal. The hyphal tips no longer disintegrate if exposed to osmotic shock. Robertson accounts for these results by supposing that in normal walls the osmotic strain is imposed almost wholly on the tip, which is the only plastic region, the rest becoming rigidified through 'wall-setting'. This latter process is inhibited in the enzyme-grown colonies and so the osmotic shock is dissipated over a wider plastic area and the hyphae do not burst. Robertson infers, therefore, that there is an inner extensible wall region and an outer more rigid region. However, this particular orientation does not seem to be a necessary assumption to account for his results.

It is now desirable to consider two problems raised by Robertson's specula-

Plate III. Wall and cytoplasmic structures in fungi (continued from Plate II)

Figs. 1 and 2. Dolipore septum of *Coprinus lagopus*. In Fig. 1 the septum is intact and the pore cap of endoplasmic reticulum can be readily seen. In Fig. 2 a nucleus is seen migrating through an apparently simple pore which is believed to have arisen from a dolipore by breakdown and disorganization. (×21 000)

Fig. 3. Inner surface of nuclear membrane of *Saccharomyces cerevisiae* after freeze-etching treatment. Note characteristic circular pores in membrane some of which seem to be open, others closed. (×28,000)

Fig. 4. Outer surface of nucleus of *S. cerevisiae* after freeze-etching treatment. Note ribosomes adhering to regions surrounding nuclear pores. It is not possible to tell whether these are polyribosomic in form. (×45,000)

[Figs. 1 and 2 by P. R. Day, from Giesey and Day, 1965 ; Figs. 3 and 4 by Mühlethaler from Moor and Mühlethaler, 1963.]

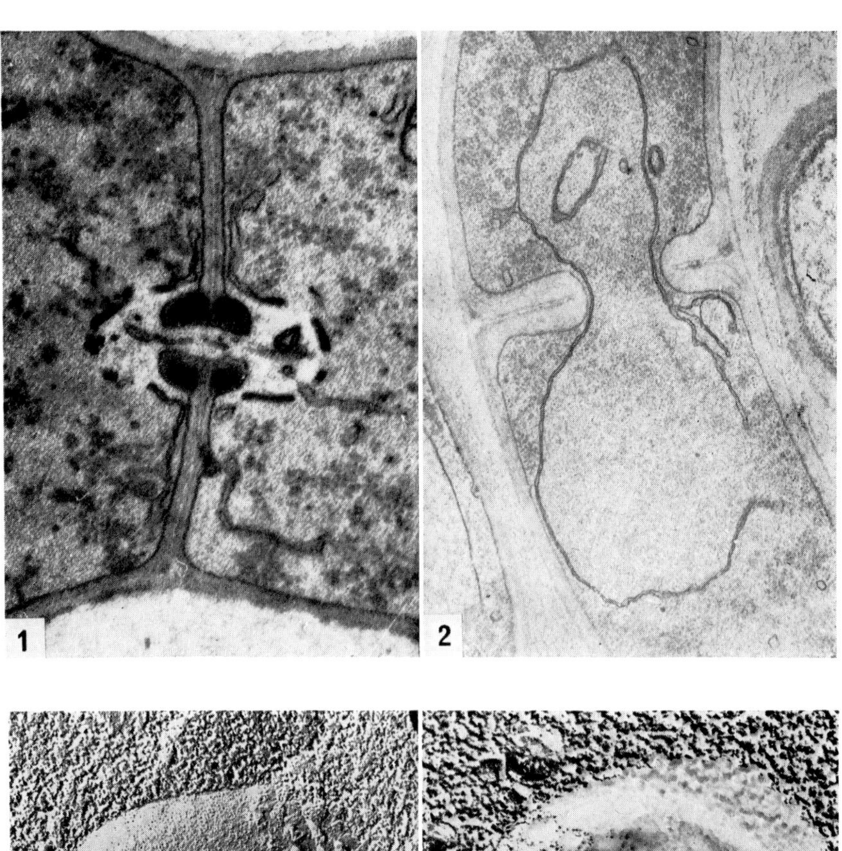
1
2

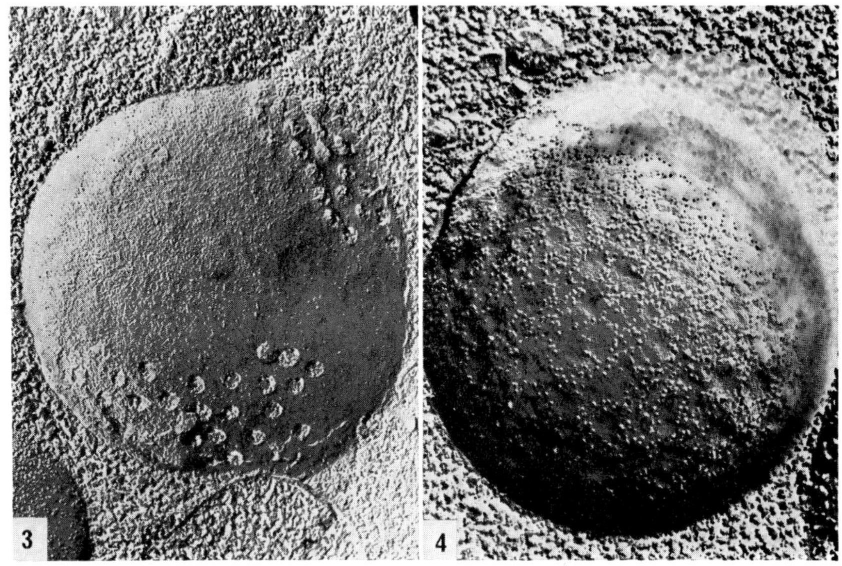
3
4

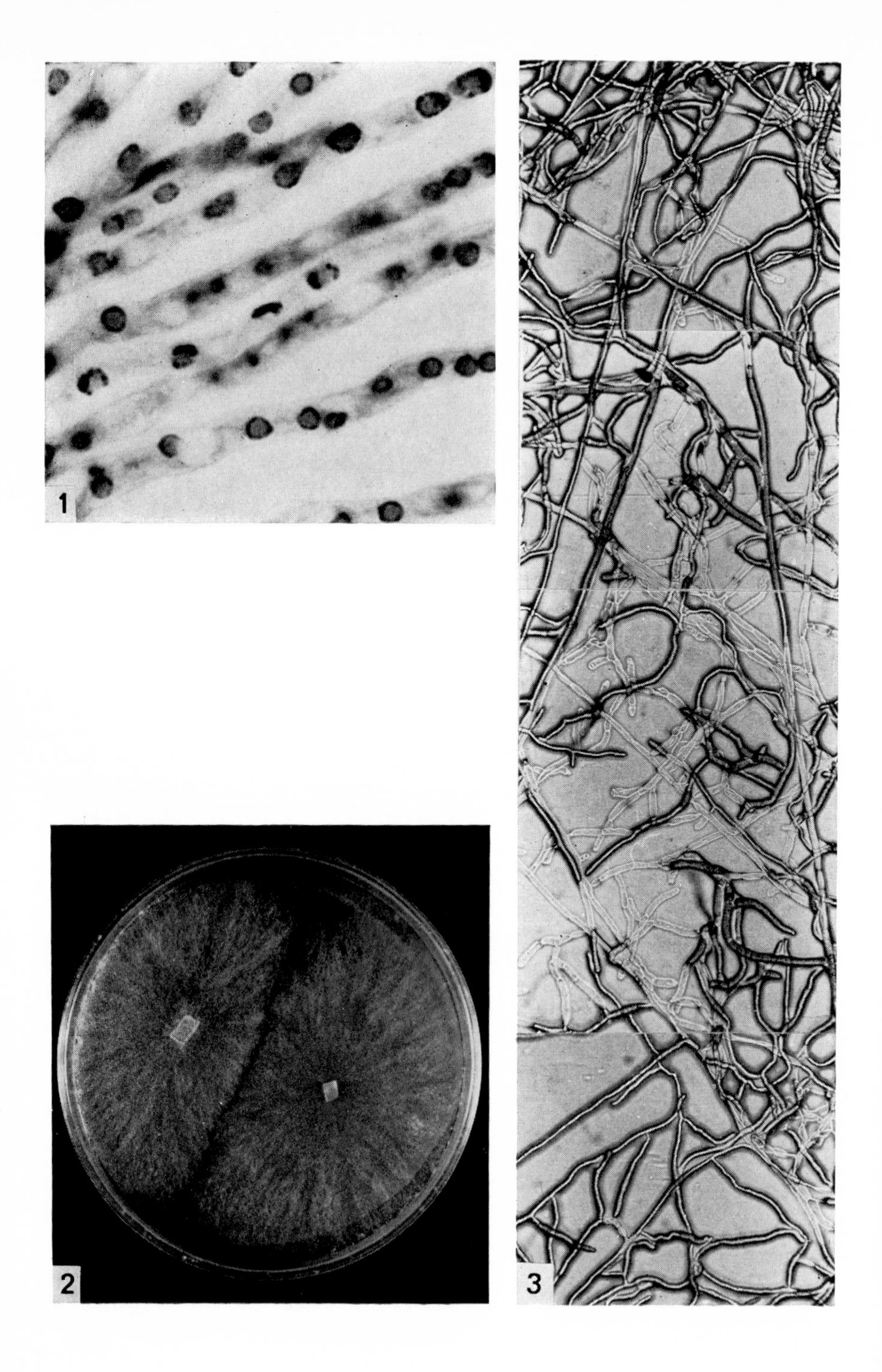
1
2
3

tions: the first is the evidence for zones in the apical region which differ in their degrees of extensibility; the second is the evidence in connexion with the physical organization of the walls of the hyphal apices.

WALL ORGANIZATION AND BEHAVIOUR IN APICAL AND SUB-APICAL GROWTH ZONES

The only direct evidence concerning the extensibility of growing regions comes from the sporangiophore of *Phycomyces*. It is necessary to describe the growth of this structure and four stages have been designated (CASTLE, 1942 and Fig. 3.8). In Stage 1 the sporangiophore grows upwards with an S-spiral* rotation. Both processes cease in Stage 2 and the tip inflates to form a sporangium which begins to mature; its contents are still yellow but later darken. Stage 3 is a period of apparent inactivity. Just before the sporangium darkens the sporangiophore recommences growth in a 2 mm zone immediately below the point of junction of the hypha and sporangium. At first, rotation is in an Z-helix but the rates of growth and rotation decline gradually after about $2\frac{1}{2}$ hr and then halt momentarily. This is the end of Stage 4a. Rotation, this time in an S-helix, and elongation then recommences in the same 2 mm growth zone; the rates pick up and elongation at 25°C may be maintained constant at 3 mm/hr (50 μm/min) for many hours. It is for this reason that many investigations have been made of growth phenomena in Stage 4b.

CASTLE (1937) showed that the 2 mm growth zone was not uniform in intensity of growth in length; it reached a maximum at about 0·4 mm below the sporangial swelling and declined toward zero at the lower end of the zone (Fig. 3.9). THIMANN and GRUEN (1960) applied minute drops of water, *c*. 0·015 mm^3, to specific regions of the growth zone of isolated sporangiophores. [Growth continues for as much as 100 hr provided isolated sporangiophores are undamaged and are stood in water or culture solutions (GRUEN,

* The terms 'left-' and 'right-handed' spiral can cause confusion and are best replaced by Z- and S-helix (or spiral). When a Z-helix is viewed ascending vertically the part nearest the observer is parallel to the oblique stroke of the Z: the mirror image of such an helix is an S-helix (ROELOFSEN, 1959).

Plate IV. Hyphal features

Fig. 1. The multinucleate condition of hyphae of a species of *Penicillium*. Some of the nuclei are showing stages of nuclear division but this is usually synchronous. (×1760)

Fig. 2. Interaction between single spore cultures of *Rhizoctonia solani* (*Thanatephorus cucumeris*) showing a clear zone formed between the anastomosing hyphae leading to their death in 24–72 hr. The cultures are grown on cellophane over-lying soil extract agar.

Fig. 3. The zone of aversion showing the killing reaction which involves the fusion cells of up to 6 or 8 cells on either side. The photograph was taken by incident light in which the darker contents of the living cells show up clearly. (×c. 180)

[Unpublished photographs supplied by C. F. Robinow (1) and N. T. Flentje (2).]

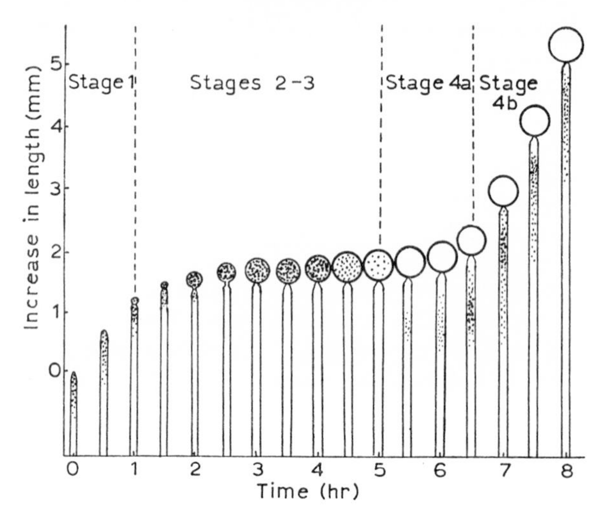

Fig. 3.8. Diagram of stages in the growth of the sporangiophore of *Phycomyces*. The darkest regions are the zones of most active growth; the stages are described in the text. (After Castle)

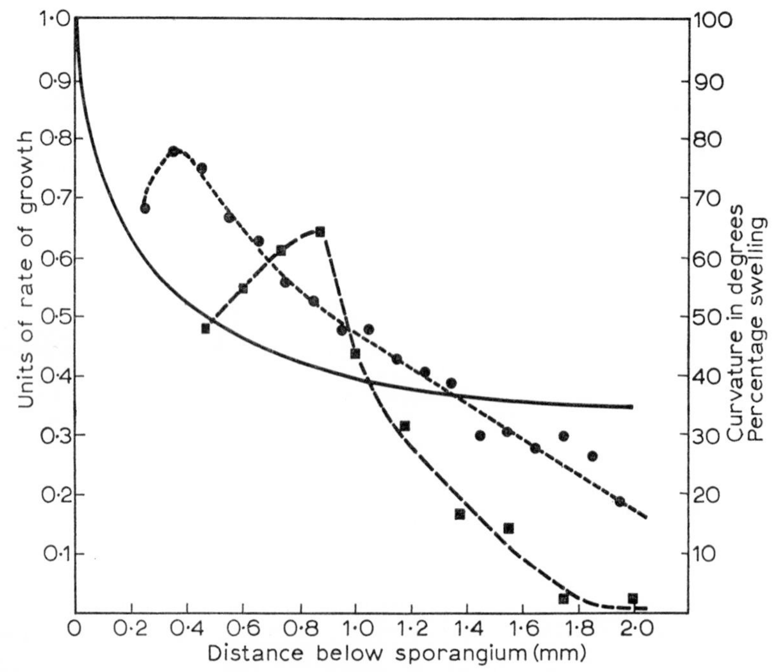

Fig. 3.9. Diagram to illustrate various features of the sub-terminal growth zone of the sporangiophore of *Phycomyces*. Key: --●-- intensity of growth; —■— degree of extensibility; ——— degree of extensibility under pressure. (From data of Castle, Thimann and Gruen, and Roelofsen)

1959a).] The water drops were carried up by the growth of the region to which they adhered but they also induced a curvature of the sporangiophore. Thus both the intensity of growth and the elasticity of the wall in the area covered by the drop of water could be found and these quantities recorded for all regions of the growth zone. They confirmed Castle's measurements of the intensity of elongation growth but found that the region of maximum extensibility, as measured by maximum curvature, was between 0·6–0·8 mm from the top of the zone (Fig. 3.9). A different kind of measurement of extensibility, mechanical extensibility, had been made earlier by ROELOFSEN (1950), who subjected detached sporangiophores to different internal pressures by enclosing them in an 'iron-lung' device. Maximum extension, under pressure, was at the junction of the growth zone and sporangial swelling and reduced asymptotically to 35% at the base of the growth zone (Fig. 3.9).

These data, considered in conjunction with those adduced by Robertson, enable speculations to be made concerning the situation at an hyphal apex. The extreme tip is zone α, a cap-like non-extensible region followed by one of maximum mechanical weakness; behind this lies zone β, that of maximum intussusception and synthesis of wall materials. Zone γ, further back, is highly labile and here extension growth may occur while behind this is zone δ where rigidification sets in with increasing tempo (Fig. 3.10). Lateral growth would originate in zone γ and disruption would occur at either of the

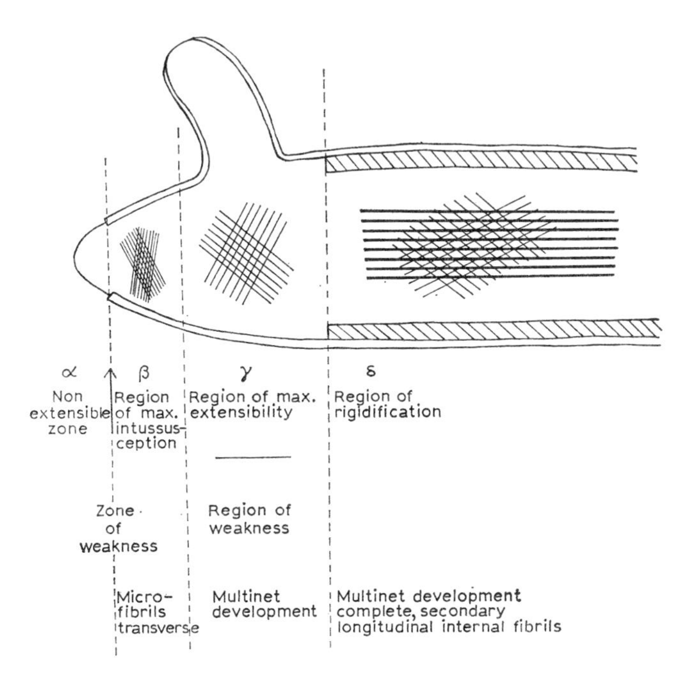

Fig. 3.10. Diagram to illustrate the organization of the hyphal apex proposed in the text

junctions of zones α and β or γ and δ. These speculations, needless to say, require to be tested by experiment.

Most of the detailed information available concerning the physical organization of hyphal apices comes from *Phycomyces*. This is fortunate, in that it enables a basis to be sought for some of the previous speculations, but unfortunate in that this is clearly a very specialized, tough, xerophytic kind of hyphal structure. The information available concerns the dispositions of microfibrillar elements left after fairly drastic chemical 'cleaning' of pieces of wall material. Although not the earliest data obtained, the appearance of shadowed preparations of hyphal tips of certain Phycomycetes in electronmicrographs, provides the clearest information on the orientation of microfibrils. They are, presumably, of chitin since the 'cleaning' treatments are likely to have removed soluble glucans and the diameter of the microfibrils is of the order of 200–300 10^{-10}m. In some cases a crossed network appears to predominate at the tip, e.g. *Gonapodya* or *Phycomyces* (ARONSON and PRESTON, 1960b; HOUWINK in ROELOFSEN, 1959) (Plate I, Fig. 1). This seems to be the case in *Allomyces* also but ARONSON and PRESTON (1960a) believe that the microfibrils at the tip show a tendency to axial orientation, i.e. parallel to the long axis of the hypha. ROELOFSEN (1959) interprets the crossed arrangement in *Phycomyces* as due to an outer region of longitudinally orientated microfibrils and an inner region orientated transversely. This is also clearly seen from preparations of the inner and outer faces of the wall of the growth zone of a Stage 4b sporangiophore. Support for this observation comes from the interpretation of x-ray diffraction patterns produced by a stack of carefully dried and aligned sporangiophores laid on top of each other (MIDDLEBROOK and PRESTON, 1952a). From the base of the growth zone of *Phycomyces* and behind, a well defined inner layer of microfibrils develops orientated in a predominantly longitudinal direction. This is presumed to represent the secondary thickening of the wall. A similar appearance is shown by *Allomyces* where an outer layer of microfibrils arranged more or less randomly overlies an inner layer of predominantly longitudinally running ones (Plate II, Fig. 1).

In non-hyphal forms, e.g. *Candida tropicalis*, microfibrils occur arranged at 90° to one another save in the region of bud formation, where they are arranged in a circular fashion around the potential point of abstriction of the bud, or bud-scar (HOUWINK and KREGER, 1953). In this fungus the microfibrils must presumably be insoluble glucans since the chitin content is so low (cf. Table 2.3, p. 22).

Observations such as these can be related to laminar appearances described earlier for some fungi (see p. 18) and can be related to the situation described in *Rhizoctonia*. Here the tip appears to be one layer thick but a series of lamellae develop on the inner side further back from the apex. The tip layer is presumably the primary wall, the lamellae representing the secondary wall.

The way in which microfibrils are formed and orientated is not understood, although speculations have been made. The situation in life is very different from that seen in electronmicrographs. CASTLE (1953) expressed this vividly when he wrote, 'The primary, growing wall of the plant cell is a structured but heterogeneous system, drenched with protoplasm, in which microfibrils are being formed and interconnected by processes ultimately geared to cellular metabolism'. Attention has already been drawn to the occurrence of special regions of the cytoplasmic membrane in yeast, thought to be involved in microfibril production (see p. 20 and Plate I, Fig. 3). The actual form of the microfibril may, of course, reflect the physical configurations imposed upon the sugar, or amino-sugar polymers, by the properties of their constituent monomers. Polysaccharide-protein complexes may well reflect the shape of the protein moiety. Little is known concerning these matters but principles such as these are well established for nucleic acids and proteins and need to be applied to polysaccharides as found in living cells. The fact that the conversion of *N*-polyacetylglucosamine into chitin can only be affected if some insoluble chitin is already present, even in a cell free situation, suggests the operation of some kind of template mechanism.

The fact that in a number of cylindrical cells, including hyphae, the innermost microfibrillar layer is orientated predominantly in a direction at right angles to the long axis has led to a purely mechanical explanation (CASTLE, 1937). This orientation is claimed to be an adaptive reaction, enabling the wall to resist the internal turgor pressure of the hypha. The transverse wall tension (T_t) is at least twice the longitudinal (T_l). Assume that the hypha is a cylinder, of radius r and length l, whose walls of thickness d, are exposed to a uniform internal pressure, P, due to its fluid contents. Then:

$$T_t = \frac{P\pi r^2}{2\pi rd} = \frac{Pr}{2d}$$

and

$$T_l = \frac{P2rl}{2ld} = \frac{Pr}{d}$$

hence

$$T_t/T_l = \frac{Pr}{2d}\Big/\frac{Pr}{d} = \frac{2}{1}$$

That the transverse tension is, indeed, greater than the longitudinal tension was shown by Roelofsen, who increased the internal pressures in detached *Phycomyces* sporangiophores until they burst by means of a longitudinal split, i.e. because of the increased transverse tension. Objections have been raised to this simple mechanical explanation by FREY-WYSSLING (1953) and PRESTON (1952). They point out that tubular cells are known in which transverse and axially orientated microfibrillar systems alternate. Most components of secondary walls do show some orientation other than transverse although the longitudinal and transverse mechanical tensions still have the same relationship as in primary walls and, finally, that there is little experimental evidence for an association of orientation of microfibrillar components and stress. These are formidable criticisms yet it still remains a remarkable, if unexplained, fact that the inner primary layer of tubular cells is predominantly orientated in a transverse manner.

A second, unresolved problem is how the outer microfibrillar strands come to be orientated differently from those in the inner layers. ROELOFSEN (1959) has proposed a multi-net theory to account for this. In essence, this suggests that the alteration in orientation arises from a stretching of each microfibrillar layer in an axial direction due to cell elongation, as it is shifted outwards by the formation of new, transversely orientated microfibrils on the inner, cytoplasmic side of the wall. Thus there is a gradual transition in the orientation of microfibrils from an inner transverse arrangement through an intermediate array of crossed microfibrils to a final outer longitudinal alignment. It is not possible to discuss this hypothesis here in detail but reference may be made to the books by Frey-Wyssling, Preston and Roelofsen mentioned earlier. It may be pointed out that the kind of transition of orientation required by the hypothesis is not at variance with the observed orientations of microfibrils in Phycomycete cell walls. In particular, ARONSON and PRESTON (1960b) have obtained an electronmicrograph of the development of a lateral branch of a rhizoidal hypha of *Allomyces* in which the sequence of microfibrillar orientations is in agreement with the multi-net hypothesis.

How far these observations on microfibrillar orientation can be related to the processes, discussed earlier, of intussusception and expansion and the proposed zones α–δ, is problematical. Evidently the origin of zone δ is to be related to the development of the secondary, internal, longitudinally-running fibrils. The intussusception zone, β, may be related to the origin of the predominantly transversely orientated microfibrils and the labile zone, γ, to a region where multi-net orientation is developing. A very great deal of observational work is necessary before it is possible to determine whether these suggestions can be substantiated. It is clear, too, that it is most desirable that electron microscopic investigations should be extended to the walls of other kinds of fungi than the few Phycomycetes investigated to date.

Spiral growth

The sporangiophore of *Phycomyces* has tended to become a model for studies of hyphal growth but it is not known whether it is an atypical structure or whether it merely exhibits, in an exaggerated form, processes typical of all fungal apices. In respect of its extraordinary spiral growth (Fig. 3.2), this is a matter of considerable uncertainty. Spiral growth has been detected in other Zygomycetes (GREHN, 1932) and there have been suggestions that it occurs elsewhere in tubular cells with tip growth, e.g. cotton hairs (ASTBURY and PRESTON, 1940; CASTLE, 1942). It is believed that spiral growth reflects the submicroscopic fibrillar anatomy but the way in which this is expressed is still the subject of controversy. Excellent discussions occur in the books of PRESTON (1952), ROELOFSEN (1959) and a review by CASTLE (1953).

CASTLE (1937) showed that the relationship between the upward movement of a marker (a flour particle) and its rotation was linear. He also pointed out that the diameter of the cell increased throughout the growth zone from apex to base and that both linear and lateral growth ceased at the lower end of the zone. He wrote 'This relation suggests that in spiral growth of the present type structural units are indeed being built into the wall at an oblique angle to the long axis of the cell, and that addition of each oblique unit furnishes a component of growth both along the cell axis and around it'. Thus all hypotheses start from the proposition that the initial microfibrillar elements are not strictly transverse but are inclined at an oblique angle to the transverse axis of the sporangiophore. This angle may increase in the outer region of the wall so that light particles adhering to the surface may be carried up at an angle varying from a few degrees to as much as 50° from the transverse axis of the sporangiophore (CASTLE, 1953). The angle can be altered by change of temperature or by constricting the sporangiophore and so changing the internal pressure.

Preston and his colleagues compare the behaviour of the wall with that of a coiled spring, fixed at one end and extended by a weight attached at the other. As stretching occurs the free end rotates; the one is contingent upon the other. In the living cell, the weight is analogous to the internal pressure on the hyphal tip and the spring to the spirally arranged microfibrillar elements. The fixed end is represented by zone δ, the rigidified cell wall region. In the case of a spring, the rotation of the end $\Delta\phi$ for an elongation Δl is given by:

$$\frac{\Delta\phi}{\Delta l} = \frac{\cos\alpha \sin\alpha\,(1-2n/q)}{r\,[\cos^2\alpha+(2n/q)\sin^2\alpha]}$$

where r is the radius of the spring; α, the angle which a gyre of the spring makes to the transverse axis; q, the coefficient of elasticity of the spring and n the torsional rigidity. Preston believes that his theory is supported by a number of facts. The predicted value n/q lies well within the range for chitin in the sporangiophore for appropriate rates of elongation. The torque which must be applied to just stop rotation should vary as r^3 and a quantitative agreement has been found with Castle's actual measurements (CASTLE, 1938; MIDDLEBROOK and PRESTON, 1952b). The relation of $\Delta\phi/\Delta l$ to r is also found experimentally to be as predicted from the equation, i.e. $\Delta\phi/\Delta l \propto 1/r$. The reversal of spiralling at Stage 4 can be accounted for by a change in the coefficient of elasticity of the chitin microfibrils and this could be affected by a change in their orientation consequent upon cessation of elongation. Thus $(1-2n/q)$ may be negative but, as growth picks up, q should again increase, the value should become positive once more and the direction of the spiral will change again. No measurements have been made of changes in n or q but, if the theory is correct, cessation of growth should be followed by reversal of the direction of rotation gradually diminishing after a period and, finally, the restoration of the original direction of spiralling. This, in fact, has been achieved by checking growth temporarily in Stage 4b by immersing the sporangiophores in weak detergent. The results are shown in Fig. 3.11 (MIDDLEBROOK and PRESTON, 1952a).

This hypothesis is so coherent and lucid that it is distressing to find that it has been challenged on a fundamental observation, the direction of the microfibrillar spiral. Preston's hypothesis requires the microfibrils to be orientated in an S-helix since in model experiments with springs only this orientation gave a rotation in the same sense as has been noted predominantly in growing

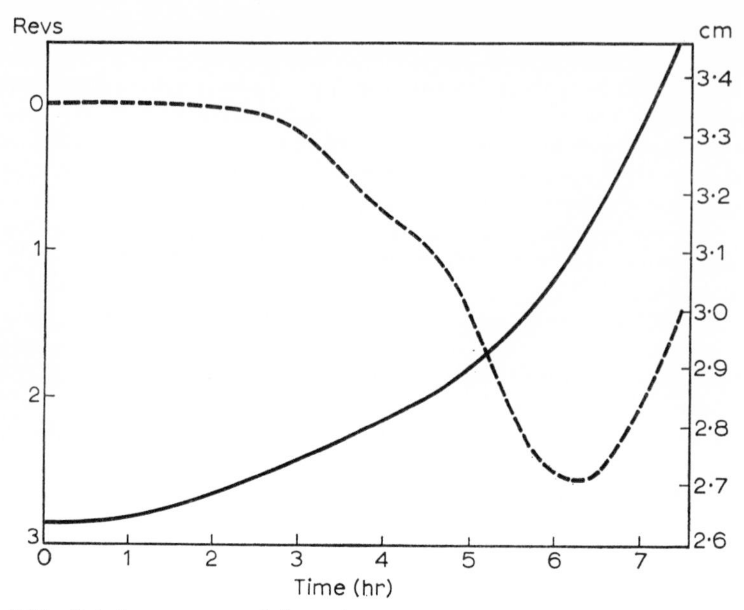

Fig. 3.11. Rotation ----- and elongation —— of a sporangiophore of *Phycomyces* after a temporary cessation of growth. Note that the rotation changes sign at the minimum value of its curve. (After Middlebrook and Preston)

Phycomyces by all observers. Roelofsen has provided optical evidence and microdissection studies which suggest that the microfibrils are predominantly in an Z-helix. He repeated Preston's observations which purported to demonstrate an S-helix arrangement in the growth zone but was unable to detect this arrangement. He suggests that Preston and his colleagues have been misled by measurements near the limit of resolution of their methods. It is only fair to say that Roelofsen's own explanation is not clear and does not include quantitative relationships such as those determined by Preston. He suggests that because the region of maximum growth does not coincide with the region of maximum wall extension, considerations other than the physical properties of the wall determine the orientation of growth. It is not clear why this should be so and a weakness of Roelofsen's hypothesis is that it provides no obvious discriminatory test. It will be apparent that neither theory provides any information as to why the microfibrils should initially be in a spiral arrangement of any kind, or of why the rotation of the sporangiophore should be predominantly an S-helix in any case.

One further remarkable group of facts remains in connexion with spiral growth. In the sporangiophore of *Phycomyces* strands of streaming protoplasm can usually be seen clearly. In the lower parts, they run axially but in the growth zone region they assume a spiral direction, identical with that shown by the rotation of the wall. When a sporangium develops (or in the mutant var. *piloboloides*, a sub-sporangial swelling), the pitch of the protoplasmic strands becomes flatter, coinciding with the reduced rate of elongation (BURGEFF, 1915; KIRCHHEIMER, 1933). The causes of these associative effects are not clear but it does suggest that ideas concerning wall growth cannot be separated from consideration of effects in the adjacent cytoplasm, a point already made elsewhere in this chapter.

Finally, it might be suggested that progress might be made if a serious examination were made of other fungi to establish how widespread and important an element is spiral growth in normal development. An alternative, and possibly more promising approach, would be to investigate the features of induced spiral growth. In the griseofulvin a useful tool is already to hand. Originally described as 'curling factor', griseofulvin will induce hyphae growing from the germinating spores of *Botrytis* to develop in a flat spiral, at concentrations as low as 0·1 μg/ml. At higher concentrations, 1–10 μg/ml. it causes un-coordinated, irregular swellings and inhibition of growth (BRIAN *et al.*, 1946, 1960). The fact that it is without effect upon Oomycetes, which, it will be recalled, possess little chitin, some cellulose and much unidentified polysaccharide material in their walls (Table 2.3, p. 22), adds to its interest (BRIAN, 1949). That it will induce curvature in the sporangiophores of *Phycomyces* indicates that it could also be used as a tool for studying wall expansion (BANBURY, 1952). It is perhaps appropriate to mention at this point that the auxins, notably indolyl-acetic acid so active in growth regulation in the rest of the plant kingdom, are apparently without such effects on the fungi (GRUEN, 1959b).

This chapter has not established many facts and it will be apparent that notions concerning apical growth are limited by the very few studies made and the limited number of organisms investigated. No apology is made for the more speculative passages. In an area of such importance and ignorance any course is worthwhile if it leads to its investigation and deeper understanding.

Chapter 4
The Fungal Colony—Vegetative Development

SHAPE AND SIZE

When a fungal propagule, such as a spore, is allowed to grow in an unrestricted manner it usually develops one or more apically extending hyphae from which lateral hyphae branch in various planes. Eventually, through some form of internal regulation, there develops a spherical colony whose outermost region consists of young, actively growing, roughly parallel hyphae. Within the colony the branching masses of main hyphae and their laterals, of various orders, may become transformed into a closed, three-dimensional network as a consequence of hyphal fusions but this does not happen in all fungi, e.g. Phycomycetes. The colony will continue to grow at a constant rate, provided that no nutritional or other restriction limits growth, and the system then appears to have reached a steady state (cf. EMERSON, 1950). The basic unit responsible for this orderly development of a colony is evidently a branching hyphal system, the tips of the hyphae showing apical growth as discussed in the previous chapter.

In practice the type of colony described is rarely seen although an approach to it can often be achieved in liquid shake-culture. Even here the morphology of the colony can vary between different species of the same genus, or within the same species grown in different media (BURKHOLDER and SINNOTT, 1945), or even in different depths of the same medium (YANAGITA and KOGANÉ, 1963a); some examples are shown in Fig. 4.1. The causes of these departures from the ideal are numerous but, in general, they arise from the imposition of some form of constraint on growth.

In plate culture a spatial constraint is imposed on colony growth so that the spherical form is replaced by the predominantly two-dimensional circular colony. This rapidly achieves a steady state, analogous to the globose colony, in which it extends radially at a constant rate. This may slow down as growth proceeds and there are two principal causes of this; growth is retarded either through the exhaustion of some essential metabolite, e.g. carbohydrate, or as a result of the accumulation of toxic metabolites, so-called 'staling products', such as organic acids or ammonia which may act temporarily or permanently. Fungi are sometimes described, therefore, as 'non-staling' or 'staling' types.

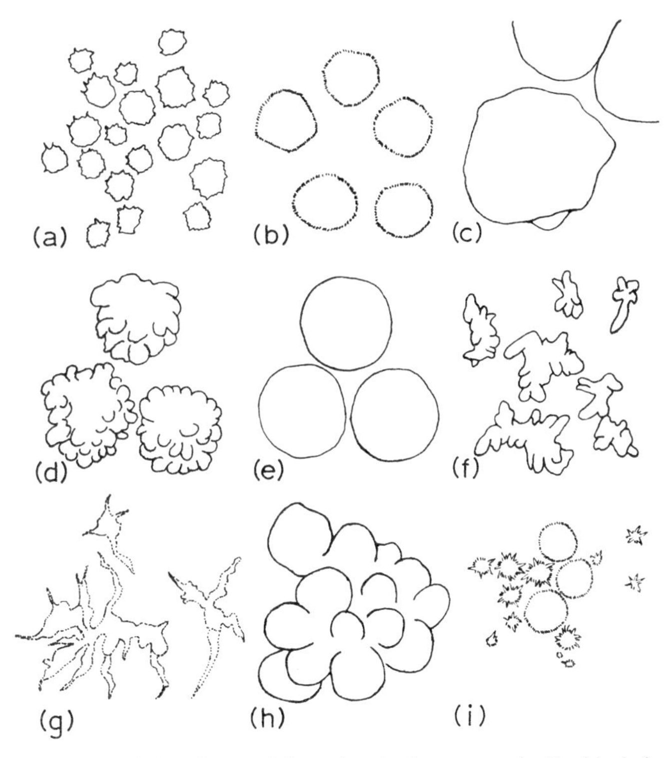

Fig. 4.1. Outlines of the shape of fungal colonies grown in liquid shake-culture. (a) *Rhizopus nigricans.* (b) *Phycomyces blakesleeanus.* (c) *Basidiobolus ranarum.* (d)–(f) *Penicillium notatum,* (d) in Czapek corn-steep medium, (e) in synthetic medium, (f) as (e) but shaking delayed. (g) *Neurospora crassa.* (h) *Alternaria solani.* (i) *Chaetomium aureum.* (After Burkholder and Sinnott)

In circular colonies on solid substrates a greater degree of hyphal differentiation is achieved than is usual in liquid shake-cultures where the differences between regions are almost entirely due to age. In plate cultures, aerial and submerged hyphae arise from the radially extending surface hyphae and they differ in their characteristic properties. Aerial hyphae may also become specialized to form simple or complex spore-bearing structures and sclerotia while submerged hyphae may develop as rhizoids, strands or rhizomorphs; these structures rarely develop in liquid shake-cultures.

Very little is known concerning the shape and size of colonies in natural habitats. This reflects, in part, the difficulty of identifying an individual mycelium under such circumstances but it also indicates lack of investigation (LARGE, 1961). In general, it may be inferred that the shape will approach the globose or circular except where mechanical, nutritional or biological features have imposed a localized constraint on apical extension growth. A superficially simple demonstration of this is the existence of 'fairy rings' whose pattern of circular growth can be distorted by topographical features, coalescence with other 'rings' or other causes. Radial extension growth can be inferred for many fungi by a study of the distribution of their reproductive structures, which are often arranged in a ring. There can be other causes of this arrangement, however, and even in such clear cut 'fairy rings' as those of *Marasmius oreades* there can be problems. In plate culture this toadstool forms a circular colony but in nature the colony is an annulus of submerged mycelium whose outer periphery extends radially but in which no mycelium can be found in the central region. It may be suggested that this situation has come about through the death and decay of the central region of the colony but there is no evidence of this from nature or from culture experiments (EVANS, unpublished). The size of fungal colonies in nature may be considerable and can certainly extend over several square yards. A simple technique is to investigate the distribution of morphological markers, fluorescent dyes or

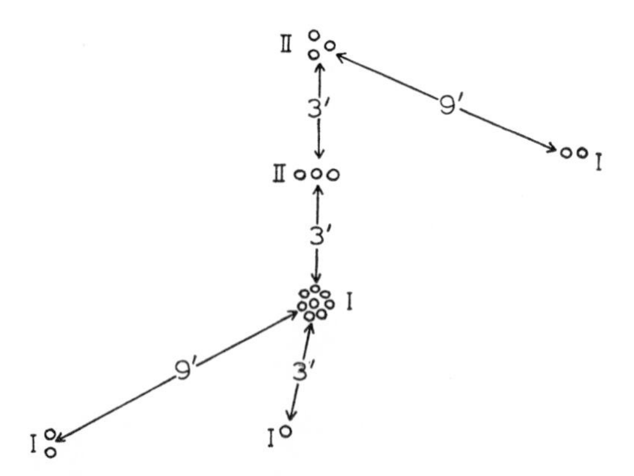

Fig. 4.2. The distribution of sporophores of *Hypholoma fasciculare* derived from the same mycelium as a guide to its distribution. Each circle represents a basidiocarp. Those labelled I were collected on 19.8.1957, those labelled II on 23.8.1957

genetic factors, in the mycelium or fruit bodies. Very little work of this nature has been done, e.g. PARKER-RHODES (1954), BURNETT and PARTINGTON (1957); an example is given in Fig. 4.2.

The hyphal branching system

When a propagule, such as a spore, germinates, its early growth in plate culture does not follow the radially expanding pattern already described. One or more hyphae grow out and may branch. At this stage their growth is likely to be limited by the amount and availability of nutrients in the immediate vicinity. Thus, around the growing hyphae, zones will rapidly become depleted of nutrients. Growth into these regions by adjacent hyphae, therefore, is bound to be restricted as a result of the limiting effect of nutrient depletion. As a result of this process growth into nutrient-rich areas will be promoted and growth into nutrient-depleted areas restricted; hence the characteristic radially expanding system soon becomes established (Fig. 4.3). In this system all main hyphae are capable of exploiting the most nutrient-rich areas and this leads to considerable competition between adjacent hyphal tips.

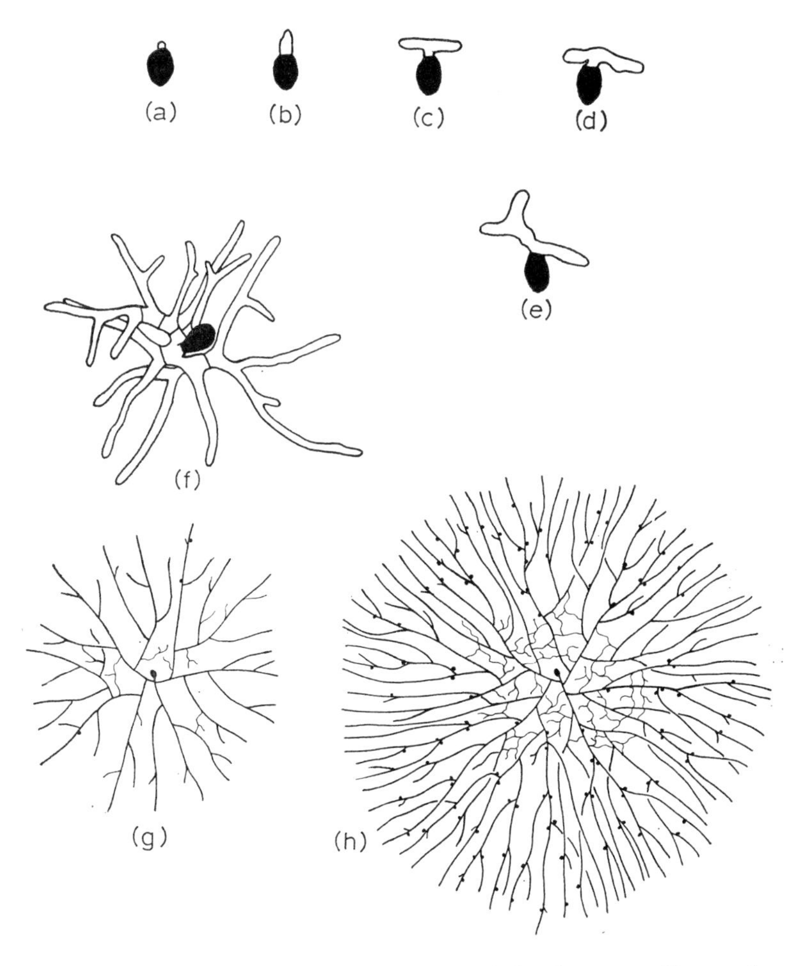

Fig. 4.3. Stages in the development of a spore of *Coprinus sterquilinus* to form a radially expanding colony in a plate culture (a–f, ×220; g, h, ×65; from Buller)

The circular form of the fungal colony is first established, therefore, as a result of competition for nutrients. Thereafter, although inter-hyphal competition continues, other internal regulating processes come into play.

Before considering the growth of the colony as a whole it is desirable to consider the production of lateral branches. This is, at present, a great enigma. Unless apical growth is in some way retarded or prevented, laterals usually do not arise until some distance behind the apex, often near the end of a segment. It may be suggested, from observations of a fungus like *Basidiobolus* for example, that a minimal volume of cytoplasm seems to be necessary before lateral branching is initiated. Until this volume is attained growth of the apical cell takes place, empty segments, previously formed, being cut off by a cross-wall. Once the critical volume is exceeded and the penultimate cell, as well as the terminal apical cell, contains cytoplasm then the former cell may give rise to a lateral branch. Observations such as these suggest that the growing apex in some way inhibits lateral formation although the mechanism of such correlative inhibition is as obscure in fungi as in higher green plants. Damage or restriction of growth at the apex enables laterals to grow out nearer to the apex than is normal (cf. p. 49) and this also suggests some mechanism of apically-determined inhibition. Primary laterals show the same type of control over the growth of laterals of higher orders.

The numbers and positions of laterals are subject to modification by environmental or genetic factors. The former is readily shown by growing almost any fungus on high and low nutrient, or water agar. The frequency of branching is generally reduced on the last two media. Recent studies of this phenomenon in *Mucor hiemalis* by PLUNKETT (1966) are of considerable value in providing an effective technique for studying the frequency of lateral branching, as measured by the lengthening or shortening of internodes. He was able to perfuse the growing mycelial margin with various complex and simple solutions and thus determine the effects of their exogenous application on hyphal branching. Arginine and glutamic acid were found to induce internode lengthening while tryptophane, cysteine or valine were found to shorten internodes, and it is suggested that α-amino nitrogen metabolism may be specifically involved in the morphogenesis of lateral hyphae. Nevertheless, so many other compounds also affect production of laterals in this and other fungi that no useful generalization can yet be made.

Genetical control was first illustrated clearly by BULLER (1933), who noted that the angle between a lateral and its parent hypha was *c.* 70°–90° in monokaryons of *Coprinus fimetarius* and *c.* 20°–40° in the case of the dikaryon. Genetical control is also illustrated by the numerous 'colonial' mutants which are known in many fungi, notably *Neurospora* (BARRETT and GARNJOBST, 1949) where they have been studied in some detail. Here laterals may be closer together, the growth habit may be dichotomous or quasi-dichotomous rather than monopodial and so on. Total or partial replacement of the normal nutrient carbohydrate by sorbose will frequently provoke phenocopies which resemble these mutants in their growth habit (TATUM *et al.*, 1949). Informative as such studies are in many ways (cf. p. 48), they have not yet provided information concerning the mechanisms responsible for the production of normal laterals.

The only thing which is quite clear from studies of both normal and abnormal branching is that the de-differentiation of a secondarily thickened cell wall and the establishment of a new apical growing region can be achieved with extraordinary rapidity, taking something of the order of 40–60 sec. It is also a general rule, in segmented forms, that the lateral arises at the end of the segment nearest to the growing apex, although exceptions do occur. The technical difficulties of studying such rapid and complex processes in such small regions need hardly be stressed.

As a colony grows outwards radially from its centre of origin it is easy to see that its margin maintains a more or less constant density of hyphal tips growing at a fairly constant rate. It is clear that the number of marginal hyphal tips must be increased proportionately as a colony expands. On the average, an increase in radius of the colony from r_1 to r_2 is associated with an increase of hyphal tips by a factor of r_2/r_1. How this is achieved has not yet been conclusively demonstrated.

For example, BUTLER (1961) has investigated this phenomenon in surface hyphae of plate cultures of *Coprinus disseminatus*, under conditions where the density of marginal hyphae was relatively low, in order that the behaviour of branching hyphae could be followed easily over three-hour periods. The length of hypha observed initially was *c.* 1000 μm long and extended to *c.* 1700 μm during the observations. Hyphal growth was strictly monopodial. Primary laterals arose at regular intervals in close association with cell division in the apex of the parent, main hypha. Secondary laterals arose regularly from primary laterals in a similar manner. The three categories of hyphae differed in their rates of growth elongation and in the diameter of their tips (Table 4.1).

Table 4.1. The tip diameter (μm) and rate of extension (μm/15 min) of main hyphae and of primary branch hyphae of *Coprinus disseminatus* at 21·5°C. (From Butler, 1961)

Hypha	Extension rate and standard deviation	Diameter and standard deviation
Main	64·6 (8·00)	5·88 (0·574)
Primary	42·2 (6·16)	4·69 (0·448)

It will be recalled (p. 42) that apical growth may be determined by many parameters including the rates at which materials are translocated to the apical region. The positive correlation between growth rate and hyphal diameter found by Butler might be thought to be a reflection of the efficiency of translocation but the relationship is not as simple as this. Main and primary branch hyphae of the same diameter showed statistically significant differences in their rates of elongation, main hyphae being the faster. Moreover, using the product 'increase in length × diameter' as a measure of total growth increment, it was shown that there is a positive correlation between the growth increments of primary branch hyphae and those of their main, parent hyphae. Growth of laterals is, therefore, a function of origin as well as of size.

During the three-hour period of observation the rates of extension of main, primary and secondary laterals remained constant in the ratio of 100:66:18. If this were always to be the case then the density of main hyphae at the growing margin should become less; this is not so. Evidently some laterals must increase their growth rates in order to 'catch up' pre-existing main hyphae, fill the gaps between them and, thereafter, slow down and maintain the characteristic, virtually constant rate. Only one observation was made which might have a possible bearing on this requirement. The cessation of growth of a lateral (for unknown reasons) appeared to result in increased growth rates of adjacent laterals although no effect was noted on main hyphae. This observation recalls the inhibitory effect of a main apical on lateral formation, already mentioned. It suggests some form of internal competition between branches, possibly for a key nutrient, although a more sophisticated regulatory mechanism could be involved. Whether the increased growth of laterals adjacent to one whose growth has ceased is a sufficient stimulus, possibly repeated, to enable one of them to exceed and eventually approach the characteristic growth rate of a main hypha is still uncertain. Further detailed observations of this type are clearly desirable.

The regulation of growth and branching is, unfortunately, not yet understood. That regulation of hyphal growth affects whole branching systems can

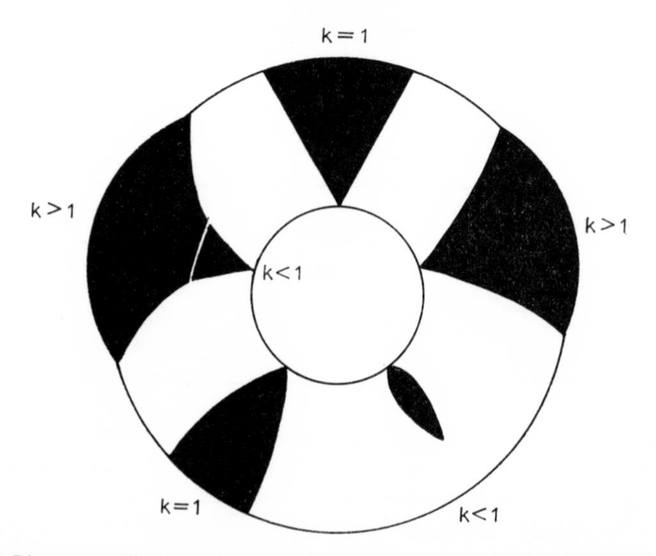

Fig. 4.4. Diagram to illustrate the commonest shapes of sectors which arise in fungal colonies in plate culture. The growth rate of each sector (k) is compared with that of the normal mycelium designated as 1. (After Pontecorvo and Gemmell)

be seen from two phenomena; the occurrence of sectors which differ in growth rate from adjacent regions of a colony and the occurrence of rhythmic growth. There is also some evidence that part of the control system of hyphal growth is located in the cytoplasm.

A sector which differs in growth rate from the rest of the colony can often be traced back to a distinct point of origin. This can be seen especially clearly if the sector also happens to differ in colour as well as in growth rate. The relationships between the shape of the sectors and their relative growth rates have been explored theoretically and in model experiments by PONTECORVO and GEMMELL (1944). The very existence of such sectors indicates that they arise from branching hyphal systems whose growth regulating properties are perpetuated independently of those in the rest of the colony. That this is the case and that their existence depends, in part, on intra-hyphal competition can be shown by inoculating plates with a spore suspension composed of two strains, in different proportions, which differ in growth rate. Such model experiments mimic natural sectors with remarkable exactitude (Fig. 4.4).

The phenomenon of rhythmic growth provides further evidence for the coordinated behaviour of complex hyphal systems. Figure 4.5 illustrates a

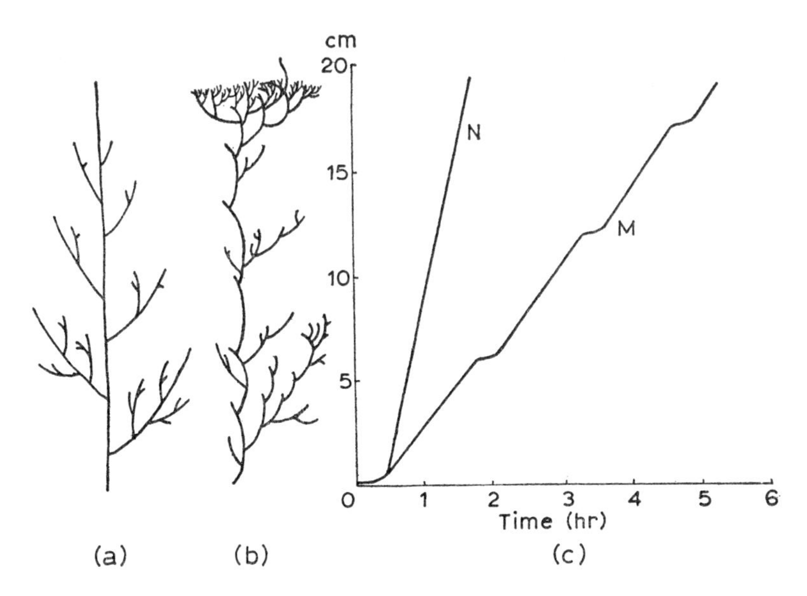

Fig. 4.5. Diagrams to illustrate the habit and growth rate of the mutant *vague* of *Ascobolus immersus*. (a) Growth habit of normal. (b) Growth habit of mutant. (c) Growth rates of normal (N) and mutant (M) compared. (From Chevaugeon)

type of rhythmic growth investigated by CHEVAUGEON (1959a, b, c) in the mutant—*vague* of *Ascobolus immersus*. Similar behaviour is shown by the mutants *patch* and *clock* in *Neurospora crassa* (BRANDT, 1953; PITTENDRIGH *et al.*, 1959; STADLER, 1959; SUSSMAN *et al.*, 1964). These mutants have in common an unusual growth pattern. Their main hyphae all grow monopodially at first but periodically extension growth is checked and they branch profusely, indeed in *clock* hyphal apices branch dichotomously. Thus a zoned appearance develops in the colony where the densely branched hyphae alternate with the normal, less densely branched hyphae. This alternation is brought about by the 'escape' of some primary laterals from the dense zone which behave as main, monopodially growing hyphae and restore the normal habit until the next branched zone develops. The periodicity of the change in habit varies; it may be circadian or longer, e.g. 40 hours in the *vague* mutant of *Ascobolus*, and its duration is affected by environmental variables such as temperature and nutrients. Chevaugeon has shown that the synchrony of this phenomenon is dependent upon the lateral continuity of the hyphae. If this is interrupted by inserting a glass slide radially in a developing colony then the zonation on opposite sides of the slide may be out of phase until lateral contact is restored in the region beyond the slide (Fig. 4.6). It seems clear

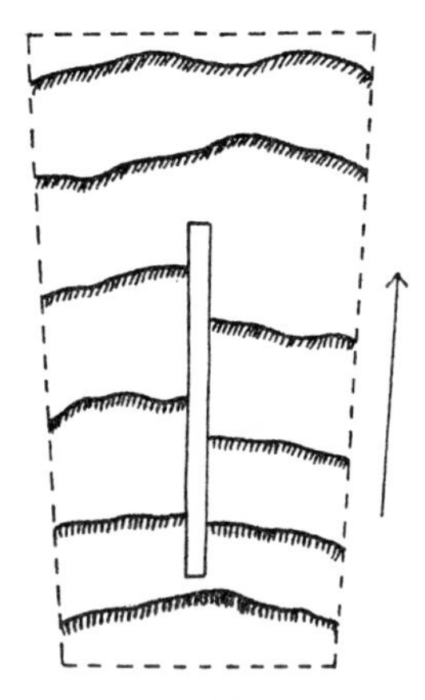

Fig. 4.6. Diagram to illustrate how the insertion of a glass plate with a radial orientation disrupts the synchrony of growth between laterally adjacent regions and how it becomes restored beyond the plate. The hatching represents the densely branched regions alternating with the regions of rapid extension growth in the mutant *vague* of *A. immersus*. (From Chevaugeon)

that these rhythmic phenomena are determined internally and that the synchronizing stimulus is propagated within the hyphae. This phenomenon differs, therefore, from the zonations often observed in plate culture due to the production and transfer to the substrate of staling substances by a fungus, which thus retard its growth. Normal growth occurs when this inhibiting zone is passed and continues until the staling phenomenon re-occurs (cf. HAWKER, 1950).

Chevaugeon's experiment with the mutant *vague* has shown the importance of cytoplasmic connexions between integrated growth regions of a mycelium. Control is exercised here by the mutant genotype but the work of Jinks and his associates has shown that growth regulation may also include a cytoplasmic component. For example, using *Aspergillus glaucus*, he has shown that statistically significant variations occur in growth rate between sub-lines derived by hyphal-tip inocula (JINKS, 1954, 1956). The design of his experiment was such as to ensure that these differences could not be ascribed to the occurrence of genetically different nuclei in the tips. He also showed that inocula containing different numbers of hyphal tips responded differently in their ability to grow (as measured by the duration of the lag period) when transferred from normal medium to medium containing 5 p.p.m. of mercuric chloride (Table 4.2).

Table 4.2. The effect of inoculum size of an homokaryotic isolate of *Aspergillus glaucus* on the mean lag period for growth on modified Czapek medium without and with 5 p.p.m. mercuric chloride. (From Jinks, 1959a)

Inoculum No. intact growing hyphal tips	Mean lag in days		Lag due to $HgCl_2$
	Czapek	Czapek + 5 p.p.m. $HgCl_2$	
4	5·6	17·0	11·4
16	2·9	12·5	9·6
32	2·1	9·6	7·5
50–200	2·0	5·3	3·3

When the results for inocula with 32 or more tips were pooled and compared with the pooled data for those with 16 or less there was a significant difference between them in respect of the frequency of deaths of inocula. Such evidence can only imply cytoplasmic differences between the vegetative hyphae of single, homokaryotic colonies in their ability to adapt to the new medium. He later demonstrated that these differences were not nuclear. Thus there seems to be clear evidence from this fungus that hyphal-tip cytoplasm can be heterogeneous in its properties and exists as a 'heteroplasmon'. Cytoplasmic segregation of various characters as well as growth rates has also been observed in *Neurospora* (PITTENGER, 1956), *Aspergillus nidulans* (ARLETT, 1957), *Nectria stenospora* (GIBSON and GRIFFIN, 1958) and *Podospora* (MARCOU and SHECROUN, 1959). It is clear, therefore, that attention will have to be paid in future to both the genotypic and cytoplasmic controls exercised on the growth and behaviour of adjacent hyphae in a single colony. How such control is exercised and integrated is not yet clear. Until much more extensive studies have been made it is hardly possible to make fruitful speculations.

The role of hyphal fusions in bringing different nuclei, or cytoplasms, or both, into the same hyphal tips is clearly of importance and, at the same time, provides a route for the transmission of coordinating stimuli, whatever their nature. This process will, therefore, be considered next.

Hyphal fusions

Hyphal fusions occur in Ascomycetes, Basidiomycetes and Deuteromycetes. They are completely lacking in nearly all Phycomycetes, although in some of these fungi fusions do take place between gametangia which in many cases appear to differ very little initially from normal hyphae, e.g. many Mucoraceae.

Little can be added to the illuminating exposition on hyphal fusions given by BULLER (1931, 1933). This author formulated an important generalization, '. . . *all hyphal fusions are essentially end-to-end ones*, i.e. that when a fusion takes

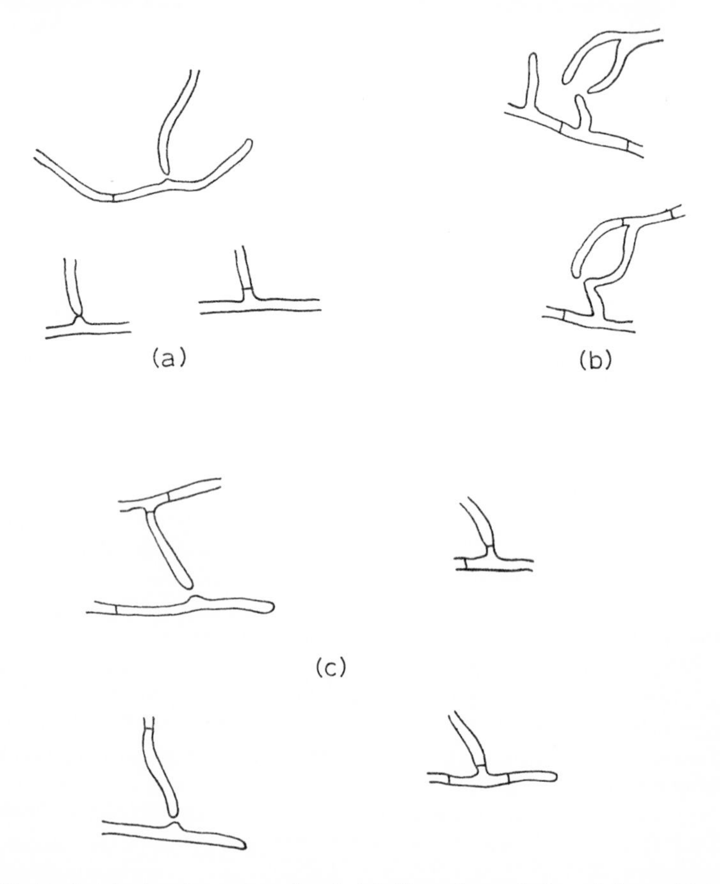

Fig. 4.7. Hyphal fusions in various fungi. (a) *Hypochnus*, fusion of tip and lateral. (b) *Hypochnus*, attraction of three tips towards each other, tip-to-tip fusion. (c) *Sclerotium*, fusion of tip and lateral with clear induction of new tip at a distance. (All ×c. 1000, after Köhler)

place, it takes place between the end of one hypha and the end of another hypha'. The generalization is illustrated in Fig. 4.7 and certain other conclusions may be drawn from these data.

The process is evidently initiated by action at a distance (telemorphosis) not exceeding 10–15 μm and this is followed by a phase in which the stimulated hyphae grow towards each other (zygotropism) and finally fuse end-to-end. Very little else is known about the process.

The telemorphotic phase can be initiated between two adjacent hyphae when lateral 'peg-hyphae' are induced mutually: thus an actual hyphal tip is not, apparently, essential to initiate the process. Observations on a number of fungi have shown that it is unusual for the tips of main hyphae to fuse. When two mycelia are apposed in plate culture, e.g. *Polystictus versicolor*, the main tips grow past each other but fusions can take place between the tips of primary, or higher order, laterals (BURNETT, unpublished). The significance of this phenomenon is not known, nor is it yet known how widespread it is.

The stimulus which initiates hyphal fusion does not seem to be completely species specific. The early work of REINHARDT (1892) and KÖHLER (1929, 1930) showed that interspecific, intergeneric and interclass fusions could occur, e.g. *Neurospora sitophila* with *N. crassa*; *N. sitophila* with *Botrytis allii*, *Mucor* sp. with *Sclerotinia sclerotiorum*. On the other hand, hyphal fusions frequently form more readily between hyphae of the same species than between different species. Buller and others have even claimed that the absence, or excessively low frequency, of fusions between two mycelia can be taken as an indication that the two forms are distinct species. The validity of this is doubtful and will be discussed later in relation to speciation (see p. 451).

For present purposes it is more significant that hyphal fusions can be non-species specific and that the process seems to take a similar course in all fungi. This suggests, as a simplest hypothesis, that there is a common underlying mechanism. Nothing is known of its nature. RAPER (1952) has suggested a theoretical system based upon a single diffusible material. This requirement seems to be essential since otherwise it is difficult to conceive of more than one substance being involved in a process which takes place between laterals derived from a common hypha. The non-acceptance of this requirement, as BULLER (1933) pointed out, can only lead to a requirement for an almost infinite number of substances inducing hyphal fusion. Raper suggests, therefore, that the hyphal tips each produce small quantities of a labile substance which diffuses outwards, setting up in each case a steep diffusion gradient around their sources of origin. It is then supposed that hyphae or hyphal tips will be induced to react over a critical and rather narrow concentration gradient of this hypothetical substance. The tips so induced will then, presumably, grow up the concentration gradient, bringing about fusion. This is the simplest possible system to postulate, it could well be that a substance is necessary to control each phase, one for telemorphosis, one for zygotropism and one for the actual fusion. A further problem is the manner in which a precise region of the wall can be induced to respond to the initial stimulus and develop a tip. This could be accounted for by the diffusion gradient hypothesis (Fig. 4.8). The wall would respond at the point in contact with the correct concentration. This might account for the occasional induction of two tips responding to a single initiator tip. Just how a cell wall is made plastic and made to act as a growing tip is an unsolved problem already discussed in relation to branch formation (see p. 64).

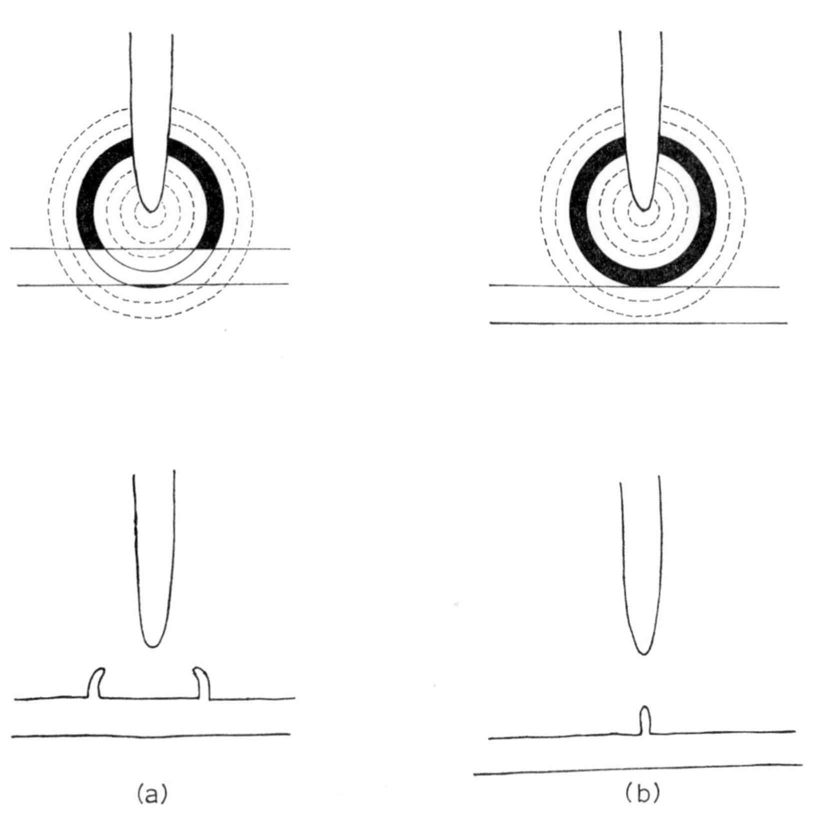

Fig. 4.8. Diagram to illustrate theory of hyphal fusions. The concentric circles represent different concentrations of a diffusible hypothetical initiator which is effective over a limited concentration range only (solid areas). (a) Initiation of tip to lateral fusion, more than one new tip initiated. (b) Initiation of tip to lateral fusion

An alternative hypothesis has been put forward based upon investigations of inhibitory staling products produced by fungi (PARK, 1961, 1963; ROBINSON and PARK, 1965; PARK and ROBINSON, 1966a). Hyphae are, in general, growing in a background of their own staling product(s), to which hyphal apices respond both by negative chemotropism (cf. p. 326) and by removing these product(s) from the medium. Thus hyphal tips are surrounded by a halo of low concentration of staling substance(s). If two apices approach each other their haloes of low concentration may overlap and they will then turn to each other and, after contact, fuse. This hypothesis is worth investigation but it neither accounts for the initiation of new hyphal tips between parallel-lying hyphae nor for the initiation of a hyphal tip in a tip-to-hypha confrontation, as readily as does Raper's hypothesis.

Considerable progress is being made in the physiological analysis of fusions between fungal cells differentiated in respect of mating ability and such studies may well provide pointers to the mechanism of hyphal fusions (see p. 410–413).

Fusion is followed by the dissolution of the wall, a process which can take place very rapidly in as little as 3 to 5 minutes. No electronmicrographs are available of the process but visual observation suggests that dissolution commences at the tips, i.e. the centre of the fused region, and spreads out to the lateral walls. This is certainly what happens in the dissolution of the fused walls of gametangia of *Phycomyces*. Here the original fusion walls are indistinguishable and the central wall substance is dissipated first. There does not seem to be any difference between the matrix and fibrillar elements in the rate of the process. During dissolution increased chitinase was detected but, if this enzyme is involved, as seems likely, it must act in a precisely localized manner (SASSEN, 1962, 1965).

Once the cell wall is penetrated cytoplasm carrying various organelles may well start to flow through the aperture. The direction of flow is often determined by the difference in osmotic pressure of the two cytoplasms when brought into contact (cf. p. 258).

In several Basidiomycetes a special consequence of hyphal fusions between compatible monokaryons (cf. p. 429) is the production of a dikaryon possessing clamp connexions. A clamp connexion is, in essence, a lateral of limited growth which grows backwards in a curve from the posterior end of a tip cell, induces a new tip at the anterior end of the penultimate cell and fuses with it (Fig. 4.9 and BULLER, 1933). The process is associated with conjugate nuclear

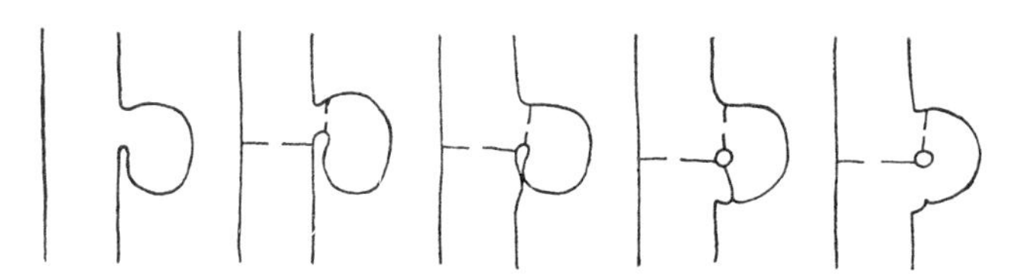

Fig. 4.9. Diagram to illustrate stages in the formation of a clamp connexion: a special form of tip-to-lateral fusion. (After Buller)

division of the two nuclei of the terminal cell and maintains this condition throughout the mycelium. This cannot be the whole function of clamp connexions, however, for conjugate divisions can occur without clamp connexions forming and the binucleate condition can still be maintained in many Basidiomycetes.

The essential consequence of hyphal fusions in all fungi, however, is that

Table 4.3. The relative numbers of hyphal fusions formed in four-day-old plate cultures of *Corticium vellereum* in relation to (**a**) nutrient concentration and pH, and (**b**) growth rate and nutrient conditions. Results are means of three replications in each case. (Adapted from Bourchier, 1957)

(a)

% Concentration of medium [1]	Mean no. hyphal fusions pH 5·1	6·7	7·1	Total
2	1·577	4·471	5·725	11·773 [2]
4	5·102	5·102	3·061	13·265 [2]
6	7·295	6·728	5·328	19·351 [2]
Total	13·974	16·301	14·114	

[1] Potato-dextrose agar.
[2] Differences in hyphal fusions were statistically significant at 5% level in relation to medium concentration but not pH.

(b)

Medium	Mean no. hyphal fusions [4]	Mean diameter colony—cm [4]
PDA [1]	64·117	2·32
CDA [2]	44·087	2·57
Malt [3]	46·941	2·01

[1] Potato-dextrose agar.
[2] Carrot-dextrose agar.
[3] 2% malt agar.
[4] Differences between numbers of hyphal fusions and between colony diameter are both statistically significant at 1% level.

hyphal branching systems which may have been competing in their exploitation of the substrate are now in contact and cooperation can replace competition. It is noticeable that hyphal fusions are more frequent on some substates and at different concentrations (Table 4.3), and this, teleologically speaking, can be regarded as an adaptation by the colony to the fullest and most economical use of the materials available. The optimum conditions for hyphal fusions do not necessarily coincide with the optimum for growth. The co-operative effect is best shown by BULLER's (1931) original example of the dung fungus, *Coprinus sterquilinus*. The spores of this Basidiomycete are voided with the faeces of horses and germinate *in situ*. Thus in a 'dung ball', for example, a large number of colonies will develop and compete for the limited materials available. If this process continued no single colony would obtain sufficient material to form a complete functional sporophore.

However, since hyphal fusions *do* occur the single mycelium so formed can exploit the whole dung ball as a single physiological unit and thus obtain sufficient material for sporophore production (Fig. 4.10).

It will be realized that this property of forming hyphal fusions makes the concept and practical limits of an individual fungal colony difficult to define. In a number of Basidiomycetes, using genetic markers, it has been possible to provide evidence that a colony can be a single physiological unit although derived from the fusion of genetically different mycelia of the same species, e.g. *Polystictus versicolor*, *Polyporus betulinus* (BURNETT and PARTINGTON, 1957; BURNETT and BELL, unpublished).

This property, enabling cooperation to replace competition, sets the fungi apart from almost all other organisms in what Buller called their 'social organization'. Its consequences have neither been fully explored nor

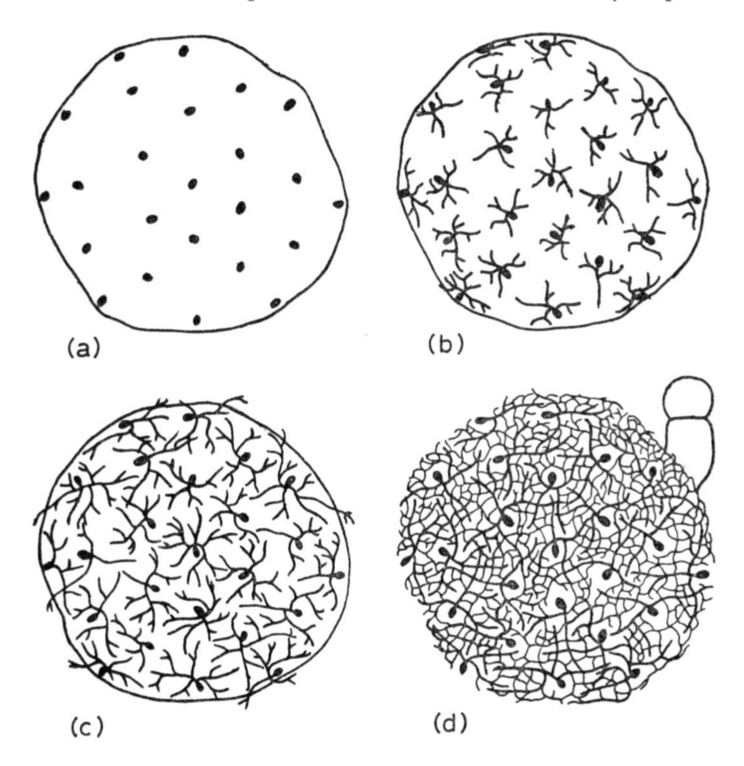

Fig. 4.10. The consequences of hyphal fusions between germinating basidiospores of a fungus such as *Coprinus sterquilinus* in exploiting a nutritionally and spatially restricted ecological niche such as a horse dung ball. (a) Spores embedded in dung ball. (b) Germ mycelium developed and competing for nutrients and space. (c) Fusions between mycelia and continuance of competition. (d) All mycelia now fused to form a single three-dimensional network capable of exploiting whole of limited environment and so producing a sporophore. This would not have been possible in other organisms in situations (b) and (c). (After Buller)

appreciated although some consideration will be given to these matters when considering recombination and speciation.

Within this growing, three-dimensional network of hyphal filaments differentiation develops at an early stage; that is, differentiation both of whole regions, as aerial and submerged mycelium, and of individual hyphal types and organs. These changes will now be considered.

DIFFERENTIATION OF THE COLONY AND HYPHAE

Hyphae immersed in substrate on nutrient agar plates are generally more highly vacuolated, have more lipid globules and may be more moribund in appearance than the aerial hyphae. These differences are borne out by differences in their biochemical activities as shown, for example, by the data of ZALOKAR (1959a). He examined the activities of certain enzymes (Table 4.4) from the surface and submerged layers of four-day-old standing cultures of *Neurospora* in liquid culture and compared them with the activity of younger mycelia and conidia.

Table 4.4. Enzyme activities at different ages and in various parts of the mycelium o *Neurospora* grown in standing liquid culture. (From Zalokar, 1959a)

	Conidia	Mycelium (age in hr)					
		8	16	96			
				Complete	Upper layer	Lower layer	Upper and lower
Succinic dehydrogenase [1]	1·92	18·7	19·6	17·8	23·4	12·5	17·9
Aldolase [2]	3·2	6·0	11·2	10·4	8·4	10·1	9·3
β-galactosidase [3]	43·2	4·0	11·4	262·0	150·0	386·0	268·0
Tryptophane synthetase [4]	0·91	1·15	1·72	1·57	1·32	1·93	1·62
Mitochondria [5]	24·0	12·0	15·4	11·9	10·3	13·6	12·0
Ribonucleic acid [6]	12·5	12·5	9·6	8·1	9·9	7·3	8·6

[1] QMB/mg protein mitochondria.
[2] mM HDP degraded/mg/hr.
[3] Colorimeter units/mg/hr.
[4] Units/mg protein.
[5] mg protein/100 mg total protein.
[6] mg/100 mgprotein.

It could be seen that in several cases there were clear differences between the upper and lower layers in, for example, succinic dehydrogenase activity. This probably reflects the environmental difference in respect of oxygen tension between aerial and submerged hyphae; other examples are known of this type of response. In *Neurospora* the oxygen-sensitive, β-carotene production is clearly restricted to the upper layer (ZALOKAR, 1954) which shows up in marked contrast to the white submerged layer. In *Aspergillus* aerial hyphae are less sensitive to carbon monoxide and cyanide but are more sensitive to lowered oxygen tension than are submerged hyphae (TAMIYA, 1942). This may reflect either differences in the effective oxidase system or modifications of the cytochrome oxidase system in different parts of the mycelium. Zalokar's data also enable a comparison to be made between the effects of ageing (8- to 16- to 96-hour-old mycelium) and differentiation (upper to lower layers). Such differences are also related to colony differentiation as shown by the histochemical studies of YANAGITA and KOGANÉ (1962) on *Aspergillus* and *Penicillium* especially in respect of basophilic substances (Fig. 4.11). It is also not uncommon for the substrate hyphae to become

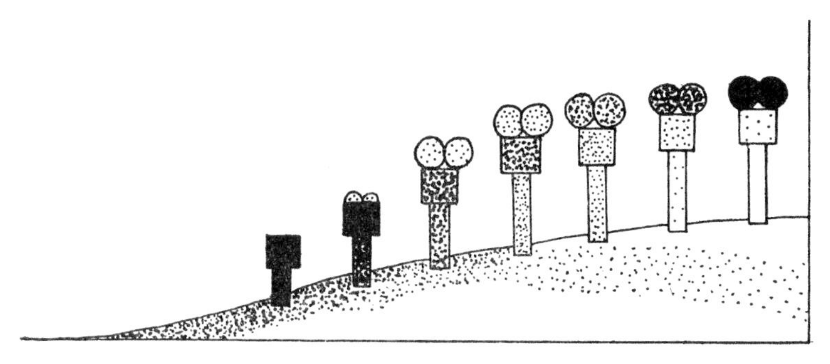

Fig. 4.11. Diagram to illustrate the unequal distribution of basophilic material in a differentiated colony of *Aspergillus*. The degree of basophily is shown by the intensity of shading. (From Yanagita and Kogané)

swollen and filled with reserve materials such as glycogen and protein, e.g. *Coprinus lagopus* (MADELIN, 1960).

Rhizoids

In addition to these general, overall differences, structural differentiation may also occur. Substrate hyphae may develop highly branched rhizoids, e.g. *Rhizopus* spp. (Fig. 4.12) which function as anchoring and, presumably, absorbing structures. In *Rhizopus* they are associated with horizontally growing, aerial, stoloniferous hyphae but this is not usually the case (see Fig. 1.2, *Rhizidiomyces*, for example). The induction of rhizoidal systems is related to the whole problem of branching and is equally obscure.

Appressoria and Haustoria

In many parasitic fungi a surface hypha develops a specialized structure, an appressorium, from which an actual penetrating infection hypha develops. The appressorium usually takes the form of a multi-nucleated, disciform

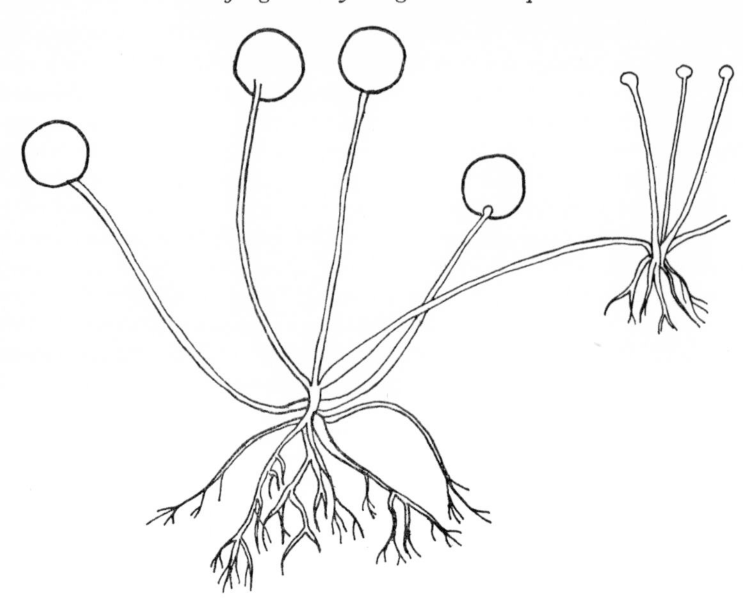

Fig. 4.12. Part of the mycelium of *Rhizopus nigricans* to show a fascicle of sporangia arising from a group of rhizoids and connected to another fascicle by means of a stolon (×100, after Atkinson)

swelling which adheres to the wall of the host cell, e.g. *Puccinia* (Fig. 4.13) but it may divide to form a small flattened plate of cells, e.g. *Botrytis*. Another kind of substrate-hyphal organ, the haustorium, develops in the host cell. This structure varies in shape and size in different fungi but it usually consists of a constricted region, where the hypha penetrates the host cell wall, and an expanded or branched region within the host cells. Recent studies with the electron microscope of the haustoria of *Puccinia*, a rust fungus (EHRLICH and EHRLICH, 1963a, b), the downy mildews *Perenospora* and *Albugo* (PEYTON and BOWEN, 1963; BERLIN and BOWEN, 1964a) and the powdery mildew *Erysiphe* (EHRLICH and EHRLICH, 1963b) have provided more detailed information and shown that there are considerable similarities between haustoria. In *Puccinia graminis* and *Erysiphe graminis* hyphal tips become cut off by a septum and mitochondria become aggregated in the region where the tip is in contact with the wall of the host cell. The tip of the haustorium and the host cell wall are said to break down so that their respective cytoplasmic membranes are in direct contact. The haustorium then penetrates the cell, enlarges, develops a wall and causes the invagination of the host cell's cytoplasmic membranes, which remain closely adpressed to the haustorium wall. The zone between differs both from the haustorium wall and the cytoplasmic membrane in appearance and its origin is disputed. In resistant wheat strains this zone persists around dead haustoria and it has been suggested that this indicates its origin from host tissue (HAWKER, 1965) but the problem is still unresolved.

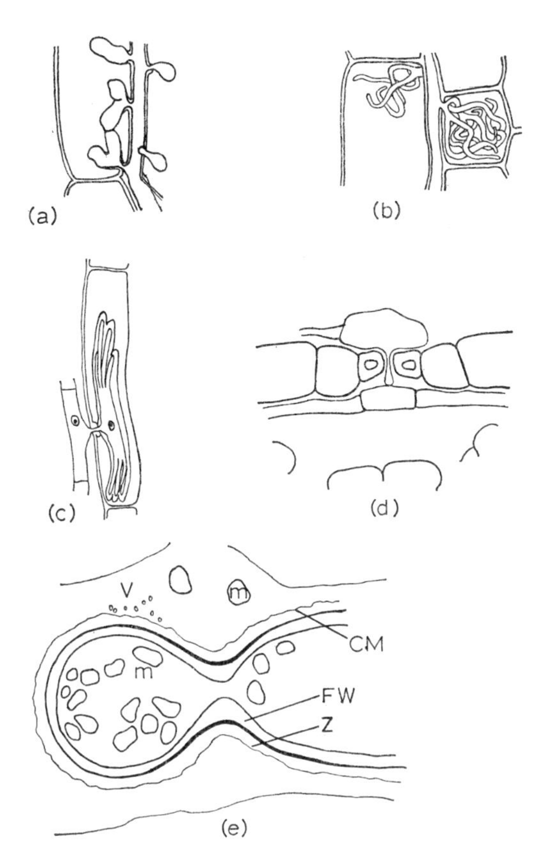

Fig. 4.13. Appressoria and haustoria of fungi. (a) Sac-like haustoria of *Perenospora parasitica* in leaf cells of *Capsella*. (b) Branched, filamentous haustoria of *Perenospora calotheca* in stem cells of *Galium*. (c) *Erysiphe graminis* penetrating epidermal cell of *Poa* by means of a fine penetration hypha enclosed in thickened cell wall of host. An elongated, branched haustorium has developed inside the host cell. (d) Penetration of stoma of *Avena* by *Puccinia coronata*. The empty germ hypha and swollen appressorium are seen above the guard cells below which is the sub-stomatal vesicle which has given rise to two lateral infection hyphae which will penetrate the mesophyll cells. (e) Tracing of an electronmicrograph of a section through an haustorium of *Puccinia graminis* in a cell of *Triticum*. The haustorium contains many mitochondria (m) and is enclosed by a wall (FW) and between this and the unbroken cytoplasmic membrane (CM) of the host cell is a zone of apposition (Z) whose origin, host or parasite, is disputed. The origins of small vesicles (V) in the cytoplasm around the haustorium are similarly disputed and could either be material from the parasite via the zone of apposition or due to the secretory activity of the host cytoplasm. ((a) ×120, after Fraymouth; (b) ×210, after de Bary; (c) ×175, after Smith; (d) ×450, after Ruttle and Fraser; (e) ×5500, after Ehrlich and Ehrlich)

The haustoria of *Perenospora* and *Albugo* are similar but it is not known whether there is a phase in early penetration when the cytoplasmic membranes of host and parasite are in direct contact.

Little is known concerning the way in which these structures arise but DICKINSON (1949) has provided experimental evidence that the appressoria of certain rust fungi of the genus *Puccinia* arise in response to a contact stimulus. When the germ hyphae from spores of these rusts were brought in contact with artificial, paraffin wax-colloidon membranes or cell walls of their hosts, nuclear division was initiated, carotenoids accumulated and side branches of limited length were put out through the continual development of new growing points in the penultimate part of the germ hypha. When cell material from leaves was added, the branching was replaced by the development of an adhesive swelling similar to an appressorium provided that the artificial colloidon membrane included waxes of high, not low, congealing points. These results suggest the action of a contact stimulus, depending on the hardness of the material. A non-specific chemical stimulus may also be involved as suggested by the effects of adding leaf material to the artificial membranes.

Hyphal traps

A number of modifications (Fig. 4.14) are developed by hyphae of the so-called 'predacious fungi' in the presence of eelworms or materials which have been in contact with eelworms. In several of them a sticky fluid is produced, especially after contact with an eelworm. In fungi, such as *Stylopage grandis*, the whole mycelium is sticky and the secretion of further fluid is localized in those regions where contact occurs. In fungi such as *Arthrobotrys* spp., *Trichothecium cystosporium* or *Dactylella* spp. the sticky regions are localized and morphologically specialized. In the first two, lateral branches form networks of hyphal loops in which the loops tend to develop at right angles to one another. The loops are covered in a viscid fluid and eelworms become entangled in the loops and stick to them. In many species of *Dactylella* the loops are replaced by short lateral branches either consisting of, or terminating in, a sticky knob-like cell. Since these branches are developed close together, the movement of an eelworm, once caught, often results in it becoming stuck to adjacent branches. A further trapping device is the development of rings both passive and constricting. In the former case, a ring of three cells is developed from a lateral stalk and eelworms accidentally push their heads through. As the eelworm struggles it becomes more tightly wedged and, although it may tear the ring from the mycelium and even pick up other rings in the same way, it can rarely dislodge the ring or rings. Constricting rings appear to be similar in form although they possess shorter and thicker stalk cells and are more robust than passive rings. The inner surfaces of the ring cells are sensitive to rubbing. If an eelworm inadvertently enters a ring the stimulus is followed in about 0·1 sec by an enormous distension of the ring cells, the stimulated cell fractionally preceding the others in its swelling.

In all cases adhesion evidently stimulates hyphal development and the eelworm is penetrated at the points of contact by fine infection hyphae. Just

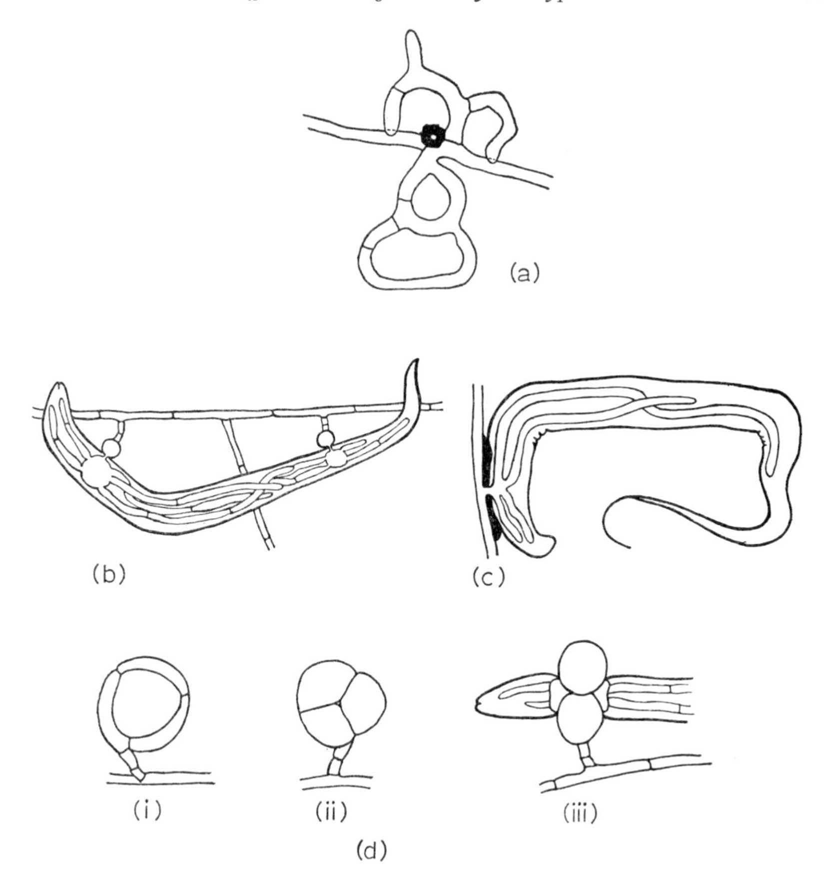

Fig. 4.14. Hyphal traps. (a) Networks of loops of *Arthrobotrys cladodes* var. *macroides* beginning to develop. (b) Sticky knobs of *Dactylella ellipsospora* adhering to eelworm in which haustorial hyphae have developed. (c) An eelworm captured by the sticky mycelium of *Stylopage grandis* (sticky fluid shown black). (d) The constricting rings of *D. bembicoides*. The untriggered ring (di) is compared with a stimulated ring which has reacted (dii) and an eelworm is shown trapped by this method and invaded by haustorial hyphae (diii). (All × *c*. 290, after Duddington)

within the cuticle of the eelworm these hyphae swell to form globular infection bulbs and from these numerous hyphae grow out and ramify in the body of the animal. This latter phase of development evidently resembles the formation of haustoria in fungi parastic on plants.

Nothing is known of the fine structure of the traps or of the subsequent infection phase nor is anything known of the biochemistry of the adhesive materials. This latter problem is especially important for, as will be described later, adhesion plays an important role in the development of various fungal structures such as sclerotia and strands.

Stromata

Many septate fungi develop aggregations of hyphae. These may be regularly, or irregularly, inflated and give the impression of forming a pseudoparenchymatous tissue or plectenchyma. This may take the form of flat plates or solid masses of tissue which are then known as stromata. The hyphae are intertwined and adhere together laterally, presumably as a result of the modification of their cell walls. The process is quite distinct from the tip fusions described earlier and in a stroma the cell contents remain distinct. Upon or within such massive stromata various kinds of vegetative and reproductive structures may develop (see Chapter 5).

Sclerotia

Aggregation and adhesion of hyphae may result in the development of more or less globose structures termed sclerotia. The largest known is formed underground by *Polyporus mylittae* in Australia. It can be as large as a man's head, weigh 15 kg, is eaten by the Aborigines and is known as 'Black fellow's bread'. More usually sclerotia range from the size of a pin-head to 1 cm in diameter but size is characteristic of the species and strain.

TOWNSEND and WILLETTS (1954) have recognized three developmental patterns. *Rhizoctonia solani*, the cause of 'black scurf' of potatoes, develops its sclerotia through the localized irregular branching and increased septation of adjacent cells which become barrel shaped. Thus a loosely compacted mass of cells arises. Species of *Botrytis* and *Sclerotium* represent a second type. They develop as a consequence of repeated dichotomous branching of a hyphal tip and increased septation of the cells so produced. This is followed by the adhesion of adjacent cells to form a solid, darkened plectenchymatous mass. An outer rind of rounded thickened cells develops and encloses a cortex of thin walled plectenchyma. The central medulla may be of less tightly convoluted and adherent filamentous cells, e.g. *Botrytis allii*, or of compacted, darkened plectenchyma, in which the cell outlines are difficult to make out, e.g. *S. cepivorum*. A third type of development shown, for example, by *Sclerotinia gladioli* has been described as the 'strand' type. Numerous small laterals arise in a localized region from one or more parallel main hyphae. As in other types septation and adhesion rapidly follow and convert the whole region into a somewhat elongated or spherical fused mass. An outer rind, cortex and medulla are often developed, their degree of differentiation being characteristic of different fungi. Amongst the most complex are those of *Phymatotrichum omnivorum* the causal fungus of the economically important root rot of cotton: here the rind is convoluted and has a bristly appearance due to the outward protrusion of pointed hyphae and strands. The innermost cells are largest and are surrounded by an irregular arrangement of large and small cells of variable size. The outermost cells are smaller and densely packed. In all these structures, however, the basic processes are common. Elongation is superseded by branching, of one kind or another; multiple septation and adhesion of adjacent cells follows rapidly. In addition, materials such as reserve carbohydrates and lipids are usually translocated to, and stored in, cells of developing sclerotia. The development of a compact, thick-walled outer rind results in a restriction of water loss and sclerotia can retain their viability for long periods. Thus, ISAAC (1946) has shown that strains of *Verticillium dahliae* can survive for five months in soil if they produce sclerotia, whereas non-sclerotial strains are short-lived.

Little is known of the factors which cause sclerotial development. There is evidence (TOWNSEND, 1957) that sclerotial initiation is similar in its nutritional requirements in respect of carbon and nitrogen to that for good mycelial growth. On the other hand maturation of sclerotia apparently does not commence until mycelial growth has been checked, when a proportion of sclerotial initials can mature. This could be due to some form of internal regulation whereby more mature initials prevent the development of younger initials. Evidence for such a regulatory mechanism is available in rhizomorph formation, a process having many similarities (GARRETT, 1953) and which will now be considered.

Strands and rhizomorphs

There is a tendency in many fungi for aerial and submerged hyphae to aggregate and form strands. The tendency may be ephemeral as in some Mucoraceae or well developed and, in the latter event, the strands may show a highly organized and differentiated structure. If the structure is a fully autonomous, apically, growing structure, it is termed a rhizomorph. Certain striking examples, the strands of the dry rot organism, *Merulius lacrymans*, or the rhizomorphs ('boot-laces') of the honey fungus, *Armillariella mellea*, early excited attention and their structure has been known for many years (e.g. HARTIG, 1874; FALCK, 1912). Strands of aerial hyphae are less common than subterranean ones. Amongst the best known of aerial strands is the so-called *Ozonium* stage of several *Coprinus* spp. This appears as orange-red tufts of coarse hairs and is composed of masses of thick-walled hyphae loosely bound together by anastomoses. Its function is obscure.

Subterranean strands have been studied in more detail. A convenient comparative account has recently been published by TOWNSEND (1954), although some of her observations do not wholly agree with those of other investigators (e.g. FALCK, 1912, and BUTLER, 1957, 1958 for *Merulius*; MACDONALD and CARTTER, 1961, for *Marasmius androsaceus*).

In the simplest strands, parallel-running hyphae become interwoven and numerous fusions develop between them. The strands grow in thickness by the development of further hyphae from the base so that the strand is broader at the base than at the apex, e.g. *Helicobasidium purpureum* (GARRETT, 1946). In others there may be a tendency for the outermost strands to be more closely interwoven than the central ones, e.g. *Hymenogaster luteus* (TOWNSEND, 1954) and in yet others, such as *Merulius* and *Phymatotrichum omnivorum*, a considerable degree of differentiation has been described and the mode of development investigated (ROGERS and WATKINS, 1938; BUTLER, 1957, 1958).

The development of strands of *Merulius lacrymans* is now fairly well understood. Their development from a wood-block 'food-base' on to the surface of non-nutrient, moist pot-chippings contained in glass tubes 25 or 38 cm long and *c.* 2 cm wide was observed directly. All strands more than 15 μm in diameter were studied (Fig. 4.15). It was shown that the number of strands developed

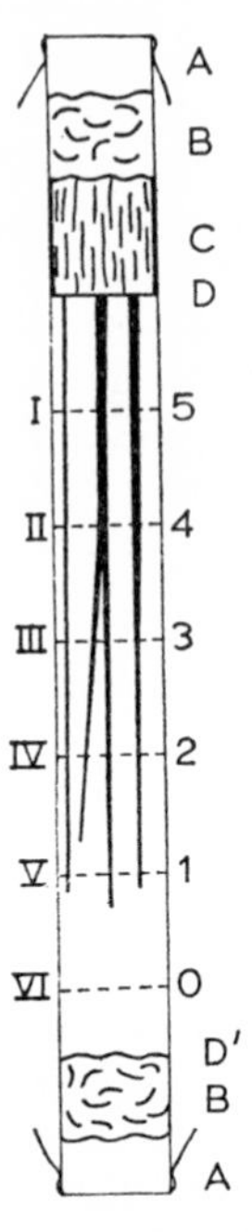

Fig. 4.15. Diagram of tube used to investigate the growth of mycelial strands of *Merulius* after 6 weeks. Key: A, Cellophane caps; B, nylon-wool plugs; C, wood-block foodbase and inoculum; D–D′, column of moist porous-pot chippings of little or no nutrient value. Three strands, one branched, are shown growing through this region; I–VI, successive weekly margins of mycelial growth; 0–5, age of original mycelium in weeks

increased in the first 2 to 3 weeks but, thereafter, remained effectively constant although the thickness of some of the strands increased throughout the period of observation (Table 4.5). Thus initiation of strands is confined to a rather narrow physiological age of mycelium. Studies of strand initiation were carried out on the first 8–10 internodes of growth from inocula on to sterile slides in a humid atmosphere. Main hyphae gave rise to increasing numbers of laterals up to about the fifth node where 2, 3 or more were present. The laterals were of two kinds. Primary laterals, 3–4 μm wide, arose at an angle of 30°–60° to the main apex from the first and second nodes only; their growth rate was equal to, or less than, that of the main hypha. Secondary laterals, only 2 μm in diameter, arose further back, from the third and later nodes, and frequently ran along the parent hypha both forwards and backwards, although they did not adhere to them; they can be described as tendril hyphae. The primary branches, although potentially of unlimited growth, showed two

Table 4.5. Strand development by *Merulius lacrymans*, 7·5 cm from food-base in tubes containing pot-chippings. (From Butler, 1957)

(a) Changes with time in mean strand number

Treatment	Strand diameter μm	Age (weeks) 1	2	3	4	5	6	7	8
Moisture content maintained	15–30	0·3	9	8	7	6	5	6	—
	45–90	0	2	4	6	5	6	5	—
	105–210	0	0	0	0·3	1	1	2	—
	225–450	0	0	0	0	0	0	0·1	—
	Total	0·3	11	12	13·3	12	12	13·1	—
Moisture content not maintained	15–30	0	9	12	10	9	8	5	7
	45–90	0	1	4	6	6	7	8	6
	105–210	0	0	0·3	1	3	2	2	2
	225–450	0	0	0	0	0	0·3	1	1
	Total	0	10	16·3	17	18	17·3	16	16

(b) Changes in diameter (μm) of individual strands as shown by successive transects along the same line round a tube

Strand	Age (weeks) 1	2	3	4	6	8
A	<15	45	75	165	195	270
B	<15	15	90	120	180	180
C	<15	15	30	30	30	30
D	<15	15	120	135	135	135
E	<15	45	120	180	195	270
F	<15	15	30	30	30	30
G	<15	30	15	15	15	15
H	<15	30	30	30	30	30
I	<15	15	15	<15	15	15
J	<15	<15	<15	<15	15	15
Total strands more than 15 μm in diameter	0	9	9	8	10	10

kinds of restriction; in some the diameter gradually tapered as growth proceeded; in others internode length and sub-branch development was restricted close to their origin as compared with more distal parts. Thus the main tendency was for parental hyphae to become enclosed within a mass of growing and branching, tendril hyphae and thus build up a strand. Strand formation was largely confined to main parental hyphae because of the restrictions on growth of the primary laterals already described. However, not all strands initiated grew and Butler has attributed this to competition between them for nutrients. There is no direct evidence that this is the case but it is suggestive that *Merulius* is a translocating fungus and that the most successful strand formers were those with large, central main hyphae, presumably efficient at translocation. That *Merulius* does translocate material was shown by experiments in which the food-base was removed, after which the mean weekly growth rate at the tip declined. A reduction in the thickness and number of strands at 10 cm from the position of the original food-base also occurred (Table 4.6). This suggests that competition for nutrients was taking place between the strands and is in line with Butler's hypothesis of competition for nutrients at strand initiation.

Table 4.6. Effect of altering food-base on strands of *Merulius lacrymans*. (From Butler, 1957)

(a) Mean weekly growth rate (mm)

Treatment	Time from inoculation (weeks)						
	4	5	6	7	8	9	10
Inoculum intact	31	32	39	31	Overgrown		
Inoculum removed (↓)	29	29↓	5	0	0	—	—

(b) Changes in mean strand number with time at 10 cm from inoculum

Treatment	Age (weeks)				
	1	2	3	4	5
Inoculum intact	2	12	12	9	10
Inoculum removed (↓)	↓0	0	0	0	0

(c) Changes in mean diameter (μm) of thickest strand/transect at 10 cm from inoculum

Treatment	Age (weeks)				
	1	2	3	4	5
Inoculum intact	15	45	72	96	99
Inoculum removed (↓)	↓0	0	0	0	0

The subsequent differentiation of the vascular hyphae, fibrous hyphae and structural hyphae (Fig. 4.16) was not followed. Structural hyphae are probably derived from the secondary branches or tendril hyphae, the main and primary branches may give rise to the other differentiated hyphae. It has been supposed that the principal conducting elements are the vascular hyphae with swollen cell cavities, walls with annular and irregularly spiral thickenings

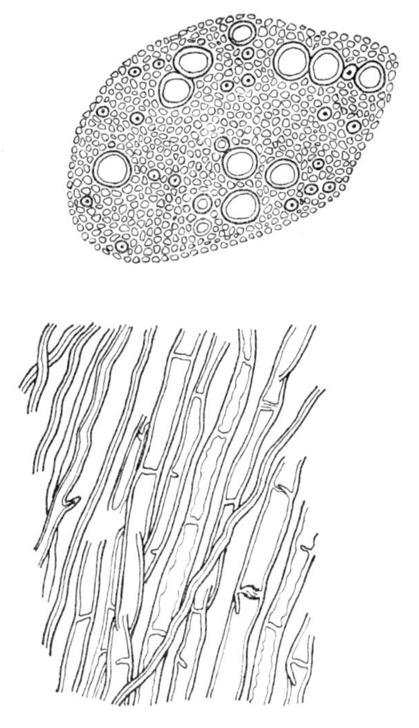

Fig. 4.16. Mycelial strand of *Merulius lacrymans* in transverse (upper) and longitudinal (lower) sections. Vascular hyphae can be recognized by their large lumens, unequally thickened main walls and cross walls broken down in part. Fibrous hyphae are narrow, thick-walled and with a small lumen. Both types are embedded in a matrix of more or less normal hyphae often termed structural hyphae. (×100, after Falck)

and cross-walls broken down so that the cells form open tubes. The fibrous hyphae supply mechanical strength and protection. At present, however, the attribution of these functions to the various cell types is speculative and evidence for the path of conduction in such strands is lacking.

Rhizomorphs are usually more highly organized and grow apically. Moreover, apical dominance may be exerted by the growing tip, so preventing development of lateral branches. TOWNSEND (1954) has described the structure of several which show a zoned structure. The rhizomorph studied most fully is that of *Armillariella mellea*. Two extreme types exist and were earlier described as *Rhizomorpha subterranea* and *R. subcorticalis*. The former consist of white flattened strands formed in the wood of infected trees or in plate culture, the latter are the hard, black flattened strands which can be found just below the bark of infected trees or in the soil around the base. Transition stages from one form to the other can be found.

The rhizomorph is made up of several thousand unbranched parallel hyphae which grow in a coordinated manner some 5 to 6 times faster than the normal hyphae, as measured in plate culture (GARRETT, 1953). Its apex

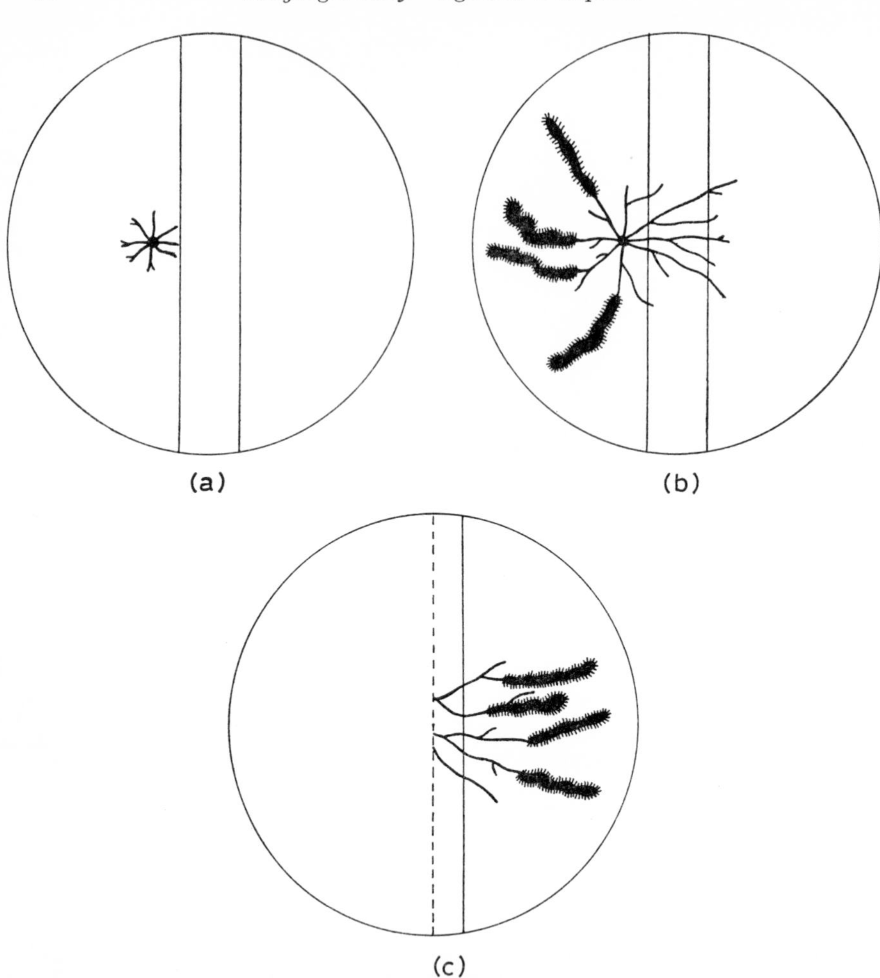

Fig. 4.17. Diagrams to illustrate experiments on the control of rhizomorph initiation and growth in *Armillariella mellea*. (a) The initial stage, a 2 mm trench has been cut out and an inoculum placed on one side. (b) On the inoculum side of the trench the mycelium has produced rhizomorphs which are now outgrowing the normal vegetative hyphae but this has not occurred to those hyphae which have crossed the trench and are now growing on the agar on the other side. (c) The mycelium has been cut along the mid-line of the trench and it and the inoculated agar removed. Now that the mycelium on the remaining agar does not have to translocate nutrients to the first-formed rhizomorphs it has differentiated rhizomorphs of its own

consists of interwoven hyphae becoming denser at its core and enclosed in a mantle of loose hyphae from about 5 mm to 10 mm behind the apex when it is replaced by a firm rind. This fringing mantle makes a rhizomorph look like a root covered in root hairs and an analogous function may well be performed by the mantle.

Table 4.7. The effect of severing the mycelial connexion between rhizomorph-forming colonies and their daughter colonies of *Armillariella mellea* in trench plate experiments. (From Garrett, 1953)

Mycelial connexion	Effects on daughter colonies		
	No. developing rhizomorphs (%)	Mean no. rhizomorphs in colonies producing them [1]	Mean no. rhizomorphs in all colonies [2]
Intact	38 (5/13)	$6 \pm 2{\cdot}06$	$2 \pm 1{\cdot}04$
Severed	93 (14/15)	$15 \pm 3{\cdot}50$	14·341

Statistically significant differences [1], $P = 0{\cdot}05$; [2], $P < 0{\cdot}0001$

The initiation of rhizomorphs, like mycelial strands and sclerotia, is partly determined by nutritional factors. Low carbohydrate and supra-optimal nitrogen (supplied as peptone) reduced the number of initials produced; as carbohydrate was increased initiation was less affected but subsequent growth of the rhizomorphs was reduced. In starvation conditions more initials were produced than could grow. GARRETT (1953) suggested that these observations indicated nutritional control of initiation. He postulated that, 'A certain threshold nutrient status of the substrate is essential for rhizomorph initiation; once a rhizomorph has started growth upon an agar plate, inception of independent initials in its neighbourhood is inhibited by the translocation into the rhizomorph of nutrients absorbed by the fringing mycelium, beyond which lies a zone of agar progressively depleted of nutrients as a result of diffusion'. He supported his views by ingenious experiments. In these he prepared plates and then cut out a 2 mm trench across a diameter. An inoculum of undifferentiated *Armillariella* mycelium was placed 1–2 mm on one side of the trench in the centre of the plate (Fig. 4.17). The mycelium grew out radially, that growing away from the trench produced rhizomorph initials, that growing towards the trench crossed it and grew on the agar on the opposite side. Only in a few cases were initials formed in these daughter colonies and then only a small number of initials (Table 4.7). When the mycelial connexion was severed in the trench and the half carrying the original inoculum removed, a quite different result was obtained. Nearly all daughter colonies produced initials and more were formed by each colony. These experiments show that an inward diffusion of nutrients must be readily available to enable rhizomorphs to be initiated. If this is prevented, as by a trench, then initiation is prevented. Moreover, once initiation has commenced it is inhibited in other parts of the mycelium unless the connexions between the regions are severed. Diffusion is thus canalized within the mycelium to existing initials and only if this is interrupted, as by the severance of the mycelium, can new initials arise in a relatively undepleted medium. The similarities between this situation and the nutritional competition proposed by Butler in connexion with strand formation in *Merulius* will be obvious. On the other hand, much less is known of the actual mechanics of rhizomorph initiation and it is not at all clear why its growth rate should be so much greater than that of individual hyphae. Garrett has suggested that it is because its growth is wholly elongation growth and is not restricted by diversion of nutrients into lateral formation. This is an ingenious but unproven speculation.

The subsequent growth of rhizomorphs is controlled in part by the availability of material in the original food-base from which it has developed. Growth is also reduced as the result of lateral formation but the precise relationships have not been fully investigated (GARRETT, 1957, 1960).

In all these modes of vegetative differentiation of the fungal colony certain themes and modes of behaviour recur. Synchronization of the behaviour of adjacent hyphal tips is clearly involved. The balance between extension growth and lateral formation is another important parameter. Internal regulation of localized growth, whether extension growth or branching, through competition for nutrients is a third factor implicated in several, perhaps all, cases. Thus the time, place and direction of translocation and its mechanism appear to be of vital importance for many growth phenomena, yet very little is known concerning the phenomenon at present (see Chapter 9).

Largely invisible, little studied, the vegetative mycelium of fungi provides an almost endless series of problems whose investigation is long overdue.

Chapter 5

The Fungal Colony—Reproductive Structures

Differentiation of the aerial parts of a fungal colony is principally associated with the production of spore-bearing structures, both for asexual reproduction and in connexion with spore formation subsequent to sexual reproduction.

In the Phycomycetes sporangia and gametangia are the most prominent differentiated structures. In Ascomycetes and Fungi Imperfecti a vast range of conidia are developed which differ in their sizes, shapes, modes of origin and the structures on which they are borne. Amongst Basidiomycetes, excepting Uredinales, the variety of spore types is greatly reduced and conidia are of less frequent occurrence. In both Ascomycetes and Basidiomycetes the 'fruit bodies' or carpophores, which bear ascospores and basidiospores respectively, show a great range of form and include some of the largest individual structures known in the fungi.

ASEXUAL REPRODUCTIVE STRUCTURES

Sporangia and sporangiola

In Phycomycetes motile or non-motile spores are usually formed within sporangia. In relatively undifferentiated forms, e.g. *Rhizidiomyces*, the whole individual apart from the rhizoidal system becomes converted into a sporangium. An example of this type of development will be considered in detail later (p. 100). There are considerable differences between such species in the mode of liberation of their spores. The production of localized or irregular holes in the wall or of 'manhole-like' sporangial lids are some of the methods employed. Nothing is known of the processes involved.

In most hyphal Phycomycetes the sporangia are confined to localized regions of the mycelium and separated from the rest of the coenocyte by a septum. A genus like *Pythium*, however, shows a wide range of differentiation which is paralleled, in part, by other species. In a species such as *Pythium gracile* and its allies any region of the mycelium may become converted into zoospores without that region even being delimited by a septum (BUTLER, 1907). The sporangium may be simple or considerably branched but, otherwise, it is hardly separable from a normal vegetative hypha. In other species zoospore development occurs in a tightly packed, branched region of the mycelium, the whole region being cut off by a septum, e.g. *P. aphanidermatum*. In the 'damping off' organism, *P. debaryanum*, definite terminal globular sporangia are developed on hyphae above the substrate. A sporangium may develop in two ways after it has put out a germ hypha; either the end swells to form a vesicle into which the sporangial contents pour and develop as zoospores to be liberated by the rupture of the vesicle, or it simply gives rise to a mycelium. The former developmental pattern occurs in wet situations, the latter under drier conditions. This pattern is followed by other Phycomycetes. Aquatic forms usually develop zoospores, although not always, by the vesicle formation characteristic of *Pythium*, while aerial, or terrestrial, forms produce non-motile spores each of which germinates directly to give a mycelium.

In many of the more highly differentiated and larger terrestrial Phycomycetes, both parasitic and saprophytic, the spores are developed on specialized sporangiophores. These frequently assume characteristic and remarkable shapes. An apparently simple type is that characteristic of mucoraceous fungi. It consists of a linear, aerial hypha often developed from a specialized cell of the mycelium (a trophocyst, e.g. *Pilobolus*) and bears a single terminal sporangium. The more or less globular sporangium is separated from the sporangiophore by a terminal septum, which often develops in such a way as to protrude into the sporangial cavity and form a dome-shaped columella. It is implied in many accounts that the terminal region is multinucleate from the outset although this matter needs to be carefully studied. The spores arise from a process of vacuolar cleavage, the details of which are complex and are only just beginning to be described precisely, e.g. in *Gilbertella persicaria* (BRACKER, 1966). The differentiated spores are generally multinucleate but, once again, it is not always clear whether this is so at the time of delimitation (see, for example, HARPER, 1899; GREEN, 1927). The genetical consequences of producing multinucleated spores may be of considerable importance and will be affected by the mode of origin of the spores, i.e. whether uni- or multinucleate when initiated (cf. pp. 380–382).

In other Phycomycetes the sporangiophores bear whorls of laterals, variously branched, or the sporangiophore itself branches in a variety of ways. The lateral sporangia borne on such sporangiophores are often smaller than the terminal ones and contain fewer spores, possibly only one. They may differ in other ways also. For example, the terminal sporangium of *Thamnidium elegans* contains several thousands of spores and has a columella and walls which deliquesce when the sporangium is mature. Its laterally borne sporangia contain far fewer spores, lack a columella and do not have deliquescent walls.

Small sporangia like these are called sporangiola. Not all sporangiola are globose; some are cylindrical and contain a linear file of spores, e.g. *Syncephalastrum*; these are often termed merosporangia. The merosporangiferous fungi include some of those with the most wonderfully branched sporangiophores known. Their patterns of branching are complex and the branches terminate in inflated tips (sporocladia) on which the merosporangia are developed, e.g. *Coemansia, Kickxella* (Fig. 5.1). The analysis of the development of such structures from an initial single hyphal tip is quite impossible at present. Detailed descriptions have been given of their development (BENJAMIN, 1958, 1959) and these provide essential material for further morphogenetic analysis. It must be admitted that the functional significance of some of these bizarre structures is obscure at present.

When the contents of a sporangiolum form a single spore the distinction between a sporangium and a conidium becomes blurred. In *Cunninghamella*, for example, the sporangiola are single-spored and indehiscent, while in the related *Choanephora* the wall is only partially separable from its contained spore. It is of some relevance, therefore, that electronmicrographs of *Cunninghamella* reveal a distinctly two-layered wall. The outer wall has been interpreted as that of the sporangiolum, the inner wall as that of the spore (HAWKER and HENDRY, cited in HAWKER, 1965). This type of observation is desirable in these forms where only a single wall appears to be present, e.g. *Kickxella* (BENJAMIN, 1958).

A beginning has been made in the analysis of the induction and development of some of the simpler sporangia-bearing structures and some of these will be considered later (p. 102).

Conidia and conidiophores

All other spore types in the fungi, with the exception of ascospores and basidiospores, may be described as conidia since they are not enclosed within a sporangium. It must be realized that this single term is used to cover a multiplicity of asexual spores produced in many different ways from an enormous range of spore-bearing structures, the conidiophores. Much attention has been devoted to the study of the kinds, origin and position of conidia due to the need to classify the Fungi Imperfecti. Different techniques have been employed, some mycologists have described the mature structures while others have sought to elucidate developmental details. Corresponding to these different approaches are different terminologies and no general agreement has yet been reached. The notions of VUILLEMIN (1911, 1912) provide the most promising basis. He recognized two distinct modes of origin and hence distinguished thallospores from conidiospores proper.

Thallospores originate by the transformation of pre-existing hyphal elements and are not, in general, detached easily. An excellent example of a thallospore is the chlamydospore, a spore type referred to and illustrated earlier (pp. 2, 10). Amongst those thallospores which are more readily separated from the mycelium are the so-called arthrospores formed by hyphal fragmentation, e.g. *Geotrichum candidum*, or from specialized regions of hyphae,

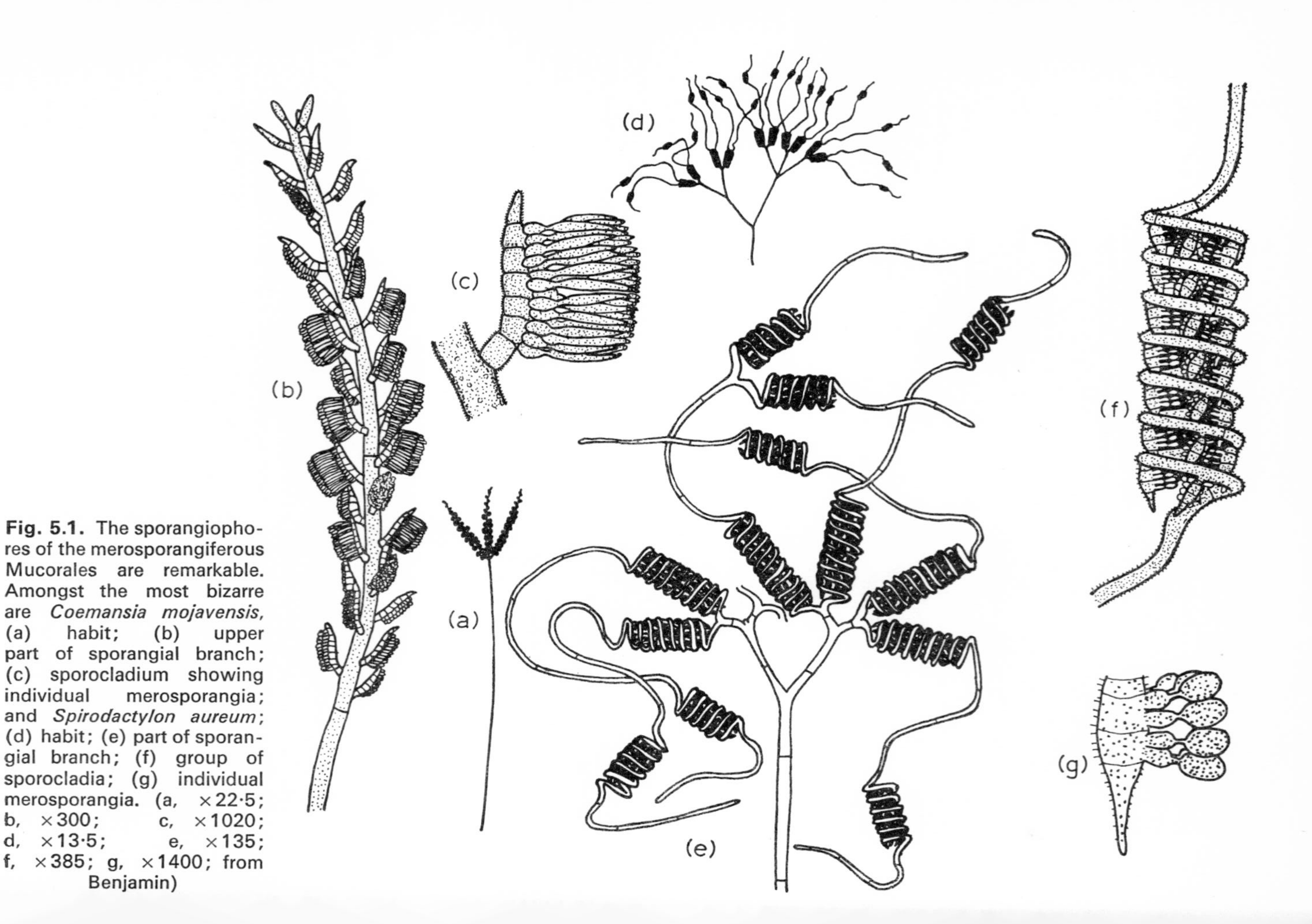

Fig. 5.1. The sporangiophores of the merosporangiferous Mucorales are remarkable. Amongst the most bizarre are *Coemansia mojavensis*, (a) habit; (b) upper part of sporangial branch; (c) sporocladium showing individual merosporangia; and *Spirodactylon aureum*; (d) habit; (e) part of sporangial branch; (f) group of sporocladia; (g) individual merosporangia. (a, ×22·5; b, ×300; c, ×1020; d, ×13·5; e, ×135; f, ×385; g, ×1400; from Benjamin)

e.g. oidia of *Coprinus* spp. (see Fig. 1.12). The buds, or blastospores, of yeasts are also thallospores as well as the multicellular dictyospores produced by the swelling and longitudinal and lateral septation of the terminal regions of a hypha or even of whole hyphae, e.g. *Alternaria tenuis*.

Vuillemin's 'true' conidiospores (*Conidium verum*) arise externally to the hyphae; they do not remain attached and are separated from their parent hyphae through the activity of the fungus. They may arise at almost any point on a hypha from a small projection which may, or may not, be readily detectable. Alternatively, they may be borne upon specialized conidiophores and arise from terminal or lateral projections, e.g. *Botrytis*, or from specialized, terminal cells, usually swollen at the base and tapering to an open-ended tip and termed phialides. The conidia are borne at, or within, the tip of the phialide and become separated as formed, e.g. *Aspergillus*, *Cephalosporium*. HUGHES (1953) has made detailed studies of spore production and recognized various types. Subsequent workers (e.g. TUBAKI, 1958) have added to, or disagreed with, Hughes' classification and concepts and, at present, the position is confused. Until there is complete agreement on the description of the developmental stages further morphogenetic analysis cannot proceed. For example, blastospores are included by TUBAKI (1958) in a group where the spore originates from a blown-out vesicle and are distinguished from arthrospores; chlamydospores are not included at all because of their lack of taxonomic significance. A series of recognized spore types is shown in Fig. 5.2. Meanwhile, some general suggestions concerning the developmental processes involved may be made.

Thallospores are evidently formed by a variety of localized processes related to wall formation. It may be supposed that increased knowledge of septum formation and the ultrastructure of hyphal walls will enable hypotheses to be made concerning the formation of arthrospores and dictyospores.

Many conidia and blastospores (although defined as thallospores by Vuillemin) have in common the formation of a terminal or lateral 'blown-out' protrusion of an undifferentiated or differentiated hyphal wall. The actual formation of conidia by phialides poses a further host of problems since budding, septation and wall separation of various kinds may be involved. It may be surmised that when more is known concerning the mechanics of formation of normal lateral branches and septa it should be possible to apply such knowledge to the problems of development of what are, in essence, tiny lateral branches. Such studies should, of course, enable some progress to be made in the developmental analysis of the bewildering variety of conidiophore forms.

The structure of conidiophores is almost as varied as that of the conidia and their modes of origin. Conidiophores may be simple or branched in various ways; an excellent account with clear illustrations is given by LANGERON and VANBREUSEGHEM (1952; 1962, English edition). It will suffice to say that the branching follows the patterns well known from the insertion of flower-heads on stems (Fig. 5.3). The origin of these modes of branching is obscure but

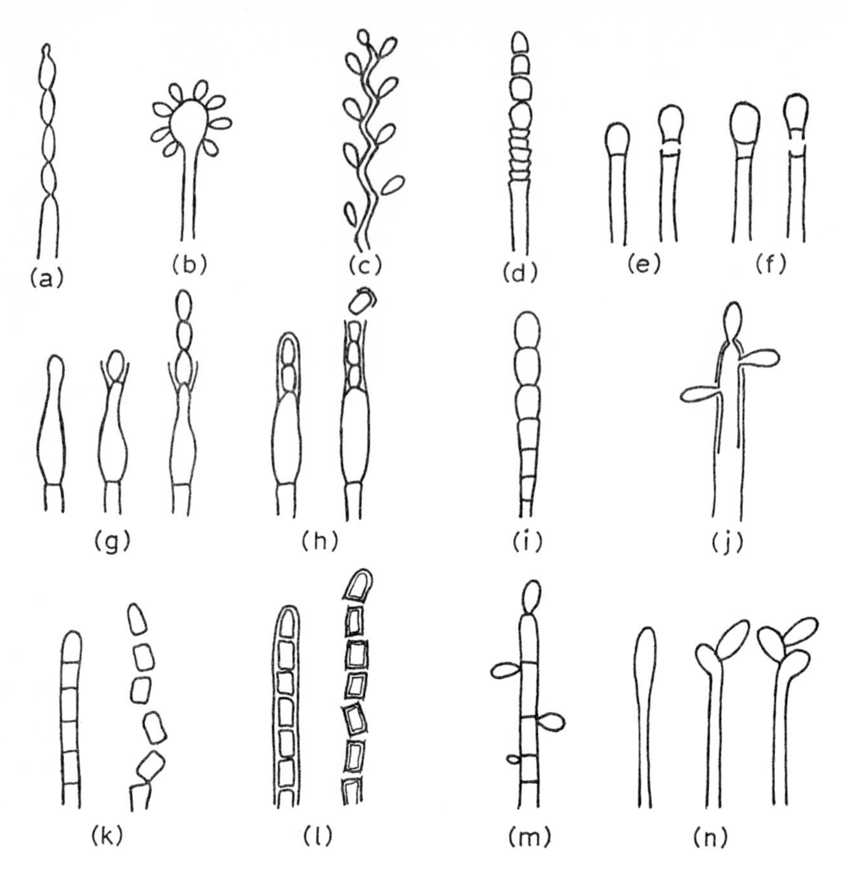

Fig. 5.2. Diagram to illustrate the origins of conidia in the Hyphomycetes. (a), (b) blastospores; (c) sympodially proliferating conidiophores with terminal conidia (sympodulae); (d), (e), (f) conidia formed by inflation of whole apex and growing by proliferation (d), or without (aleuriospore), sometimes separating by circumscissile rupture (f); (g), (h) phialospores, the latter wholly endogenous; (i) basipetal chain of arthrospores; (j) porospores, blown out through minute pores; (k), (l) exogenous and endogenous arthrospores; (m) conidiophore with basal meristem and blastospores; (n) basipetal, alternately obliquely formed blastospores. (After Madelin)

Robertson's experiments on hyphal apices will be recalled and it is clear from these that checks on the growth of the hyphal apex are frequently followed by swelling and/or branching. Thus a series of internal checks on the apical growth of successive branches could account for many of the branching patterns found in conidiophores (ROBERTSON, 1959). PARK's (1963) demonstration that the 'staling' products of many fungi induce a growth inhibition acting on the hyphal tip which is frequently followed by development of asexual reproductive structures is another example of the same principle.

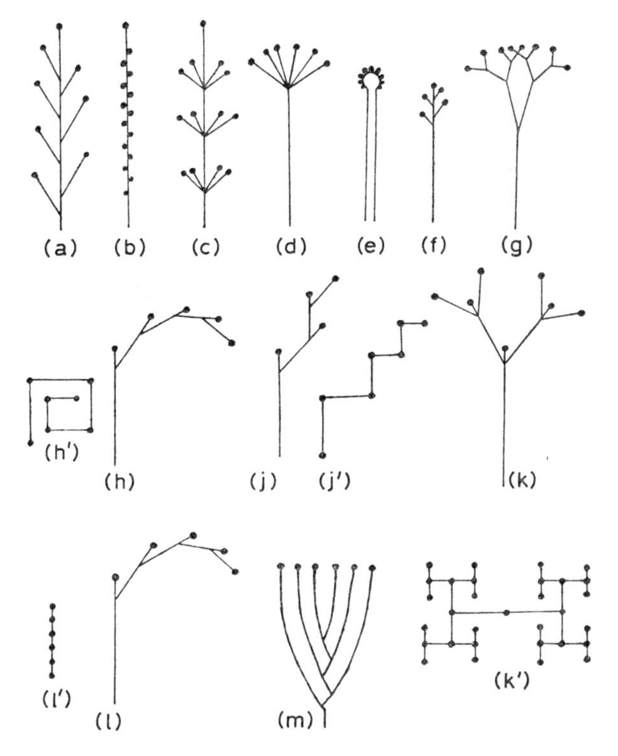

Fig. 5.3. Arrangements of conidia and branching of conidiophores. (a) cluster; (b) spike; (c) verticillium; (d) umbel; (e) head; (f) monopodial, basipetally branched; (g) dichotomous branching; (h′), (h) monochasium; (j), (j′), helicoid cyme; (k) dichasium; (l′), (l) scorpioid cyme; (m) rhipidium; (h)–(m) are forms of sympodial branching. (From Langeron, after Zopf)

In addition to the simple branching systems, many fungi produce aggregated, conidial-bearing structures. The simplest form is shown by the development of numerous adjacent conidiophores which become loosely or tightly intermingled to form a compound structure termed a coremium (Fig. 5.4). An excellent range of these structures is shown in the genus *Penicillium* and has been studied recently by Carlile (CARLILE, *et al.*, 1961, 1962a, b). In the most complex forms, such as *P. claviforme*, the growth of the whole structure is regulated and that of its individual hyphae synchronized; hypha-to-peg and peg-to-peg anastomoses are common. Experimental decapitation is followed by regeneration of the apex as a whole. The final phase of development is the synchronized production of conidia at the apex. In the related *P. isariiforme*, growth is indeterminate and apical growth continues simultaneously with sporulation, about 1 cm below the apex. Anastomoses do occur but parallel orientation is largely determined by unidirectional illumination; light is, indeed, essential for coremium formation. In diffuse light coremia develop as a 'feathery brush', presumably due to the failure of the phototropic orientating mechanism. In other species, e.g. *Stysanus stemonites*, the component hyphae become laterally adherent in a manner reminiscent of that in sclerotia or rhizomorphs. A large group of Fungi Imperfecti, the Stilbaceae, have in common the possession of coremia and many of the characteristic morphological coremial forms are given names such as *Isaria*, *Graphium*, etc. However, it is frequently found that different forms do not

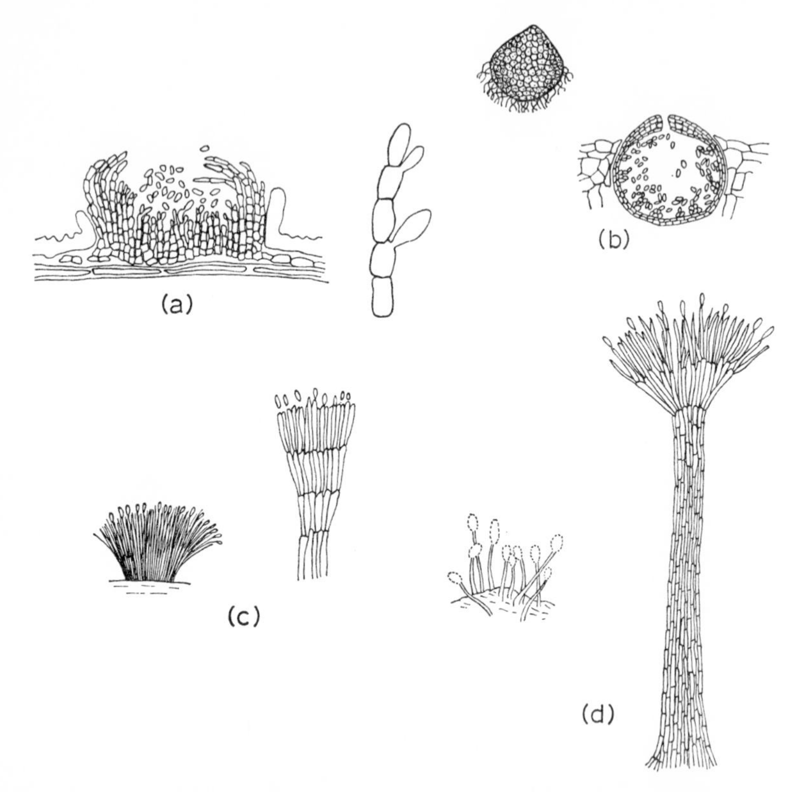

Fig. 5.4. Aggregated conidial bearing structures. (a) Acervulus of *Catenophora pruni* with enlarged conidiophore; (b) pycnidia of *Phoma* spp., note conidia in sectional drawing; (c) and (d) synnemata of *Myriothecium rorida* and *Didymostilbe* sp., respectively, note terminal conidia but rather different hyphal construction of stalk. (All redrawn from Barnett)

produce their spores in the same way. Accordingly, these are classified according to their method of sporulation and it is clear that a coremium-like form is a particular biological solution to the problem of elevating large numbers of spores in close proximity. The different ways in which the spores are borne are also related to the type of growth of the aggregation, either determinate, e.g. *Graphium*, *Stysanus*, or indeterminate, e.g. *Isaria*.

Three other types of aggregated conidiophores occur, the sporodochium, acervulus and pycnidium. In the first of these the conidiophores are loosely massed together and arise from a mass of aggregated hyphae or stroma. Thus in appearance sporodochia appear to form crusts, discs or cushions of tissue, e.g. *Nectria*, *Hypoxylon*, and they may be fleshy, woody or carbonaceous in texture depending upon the numbers and distribution of hyphae with greatly thickened walls in the stromata. The associated conidiophores each possess their own characteristic features. An acervulus is similar but the mass of conidia arise without the development of a stroma. Acervuli are often found in parasitic fungi where closely-packed, parallel-orientated conidiophores develop just below the epidermis of the host and then burst through as an erumpent mass, e.g. uredo-, aecidio- and teleutosori of rust fungi. Acervuli are formed by Oomycetes, Ascomycetes, Fungi Imperfecti and Uredinales. Perhaps the most complex asexual reproductive structures are pycnidia. They occur in Ascomycetes and Fungi Imperfecti and as the spermogonia of rust fungi. They are globose or flask-shaped and hollow. Their inner wall is lined with a layer of conidia, which are eventually liberated through an orifice in the wall or at the end of a protruding neck. They are nearly always sessile and arise sometimes from mycelial hyphae or, more commonly, from a stroma in which they are usually embedded, at least in part. The development of numerous pycnidia was described in great detail by the nineteenth-century mycologists and summarized, with further original observations, by KEMPTON (1919). He recognized three patterns of development which are extremely similar initially to those already described for the development of sclerotia. Simple meristogenous development occurs when adjacent cells of a hypha become transversely septate. Septa then develop at right angles to these and in the third plane so that the originally adjacent cells become converted into a young, three-dimensional aggregation of cells. In some cases two or more adjacent hyphae may undergo this kind of localized division in concert; this is the compound meristogenous mode of development, e.g. *Phoma pirina*. In other cases, however, the cell aggregation arises as the result of several hyphae going towards a common centre and becoming interwoven, their cells divide transversely and the products frequently become fused together. This is symphogenous development and is shown by *Zythia fragariae*. The cells in these masses, however formed, increase their volume and continue to divide and incorporate adherent cells. Eventually the central cells become separated or disappear by resorption so that a number of cavities develop. Each cavity is then lined with conidia or conidiophores either over the whole surface, which may be irregular, or at the base alone. An ostiole usually appears towards the end of the developmental period. Details are variable and not clear. In some cases hyphae curve away from each other in a particular region leaving an orifice, or become arranged in other patterns. The actual opening may be brought about by absorption of water by pycnidial contents, their swelling and consequent stretching of the wall adjacent to the ostiole. It is not uncommon for the spores to be extruded during wet weather embedded in a long, mucilaginous filament.

Many detailed descriptions of the development of these complex asexual reproductive organs exist but there is little experimental work on the causes of their development. Nothing is known, for example, of the stimuli which lead to the localization of stroma development, at a particular region in the thallus. Nutritional factors are undoubtedly involved since many fungi will only produce pycnidia in culture if cellulose is incorporated as a source of carbohydrate. It is clear that the production of quite massive structures such as these, for many are visible by unaided vision, must involve considerable nutritional demands so that a prior period of growth and storage is almost certainly a necessary prerequisite. In view of what is now known concerning the initiation and morphogenesis of simpler spore-bearing structures (p. 104) and of larger ones, such as basidiocarps (p. 133), the present time would seem opportune for the experimental investigation of the factors involved in the production of complex, spore-bearing structures.

Morphogenetic studies on simple spore-bearing structures

A beginning has been made to the study of factors which specifically initiate the development of simpler spore-bearing structures. A valuable review of this topic is included in HAWKER (1957). She has stressed the truth of KLEBS' view (1898–1900) that development of an asexual reproductive structure is usually favoured by the same general conditions which favour mycelial growth but usually within a narrower range of environmental conditions. It is also often the case that asexual reproduction is less dependent upon exogenous nutrients and this implies a dependence upon endogenous metabolites produced by the diverse biochemical pathways developed during differentiation of the mycelium. Once the appropriate internal balance has been achieved then a 'trigger' reaction or stimulus may act as an apparently specific stimulus to initiate a particular development.

BLASTOCLADIELLA

There are now some reasonably well documented examples of such situations. One such case is the water mould *Blastocladiella emersonii*, which is being studied intensively by Cantino and his colleagues who have recently summarized their work (e.g. CANTINO and LOVETT, 1964; CANTINO, 1966).

Motile zoospores of the fungus settle down, retract their flagellum and develop a uninucleate germ tube. This grows into a rhizoidal system embedded in the substrate matrix and virtually devoid of cell contents. An upper, extra-matrical cell develops into a single multinucleated cell which is wholly transformed into either an ordinary colourless sporangium (OC plant) or a thick-walled pigmented resistant sporangium (RS cell). The former liberates zoospores when mature, the latter remains dormant until induced to release its zoospores. Once released from either type of sporangium, the zoospores initiate a new cell generation. Much of Cantino's success has sprung from his use of submerged, synchronized, single-generation cultures of some 10^8–10^9 individuals. Any phase can be produced at will in quantity so enabling analysis or treatments to be readily undertaken (LOVETT and CANTINO, 1960; GOLDSTEIN and CANTINO, 1962).

Cantino and his associates have been able to describe the metabolism and composition of the developing thallus and to detect a 'trigger' reaction which shifts development from the OC to the RS pathway. They have begun to analyse the intracellular operation of this trigger reaction.

During the early phase of development of the extra-matrical region, growth is exponential and the biochemical composition is changing continuously. Dry weight, soluble polysaccharide and soluble protein all increase exponentially, the first two exceeding the last by some 20% in the middle third of the plants' generation time. The rate of polysaccharide synthesis and generation time are increased by exposure to white light but no change is induced in the final OC development. However, if the concentration of bicarbonate in the medium (or the CO_2 tension) is increased before three-fifths of the plants' generation time has elapsed then a profound metabolic shift is induced and this is followed by morphological changes to give RS plants. The metabolic changes detected are largely quantitative but also include novel developments such as melanin deposition in the walls and carotene formation in the cytoplasm. Quantitative changes involve decreased growth rate, Q_{O_2} and glucose utilization and increased synthesis of chitin, lipid and soluble polysaccharides; major changes in the composition of the soluble proteins a reduction in the activity of α-ketoglutaric dehydrogenase, succinic oxidase and cytochrome oxidase with an enhanced activity of isocitric dehydrogenase. These last changes are believed to be the 'trigger' reactions which determine the morphogenetic response. Bicarbonate is thought to induce a block in the tricarboxylic acid cycle (TCA cycle, see p. 269). In OC cells the activities of α-ketoglutaric dehydrogenase and isocitric dehydrogenase are approximately the same but the consequence of the bicarbonate 'trigger' is to shift the balance from unity to 1:10 in favour of isocitric dehydrogenase. Thus oxidative decarboxylation of isocitrate to α-ketoglutarate is replaced by reductive carboxylation of α-ketoglutarate, viz:

$$\text{Isocitric acid} \underset{+HCO_3^- \text{ or } CO_2}{\overset{-HCO_3^- \text{ or } CO_2}{\rightleftharpoons}} \alpha\text{-ketoglutaric acid}$$

It is not yet clear how all the other changes follow from this weakening of the TCA cycle but there seems little doubt that this is the key process. The evidence is largely circumstantial. Interference with the TCA cycle might be expected to act as a rate-limiting factor, so reducing O_2 and glucose consumption and hence growth; such changes are observed. Moreover, since metabolic routes have been altered changes in synthetic activity might be expected and these are in fact observed for polysaccharide, protein, melanin and carotene. It is known that the isocitrate produced is converted to succinate and glyoxylate and the latter transaminated to glycine. It has been suggested, by comparison with other organisms, that the succinate and glycine could then act as intermediates in carotene and purine synthesis. It is clear that the detailed elucidation of the biochemical changes is likely to take some time but at least a start has been achieved. Reversal of RS development is possible in cells up to about 45% of the generation time. As might be expected, it is associated with a restoration in the balance of the activities of the two dehydrogenases. The point of no return in RS cells is achieved in a very short (2 hr) critical period associated with profound changes in the composition of cellular RNA and, in particular, decrease in soluble RNA and an increase in NaCl-insoluble RNA.

Finally, Cantino's research group have recently examined changes in the last 5% of the generation time, which is associated with the final phase of sporangial development, spore formation (CANTINO, 1965; DOMNAS and CANTINO, 1965a, b). This work has been done with OC plants. There appears to be an increased requirement for arginine and although its fate is unknown, it has been speculated that it could be involved in the formation of the characteristic nuclear cap. Glutamic acid was also consumed rapidly in the last stage of development and incorporated into proteins. A *b*-type cytochrome also seems to occur in a soluble condition just before sporogenesis.

The significance of all these observations is not yet clear but they have been described briefly because this work represents the most detailed biochemical description available in a fungus for any morphogenetic changes associated with reproduction. The examples now to be described are more concerned with the recognition and primary analysis of 'trigger' mechanisms.

PHOTOMORPHOGENETIC EFFECTS ON MUCORACEOUS FUNGI

Light is known to have profound effects on the initiation of asexual reproductive structures in several mucoraceous fungi (CARLILE, 1965). A particularly popular genus for investigation has been *Pilobolus*. This dung fungus can be grown on a medium of defined composition and its sporangia show diurnal periodicity, both of development and discharge which are related to the diurnal rhythm of light and dark (KLEIN, 1948). PAGE (1952, 1956, 1959) has shown that exposure to light triggers off trophocyst formation in *P. kleinii* once it has been growing for some time. Two processes are involved, the flow of protoplasm to the site of the initial, with a consequent localized swelling of the hypha, and its delimitation by transverse walls. The action spectrum for initiation showed the range 3800–5100 10^{-10}m to be effective with a fairly sharp peak at 4800 10^{-10}m. This suggested that either carotenes or flavins might be acting as photoreceptors and experiments were done with inhibitors of carotene formation such as diphenylamine and the supposed competitive inhibitor of riboflavin, lyxoflavine. This latter substance inhibited trophocyst formation (and growth also) but this could be relieved by the simultaneous addition of riboflavin (Fig. 5.5). Diphenylamine inhibited, but did not prevent, carotenogenesis and had no effect on trophocyst production. Page, therefore, postulated a flavin-mediated light absorption as the first stage in trophocyst formation; thereafter they develop into sporangiophores in darkness. Indeed, if cultures are in continuous light or darkness only a few sporangiophores develop. Optimum development of both kinds of structures was obtained by exposure of the initial culture to 2 min white light followed by 19 hr of darkness and then a second exposure of 64 min to light. Thus the trophocysts have to 'mature' in the dark until a second photomorphogenic 'trigger' initiates sporangiophore formation. Page has also shown that the subsequent elongation of the sporangiophore is dependent upon an exogenous supply of thiazole. Different species of *Pilobolus* differ in their light reactions. *Pilobolus*

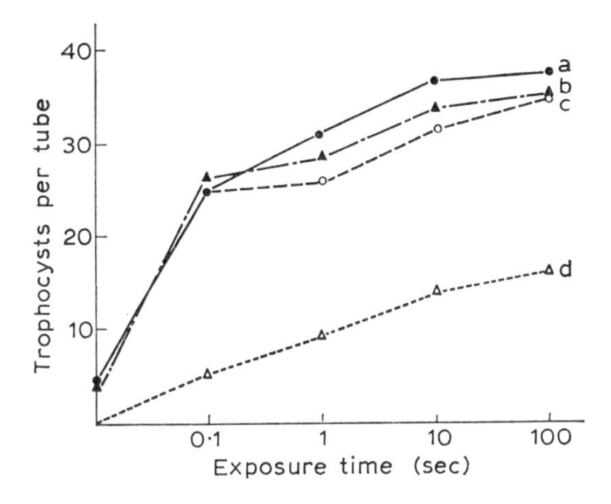

Fig. 5.5. Effects of lyxoflavine and riboflavin on trophocyst formation by *Pilobolus*. (a) Basal medium alone; (b) basal + lyxoflavine (100 μg/ml.) + riboflavin (10 μg/ml.); (c) basal + riboflavin; (d) basal + lyxoflavine. All cultures grown 4 days in dark then exposed to either 0·1, 1, 10 or 100 sec of white light. (From Page)

longipes and *P. sphaerosporus* have no light requirement whereas *P. crystallinus* and *P. umbonatus* require it only for sporangium formation. However, light-induced periodicity of sporangium discharge is known in *P. sphaerosporus* but this will be considered with other mechanisms of spore discharge (see Chapter 6).

Alternating light and dark periods, that is photoperiodic effects, act as morphogenetic 'trigger' reactions in other fungi. In *Choanephora cucurbitarum* and *Thamnidium elegans* photoperiodic induction effects determine the presence or absence of conidia in the former and the relative balance of sporangia and sporangiola in the latter. BARNETT and LILLY (1950) have shown that only the order light/dark periods will induce conidial formation at an intensity of 699·4 lux; continuous light or darkness, or dark/light periods are ineffective. Continuous illumination at low intensities such as 10·76 lux will, however, initiate conidial development. They postulate a two-step inductive process. The difficulties inherent in generalizing about such effects are well shown in this species by CHRISTENBERRY's (1938) report on another strain of the fungus. Here, although maximum conidial production was achieved with light/dark periods, especially with long wavelengths, he was also able to induce conidial development in total darkness. However, his results may be accounted for as a consequence of his poor temperature control for this too is known to affect conidial and sporangial production (BARNETT and LILLY, 1955). A similar but more complex situation has been described by LYTHGOE (1961, 1962) in *T. elegans*. The sporangiophores develop in an analogous sequence to that already described for *Phycomyces* (cf. p. 51). Continuous illumination above about 15°C results in:

(a) transient decreases in the growth rates at specific times in Stage 1 and 4 sporangiophores,
(b) alteration of the total duration of Stage 1,
(c) the induction of sporangium formation during Stage 1 (although this shows a transient decrease at a specific time) and
(d) formation of sporangia rather than sporangiola.

With the same total dosage of light, intermittent 30-second periods of light followed by dark removed the transient check to sporangium formation without greatly altering the total proportion of sporangia produced. Lythgoe provides evidence that the transient checks both to sporangiophore growth and sporangial induction in Stage 1 under continuous illumination are either the same or intimately linked. He also suggests that two processes are involved in the induction of sporangia. An hypothetical activator molecule is produced but cannot become incorporated in the appropriate metabolic sequence because a light-activated inhibitor is also produced. In intermittent illumination, of course, the dark period either results in inactivation of the inhibitor or its elimination, so that the activator can become incorporated in the metabolic sequence leading to sporangium induction. In continuous light the inhibitor must eventually become inactive and the transient inhibition is thus removed. Nothing is yet known of the biochemical processes involved with these three fungi.

COREMIUM FORMATION IN PENICILLIUM

CARLILE (1965; CARLILE *et al.*, 1961, 1962a, b) has investigated the action of light as a stimulus to coremium formation in five species of *Penicillium* in the subsection Asymmetrica-Fasciculata. All species are sensitive to illumination. Some such as *P. clavigerum* show reduced growth rate in continuous illumination but the production of coremia is not influenced by light. *Penicillium expansum* shows a phototropic response of its hyphae to unilateral illumination and laterals which develop into conidiophores arise on the illuminated side. In *P. isariiforme* normal coremium formation depends upon unilateral illumination and both the individual hyphae and the coremium itself are positively phototropic. The behaviour of the whole structure appears to be explicable in terms of the phototropic orientation of the component hyphae. Transfer to a 24-hr dark period results in a cessation of growth but otherwise the rate of growth is not affected by light and dark periods. Sporulation is dependent upon light. The most complex situation is found in *P. claviforme*. While coremium initiation is affected both by light and nutrition it can occur independently of the former. However, subsequent development is determined by a brief, photosensitive phase during the first 2 mm of elongation of the coremium. The blue end of the spectrum is most effective. Transfer to darkness in this sensitive period results in the development of shorter, more branched coremia. Growth after the first 2 mm and the production of conidia at the apex when growth ceases are not photosensitive. Decapitation of a coremium is followed by an 18 to 30-hr quiescent period followed by regeneration. From the first 1·5 mm of the growth the regenerated coremium is again photosensitive.

It is to be hoped that the identification of these 'trigger' reactions will provide a starting point for a complete biochemical and morphogenetic analysis of the differentiation of asexual reproductive structures.

CARPOPHORE CONSTRUCTION

Many Ascomycetes and most Basidiomycetes, with the exception of the rust and smut fungi, develop large, readily visible, spore-bearing structures termed ascocarps and basidiocarps respectively. The largest fungal carpophores include the South American *Geopyxis cacabus* with a stalked apothecium 1 m high and 50 cm across, but the majority of Discomycetes are sessile and but a few millimetres in diameter at the most. Basidiocarps can be even larger. In the North Temperate region the giant puff-ball, *Calvatia gigantea*, has been recorded as 1·6 m long by 1·35 m broad and 24 cm high and a bracket fungus, *Polyporus squamosus*, as 2 m in diameter and weighing 31·5 kg, while a toadstool-like fungus from the Malagasi Republic, *Boletus colossus*, was some 30 cm high with a stalk 22 cm in diameter, a cap 60 cm broad and 4–6 cm thick, and the whole weighing 6 kg.

A great deal of descriptive information is available concerning these structures but remarkably little experimental work has been done and much of it has not proved very rewarding in the sense of clarifying the developmental processes. Some of the most provocative descriptions have been those of Corner, who has attempted to analyse the form-factors involved in carpophore structure in Discomycetes, agarics, polypores and Clavariaceae (e.g. CORNER, 1929–1931; 1948a; 1932a, b, 1953; 1950; 1961; 1966). His method has consisted of tracing the directions in which hyphae run and their modes of branching and differentiation and so inferring the growth processes involved. Such an analysis provides an essential foundation for the experimental investigation of the hypotheses to which they give rise.

Certain features are shown by all carpophores. The spore-bearing structures, asci or basidia, are usually arranged in a regular parallel manner to form an hymenium amongst which other cell types are interspersed. The hymenium, composed of terminal apical cells, is usually enclosed, supported by, or carried aloft on sterile tissue and it is the range of form of this sterile tissue that gives the diversity of form to the carpophore. Hyphal differentiation within the carpophore appears to serve one of two functions: it either provides supporting cells to maintain the mechanical rigidity of the

Fig. 5.6. Diagram to illustrate various types of carpophores in section. The hymenial layer is represented by a darker line or area in each case. (a)–(k) ascocarps; (l)–(w) basidiocarps. (a)–(e) Discomycetes such as *Peziza, Helvella, Mitrula, Morchella, Sarcoscypha*; (f)–(h) Pyrenomycetes such as *Sordaria, Daldinia*, sclerotia and stromata of *Claviceps*; (i)–(k) hypogaeus forms such as *Genea, Tuber, Balsamia*; (l), (m) pore fungi such as *Polyporus* and *Boletus*; (n) club fungus, *Clavaria*; (o) gill fungus, *Psalliota*; (p) spine fungus, *Hydnum*; (q) crust fungus, *Stereum*; (r) stink-horn, *Phallus*; (s) puff-ball, *Lycoperdon*; (t) earth-star, *Geaster*; (u) bird's-nest fungus *Cyathus* with peridiolum; (u′) single peridiolum highly magnified in section; (v) *Sphaerobolus* with projectile v′ containing basidiospores; (w) hypogaeus form such as *Hymenogaster*

carpophore or it is concerned with the spore liberation mechanism. Some examples of carpophore shape in relation to the position of the hymenium are given in Fig. 5.6.

Ascocarp construction and differentiation

Ascocarps, whatever their final form, originate in much the same way from an ascogonium. This is a specialized lateral branch of a hypha which may be associated with a definite sexual reproductive organ, the oogonium, or may merely be a convoluted spirally twisted hypha. It is not clear why so many ascogonia possess this spirally twisted form. Once sexual reproduction has been initiated, through the coming together of compatible nuclei in the ascogonium, further development commences. Hyphae develop from the outside of the ascogonium and from immediately adjacent regions and invest and enclose the structure, e.g. *Aspergillus* (Fig. 1.8, p. 12). In some Pyrenomycetes this happens prior to the association of compatible nuclei and the small structures so formed are known as protoperithecia, e.g. *Neurospora* (Fig. 1.9, p. 13). The ways in which sexual reproduction is initiated are discussed later (Chapter 15).

APOTHECIAL DEVELOPMENT

In the development of apothecia the subsequent development of this loose weft of enveloping hyphae can follow one of three patterns (Fig. 5.7). In angiocarpic forms, e.g. *Ascobolus stercorarius*, the hyphae branch within the weft, probably in a sympodial manner, and grow in all directions to give a ball of interwoven hyphae. Cells in the outer layers enlarge to form a pseudo-parenchymatous cortex. The walls of these cells may also thicken. Within the cortex a mucilage cavity now develops in the upper half of the mass. Hyphae from cells lining the cavity form a regular palisade layer and those on the floor of the cavity become intercalated with binucleated ascogenous hyphae which have developed from the dikaryotic ascogonium. Palisade hyphae become arrested in their upward growth and may swell towards their tips, a common result of a check on hyphal growth (cf. p. 49), to form paraphyses (sterile hairs) while the ascogenous hyphae develop into asci. In this way the tissue lining the floor of the cavity becomes transformed into a hymenium. The expansion of the internal tissue ruptures the cortical layers overlying the mucilage cavity and so exposes the hymenium. At this stage the hyphae around the rim of the mucilage cavity form a marginal growing point and a secondary phase of expansion growth may take place. In this the apices of the marginal growth zone grow and divide sympodially. The laterals produced inwards become orientated parallel to the walls of the hymenium and become transformed into paraphyses; those produced outwards enlarge and contribute to the cortical pseudo-parenchymatous tissue.

Gymnocarpic development, e.g. *Ascobolus magnificus*, differs in that in the earliest development the hyphae investing the ascogonium do not form a closed sheath but continue to grow upwards in a well-defined palisade layer. Eventually this upward growth is checked from the centre outwards and so

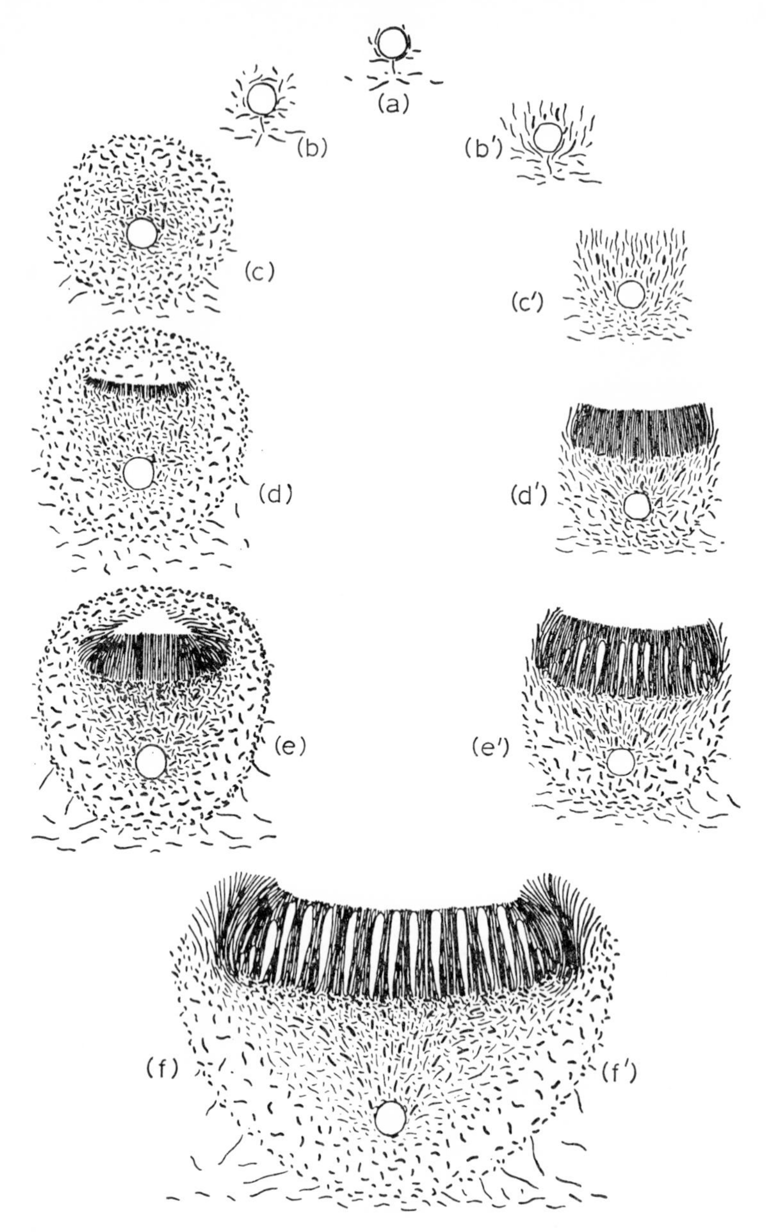

Fig. 5.7. Diagrams to illustrate the development of different types of apothecia. The archicarp is indicated by a hollow circle in the non-pileate forms. (a)–(f) angiocarpic development; (a), (b′)–(f′) gymnocarpic development. (After Corner)

the hymenium is initiated, while at about the same time the cells of the lower part of the ascocarp commence to enlarge. Thus the hymenium develops on a mass of pseudo-parenchymatous cortical tissue. The development of the ascogenous hyphae and asci and the initiation of a secondary phase of marginal growth follow the same course as that described for the angiocarpic apothecium.

Hemi-angiocarpic development as shown by *Peziza aurantia*, for example, is intermediate. Hyphae arch over the ascogonium. They do not form a closed sheath but become converted into cortical hyphae; otherwise the development more resembles that of gymnocarpic forms.

The essential processes in such developments are (a) the transfer of growth from one region to another, as in the development of the marginal growth zone, (b) the inflation of cells often associated with wall thickening, (c) the change-over from monopodial to sympodial growth and (d) the development of mucilage. Processes (a)–(c) have all been described before in connexion with the hyphal differentiation of vegetative structures (Chapter 4). However, the formation of gels and mucilages frequently accompanies the differentiation of spore-bearing regions or structures and has not been considered before. MOORE (1965) has recently made a careful study of the ontogenesis of gel tissues in several Ascomycetes and Basidiomycetes. She has shown that there are two distinct processes. In the first, the mucilage is developed by the disintegration of the hyphae and may, or may not, be preceded by the swelling of their walls. This kind of development certainly occurs in some Ascomycetes (CORNER, 1929). Alternatively, mucilage detected by selective staining with mucicarmine can be seen to develop within the hyphae and it is then secreted through minute pores in the wall. Once secreted it often undergoes a rapid change in viscosity to form a gelatinous mass which may be inter-hyphal or, if there is a high production of mucilage, a complete gel matrix. A region where there is intensive production of mucilage or fungal gel has been termed a gliatope by Moore. This kind of development is seen in the gelatinous species of *Bulgaria* or the gelatinous heads of genera such as *Leotia*. Very little is known about the nature of fungal gels but there is evidence that some of them are uronic acid derivatives, e.g. manno-arabino-uronic and polyglucuronic acids (QUILLER, 1942, cited by MOORE, 1965).

Leotia is of interest because it is an example of a stalked apothecial type. A number of such forms are known, including the edible Morels, *Morchella* spp. which may stand 20 cm or more high. CORNER (1929–31) has shown that the stalked forms arise by an intercalated stage of upward growth of the hyphae investing the ascogonium so that it is carried up on a 'shaft of hyphae'. Quite early in development the hymenium appears and a marginal growing zone is either established, e.g. in *Helvella elastica*, or not, e.g. *Mitrula pusilla*. Intercalary growth in the former leads to a reflexing of the cup-shaped hymenium to give the characteristic helvelloid appearance; in the latter a cap-like development results since the phase of marginal growth is absent. In essence the whole ascocarp of *H. elastica* is differentiated when it is about 1 mm high although its final size may be 4–6 cm. This phase of expansion is due to a 3 to 4 times increase in cell size (3–5 μm to 10–20 μm) of the cortical cells but there must also be intercalated growth after the stem has been laid down to account for the, approximately, fifty-fold expansion. This kind of development by expansion is characteristic of the larger carpophores and will be considered in more detail in relation to the development of the basidiocarp (pp. 125–128).

PERITHECIAL DEVELOPMENT

The early development of the perithecium is similar to that of the angiocarpic apothecium. A globose or piriform mass of hyphae develops, the outer cells of which become somewhat inflated to form the wall of the perithecium, or in some cases, the cleistothecium. In the former a pore, the ostiole, is formed by the splitting of hyphae and this may terminate in an elongated neck-like protrusion or a short papilla. In any event the ostiole is lined within by short hair-like growths or periphyses. All the other tissues within the perithecial wall have been defined as the centrum and they may develop in a number of ways. These have been described by several mycologists, notably LUTTRELL (1951) who has devised a scheme of classification based on the types of centrum development. In all cases a perithecial cavity arises and its floor and possibly the walls become lined with hymenium. It is hardly possible to detail the various developmental patterns here. The kind of distinction drawn is to be seen by comparing a genus such as *Sordaria* with one such as *Nectria* (the coral-spot fungus of dead wood). In the former, the perithecia are separate with ostiolar epiphyses and long club-shaped asci interspersed with evanescent, basal paraphyses so that none are visible when the ascospores are ripe. In *Nectria* the perithecium and ostiole resemble those of *Sordaria*, the asci are clavate or cylindric but true basal paraphyses are completely lacking. Instead, pseudo-paraphyses arise as sterile threads from the roof of the perithecial cavity, grow down to the cavity floor and there become attached amongst the asci. Thus in both genera the hymenium may appear to be similar superficially but, in fact, the origins of the sterile cells between the asci is quite different in the two cases.

There is little differentiation of hyphal types other than those already described for apothecial forms. Ascogenous hyphae and asci develop in a similar manner although there are variations in the number of wall layers and the mode of opening of the asci (see Chapter 6). In several taxonomic groups the perithecia develop in extensive stromatic tissues; this is of the usual pseudo-parenchymatous type but the component hyphae are characterized by having thick darkly pigmented walls. The nature of these cell wall constituents is obscure. There is no doubt that acid-resistant aromatic materials do occur in cell walls and these have been referred to as melanins and fungal 'lignin', although it is unclear what exactly is meant by the latter term (BU'LOCK and SMITH, 1961). A considerable number of Ascomycetes are separated taxonomically from stromatic Pyrenomycetes as the Loculoascomycetes (LUTTRELL, 1955). In these fungi the asci are bitunicate and develop in cavities within a stroma. If a single ascus develops, a uniloculate ascostroma, it is termed a pseudo-perithecium since, unlike a true perithecium, there is no special wall around the centrum. It should be realized that often this difference is not easily recognized in a mature ascocarp and may only be detected after a study of the developmental processes.

SUBTERRANEAN ASCOCARPS

A further kind of ascocarp is the subterranean one of the Tuberales or truffles. These gastronomic prizes, together with the false truffles (Elaphomycetales), consist of an irregularly shaped, tough peridium enclosing a mass of tissue which can be regarded as being made up of convoluted folds on the surface of which globose asci enclosing 2 to 8 ascospores develop. Because of their subterranean habit little or nothing is known of their development but it will presumably be found to be not unlike that of some of the sclerotia described earlier (p. 82, Chapter 4).

ASCOSPORE FORMATION

Ascospores arise within the cytoplasm of the ascus by free-cell formation after meiosis. Early observers such as HARPER (1905) and FAULL (1905) gave rather different accounts of the process. Harper claimed that each centriole at the tip of a beaked nucleus gave rise to a system of astral rays which elongated and fused together to form a hemi-ovoid membrane whose margin extended and eventually enclosed a nucleus and some cytoplasm (Fig. 5.8). Faull, on the

Fig. 5.8. Ascospore delimitation. Astral rays emanating from a beaked nucleus appear to delimit the cytoplasm around the nucleus. ($\times$ 670, after Harper)

other hand, could not detect the fusion of astral rays although he described how a more or less hyaline, limiting layer developed in the region of each centriole and eventually delimited the spore. It is possible, as HAYMAN (1964) has suggested, that both mechanisms may occur. So far, electron microscope studies have not given a clear picture of what is taking place (BRACKER, 1967).

Basidiocarp construction and differentiation

Basidiocarps exceed all other carpophores in their range of size, complexity of construction and degree of tissue and hyphal differentiation. Despite this complexity the underlying processes, already described, are still the same, viz hyphal aggregation and compaction, localized branching and swelling, wall thickening and gelatinization. The 'toadstool' type of structure is that which has received most detailed study but even here development is far from adequately understood. Attention will, however, be focused largely on this type in this account. The occurrence of similar kinds of hyphal and tissue construction in boletes, polypores, spine fungi (Hydnaceae) and even in Gasteromycetes, superficially so different, suggests that common principles of development will eventually be discovered.

The origin of a basidiocarp may be a single cell, e.g. *Coprinus stercorarius* (BULLER, 1933, and Fig. 5.9). Here lateral branches divide rapidly and more

Fig. 5.9. The origin of a basidiocarp from a single dikaryotic cell in *Coprinus stercorarius*. (×325, ×170, ×260, from Buller, after Brefeld)

or less irregularly to form a tight cluster of hyphae whose innermost cells rapidly become compacted to form a pseudo-parenchymatous mass of tissue. Differentiation of the adult tissues and organs is achieved rapidly in these minute primordia which may remain checked at this stage. Subsequent development consists largely of their sudden, rapid expansion to full size and the maturation of the hymenium. How far it is really the case that a basidiocarp can arise from a single cell is disputed. REIJNDERS (1963) has claimed that development almost certainly occurs from more than a single cell in many species of toadstool. It seems probable that further studies will reveal that basidiocarp primordia arise in much the same ways as other aggregated structures such as sclerotia or pycnidia (see pp. 82; 99). The subsequent phases of an early differentiation, expansion and final maturation do, however, seem to be universal.

ANCHORING STRUCTURES

Many basidiocarps are bulky and, therefore, require a considerable reserve or source of material on which to draw during development. As a consequence basidiocarps do not usually arise until a fairly extensive vegetative mycelium has developed and are frequently found to be attached to appropriate food-bases by well developed strands or rhizomorphs, e.g. *Phallus*, *Armillariella* or, in some cases, even more bulky structures called pseudorhiza. The basidiocarp primordia of such toadstools often develop a mycelium on the food-base of submerged wood or roots. The base of the primordium then elongates by intercalary growth, so developing the pseudorhizal structure, and carries the basidiocarp to the surface of the ground. Growth in length of the pseudorhiza then ceases although it frequently increases to its maximum girth just below ground level. Pseudorhiza may be annual and unbranched, e.g. *Collybia radicata* 15–16 cm long, or branched, e.g. *Coprinus macrorhizus*, or, most exceptionally, perennial and branched structures, e.g. *Collybia fusipes*, *c.* 30 cm long (BULLER, 1934).

PRIMORDIUM STRUCTURE AND DIFFERENTIATION

The primordia are of two main types. In one type characteristic of basidiocarps with stalks, the primordial shaft consists of longitudinally-running, interwoven hyphae forming a compact sub-cylindrical or sub-conical structure. REIJNDERS (1963) has contrasted this type with what he terms the 'diffuse type', common in Aphyllophorales. Here the hyphae spread out from a growth centre to form a hemisphere or an inverted fan-like segment. In the compact type, apical synchronous growth is shown by the hyphae at the distal end of the primordial shaft so that a stipe (stalk) develops. From the growing end the pileus (cap) develops by exogenous or endogenous growth. In the former, peripheral main hyphae grow outwards and this is often associated with a check in the growth rate of more centrally placed hyphae. Thus a marginal growth zone develops and this may become curved towards the stipe, as a consequence of the more rapid inflation of hyphae on the upper side. By intercalary growth gills arise on this smooth uncurved surface as a series of radiating, parallel anastomosing or dichotomizing ridges. Such a development, in which the gills arise on the surface and are never enclosed

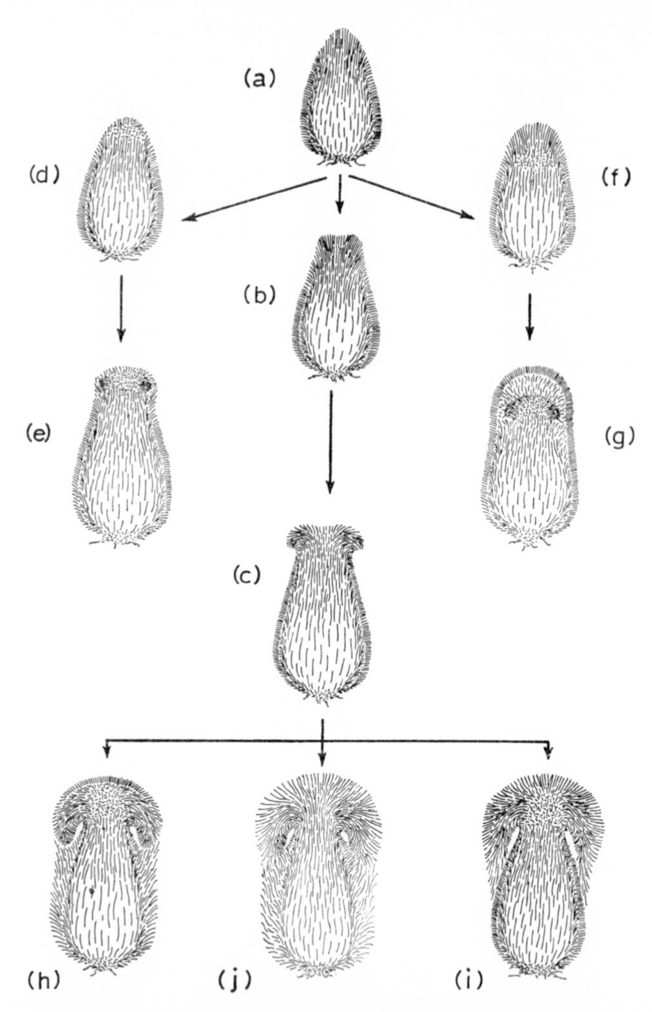

Fig. 5.10. Diagrams to illustrate development of different types of basidiocarp and of the veils. (a) The primordial shaft, corticated and with the medullary hyphae beginning to enlarge; (b) and (c) a pileus developing exogenously; (d) and (e) a pileus developing slightly endogenously; (f) and (g) a deeply endogenous pileus developing; (h-i) outgrowths from the surface of the stem (h), from the margin of the pileus (i), and from both regions (j) of a primarily gymnocarpic form.
Thus (a), (b), (c) represent development of a gymnocarpic type such as *Cantharel us*; (a), (d), (e) and (a), (f), (g) represent slightly angiocarpic types such as *Collybia* and fully angiocarpic development, e.g. *Amanita*; (h), (i) and (j) show how originally gymnocarpic types can become enclosed by partial or complete veils and so appear, eventually, to be angiocarpic or hemi-angiocarpic in the mature basidiocarp. (After Corner)

in a sheet or sheets of hyphae, is termed gymnocarpic, e.g. *Cantharellus* (Fig. 5.10b, c).

However, when the pileus develops endogenously, sub-terminal lateral branches grow outwards a short distance below the apex of the primordial shaft. The main growth zone thus tends to become displaced from the apex of the primordial shaft to the annular margin of the outwardly directed laterals. The pileus develops more (Fig. 5.10f, g) or less (Fig. 5.10d, e) enclosed within the original and still growing hyphae of the apical region of the primordial shaft. The gill folds still arise on the under surface of the pileus and this is either preceded by, or is simultaneous with, the development of an annular cavity which surrounds the stalk below the pileus. This gill cavity is, of course, enclosed from the moment of its initiation by the peripheral layers of hyphae of the primordial shaft. Such a development is said to be angiocarpic and the layer, or layers, of tissue which enclose the basidiocarp form a veil. A universal veil totally encloses a basidiocarp, e.g. *Amanita* spp. When the stipe eventually elongates, the veil may become stretched at the margin of the gill cavity and ultimately torn. The lower part forms a cup-like volva at the base of the stipe, while the upper part becomes a series of scales on the pileus. In other forms this does not occur due to changes in the component hyphae so that they may become converted either into a mealy layer on the pileus and a granular ring on the lower part of the stipe, e.g. *Amanita excelsa*, or even to a slimy coating, e.g. *Panaeolus campanulatus*. A partial veil only encloses the region between the margin of the pileus and the stipe and, on being ruptured, may leave a ring of tissue on the stipe, e.g. *Lepiota* spp. and sometimes teeth-like remnants around the margin of the pileus, e.g. *Psathyrella candolleana*. The genus *Cortinarius* can be recognized by the cobweb-like filaments, or cortina, which cover the gill chambers of young basidiocarps and often persist as fibrils, at the margin of the pileus. It is the partial veil which ruptures to form the cortina.

The veils do not originate exclusively as a consequence of the endogenous origin of the pileus. CORNER (1934) has shown clearly that a further process is involved, namely the outgrowth of hyphae from either the margin of the pileus or the surface of the stem or from both these regions (Fig. 5.10g, h, i). Such outgrowths can result in the gills becoming enclosed in an originally gymnocarpic form or can increase the degree of angiocarpy in forms with endogenous pileus development. For example, in the gymnocarpic species *Lactarius torminosus* there is a noticeable outgrowth of marginal hyphae from the pileus and there is considerable growth from the stipe of *Collybia radicata*. In *Pluteus admirabilis*, hyphae from the incurved margin mingle loosely with those from the stem and just fail to enclose the potential gill region within a partial veil. A rudimentary marginal veil is developed by such outgrowths in *Collybia apalosarca* where the pileus is slightly endogenous (hemi-angiocarpic). In *Psalliota* species the angiocarpic development arises from the deeply endogenous development of the pileus but this is enhanced by considerable marginal outgrowth from the pileus.

Corner's analysis is important for it draws attention to the three variables which determine the degree of enclosure of the basidiocarp, viz (a) the exogenous or endogenous origin of the pileus, (b) the amount of additional hyphal outgrowth from the pileus margin, the stipe or both regions and (c) the extent and effectiveness of the intermingling and fusion of these hyphal outgrowths. A universal veil is determined largely by the place of origin of the pileus, a partial veil more by the extent of hyphal outgrowth, intermingling and fusion. Angiocarpy, hemi-angiocarpy and gymnocarpy are thus derived conditions and it is not surprising that these terms have caused some confusion since intermediate conditions can obviously arise through the interplay of the three basic variables.

GILL STRUCTURE

While there is now some understanding of the degree of enclosure of the basidiocarp, the origin of the gill cavity and of the gills is still far from being understood. Gills usually arise as more or less symmetrically radiating ridges which grow out centripetally on the underside of the pileus. They may extend from the stipe to the margin of the pileus or for only part of this distance and they may be free from each other or show varying degrees of fusion or branching. As the gills diverge, secondary gills arise at a certain distance in the gaps between the primaries and this process may be repeated to give tertiary or even quaternary gills. The primary gills are attached for a greater or lesser distance down the stipe from its junction with the pileus; the degree of attachment may be altered during development. The various types of gill attachment, e.g. free, adnate, decurrent, etc., are of considerable diagnostic value.

The regularity of gill development and of gill spacing suggests some kind of regulating mechanism which may be termed a gill field and, indeed, there is some experimental evidence in favour of this view (see p. 141). The number and order of gills in a fruit body seem to bear some relation to the width of the stem apex and of the pileus, to judge from CORNER's (1934) data on *Collybia apalosarca* (Table 5.1). Such data must be viewed with caution, however, for they are derived from measurements on mature basidiocarps whereas the form factors responsible must operate in the primordial stage prior to the phase of rapid expansion.

At least three patterns of gill origin have been described. In genera of the gymnocarpic cantharelloid alliance the gill folds arise as 'lines of excessive intercalary growth of the initially smooth hymenium', so that, 'it is thrown, thereby into obtuse ridges' (CORNER, 1966). The process is evidently not a purely mechanical folding. The gills consist of a hymenium of terminal basidia, derived from the sub-hymenial layer, which continues to intercalate more basidia during growth. The gill fold lacks any well developed central tissue, or trama, the hyphae within the fold being pulled apart as their terminal regions develop into basidia. The whole surface of the gill fold, including its margin, is fertile from the first and the fold continues to thicken by the continuous outgrowth of new basidia which may enclose and protrude beyond those which were formed first. In such species, e.g. *Cantharellus infundibuliformis*, the thick, obtuse gill folds so developed may come to resemble those

Table 5.1. The number of primary gills, the ranks of gills and the widths of the stipe apex (mm) and pileus (mm) for 25 basidiocarps *Collybia apalosarca*. (From Corner, 1934)

(a) Gill ranks, pileus width and primaries

Gill ranks	Pileus width	No. of primaries
	4	12
	5	11
	6	17
1–2 or 2	8	11
	14	15
	17	20
	21	15
	10·5	14
	18	19
	18	20
2–3 or 3	19	19
	20	18
	20	13
	22	19
	28	18
	36	31
	42	23
	45	25
	65	30
3–4 or 4	72	30
	74	31
	76	28
	86	32
	89	26
	93	29

(b) Width of stipe apex, pileus and primaries

Stipe apex	No. of primaries	Pileus width
	11	5
0·5–0·1	11	8
	12	4
	13	20
	14	10·5
1·1–1·9	15	14
	15	21
	17	6
	18	20
	18	28
	19	19
2–3	19	22
	20	17
	20	18
	23	42
4–6	25	45
	31	36
	31	74
	26	89
	28	76
7–8	29	93
	30	65
	30	72
	32	86

produced by *Hygrophorus* spp. but these latter arise in a different manner, typical of most of the Agaricaceae (Fig. 5.11). In this pattern, each gill is derived from a distinct, downwardly-directed outgrowth of hyphae, in the form of a flat plate, from the underside of, and at right angles to, the pileus. The main growth region is the edge of the gill which, initially, is sterile. Thus a well developed trama is produced. Peripheral hyphae tend to diverge somewhat from the parallel orientation and become compacted at their extremities to form the sub-hymenium from which the outwardly directed palisade hymenium is developed. The thickening of the gill may be brought about by the basifugal inflation of the tramal cells. This divergent pattern may be apparent in the mature gill which is then termed bilateral, a term used by FAYOD (1889) who first systematized descriptions of gill structure. It is now clear, however, that a divergent phase is an ontogenetic stage in the development of other types of agaric gills (REIJNDERS, 1963).

A readily derived pattern, the regular type, arises through an increase in the number of central parallel hyphae and a masking of the originally divergent hyphae so that the sub-hymenium is compressed and the hymenium appears to arise abruptly at right angles to the tramal hyphae. In the irregular type the further ramifications of the tramal hyphae result in the hyphae

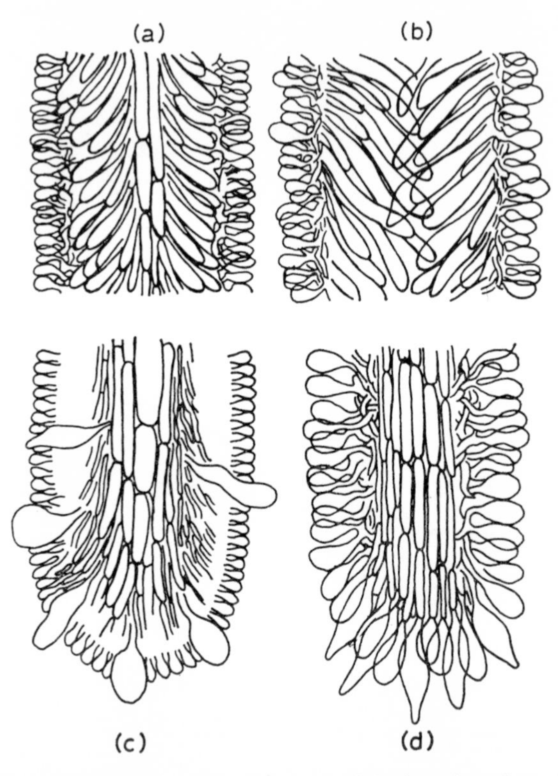

Fig. 5.11. The arrangement of cells in the trama of the lamellae of agarics as seen in tangential section. (a) bilateral trama as in *Amanita*; (b) inverse trama as in *Pluteus*; (c) regular trama; (d) regular to bilateral trama as occurs in many agarics. (From Heim)

appearing to be interwoven. The most remarkable pattern, characteristic of *Pluteus* and *Volvariella* spp., is the inverse trama. Here the tramal hyphae consist of inflated downwardly directed cells whose free ends are at the mid-line of the gill as seen in tangential section. In development, however, the gills of these fungi pass through a brief divergent phase to become regular. Then hyphae from the sub-hymenium develop, become inflated and replace the original tramal cells, so giving the inverse pattern.

Numerous intermediates exist, as might be expected, the same patterns can arise in different ways and there are other possible modes of construction.

For example, *Amanita* spp. have typically a bilateral trama but this arises secondarily from a regular trama, the inflated cells directed obliquely outwards arising from the central hyphae of the trama and eventually replacing them. CORNER (1966) has also suggested a further mode of development in the genera *Craterellus* and *Pterygellus*. Here the gill folds arise in a manner similar to that in most cantharelloid fungi but 'the folded hymenium is reinforced by a radiating strand of hyphae parallel to those of the pileus, not descending at right angles as in agarics'.

The factors which cause the hyphae to change their direction of growth in these various ways is unknown. The existence of certain agarics such as *Asterodon* spp. (CORNER, 1948b), in which conspicuously thickened hyphae can be traced through the basidiocarp to the hymenium, demonstrates the reality of a continuing regulatory process throughout primordium development.

DIFFERENTIATION IN APHYLLOPHORALES

In other Homobasidiomycetes with similar shaped basidiocarps the shape of the hymenial surface differs. Nevertheless similar processes are involved In *Polystictus xanthopus* CORNER (1934) has pointed out that the pores arise acropetally behind the pileus margin in a regular manner and are as crowded as possible. In the pore area the hyphae are interwoven but in the regions of the dissepiments, hyphae grow outwards and downwards, delimiting the final pore size from the outset, and ceasing growth after some 120–140 μm when they lose their polarity and become lobed or shortly branched and so partially occlude the mouths of the pores. Such regular and ordered differentiation suggests the existence of a pore field, analogous to a gill field, and presumably similar processes determine pore development in Boletaceae and spine development in Hydnaceae. There is evidently a rich field of morphogenetic study awaiting investigation in these fungi.

DIFFERENTIATION IN GASTEROMYCETES

The development of gasteromyceteous basidiocarps is not well understood and this reflects in part their complicated and heterogeneous structure and in part their frequently inaccessible and subterranean habit. FISCHER (1933) has attempted to bring some order into the patterns of development which occur (Fig. 5.12). In essence the development is either angiocarpic or effectively so, since in forms such as *Hemigaster* or *Elasmomyces* the potentially hymenial surfaces become enclosed at an extremely early stage. Thus, essentially, gasteromycete basidiocarps can be regarded as being enclosed within a peridium, one or more layers thick. Inside, one or more hymenium-lined cavities develop from potentially sporogenous tissue, the gleba. Once basidiospores have developed the gleba may undergo autodigestion or conversion into dried-up, thread-like strands. The tissue between these cavities constitutes the trama and serves, in different species, for conduction, mechanical support or even dehiscence in some cases (cf. p. 153). The variation arises primarily in the origin of the glebal tissue. Four types have been recognized:

(a) The lacunar type where, in the extreme case, a single cavity or, more

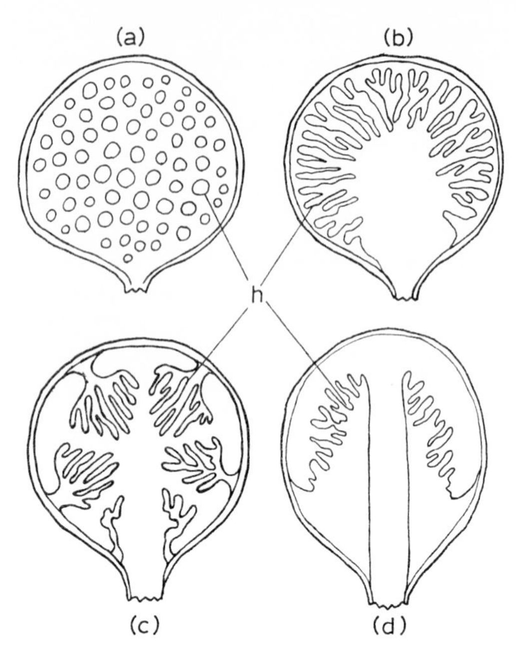

Fig. 5.12. The basic forms of construction in Gasteromycetes. (a) The lacunar type, cavities lined with hymenium, e.g. *Scleroderma*; (b) The coralloid type, an irregular cavity arises in the upper half covered by hymenium, e.g. *Hysterangium*; (c) the multipileate type, several irregular cavities, e.g. *Clathrus*; and (d) the unipileate type, the hymenium on the underside of the pileus-like structure, e.g. *Phallus* (h = hymenium). (From Fischer)

usually, cavities, lined with an hymenium, arise within the basidiocarp, e.g. Nidulariaceae, the bird's-nest fungi, several cavities; *Protogaster*, a single cavity.

(b) The coralloid type, in which a single cavity arises in the upper half of the basidiocarp. This is enclosed above by tramal tissue which may unite with the inner side of the peridium but, at the base of the cavity, branching masses of hyphae grow upwards and outwards to form 'coral-like' branches covered by a continuous hymenium, e.g. *Hysterangium* spp.

(c) The multipileate type is one in which a number of tramal branches grow rapidly, reach the inner surface of the peridium and there spread out over it. Coral-like branches then grow from these main branches into the spaces into which the basidiocarp has been divided, e.g. *Clathrus* spp.

(d) The unipileate type resembles the multipileate but here a single tramal branch develops as a column of tissue from base to apex and spreads out in a pileus-like manner, on the inner surface of the peridium. Hymenium develops on coralloid branches on the underside of the pileus-like structure, e.g. *Secotium* spp.; *Phallus* spp., the stink-horn.

It will be seen that development in these fungi is complex and, at present, comparison with developments in other basidiocarps is difficult. Nevertheless, there are some similarities. The origins of cavities in mycelial tissue is a problem superficially similar in the gleba and in the gill cavity of Agaricaceae and the divergent and regular patterns of tramal differentiation in the gills of these fungi have their parallels in Gasteromycetes. Comparisons with the patterns of development in the subterranean Ascomycetes, such as the truffles, might also be rewarding. At present there is a great need for developmental studies in the Gasteromycetes, especially of the tropical and subterranean forms which are virtually unknown.

HYPHAL DIFFERENTIATION

The primary phase of development, in which the basic structure of the basidiocarp is laid down, is accompanied or followed by hyphal differentiation of the outer surfaces, the context (or flesh) and of the hymenium itself.

CORNER's (1932a, b; 1948b, 1950, 1953, 1961, 1966) pioneering studies have

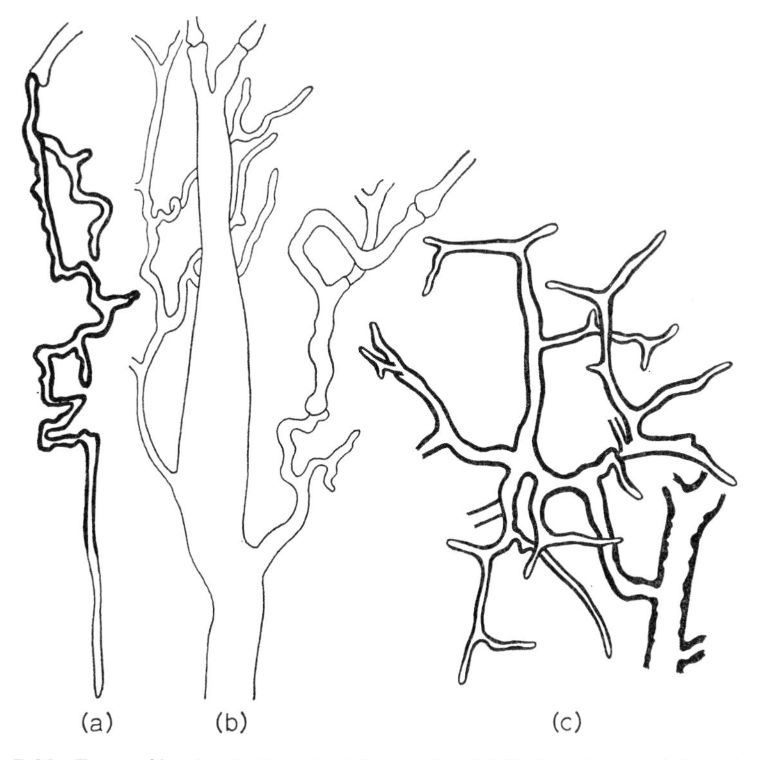

Fig. 5.13. Types of hyphae in Homobasidiomycetes. (a) Skeletal hypha of dissepiment of *Polyporus betulinus* arising from generative hypha; (b) inflated and elongated skeletal hypha with tapering ends of *Trogia stereoides* and generative hyphae; (c) branched binding hypha of *Polyporus sulphureus* arising from broader generative hypha. (a, ×570; b, ×310; c, ×285. After Corner)

provided a rational basis for the description and, ultimately the analysis, of hyphal differentiation in Agaricales and Aphyllophorales. He has recognized three basic types of hyphae (Fig. 5.13):

(a) Generative hyphae: thin-walled, branched, usually septate, with or without clamp connexions. They may become inflated, especially in Agaricales. They give rise to the other hyphal types.
(b) Skeletal hyphae: thick-walled, typically unbranched, commonly aseptate, straight or slightly flexuous constructional hyphae of the first order.
(c) Binding hyphae: narrow, thick-walled, much branched, rarely septate hyphae of intricate and limited growth.

In the most complex basidiocarps generative hyphae provide the ground plan and skeletal hyphae the firm constructional framework which is locked together firmly by the binding hyphae. Corner has recognized five principal constructional types.

(a) Monomitic, where the hyphae are all of one type, generative, although this does not preclude the presence of hyphae of different sizes and shapes, e.g. inflated and uninflated.
(b) Dimitic, with two hyphal types present. Two kinds of dimitic forms occur, those with generative and skeletal hyphae, e.g. *Polyporus betulinus*, and those with generative and binding hyphae, e.g. *P. squamosus*.
(c) Sarcodimitic, with generative hyphae and thick-walled, very long, fusiform, inflated cells whose ends taper to relatively narrow septa. These cells replace the skeletal cells of other fungi. Examples are *Mycenella* spp., *Rimbachia paradoxa* and *Trogia* spp. which are predominantly tropical.
(d) Trimitic, with generative, skeletal and binding hyphae as in the common European polypore of dead standing wood, *Polystictus versicolor*.
(e) Sarcotrimitic, where generative hyphae give rise not only to inflated, fusiform skeletal cells (cf. sarcodimitic) but also to narrow, thick-walled hyphae resembling binding hyphae in structure and function but differing from them in being septate, e.g. *Trogia* spp.

The enormous strength which can be imparted to basidiocarps by these complex types of construction can readily be discovered if an attempt is made to cut or dissect the context of almost any polypore and compare it with that of a monomitic agaric.

It is not always easy to recognize the constructional system since, for example, generative hyphae may rapidly give rise to specialized hyphae so that the former can only be found with difficulty at the margin of basidiocarps that are still growing, e.g. *Elfvingia applanata*. A further feature that may be of importance is the modification of the clamp connexions as a consequence of hyphal differentiation. In general, only generative hyphae bear clamps and not all of these nor at every cross-septum. The typical clamp connexion often becomes modified to a 'medaillon type'. The clamp is open, the side branch and main hypha are of much the same diameter and more or less symmetrically developed. ZELLER (1916) has described various transitional stages in *Lenzites sepiaria*.

Other types of differentiation occur and it is possible that other patterns of construction may be recognized. For example, CUNNINGHAM (1946) recognized '*Bovista*-hyphae' but Corner has interpreted these as inflated binding hyphae. Even generative hyphae are often differentiated by their size and behaviour into two classes. The 'ground hyphae' (HEIM, 1931) of Agaricales are usually wide (up to 15–20 μm) but give rise to slender 'connective hyphae' (1·5 μm) which become interwoven amongst the ground hyphae. Thus they perform an analogous function to binding hyphae but differ in being neither thick-walled nor of intricate, limited growth. An even more characteristic cell type is the sphaerocyst, characteristic of the pilei of the Lactario-Russula alliance. Here clusters of spherical, inflated cells, rather like bunches of grapes, arise from the generative hyphae.

A further class of differentiated hyphae are those supposed to be involved in transport or storage. Translocating hyphae have been recognized in the centre of the stipe of *Lentinus tigrinus* (LITTLEFIELD *et al.*, 1965) and have also been described in fungal strands. Laticiferous hyphae, usually extremely long, with relatively thick walls and sometimes septate, are known from many fungi, notably the genus *Lactarius*. Here the latex, whose function is quite unknown, is abundant, coloured or opaque, and a cut surface 'bleeds' readily. Another hyphal type, derived from connective hyphae and termed vascular (VAN BAMBEKE, 1892, 1894) or vasiform (VUILLEMIN, 1912) is long, sinuous, thin-walled and aseptate with infrequent globular swellings and an irregular diameter. The contents are always highly refractive and probably contain lipids or essential oils. Oleiferous hyphae would seem to describe them more precisely but their function is unknown.

Further types of differentiation occur which resemble those of other fungi. For example, the characteristic gelatinous texture of the basidiocarps of the Tremellales, or jelly fungi, is due to the gelatinization of the hyphal walls without their disintegration (MOORE, 1965). Presumably other developments, like the jelly which encloses young basidiocarps, of the stink-horns, e.g. *Phallus*, *Clathrus*, will be found to follow one or other of the general modes of gel formation described for other fungi.

The outer covering of a basidiocarp will clearly reflect, to some extent, the degree of hyphal differentiation within but the systematized knowledge concerning such regions is largely based upon the appearance of mature basidiocarps, although comparisons have been made with similar regions in ascocarps. LOHWAG (1941) has recognized four types of outer covering:

(a) A derm is where the outermost hyphae are more or less perpendicular to the surface of the basidiocarp. Thus if the dermal elements branch and are hair-like (a trichoderm) the surface will be silky or velvety, e.g. *P. versicolor*. In others parallel hyphae may be closely or loosely packed together or even form a compacted pseudo-parenchymatous tissue.

(b) A cutis is developed when the floccose hyphae run parallel to the surface. They may be cemented together or held by mucilage derived from the hyphal walls, e.g. *Mycena vitilis*.

(c) When the outer layers are not clearly delimited but merely consist of more closely packed hyphae, the term cortex is used.

(d) An equally artificial term, crust, is employed to describe any hard, contrasting surface layer regardless of its origin, e.g. *Fomes* spp.

It is clear that the application of the type of hyphal analysis developed for the context by Corner and others will enable a more precise analysis to be made of the origin and kinds of outer covering of basidiocarps.

THE HYMENIUM

A special type of outer layer, covering the gills, spores, spines or other surfaces is, of course, the hymenium. It is characterized by the parallel arrangement of the elements and the presence of basidia. Great attention has been paid by taxonomists to the cell types of the hymenium. These include the basidia, the basidiospores, cystidia and undifferentiated hyphae.

Basidia are important taxonomically and a general distinction is drawn between those which are septate, the Heterobasidomycetes and the unseptate Homobasidiomycetes. The principal types of heterobasidium which occur in the basidiocarps of the Tremellales are well known but their terminology has been greatly confused as a result of the phylogenetic speculations which have surrounded them (Fig. 5.14 and cf. AINSWORTH 1961 for bibliography). In

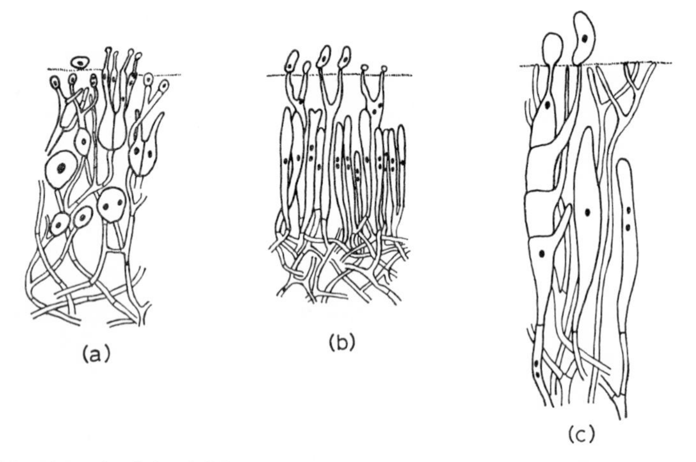

Fig. 5.14. Heterobasidia of (a) *Tremella mesenterica*; (b) *Dacromyces deliquescens*; and (c) *Auricularia auricula-judae*. In each case mature and immature basidia are embedded in a gelatinous matrix. (All ×400. From Gäumann, after Dangeard and Sappin-Trouffy)

essence, the basidia are divided either transversely or vertically. However, heterobasidia are not confined exclusively to the Tremellales and basidiocarps looking like typical members of the Thelephoraceae, Polyporaceae, Hydnaceae and Clavariaceae are known with heterobasidia, e.g. *Sebacina,* cf. with

Corticum; *Aporpium*, cf. with *Poria*; *Pseudohydnum*, cf. with *Auriscalpium*; *Tremellodendron*, cf. with various Clavariaceae. There are even genera such as *Hyaloria* and *Phleogena* whose wholly angiocarpic development suggests comparison with Gasteromycetes. There are also genera, e.g. *Pseudotremellodendron* (REID, 1957), where the basidia are only partially septate and, therefore, fall between the two principal types. Homobasidia, by contrast, are remarkably uniform, ranging from the cylindrical to the clavate in shape, the latter being the predominant form. Cylindrical basidia often have their meiotic spindles arranged longitudinally at different levels—the stichobasidial type—whereas clavate basidia may have transverse spindles at the same level—the chiastobasidial type. However, these distinctions are far from absolute and spindle orientation appears to be essentially a packing problem (cf. p. 418). The other variable which is often a taxonomic characteristic is the number of sterigmata produced. In the vast majority of Agaricales and Aphyllophorales the number is either 2 or 4 save for occasional, and obvious, aberrant basidia. In Gasteromycetes, however, there are frequently more than 4, e.g. *Cyathus* spp., or there may be none at all, e.g. *Calostoma* spp., although here some 9 to 12 sessile basidiospores develop. In general, each sterigma bears one basidiospore asymmetrically.

Basidiospores are usually unicellular and dorsi-ventrally symmetrical, with a basal hilum where they were attached to the sterigmata. They are variable in size, shape and ornamentation, and in the thickness and number of layers of their walls. They may be variously pigmented although the colours used in taxonomic works nearly always describe the spore mass rather than that of an individual spore. Colour is of diagnostic value in the Agaricales.

In addition to the basidia, cystidia, sterile cells of various but characteristic shapes and sizes may also occur in the hymenia of Agaricales and Aphyllophorales but less often in Gasteromycetes and Tremellales (Fig. 5.15). A true cystidium is said to originate in the trama, and hyphal elements which resemble them but are, in fact, terminal prolongations of laticiferous or oleiferous hyphae are termed pseudo-cystidia. How far this statement is meaningful is uncertain for CORNER (1953) has shown clearly how in *Polystictus microcyclus* the terminal cell of an hymenial hypha differentiates into a thick-walled pointed seta, while basidia develop sympodially behind. Thus the seta eventually becomes embedded amongst basidia through which it appears to have grown. A complex nomenclature has grown up to describe cystidia (cf. ROMAGNESI, 1944) but many authors use the simple descriptive prefixes pilo-, cheilo-, pleuro- and caulo- to describe their positions on the surface of the pileus, the edge or side of the gill and on the stipe, respectively. Their function is quite unknown save in the special case of the large, stout pleurocystidia of certain *Coprinus* spp., e.g. *C. atramentarius*, which stretch between the thin, closely packed gills and hold them some 0·12–0·17 mm apart.

LATER DEVELOPMENT, THE MATURE BASIDIOCARP

In many Holobasidiomycetes initiation is followed by a pause and there is then a sudden, final, rapid phase of expansion and maturation. This is shown

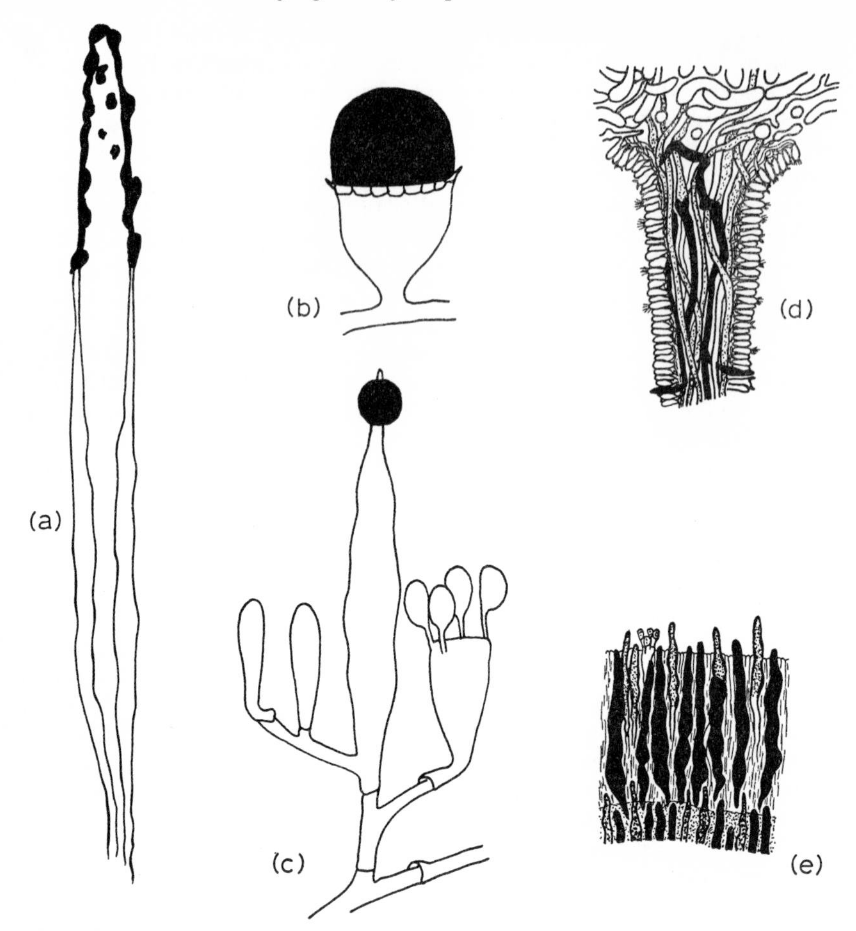

Fig. 5.15. Structures in the hymenium. (a) Cystidium of *Peniophora* sp. (× 370); (b) stephanocyst of *Gloeocystidium* sp. the upper half filled with oil (× 1280); (c) cystidium with oil droplet and basidia of *Peniophora* sp. (× 1000); (d) part of the gill of *Collybia lacrimosa* showing latex containing cells (× 125); (e) cystidia and gloeocystidia in hymenium of *Peniophora lundelii* (× 215). (After Lohwag)

clearly in Fig. 5.16 which is based on the work of BONNER *et al.* (1956) with the common mushroom. The region of elongation of the stipe is greatest just below the pileus, a general phenomenon in agarics. After Stage b the pileus expands in a gradient, there being virtually no growth in the centre and the maximum expansion being at the margin. Since the early claim of DE BARY (1887) it has been held that there is no increase in hyphal number in the stipe and no further cell formation, as a consequence of transverse divisions of the hyphae, during elongation. Precise data are not readily available but measurements of hyphae, as seen in transverse section of a stipe, seem to bear out this contention. The dry weight/wet weight plot of whole

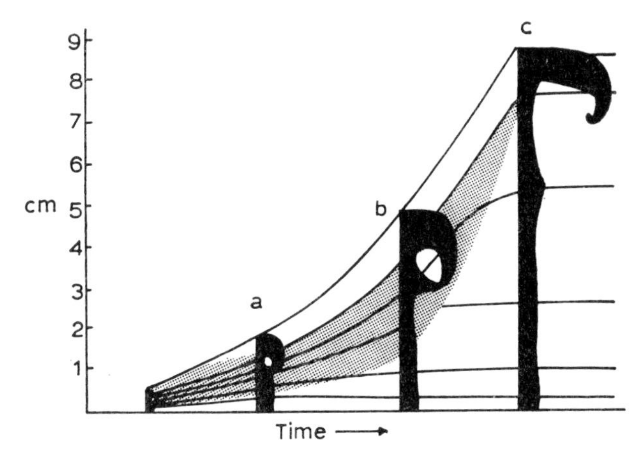

Fig. 5.16. Diagram to illustrate successive stages in the enlargement of the agaricaceous basidiocarp of the cultivated mushroom. It can be seen that extension growth is reduced eventually to a region at the apex of the stipe. Stippled areas represent the zone of elongation. The pileus expands at the margin after stage b. (After Bonner *et al.*)

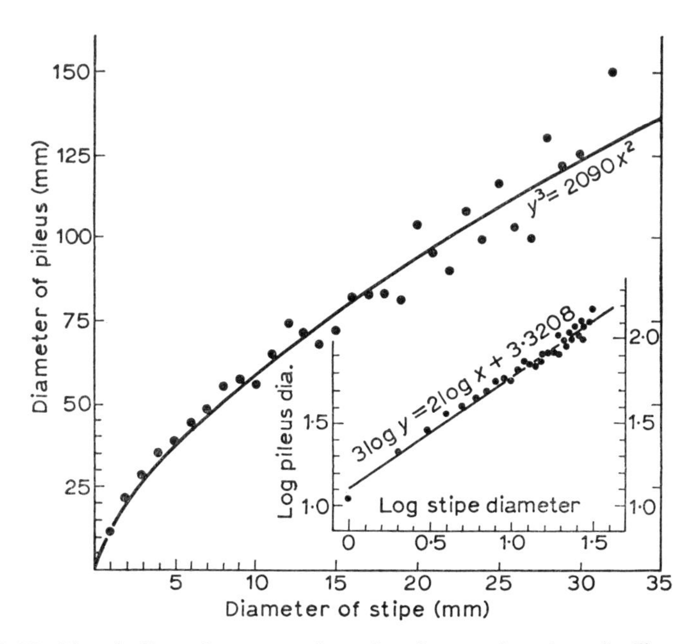

Fig. 5.17. Plot of pileus diameter against stipe diameter based on the illustrations in Lange's *Flora Agaricina Danica*. It can be seen that the plot of the cube of the diameter of the pileus (≡ to the weight) against that of the square of the stipe (≡ to cross-section) fits reasonably well with the predicted lines. (From Bond)

basidiocarps was linear indicating that the increase in size was due to translocation both of materials and water into the primordium. It seems probable that similar behaviour is shown by other fungi (cf. p. 134).

INGOLD (1946) has drawn attention to the fact that the size and form of the mature agaric basidiocarp are interrelated. Smaller basidiocarps have relatively slender stalks, larger ones relatively thick stalks. This is to be expected if thickness of the stipe is just sufficient to support the pileus. When stipe diameters were plotted against pileus diameters the best fit was obtained on the assumption that the diameter of the stipe will vary to maintain a constant weight per unit cross-section, as the volume (and hence the weight) of the pileus varies. BOND (1952) extended these calculations and replotted the data in a mathematically more acceptable form which is reproduced here (Fig. 5.17).

In Gasteromycetes, so far as can be discovered, differentiation and increase in size largely go together save in certain exceptional cases such as the expansion of the stalk of the stink-horns, *Phallus* spp. Whether this process involves increase in cell number, as well as of cell size, is not known, nor is it known whether expansion involves the transfer of materials other than water. This may well not be so since 'eggs' (young, unexpanded fructifications) will rapidly elongate if kept in water, on moist cotton wool or even in drier substrates. It is possible that water and materials may be absorbed from the distended volva gel which lines the peridium (INGOLD, 1959).

BASIDIOSPORE FORMATION

In Homobasidiomycetes the basidiospores appear to arise by a process called by CORNER (1948a) the 'ampoule-effect'. After a series of changes during the growth of the basidium involving vacuolation and the translocation of cytoplasm into the developing structure a final phase of vacuolation sets in. Smaller vacuoles coalesce to form a single vacuole which increases in size forcing the cytoplasm to the distal end of the basidium. At the sites of the sterigmata the wall must become plastic and the sterigmata are forced out. The continuing process of vacuolation forces the cytoplasm into the sterigmata and it appears to be extruded at the tip under pressure. Wall formation and wall 'setting' occur and the timing and pattern of the processes results in the various shapes which different basidiospores eventually assume (Fig. 5.18). The correlation between the vacuolation process described by Corner can be seen clearly in the much earlier drawings of basidiospore development in the yeast-like *Sporobolomyces* (BULLER, 1934). A recent study by WELLS (1965) of development in *Schizophyllum commune* suggests that the junction of the point of attachment of a basidiospore to a sterigma is extremely complex but details are not yet clear.

Morphogenetic studies on complex spore-bearing structures

As with studies on simpler spore-bearing structures referred to earlier (p. 100) there is a huge body of knowledge concerning the effects of various environmental and nutritional factors on the production of carpophores of various kinds. HAWKER'S (1957) review is a brave attempt to bring some order to such

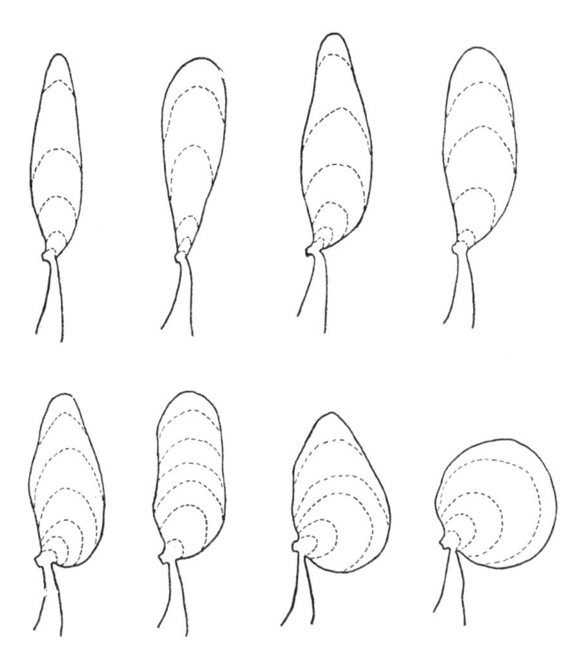

Fig. 5.18. The shapes which may be assumed by developing basidiospores depending upon the relative rates of wall formation and wall setting as a basidiospore is 'extruded' from a sterigma. The dotted lines represent successive shapes during growth. (From Corner)

observations but no very useful generalizations have yet emerged. In the earlier discussion (pp. 100, 102) attention was focused on situations where some kind of a specific 'trigger' reaction appeared to be involved. Here, attention will be directed rather to different types of investigation (nutritional, genetical, physiological) as examples of modes of attack on these somewhat intractable but stimulating problems.

NUTRITIONAL STUDIES WITH PYRENOMYCETES

There are many published accounts of the effects of nutrition on perithecium production but those which have provided the most useful general notions have been by HAWKER and her associates (ASTHANA and HAWKER, 1936; HAWKER, 1939, 1947, 1948; HAWKER and CHAUDHURI, 1946). The most valuable studies were concerned with the effect of different carbon sources on the number of perithecia produced in plate culture, initially in *Sordaria (Melanospora) destruens* but later in other Ascomycetes. Hawker showed that nutrient concentration and the nature of the carbon source affected vegetative growth differently from reproductive growth. The latter is highly correlated with the ease with which the carbon source is assimilated (Fig. 5.19). M and F represent the mycelial and reproductive responses respectively. M_1 and F_1 are the responses with simple readily assimilated sugars such as hexoses; M_2 and

F_2 illustrate the responses of more complex sugars which can be readily hydrolysed and assimilated, e.g. sucrose for *S. destruens* or lactose for *Podospora* sp.; M_3 and F_3 are the opposite extremes where hydrolysation of the sugar is so ineffective that 'starvation growth' alone is achieved. The effects are thus concentration responses and differences between fungi reflect differences in their hydrolysing ability. For example *S. destruens* and *Melanospora zamiae* both respond effectively to sucrose but the former requires a ten times greater concentration of sucrose than the latter (5·0:0·5%) to achieve the same result. A sucrose assay, however, showed that *M. zamiae* can utilize

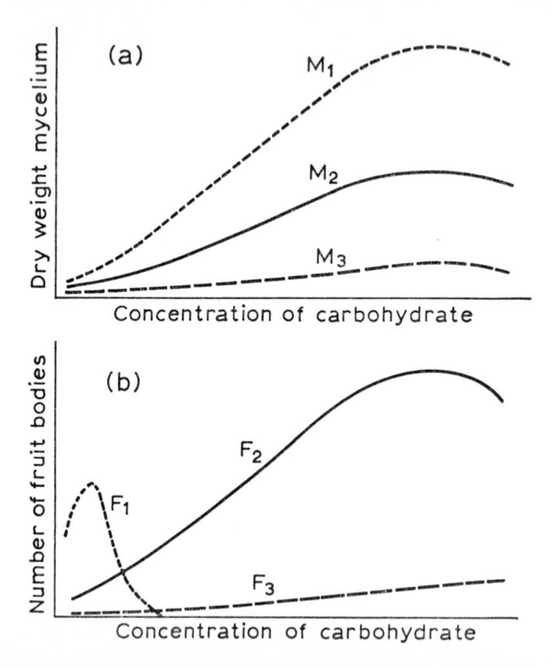

Fig. 5.19. Diagrams to illustrate the effect of different concentrations of carbohydrates on (a) mycelial dry weight and (b) numbers of ascocarps. M_1, F_1 represent mycelial and fruiting responses, respectively to hexoses; M_2, F_2 to a sugar which requires steady breaking down, such as sucrose and M_3, F_3 the case where the sugar is not readily broken down and starvation growth ensues. (From Hawker)

sucrose far more effectively than *S. destruens*. Indeed, it can be shown that any factors which hydrolyse sucrose in the medium also increase the perithecial production of *S. destruens*, e.g. increase in temperature of incubation or contamination by an organism producing an active extracellular sucrase. This effect is not simply due to the greater availability of hexoses. A comparison between the number of perithecia produced on a medium containing glucose and fructose at the same concentrations as those produced by hydrolysis on a sucrose plate shows that more perithecia are produced in the latter case. Hawker suggested that a significant factor involved was the ease of phosphorylation of intermediates. When small quantities of fructose diphosphate, glucose-1-phosphate (glucose-1-P) or a mixture of hexose monophosphates (HMP) was added to fructose or hexose media an increase was observed in perithecium production. Glucose-1-P added to sucrose medium also resulted in an increased rate of respiration, a feature often associated with increased efficiency of reproduction. Some support for Hawker's suggestions comes from work with *Chaetomium globosum*, another Pyrenomycete (BUSTON *et al.*, 1953; BUSTON and KHAN, 1956; BUSTON and RICKARD, 1956). They showed that addition of jute extract to a sugar/salts medium increased production of perithecia, that the extract contained traces of glucose-1-P and fructose-1:6-P and that addition of these alone was also effective. Enhanced production of perithecia by *C. globosum* when contaminated by an *Aspergillus* was traced to the production of organic phosphate esters and these too, when added alone, were found to cause the response. These experiments are suggestive but it is not clear how phosphorylated esters could increase perithecium production and the work needs to be reinvestigated and extended.

While Hawker's work has provided a more rational explanation of how sugars affect reproduction in Pyrenomycetes, it must be confessed that the experiments throw no light on the different effects of any carbon source on vegetative versus reproductive growth.

MORPHOLOGICAL AND BIOCHEMICAL MUTANTS IN PYRENOMYCETES

The very great use made of Ascomycetes, notably *Neurospora*, by geneticists because of the relative ease of tetrad analysis in these organisms has resulted in the production of many mutants. In a number of genera, notably *Glomerella*, *Sordaria* and *Podospora*, mutants are known which are blocked at various stages of development of the reproductive structures; some of these are listed in Table 5.2. The most fully investigated case is that of *Sordaria macrospora* by ESSER and STRAUB (1958). Five classes of genes, all non-linked, but each often having a number of multiple allelomorphs, were found. Some produced no reproductive organs at all, others ascogonia but no further development. Yet others produced protoperithecia only and, finally, the two classes which produced perithecia differed in that one was quite sterile while the other produced ascospores (but was incapable of discharging them). In most cases the mode and rate of hyphal growth was also modified so that analysis in terms of hyphal behaviour is a definite future possibility. A similar range was found in *S. fimicola* but there were fewer in the other genera examined. The biochemical bases of these mutants is not understood but from *S. macrospora* and *P. anserina* it is clear that associated changes occur in the melanin pigments and/or the phenoloxidase systems of the fungi. This is of particular interest since earlier somewhat equivocal claims had been made by HIRSCH (1954; WESTERGAARD and HIRSCH, 1954) that protoperithecium production and fertility in *Neurospora crassa* were associated with tyrosinase production. At present no clear correlation can be seen but study of the comparative biochemistry of these or similar mutants might well prove rewarding. In addition, the ability of these fungi to form heterokaryons provides a further

Table 5.2. The effects of genes on developmental stages in perithecium production in various Pyrenomycetes. (From Esser and Straub, 1958; Carr and Olive, 1959; Wheeler and McGahen, 1952; Esser, 1956)

Effect	Species: *Sordaria macrospora*	*S. fimicola*	*Glomerella cingulata*	*Podospora anserina*
No reproductive organs	*c* *r*	*st*-2	A^1	z
Ascogonia but no protoperithecia	*cit* *spd* *p*	—	F^1 *st* 1	*pa* *sp* *al* *fu*
Protoperithecia but no perithecia	*pl* *f* *l*	*st*-1	—	*i* *lg*
Sterile perithecia	*s* *min* *pa*	a^3	B^2dw1	—
Fertile perithecia but no spore discharge	*n* *in* *m*	*st*-5	—	—

means of analysis of the processes involved. For example when a number of Esser's mutants of *S. macrospora*, e.g. *r*, *spd*, *f*, were mated, crossed perithecia were produced but also normal, selfed perithecia on the mutant mycelia which are normally incapable of such development. Tests were made to see if this perithecial induction effect could be brought about by adding filtrates of one mutant to another, e.g. *s*1 filtrate on *f*1 or by growing them separated by a dialysis membrane or sintered glass filter. No evidence was obtained that extracellular diffusible agents were involved so the basis of such development must be intracellular. This is of particular interest for MARKERT (1949) had claimed earlier that in *G. cingulata* diffusible agents affected perithecial development.

The relative ease with which such morphologically blocked mutants can be obtained in Ascomycetes suggests that they are potentially a most valuable tool for the experimental analysis of the development of reproductive structures.

PHYSIOLOGICAL AND BIOCHEMICAL STUDIES WITH HOLOBASIDIOMYCETES

Progress in these studies has occurred in two ways, in the study of the general physiology and biochemistry of the basidiocarp and the specific control of stipe elongation and expansion of the pileus.

Table 5.3. The distribution of dry weight of fungus tissue between mycelium and basidiocarps in a developing plate culture of *Coprinus lagopus*. (From Madelin, 1956a)

Stage of development	Age (days)	Mean dry mycelium (mg)	Weight (4 replicates) basidiocarps (mg)	Total (mg)	
Initiation of primordia	8·9	26	0·2	26·2	
	9·7	30·1	0	30·1	
Enlargement of primordia	10·9	31·6	3·9	35·5	
	11·7	**37·2**	**4·5**	**41·7**	
Maturation of primordia	12·7	**22·0**	**19·3**	**41·3**	
	13·7	30·9	17·5	48·4	Deliquescence of basidiocarps (48·4–62·8)
	15·0	30·2	29·9	60·1	
New crop initiated	16·1	39·3	13·7	53·0	
	16·7	43·0	19·8	62·8	
Significant differences between means (mg) $P = 0{\cdot}05$		7·2	5·7	7·3	
$P = 0{\cdot}01$		9·7	7·7	9·9	

Note that over 12 to 13 days the total dry weight remained almost stationary and the loss in mycelial weight (15·2) almost exactly equals the gain in weight of basidiocarps (14·8). Just after 15 days the loss in total dry weight is almost exactly balanced by the loss in weight of basidiocarps through deliquescence. The former comparison is statistically significant, the latter just falls short.

The construction of a large basidiocarp clearly involves the transfer of much material and this has recently been re-examined by MADELIN (1956a, b; 1960) in *Caprinus lagopus*. He has shown that, in plate culture, hyphae immersed in the agar swell and stain heavily with I/KI solution. Once basidiocarps begin to develop the swollen cells vacuolate and lose their dense contents. He interprets this as suggestive evidence for storage of glycogen followed by its transfer to the basidiocarp. He demonstrated a direct transfer of materials by following the changes, in dry weight, of the mycelium and basidiocarps over a 17-day period, during which a crop developed, deliquesced and a new crop began (Table 5.3). The significant data are over the period days 12–13. Here the total dry weight was virtually the same but an increase in the basidiocarp weight (14·8 mg) was almost exactly equalled by a loss in the weight of vegetative mycelium (15·2 mg). This is in keeping with Bonner's observation that the linearity of a dry wt./wet wt. plot during the elongation of *Agaricus campestris* represents translocation both of water and materials (Fig. 5.20 and BONNER *et al.*, 1956). Madelin also investigated the

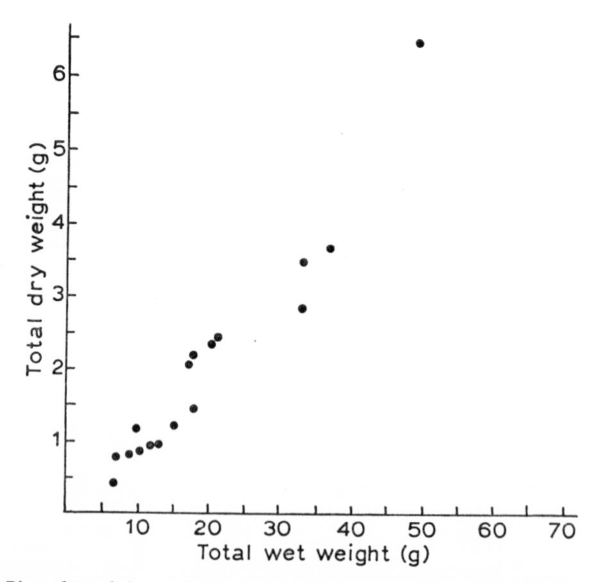

Fig. 5.20. Plot of total dry weight against total wet weight of developing basidiocarps of the common mushroom during the phase of elongation. Translocation both of water and materials is evidently involved. (From Bonner *et al.*)

factors which governed the distribution of basidiocarps on a plate colony. He noted that primordia were usually dispersed uniformally within an annular zone of the colony but that those that developed further were generally distributed unevenly. He adopted a suggestion of HEIN's (1930) concerning the abortion of basidiocarps in mushroom beds, namely that some primordia would be more favourably placed than others through slight differences in their times of initiation for instance, to initiate a flow of nutrients in their direction and, therefore, away from the other primordia. Thus, if a culture were halved neither half would be able to draw on the intra-hyphal reserves in the other half and so they should produce similar yields of basidiocarps. Two sets of plate cultures were set up; in one set each culture was bisected in a sterile manner and one half was transferred to another sterile petri plate, the other set was left intact (Table 5.4). The yields of mycelium were similar in both sets. However, in the intact plates there were great differences in basidiocarp yield between the two halves whereas the yields in the two halves of the bisected plates were virtually equal. This is in accordance with the hypothesis proposed. It should also be noted that this is not a uniform nutrient depletion throughout the colony for, if that were the case, it would be expected that all primordia would develop to a size in accord with the nutrients available in the agar medium. This is not the case. Nevertheless, the similarity in dry weights of the vegetative mycelia from each half of the plates, both intact and bisected, shows clearly that material is drawn uniformly from all parts of the mycelium but only transported to certain basidiocarps.

Table 5.4. The effect of dividing plate cultures of *Coprinus lagopus* (at the time of initiation of primordia, 10 days) on the distribution of material in basidiocarps developed in the next 6 days compared with intact cultures. (From Madelin, 1956a)

Distribution of Material	Mean dry weight (5 replicates) Intact cultures (mg)	Halved cultures (mg)
Basidiocarps		
More heavily fruiting half	27·4+8·4	22·5+2·5
Less heavily fruiting half	3·6+1·8	18·8+2·5
Difference	23·8	3·7
Significance of difference (*P*)	(0·03)	(0·3–0·4)
Vegetative mycelium		
More heavily fruiting half	36·3+7·3	29·4+6·7
Less heavily fruiting half	41·4+4·4	31·3+4·5
Difference	5·1	1·9
Significance of difference (*P*)	(0·5–0·6)	(0·8–0·9)

The halves in the intact culture were determined by a diameter drawn on the base of the petri plate at the start of the experiment. Note the almost equal yields of vegetative mycelium in divided and control plates and the unequal distribution of basidiocarps in intact plates compared with the divided plates.

Transport has been shown to be involved in other agarics, e.g. *Collybia velutipes* (PLUNKETT, 1951) and even in the stipitate polypore, *P. brumalis* where, by contrast, growth is mainly due to formation of new hyphae at the marginal growth zone of the pileus (PLUNKETT, 1958).

A more recent study of the biochemical morphogenesis of *Schizophyllum commune* in plate culture has not only confirmed this phenomenon but has extended understanding to the kinds and location of materials involved (WESSELS, 1965). Four developmental phases were recognized:

(a) Undifferentiated growth of submerged vegetative hyphae.
(b) Initiation of primordia amongst hyphae on the surface through their aggregation.
(c) Growth of the primordium to form a globose structure which develops a concave apical cavity. A hymenium develops and lines this cavity whose margin proliferates to develop a cup-shaped structure. A mutant, K.35, employed in this work ceases to differentiate further. The hymenium develops gills.
(d) Pileus formation is accompanied by the indeterminate growth of one side of the cup to form a sessile fan-shaped basidiocarp, whose gills appear to radiate out from its point of attachment.

The total nitrogen, total carbohydrate, principal polysaccharides, R.Q., Q_{O_2} and Q_{CO_2} in nitrogen, were determined in different regions during development. In addition, the effect of thiamin on primordium initiation and concomitant changes in nucleic acids and protein metabolism were studied. Wessels was able to achieve a fair degree of synchrony in the developmental sequence between primordia in the same culture but this was lost in the final phase. Here, as described earlier for *C. lagopus*, only certain primordia developed pilei and the rest were either arrested in development or degenerated. Results are summarized in Fig. 5.21 and Table 5.5.

Undifferentiated mycelia have a low growth rate and lower requirements for oxygen and thiamin than differentiated hyphae. In fact, undifferentiated hyphae could be grown without thiamin and this provided a fairly uniform mycelium with a low basal metabolic rate for establishing the synchronous cultures.

Primordium initiation is associated with a change in the hyphal walls, which become stainable with thionin, but whether this represents a change in the structure of hyphal walls associated with their ability to aggregate is not known. Exogenous nitrogen must be available and initiation does not seem to be possible in the absence of net protein synthesis. Exogenous glucose is also essential and so is thiamin. Initiation can be inhibited by absence of any of these nutrients, by high CO_2 concentration or temperatures above 25°C. During the transition the Q_{O_2} increases, the R.Q. is up to values of 2 and the fermentative ability is also high.

Once the primordia are formed the R.Q. declines rapidly and settles down at about 1. The Q_{O_2} continues at a high level until the exogenous glucose has been utilized, when it drops dramatically, although still at a high rate, and then slowly declines. During growth of the primordia exogenous glucose is necessary but exogenous nitrogen is not, the requirements of the primordia being met by translocation of the bulk of the nitrogen compounds from the submerged mycelium. Most of the glucose is converted into cellular carbohydrate, chiefly in the walls. Over 80% of wall carbohydrate is glucan and

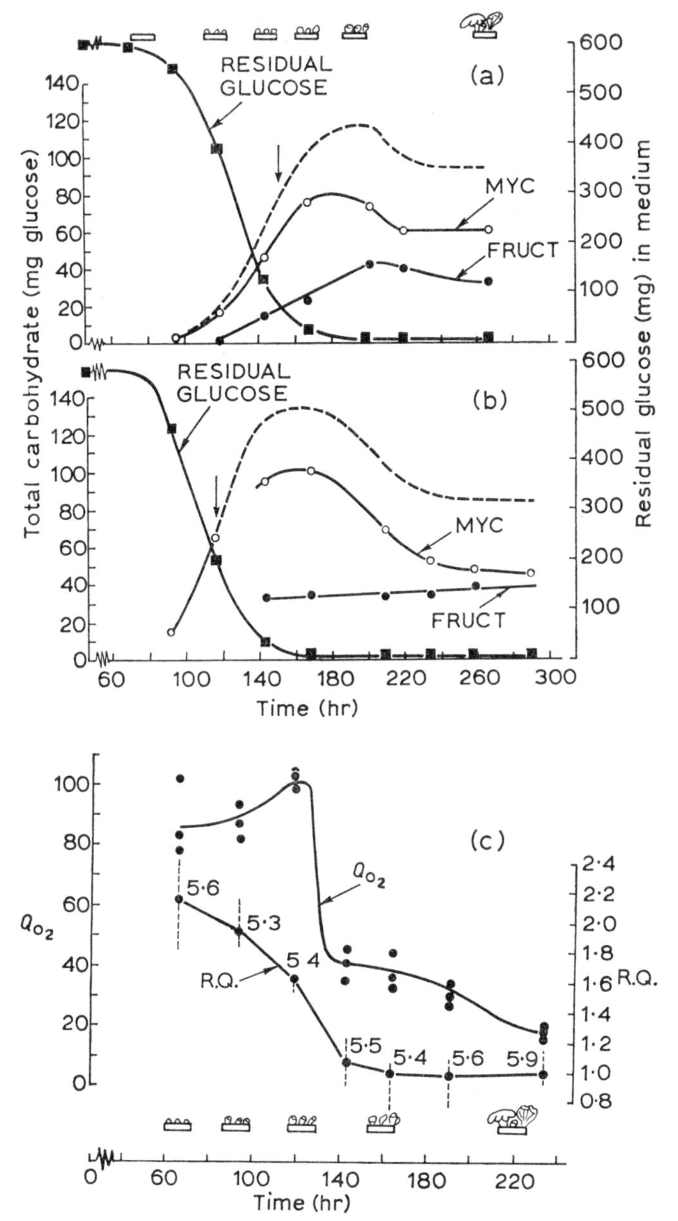

Fig. 5.21. Graphs to illustrate changes in various components during the development of the basidiocarp of *Schizophyllum commune*. The diagrams at the top and bottom illustrate the approximate gross appearance of normal developing basidiocarps. (a) and (b) represent changes in the total carbohydrates in the mycelium and basidiocarps and the exogenous glucose supplied. (a) is the curve for the mutant K.35 which never develops proper pilei, the *cup* mutant; (b) is that for a normal dikaryon, K.8. In both cases the vertical arrow indicates when the whole culture achieves its maximum nitrogen content. (c) represents the Q_{O_2} and R.Q. of normal cultures of K.8. Note that the scale of (c) is slightly different from (a) and (b). In all graphs, MYC = mycelium; FRUCT = basidiocarps; RES. GLUCOSE = residual (exogenous) glucose; ----- = total carbohydrate. (From Wessels)

Table 5.5. Changes in S- and R-glucan during development of cultures of *Schizophyllum commune* under different conditions. (From Wessels, 1965)

(**a**) Changes in 7- and 12-day cultures at 25°C under conditions of carbon and nitrogen starvation.

Values as mg/30 ml. culture
M = mycelium F = all fructifications

Glucan	7-day culture			12-day culture		
	F	M	M+F	F	M	M+F
S	30·1	51·2	81·3	36·3	28·9	65·4
R	10·7	19·6	30·3	5·7	3·0	8·7
S/R	2·8	2·6	2·7	6·4	7·6	7·5

Note relatively greater loss of R-glucan between 7 and 12 days and gain in S-glucan by fructifications after 12 days.

(**b**) Relative distribution of glucans in various structures after 16 days in low-phosphate 4% glucose: 0·15% asparagine medium.

Values as mg/20 ml. culture

Glucan	S	R	S/R
Mature basidiocarp	8·86	3·80	2·33
Stunted basidiocarp	72·8	3·18	22·89
Stroma	64·06	4·21	15·22
Undifferentiated mycelium	1·06	0·19	5·58

Note differences in ratio between mature and stunted basidiocarps due to degradation of R-glucan and transfer of products to mature basidiocarps.

there are two principal components, the alkali-insoluble R-glucan (cf. yeast glucan) and an alkali-soluble S-glucan.

In the final phase of pileus formation neither exogenous nitrogen nor carbon is necessary; indeed, an exogenous carbon supply prevents pileus formation. Analysis of the carbohydrates of those basidiocarps which mature, those which remain stunted and the stroma on which the primordia develop, shows a dramatic degradation of R-glucan in the last two manifested by a high S/R ratio (15–23:1); normal basidiocarps have an S/R ratio of 2–3:1. There is little doubt that pileus formation is dependent upon a low but continuous supply of glucose which is normally supplied by hydrolysis of the R-glucan and transport of the glucose so formed to the basidiocarps which then mature. This latter phase is analogous to that described for *C. lagopus* earlier, although there the material appears to be derived from cytoplasmic polysaccharide in the submerged, inflated storage cells. Comparison with the general exogenous carbon requirements for effective reproduction in Pyrenomycetes described earlier, is also striking (cf. p. 120). The most telling evidence in favour of this reutilization of polysaccharide material in *S. commune* comes from the fact that pilei can be induced in the mutant K.35 by feeding with a continuous low concentration of glucose, while pilei can be inhibited in the normal form by transfer to a high concentration of glucose. It was shown that K.35 cannot achieve net degradation of R-glucan, S-glucan or chitin, thus it is incapable of supplying glucose at the appropriate time; an exogenous supply makes good this deficiency. In the normal form, the maintenance of a high exogenous concentration of glucose results in greater primordial growth and thicker cell walls; the glucan degradation process is inhibited, hence no glucose is available and so no pilei develop.

This meticulous and elegant study of Wessels is the most detailed biochemical investigation of basidiocarp morphogenesis available. Its most striking feature can best be described in the investigator's own words, '. . . growth of the primordia can occur in the absence of a nitrogen supply in the medium and formation of the pileus occurs even in the absence of both an exogenous carbon and a nitrogen supply. Thus, the whole morphogenetic system may be regarded as a system in which particular structures are produced in sequence, each structure providing the substrate for the following. The vegetative mycelium supplies the nitrogenous compounds for the growth of the primordia and the expanding pilei draw completely upon the stunted fruiting bodies and the stroma. . . . The distinction between reserve and structural constituents has now become rather meaningless in the system studied here'.

That cell wall constituents may even be involved in supplying material for spore production is suggested by CORNER's (1934) observation of the 'corrosion' of the walls of hyphae above the dissepiments in *Polystictus xanthopus*.

GROWTH REGULATION IN AGARICS

Analyses such as those of Wessels provide information concerning the kinds and sequences of processes involved but do not, as yet, provide much information concerning the regulation of growth. This topic has been approached in agarics mainly by the use of operations on developing basidiocarps.

As early as 1842 SCHMITZ noticed that removal of part of the pileus did not prevent elongation of the stipe but that total removal inhibited growth. Several other mycologists noticed this type of effect but it was BORRISS (1934a) who first demonstrated, with *Coprinus lagopus*, that if a few gill lamellae attached to a slice of the pileus were left on the stipe it would continue to elongate and show weak curvature. He suggested that the gills produced a hormone which acted on the stipe prior to its phase of rapid elongation. Hagimoto has confirmed that the gills are the site of growth regulation in a number of agarics (HAGIMOTO and KONISHI, 1959, 1960; HAGIMOTO, 1963a, b; KONISHI and HAGIMOTO, 1962) and has made claims of an active extract in water, ethanol or organic solvents. The most fully documented study, however, that of GRUEN (1963) was with the common mushroom. He provided accurate data of growth in basidiocarp length and pileus diameter from primordia of initial length 1–1·5 cm and 0·8–1·7 cm initial diameter, respectively, up to 10 cm or more. He then carried out a number of operative procedures on the pileus and observed the subsequent growth and, in some cases, curvature of the stipe

Table 5.6. Mean total growth increments of intact basidiocarps of *Psalliota hortensis* var. *bisporus* and, after various operations, of bilaterally symmetrical and initially parallel-sided cap slices made from primordia of initial length 2·6–3·0 cm and average width 2·8 cm. Standard error of means and numbers of specimens (in parentheses) are given. (From Gruen, 1963)

Operation	Basidiocarps. Total growth (cm)	Pileus slices. Increase in diameter (cm)
Intact	5·7+0·20 (30)	7·6+0·38 (24)
Cap slice+gills	4·9+0·28 (11)	9·2+0·77 (11)
Cap stump+gills	5·3+0·49 (11)	
Cap stump+gills at outer edge	2·8+0·43 (7)	10·6+1·46 (7)
Cap slice−gills	1·8+0·19 (12)	2·7+0·35 (11)
Decapitated stipe	1·1+0·23 (14)	
Stipe+cap centre	1·5+0·27 (14)	
	Stipe middle	Stipe edges
Gills	3·5+0·51 (10)	4·3+0·38 (10)
Gills removed after 2 days	3·0+0·32 (9)	3·1+0·21 (9)
Gills+1–2 m pileus	3·8+0·25 (6)	4·6+0·28 (6)

and expansion of the pileus (Table 5.6). He showed that although there is some residual growth after decapitation the stipe's growth rate declines, ceases and fails to recover, whatever the size of the original basidiocarp (in the range 1·3–5 cm). This was true even if the centre of the pileus was left intact on the cap, although the onset of the decline in growth rate was later in primordia 3–5 cm high. On the other hand, if gill tissue was left on the pileus, even as little as one eleventh of the normal, then the growth rate continued to increase although, depending on the severity of the operation, not necessarily as much as in an intact basidiocarp. The growth-promoting region was located; it was in the gills. Complete removal of gills was followed by a decline in growth, complete removal of pileus save for a thin basal layer holding the gills was followed by a temporary check and then a recovery. If gills were removed save only at the outer edges there was a decline in growth similar to that obtained when the lamellae were completely removed but the cap tissue expanded normally. Unilateral removal of pileus and gill tissue was followed by a curvature and elongation of the stipe away from the untreated region (negative curvature). Gruen concluded, therefore, that the lamellae were, as Borriss had suggested, the regulatory centre for stipe elongation and pileus expansion. The transmission of the, presumably, hormonal stimulus to the stipe was either through gill tissue or hyphae in the pileus immediately adjacent to the base of the gills. Since pileus material of primordia 3 cm or larger did promote stipe growth some residual growth-promoting material was presumably present but *P. hortensis* f. *bisporus* is incapable of regenerating a new and physiologically effective growth-controlling centre. Thus many of

Hagimoto and Konishi's earlier observations were substantiated but there is, as yet, no confirmation of their claims concerning the isolation of the effective growth regulator(s), or that gill diffusate and pieces of stipe or cap can induce growth curvatures. It is clearly desirable that these claims should be investigated and an attempt made to isolate the growth regulator(s) involved.

Whether this type of regulation will be found throughout agaric basidiocarps is not clear. For example HAWKER (1950) claimed that unilateral decapitation induced curvatures in basidiocarps of *Collybia velutipes*. This could not be confirmed by BANBURY (1959).

Other types of regulation are possible. BEVAN and KEMP (1958) observed that basidiocarps develop more rapidly from pieces of cut stipe than from mycelium and suggest that this indicates a basidiocarp-inducing substance. Such a substance has been isolated from *Schizophyllum* (LEONARD and DICK, 1968). Another interesting type of regulation was observed in the early surgical experiments of MAGNUS (1906). If the potential gill area is removed from developing basidiocarps, irregular outgrowths may arise from any part of the remaining surface and form anastomosing cerebelloid-like gill structures. Even so these outgrowths appear to be arranged at a constant distance from each other. Whether this is due to a zone of inhibition around each developing outgrowth is not known. Clearly, there is here evidence for regulation of packing or of a morphogenetic field. The development of gills, therefore, involves both regulation of gill initiation and of spacing but whether these are manifestations of the same or of different processes is unknown. There is an urgent need for further investigations into all these types of regulation of morphogenesis. The small size of the primordia and of the angiocarpic or pseudoangiocarpic development of many agarics and most Gasteromycetes suggests that regulation of hymenial patterns will best be investigated initially by concentrating on gymnocarpic hydnaceous or poroid forms. Stipitate basidiocarps, however, are likely to provide the most rewarding material for investigations of elongation phenomena.

EXTERNAL DETERMINANTS OF BASIDIOCARP FORM

Because of the difficulties of investigating small primordia and because of the striking effects on development of such factors as light and gravity, particular attention has been paid to them. No general hypothesis has yet been formulated in respect of the action of either of these factors but some progress has been made in particular examples.

Basidiomycetes seem to fall into two groups, those which do not require light for normal basidiocarp formation (although a dark period may or may not be essential) and those which require light for all or part of the period of development. In the former group light may often affect development without being an absolute requirement. For example, MADELIN (1956a) found that *Coprinus lagopus* would develop primordia after 15–16 days at 25°C in plate culture but that the process was greatly accelerated by exposures as brief as 1 second at 1690 lux or 5 seconds at 1·076 lux. In such facultative light requirers or in obligate light-requiring fungi two distinct processes are involved. Light may be necessary for initiation of primordia or it may affect subsequent stages of differentiation, viz:

Initiation, e.g.	*Coprinus nythemerus* *Pleurotus ostreatus* *Fomes roseus: texanus; rimosus* *Stereum umbrinum: veriforme*
Stipe inhibition, e.g.	*Coprinus bisporus: lagopus: stercoreus* *Polyporus agariceus*
Pileus formation, e.g.	*Polyporus squamosus; brumalis* *Polystictus versicolor* *Lentinus lepidus; tuber-regium*
For hymenium and spore formation	*Polyporus pargamenus* *Poria ambigua.*

Many of these light-induced responses are sensitive to the blue end of the spectrum, 4000–5000 10^{-10}m (MARSH *et al.*, 1959), a fact first elucidated with precision by BORRISS (1934b) in his studies on *C. lagopus*. Red and orange light, in general, are ineffective and simulate total darkness. The inducing wavelengths are the same as those effective in initiating phototropic and light growth responses but the nature of the photoreceptor is unknown, although providing a fertile field for speculation (see p. 320). Very little progress has been made to date with this problem and it is, perhaps, of more value to consider the interaction of certain external factors as a guide to the kind of complex situation that must exist in nature. An excellent example is the work of PLUNKETT (1956; 1961) on the development of the stipitate polypore, *Polyporus brumalis*. In complete darkness some primordia did develop but their number was substantially increased in light intensities as low as 430·4 lux and, with increased light intensity, an earlier and numerically greater response was shown. Development of the pileus was totally inhibited if stipitate primordia were returned to total darkness. Pileus formation was also found to be sensitive to low transpirational water loss. This is because the rate of translocation in this fungus is largely determined by the rate of transpiration from the apical regions in developing basidiocarps. In still air, pilei are either absent, reduced or modified and the stipes tend to be longer. The effect of light and transpiration loss are additive. Once a pileus is formed it is positively phototropic as is an epileate stipe. The growth of the stipe was confined to the terminal 5 mm. Stipes, when illuminated unilaterally, curved to the light and often, if exposed to a stream of dry air, developed pilei in about 2 days which expanded at right angles to the long axis of the stipe. Stipe elongation continued unchanged for a further 2–3 days and, thereafter, the stipes turned upwards and within 2 days were growing vertically upright, notwithstanding the unilateral illumination. Additional experiments showed that epileate stipes were negatively geotropic in the absence of light. By changing the position of the unilateral source after pileus initiation it was possible to show that such pileate stipes could respond to a strong phototropic stimulus (Fig. 5.22). Plunkett believed that the pileus, when developed, acted as a light screen which masked the response of the stipe in the initial experiment so that the final negative geotropic response became dominant. By attaching black paper discs to epileate stipes, so mimicking pilei, he was able to reproduce the behaviour of normal pileate basidiocarps. He then followed pileus development on a klinostat with incident light from different directions and showed

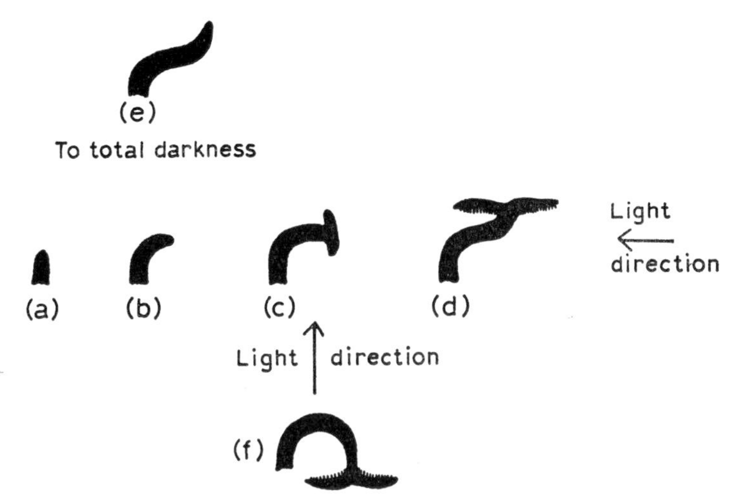

Fig. 5.22. Behaviour of basidiocarp of *Polyporus brumalis* in response to unilateral illumination. (a)–(d) show successive curvatures towards a unilateral light source until the stipe apex is shaded by the expanding pileus when the stipe curves upwards. The same result is achieved by transferring stipes at stage b to total darkness (e). If a stage c basidiocarp is illuminated from below it behaves, in fact, as in d but grows downwards because of the changed direction of the light (f). (After Plunkett)

that pileus initiation is unaffected by directional stimuli both of light and gravity. On the other hand, the morphologically upper surface of the pileus was always found to be directed towards the highest light intensities, the hymenium on the other side. This was even true of pilei which developed in an inverted position. Here then is clear evidence of a series of complicated interactions of light, gravity, humidity and wind velocity, all of which will play important morphogenetic roles in the development of basidiocarps in natural environments. The situations described here have their probable parallels in other fungi and, no doubt, other kinds of interactions of physical environmental variables can and do occur.

This chapter has attempted to outline the general kinds of construction found in the spore-bearing structures of the fungi. It has, of necessity, been a very brief outline and the treatment of experimental investigations of morphogenesis has been even more cursory. Nevertheless, certain features of general interest have emerged. Basically, despite the fact that morphogenetic processes appear to transcend the limits of hyphae and apparently affect, in large structures at least, entire hyphal aggregations, their origins lie in the facts of hyphal behaviour. Growth, branching and aggregation are the fundamental processes, together with the control of direction of growth. Sufficient examples have been given to show that valuable information can be obtained by investigations of the biochemical changes associated with the onset and development of the reproductive phase, of the 'trigger' reactions or stimuli which initiate development or switch its direction, and of the kinds of regulatory process which control the morphogenesis of spore-bearing structures. Yet, despite the value of work of this nature the fundamental processes involved in morphogenesis at the hyphal level both elude study and, relative to other types of enquiry, are neglected by investigators. It is to these processes that attention should be given if a more coherent and analytical approach is to be brought to the study of the construction of reproductive structures of fungi.

Chapter 6
Spore Liberation, Dispersal and Germination

The topics of this chapter have attracted considerable attention and are especially well documented in a number of books which can be read with both profit and pleasure. Spore liberation and dispersal are recurring themes in Buller's seven great volumes of *Researches on Fungi* (1909–1950 and Kew MS). INGOLD has summarized more recent literature including many of his own researches in *Dispersal in Fungi* (1953) and *Spore Liberation* (1965), while GREGORY has done much the same for air-borne dispersal in *The Microbiology of the Atmosphere* (1961). SUSSMAN (1965) has recently reviewed most adequately the topic of spore germination. The account given here rests heavily upon these works and will, therefore, be largely an outline of the subjects involved.

SPORE LIBERATION

The distinction between spore liberation, the actual process of detachment from the spore-bearing structure, and spore dispersal, the subsequent movement of the spore before coming to rest on a substrate, is somewhat artificial but convenient.

Spore liberation may be a passive or an active process. Amongst the former are a number of different methods depending on whether the spores are dry or slimy, enclosed by a membrane or associated with material which attracts other organisms to it. Actively liberated spores are projected a greater or lesser distance by a variety of syringe-like devices generally based upon the hydrostatic pressure of the spore-bearing structures or upon devices triggered off through changes in turgidity or the development of negative pressures in cells adjacent to the spores or in the spores themselves.

Table 6.1. The mean number of conidia of *Trichothecium roseum* caught by an impactor following the passage of air streams of different velocities and relative humidities over the fungus. (From Zoberi, 1961)

(a) Catch after flows of constant volume and relative humidity (31 %) but different speeds.

Speed of air flow (m/sec)	Mean catch (6 experiments) spores
1·7	132
2·5	399
3·3	1234
5·0	2358
6·7	5879
10·0	12375

(b) Catch in successive 1-minute intervals at a constant velocity (5 m/sec) but two different humidities.

Intervals (1 min)	Mean catch * (6 experiments) spores	
	R.H. 50·5 %	R.H. 95 %
1	3000	410
2	850	51
3	488	26
4	282	13
5	231	13
6	196	7

* Data rounded off, since they were read off a graph.

The liberation of dry spores

Some of the most important epidemic-producing pathogenic fungi such as the rust fungi (Uredinales), the downy and powdery mildews (Perenosporales, Erysiphales) and the potato blight organism (*Phytophthora*) have dry spores. These are detached and dispersed by the wind. This is also true of many of the common moulds, e.g. grey mould (*Botrytis*), pin moulds (*Rhizopus* spp.), grey and blue moulds (*Penicillium*, *Aspergillus*). Although in some cases the force of the wind is the agent detaching the spore, in many cases there are ancillary structures which either lead to release but not the liberation of the spores, or provide a weak connexion which can readily be broken by any shock. ZOBERI (1961; INGOLD and ZOBERI, 1963) has made a study of the take-off of mould spores in relation to the speed and humidity of the wind. Dry-spored types were found to be most effectively liberated at low relative humidities and high wind-speeds, e.g. *Trichothecium roseum* (Table 6.1). Separation devices include various modifications of the wall between the spore and hypha, intercalary cells and disjunctors (Fig. 6.1). Frequently the septum between adjacent conidia, e.g. *Albugo*, *Aspergillus*, which is simple differentiates into three lamellae, the central one of which gelatinizes and is soluble in water, so ensuring separation of the conidia (ZALEVSKI, 1883). Between the aecidiospores of many rusts there are actual intercalary cells which function in an analogous manner, separating the spores by their

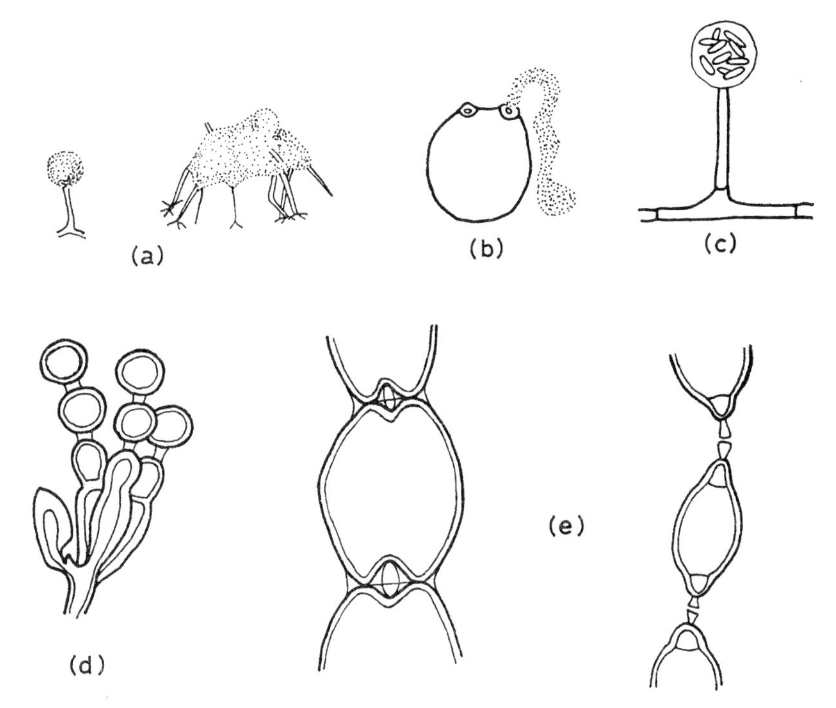

Fig. 6.1. 'Slime' and 'dry' spore types. (a) *Haplographium bicolor*, showing spore mass enveloped in mucilage (×100); (b) *Dendrophoma faginea*, pycnidium releasing spores in a mucous strand (×*c*.75); (c) *Coprinus fimetarius*, oidia held in a drop of sugary secretion attractive to insects (×700); (d) *Albugo candida*, spores separated by disjunctors (×550); (e) *Sclerotinia vaccinii*, similar to (d) but disjunctor more highly differentiated (×900, ×450). (From Langeron, after various authors)

gelatinization. Species of *Sclerotinia* have a more elaborate mechanism for spore separation and detachment. The transverse septum becomes double although all the conidia are still enclosed in the primary membrane. From each new membrane a central body is produced and these fuse to form a single elongated disjunctor. The pressure exerted by the intercalation of this disjunctor leads to a rupture of the primary membrane in the plane of the septum and the conidia are then held together only loosely by the disjunctor (WORONIN, 1888). In some cases small hygroscopic movements of the conidiophore may be sufficient to detach spores but not to liberate them. An example is grey mould (*Botrytis cinerea*) where hygroscopic movements detach the spores which may then be blown away or dispersed as a consequence of rainsplash. This latter effect is considered later (p. 151). Another way in which water affects passive liberation of spores is through mist pick-up. Drifting droplets of water in mist may come into contact with spores which will be attracted and held to them by surface tension and so carried away. How far this is really effective is unknown but such claims have been made in respect of *Cladosporium*, for example (DAVIES, 1959).

The liberation of slime spores

In many of these the spores are liberated as a consequence of the gelatinization of their walls or through the autolysis of adjacent structures or cells. Good examples are to be found in the Mucoraceae where the sporangium wall dissolves, leaving the spores held together in a drop of mucilage containing water, e.g. *Mucor hiemalis*, *Phycomyces* spp., *Pilaira anomala* (DOBBS, 1939). A similar kind of behaviour is shown by many conidia-producing fungi where the spores form in a similar drop at the ends of the conidiophores, e.g. *Graphium*, *Cephalosporium*, *Verticillium* spp. (Fig. 6.1). These drops are not blown away by the wind and, in a dry atmosphere, the spores involved are left sticking firmly to each other. While sticky they may be liberated through contact with a passing animal, or by spreading out, like oil, over a film of water. When dry they may become detached as a consequence of some physical shock, although this must be considerable. ZOBERI (1961), for example, was unable to detach any slime spores at all in his experiments with wind speeds up to 16·6 m/sec over a range of 20–90% R.H. At an R.H. of 45·5% and a wind velocity of 8·8 m/sec he was able to detach the slime spores of *Trichoderma viride* but this was only one-third as effective at 91% R.H. It is possible that this reflects the different susceptibility to detachment of dried down and still sticky conidia.

Many pycnidia also produce drops, as in the pycnidiospores of rust fungi, but in other cases spore formation is associated with the production of mucilagenous material, usually by the autolysis of sterile cells in the pycnidium. A similar kind of behaviour is shown by several Pyrenomycetes. In all such cases the spores are exuded through the ostiole, embedded in slime, as a slime tendril which may be long or short, e.g. Sphaeropsidales and conidial stages of Ascomycetes such as *Diatrype stigma* (several centimetres long); perithecia of *Chaetomium* and *Ceratostomella* spp. (Fig. 6.1). A somewhat similar appearance is given by the gelatinous finger or tongue-like tendrils of teleutospores produced by the rust fungus *Gymnosporangium*, often on *Juniperus*. Here, however, it is the walls of the teleutospores themselves which swell up and become gelatinous when wet and the mass is then exuded.

A rather different form of spore liberation is achieved in those cases where the spore-drop or associated slime is attractive to animals, especially insects. The spores become detached in the attractive material which is then removed by the animal. Examples are the sugar-containing droplets exuded by the pycnidia of many rusts, the oidia of agarics and the *Sphacelia* stage of the ergot organism (*Claviceps purpurea*), or the sweet to nauseating slimes of Phallales. The best known examples of these are the stink-horns of various types (*Phallus, Dictyophora, Mutinus*) and the hollow, spherical lattice-wall of *Clathrus* and the Australian *Aseröe*. Spore liberation is here dependent upon a complex series of processes. The basidia in all cases are embedded in a dark-green sugary gluten which encloses the hymenial surfaces and is itself encased in a gelatinous volva and an outer peridium, this being spherical or egg-shaped. The 'eggs' are just subterranean and evidently take up water so that the volva gel swells. When this happens the characteristic sickly-sweet odour which is attractive to flies and slugs develops. Slugs are said to be able to detect phalloids some 2 m away! (RAMSBOTTOM, 1953). The cells of the stalk or lattice work, as the case may be, now enlarge, the peridium splits and the hymenium, with its covering of gluten, is carried above the ground where the material, including the spores, is rapidly removed by flies and slugs. For example COBB (cited in RAMSBOTTOM, 1953) has found $22{\cdot}4 \times 10^6$ spores in a single 'fly speck' voided after feeding on a *Phallus* sp. in Hawaii. How and when the spores become detached from the basidia is unknown. *Clathrus* and *Aseröe* are also visually attractive to human eyes, the former being crimson, the latter having an apical orbicular disc from which protrude 5–9 horizontal bifid rays, the whole being bright red. It seems doubtful whether this is a means of attracting flies because most of them are relatively insensitive to colour. The nature of the glutens and the odours of Phallales are unknown. Another form of insect- and mammal- (especially rodent) attracting mechanism is found in hypogaeus fungi, especially truffles. These are detected by their odour and are attacked by flies and beetles who lay their eggs in them, or by rodents who grub them up for food. This, at least, hastens the breakdown of the peridium and so the enclosed asco- or basidiospores, which are presumably already liberated within the carpophore, can escape. Very little accurate information on these matters is available at present.

Spore liberation by impaction

The best known examples of this type of spore liberation are the puff-balls and their allies (Lycoperdales), the various types of 'splash-cups' such as those of the bird's nest fungi (Nidulariales) and other less specialized fungi (Fig. 6.2).

In the puff-balls and their allies the contents of the basidiocarp are a mass of dry spores interspersed with fine dried-out hyphae, the capillitial threads. The basidiocarp opens by a simple (*Lycoperdon*) or complex (*Geaster*) ostiole. Spores may be blown out by the suction of the wind across the ostiole but, by means of high speed photography, GREGORY (1949) has shown they can also be forced out when the basidiocarp is struck by raindrops. The peridium is unwettable, a raindrop compresses it and air is expelled with spores; the peridium then springs back to shape. Any other form of impaction such as a kick, is as effective and INGOLD (1965) has made the interesting suggestion that desert fungi with similar spore-releasing mechanisms may be operated by impacting wind-blown sand-grains, e.g. *Tulostoma* spp. (*T. brumale* is an

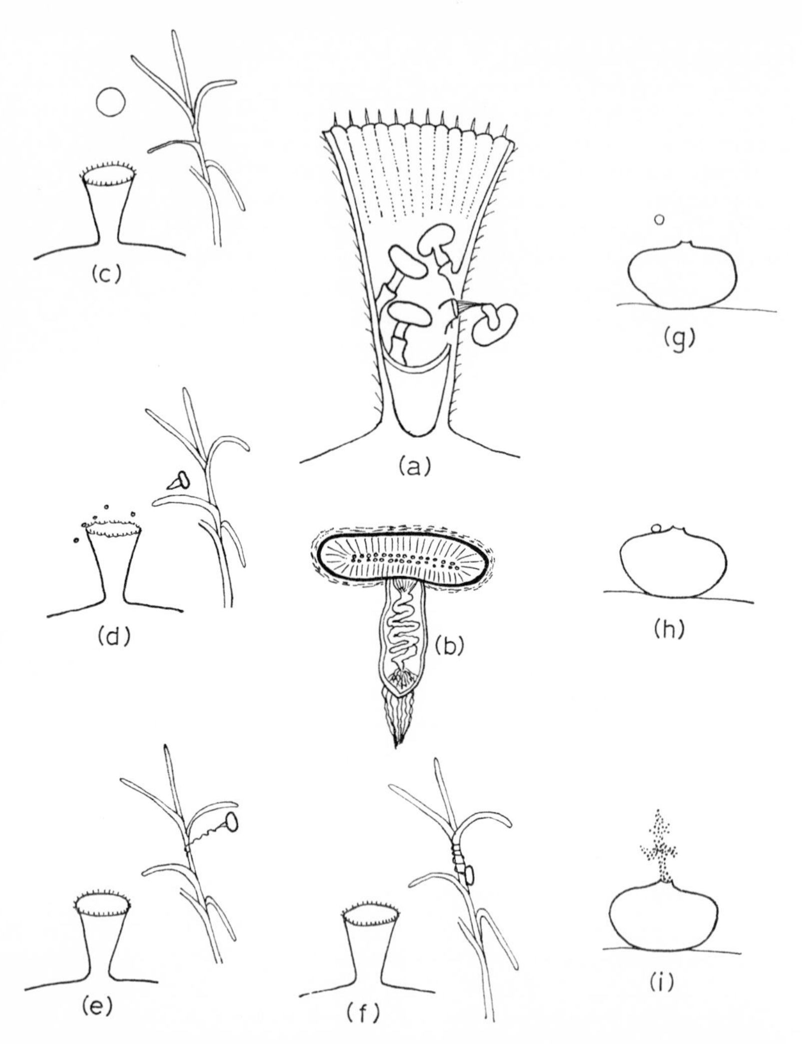

Fig. 6.2. Diagrams to illustrate liberation by impaction. (a) Splash-cup dispersal by basidiocarp of *Cyathus striatus*, a bird's-nest fungus, showing cup-shaped peridium containing peridiola attached to wall; (b) section through a peridiolum. The terminal hymenial layer enclosed in a tunica, the funicular cord attached at its base by an hapteron to the middle piece, all enclosed in the purse; (c)–(f) successive stages of dispersal consequent upon a raindrop falling into the peridium. In (e) and (f) the hapteron catches on a plant stem, the funicular cord extends and the resultant jerk when it reaches full extension winds it around the stem (after Brodie). (g)–(i) consequences of a raindrop falling on the outside of the peridium of a puff-ball (*Lycoperdon*). The puff-ball is momentarily compressed and a puff of spores is expelled. (After Gregory)

uncommon British species of sand dunes). In other Gasteromycetes, some of them allied to the puff-balls, liberation is effected by more drastic movement of the capillitial threads (see below).

BRODIE (1951) has studied splash-cup dispersal mechanism in the Nidulariales. The basidiocarps are either cup-shaped structures containing peridiola, e.g. *Crucibulum*, *Cyathus*, or small heaps of peridiola. Peridiola are ovoid, flattened structures within which the basidiospores develop and become detached. In *Cyathus* and *Crucibulum*, peridiola are attached by a spirally coiled and enclosed funicular cord with a basal hapteron, to the wall of the cup, but this structure is lacking in other genera. If a raindrop falls into the basidiocarp it breaks up, is reflected from the splash-cup and carries the peridiola out with it. If the peridiolum of *Cyathus* or *Crucibulum* strikes an object the hapteron may adhere to it as its component hyphae have spread out; the funicular cord tends to wind round the object, rather like a bolas, so anchoring the peridiolum.

Splash dispersal is probably more widespread than these rather specialized examples suggest. For example, the impact of large drops of water (5 mm) from a height on to a twig covered in conidial stromata of the coral spot fungus (*Nectria cinnabarina*) caused the drop to break into several thousand droplets all containing spores (GREGORY *et al.*, 1959). This is a fungus with slime spores and since splash dispersal has already been referred to in connexion with dry-spored forms (p. 148), the phenomenon may be extremely widespread.

Liberation of spores by hygroscopic movements

This mode of liberation occurs in dry-spored fungi and is considered here because it is a quasi-active mode of liberation. An often-quoted case is that of *Perenospora tabacina* (PINCKARD, 1942) where the dichotomously branched conidiophore twists, dislodging the spores, through hygroscopic movements consequent upon drying out. It seems possible, however, that such spores could be blown off by wind action. In a number of Gasteromycetes there may be a more highly developed process. In *Podaxis* and *Battarraea* spp. (two stipitate genera) the sporogenous tissue consists of spores intermingled with well developed capillitial threads or elaters. In *Podaxis* the capillitia are coiled threads, 2–3 cm long, arising from the central stipe and passing through the spore masses to the outer peridium. In *Battarraea* the spores are interspersed with coarse, stiff capillitial threads attached to the stipe and free, elongated spirally thickened cells similar to the elaters of liverworts. It is not known with certainty whether the hygroscopic movements of these capillitial threads and elaters lead to spores being released when the peridium separates from the stipe but the occurrence of similar structures in the sporangia of slime moulds and liverworts strongly suggests a functional analogy.

Spore liberation by trigger devices

The observations of MEREDITH (1961, 1962, 1963) have shown that in some cases, dry spores are liberated by an active discharge mechanism. The cell concerned loses water by evaporation so that it develops a negative pressure and tends to be distorted. As the process continues water molecules in the cells tend to cohere both amongst themselves and to the walls. Thus the water is exposed to an increasing tension until a point is reached where the tension is released by the water expanding into a gas phase and so the distorted walls recover. This rapid process imparts a jerk to the cell concerned and to its neighbours. In some cases the cell concerned is the conidiophore and the jerk, at recovery from tension, discharges one or more conidia into the air, e.g. *Deightoniella torulosa* (Fig. 6.3), *Zygophiala* spp.; in others, however, gas

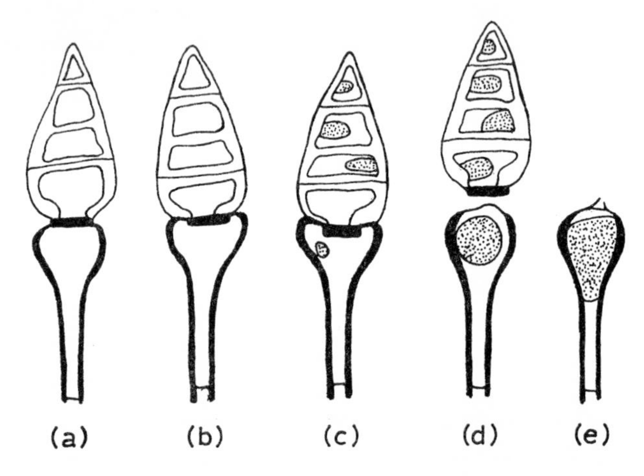

Fig. 6.3. *Deightoniella torulosa* (×600). (a)–(e) Stages in a conidium being jerked from a conidiophore through a sudden reversal of increased tension on the walls being released by water expanding into a gas-phase bubble (d). (From Meredith)

bubbles develop in one or more of the conidia immediately prior to discharge, e.g. *Curvularia* spp., *Alternaria tenuis*. In all cases the jerk of recovery is the affective agency of liberation but in some cases the process is, perhaps, assisted by hygroscopic movements of the conidiophore similar to those described for *P. tabacina* earlier, e.g. *Memnoniella subsimplex*. It is of interest that INGOLD (1956) had previously predicted this type of negative-pressure mechanism when discussing the occurrence of a gas-phase in fungal spores. However, it is clear that while such gas bubbles may not be uncommon in fungal cells, they are not necessarily associated with spore liberation.

In other fungi discharge is triggered off by the development of high positive turgor pressures in the cell which cause them to round off. Examples are known from Perenosporales, Entomophthorales and Uredinales. In *Entomophthora coronata* (Fig. 6.5a, b), a single terminal conidium is borne on a simple conidiophore, each has its own enclosing membrane but both are enclosed by a common outer membrane. This ruptures around the point of contact, the conidiophore suddenly swells up, its terminal apex bulges out and the conidium is discharged. A similar rounding off occurs in *Sclerospora philippinensis* but here several conidia are discharged simultaneously from the beaked tips at the end of the conidiophore. In the aecidia of many rusts the aecidiospores are packed closely together and each assumes a polyhedral form. Once they are separated by the gelatinization of the intercalary cells they continue to adhere but swell up by absorbing water. Thus each spore tends to round off and the outermost spores are shot off, singly or in groups, as a consequence of the cumulative pressures developed. They may be discharged as much as 7–8 mm, although 3–4 mm is more usual.

The most striking discharge mechanism depending upon the rounding off of cells due to their increased turgor pressure is shown by the Gasteromycete, *Sphaerobolus* (Fig. 6.4). An excellent account of this mechanism is given for

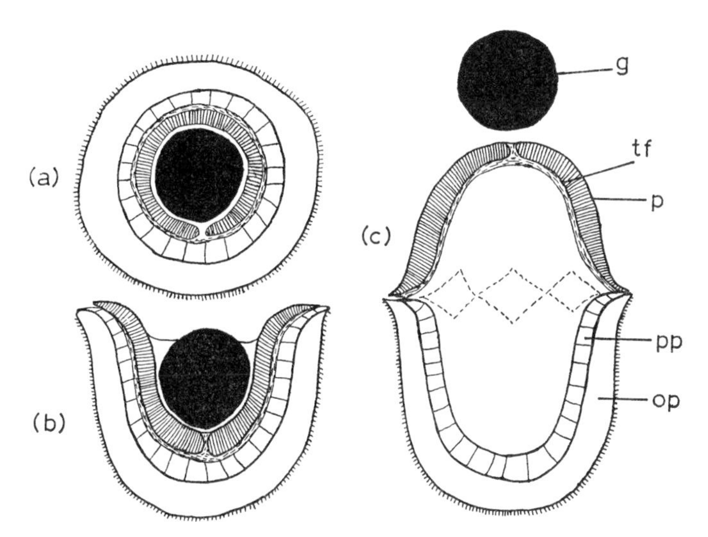

Fig. 6.4. *Sphaerobolus stellatus*, Median L.S. diagrams to illustrate discharge of basidiospore containing projectile. (a) young basidiocarp; (b) mature basidiocarp which has split open; (c) basidiocarp after inversion of inner, palisade layer. Key: op, outer peridial layers, outer fibrous and inner gelatinous; pp, pseudo-parenchymatous layer; tf, layer of tangentially arranged fibres; p, palisade layer, usually orange coloured; g, glebal mass

S. stellatus by BULLER (1933). The basidiocarp is a minute sphere some 2 mm in diameter. Its core is composed of a spherical glebal mass of basidiospores, gemmae and cystidia being embedded in a gelatinous matrix. This is surrounded by six histologically distinct layers which comprise the peridium. The peridium splits at the apex to give six to eight teeth and, at the same time, a split develops between the outer three and the inner layers except at the apices of the teeth. The innermost layer gelatinizes so that the glebal mass rests in a drop of fluid held in an inner and an outer cup. The innermost layer of the inner cup is composed of large parenchyma-like cells which, initially, are full of glycogen. This becomes hydrolysed to glucose (WALKER and ANDERSON, 1925; ENGEL and SCHNEIDER, 1963), hence its suction pressure increases and it takes up water. The cells tend to expand but the outer layer of the inner cup is composed of rigid, tangentially running interwoven fibres which resist expansion. The tension so set up is finally released by a sudden inversion of the inner cup, accompanied by a minute detonation, which thereby projects the glebal mass some 2 m high and 4–5 m horizontally. The everted cup remains like a little balloon still attached to the outer cup tissues at the apices of the teeth. Light is necessary for initiation and development of the basidiocarp and both young and unopened, mature basidiocarps show a directional response to light (ALASOADURA, 1963), although the mechanism of phototropism is not understood.

Spore liberation by syringe devices

The essential processes involved in this type of spore liberation are the development of a high hydrostatic pressure within a cell and its release by extrusion of the cell contents as a jet through a fine orifice. Exogenous spores are thereby shot off, endogenous spores are either projected through the orifice by the pressure or ejected with the fine jet. This type of spore liberation is found in a few Phycomycetes, e.g. *Empusa*, *Basidiobolus*, *Pilobolus* and Fungi Imperfecti, e.g. *Nigrospora sphaerica*; the majority of Ascomycetes, both Discomycetes and Pyrenomycetes in respect of the ascospores, and in the discharge of the basidiospores of all Basidiomycetes (except in Gasteromycetes where little is known of how basidiospores become separated from basidia). Since there are differences, largely imposed by the morphology of the spore-bearing structures, between these groups, it will be convenient to describe them on a taxonomic basis.

In *Pilobolus* (Fig. 6.5c, d) and *Empusa* the walls of the apices of the sporangiophores and conidiophores, respectively, are extremely elastic and have a circular transverse line of weakness around their apices. The hydrostatic pressure of the sporangiophore (or conidiophore) increases until it ruptures, then it contracts violently, squirts out a jet of sap and so projects the sporangium (or conidium). In *P. kleinii* the distance to which the sporangium is discharged is about 2 m. Particular interest applies to *Pilobolus* because of its phototropic response which results in the tip of the sporangium being directed towards the maximum light intensity (see Fig. 12.2, p. 322). The discharge of conidia by *Basidiobolus ranarum*, commonly found on frog dung, is somewhat different. The terminal conidium is borne on a swollen conidiophore but the line of dehiscence is here towards the base of the swelling. At rupture the conidium adheres to the upper part of the swollen conidiophore which contracts, squirting sap out and so driving the conidium forward some 1–2 cm. The two parts usually separate in flight.

The only Fungus Imperfectus known to have an active syringe-discharge mechanism is *Nigrospora sphaerica*. It is essentially like that described for *Empusa*, rather than the *Basidiobolus* type (WEBSTER, 1952).

The active discharge of basidiospores may be similar to that of *Pilobolus* or

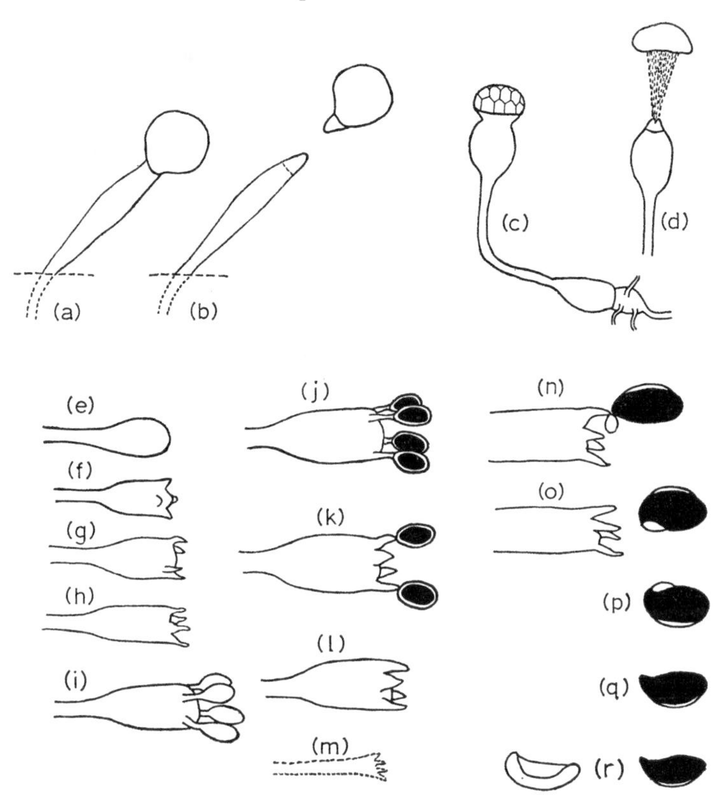

Fig. 6.5. Discharge by rounding-off and syringe-devices. (a), (b) discharge of conidium by *Entomophthora coronata*. The columella can be seen bulging into the conidium. At discharge the incurved, conidial wall can be seen to have bulged out (×340); (c), (d) discharge of whole sporangium of *Pilobolus* sp. The jet is directed on the base of the sporangium (×13·5); (e)–(m) diagrams of successive stages in development and discharge of basidiospores; (n)–(r) diagrams of discharge of basidiospore of *Coprinus sterquilinus*. Note drop exuded at hilum; its spread over the spore surface after discharge and the concave contraction of the spore as it falls (shown at (r) in section). (After Ingold, Zopf and Buller)

Empusa but certain aspects are somewhat different. In particular, basidiospores are borne asymmetrically at the tip of fine sterigmata and their discharge is associated with the secretion of a small drop of liquid. Such spores are often termed ballistospores. A typical Holobasidiomycete basidium is a terminal clavate cylinder which develops sub-apically, four incurved projections, the sterigmata. From the tip of each of these, a spherical to ovoid swelling develops to form the basidiospores. Each is attached to a sterigma by a minute projection, the hilum, situated at the base of the, usually, concave adaxial face of the basidiospore (cf. Fig. 5.18, p. 129). The four spores are thus held asymmetrically facing outwards from the apex of the basidium.

Buller made the critical observation that some 5–40 sec before the spores are projected a drop of fluid emerges on the adaxial side at the base of the hilum. It enlarges and may achieve a diameter half that of the basidiospore. When it has reached its maximum diameter the basidiospore is suddenly and violently projected and can then be seen to have the drop adhering to, and spread over, its concave, adaxial face. One basidiospore only is projected at a time but all four follow each other successively, usually in opposite pairs. After discharge there is no detectable shrinkage of the sterigmata and only exceedingly rarely is a drop of liquid to be seen adhering to any of them (Fig. 6.5n to r).

Before these critical observations there were two hypotheses. BREFELD (1877) (who thought the spores were discharged simultaneously) believed that they were projected by tiny jets consequent on the bursting of the tips of the sterigmata, although these would have had to become re-sealed extremely rapidly. BULLER (1909), on the other hand, thought that basidiospore discharge was due to a rounding off process analogous to that already described for *Entomophthora* or *Sclerospora*. Both theories have difficulties. The latter neither provides an explanation of the invariable presence of the exudation droplet, nor does it account for the successive discharge of the basidiospores. The former theory does not account for this last point, there is no evidence of contraction of the sterigmata and their proposed rapid sealing would have to be exceedingly efficient if it were never to fail. Further studies have done little to resolve these difficulties. Support for Brefeld's view has come from MÜLLER and CORNER, for Buller's 1909 view from PRINCE (1943) and BEGA (in INGOLD, 1965).

MÜLLER (1954) studied discharge in the basidiospore-producing yeast *Sporobolomyces*, a genus already studied by Buller and remarkable for the ability of the same sterigma to produce successively up to four or more basidiospores. He observed cases where the drop disappeared suddenly but the spore remained attached. He supposed that the drop was squirted away by the hydrostatic pressure of the basidium and normally adhered to the basidiospore so carrying it away as well. This was supposed not to have occurred when the spore was left and the drop disappeared. CORNER (1948a) confirmed that the hydrostatic pressure is developed through the enlargement of the basal vacuole so that cytoplasm is forced into the sterigmata and he claims that this, not water as Buller had supposed, constitutes the material of the drop. This also explains the sticky surface of the discharged basidiospore, a fact well known and commented upon earlier by Buller. He too supposes that after the cytoplasm has exuded from the stretched wall, the tip breaks and a minute jet blows off the spore. The drop is exuded at the hilum region because this is the least hardened and most stretched region of the basidiospore.

Support for the opposing theory has come from the observations of Prince and Bega on the rust fungi *Gymnosporangium nidus-avis* and *Cronartium ribicola*, respectively. In these fungi there seems to be a distinct cross-wall separating the basidiospore hilum from the sterigma tip so that discharge could be due to the rounding off of the wall of the latter. The drawing of the (unpublished) electronmicrograph of *C. ribicola* shows the projecting hilum and a distinct depression at the apex of the sterigma into which the hilum could fit like a peg in a socket. A similar interpretation can be made of electronmicrographs of sterigmata of *Schizophyllum* (WELLS, 1965). These observations certainly make a rounding-off hypothesis possible but do not in any way provide an explanation of the formation of the drop.

The remaining two hypotheses have little to support them. INGOLD (1939) developed a suggestion of Buller's that in some way the force of surface tension developed on the drop is responsible for discharge. He showed that the surface energy of the drop at maximum size was of the order of 12×10^{-13}J and that of the drop adhering to the spore after discharge $9{\cdot}5 \times 10^{-13}$J thus a maximum of $2{\cdot}5 \times 10^{-13}$J is available for work. Some $0{\cdot}34 \times 10^{-13}$J is required to account for the initial velocity of the spore and this is but one-seventh of that available. These calculations make no allowance for the work done in the change of shape of the drop, its adherence to the spore wall, or the fact that it is probably not pure water. But even if these assumptions were correct it is not clear how the energy available could be utilized to project the spore. OLIVE's (1964) novel hypothesis does not meet the fact repeatedly observed, that the drop adheres to the discharged spore. The drop is supposed to be due to accumulation of gas, perhaps CO_2, which blows a bubble between the double wall of spore and sterigma and dislodges the spore when it bursts violently!

In fact, the mechanism of ballistospore discharge is unknown. Since most of the process can readily be observed there seems to be more need for an ultra high-speed record of the actual process of projection rather than further speculations.

The mechanisms of syringe liberation are better understood in the Ascomycetes. In the cleistothecial forms, such as the powdery mildews (Erysiphales), the asci within the cleistocarp take up water and swell. The cleistocarp is ruptured and the ascus or asci within protrude or may even squeeze out and finally be violently ejected. Continued swelling of the contained or ejected asci results in their bursting and scattering the enclosed ascospores in all directions, e.g. *Sphaerotheca*, *Erysiphe* (Fig. 6.6).

In the Discomycetes the exposed hymenium of the apothecium permits the simultaneous discharge of ascospores and this gives rise to the phenomenon of 'puffing', i.e. synchronized discharge of mature asci. An ascus discharges its spores as follows. It becomes turgid as the result of developing a high, positive hydrostatic pressure and this is often associated with the development of a vacuole which forces much of the cytoplasm and the ascospores to the apex of the ascus. The ascus frequently enlarges and may protrude from the rest of the hymenium in a state of high tension. This tension is released when the operculum, a small, hinged apical lid, gives way and the elastic walls of the ascus contract, driving out the upper half of the contents as a cylindrical jet which dissipates into a cloud of ascospore-carrying droplets. The distance of discharge depends on the species and ranges from 0·5–17·0 cm.

The basic process is aided in a variety of ways in Discomycetes. The ascospores are often attached to each other and to the operculum, e.g. *Sarcoscypha protracta*, so that they are discharged together, although presumably in rapid succession since the diameter of an ascospore is often similar to that of the apical opening. The asci are held upright by mutual compression of the elements of the hymenium but, in addition, many are positively phototropic in some respect. In *Ascobolus stercorarius* the ascus tips can bend in response to unilateral light in that part protruding above the hymenium while, at the

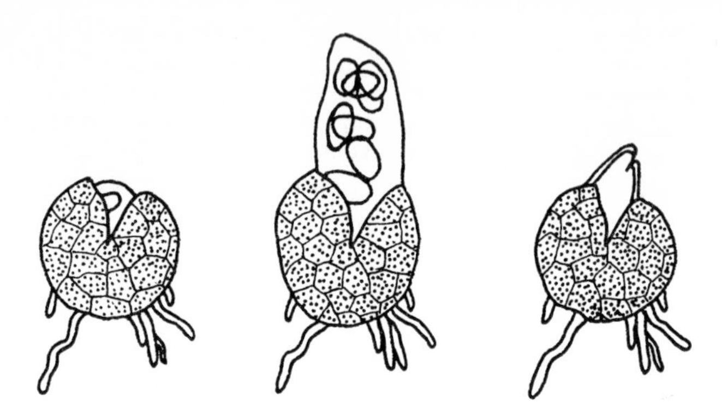

Fig. 6.6. *Sphaerotheca mors-uvae.* Discharge of ascus from cleistocarp due to its swelling followed by release of ascospores through an apical slit in the ascus. (× *c.* 200, after Salmon)

other extreme, *S. protracta* can only react by a slight displacement of the operculum towards the light. In many cases also, the trigger which leads to the synchronized ejection by all mature asci is a mechanical shock such as touch or wind displacement, and the discharge is then accompanied by an upthrust of air due to the many thousands of jets and spore drops being directed at the same time in the same direction. This puff of air probably assists in carrying them further than if they were ejected in smaller numbers (BULLER, 1934). The larger stipitate Discomycetes do not show this tactile response and do not usually 'puff' e.g. *Morchella.*

Some Discomycetes, the inoperculate forms, have asci which only dehisce by an irregular apical split, e.g. *Geoglossum, Trichoglossum.* In these genera the ascospores are filiform and septate and after dehiscence one is pushed through the apical pore, slowly at first but with increasing velocity until finally it is discharged. Before the ascus can contract, another ascospore blocks the pore and the process is repeated until the last spore has been ejected; the ascus then collapses (INGOLD, 1939). Not all inoperculate Discomycetes behave like this, however, and some are capable of effective 'puffing', e.g. *Helotium, Sclerotinia* and *Rhytisma acerinum*, the tar-spot fungus of sycamore leaves (BULLER, 1934; WOLF and WOLF, 1947).

It is not entirely clear how the asci do in fact rupture or open by their operculi. Studies such as those of CHADEFAUD (1942) have revealed the enormous anatomical complexity of the apical regions of asci during development and it is clear that the end-result, dehiscence, may come about in a variety of ways.

Pyrenomycetes never exhibit 'puffing' for the ostiole of the perithecium, enclosing the asci, is usually so narrow that only one ascus can pass through it at a time. Three types of active discharge have been described (Fig. 6.7). The first is illustrated by *Sordaria fimicola* and is probably the commonest method amongst coprophilous Pyrenomycetes (INGOLD, 1953). The asci remain attached to the base of the perithecium, their walls consist of a single layer and are highly extensible. The asci elongate up the neck canal, one at a time, until the tip is forced through the ostiole. Increasing internal hydrostatic pressure leads to the rupture of the ascus and the spores are forced out singly or in groups. As the pore is about 4 μm in diameter and the ascospores about 13 μm there is an alternate extension and contraction of the ascus pore and the latter movement may tend to break the mucilaginous thread which holds the eight ascospores together in the undischarged ascus. In some cases, all the ascospores are ejected in a mass, together with the apical part of the ascus which has dehisced by a circumsessile slit, e.g. *S. curvula*.

A second type first described by PRINGSHEIM (1858) is *Pleospora scirpicola* and there are many other species like it. It is similar to the *Sordaria*-type but the ascus wall is double, the outer ectoascus being thick and rigid, the inner endoascus thinner and extensible. As the ascus imbibes water it tends to swell but this is resisted by the ectoascus until it is finally ruptured by the high internal pressure. The endoascus then elongates suddenly to some two to three times its original length and penetrates the ostiole. The uppermost of the ascospore, which by now has been displaced upwards from the base of the ascus, is violently expelled through an apical pore. There is a retraction of about half an ascospore length by the ascus and the next spore touches the pore and the discharge is repeated and followed by a second retraction and so on. Finally the empty exoascus contracts, its wall swells and it disintegrates. There are several variants; in some all the ascospores are ejected *en masse*, e.g. *Sporomia intermedia*; in others the ectoascus dehisces in a circumsessile manner leaving a thimble-like cap over the apex of the endoascus which is pushed off at the first discharge, e.g. *Lecanidion atratum*. A further variant is found in *Glomerella* spp. where the ectoascus remains intact save for a papillate pore. The endoascus perforates and ascospores are squeezed through the ectoascal pore which snaps shut between successive ejections.

The third type is found in Pyrenomycetes with long perithecial necks such as *Ceratostomella* or *Ophiobolus* spp. The asci become detached and lie free within the perithecium. As more asci are formed basally they pack the perithecial cavity and are forced into the neck. Finally, one protrudes, it swells but its lower half is held firmly by the ostiole so that it bursts apically, liberating the ascospores, and is itself extruded when empty, its place being taken at the ostiole by the next ascus. In some species, e.g. *Linospora gleditsiae*, the asci collect in a mucoid drop and are disseminated by water before

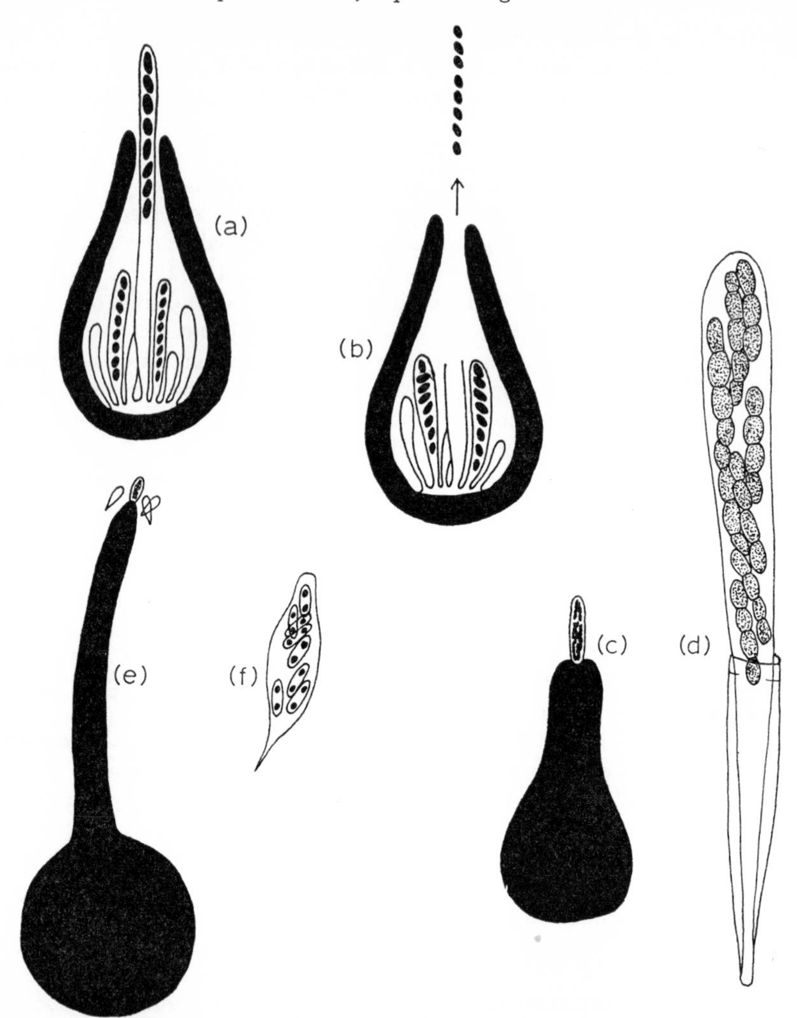

Fig. 6.7. The three modes of spore liberation in Pyrenomycetes. (a), (b) spore discharge (diagrammatic) by *Sordaria*. The asci elongate and, one at a time, protrude through the ostiole, discharge their spores and retract; (c), (d) *Sporomia intermedia*. An ascus protrudes from the perithecium (c, ×85), its outer walls have stretched and split allowing the inner walls to elongate and discharge the spores (d, ×470); (e), (f) *Ceratostomella ampullacea*. A perithecium discharging its spores, an acus being held in the ostiole before bursting, three empty asci extruded nearby (e, ×40). A mature detached ascus is also shown (f, ×330). (From Ingold)

swelling and bursting. They evidently approach the spore-tendril type found in other species of *Ceratostomella* and *Ophiostoma*.

One problem which has hardly been touched on in this and the previous section is the production of high internal hydrostatic pressures. There is extremely little information on this topic, in part because it has proved difficult to make osmotic determinations with fungal cells, in part because many of the cells are inaccessible at the appropriate stage of development, but also because little attention has been paid to the problem. There are three ways in which such pressures could arise. More cytoplasm including vacuoles could be translocated into the cell; the osmotic properties of the cell could alter by a change, for example, of glycogen to hexose so that water was taken up exogenously; or there could be an increase in endogenous water leading to an increased volume of vacuole and hence increased pressure within the cell. Any one or all three processes could be effective at any one time in development, as for example in the basidium where cytoplasmic translocation and vacuolar enlargement take place at different ontogenetic stages.

Periodicity of spore liberation

Spore-trapping in the immediate vicinity of spore-producing structures and in the air generally has revealed a surprising number of periodic patterns. Many of these are related to the diurnal rhythm of alternating day and night, others are related to the water relations of the reproductive structures, and yet others reflect changes in climatic cycles. Information on these periodical changes is summarized by INGOLD (1965) and GREGORY (1961).

Periodical discharges in relation to availability of water have been frequently studied and are, perhaps, most easily understood. Many Ascomycetes are drought-enduring xerophytes and their spore discharges are correlated with periods of effective wetting by rain. This often has important consequences for the epidemiology of pathogenic fungi and for appropriate control measures. Lack of periodicity is shown by many pathogenic stromatoid Ascomycetes and Basidiomycetes, especially polypores. By supplying or withholding water in various ways both from attached and detached carpophores, Ingold has been able to show that continuous spore liberation may be due to the availability of water from the host or from tissues (especially gels) in the fungus.

Temperature and rainfall, the two external factors which greatly affect the development of reproductive structures, themselves appear to be the prime determinants in long term periodic fluctuations, e.g. annual cycles of air spora.

Temperature, humidity and light are often determinants of diurnal rhythms. It has been suggested that the day-maximum periodicity of *Erysiphe* spores is related to light, that of *Fomes annosus* basidiospores to temperature and the pre-dawn maximum of *Sporobolomyces* ballistospores to the high humidity which is necessary for their discharge. Sometimes the causes are not immediately obvious, for example, the mid-day maximum of *Ustilago nuda* (loose smut of barley) chlamydospores. This was eventually correlated with wind velocity which operates by affecting infected plants since they are taller than healthy plants, remain upright in the wind and so the dry powdery spores are readily dispersed. In a number of experimental studies endogenous diurnal rhythms have been found which persist even when the fungi are

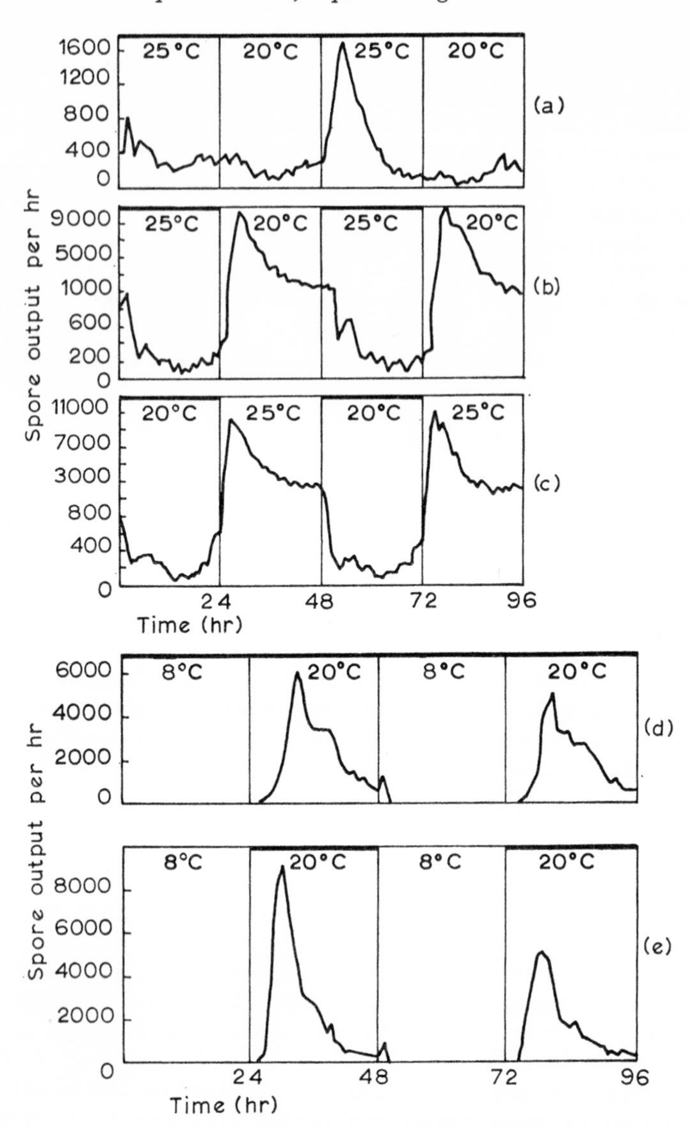

Fig. 6.8. Periodicity of ascospore discharge by *Sordaria fimicola* in relation to temperature, light and darkness. Dark periods with solid bar above them. (a) culture grown in continuous darkness with alternating 24-hr periods at 25° and 20°C. Note development of temperature-regulated periodicity. (b), (c) cultures grown in light and then transferred to alternate 24-hr dark and light periods and alternate 25°/20°C regimes in opposite phase. Note dominating influence of light-induced periodicity over temperature effect shown in (a); (d), (e) 8° and 20°C 24-hr alternating periods in total darkness (d) or in alternate light and dark periods, the light period coinciding with the low period (e). Note that in these conditions, temperature-induced periodicity dominates over light effects as shown in (b) (c). (From Ingold)

exposed to continuous darkness or light (Fig. 6.8). While descriptions of such rhythms are not uncommon it must be confessed that the elucidation of their causal factors is proving extremely intractable. As with many other organisms, fungi such as *Daldinia* and *Pilobolus* (INGOLD and COX, 1955; SCHMIDLE, 1951; UEBELMESSER, 1954) appear to have an internal 'biological clock' which can be set by light or temperature but which will not depart from a 24-hour period and will continue for some time without re-setting.

SPORE DISPERSAL

Spore dispersal is essentially a passive process determined by physical agencies such as wind or water or by living organisms. However, it is convenient to consider first the ballistic problems involved in active discharge mechanisms.

The ballistics of spore discharge

Spores which are projected violently through the air follow a quasi-parabolic trajectory like most other projectiles. If they have not reached the ground by the time the energy of discharge has been dissipated then they will fall freely under gravity. Indeed, it was this latter property which was first investigated by Buller in an attempt to see whether the spores of agarics and polypores follow Stokes' law.

The most important quantities which determine the horizontal or vertical range of a projectile are its initial velocity, its angle of elevation and the resistance it encounters in its passage through the air, although this last is usually related in some way to the speed. Unfortunately there are hardly any measurements or estimates for most of these quantities. Moreover, most of the theoretical treatments of such problems deal with perfect, smooth spheres and this is an extremely unusual shape for a fungal-spore projectile.

Most measurements have been relatively simple and concerned with the horizontal range, more rarely the vertical range, and the rate of fall in still air under gravity.

The fungus projectile for which most data is available is *Sphaerobolus*. BULLER (1933) has summarized his findings and those of Miss Walker on the horizontal and vertical ranges achieved. In this fungus, the projectile is reasonably spherical and about 1–1·25 mm in diameter. They were able to show that the horizontal range of the glebal mass was affected by the angle at which it was projected (this being greatest at 45°, less at 40° and zero at 90° to the horizontal) and that the maximum horizontal range was generally about twice the vertical. It is true that the maximum ranges achieved by the Nebraska isolation of *Sphaerobolus* have a ratio of about 1·2:1 but it seems probable that the potential horizontal range is greater than that measured. These measurements suggest that a fairly normal paraboloid trajectory is followed by spore masses of *Sphaerobolus*. It contrasts with INGOLD's (1960) claims that the ratio is about 1:1 for projected ascospores of *Sordaria fimicola* although under his experimental conditions the ascospores were intended to be discharged either strictly vertically or horizontally. Thus the two sets of data are not directly comparable. INGOLD (summarized 1965) has investigated another relationship in Ascomycetes chiefly amongst *Sordaria* spp. and their allies. This is:

$$D_x = \frac{u_x v}{g} \quad \text{but} \quad v = \frac{2}{9}\frac{\sigma-\rho}{\eta}gr^2$$

hence

$$D_x = \frac{2(\sigma-\rho)u_x}{9\eta}r^2 = kr^2$$

where D_x = horizontal range; u_x = initial horizontal velocity; v = terminal velocity; g = acceleration due to gravity; σ = density of projectile; ρ = density of air; η = viscosity of air; r = radius of projectile. (Formulae derived by BARLOW, in BULLER, 1909)

The relationship $D_x \propto r^2$ is only true for spherical projectiles of the same initial velocity and density discharged in the same medium. Most of these conditions have either not been checked or do not obtain. The best experimental agreement with the relationship was found with *Entomophthora coronata* where the conidium approximates to a sphere but varies in size (Fig. 6.9). In

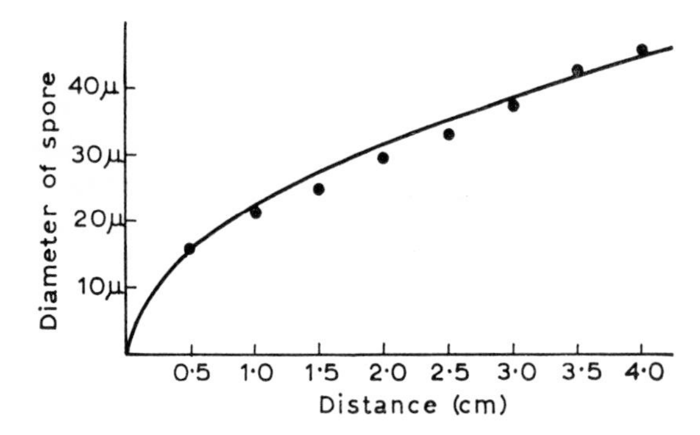

Fig. 6.9. The relationship between the diameter of the conidium of *Entomophthora coronata* and the distance to which it is discharged. The line represents the relationship $d = kr^2$. (After Ingold)

other cases, notably Ascomycetes (e.g. INGOLD, 1961; WALKEY and HARVEY, 1966), this relationship was frequently found not to hold. This is not entirely unexpected. Hardly any of the ascospores were spherical, their densities and initial velocities were unknown and, in many cases, the spores were projected adhering to each other and breaking up into larger or smaller units. It is of interest that plots of D_x against V (where $V^{2/3}$ is the volume of the spore or spore mass) did show an approach to a linear relationship when comparisons were made between closely related members of the same family, e.g. Sordariaceae, provided that their spore structure and mode of projection were similar and hence their initial velocities might be expected to be similar. It was also found that the approach to a linear relationship was only found in species where the ascospores were well separated into groups of different numbers. The process of break-up of the ascus jet has been studied by both BULLER (1909) and INGOLD and HADLAND (1959). Buller drew attention to the fact that a free cylinder of liquid will split into a number of distinct spherical drops where

$$n = \frac{l}{\pi d}$$

where n = number of drops: l and d = length and diameter of cylinder. Thus if a column is *c.* 25 times longer than the diameter it should separate into eight drops. No direct observations are available unfortunately concerning the dimensions of ascus jets and only a few measurements of the mucilage tail which holds some together, e.g. *Pleurage taenioides*. The jets of *Sordaria fimicola* are of the order of 400 μm long, those of *P. taenioides*, 1730 μm. Ingold has shown for the former that a break in the jet is equally likely between any two spores and the numbers of projectiles with different numbers of spores in them agree closely with theoretical expectations. It was also shown that there was a greater tendency for spores to adhere at low temperatures than at high temperatures presumably due to changes in the viscosity of the mucilage. There is no direct observational evidence, however, that the ascus jets or spore masses round off to form spherical projectiles as Ingold has suggested.

The relationship discussed on p. 164 was derived by BARLOW (BULLER, 1909) in an attempt to provide a formula for the trajectory of agaric or polypore basidiospores. He regarded it as compounded of independent horizontal and vertical motions and considered them separately, the former by the equation of motion, the latter by incorporating Stokes' expression for the resisting force on a moving sphere. He obtained the expression:

$$y = \frac{v^2}{g}\left[-\log_e\left(1 - \frac{x}{D_x}\right) - \frac{x}{D_x}\right]$$

where y = vertical distance of a point on the trajectory below the point of liberation; v = terminal velocity; g = acceleration due to gravity; D_x = maximum horizontal range; x = horizontal distance from the point of liberation.

This expression is not wholly satisfactory and its application is difficult. Buller was able to measure D_x with some accuracy by seeing at what distance from the surface of a gill or pore the spore deposits formed. Since the basidiospores are sticky they do not move appreciably on landing. In agarics and polypores D_x is extremely short, not exceeding 0·2 mm, as compared with 0·85 or 1·2–1·4 mm in rusts and *Tilletia tritici* respectively or 1–2·5 cm in *Sporobolomyces*. Although all these are ballistospores, the differences appear to be correlated with the fact that too great a projection in agarics and polypores would merely deposit the spore on the face of the next gill or the other side of the pore, whereas in the others the trajectory is into free air. Thus natural selection has evidently operated on those factors which control the initial velocity of ballistospores. Buller also attempted to measure v, originally as an experimental test of Stokes' law. Since Stokes' treatment applies to a perfect sphere and the density of the sphere and the density and viscosity of the medium (air) must be known, he initially selected *Amanita vaginata.* He was able to determine accurately the density (1·02) and diameter (*c.* 10 μm) of its basidiospores and, apart from a minute hilum, each spore was smooth and spherical. These latter attributes are rare in fungi and the variants of Stokes' law developed to cover certain shapes, e.g. ellipsoids where

$$v = \frac{2}{9}\frac{\sigma - \rho}{\eta} g.b. \sqrt[3]{a}\,\sqrt[3]{b^2},$$

a and *b* being the axes (FALCK, 1927), do not cover all spore shapes.

Buller also found that the basidiospores had a higher velocity, when first observed just below the gills, than predicted by Stokes' law, and, as they fell, their velocity was reduced to something less than that expected. Moreover, humidity affected the velocity, the drier the atmosphere the slower the fall. These observations applied to other non-spherical basidiospores of agarics or polypores and have since been found to apply to those of *Puccinia graminis tritici* (WEINHOLD, 1955) and other spores. The causes of these discrepancies are not fully understood. Some of the possibilities are:

(i) Experimental errors of which the most probable is the inability to obtain aerodynamically still air or to avoid drag at the boundary of the container in which the fall is studied (BULLER, 1909; YARWOOD and HAZEN, 1942).
(ii) Surface drag on the spores due to protrusions or ornamentations of the spore wall; this has never been properly assessed (MCGUBBIN, 1944).
(iii) Changes in the mass and shape of the spores as they fall. There is clear evidence of the deformation of ballistospores as they fall, e.g. *Coprinus sterquilinus* (Fig. 6.5). BULLER and HANNA (1924) claimed that this was due to loss of water during fall but no quantitative measurements have been made and the evidence is circumstantial.
(iv) Spores usually carry an electrostatic charge and GREGORY (1957) showed that this could increase the terminal velocity of spores of *Ganoderma applanatum* by about 0·05%, through interaction of the spore's positive charge and the negative charge of the flat ground in fine weather.

In view of these uncertainties concerning the experimental determination of the terminal velocity of falling spores it is clear that the use of such values to determine the initial velocity can only be rough approximations.

While the study of falling spores in still air is of interest at certain stages of aerial spore dispersal, the major process is dependent upon turbulent conditions and these will now be considered.

Dispersion and deposition of aerial spores

Spores once liberated are either below the boundary layer or above it. In the former case the air is still and the spores fall to the ground in conditions approaching those discussed at the end of the previous section. In the latter case, however, they are airborne in the turbulent layers of the atmosphere and become dispersed by the eddies and convection currents of the atmosphere. The boundary layer is not sharply defined, being broken by local eddies. In general, it is nearest to the ground on a sunny day (1 mm) highest in still, clear night conditions (10 m) and intermediate in cloudy conditions becoming lower as wind velocity increases (Fig. 6.10). Admirable accounts of these

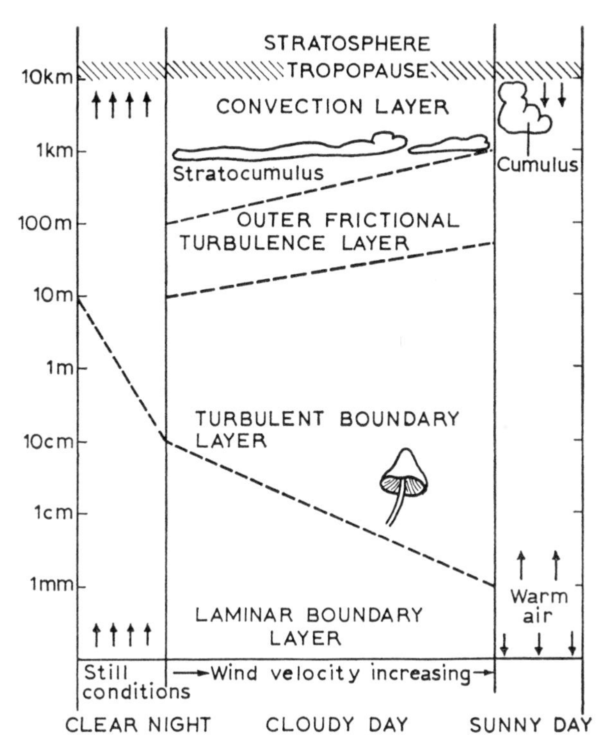

Fig. 6.10. Diagram to illustrate the effects of atmospheric conditions on the layers of the atmosphere. (After Gregory)

atmospheric patterns are given by GEIGER (1965) or, in a context relevant to dispersal, by GREGORY (1961). Once in the turbulent layers the dispersion of spores has been found to be similar to that of other particles such as smoke particles and the methods employed by meteorologists have been found to be applicable to spore clouds.

The existence of a large number of mechanisms to project small and light spores relatively small distances from the reproductive structure can now be rationalized. In essence, active spore liberation must ensure that the spore is projected beyond the surface boundary layer into the turbulent region where eddy diffusion will bring about further transport. The structure of stipitate fungi, particularly agarics, is also related to this property. Toadstools discharge their spores outwards from the gills in a more or less horizontal trajectory followed by a vertical fall but, if the pileus is high enough, it will penetrate the boundary layer. The falling spores, once clear of the gills, will then be dispersed in the turbulent layers of the atmosphere. INGOLD (1946) and BOND (1952) have both shown that with decreasing size, the height of the basidiocarp also decreases to a minimum of about 34 mm (or, if the average value between the two plots is taken, about 20 mm) (Fig. 6.11). Selection may

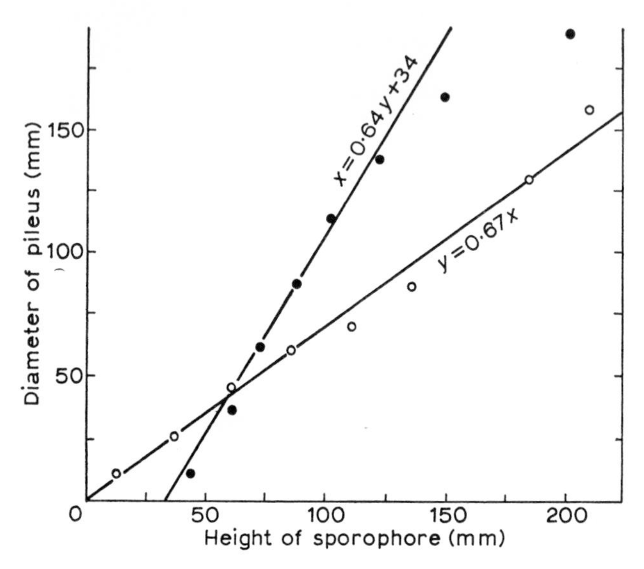

Fig. 6.11. The relationship between the height of the basidiocarp and the diameter of the pileus. Two plots are shown, the regression of height on diameter and vice versa, the data are based on the drawings of agarics in Lange's *Flora Agaricina Danica*. Note the minimal basidiocarp height is 34 mm or, if the average of the two regressions is taken, *c.* 20 mm. (From Bond)

be supposed to have operated to maintain a minimal height. Below this, eddy diffusion would be greatly reduced and liberated spores would, too often, be below the surface boundary layer.

Diffusion in the turbulent layer is affected by scatter, wind velocity and direction. Scatter around a source occurs in all planes and, as a cloud of spores blows along in an ever diverging mass, some of its component spores at the lower side will be deposited. It is clearly important to discover how far the spores travel and what proportion are deposited. Experimental studies, such as those of STEPANOV (1935) with spores of *Tilletia* and *Bovista*, show clearly that the majority of spores are deposited near to the source (Fig. 6.12). Indeed some 99% of all the spores will have been deposited within 100 m of the point of liberation. Nevertheless, long-range dispersal can and does occur over far longer distances as is shown in Table 6.2 based on the studies of STAKMAN and HAMILTON (1939) on *Puccinia graminis*. Here, the infected wheat plants in Southern U.S.A. liberate their uredospores as they ripen before those to the north and at a time when the prevailing wind is southerly. Thus spores trapped in more northerly areas can only have come from these southern sources. A similar kind of situation in India has provided evidence

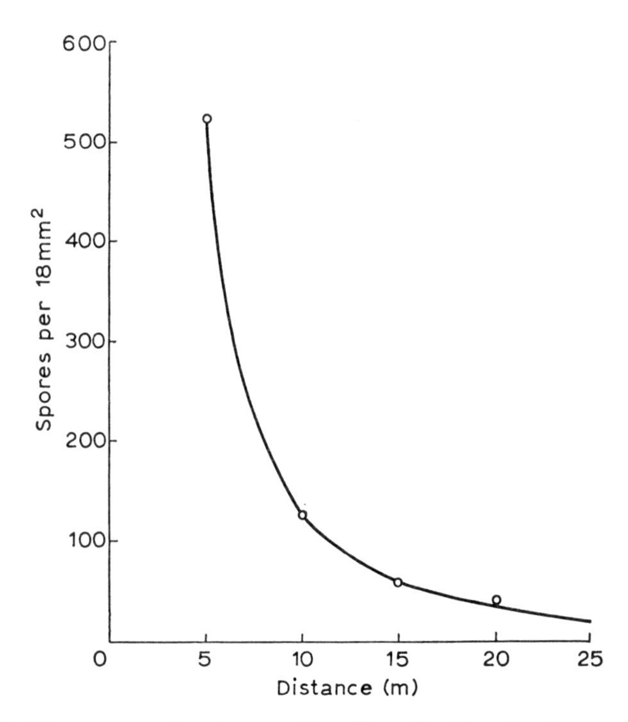

Fig. 6.12. The steep gradient in spore deposition after liberation from a point source. The curve is for *Tilletia* brand spores. (after Stepanov)

Table 6.2. Deposition of uredospores of *Puccinia graminis tritici* on 24/25 May, 1938, in area of the infected source and places to the north in the U.S.A. Deposition assumed to be due to spore cloud borne on southerly winds. (From Stakman and Hamilton (1939), after Gregory, 1961)

Place	Distance from source (km)	Uredospores per 0·09 m^2 per 48 hr
Dallas, Texas	0	129,216
Oklahoma	300	6288
Falls City, Nebraska	560	7680
Beatrice, Nebraska	840	1968
Madison, Wisconsin	970	192

for dispersal over about 2000 km (MEHRTA, 1940, 1952). This long-range dispersal comes about when the source of liberation is not a point but an area, so that the 1% of spores which get beyond 100 m are an appreciable absolute number, and in conditions of high atmospheric turbulence. Spores have been isolated from the troposphere, although above 2500–3000 m the numbers decrease appreciably, and from those regions far out over the oceans or the polar regions, i.e. well away from immediate sources of spores (GREGORY, 1961).

The study of spore dispersal is one of the branches of theoretical mycology which is most advanced. GREGORY (1945) was able to apply the treatment of particulate clouds elaborated earlier by SUTTON (1932) to air spora. The Gregorys (GREGORY, 1945) derived expressions to describe the initial scattering process and the number of spores deposited per unit area at given distances from a variety of sources such as a point source, a line source and so on.

That for a point source is

$$d_w = \frac{p2Q_x}{\pi C^2 x^m}$$

where d_w = no. of spores deposited; x = downwind distance travelled; Q_x = no. of spores in cloud after travelling x; p = deposition coefficient; C = coefficient of diffusion; m = a parameter indicating the degree of turbulence: it is 0 with no turbulence.

p, C, m are determined empirically, Q_x can be found from

$$Q_x = Q_o \exp\left[-\frac{2px^{(1-1/2m)}}{\sqrt{\pi}\, C(1-\frac{1}{2}m)}\right]$$

Stepanov's data showed good agreement with the theoretical predictions for normal turbulence where $m = 1{\cdot}75$ and C and p were calculated from his data (Fig. 6.13). Recently GREGORY (1961) has provided a series of graphs from which appropriate values may be read off and compared with experimental data (Fig. 6.14). The predictive value of such treatments is certainly improving and their significance for forecasting the spread of air-borne pathogens is of increasing importance (GREGORY, 1961; HORSFALL and DIMOND, 1960).

Deposition occurs in one of four ways:

(a) The spores may pass the boundary between the turbulent layer and the still stream of air and then sediment under gravity.

(b) They may enter the still layer by turbulent deposition which is related neither to gravity nor to wind direction. This process, studied in wind-tunnel experiments and artificial spore traps in the open, results in spores being deposited more rapidly than would be expected from sedimentation due to gravity and on under as well as upper surfaces. Presumably local turbulence breaks through the skin of still air and the spores are deposited directly on the surface.

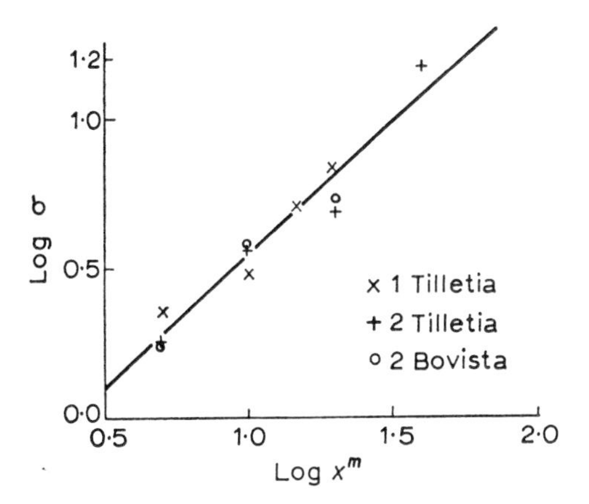

Fig. 6.13. The relation between the standard deviation of spores deposited at various distances from their mean position and the distance from the source. The experimental data are those of Stepanov who liberated spores from a point; the line is the regression based on Sutton's eddy-diffusion theory where $\sigma^2 = \frac{1}{2}C^2x^m$. C = coefficient of diffusion, x = distance travelled and m = measure of degree of turbulence. Note close agreement of experimental data with theoretical expectation. (From Gregory)

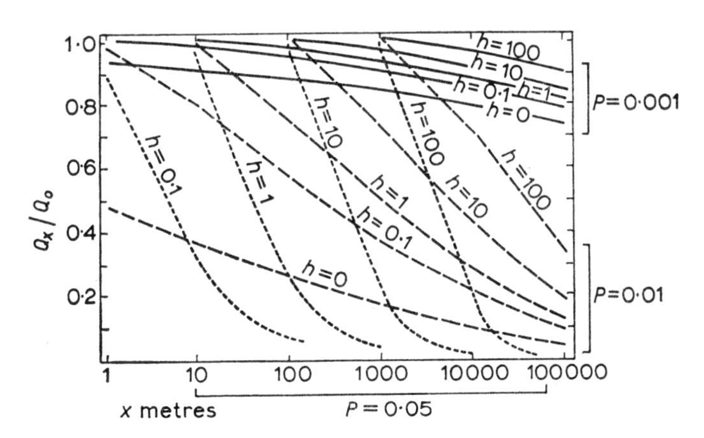

Fig. 6.14. Theoretical curves of Q_x/Q_0 against distance (x) for different heights of liberation (h), confidence limits at P = 0·05, 0·1 and 0·01 are given. Such curves can be used to test predictions concerning spore dispersal. (From Gregory)

(c) In a similar manner spores are deposited by impaction on the windward side of small projecting areas such as twigs or leaves. Impaction is most effective when large spores are blown by high winds on to small areas; turbulent deposition is efficient at wind speeds of 5–10 m/sec and increases with increasing wind speed, although unrelated to its direction or the extent of the potential area of deposition.

(d) Finally, spores can be deposited by being 'washed out' in precipitated water as rain, hail or snow. This process is most effective on clouds of small spores widely dispersed both horizontally and vertically. The upper atmosphere is cleared of spores but the lower atmosphere may show an increased air spora if the 'washing-out' agency is rain, due to its effectiveness as an agent of spore liberation or in promoting conditions suitable to spore proliferation. The continuous recording of air-spora by a variety of devices is now commonly practised and a variety of diurnal and seasonal rhythms, which may be modified by local physiographic or meteorological features, have been described. They are summarized by GREGORY (1961).

Dispersal by animals

In some cases the spores are borne in a material which attracts the animal responsible for dispersal, in other cases the dispersal is fortuitous and arises through the spores being swallowed or adhering in some way to the animal. Examples of the former situation are common and involve especially the insects as agents of dispersal. The Gasteromycetes are a group with particular adaptations to this mode of dispersal as in the stink-horns and their allies. No systematic study seems to have been made of the insect-attractant principles of the slimes but, to the human nose at least, they all seem to be similar. There is no doubt of their functional effectiveness in attracting flies of various kinds. Another common kind of attractant is the production of nectar-like secretions as in *Claviceps purpurea* (conidial stage), the pycnidial stage of many rust fungi and, in some cases, their aecidial stages. In most cases investigated the predominant constituent of the nectar is sucrose but, in addition, unidentified odoriferous compounds are also present, especially in pycnidial drops.

In examples such as these, the insect visits the fungus primarily to feed on the attractive material and this principle also obtains in hypogaeous forms where it is often larger animals especially rodents, which feed on the carpophore. Slugs too are frequent feeders on larger fleshy fungi and have been shown to swallow and transmit spores unharmed (BULLER, 1933). Here the attraction appears to be the basidiocarps as a whole.

In many cases the dispersal process is quite fortuitous. For example many of the dung fungi produce spore-bearing structures from which the spores are ejected violently and come to rest on surrounding vegetation or material which is later eaten by the appropriate herbivore. It is notable that the spores or spore masses of such fungi are often sticky. Another and very different example is the peridiola of the bird's-nest fungi (cf. Fig. 6.2). Once dispersed primarily by the splash-cup mechanism they may undergo a second phase of dispersal after ingestion by an herbivore. Dispersal of the fungi by slime spores is also fortuitous, save that the sticky slime is likely to promote adherence to passing organisms. It seems not improbable that the numerous slime fungi in the soil are dispersed in this way and also the numerous fungi which grow in the galleries made by the various kinds of boring insects. The most notable examples here are perhaps the Dutch elm disease fungus, *Ceratostomella* (*Graphium*) *ulmi*, transmitted by beetles of the genus *Scolytus*, and the 'fungus growing' ambrosia beetles, ants and termites. In all these cases the transmission of fungal material, usually spores, is ensured by such means as simple adherence or by more specialized modes of ingestion which may involve the development of specialized organs. These latter, which are found at their most elaborate in scolytid beetles as well as in platypodid beetles consist of invaginations or pits of the outer integument. The cavities may secrete an oily liquid from a lining of specialized cells in which the fungus is maintained in a toruloid condition (see BAKER, 1963; and BUCHNER, 1953, and Chapter 12, pp. 356–357).

Hemiptera and biting insects may also act as passive agents of dispersal as well as pollinating insects. In the former case the spores adhere to the sucking or biting mouth parts and in this way, for example, the common brown-rot of apples and plums (*Sclerotinia fructigena*) is transmitted from fruit to fruit by wasps in early autumn. In those fungi which infect anthers, e.g. *Ustilago violacea* in Campion (*Silene* spp.), *Botrytis athophila* in red clover, the pollinating insect picks up fungal spores with the pollen and so disperses the spores.

There is hardly any information concerning the distance to which spores can be dispersed by animals. In the case of inadvertent dispersal by pollinating insects it may be supposed that the pattern of dispersal is similar to that of the pollen. BATEMAN (1947) has studied this in relation to contamination of seed crops by 'foreign' pollen carried by bees and other foraging insects. He has derived an expression from the results of both experiments and theoretical treatments which is remarkably similar to that due to contamination by wind-borne pollen. In other words, the pattern of spore dispersal by insects is very similar to that discussed in the previous section for short-range, wind-dispersed spores. How far such behaviour is applicable to the cases of ingested spores by other insects is unclear. Equally unclear is the pattern of distribution by larger animals. All that can be said usefully at this stage is that it is now clear that all animals appear to be confined to larger or smaller territories and that animal spore dispersal is unlikely to be greater than the greatest distance apart within a particular animal's territory.

Dispersion and deposition of aquatic spores

Ingold and others have described a submerged aquatic mycoflora which produces non-motile spores. A great number of these are tetra-radiate to a greater or lesser extent (INGOLD, 1966). They occur in all the main groups of fungi, save Phycomycetes, although the majority are Fungi Imperfecti. Other shapes are sigmoid, crescent-shaped, cylindrical or sub-spherical (Fig. 6.15). WEBSTER (1959) has carried out experiments on the rates of sedimentation and deposition by impaction of aquatic spores. The former class of experiments shared the same difficulty as did those of Buller on air-borne spores; it was difficult to avoid convection effects. No clear picture emerged of a correlation between rates of sedimentation and spore structure. The experiments on impaction were more rewarding. The spores, at different

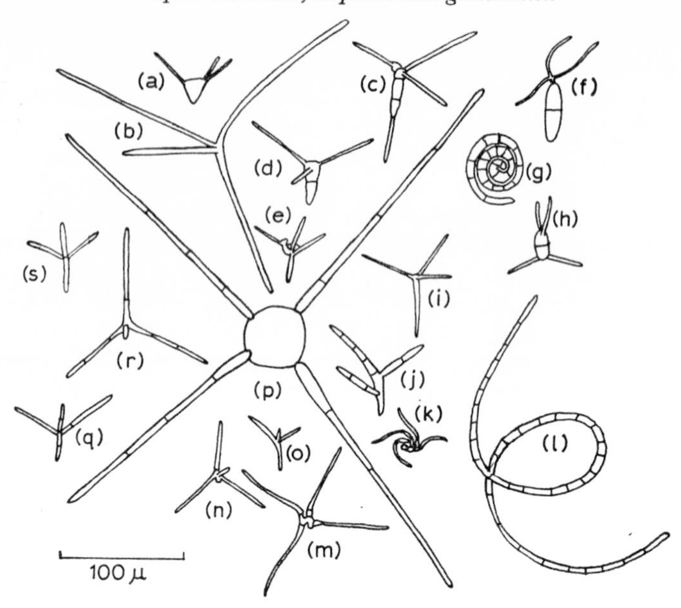

Fig. 6.15. A selection of tetra-radiate and sickle-shaped spores of aquatic fungi; f is an ascospore, s, a basidiospore, the rest conidia. (a) *Clavariopsis* sp.; (b) *Tetrachaetum elegans*; (c) *Culicidospora aquatica*; (d) *Clavariopsis aquatica*; (e) *Tetracladium marchalianum*; (f) *Robillarda* sp.; (g) *Helicomyces* sp.; (h) *Halosphaeria quadricornuta*; (i) *Heliscus tentaculus*; (j) *Tricladium splendens*; (k) *Ingoldia craginiformis*; (l) *Anguillospora* sp.; (m) *Campylospora chaetocladia*; (n) *Tricelophorus monosporus*; (o) *Alatospora acuminata*; (p) *Actinospora megalospora*; (q) *Articulospora tetracladia*; (r) *Lemonniera aquatica*; (s) *Digitatispora marina*. (After Ingold)

concentrations and different water speeds, were passed along a cylindrical tube past a trap consisting of a vertical glass rod coated with collodion. The majority of spores were deposited on the centre-line of the 'trap' facing upstream and numbers fell off on either side. Seventeen spore types were tested and the trapping efficiency, defined as the number of spores trapped per unit area as a percentage of those which would have passed through the area if the trap had not been there, was determined. The 10 most efficient were tetra-radiate spore types, then 3 sigmoid types and the rest were more or less cylindrical or ellipsoidal. One species, *Tetrachaetum elegans*, in particular, was far more effectively trapped than any other at all water speeds from 5 cm/sec to 45 cm/sec but the reason for this is obscure. The greater efficiency of trapping of the tetra-radiate spore was confirmed by exposing equal concentrations of mixed spore-type suspensions. In general more tetra-radiate spores were retained than any other type (Table 6.3). Webster has suggested that the advantage of this spore type lies in its ability to be deposited with a three-point contact on the substrate. They form appressoria at the tips of the three arms in contact with the substrate and this enables them to remain most effectively.

Table 6.3. Trapping efficiency (per cent) for various aquatic hyphomycete spores at five water speeds (mean values from three experiments). Calculations based on spore deposits in row 1 only. (From Webster, 1959)

Species	5 cm/sec	15 cm/sec	25 cm/sec	35 cm/sec	45 cm/sec	Mean of all observations
Tetrachaetum elegans	12·8 ± 11·05	3·94 ± 1·56	2·69 ± 1·09	2·1 ± 1·22	1·31 ± 0·25	4·56
Articulospora tetracladia	0·9 ± 0·13	0·35 ± 0·002	0·51 ± 0·002	0·33 ± 0·001	0·30 ± 0·021	0·477
Tricladium splendens	0·69 ± 0·28	0·32 ± 0·06	0·39 ± 0·28	0·21 ± 0·13	0·10 ± 0·07	0·344
Tetracladium marchalianum	0·63 ± 0·22	0·28 ± 0·024	0·37 ± 0·029	0·27 ± 0·02	0·15 ± 0·002	0·339
Clavariopsis aquatica	0·23 ± 0·28	0·3 ± 0·17	0·4 ± 0·33	0·3 ± 0·15	0·28 ± 0·24	0·301
Lemonniera brachycladia	0·19 ± 0·05	0·27 ± 0·06	0·46 ± 0·05	0·36 ± 0·08	0·18 ± 0·03	0·293
Tetracladium setigerum	0·56 ± 0·11	0·24 ± 0·022	0·30 ± 0·04	0·19 ± 0·04	0·12 ± 0·05	0·281
Lemonniera aquatica	0·61 ± 0·45	0·35 ± 0·09	0·13 ± 0·033	0·04 ± 0·014	0·039 ± 0·18	0·232
Tricladium gracile	0·32 ± 0·15	0·39 ± 0·14	0·23 ± 0·01	0·1 ± 0·05	0·07 ± 0·13	0·222
Varicosporium elodeae	0·14 ± 0·1	0·16 ± 0·05	0·15 ± 0·06	0·12 ± 0·05	0·1 ± 0·04	0·133
Anguillospora longissima	0·1 ± 0·066	0·11 ± 0·07	0·1 ± 0·06	0·1 ± 0·04	0·06 ± 0·01	0·096
Flagellospora curvula	0·049 ± 0·03	0·06 ± 0·026	0·136 ± 0·07	0·14 ± 0·05	0·08 ± 0·054	0·095
Flagellospora penicillioides	0·085 ± 0·10	0·023 ± 0·016	0·023 ± 0·01	0·013 ± 0·09	0·063 ± 0·002	0·023
Heliscus lugdunensis	0·01 ± 0·02	0·004 ± 0·005	0·009 ± 0·008	0·007 ± 0·003	0·006 ± 0·003	0·008
Dactylella aquatica	0	0·004 ± 0·002	0·009 ± 0·036	0·008 ± 0·0002	0·007 ± 0·002	0·006
Glomerella cingulata conidia	0·0013 ± 0·002	0·0006 ± 0·009	0·002 ± 0·003	0·0012 ± 0·005	0·0023 ± 0·002	0·0025
Tricellula aquatica	0·0006 ± 0·00009	0·0023 ± 0·004	0·0026 ± 0·001	0·0029 ± 0·001	0·0032 ± 0·0027	0·0024

To investigate this *T. elegans* spores were deposited and then subjected in one series to a water flow of 31 cm/sec for 1 hr immediately after deposition and, in a second series of experiments, after 12 hr. Seventy-three per cent of the spores were washed off in the first series but only 19% in the second (299/420: 78/431). Moreover, deposition is more effective on a rough surface than on a smooth and it is clear, therefore, that deposition of tetra-radiate spores will rapidly increase the roughness of the surface and hence the efficiency of the trapping.

Amongst the Phycomycetes the aquatic species produce motile zoospores. Their movement has been little studied. COUCH (1941) has given the best account but many more observations using the more efficient optical and photographic methods now available are desirable. The structure of flagella was described in Chapter 2 (pp. 34–37) and it will be recollected that four types of zoospore are known:

No. of Flagella	Position	Type
Uniflagellate	Anterior	Tinsel (Hyphochytridiomycetes)
	Posterior	Whiplash (Chytridiomycetes)
Biflagellate	Forward Backward	Tinsel Whiplash (Oomycetes)
	Forward Backward	Blunt Whiplash (Plasmodiophoromycetes)

There seems to be no detailed account of the way in which biflagellate zoospores of Plasmodiophoromycetes move and only Couch has described the movement of *Rhizidiomyces apophysatus*, a member of the Hyphochytridiomycetes. The cell rotates as it moves, flagellum forward, and travels in a wide spiral path. The flagellum shows alternating single and double images which suggests that it is beating in one plane. Couch was not certain how the movement of the flagellum developed but believed that undulations began just behind the tip and travelled back to the base of the flagellum. In this respect these fungi resemble the Trypanosomatidae (HOLWILL, 1964) rather than the better known forms such as *Euglena*.

The behaviour of the posteriorly flagellated zoospores appears to be similar to that described for spermatozoa of various kinds (SLEIGH, 1962). Couch has described four patterns of movement: (a) a nearly straight or circular path without rotation, (b) a nearly straight path with clockwise rotation, (c) a spiral path with clockwise rotation and (d) an irregular path involving 'hopping and darting about'. The first two are the most common and are probably associated with undulations which develop at the base of the flagellum and travel to its tip. The beating of the flagellum, unlike that of spermatozoa, is probably in one plane or slightly three-dimensional, but the observations are not really sufficiently critical to decide this matter (COUCH, 1941; KOCH, 1959, 1961). According to Koch the irregular motion is due to frequent starting and stopping associated with changes in direction. This may arise as a consequence of stopping abruptly for this is usually associated with a jerk of the flagellum which pivots the zoospore round. More attention has been given to the biflagellated zoospores of the Oomycetes. Couch made the novel suggestion that both flagella were active. Care has to be taken in comparing observations of flagellar behaviour in the Saprolegniales, a favourite group for study, since there are two motile stages separated by a non-motile phase (diplanetism). In the first stage in *Saprolegnia*, the zoospores perform slow broad spirals, rotating as they move, one forwardly-directed flagellum undulates in one plane, the other, posteriorly-directed flagellum undulates in the plane at right angles to the other flagellum. The movement of the former may be much less active than that of the latter, when the whole zoospore moves in a narrow circle. In abnormal zoospores lacking one flagellum, those without a forwardly-directed one swam smoothly in a straight line and did not rotate; those without a backwardly-directed flagellum swam in an awkward, irregular spiral and rotated. The second swimming stage lasts much longer and the zoospores are more active. Their movement is similar to that of other biflagellated species which do not show diplanetism, e.g. *Pythium*, *Phytophthora*. They follow a drawn out spiral path, rotate vigorously at all times and the flagella undulate in planes at right angles to each other. In some cases, e.g. *Allomyces*, the backwardly-directed flagellum appears as straight structure stretched out behind the zoospore but Couch believes that, in fact, it is undulating in the vertical plane and the zoospore is not rotating, hence its movement cannot be detected. It was known that the backwardly-directed flagellum of the second-stage zoospores of *Saprolegnia* could be twice as long as the forwardly-directed flagellum but MANTON *et al.* (1951) also drew attention to the possibility of its sheath being flattened to form a fin-like expansion along two sides of the axis. MCKEEN (1962) has suggested that it is the backwardly-directed flagellum which propels the zoospores and that this is associated with the more elaborate structural development. In his view the other flagellum acts as a rudder. The claim that in *Plasmodiophora* the short blunt flagellum projects forward moving regularly and slowly while the long whiplash projects backwards undulating irregularly suggests that here the anterior flagellum is the primary tractile agent, despite the discrepancy in size (KOLE, 1965). If the two stages of *Saprolegnia* are comparable it is also difficult to relate McKeen's view to the propulsive action of the anteriorly-directed flagellum of the abnormal uniflagellated zoospores. At present it is clear that the propulsive mechanism of biflagellated zoospores is unresolved.

There seems no reason to doubt, from both the structural and hydrodynamical aspects, that once the movements of flagella have been accurately described they will be found to function in a manner analogous to that of better understood flagella such as spermatozoa but it would be of interest to make observations to see if fungal flagella behaved according to theories developed for these structures in other organisms (e.g. GRAY and HANCOCK, 1955; BURGE and HOLWILL, 1965).

Virtually nothing is known of the range of dispersal of motile zoospores nor is much known of the kinds of aquatic conditions which they may face on release from a sporangium. MANTON *et al.* (1951) have shown that the interpolated non-motile stage in *Saprolegnia* is associated with the development of an enclosing cyst bearing little anchor-like appendages. It may be that this stage is that in which long-range dispersal is achieved and once transported and anchored, the enclosed contents produce a motile stage which is responsible for deposition over a short distance.

It has been known for many years that zoospores accumulate in response to chemical attraction (PFEFFER, 1884; FÏSCHER and WERNER, 1958) and this is important in infection by motile zoospores of root parasites, for example. Recently ROYLE and HICKMAN (1964a, b) have studied the factors involved in the accumulation of biflagellated zoospores of *Pythium aphanidermatum* on pea roots. They showed that accumulation was promoted by, and directed towards, punctures made in the epidermis, root exudates and extracts diffusing from agar in capillary tubes and a variety of mixtures or single compounds. The most effective of these last in replacing root extracts were weak base-adjusted glutamic acid and a mixture of casein hydrolysate+glucose+fructose+sucrose (all 1%). Most workers with other fungi have found

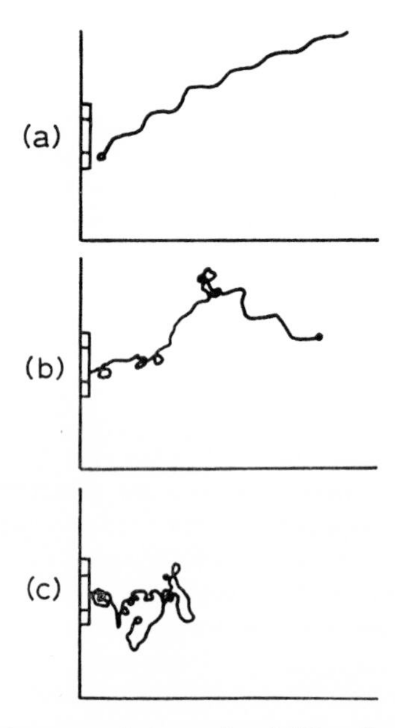

Fig. 6.16. Dark ground traces of the path of biflagellated zoospores of *Pythium aphanidermatum* in response to various diffusates in a capillary tube sealed with a permeable cap (on the left-hand side). (a) with plain water agar in capillary tube; random movement, the zoospore will pass by; (b), (c) with *cv.* Alaska pea root-exudate agar. Note jerky motion with looping which keeps zoospore near source of diffusate, and becomes emphasized as it approaches the source of diffusate (c). (From Royle and Hickman)

mixtures of amino acids with or without sugars or inorganic salts to be effective in promoting zoospore accumulation. The mechanisms of such processes are by no means clear as yet (cf. ROTHSCHILD, 1956). More important in the *Pythium* work was an elucidation of the behavioural response leading to zoospore accumulation. Zoospores moved at random in the manner already described until they entered the zone of attraction, they then changed their direction towards the source of diffusing material. This was shown by measuring the angle between the track of the zoospore and the source of diffusate, either water agar or root-extract agar. The mean angles were $40{\cdot}6° \pm 26{\cdot}2°$ and $14{\cdot}5° \pm 19{\cdot}8°$ respectively; a statistically significant difference. As the source of diffusate was approached the type and rate of movement changed and became a series of irregular, jerky loops in no particular direction but within a closely circumscribed area around the mouth of the capillary containing the diffusate. The nearer the source was approached the more angular the motion and the greater the frequency of looping. The rate of movement was almost halved. In this way large clusters of zoospores rapidly built up at particular regions and this was frequently followed by their settling and encystment. No data are given but from their figures it appears that the attraction can operate over distances as great as 200–300 μm (Fig. 6.16). There is no clear evidence that this two-step type of approach operates in other organisms attracted by diffusing chemicals; *Saprolegnia*, for example, is only said to show a change in direction (MÜLLER, 1911) but this may reflect inadequate observation rather than a real difference in behaviour. It will be of interest to know how far such mechanisms of attraction operate in nature to bring about the settlement of motile zoospores.

SPORE GERMINATION

Since spores are so diverse in their nature and functions it is not surprising that a simple phrase, 'spore germination', denotes a wide range of processes. The essential process is the restoration of normal metabolic and physiological activity after a period, however brief or long, in which these processes have been checked, changed or have all but ceased. Thus germination cannot be considered apart from dormancy and the germination of a spore highly adapted to the dormant condition will inevitably be very different to that of a small portion of abstricted hypha, as in an arthrospore, which is capable of recommencing growth almost at once.

The immediate consequence of germination in most fungi is the production of one or more germ tubes, i.e. short, apically-growing hyphae, which give rise either to the normal vegetative mycelium or, very rapidly to another sporangial phase, e.g. zygospores of Mucoraceae. Exceptions to this sequence occur. For example, *Phytophthora* spores may either develop a germ tube or give rise to a vesicle containing motile zoospores, the alternatives being determined by the environmental conditions, notably humidity. Basidiospores of *Tilletia* spp. can give rise to a succession of secondary spores or to a germ tube and mycelium. Attention will be concentrated here, however, on the commonest development, the production of a germ tube.

Morphological and cytological changes

In many spores, save those with unduly thick walls, the first observable change is in the size of the spore, which swells, and often in its refractive index, which falls for a variety of reasons. Thus the spore wall becomes extended, presumably by stretching rather than by the formation of new material, although there is no definite evidence on this point. In many mucoraceous fungi, e.g. *Rhizopus*, *Cunninghamella*, *Gilbertella* (HAWKER and ABBOTT, 1963c; HAWKER, 1966; FULLER, 1966) an inner wall develops within and, on the rupture of the original wall, the new wall grows out to form the germ tube. A similar process occurs in *Aspergillus* and *Fusarium* (HAWKER, 1966; MARCHANT, 1966a; MARCHANT and WHITE, 1966) and this is borne out, in the latter fungus, by studies using the fluorescent antibody technique where, it will be recalled, the spore wall did not apparently contribute to the wall of the germ tube (cf. Fig. 3.3, p. 41). How the spore wall is ruptured, whether mechanically, enzymatically, or by a combination of both, is unknown. In some spores, e.g. *Botrytis*, the wall is double from the outset and here the inner region develops as the germ-tube wall. It would clearly be of interest to apply the fluorescent antibody technique to such cases.

The changes in refractive index may represent increased hydration or reflect other metabolic changes. There is certainly suggestive evidence of changes in the shape, structure and number of several cell organelles. The endoplasmic reticulum is said to be poorly developed in many spores, e.g. sporangiospores of *Rhizopus*, conidia and ascospores of *Neurospora*, uredospores of *Puccinia* and basidiospores of *Schizophyllum* (HAWKER and ABBOTT, 1963c; WEISS, 1965; SUSSMAN, 1966; WILLIAMS and LEDINGHAM, 1964; VOELZ and NIEDERPRUEM, 1964). The mitochondria are also said to alter and, in general, to increase in numbers. TURIAN (1962) has claimed that the numbers of cristae within the mitochondria increase when the stationary zoospore of *Blastocladiella* develops a germ tube and a similar claim is made for *Puccinia*. The difficulty in assessing such claims is that fixatives do not always penetrate spore walls uniformly or effectively and may cause damage or distortion to already fairly dehydrated cytoplasm. Thus some of the changes may really be the consequence of fixation artefacts. Hydration, when it occurs, is not reflected by increased vacuolation at the ultrastructural level, to judge by present evidence, so that the water is likely to be bound rather than free.

It will be clear that there are still far too few studies of the morphological changes at the ultrastructural level for useful generalizations to be made.

Biochemical and metabolic changes

While the visible events at germination have a certain similarity the metabolic changes underlying them are more varied. Spores vary in two principal respects, the degree of dormancy imposed upon them and their nutrient content.

SUSSMAN (1965) has provided an encyclopaedic account of dormancy in fungal spores but, in essence, two contrasting conditions are recognized, constitutive and exogenous dormancy. The former block to germination is due to innate properties of the spore, the latter block arises from environmental factors. The borderline between these two kinds of dormancy can be narrow and the conditions are evidently associated with the biological function of the spores concerned.

Thick and apparently impermeable walls, e.g. ascospores of *Neurospora tetrasperma*, are often associated with constitutive dormancy and in this species it has been possible to locate the block in the exospore wall, for some germination occurs if it is removed (LOWRY and SUSSMAN, 1958). Internal metabolic blocks or deficiencies may also be involved and, in the same species for example, the trehalose utilized during development is somehow separated from the trehalase which initiates its utilization at germination (BUDD, SUSSMAN and EILERS, 1966). Another cause of constitutive dormancy is the presence of internal inhibitors within the spores, as in the uredospores of *Puccinia graminis tritici*. This condition is only marginally different from that in which an inhibitor produced within the spores is released and inhibits adjacent spores from germinating. Inhibition may also be imposed by the adjacent mycelium of the same species as in *Botrytis cinerea* (CARLILE and SELLIN, 1963) where total inhibition of conidial germination was brought about by the older parts of the mycelium after 6½ hours. This may appear to be an example of innate inhibition unless the spore is removed from the mycelium when germination can occur without hindrance.

Exogenous dormancy may well be widespread in nature; one example is the widespread fungistasis of many soils (DOBBS and HINSON, 1953). The causes of this phenomenon may be similar to those described for *Botrytis* but several other explanations have been suggested (see various hypotheses in BAKER and SNYDER, 1965). Factors of obvious significance in inducing exogenous dormancy include relative humidity, temperature, pH, toxic and inhibitory materials including 'staling substances', CO_2 and a host of others. Descriptions of particular examples are available in several general texts of fungal physiology (e.g. HAWKER, 1950; COCHRANE, 1958) and reviews of spore germination (e.g. ALLEN, 1965; GOTTLIEB, 1950; COCHRANE, 1960; SUSSMAN, 1965).

The presence or absence of sufficient endogenous nutrients for spore germination also confuses the designation of the types of dormancy shown by any fungus. When species require specific, or even general nutrients to germinate, e.g. biotin for *Memnoniella echinata* (PERLMAN, 1951) or C-, N-, P- and S-containing compounds for *Glomerella* (LIN, 1945), they may appear to show exogenous dormancy if the nutrients are not available in the immediate vicinity of the spore. Thus the study of dormancy and nutrient requirements of particular fungi cannot readily lead to useful generalizations and three examples will be discussed later.

However, with the exception of the conidia of some of the powdery mildews

(Erysiphales), spores have a low water content and hydration, with or without swelling, is an essential step at the onset of germination. In most cases the water taken up can be present as liquid water or in the gas phase in the atmosphere but there are exceptions. The spores of the downy mildews (Perenosporales) are said to germinate only in the presence of liquid water while free water is inimical to conidia of powdery mildews although quite specific requirements may exist for a particular range of relative humidities of the atmosphere (WALKER, 1957; SCHNATHORST, 1965).

The process of water uptake is probably complex and there is clear evidence of this in some Mucoraceae, *Aspergillus* and *Fusarium* species. A primary requirement appears to be a change in the permeability of the spore wall although how this comes about is unknown. In *Fusarium* and *Aspergillus* there follows a non-metabolic phase in which water is taken up by the outer sheath of the spore in the former or a stretching of the wall in the latter (MARCHANT and WHITE, 1966; YANAGITA, 1957). WOOD-BAKER (1955) has claimed that a primary phase of spore swelling in mucoraceous fungi is less sensitive to anaerobiosis than later stages but, in general, this is not so and glucose appears to be necessary for swelling in these fungi (EKUNDAYO and CARLILE, 1964; EKUNDAYO, 1966). The process is independent of the osmotic pressure of the glucose solution and this suggests that a non-osmotic, energy-requiring process may be involved. A similar requirement for both glucose and a nitrogen source to ensure maximum swelling in *Fusarium* macroconidia is not opposed to this view (MARCHANT and WHITE, 1966). Not surprisingly, later stages such as germ tube development are metabolism dependent for biosynthetic processes are clearly implicated. The supposed active uptake of water by spores, however, does suggest that some metabolic changes have occurred primarily within the spore.

Attention will now be focused on the subsequent changes and, since each spore is almost a special case, three examples have been selected for description. The first is the ascospore of *Neurospora tetrasperma*, a spore showing typical constitutive dormancy; the second is the uredospore of the wheat-stem rusts (*Puccinia graminis tritici*), a somewhat specialized spore of an obligate parasite; the third is the conidium of *Aspergillus niger*, fairly typical of the numerous kinds of conidia. Unfortunately, studies made of these three spore types are concerned with different aspects so that a truly comparative account cannot be presented. This, indeed, is almost impossible save for *Neurospora* conidia and ascospores where SUSSMAN (1966) has provided the best comparative account available.

Particular attention has been paid to the activation processes involved in breaking the constitutive dormancy of ascospores. Heat shocks or treatments with chemicals such as furfural or other heterocyclic compounds and organic solvents are effective (GODDARD, 1935; SUSSMAN, 1953; SUSSMAN, LOWRY and TYRRELL, 1959). The permeability of the cell wall to anions and cations is increased greatly and, thereafter, germination can proceed in an hour or so in distilled water in the presence of oxygen utilizing the endogenous reserves of the ascospore. Lipids constitute about 27% of the dry weight of the spore and carbohydrates about 33%. Some 70% of the lipid fraction is in the form of esters and over 40% of the carbohydrates is trehalose. Some of the changes associated with germination are set out in Table 6.4.

Dormant spores appear to utilize lipids as the main source of energy but, on activation, trehalose is preferentially metabolized. Pyruvate, ethanol and acetaldehyde accumulate and the first two decline after some two hours as the utilization of trehalose declines. Coincident with this is an increase in TCA cycle acids such as α-ketoglutarate and fumarate and the metabolism shifts from being predominantly glycolytic to become more oxidative. It is at this point that germination can be observed. An exogenous supply of hexose is necessary for further growth and if it is not available the ascospore reverts to utilizing its endogenous lipids. There is a marked rise in biosynthetic processes at germination. Carbon from exogenously supplied [^{14}C]-glucose is increasingly incorporated into amino acids and, indeed, the protein content of the ascospore doubles within two hours of activation. It is significant that HENNEY and STORCK (1964) have shown that ribosomes present in dormant ascospores of the related *N. crassa* are replaced by polyribosomes in germinating ascospores. No data are yet available about the synthesis of wall materials.

A puzzling feature of activation and germination is that all the necessary enzymes for the metabolic pathways appear to be present in the dormant ascospore, notably trehalase. Thus it has been suggested that activation involves not only changes in wall permeability but also the association of trehalase and its substrate. This could be by the removal of an inhibitor of the enzyme, the breakdown of some form of internal compartmentalization between sugar and trehalase or even other possibilities (HILL and SUSSMAN, 1965). Activation is evidently a complex process.

Uredospores have fairly thick walls although not as thick as those of *N. tetrasperma* ascospores. However, they do seem to be freely permeable and suffer considerable and rapid losses by leaching of organic acids, amino acids and soluble carbohydrates when suspended or shaken in water or nutrient solutions. Germination can occur in pure water or in atmospheres of high relative humidity but it is frequently erratic. ALLEN (1955, 1957) demonstrated that this was due, in part, to the presence of inhibitors which were soluble in water or even volatile. Trimethylethylene has been suggested as the effective agent (FORSYTH, 1955). The situation is complicated further by the presence of germination stimulators which are released by autolysed uredospores and natural collections often include such spores. Pelargonaldehyde has been implicated and certainly parallels the behaviour of the natural stimulator but other aldehydes show similar behaviour (FRENCH *et al.*, 1957; FRENCH, 1962). Thus dormancy in uredospores is on the borderline between the constitutively and exogenously determined states. A further check to germination may be imposed by light of intensity 2152 lux or more, which checks germ tube growth but this can be overcome by a subsequent dark period.

Germination can take place after hydration in an aerobic atmosphere on endogenous reserves alone. These are largely lipids and they are utilized both

Table 6.4. Changes in various metabolites and respiratory measurements with time during the germination of ascospores of *Neurospora tetrasperma*. (From Sussman, 1954, 1961; Sussman, Distler and Krakow, 1956; Lingappa and Sussman, 1959; Hill and Sussman, 1965)

Phase	Time (hr)	Respiratory measurements Q_{O_2}	Q_{CO_2}	R.Q.	Metabolites Pyruvate	Ethanol	Alcohol-Sol CHO	Trehalose	Lipids
Dormant	0	0·21–0·59	0·13–0·36	0·6	1	0	14	2·0	26·6
Activated	1	4·5	—	0·8–1·2	8·5	83	8·9	3·5	25·7
	2	10·9	—	—	12·0	103	6·5	6·0	*c.* 24·4
Germinating	5	16·4–24·2	10·0–13·8	0·6	6·2	—	3·3	—	*c.* 22·8

Units Q_{O_2}: mm^3 O_2/hr/mg dry wt.
Q_{CO_2}: mm^3 CO_2/hr/mg dry wt.
Pyruvate: μg/mg N
Ethanol: μg/100 mg dry wt.
Alc.-sol carbohydrate: mg/100 mg dry wt.
Trehalase activity: μgglucose $\times 10^{-3}$
Lipids: mg/100 mg dry wt.

in dormant and germinating uredospores. There seems to be good evidence for the enzymes of the EMP glycolytic pathway, the pentose phosphate pathway, the TCA cycle and the glyoxylate cycle (cf. review by STAPLES and WYNN, 1965). When [^{14}C-] acetate or fatty acids were fed to germinating uredospores it was evident that they were directed into the same metabolic pathways and via transaminations into amino acids and proteins. Such observations bear out the limited observations made of changes in chemical composition of uredospores on germination. Fatty acids declined, although non-saponifiable lipids increased, and there was no net gain in lipids. Somewhat unexpectedly, there was no net gain in protein although considerable turnover occurred, and, indeed, in wheat-stem rust there is a slight initial fall. Basically, however, none of these changes seemed to differentiate dormant from germinating uredospores save the more rapid conversion of lipid to soluble carbohydrates. The striking feature of the germination of these spores is the relatively small rise and change in metabolic tempo and direction once germination has occurred. It may be suggested, therefore, that the clue to their further development, which has not been achieved unequivocally apart from a living host, must lie in the tissues and metabolism of the latter.

By contrast with these examples, the germination and subsequent development of the conidia of *A. niger* is relatively straightforward. The check to germination seems to reside largely in the degree of hydration of the spores and their insufficient endogenous nutrients. There is some evidence that germination is checked by a diffusible material from the parent mycelium (CARLILE and SELLIN, 1963) but once free of the latter, water and exogenous C- and N-compounds are all that are necessary for germination. Hydration and swelling have received some attention and YANAGITA (1957) has described a two-stage process prior to the emergence of the germ tube. The first phase is unaffected by temperature and nutrients and is, perhaps, a purely physical phase; the second phase is temperature sensitive and involves the incorporation of CO_2; it is, therefore, clearly a metabolic phase. Thereafter, as germination proceeds there are striking changes in the enzyme content of the conidium. Enzymes of the EMP pathway, pyro- and glycero-phosphatases increase and only then does the germ tube develop. Protein synthesis also develops rapidly at germination. Choline sulphate, deposited in the conidium during its formation, is utilized in the synthesis of sulphur-containing protein. Carbon from [^{14}C]-$NaHCO_3$ is incorporated into nucleic acids and new RNA is synthesized 30 min from germination while the conidia are still swelling. Nucleotides synthesized in this early phase incorporate phosphate from endogenous polyphosphate. Thus the whole apparatus of protein synthesis seems to develop during germination and this may also be true of the wall forming system for it has been claimed that DNA, RNA and protein are incorporated into the spore wall before germ tube development. Apart from this little is known of wall synthesis in conidia of this fungus. Some relevant data concerning germinating conidia of *A. niger* are set out in Table 6.5.

These three patterns of germination highlight the different functions and potentialities of the spore types concerned.

Table 6.5. Changes in various metabolites, enzymes, uptake processes and respiratory measurements with phase of germination in conidia of *Aspergillus niger.*

Time or phase (hr)	Dry Wt.	Total N	Protein N	Alanine synthesis	O_2 uptake	Glucose uptake	Phosphorus uptake	Catalase	Hexokinase	Phosphoglucomutase	Phosphohexoisomerase	Aldolase
0–3 Swelling [1]	0	0	0	0·12	0·49	1·0	0·43	0·34	0	0	0	0
3–6 Swelling [1]	16	0·72	0	0·5	1·9	1·9	1·0	—	0	0	0	0
12 Swelling [2]												
6–9 Germ tubes [1]	45	1·94	1·72	1·1	8·8	8·6	1·6	—	—	—	—	—
18 Germ tubes [2]	—	—	—	—	—	—	—	0·0082	0	15	41	1·51
24 Long germ tubes and hyphae [2]	—	—	—	—	—	—	—	—	—	58	—	11·88

Units Dry wt.: μg 10^{-7}/hr/spore
Total and Protein N: μg 10^{-8}/hr/spore
Alanine: O_2 and glucose uptake: μg 10^{-7}/hr/spore
Phosphorus uptake: μg 10^{-8}/hr/spore
Hexokinase, phosphoglucomutase, Phosphohexoisomerase, aldolase: Activity/40 mg spore material
Catalase: K/ml. spore suspension after 3 min

[1] Data of Yanagita (1957), Strain 1617; 1 mg spores (*c.* 7×10^{-7})/10 ml. medium at 30°C
[2] Data of Bhatnagar and Krishnan (1960a, b), Strain NRRL 599; 1 g spores/20 ml. medium at 27–30°C

Neurospora ascospores are able to increase the longevity of the reproductive units because of the complex requirements for activation prior to germination. It is not known how far their metabolism at germination is adapted to the conditions which they are likely to meet on germinating in nature. Uredospores of wheat rusts have mechanisms, readily removed because of their soluble and volatile nature, which defer germination until the spores are somewhat dispersed and diluted. Thereafter, they have enough material to exploit slowly and economically their immediate surroundings and to develop the necessary organs of penetration. Their subsequent fate, however, is seemingly dependent upon the host. The conidia of *Aspergillus* differ from both these spore types in being adapted to rapid exploitation of a favourable habitat. Provided that the environment has material which can be incorporated readily into the conidium then the latter has the cellular machinery and necessary endogenous raw materials to produce the adaptive and synthetic enzymes systems necessary for growth.

The study of spore germination can evidently be conducted as an interesting biochemical and physiological exercise in the laboratory but, to be truly meaningful, it must be related ultimately both to the natural environment in which germination actually occurs and to the biology of the parent fungus as a whole.

Chapter 7
General Aspects of Fungal Nutrition and Metabolism

Fungi, like all other living organisms, take up external materials through their membranes, transport them to the active metabolic sites within the mycelium and there transform them to provide energy for maintenance and biosynthetic processes; during this sequence various materials may also accumulate in or be lost from the fungus. These processes will be dealt with in subsequent chapters; here, attention will be directed to the most general features of fungal metabolism, modes of nutrition, the nutrients fungi utilize and the efficiency with which they convert nutrients to their own substance.

Modes of nutrition

All fungi are heterotrophic for carbon compounds and many are heterotrophic for other materials as well, e.g. vitamins. They obtain these essential materials as saprophytes, as facultative or obligate parasites and as facultative or obligate symbionts with green plants or insects. It might be supposed that in cases of obligate parasitism or symbiosis the heterotrophic requirements of the fungi concerned would be greater than if they were saprophytes or only facultative in their parasitism or symbiosis. This may well be so but at the present time the nature of the nutritional requirements for obligate parasites is unknown. It is clear that some growth of obligate parasites can be achieved with a supply of fairly usual nutrients, utilized apparently in a normal way as, for example, in the germinating uredospores of rust fungi. Nevertheless, this growth ultimately becomes restricted for reasons which are not known. A similar situation occurs in other fungi not so obviously demanding in their requirements. An example is *Ustilago maydis*, the causal organism of smut disease of maize. Monokaryons of this fungus will grow readily as saprophytes in plate or liquid culture and have no especial nutritional requirements. A dikaryon between compatible monokaryons will not grow vigorously on a synthetic medium or produce the characteristic brandspores but this is achieved if a susceptible maize plant is infected with a dikaryon. More surprising is the finding that a diploid derived from two compatible monokaryons will, however, grow readily in a synthetic medium in a yeast-like manner reminiscent of the normal haploid strains (HOLLIDAY, 1961a, b). The explanation of this behaviour must lie in the genetic control exercised on metabolic processes or in the spatial location of certain key processes within the mycelium.

Sometimes the requirement for growth can be met in synthetic culture as in the case of *Phytophthora* spp., the majority of which can be grown readily in defined media, but not in a natural environment. These species often occur as root parasites of higher green plants but outside their hosts they may never exist in the mycelial phase but only as dormant chlamydospores or oospores (HICKMAN, 1958; VUJIČIĆ and PARK, 1964). In natural conditions there also exist 'ecologically obligate parasites' such as *Ophiobolus graminis*. These are fungi which are saprophytic in culture but are unable to compete as such in a natural environment. The boundary between obligate and facultative parasitism is, therefore, not a purely nutritional one but will be determined also by the ecological situation of the fungus concerned. Similar considerations apply to facultative parasites compared with saprophytes and to comparisons of symbiotic and non-symbiotic forms. For example, *Armillariella mellea* is capable of forming an endotrophic mycorrhizal association with the orchids *Gastrodia elata* (KUSANO, 1911) and *Galeola septentrionalis* (HAMADA, 1940a, b); of acting as a most serious and distinctive facultative root parasite or of persisting saprophytically (GARRETT, 1960). There does not, therefore, seem to be a close degree of correlation between the degree of heterotrophy and the mode of nutrition.

Nutrients for fungi

A remarkable range of carbon compounds is utilized heterotrophically by fungi. Carbon dioxide can be fixed by many fungi but cannot be used as an exclusive source of carbon for metabolism. Some fungi can use almost all carbon compounds, albeit with varying degrees of efficiency. Inability to utilize a particular carbon compound may reflect the inability of the compound to be taken up for genetic or environmental reasons. The most important of the latter is the pH of the medium. For example, MOSES (1955) found that cells of *Zygorhynchus moelleri* were permeable to glucose and acetate only at pH 6·8, whereas at pH 3·4 the cells were permeable to all the other TCA cycle intermediates but not to acetate. Such pH effects may arise from the presence or absence of other non-carbohydrate nutrients in the medium. Different strains of the same fungus may vary in their ability to utilize certain classes of carbon compounds or to use the same compounds in quantitatively different ways. When presented with a mixture of carbon sources there is often preferential utilization of one form over others. Unfortunately, although this situation may often occur in nature, very little experimental work has been done in controlled conditions, principally because the single most effective carbon source is usually supplied for effective culture (LILLY and BARNETT, 1953). In studies on penicillin production, however, some data became available for *P. chrysogenum*. Here, acetate is utilized before lactate plus glucose and lastly, lactose in a mixture of these four carbon sources (JARVIS and JOHNSON, 1947). Sometimes utilization is more effective in mixtures than when carbon sources are supplied singly; an example of such interaction is the data on glucose and/or galactose utilization by *A. niger* (Table 7.1). The causes of such interactions may differ in different fungi and with different carbon sources. In much culture work the amount of carbon made available to the fungus must be far in excess of that likely to be available for most of the time in nature; indeed, it may well be in excess of anything to which the fungus is ever likely to be exposed. There is a great need for studies both on the availability of carbon and other sources in the natural environment, and on the utilization of mixtures, or single carbon compounds at such natural concentrations in culture. Until then it will not be possible to assess adequately or to account for the growth of fungi in nature.

Table 7.1. The effect of glucose and galactose, singly and in combination, upon the mycelial yield of *Aspergillus niger* in liquid culture at 20°C after 7 days. (From Horr, 1936)

Carbon source (g/l.)		Yield (mg dry wt)
Galactose	10	45·1
	18	42·4
	20	44·3
Glucose	2	145·6
	10	411·0
Galactose + glucose	18+2	577·0
	10+10	1151·6

Certain groups of fungi show clear specializations in respect of certain carbon sources. For example, *Leptomitus lacteus*, *Apodachlya brachynema* and perhaps other Leptomitales are incapable of utilizing sugars and grow on fatty acids. In most cases, however, the nutritional restriction is not confined to a taxonomic group; examples are the saprophytic 'sugar' fungi and the 'brown and white rot' fungi. The first of these is a well characterized ecological group of soil fungi. They are fast growing and capable of rapidly utilizing sugars and simple carbon compounds during a brief vegetative period before sporing and passing to a dormant phase. They are frequently incapable of utilizing polysaccharides, especially cellulose, and are mostly Phycomycetes. The last two groups are wood-rotting fungi. The first can utilize cellulose but not lignin, the latter can utilize both classes of compound, e.g. *Polyporus betulinus* and *Polystictus versicolor* respectively.

Heterotrophy for nitrogen has also been investigated but the position is still rather obscure (ROBBINS, 1937; LILLY and BARNETT, 1951). There is still disagreement as to whether fungi can, or cannot, fix atmospheric nitrogen although it is quite certain that it is not a widespread ability. Species of soil-inhibiting *Rhodotorula* and the yeast-like *Pullularia pullans* have been shown to fix ^{15}N using mass-spectrometric methods (METCALFE and CHAYEN, 1954). In general, however, fungi utilize inorganic or organic sources of nitrogen. A number of fungi are known to be capable of utilizing both nitrate and ammonia, others are incapable of utilizing nitrate but use ammonia, e.g. *Absidia* spp., *Mucor hiemalis*, *Lenzites trabea*, *Marasmius* spp. (ROBBINS, 1937; LINDEBERG, 1944). Certain groups of water moulds, the Saprolegniaceae and Blastocladiales, include a number of species which are said to grow only with organic nitrogen as, for example, amino acids. This is always difficult to demonstrate for inability to grow may reflect an effect of low pH rather than inability to utilize ammonia, e.g. as in *Allomyces javanicus* (MACHLIS, 1953). In other fungi this pH effect has been excluded and in two species of *Thraustochytrium* adequate growth can occur only with a mixture of amino acids or any one of glutamate, L-aspartic acid and L-asparagine. Most, if not all fungi can dispense with inorganic sources if amino acids are available. However, fungi usually show a fairly rigid specificity to the isomeric form, L- or D-rotatory, in which different amino acids are supplied. Ability to utilize the D- form of one amino acid does not indicate that the D- forms of other amino acids will also be readily utilized. A very few have a specific complete or partial requirement for an amino acid in addition to an inorganic source, e.g. *Cenococcum graniforme*, a mycorrhizal form, requires histidine; *Trichophyton* spp., histidine or arginine. This paucity of fungi heterotrophic for amino acids is perhaps surprising since amino acid-requiring mutants of most fungi can readily be produced experimentally. Presumably such mutants are selected against very heavily in nature.

Just as with the carbon sources, mixtures of nitrogen sources show various kinds of effects. Ammonia is usually preferentially taken up but amino acids, especially, show complex interactions which are not readily predictable and are not always understood.

Sulphur and phosphorus are required by fungi much as in other organisms but they can usually be utilized from simple inorganic forms such as sulphates for sulphur or phosphates for phosphorus. It is clear, however, that S-containing amino acids can act as alternative sources of sulphur. Exceptionally, in the case of the Saprolegniaceae, growth is not supported by sulphate alone but is so supported by S-containing amino acids, sulphide or thiourea. In some cases the requirements are much more specific, e.g. *Allomyces arbuscula* requires methionine specifically as its sulphur requirement (INGRAHAM and EMERSON, 1954).

The major and minor metallic element requirements of fungi are very similar to those of other organisms and the form in which the element can be utilized is the usual anion. Numerous studies have been made of the effects of mineral nutrition on fungal growth and the precise work of STEINBERG (1939) showed, for *A. niger*, how a proper balance must be achieved between major and minor nutrients if maximum growth is to result. Various experimental procedures exist for the investigation of the 'salt balance' of a medium. These include the triangle method for combinations of three salts (HAENSELER, 1921) and more elaborate factorial and statistical schemes to encompass a larger number of nutrients (e.g. TALLEY and BLANK, 1941).

One of the best known and fully investigated features of fungal heterotrophy is their requirements for vitamins and growth factors. Several good reviews of this topic, both general and particular, exist (SCHOPFER, 1943; ROBBINS and KAVANAGH, 1942; FRIES, 1961, 1965). The requirement arises from the inability of particular fungi to synthesize the appropriate growth factor. The widespread occurrence in fungi of such deficiencies for vitamins contrasts greatly with the rare occurrence of requirements for amino acids. This is especially striking since both kinds of deficiency can be readily induced experimentally in mutant strains. Vitamins are normally required in the concentration range 0·01–1 ng/ml. whereas other growth factors are effective in the range 10–1000 ng/ml. or even as high as at 1 μg/ml. Amino acids, on the other hand, are usually required in greater quantities and this may be the basis of a possible differential selective effect on deficiency mutants; amino acid deficient mutants being, in the wild, more lethal.

The principal vitamins required are thiamin (B_1), and biotin, and, less frequently, pyridoxine (B_6). The vast majority of fungi which require thiamin can synthesize thiazole but require pyrimidine which combines in a manner not yet understood to form the effective vitamin. Why pyrimidine deficiency should be the commonest growth factor requirement in the fungi is quite obscure. Requirements for nicotinic acid and pantothenic acid are relatively uncommon save in yeasts where the latter requirement is fairly common. There are but single reports of a requirement for riboflavin (B_2) by *Poria vaillantii* and for *p*-amino-benzoic acid by *Rhodotorula aurantiaca* (ROBBINS and MA, 1944).

In some cases, notably the yeasts, multiple vitamin requirements are not uncommon. For example, *Kloeckera brevis* requires thiamin, biotin, pyridoxine, nicotinic acid, pantothenate and, in addition, inositol, although this is required at about 1 μg/ml. medium.

Vitamin requirements are usually characteristic of all the strains of a particular species but they may only become apparent in certain situations. Spores of *Myrothecium verrucaria*, for example, can grow exceedingly slowly in the absence of biotin but addition of the vitamin just after germination increases their growth rate enormously. The mycelium, however, requires no exogenous biotin for normal growth (MANDELS, 1955). A similar situation occurs in *Memnoniella echinata* where the spore requires biotin and is apparently incapable of synthesizing it from desthiobiotin (its precursor) although this can be done by the mycelium (PERLMAN, 1951). Environmental factors or medium constituents may also affect the response. Many fungi become more exacting as they reach the limits of their temperature tolerance, presumably because synthetic mechanisms are impaired, e.g. *Coprinus fimetarius* requires methionine at or above 40°C. The balance of the mineral constituents of the medium determines a thiamin requirement in *Pythium butleri* (ROBBINS and KAVANAGH, 1938) while the replacement of glucose by fructose as carbon source eliminates the biotin heterotrophy of *Neurospora crassa* (STRIGINI and MORPUGO, 1961).

A number of unknown or poorly understood growth factor requirements still await investigation. It seems possible that Vitamin B_{12} may prove to be required by some fungi and it is claimed that it is so for a marine saprolegniaceous form (ADAIR and VISHNIAC, 1958). The dung fungus *Pilobolus* will only grow effectively if either natural materials or dung decoction are added to a nutrient agar. This is because dung apparently contains a haematin compound essential for normal growth. Its chemical constitution is not yet known (HESSELTINE *et al.*, 1953).

There appears to be no correlation between vitamin heterotrophy and the mode of life of the fungus. Parasites and saprophytes are apparently equally likely to show the condition. The implication of this is that vitamins must be relatively common in the environmental niches occupied by most fungi. This is also in line with the reason suggested earlier for the more frequent occurrence of vitamin heterotrophy than of amino acid heterotrophy.

The efficiency of nutrient utilization by fungi

It will be appreciated by now that it is exceedingly difficult to assess the nutrient status of the environment and its effects on metabolic processes in any fungus. Indeed, all that can be done at present is to define particular situations involving defined levels of carbon and nitrogen nutrition, salt concentrations and external environmental variables, and to measure the efficiency of utilization of such materials in these circumstances. In fact, the only data available are concerned almost entirely with the measurement of carbon utilization. The essential techniques employed are to determine more or less fully a carbon balance sheet and to derive from such data some measure, usually a ratio, of material supplied to that retained or lost. Several such measures exist. A common and readily determined ratio is the 'economic coefficient':

$$\frac{\text{Dry wt. mycelium} + \text{spores}}{\text{Wt. carbohydrate consumed}} \quad \text{as a percentage}$$

A more sophisticated measure is that attempted by Terroine and his associates (TERROINE and WURMSER, 1921, 1922a, b, c; TERROINE and BONNET, 1930) who tried to estimate the thermodynamic efficiency of carbon utilization. They estimated the Rubner coefficient:

$$\frac{\text{Heat of combustion of mycelium} + \text{spores}}{\text{Heat of combustion of total carbon source utilized}}$$

The Rubner coefficient is always appreciably higher than the economic coefficient because the heat of combustion of 1 g of fungal mycelium is higher than that of the carbohydrate from which it is derived. This is because the mycelium includes materials such as fats with a higher specific calorie content than, say, glucose, i.e. 38·9 kJ/g to 15·7 kJ/g, so that the average figure usually adopted for mycelium is 20·1 kJ/g. For example, *A. niger* grown on a glucose medium has an economic coefficient of 44% but a Rubner coefficient of 56% (TERROINE and WURMSER, 1922a; corrected to 59·4% by FOSTER, 1949, since not all the glucose was converted into mycelium and CO_2).

A further measure is the 'Absolute efficiency'. This is the proportion of energy directed into synthesis only and leaves out of account that utilized in maintaining fungal material already formed. Since such maintenance is an integral requirement for normal growth there seems to be little real significance in such a measure for most purposes.

Raistrick and his collaborators (RAISTRICK *et al.*, 1931) have provided a notable series of examples of carbon-balance sheets for some 216 strains of fungi, mostly Deuteromycetes and Ascomycetes (Table 7.2). The principal deductions to be made from such data are that most fungi lose more carbon as CO_2 than they transform into cellular carbon and that, depending on conditions, a good deal of carbon is in material which may accumulate in, or be lost from, the mycelium such as organic acids or polyhydric alcohols. In general, high CO_2 production is associated with a high carbon content in the 'volatile neutral compounds', i.e. ethanol or in the 'carbon not accounted for', usually polyhydric alcohols, glycerol or polysaccharides. The Basidiomycete data suggest that a rather more efficient conversion of carbon occurs but they are too few to make valid generalizations.

The efficiency of utilization of carbon compounds can be altered by changes in the concentration of the principal source of carbohydrate, by variation of the composition of the medium including most notably the presence or absence of trace elements, by the duration of the experiment, or by the strain of fungus employed. Examples are given in Table 7.3, mostly from experiments appertaining to industrial fermentations for it is under such conditions that an effective yield of product is of the greatest importance. The alterations brought about in yields are due to a changed balance between carbon incorporated into mycelial material and that in soluble metabolic products. The better balanced a medium is, the more efficient is the conversion of carbon into mycelial material and the better the growth achieved. A low economic coefficient is frequently a sign of incomplete utilization of carbohydrate and indicates the accumulation of 'waste' products, i.e. materials not rapidly utilized to CO_2 with concomitant release of energy.

The data of Terroine and his colleagues indicate that, in general, the efficient utilization of carbohydrate is related to the number of carbon atoms in the substrate. This presumably reflects the efficiency with which the carbon compounds can be converted into intermediates suitable for assimilation.

The difficulties experienced in measuring the efficiency of conversion of nutrients under controlled cultural conditions are great and their interpretation

Table 7.2. Summaries of carbon balance sheets for various fungi. (After Raistrick *et al.*, 1931)

Fungus	Titratable by N/1 acid (ml.)	C Non-volatile acids (g)	C Volatile acids (g)	C Volatile neutral (g)	C Unaccounted (g)	C In mycelium (%)	Resp. coeff.	Main product
Aspergillus niger	10·2	0·308	0·007	0·792	0·349	49·3	1·58	Alcohol, TCA acids
A. wentii	7·8	0·474	0·001	0·009	0·709	50·0	1·12	Gluconic acid
Penicillium terrestre	0·7	0·087	0·10	1·797	0·306	47·5	2·26	Ethanol
Fusarium solani var.	4·8	0·067	0·212	0·488	0·311	52·2	1·45	Ethanol, fatty acids
Helminthosporium geniculatum	0·3	0·094	0·029	0·818	0·622	57·4	1·56	Ethanol, Mannitol
Chaetomium sp.	0·8	0·130	—	0·001	0·069	53·1	1·22	CO_2
Ustilago avenae	0·9	0·170	0	0	0·161	52·0	1·01	CO_2

All fungi grown on (g) glucose 50: $NaNO_3$ 2 ; KH_2PO_4 1; KCl 0·5 ; $MgSO_4.7H_2O$ 0·5 ; $FeSO_4.7H_2O$ 0·01 ; H_2O—1000 ml.

Table 7.3. The effects of carbohydrate concentration, micro-nutrients and strain specificity on the efficiency of utilization of carbon compounds or of growth by various fungi.

(a) Carbohydrate concentration and zinc in *Rhizopus nigricans*. (After Foster and Waksman, 1939b)

Initial concentration glucose (g/10 ml.)	Glucose used (g/10 ml.)		Economic coefficient (g mycelium/g glucose used ×100)	
	+Zn	−Zn	+Zn	−Zn
2·5	1·7	2·07	14·1	40
5·0	2·65	3·89	13·9	21·3
10·0	2·54	6·49	16·0	12·5

(b) Economic coefficients of different strains of *Aspergillus versicolor*. (After Raistrick *et al.*, 1931)

Strain designation	Period (days)	Initial carbon as glucose (g)	Carbon utilized (g)	Mycelium dry wt. (g)	Economic coefficient
AC7	16	5·020	2·387	2·121	88·8
AC8	22	5·094	2·419	1·846	76·3
AC28	15	5·094	2·173	1·872	86·4
AC33	33	5·094	1·145	0·938	82·0
AC47	32	5·001	1·541	1·138	73·6
AC25	36	5·094	1·620	1·111	68·8
AC30	43	5·094	1·543	1·158	75·0
AC18	39	4·834	1·324	0·800	60·4

difficult. The difficulties are vastly increased if attempts are made to make such measurements under natural conditions. However, it seems desirable to make such attempts if the significance of fungi as energy-transferring agents in ecosystems is to be assessed. Such data are particularly pertinent, for example, to the role of fungi in decomposition processes in the soil, in biogeochemical cycles and in the assessment of the biological efficiency, rather than the more usually quoted economic losses achieved by parasites.

Considerable bodies of data already exist concerning the amount of fungal material in the soil although it is usually in the form of spore counts or lengths of mycelium, e.g. 38 m of hyphae/g air-dried soil from unmanured plots at Broadbalk, Rothamstead (cited in RUSSELL, 1950). There are also some estimates of the efficiency of litter decomposition by Basidiomycetes in pure culture. It is usually of the order of 40–65% (GRAY and BUSHNELL, 1955). Economic coefficients of this order are comparable to those obtained by Raistrick and co-workers for common saprophytic moulds. Nevertheless, such data are very far from being applicable to the natural environment. The substrates are heterogeneous and efficiency of utilization is likely to be affected, in addition, by adjacent fungi and other organisms as well as by the fungus itself. For example, although soil Basidiomycetes and common moulds may appear to have a similar economic coefficient in laboratory experiments, their significance in the soil environment must be very different. BURGES (1960) has described a number of typical growth patterns including that of 'Penicillium' and the 'Basidiomycete patterns'. In the former pattern a highly localized, specialized piece of substrate is densely overgrown and heavy spore production ensues rapidly with little or no extension growth; the spores then remain as dormant propagules. By contrast, the 'Basidiomycete pattern' involves production of a perennial mycelium which builds up at the point of colonization on a substrate and then grows out over considerable distances to other substrates by means of strands of greater or lesser complexity. Thus in the 'Penicillium pattern' substrate utilization is a periodical phenomenon of alternating phases of rapid exploitation and dormancy. The 'Basidiomycete pattern' may involve a much slower exploitation but this lasts for a far longer period. The formation of propagules, basidiospores or sclerotia, e.g. *Waitea circinata* (WARCUP, 1959), occurs throughout this period and, in their turn, they may germinate and metabolize effectively during the period of exploitation of the primary mycelium from which they have been derived.

Similar difficulties and complications arise in attempting to measure the biological efficiency of parasites. Here data exist concerning the losses sustained in substrate materials but very few concerning the gain in materials by parasites. It should not be too difficult to obtain appropriate data in some cases. Ectoparasites such as downy mildews could, perhaps, provide systems where the gain in mycelial matter could be compared with the loss of material in infected leaves and where appropriate respiration measurements could be made. Similarly, with fungi such as cellulose-destroying Basidiomycetes, where the principal carbon source in nature is almost certainly cellulose, model laboratory experiments should be possible; indeed, pertinent data concerning loss in weight of substrate are already available.

The study of fungal nutrition and metabolism and the efficiency of these processes is an almost endless task. The accumulated data on fungal nutrition is already enormous but sufficient has perhaps been said to draw attention to the fact that much of such data has little bearing on growth in natural conditions. FOSTER (1949) considering soil fungi has written, 'Due to the rather frugal and precarious nutritional environment prevailing in this natural habitat of molds, it appears not unreasonable that they have become adapted to survival and existence under threshold nutritional conditions by their high efficiency of utilization of the limited energy source available'. Very few experiments, however, have ever been done on fungal metabolism under such threshold conditions; Foster's view cannot be validated or assessed until they have been done. There seems to be general support for such a view since fungi are primarily aerobic organisms. It is true that many can persist under anaerobic conditions for some time, notably yeasts, but it is a striking fact that only a single, obligatorily-anaerobic fungus has so far been discovered (EMERSON and

WESTON, 1967). Under conditions of limiting carbohydrate nutrition an aerobic organism will be more efficient at utilizing carbohydrate than an anaerobic one since, in the former, the ultimate products are cell material and CO_2, whereas, in the latter, complete conversion of the substrate does not occur. Production of large amounts of other metabolic products in conditions of high carbohydrate nutrition, which occurs in various industrial fermentations, must be regarded as inefficient metabolic utilization on the part of the fungus. Some have gone further, '. . . the metabolism of the organism becomes deranged. It becomes, so to speak, pathological. This pathological behaviour is a direct result of the influence of environmental conditions' (FOSTER, 1949). This extreme view is debatable. It is difficult to suppose that the enormous range of known fungal metabolites and their accompanying biosynthetic mechanisms are solely a reflection of 'pathological' responses but it must be admitted that in most of these cases no functions can yet be attributed to the metabolites.

Because it is simply not possible to compass all the nutritional requirements or metabolic activities of fungi in a book such as this, limited attention will be paid to such processes in subsequent chapters but a number of valuable summaries exist to which reference may be made (RAISTRICK *et al.*, 1931; FOSTER, 1949; COCHRANE, 1958; BIRKINSHAW, 1965). Attention will now be focused on those general processes common to all metabolic activities and it will be seen that, even in this area, knowledge of some fundamental processes is woefully inadequate.

Chapter 8
Transport Processes in Fungi

All nutrients must be transported through the two outer membranes of a fungus, its cell wall and cytoplasmic membrane, before they can be utilized metabolically. Extracellular products must similarly pass outwards. Transport processes have not been studied in a wide range of fungi and, indeed, this is true of many aspects of metabolism. A good deal of information is available for yeasts but less for filamentous fungi. This is unfortunate if the cell wall affects transport, because of the wide variations of wall composition reported in different fungi (cf. p. 22). However, it seems probable that transport processes are basically similar in all organisms so that this limited survey may not prove to be too serious a restriction. Naturally, however, information about transport in a unicellular organism gives little information about transport across hyphal membranes in different parts of a complex colony or in a large, vegetatively and reproductively differentiated mycelium such as an agaric. The process about which least is known in fungi is the gain and loss of water by hyphae.

In this account the permeability of the outer membranes will be considered first, then the general problems of transport followed by the consideration of particular mechanisms for water, electrolyte and non-electrolyte transport and finally the problems involved in the release of extracellular products. Mention may also be made conveniently at this point of the use of the terms transport, uptake, accumulation and assimilation. Transport is employed to designate the influx into, or efflux out of a cell of a particular solute. An apparent net gain by the organism of a solute as compared with its external concentration is described as uptake, absorption or accumulation and, if the solutes so gained by the cell are metabolized, they are said to be assimilated. Sometimes utilization of solutes is so rapid, commencing upon entry into the cell, that the transport processes cannot readily be separated from the assimilatory processes; uptake will then have to be used to cover both phenomena.

PERMEABILITY OF THE CELL WALL AND CYTOPLASMIC MEMBRANE

It is not easy to separate the two outer membranes of fungi (see pp. 19, 20) and it is doubtful whether they are separable in the living condition. Consequently not many studies have been made on the permeability of the two components separately. It will be convenient to consider three regions, however, the cell wall, the cytoplasmic membrane and the region between them.

Cell wall permeability

There is clear evidence from the studies of Sussman and his collaborators that a variety of substances, including cations and anions, can be bound to the walls of ascospores of *Neurospora tetrasperma* (SUSSMAN, 1954; SUSSMAN and LOWRY, 1955; LOWRY, SUSSMAN and VON BÖVENTER, 1957; SUSSMAN, VON BÖVENTER-HEIDENHEIM and LOWRY, 1957; SUSSMAN, HALTON and VON BÖVENTER-HEIDENHEIM, 1958). Using dead or dormant ascospores or cytoplasm-free fragments of walls, it was shown that Ag^+, UO_2^{2+}, Cu^{2+}, Th^{4+}, Ce^{3+}, Ac^{3+}, the cationic dye methylene blue and antibiotics such as polymyxin B sulphate became bound to wall material. The process followed a typical adsorption isotherm and the cations could be readily eluted by chelating compounds or sulphydryl compounds. The adsorption was site-specific and prior treatment by polymyxin B or hexol nitrate could block or reduce binding of cations (Table 8.1). Similar phenomena are known for cations and nystatin in yeast and with heavy metal cations for conidia of *N. sitophila*, *Alternaria tenuis* and *Penicillium italicum* (ROTHSTEIN and HAYES, 1956; LAMPEN *et al.*, 1962; OWENS and MILLER, 1957; MÜLLER and BIEDERMANN, 1952; SOMERS, 1963). In these examples the amount of bound cation is appreciably lower than in *Neurospora* ascospores, e.g. 2% of total cellular cations in yeast compared with 3–11% in *Neurospora* ascospores. These differences are likely to be attributable to the much thicker walls of the ascospores.

Table 8.1. Methylene blue uptake by fragments of walls of *Neurospora tetrasperma* prepared by sulphuric acid treatment without or with prior treatment by polymyxin. (From Lowry and Sussman, 1956)

	Starting concentration of dye 2×10^{-5} M			
	Untreated		Polymixin treated	
Time	Dye uptake $\times 10^{-5}$ M	% uptake	Dye uptake $\times 10^{-5}$ M	% uptake
10 min	1·92	96	0·52	26
2 hr	—	—	1·35	67
3 hr	—	—	1·54	77
30 hr	—	—	1·68	84

However, this adsorption phenomenon is unconnected with transport. If ascospores are 'activated', rapid uptake occurs of Ag^+, for example and the cation penetrates the spore and impairs respiration. If ascospores are pretreated with Ag^+ so that it is heavily bound to the walls and then activated, the penetration phase and inhibition are not accelerated. Moreover, if Ag^+ binding is inhibited by pretreatment with a blocking agent it in no way affects Ag^+ uptake, penetration and respiratory inhibition when the ascospores are activated.

SUSSMAN (1961) has also been able to show that dormant ascospores can be reversibly dehydrated with glycerol, causing shrinkage of the protoplasm from the spore wall so that the latter is presumably fully permeable to water.

Cell walls may, therefore, retain cations and other compounds but this does not affect their permeability to materials undergoing transport into the cell. The significance of this phenomenon for the life of fungi is obscure. Clearly the retention of materials in walls could enable them to be made available at germination, which might occur in the absence of such materials in the immediate environment. A similar kind of situation will, of course, arise in those fungi whose outermost layer is sticky and gel-like and here the larger adsorbing region may hold up the passage of materials for a short time, e.g. *Fusarium culmorum* (MARCHANT, 1966b).

Recently attempts have been made to study the permeability of isolated cell walls of *S. cerevisiae* and *N. crassa* to a series of metabolically inactive polyethylene glycols and dextrans of different molecular weights. *Saccharomyces cerevisiae*, did not take up polymers of molecular weight above 4500 (GERHARDT and JUDGE, 1964) and this compares remarkably well with a figure of 4750 for normal *Neurospora* (TREVITHICK and METZENBERG, 1966b). However, an *osmotic* mutant of *Neurospora* was permeable to polymers of molecular weight 18,500. The walls of this mutant appear to be thinner and are readily broken to release protoplasts. Their galactosamine/glucosamine ratio was some 30-fold higher than that of wild-type hyphae. The interpretation of the observations proved difficult but it was suggested that there was a continuous range of pore sizes in the wall, including a few very large ones. In the mutant *osmotic* these latter, large pores were thought to have been increased greatly. These findings were compared with studies of 'molecular sieving' of *Neurospora* invertase during secretion. METZENBERG (1964) showed that invertase can exist as an active monomer (light invertase) or as an aggregated active form (heavy invertase), the transition being reversible. TREVITHICK and METZENBERG (1966a) showed that more invertase was secreted into the medium with a higher percentage of light invertase by mutants such as *osmotic* than in the case of wild-type hyphae (Fig. 8.1). They suggested that this too was related to an increase in the number of larger pores (although the molecular weight of light invertase exceeds 18,500) and demonstrated that it was not just due to leakage through cytolysis. They refer to the possible existence of branching, three-dimensional micro-channels of diameters 40–70 10^{-10}m discovered by a combination of electron microscopy and enzyme studies. The investigators, MANOCHA and COLVIN, suggest that 'they serve as conduits for movement of

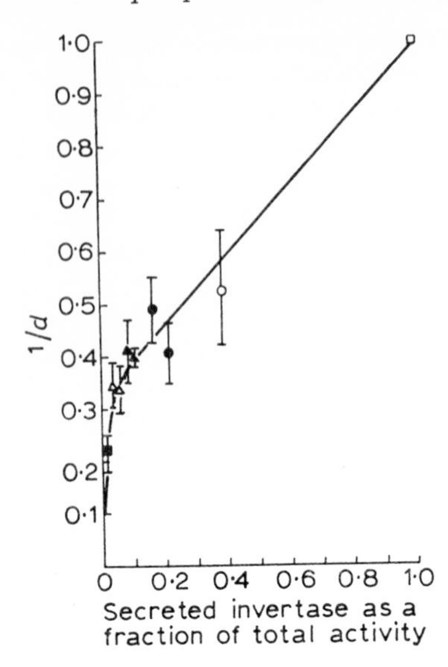

Fig. 8.1. Plot of the amount of invertase(s) secreted into the medium in 3 hr by wild type and mutants of *Neurospora crassa* against the reciprocal of the degree of fractionation (d) (d = ('light' invertase in medium/total invertase)/('light' invertase in bound fraction/total in bound fraction), see text). If cytolysis contributes significantly to release of enzyme then $1/d = 1{\cdot}0$. Thus different strains show differing sieving. Symbols: △, wild type; ▲ *crisp*; ●, *crisp osmotic*; ○ *osmotic*; □ wild-type protoplasts; ■ wild type + cycloheximide 10 μg/ml.
(From Trevithick and Metzenberg)

macromolecular substances, perhaps in both directions'. Whether these micro-channels are related either to the hypothetical 'pores' postulated in relation to the permeability studies or, indeed, to movement of macromolecules must await further detailed, studies like those of MANOCHA and COLVIN (1967). The work published so far does, however, indicate that wall structure can affect the transport of at least one kind of molecule, invertase.

Permeability of the cytoplasmic membrane

The implication of Sussman's studies is that it is the cytoplasmic membrane which regulates the transport of materials into the cell. This is in keeping with findings in cells of other organisms, notably the walled cells of green plants. Some of the most striking evidence in support of this view is available from studies on the effects of polyene antibiotics on fungi (LAMPEN, 1966) and the properties of protoplasts, i.e. fungal cell contents free, or effectively free of the cell wall (VILLANEUVA, 1966).

Polyene antibiotics such as nystatin kill yeasts and hyphal fungi such as *Neurospora* by inducing leakage of ions such as K^+, NH_4^+, carboxylic acids and sugars, or by impairing concentration mechanisms and eventually, as a consequence of increasing internal acidification, by autolysis of cell contents. It has been shown that nystatin is bound on to cytoplasmic membranes of yeast almost instantaneously, even at 0°C. Using protoplasts of *N. crassa* a similar effect was observed but the nature of the binding process was also elucidated. Here treatment with organic solvents releases the nystatin but binding can be restored very considerably if the membranes are exposed to ergosterol. The nystatin is thought, therefore, to be bound to the steroid portion of the cytoplasmic membrane.

Table 8.2. Chemical analysis of protoplast membranes of *Candida utilis* and *Saccharomyces cerevisiae*. Data as % of dry wt. of whole protoplast. (Data of Garcia Mendoza and Villaneuva, 1965; Boulton and Eddy, 1962; Boulton, 1965)

	Candida	*Saccharomyces*
Carbohydrate	5	4 (Hexose)
Hexosamine	0	—
Protein	37·5	40
Lipid	40·0	40
RNA	1·0	5·0
DNA	—	0·8
Total N	5·3	—
Total P	0·7	—

Studies of cytoplasmic membranes in intact fungal cells and protoplasts have shown that it is a typical 'unit membrane' which seems to surround all cells. Chemical analyses have been made of the cytoplasmic membranes of protoplasts of *Candida albicans* and *Saccharomyces cerevisiae* (Table 8.2; GARCIA MENDOZA and VILLANEUVA, 1965; BOULTON and EDDY, 1962; BOULTON, 1965) and electronmicrographs show them to have an outer unit membrane in *Candida* and *Polystictus versicolor* (STRUNK, 1964). The high lipid and protein content suggest that fungal membranes resemble, chemically, the cytoplasmic membranes of other organisms. They may well have a structure such as that hypothesized by DANIELLI (1954) in which protein coats an internally arranged bimolecular lipid plate broken in places by polar pores coated with protein and, therefore, permeable to water (Fig. 8.2). The action of antibiotics such as nystatin on such a membrane is thought to result from the insertion of antibiotic molecules within the lipid (sterol) layer, leading to molecular re-orientation and loss in mechanical rigidity and hence mechanical failure of the membrane (LAMPEN, 1966). Adoption of such a model also suggests that penetration of most materials will be related to their lipid solubility since the 'polar pores' are too small to allow entry of ions or larger molecules other than water. Unfortunately no systematic data exist concerning penetration of substances in relation to lipid solubility in fungi. However, HOLTER and OTTOLENGHI (1960) have compared the permeability of intact cells of *Saccharomyces* with that of derived protoplasts. The properties shown by the cells were retained precisely by the protoplasts, e.g. impermeability to sucrose and melibiose. In the case of glucose, fructose and mannose, uptake occurred and was

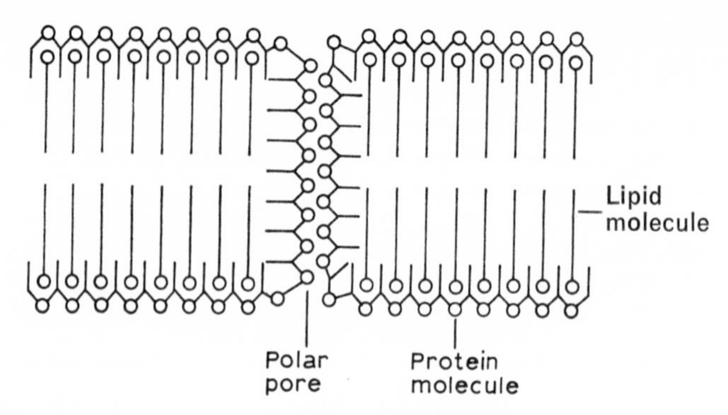

Fig. 8.2. Diagram of structure of protoplasmic membrane proposed by Danielli

mediated by a common transport system as in intact cells (HEREDIA *et al.*, 1963, and see p. 222).

Fungal cells are remarkably resistant to osmotic changes in the sense that plasmolysis is not readily achieved and although apical growth may be checked it is soon resumed (ROBERTSON, 1958, 1959; PARK and ROBINSON, 1966b; see also pp. 49, 208). Protoplasts, by contrast are osmotically sensitive. They burst easily in solutions of low osmotic pressure or water and contract in hypertonic solutions (SVILHA *et al.*, 1961). OTTOLENGHI and LILLEHOJ (1966) have shown that yeast cells can adapt to 3 M sorbitol or 2 M KCl and that protoplasts derived from them are then stable in solutions of the same molarity although 0·5–1 M KCl are the usual concentrations in which yeast protoplasts are stable. The difficulty frequently experienced in plasmolysing the cytoplasm away from the walls of intact hyphae exposed to hypertonic solutions reflects the close connexion between wall and cytoplasmic membrane in the living fungus.

Free space in fungi

There is some evidence that substances such as electrolytes or enzymes can exist in a region physiologically outside the permeability barrier although the precise spatial location of this region of 'free space' is not clear. 'Free space' has been defined as 'the phase in the cell or tissue into which solutes move relatively freely, and was distinguished from the osmotic volume into which solutes, but not the solvent, penetrate relatively slowly' (BRIGGS, HOPE and ROBERTSON, 1961). It has been pointed out that the term 'apparent free space' (AFS) is preferable since the method of measurement and assumptions concerning the properties of 'free space' affect the estimated value of the parameter. Estimates of the free space of yeast cells were made by CONWAY and DOWNEY (1950). They showed that materials such as inulin, gelatine and peptone would not enter cells and occupied 26% of a centrifuged cell suspension, a volume close to that expected between closely packed spheres of the same dimensions as yeast cells. Sodium and potassium chloride, succinate, arabinose and galactose, however, occupied some 33–34% of the cell mass under the same conditions, i.e. without uptake. Thus some 10–12% of the cell volume in the outer region of yeast cells can be occupied without entry into the cell. Unfortunately, no other data are available for fungi.

A different kind of observation suggests that substances exist in a region outside the cytoplasmic membrane. This is the observation that enzymes exist 'on the surface' of cells. In *Saccharomyces* spp. invertase and melibiase activity is associated with walls (BURGER, BACON and BACON, 1958, 1961; FRIIS and OTTOLENGHI, 1959a, b) and this is true of invertase in the spores of many fungi (MANDELS, 1953; METZENBERG, 1963). If the walls are disintegrated the enzyme activity is lost but intact protoplasts could then produce and liberate the enzyme (TREVITHICK and METZENBERG, 1964). It is perhaps significant that extracellular yeast invertase has a high mannan content which appears to act as a stabilizer and that mannan has been recorded from yeast cell walls (N. NEUMANN, in LAMPEN, 1965; NORTHCOTE and HORN, 1952). Metzenberg's careful comparison of the behaviour of *Neurospora* conidia and purified invertase from *Neurospora* showed their identity of behaviour in relation to pH, their Michaelis constants and response to inhibition by aniline and 2 N HCl. Some 9–19% of enzyme was freely extractable from conidia with cold water; sonic disruption over intervals up to ten minutes resulted in increasing amounts being liberated although the total activity remained the same. This is mostly easily interpreted as indicating that 80–90% of the invertase is somehow associated with the region external to the cytoplasmic membranes but within the outermost regions of the wall. How far this region can be equated to the free space discussed earlier is uncertain.

It may be concluded then that the cell wall and cytoplasmic membranes both play important roles in transport phenomena but it is the latter which is principally responsible for the regulation of transport by the cell. The cell wall supplies rigidity and shape to the cell and enables cells to accumulate substances so that great osmotic pressure differences may be maintained between the inside and outside of the fungus. Size, therefore, is neither determined by the osmotic pressure of the cell contents nor changes consequent to uptake and leakage, as it is in unwalled cells (TOSTESON and HOFFMAN, 1960). The wall in association with the cytoplasmic membrane also delimits a region in which substances may accumulate passively in close association with the permeable membrane and into which enzymes and materials may be liberated yet retained adjacent to the cell surface. Some 'molecular sieving' of macromolecules may occur at this phase.

Transport phenomena are, therefore, largely concerned with the cytoplasmic membrane and attention will now be directed towards the processes involved.

POSSIBLE MECHANISMS FOR TRANSPORT

The mechanisms available will, of course, vary to some extent with the nature of the substance being transported. In general, two kinds of processes may be involved; passive processes, which involve diffusion of some kind or other, and active transport, in which there is at least one phase dependent upon metabolic energy and a consequent accumulation against a concentration difference. Both passive and active transport processes may be involved in the transport of any particular substance. Knowledge of the processes, their attributes and their demonstration is at least as meagre in fungi as in green plants and only a limited number of species, notably yeasts, have been studied. Several detailed discussions of transport processes in plants are available, viz, DAVSON and DANIELLI (1952); BRIGGS, HOPE and ROBERTSON (1961); SUTCLIFFE (1962); JENNINGS (1963); DAINTY (1964); BRIGGS (1967); but only Jennings provides much information concerning fungi. CIRILLO (1961a), ROTHSTEIN (1965) and NICHOLAS (1965) deal with particular problems in fungi in some detail.

Passive transport processes

The simplest example of a passive transport process is the movement of a solute both due to and proportional to a difference in concentration, i.e. free diffusion, and the rate of transport of a solute is then proportional to the concentration difference, its molecular size and lipid solubility. However, early studies of sugar transport into mammalian erythrocytes showed that it did not follow this relationship. The rate of transport did not increase proportionally with increasing external concentrations but approached saturation and the rate was greater than expected for substances of particular molecular weights and lipid solubilities. Moreover, structurally similar substances could show competitive inhibition and uptake was specific for isomers. This process, which does not involve metabolic energy, was termed 'facilitated diffusion'. Its mechanism is not understood. The diffusing substances which show this behaviour are polar and are supposed to become attached to a carrier or enzyme while being transported. This could account for the specificity of the process and its saturation kinetics but whether the carrier (or enzyme) is static or mobile is not clear.

For electrolytes, passive transport can be detected by either measuring the electrical potential difference across the membrane or the ratio of the influx to the efflux of ions and, in either case, comparing the experimental values with those theoretically expected from the equations of Nernst or Ussing. A departure from these is suggestive of an active process but under certain conditions, notably when ions are moving in files through long narrow pores, measurements may be misleading (cf. BRIGGS, HOPE and ROBERTSON, 1961, for full discussion).

Electrolyte transfer may also be modified by a situation in which certain ions are immobilized or restricted in their movements. This may come about as a consequence of the binding of an ion to an insoluble structure or its restriction by an impermeable membrane. In general, in plants, anion immobility within the cell is the commonest phenomenon of this type and hence cations tend to accumulate. The Donnan equilibrium which results appears to be a phenomenon affecting the distribution of ions in the free space of green plant cells but nothing is known with certainty concerning the situation in fungi.

Active transport processes

When solutes are accumulated against a concentration difference and when metabolic energy appears to be necessary for the process then an active process may be suspected. On the other hand these criteria are not always easy to apply. If the substance transported is metabolized on entry into the cell, e.g. sugars, then the difference between active transport and facilitated diffusion becomes marginal. The ambiguities in interpreting departures from expected flux ratios for ions, which have already been referred to, make the recognition of active transport difficult with electrolytes. A link with oxidative respiration is often regarded as providing circumstantial evidence for active transport but the nature of the interaction of these processes is still obscure. DAINTY (1964) has pointed out in connexion with green plants that two schools of thought exist, 'One school . . . considers that redox energy available in the electron-transport chains is used directly. . . . Basically, ion transport is looked upon as a consequence of a primary separation of positive and negative charges during respiratory electron transfer. . . . The other, perhaps more numerous school, takes the apparently more conventional view that ATP . . . is the immediate source of energy for driving ion pumps'. The same considerations could apply to fungi.

Some form of carrier molecule is often thought to be involved in active transport since the kinetics of the process often resemble enzyme kinetics but its, or their, nature is not known. Finally, it is suggested that active transport may be mediated through the action of a contractile protein or by incorporation of substances in ultra-microscopic invaginations of the membranes which become closed to form vesicles during transport, disintegrating once they are within the membrane. This vesiculation phenomenon, pinocytosis, certainly occurs in some animal and higher green plant cells but a search for it in yeast, using a fluorescent-labelled antibody technique to detect ingested protein, failed to reveal its occurrence (HOLTER and OTTOLENGHI, 1960).

TRANSPORT OF WATER

There is no doubt that the osmotic pressures of fungal hyphae are considerable, that they can readily adjust to changes in concentrations of external solutions and that fungi can withstand, and grow under extremely high external osmotic pressures. Thus KLAUS (1943) measured the growth of cultures of *Alternaria solani* after 10 days at 17–18°C in media of different osmotic pressures and showed that growth began to fall off at *c.* 35 bar but was not suppressed until *c.* 132 bar. Moreover, there are halophytic fungi and osmophilic yeasts which live in environments of high osmotic pressure such as brine, nectar, honey and fruit juices. Most fungi, however, are not usually resistant to much more than about 2 M sucrose (*c.* 50 bar). There are not many determinations of the osmotic pressure of fungal cells on which reliance may be placed. The reason for this is, in part, methodological. Although vacuoles do occur in fungal cells they are lacking from the hyphal tip and sometimes from cells adjacent to it. Thus the normal plasmolytic methods used by botanists are not readily applied and, in any case, provide no information concerning the actively growing apical regions. Cryoscopic methods have been used and give a range of values but these must represent the average for hyphal fungi and different values may occur at their apices, e.g. 2·33 and 1·3 atm in *Neurospora*, although the method employed was plasmolytic. PARK and ROBINSON (1966b) have argued that extension growth at the hyphal apex depends on the internal pressure so that application of substances which arrests its growth or cause it to burst can be used to estimate the internal pressure. They applied sucrose solutions of different molarities and either, determined the time to resumption of normal extension growth, or treated the hyphae with 0·5% acetic acid, to cause apical extrusion of contents, at graded intervals from 5 sec to 4 min; the number of tips which burst in one minute after this treatment was recorded. Table 8.3 shows some of their results and it is clear that hyphal apices of *A. niger* are capable of rapid osmotic changes after which growth could recommence.

Table 8.3. The effect on hyphae of *Aspergillus niger* of treating them with sucrose solutions alone or followed by 0·5% acetic acid (v/v). In the first case the time taken to resume normal growth was determined; in the latter, the time to the first acetic acid treatment (given at 5-sec intervals) to cause bursting 1 min after treatment. (From Park and Robinson, 1966b)

Sucrose M	Growth delay (sec)	Acetic acid treatment (sec) (no. of treatments)
0	0	5 (1)
0·06	0	5 (1)
0·1	0	5 (1)
0·2	50–60	30 (6)
0·23	90	60 (12)
0·03	300–330	240 (48)

The assumption made in this type of work is that transport is essentially due to osmosis but there are suggestions that water uptake by certain spores at germination could be an active process. In *Rhizopus arrhizus* sporangiospores and *Fusarium culmorum* conidiospores (EKUNDAYO and CARLILE, 1964; EKUNDAYO, 1966; MARCHANT and WHITE, 1966) it has been demonstrated that spore swelling is dependent upon the presence of exogenous glucose. In *R. arrhizus*, rate of increase in volume and hence rate of water uptake was constant per unit surface area during a twenty-fold increase in volume. The rate was also independent of a range of 14 bar in the external medium. Swelling was inhibited by a range of respiratory inhibitors. Uptake may, therefore, be active. However, D_2O was shown to be freely exchangeable with H_2O whether or not swelling occurred so that the effect could be regulated by a change in wall plasticity, so enabling osmotic water uptake to be followed by swelling.

It will be evident that water uptake by fungi is a neglected subject. Determinations both of the osmotic equivalents of fungal cells and also of the rates of uptake of water using isotopic techniques, for example, are much needed.

TRANSPORT OF ELECTROLYTES

Very few fungi have been examined in connexion with the transport of electrolytes and surprisingly few cations have been studied. Work on this topic is largely dominated by studies on the transport of phosphate and alkali–metal ions by yeast cells.

Fresh commercial yeast cells when washed and starved for a few hours while being aerated possess very constant and reproducible properties. Their ionic content is typically like that in Table 8.4. Most of the K^+ is probably unbound and this would account for the osmotic pressure of 11–13 atm (EDDY and WILLIAMSON, 1957) but the bivalent cations are likely to be bound to inorganic polyphosphates or phosphoryl groups of nucleic acids.

Cation transport

Consideration will be given here to studies on monovalent and bivalent cations with the exception of the ammonium ion. Although illogical in one sense it seems best to consider the transport of all inorganic nitrogen compounds together since they show close interrelationships.

Table 8.4. Ionic composition of fresh, commercial *Saccharomyces cerevisiae* after storing with aeration. All values, mM/kg wet wt. (From Rothstein, 1955)

Cations		Anions	
K^+	170	HCO_3^-	50
Mg^{2+}	20	Total P	120
Ca^{2+}	3	(TCA extractable)	(40; *ortho*-P 15)
Mn^{2+}	0·2	Succinate	7
		Ether sol. acids	15

POTASSIUM AND SODIUM

The first critical observations concerning transport of potassium across the cell membrane of yeast were made by HEVESY and NIELSEN (1941) and ROTHSTEIN and ENNS (1946). The former workers examined the permeability of strongly fermenting yeast to potassium using ^{42}K as an indicator over a two-hour period. It became clear that potassium was exchanged between cells and nutrient solution at a level below that due to complete interchange. Five years later Rothstein and Enns showed that potassium could be transported against a concentration gradient and that the influx of K^+ was associated with an efflux of H^+ and some K^+ in fermenting yeast cells. The process of K^+ transfer was studied in detail by the use of extremely ingenious isotopic techniques employing yeast in the equivalent of steady-state conditions in relation to the external [K^+] by ROTHSTEIN and BRUCE (1958a, b). In K^+ free solution there is a small constant efflux of K^+-ions, *c.* 15 mM/kg cell/hr, balanced by an influx of H^+. As the external [K^+] is increased the yeast cells go through an equilibrium position, where there is neither net gain nor loss, to a situation where there is an increasing gain in K^+-ions. The influx is at first linear but is somewhat flattened at higher concentrations due probably

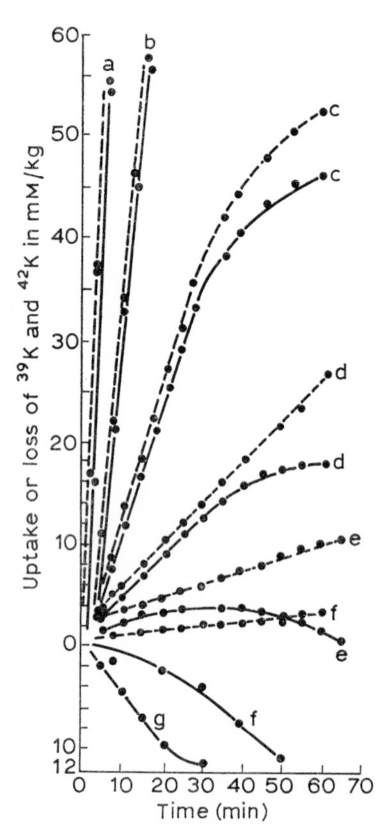

Fig. 8.3. Uptake or loss of ^{39}K (solid lines) and ^{42}K (broken lines) at various exogenous concentrations by fermenting yeast cells. Concentrations: (a) 1×10^{-2} M; (b) 2×10^{-3} M; (c) 1×10^{-3} M; (d) 3×10^{-4} M; (e) 1×10^{-4} M; (f) 3×10^{-5} M; (g) no potassium. Glucose (0·05 M) was in the medium which was at pH 4·0. (From Rothstein and Bruce)

to the efflux of absorbed K^+ (Fig. 8.3). The rate of transport increases with increasing external $[K^+]$ to a maximum in a manner analogous to that applicable to the kinetics of enzyme action. This suggests that a carrier is involved in the transport process, as might be expected since transport of H^+-ions is involved.

The process is highly sensitive to pH being greatly reduced below about pH 5 (Fig. 8.4). The process of accumulation is clearly dependent upon the

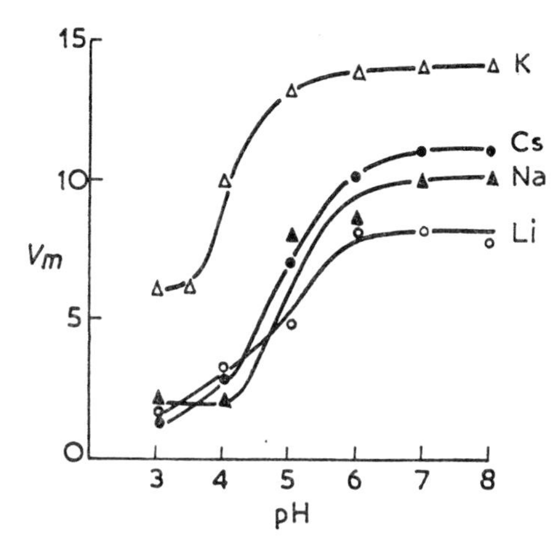

Fig. 8.4. The effect of pH on the maximal rate of transport (V_m) of alkali metal cations by yeast. (From Rothstein)

exchange of H^+ from the cell, this leads to internal accumulations of organic anions such as succinate, acetate and bicarbonate produced by the yeast's metabolism. Thus the pH of the cell rises, albeit slowly, from 5·8–6·2 (CONWAY and DOWNEY, 1950) and this tends to reduce K^+ uptake and so limits inward transport. It is also affected by the presence of other ions, by temperature, by inhibitors like sodium azide, cyanide or mercury salts and methylene blue.

Tested separately all the alkali cations, sodium, rubidium, caesium and lithium, show a similar transport pattern to that of potassium (CONWAY and DUGGAN, 1958; ROTHSTEIN, 1964) but with far lower relative affinities, viz K:Rb:Cs:Na:Li in proportions 100:42:7:3·8:0·5. Combinations of ions, for example sodium and potassium, show competitive inhibition, following the Michaelis–Menten equation, according to their relative affinities (Fig. 8.5). Since the affinity for K^+ is so much higher at all pH levels from 3–8 than that for other alkali ions, it is clear that the cell is likely to selectively favour K^+ accumulation under most conditions. Moreover, the affinities of different cations in relation to efflux enhance this effect. ROTHSTEIN and BRUCE (1958b) have shown that the influx process has a high affinity for K^+- and H^+-ions but a low one for Na^+-ions; on the other hand, the outward transport process differs and here both H^+- and Na^+-ions have a higher affinity than K^+-ions. Thus, although the yeast cells were not in steady-state conditions, as in

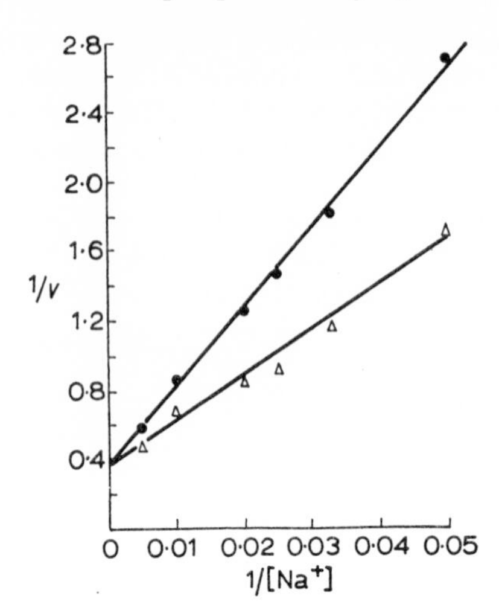

Fig. 8.5. Competitive inhibition between Na^+ and K^+ by fermenting yeast. *V* is the rate of uptake of Na^+ ions. Symbols: △, sodium acetate only; ● sodium acetate + potassium acetate. (From Conway and Duggan)

Rothstein's experiments, it has been shown that Na^+-ions are preferentially transported from yeasts with high internal concentrations of Na^+ ('sodium'—yeast, CONWAY *et al.*, 1954).

The Q_{10} for the efflux of K^+-ions lies between 3 (5–15°) and 2 (15–25°); there are no reports for other alkali ions. Sodium azide and dinitrophenol (DNP) at concentrations *c.* 10^{-3} M increase efflux initially, especially in the presence of glucose, but later the rate of efflux becomes less. Cyanide has a similar but quantitatively smaller effect than azide or DNP. Efflux of Na^+ (with succinate) is sensitive to cyanide but not azide. The effects of methylene blue and mercury ions are both more dramatic and specific. Both lead to dramatic increases in the rate of efflux and they appear to be bound irreversibly to the yeast cell. In the case of mercuric chloride the increased efflux is prevented by cysteine if present from the outset but not if it is added after the mercury salt. This has led to the suggestion that sulphydryl groups are implicated in determining the permeability of the cell membrane to cations (PASSOW *et al.*, 1959; PASSOW and ROTHSTEIN, 1960). This postulated effect is likely to be indirect since other metals, e.g. lead and zinc, will react with sulphydryl groups yet the action of mercury is quite specific in this case. Thus Rothstein's suggestion that these agents affect the redox potential at the yeast cell membrane is plausible and could be related to the observed effects of other factors such as pH or metabolism and the consequences of metabolic inhibitors. This explanation is also inadequate, however useful it may be as a working hypothesis. For example, anaerobiosis decreases the rate of efflux of potassium from yeast cells but efflux is increased by azide or cyanide which induce anaerobiosis within the cell.

Apart from studies with yeast, transport of monovalent cations has only been studied in mycorrhizal roots of beech (*Fagus sylvatica*) although rubidium has been shown to bind competitively against potassium on the cell surface of *Neurospora* (LESTER and HECHTER, 1958). It is not known if absorption and loss of K^+-, Na^+- and Rb^+-ions by intact mycorrhizas can be equated to uptake by the fungal tissue alone. HARLEY and WILSON (1959) noted that some 60% of the ^{86}Rb supplied was initially accumulated in the sheath. They pointed out that the fungal sheath behaved in the same way to rubidium as it did to phosphate, i.e. it accumulated Rb^+ more rapidly than core tissue but, eventually, to a less pronounced degree. If it is assumed that uptake by intact mycorrhizas reflect the uptake characteristics of the fungal tissue a number of general similarities are found to the process in yeast. Transport, especially loss of K^+-ions, was temperature sensitive (being greatly reduced above 20°C) and inhibited by low oxygen concentrations, particularly below 5%, as well as by anaerobiosis (Fig. 8.6), sodium azide and DNP. No

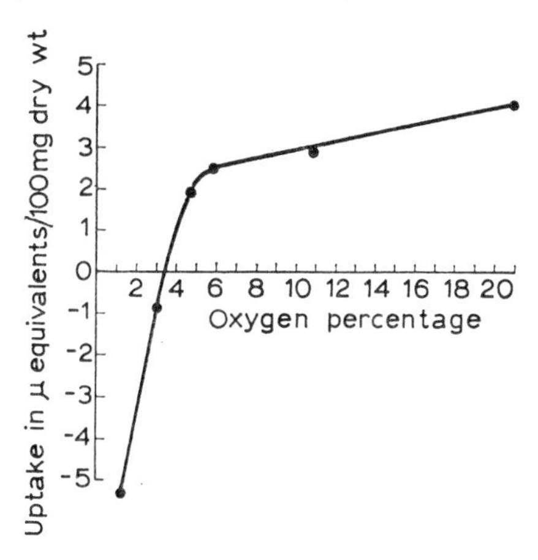

Fig. 8.6. The uptake of phosphate by mycorrhizal roots of beech in relation to oxygen concentration of the medium. (From Harley and Wilson)

measurements were made, however, on rates of influx or efflux so that exact comparisons with the work on yeast cannot be made. Potassium was most effectively absorbed from mixtures of cations (Table 8.5).

This work bears out the general observation that cation transport is associated with metabolic activity and suggests some form of active uptake.

BIVALENT CATIONS

A number of bivalent cations can be shown to be reversibly bound to specific sites at the yeast surface without actually entering the cells. A notable example is the uranyl ion, UO_2^{2+}, and two kinds of binding sites can be distinguished; one, polyphosphate-like, becomes saturated at lower concentrations of UO_2^{2+} than the other, apparently carboxyl in nature (ROTHSTEIN and MEIER, 1951). Later, ROTHSTEIN and HAYES (1956) showed that the same two kinds of site bind Mn^{2+} and other bivalent cations (Fig. 8.7). Use has been made of

Table 8.5. Absorption of K, Rb and Na from single salt solutions or from mixtures of their chlorides by beech mycorrhizas. (Data of J. M. Wilson, from Harley, 1959)

(**a**) Uptake as μg equiv/100 mg dry mycorrhizas/hr at 15°C

	KCl or RbCl 0·1 mM	Both 0·1 + 0·1 mM	KCl or RbCl 0·2 mM
K	1·50	0·82	1·78
Rb	1·61	0·44	2·01
Total Rb + K	—	1·46	—

(**b**) Uptake as μg equiv/100 mg dry mycorrhizas/24 hr at 15°C. All salts 0·1 mM

	K absorbed		Na absorbed	
	Uptake	% Control	Uptake	% Control
KCl	1·68	100	—	—
NaCl	—	—	0·41	100
KCl + NaCl	1·37	82	0·06	14

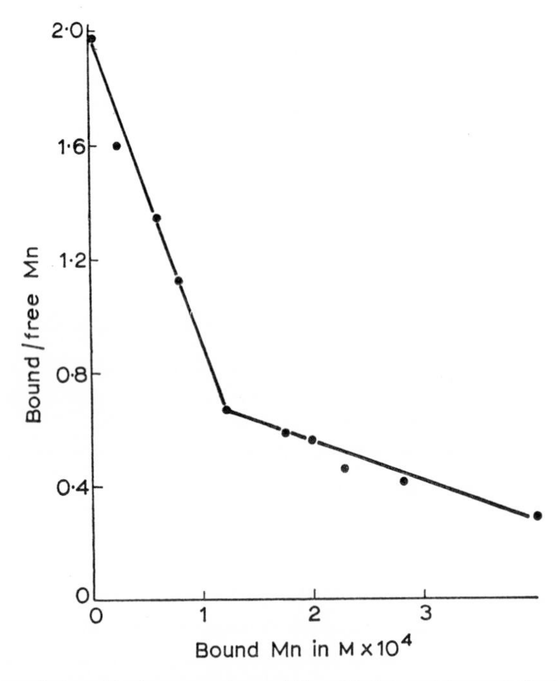

Fig. 8.7. The relationship between bound/free Mn^{2+} and bound Mn^{2+} to show that the binding process probably involves two sites. (From Jennings, after Rothstein and Hayes)

non-transportable ions such as UO_2^{2+} in studying the transport of other substances (see p. 223) but attention will now be concentrated on those bivalent cations that can enter the cell.

Conway and his associates (CONWAY and DUGGAN, 1958; CONWAY and BEARY, 1958) showed that Mg^{2+}-ions could apparently be transported by the same system, in yeast, responsible for transporting alkali ions. The relative affinity was similar to that of Li^+-ions, i.e. 0·5 (cf. K^+—100) and an exchange of Mg^{2+}- and H^+-ions was involved. There were, however, differences. Magnesium transport was sensitive to anaerobiosis, unlike K^+ transport, and DNP only inhibited Mg^{2+} transport when it was available as $MgCl_2$. Thus the associated anion affected uptake and Mg^{2+} uptake from magnesium acetate, for example, was not inhibited by DNP. Calcium was also shown to behave in a similar manner (CONWAY and DUGGAN, 1956, 1958) although a detailed study was not made. The important feature of the work of Conway and his colleagues was that there seemed to be a common cation carrier in, or at, the cell surface of yeast cells. Thus, depending upon their relative affinities and external concentrations, it might be expected that there could be interference of the uptake of one cation by any other. This, indeed, occurs although it has not received much attention in fungi.

An entirely different kind of transport system has been described by ROTHSTEIN *et al.* (1958) and JENNINGS *et al.* (1958) to account for the transport of bivalent cations. In this system only uptake is involved and there is a distinct order of affinity, thus Mg^{2+}, Mn^{2+}, Ca^{2+} and Si^{2+} compete, Mg^{2+} having the highest affinity. If yeast cells are exposed to phosphate and glucose they absorb the former, especially if K^+ is present also. Such cells can be washed and returned to an Mn^{2+}-containing solution which is free of phosphate and K^+; Mn^{2+}-ions are then rapidly absorbed. The maintenance of this transport system, which is highly specific, is contingent upon the continued availability of glucose and phosphate, which is taken up simultaneously. It is also affected to some extent by the potassium concentration of the cells; low PO_4^{3-} and high K^+ are associated with low absorption but high PO_4^{3-} and low K^+ are most effective in promoting transport. It was suggested that a carrier was involved in this uptake process; probably a phosphorylated compound formed reversibly as a direct product of the PO_4-transport system. This meets the observation that the uptake of both Mn^{2+} and PO_4^{3-} are inhibited in an analogous manner by arsenate, redox dyes and sodium acetate. No further information is available concerning this system in yeast or other fungi and it is said not to have been described in other organisms (JENNINGS, 1963).

Anion transport

PHOSPHATE AND ARSENATE

Phosphate is intimately involved in many metabolic processes which sometimes makes it difficult to distinguish its transport from its utilization. Two kinds of fungal material have been investigated in some detail, yeasts and the mycorrhizal roots of beech. In many features the transport processes in these very dissimilar structures are similar.

Phosphate transport does not appear to be possible without the provision of metabolic energy by the fungus. Thus, in yeast, uptake is effective in the presence of fermentable sugars and in aerobic and anaerobic conditions (MULLINS, 1942; GOODMAN and ROTHSTEIN, 1957) but it is greatly reduced in the presence of sodium azide even though glycolysis is unaffected (KAMEN and SPIEGELMAN, 1948). In beech mycorrhiza phosphate is absorbed in the absence of exogenous sugars but is reduced by low O_2 tensions or the presence of azide. Indeed, phosphate may actually leak out in either of these last two conditions (HARLEY *et al.*, 1953, 1956, 1958; HARLEY and BRIERLEY, 1954). The uptake process is also highly sensitive to temperature.

Phosphate could be transported as either the monovalent, the bivalent or as the trivalent ion, $H_2PO_4^-$, HPO_4^{2-} or PO_4^{3-}. Uptake is highly pH sensitive, decreasing dramatically above about pH 6·0, i.e. in the range in which phosphoric acid is shifting from a predominantly univalent to a bivalent form; at higher pH values it becomes trivalent. Thus it seems probable that phosphate is mainly taken up as $H_2PO_4^-$, both by yeast and beech mycorrhiza (GOODMAN and ROTHSTEIN, 1957; HARLEY and MCCREADY, 1950).

Once within the yeast cell it was supposed by GOODMAN and ROTHSTEIN (1957) that phosphate would become bound since it did not greatly increase the inorganic orthophosphate of the cell. HARLEY and LOUGHMAN (1963) and JENNINGS (1964a) have shown that this is not wholly true for beech mycorrhiza. Here there is evidence that phosphate below 10^{-3} M in the external solution equilibrates with a small 'pool' of orthophosphate which is associated with incorporation into a bound form, i.e. it is mostly bound. However, above 10^{-3} M phosphate, orthophosphate is transferred increasingly from the 'small' orthophosphate 'pool' into a large 'pool' (Fig. 8.8). The nature and location of these 'pools' is unclear but they evidently represent spatially separated intracellular compartments of some sort. These ideas may be summarized as

$$\text{(i)}\quad <10^{-3}\,\text{M}\ H_2PO_4^- \xrightarrow[\text{Transport}]{\text{Primary}} \begin{matrix}\text{Small}\\ \text{orthophosphate}\\ \text{pool}\end{matrix} \rightleftharpoons \begin{matrix}\text{Bound}\\ \text{phosphate}\end{matrix}$$

$$\text{(ii)}\quad >10^{-3}\,\text{M}\ H_2PO_4^- \xrightarrow[\text{Transport}]{\text{Primary}} \begin{matrix}\text{Small}\\ \text{orthophosphate}\\ \text{pool}\end{matrix} \rightleftharpoons \begin{matrix}\text{Bound}\\ \text{phosphate}\end{matrix}$$

$$\begin{matrix}\Updownarrow\\ \text{Large orthophosphate pool}\\ \Updownarrow\\ \text{Other P-compounds}\end{matrix}$$

The amount of phosphate transported is affected by a number of factors. There is little competitive inhibition by other anions in the case of yeast, even in the presence of exogenous respirable substrates such as glucose (ROTHSTEIN,

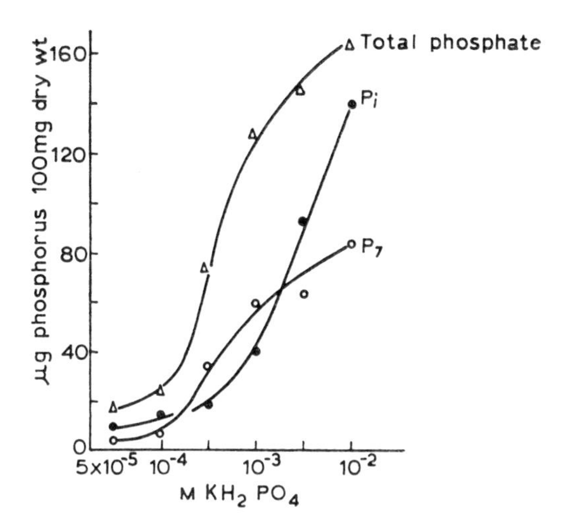

Fig. 8.8. Increase in phosphorus content of P_i and P_7 in excised tips of mycorrhizal roots of beech after exposing them to a range of phosphate concentrations for 12 hr at 23°C. P_i = orthophosphate; P_7 = compounds hydrolysed by 2N HCl in 7 min at 110°C. There was no change in non-hydrolysed compounds (P_s) or acid-insoluble phosphate. (From Jennings)

1955). The potassium content of the cell is, however, an important factor and a K^+-rich cell can absorb far more phosphate than a cell low in K^+-ions. This is thought by ROTHSTEIN (1960) to be due to the fact that the buffer capacity of the cell is greater in K^+-rich cells. Thus K^+-ions tend to shift the internal pH to the alkaline side but phosphate ions tend to the opposite extreme, for $H_2PO_4^-$ acquisition is balanced by loss of cellular OH^--ions. JENNINGS (1964) has described quite a different situation. Pretreatment with 10^{-2} M to 10^{-1} M salts of NH_4^+, K^+ or Na^+, other than phosphate, inhibits subsequent phosphate absorption by beech mycorrhizas. The effect seems to be on cell permeability which leads to an efflux of a variety of phosphorus compounds. For example, it can be prevented by the presence of Ca^{2+}-ions which are thought to be involved in ionic leakages between COO^- groups in the phosphatidic acids of the cytoplasmic membrane (Table 8.6). Jennings has pointed out that this phenomenon will affect the interpretation of data on the uptake of phosphates from salts with monovalent cations, e.g. KH_2PO_4 in the range 10^{-2} M to 10^{-1} M.

While, therefore, there is some apparent non-specific efflux under these conditions, it is still important to remember that the influx of phosphate is a specific process associated in some way with a requirement for metabolic energy.

The only other anion to which much attention has been paid is arsenate. This is taken up in yeast by a mechanism similar to that for phosphate, with which it competes. The transport system is progressively and irreversibly

Table 8.6. Effect of presence of $CaCl_2$ on loss of ^{32}P from mycorrhizal roots of beech in KCl. Roots supplied with 5×10^{-5} M phosphate (1–20 μCi ^{32}P/ml.) for 5 min at 25°C then to 10^{-1} M KCl for 30 min at 25°C with, or without, 10^{-3} M $CaCl_2$. (From Jennings, 1964b)

Treatment	Loss of radioactivity as % total radioactivity initially present before treatment (replicates)
None	1·7:1·8
KCl	2·8:2·7
KCl + $CaCl_2$	0·9:0·8

inhibited due to a direct effect of the arsenate on the transport system (ROTHSTEIN, 1963).

INORGANIC NITROGEN COMPOUNDS

The most important compounds to be considered here are 'ammonia' (used to include the ion or the undissociated molecule) and nitrate. Attention has already been drawn to the preferences, facultative or obligatory for 'ammonia' compared with nitrate in many species or groups of fungi (Chapter 7, p. 191). Considerable attention has been paid to the interaction between these two nitrogen sources and to their assimilation within the cell but rather less to their actual mode of transport.

In general 'ammonia' is transported preferentially to nitrate, this is especially striking when ammonium nitrate is the nitrogen source available. The significant observations on these processes have been made by MacMillan and Morton (MORTON and MACMILLAN, 1954; MACMILLAN, 1956a, b) on a variety of mucoraceous and imperfect fungi but particularly in *Scopulariopsis brevicaulis*.

Mycelium was grown in a synthetic medium containing potassium nitrate and glucose and transferred to a buffered medium (pH 7·0) containing ammonium sulphate and glucose. Ammonium was lost from the medium and free ammonia rose rapidly to a constant level within the mycelium. The uptake ceased after about one hour in the absence of glucose but continued unabated in its presence. If the mycelium was transferred to ammonium-free buffer there was a rapid efflux (Fig. 8.9a). The cells are evidently freely permeable to 'ammonia'. This is a finding well documented for yeast (ÄYRÄPÄÄ, 1950) and *Neocosmospora vasinfecta* (BUDD and HARLEY, 1962a, b). MACMILLAN (1956a) showed that the internal concentration of 'ammonia', over a range of about 1·0–4·0 mg 'ammonia' N/ml. in the external solution, was about the same or slightly higher. Moreover, the rate of entry increases with concentration (Fig. 8.9b). It seems reasonable to conclude therefore, that uptake is due to passive diffusion. Finally, it was shown that 'ammonia' was accumulated from a solution of constant concentration of 'ammonia' more rapidly above pH 7 and up to pH 8·5–9·0 than below. Since pH values above pH 7 increase the proportion of undissociated ammonia in the external solution, it is probable that it entered in this form. One way of testing this is to estimate the internal pH of the cells on the assumption that they contain the same concentration of undissociated molecules as does the external solution whose pH is being varied. A comparison can then be made between observed and expected changes in the internal pH. The measurement of the latter is exceedingly difficult and can only provide an averaged value even though particular localized regions may show large differences. The comparison was, however, reasonably in agreement with the assumption (Fig. 8.9c, d).

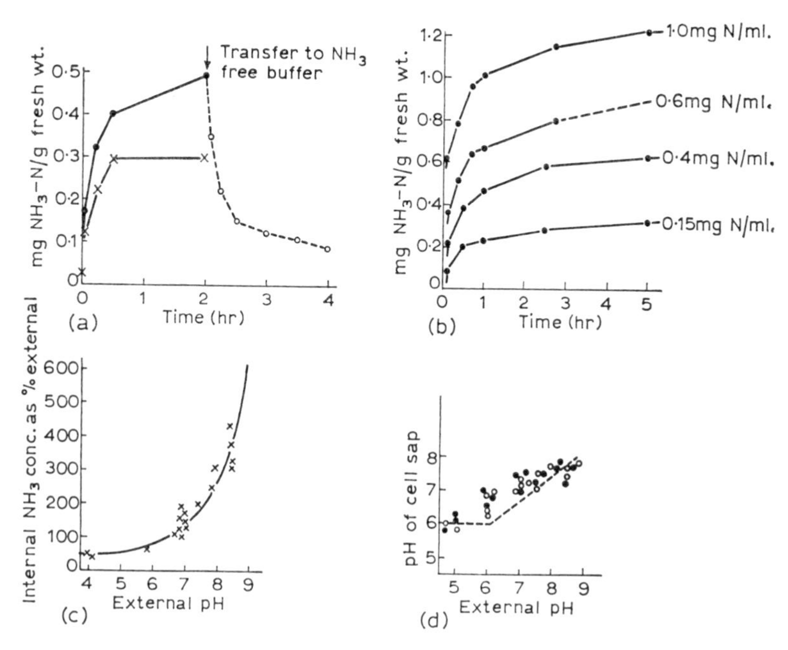

Fig. 8.9. Effects on 'ammonia' uptake in relation to presence or absence of glucose, concentration of ammonium sulphate and pH in *Scopulariopsis brevicaulis.* (a) 'Ammonia' in the mycelium shaken in phosphate buffer at pH7 with $(NH_4)_2SO_4$ (0·4 mg N/ml.) without glucose —●—, or with 1 % glucose —×—, or on transfer to ammonia-free buffer —○—; (b) effect of different concentrations of $(NH_4)_2SO_4$ on mycelial 'ammonia'; (c) the effect of standing mycelium in buffered solutions of different pH on the internal ammonia concentration; (d) as in (c) to show the effect on the pH of expressed cell sap with ammonia supplied, ●, or without, ○. The dotted line shows the calculated internal pH. (From MacMillan)

In all cases 'ammonia' once within the mycelium was rapidly assimilated by metabolic activities into organic, nitrogen-containing compounds. Further information has been obtained in *Neocosmospora*, however, about the non-assimilated 'ammonia' pool in the mycelium. MacMillan believed that this was not affected by application of respiratory inhibitors in *Scopulariopsis* but this is not so in *Neocosmospora*, although the sensitivity of the processes involved is much less than that of nitrogen assimilation (Table 8.7). The 'ammonia' in the mycelium can be separated into two components, the 'exchangeable' and the 'residual'.

Table 8.7. The influence of sodium azide, 2,4-dinitrophenol and anaerobiosis on 'ammonia' uptake (NH_3) and increase in nitrogen fractions (organic N) in *Neocosmospora vasinfecta*. (From Budd and Harley, 1962a)

Medium: 28 mM NH_4Cl + 2% glucose buffered at pH 6·2 at 25°C. Data as % of uninhibited control.

Treatment	Increase of N in	Time (min) 5	10	32	60
10^{-3} M NaN_3	Organic N	43·1	20·1	49·8	25·2
	Mycelial NH_3	71·1	61·5	61·5	73·8
	Total uptake	58·0	48·8	44·8	32·3
10^{-4} M 2,4-DNP	Organic N	61·5	70·2	76·3	75·4
	Mycelial NH_3	81·3	80·0	82·0	86·2
	Total uptake	68·1	73·4	77·3	76·6
Anaerobiosis	Organic N	18·8	62·7	44·2	31·9
	Mycelial NH_3	69·8	60·9	62·7	67·5
	Total uptake	32·1	62·0	49·0	38·0

'Exchangeable ammonia' can be extracted by shaking mycelium with 10^{-3} M solutions of alkali chlorides, especially KCl; 'residual ammonia' is not extracted by this treatment. 'Exchangeable ammonia' is itself made up of two components for, while much of it can be extracted by washing with distilled water, all of it is only obtained by KCl extraction, with exchange of 'ammonia' for K^+ ions. These components are perhaps associated with a simple aqueous phase comparable with that described earlier as 'free space' (p. 204) and with exchange sites on the wall or possibly the cytoplasm (cf. pp. 200, 213).

Increase in 'residual ammonia' is markedly affected by the external pH and by respiratory inhibitors or anaerobiosis but it is not clear where it is sited in the cell. It can also be metabolized on incubation, so that it is fairly readily available to the assimilatory machinery of the cell. These observations suggest that 'residual ammonia' must be located near to the cell surface and 'is held in the cytoplasm, either by diffusion-resistant membrane(s) or by binding to specific sites' (BUDD and HARLEY, 1962b). A decision cannot yet be made and the role of respiratory energy in the process is unclear but either of the suggestions made could involve expenditure of energy for maintenance of a membrane system or binding sites.

‘Ammonia’ transport, therefore, appears to involve a number of stages as set out below:

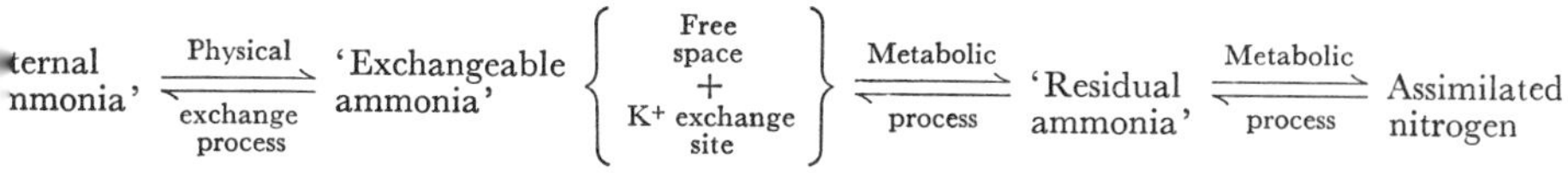

ternal nmonia’ $\underset{\text{exchange process}}{\overset{\text{Physical}}{\rightleftharpoons}}$ ‘Exchangeable ammonia’ $\left\{\begin{matrix}\text{Free space} \\ + \\ \text{K}^+ \text{ exchange site}\end{matrix}\right\}$ $\underset{\text{process}}{\overset{\text{Metabolic}}{\rightleftharpoons}}$ ‘Residual ammonia’ $\underset{\text{process}}{\overset{\text{Metabolic}}{\rightleftharpoons}}$ Assimilated nitrogen

Nitrate transport has received less attention but uptake of NO_3^--ions from KNO_3 in buffer with and without external glucose has been studied in *S. brevicaulis* by MACMILLAN (1956b). In the absence of glucose there is a primary phase of uptake which is insensitive to cyanide followed by a second phase sensitive both to cyanide and azide. In the first 3–5 hr little free NO_3^- can be detected in expressed cell sap; it is, therefore, postulated that it is bound, perhaps by adsorption. Once the sites are saturated, free NO_3^--ions accumulate. A good deal, although not all of the NO_3^- so accumulated, can be lost from the cell by washing in sodium/potassium phosphate buffer. It is not clear whether this is a phenomenon comparable to that just described in the case of exchangeable ‘ammonia’ and KCl. Subsequent accumulation from KNO_3 solution is evidently dependent upon the energy released by endogenous respiration since it is markedly depressed by respiratory inhibitors. These two phases appear to be superficially comparable to those of the acquisition of exchangeable and residual ‘ammonia’ already discussed but there is no evidence to suggest that they have any more fundamental identity, whether of mechanism or of site.

In the presence of glucose the situation is quite different, nitrate is assimilated as soon as it enters the cell and the rate of entry is 10–12 times more rapid (1 mg N/g Fresh wt./12 hr without glucose; 6 mg N/g Fresh wt./2 hr with glucose). The NO_3^- in the mycelium remains approximately constant with or without external glucose present for the first 4–5 hr and remains more or less at this level, thereafter, in the presence of glucose. This situation is also superficially comparable to that which occurs with ammonia.

Attention has also been paid to transport from external ammonium nitrate solutions in *S. brevicaulis* (MORTON and MACMILLAN, 1954; MACMILLAN, 1956b) and other fungi in the presence and absence of glucose. In the presence of glucose these fungi behave as if only ‘ammonia’ were available for a considerable time. Once, however, the concentration of ‘ammonia’ falls below about 0·002 N then nitrate assimilation commences quite suddenly and proceeds simultaneously with ammonia assimilation at approximately the same rate. Tests at various concentrations of nitrate and ammonia of the same order of magnitude as are found in natural soils suggested that this same sequence was likely to occur in a natural habitat. On the other hand, the accumulation of NO_3^--ions is unaffected by the presence of external ‘ammonia’ in the absence of glucose. This suggests that the inhibition of nitrate by ‘ammonia’ is a phenomenon associated with assimilation rather than transport.

The uptake of nitrite which can be carried out by many fungi without a prior period of adaptation, e.g. *Fusarium niveum*, *S. brevicaulis*, also has a bearing on this problem. Ammonium nitrite (made by mixing $(NH_4)_2SO_4$ and KNO_2 and sterilized by Seitz filtration) supports growth of *S. brevicaulis* without a dramatic drop in pH, an indication that both 'NO_2' and 'ammonia' are taken up equally. It was confirmed that both were utilized at about the same rate. Thus it seems probable that the inhibition of NO_3^- uptake by 'ammonia' is not a competitive inhibition since there is no such interaction with NO_2^--ions. The nature of the ammonia/NO_3^- inhibition phenomenon is still obscure and needs to be investigated further.

TRANSPORT OF NON-ELECTROLYTES

Sugars

MONOSACCHARIDES

Many sugars are readily metabolized by cells so that, as with other highly reactive compounds, it is often difficult to distinguish transport from assimilation. Nevertheless, a number of sugars are known which are either not utilized at all by fungal cells, or only after a period of adaptation, e.g. L-sorbose and L-galactose respectively so that measurements can be made before assimilation has advanced very far.

BURGER *et al.* (1959) studied the transport of L-galactose and L-arabinose by yeast cells. These sugars entered the cells but did not accumulate in excess of the external concentration, with which the intracellular concentration showed a linear relationship. When glucose was supplied it could not be detected free within the cell unless the respiratory inhibitor, 1 mM iodoacetate, was added. The combined effects of galactose (or arabinose), glucose and

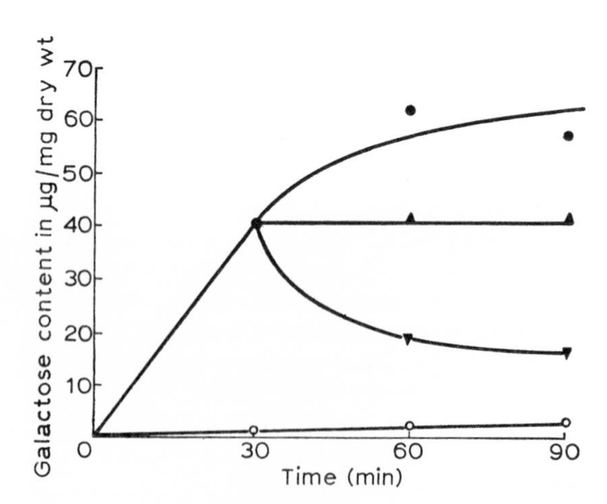

Fig. 8.10. The effect of glucose and iodoacetate on uptake of 5% galactose by yeast at 30°C. Symbols: ●, control; ▲, glucose and iodoacetate and ▼, glucose only, added after 30 min; ○, glucose and galactose added at zero time. (From Burger *et al.*)

iodoacetate are shown in Fig. 8.10. Glucose supplied simultaneously with galactose inhibits accumulation of the latter; when added after a period of accumulation it results in a loss of intracellular galactose. However, glucose + iodoacetate added after galactose accumulation leads to a stabilization of the galactose content at whatever value it has reached. Arabinose gave the same result as galactose. Glucose seems to inhibit transport, whether influx or efflux, of these sugars if it is present as the free molecule with either of them in the external or internal medium. From these experiments galactose and arabinose do not seem to be transported by an active mechanism. CIRILLO (1961b) has provided further evidence that L-sorbose, L-galactose, D-xylose, D-arabinose, L-fucose, L-arabinose and D-ribose were transported in a similar manner by yeast in that order of decreasing efficiency. Metabolic energy was not involved, sugars showed competitive inhibition and the system could be inhibited by UO_2^{2+}-ions. This last observation was of particular importance for UO_2^{2+}-ions were already known to inhibit the uptake of fermentable sugars, such as glucose and fructose by yeast cells (ROTHSTEIN, 1954). Exposed to 2×10^{-5} M uranyl ions, some 1×10^{-3} M UO_2^{2+}/l. of cells is bound and glucose uptake is inhibited about 90%. The bound ions have no effects on intracellular metabolism and they seem to be combined with sites on, or very near the surface of the yeast cell. For example, although the internal concentration of orthophosphate is 10^{-2} M, addition of as little as 2×10^{-4} M orthophosphate reduces the inhibition by two thirds, i.e. glucose uptake is only 30% inhibited (ROTHSTEIN, MEIER and HURWITZ, 1951). Inhibition is directly proportional to numbers of bound sites and a cell is 100% inhibited by taking up $4{\cdot}6 \times 10^7$ molecules; sites so detected are claimed to be polyphosphate in nature (ROTHSTEIN *et al.*, 1948; ROTHSTEIN and MEIER, 1951). In the presence of even higher exogenous concentrations of UO_2^{2+}-ions a further 3×10^7 sites per cell are found to bind the ions but in a relatively unstable manner. These sites appear to be carboxyl. The two classes of sites perhaps amount to some 2% of the yeast cell surface and are the same ones which bind cations (ROTHSTEIN and HAYES, 1956 and cf. p. 213).

These observations suggest that transport of sugars, both those metabolized and those not metabolized, follow the same initial route. It has also led to the suggestion that sugar transport is associated with some sort of interaction with phosphate, perhaps a phosphorylation. However, CIRILLO (1962) has compared D-glucose transport with that of L-sorbose under conditions which inhibit both phosphorylation and glucose metabolism. Experiments were done under N_2 plus 10^{-3} M iodoacetic acid and both sugars were taken up. There was an initial rapid phase of glucose uptake which became slower and was eliminated at 6°C. Inhibition of sorbose transport by exogenous glucose occurred and by studying a range of concentrations it appeared that the affinity for glucose compared with sorbose was of the order of 25:1. The kinetics of transport were studied at external concentrations of 50 mM and 500 mM for both sugars. At the lower concentration more glucose is transported than sorbose but the relationship is inverted at the higher concentrations. This rather puzzling observation is accounted for by a theory of carrier transport developed originally from studies with mammalian blood cells (see p. 206). In this theory:

at low concentrations $T \propto 1/K_t$

and at high concentrations $T \propto K_t$

where T = rate of transport; K_t = the dissociation constant of the hypothetical carrier-substrate complex; $1/K_t$ = is the affinity.

Thus at low concentration glucose is transported more rapidly than sorbose because of its higher affinity and vice versa at high concentrations.

Another phenomenon shown is 'counterflow', which occurs when an exchange reaction occurs between an internal sugar and an external sugar, the efflux of the former being against a concentration gradient. This was shown to occur when yeast cells in equilibrium with external sorbose were exposed to glucose, or between the two metabolically active sugars glucose and fructose which have a similar affinity ($1/K_t$) (CIRILLO, 1961b; 1962). Cells were equilibrated with fructose under N_2 plus inhibitor and then exposed to 250 mM glucose+250 mM fructose mixture. The uptake of glucose was compared with that taken into yeast cells which had not previously been exposed to fructose (Fig. 8.11). In equilibrated cells glucose transport was accelerated

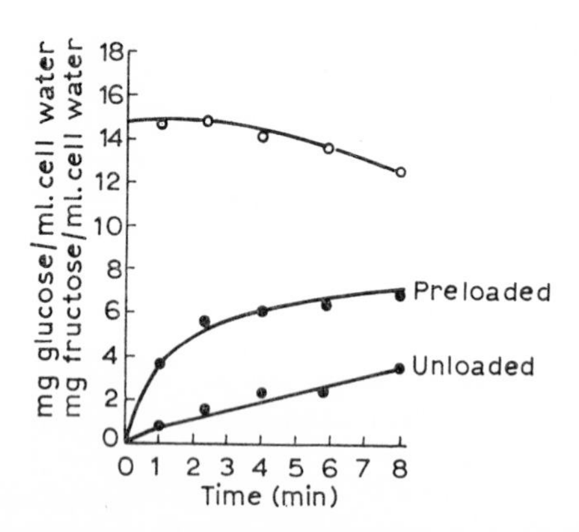

Fig. 8.11. Transport of glucose (●) into yeast cells either pretreated with fructose (preloaded) or not (unloaded) and fructose counterflow (○) out of cells. (From Cirillo)

compared with non-equilibrated cells and there was also an efflux of fructose. The internal fructose/glucose concentration was greater than unity, whereas the external ratio was unity. As glucose transport was accelerated influx of fructose was, therefore, greatly inhibited and so counterflow of fructose occurred from within the cell even though it was against a concentration gradient.

Similar kinds of observations have been made in other fungi. Uranyl ions are known to inhibit sugar uptake by *Neurospora* (COCHRANE and TULL, 1958); competitive inhibition is common in a number of fungi between various sugars, e.g. glucose and fructose in *Chaetomium globosum* and *Tricholoma nudum* (WALSH and HARLEY, 1962; REUSSER *et al.*, 1960); glucose and ribose in *Rhodotorula* (KLEINZELLER *et al.*, 1952); and mannose, glucose and xylose, in that order by the Basidiomycete, *T. nudum* (CIRILLO and GRAYSON, in CIRILLO, 1961a).

DISACCHARIDES AND POLYSACCHARIDES

Sucrose and maltose are utilized by many fungi. In some fungi sucrose is converted into the hexose moieties by enzymes on or near the cell surface (see p. 242) and then glucose or both hexoses are taken up, e.g. *Chaetomium* and yeast respectively; in other cases the sucrose is apparently transported intact, e.g. *Myrothecium verrucaria* spores (MANDELS, 1951).

Maltose uptake has been studied by ROBERTSON and HALVORSON (1957) in yeast and it can be transported intact or after cleavage. It has been claimed that transport requires metabolic energy since it is inhibited by 5×10^{-3} M DNP and 10^{-2} M NaN_2 or NaF. The DNP concentration used is high for uncoupling phosphorylation alone and CIRILLO (1961a) queries whether the inhibition may not have been due to a toxic effect rather than to inhibition of the supply of respiratory energy.

Sugars and polysaccharides larger than disaccharides are almost certainly hydrolysed to their component monosaccharides before transport takes place. This takes place at or near the cell surface (cf. pp. 201, 242) or by the production of extracellular enzymes. This topic will be considered at the end of this chapter.

Amino acids

Although the amino acid nutrition of fungi has received great attention very little study has been made of their transport into fungal cells. Most studies have been made on two yeasts, *Candida (Torulopsis) utilis* by Cowie and his associates (summarized in COWIE, 1962; HALVORSON and COWIE, 1961) and *Saccharomyces cerevisiae* (HALVORSON, FRY and SCHWEMM, 1955; HALVORSON and COHEN, 1958; SURDIN *et al.*, 1965). More limited studies have been made with the filamentous fungi, *Botrytis fabae* (JONES and WATSON, 1962; JONES, 1963) and *Neurospora crassa* (MATHIESON and CATCHESIDE, 1955; ZALOKAR, 1961; DE BUSK and DE BUSK, 1965; WILEY and MATCHETT, 1966).

Cowie and his colleagues showed that free amino acids occurred within *Candida* cells grown in steady-state conditions in two pools, the 'expandable pool' and the 'internal pool'; the former is associated with transport. The amino acid composition of the 'internal pool' is similar to that of cellular proteins. That of the 'expandable pool' is highly variable and depends upon the available, exogenous amino acids, their external concentration and the presence of other exogenous materials. Considerable but not complete exchange can readily be effected by osmotic shock or by the presence of external amino acids. For example when [^{14}C]-threonine was supplied to *Candida* it was absorbed but on transfer to water almost 50% of the threonine taken up was lost extremely rapidly. In another experiment [^{14}C]-arginine was supplied until it represented 57% of the 'expandable pool's' amino acids at equilibrium. Cells were then transferred to media containing either [^{12}C]-arginine or no arginine at all. In the first case there was a gradual loss of [^{14}C]-arginine to the medium, the 'internal pool' and to protein. In the second case there was a rapid decrease in the arginine content of the 'expandable pool' and a considerable efflux into the medium. After a time this decrease was halted and reversed somewhat, presumably when the external concentration exceeded that of the internal 'expandable pool' (COWIE and MCCLURE, 1959).

The rapid exchange of amino acids demonstrated by these experiments is not wholly unregulated. It occurs in the presence of either exogenous or endogenous fructose, free or assimilated, and there is discrimination between stereo-isomers and analogues. For example if DL-*p*-fluorophenylalanine and L-phenylalanine are supplied exogenously in a ratio of 10:1 at concentration of 10^{-3} to 10^{-4} M, their internal concentrations at equilibrium are very different (Table 8.8).

Table 8.8. The internal concentrations of DL-*p*-fluorophenylalanine (*p*FPhe) and phenylalanine (Phe) when supplied to *Candida utilis*. *p*FPhe and Phe supplied at 10^{-3} to 10^{-4} M as either [^{12}C]-Phe + [^{14}C]-*p*FPhe or [^{14}C]-Phe + [^{12}C]-*p*FPhe to cells in exponential phase of growth. (From Kempner and Cowie, 1960)

	External	Expandable pool	Internal pool	Protein
*p*FPhe	—	23·9	14·0	725
Phe	—	4·58	6·1	305
Ratio	10:1	5·2:1	2·3:1	2·38:1

Pool values as nm moles/g F.Wt.
Protein value as nm moles/Δg F.Wt. where Δ = gross net weight of cells, i.e. mass of cells grown *after* addition of tracer materials.

The studies of the Institut Pasteur group with *S. cerevisiae* have shown up some additional features. Competitive inhibition between amino acids is not markedly related to their structure, a feature generally shown by bacteria such as *E. coli* (COHEN and RICKENBERG, 1956). However, basic α-amino acids and L-forms are accumulated preferentially over others. Many amino acids had a similar pH optimum between 5 and 6·5 and the non-specific character of their uptake suggested accumulation in a common pool. Uptake can be described by the Michaelis-Menten equation as well as competitive inhibition and some K_m values were estimated. Exchange with previously accumulated amino acids was not wholly complete, however. Respiratory energy was apparently utilized for transport as well as for assimilation. Thus it was shown that 2×10^{-4} M sodium arsenate would inhibit amino acid utilization in protein synthesis but not accumulation. On the other hand, accumulation is inhibited by $10^{-3}-5\times10^{-3}$ M dinitrophenol or sodium azide, inhibitors known to affect respiratory processes.

These observations have led the French workers to postulate a specific enzyme system termed a 'permease', a concept first developed in connexion with bacteria (COHEN and MONOD, 1957), which is responsible for transport. SURDIN *et al.* (1965) have described a single gene mutant of *S. cerevisiae* which differs from the parent strain in that its ability to transport amino acids is reduced to about one tenth (Table 8.9). Although single gene mutants are

Table 8.9. Relative rates of accumulation of amino acids under similar conditions by *S. cerevisiae* Strain 4094-B, and its mutant derivative Strain 4094-B4 which differs by one gene only. (From Surdin *et al.*, 1965)

Amino acid (L-forms)	4094-B	4094-B4
Glutamine	8	1
Aspartic	0·6	1
Threonine	6	8
Methionine	13	1·8

known in *E. coli* which are claimed to result in altered permeases, the occurrence of a single gene mutant in yeast with different transport abilities from the parent strain does not necessarily support the permease concept, however suggestive the analogy may seem.

Observations on hyphal fungi confirm and extend those described for yeasts. In *Botrytis* competitive inhibition between amino acids is known, uptake is related to the external concentration of the acid(s) and is sensitive to temperature and pH. The optimum pH ranged from *c.* 3·5 for L-lysine to 4·5 for L-glutamic acid in buffered medium but the rate did not alter from that when uptake occurred from distilled water. Age of mycelium also affected the uptake and rates became slower in both species as the mycelium aged. Unsubstituted —NH_2 and —COOH groups were necessary for uptake to occur. L-histidine uptake was highly sensitive to 10^{-4} M DNP and 10^{-2} M sodium azide which inhibited uptake by 96% and 72% respectively.

Phenylalanine and tryptophane transport by *Neurospora* conidia could be accounted for by Michaelis-Menton kinetics with a notable reduction at low temperatures. In the case of phenylalanine (DE BUSK and DE BUSK, 1965) transport but not utilization was greatly reduced by 2,4-DNP and azide and, indeed, was followed by a considerable efflux. WILEY and MATCHETT (1966) used a tryptophane-requiring mutant *td-201* as well as wild type strains. Above 40°C there was a marked decrease in uptake (Fig. 8.12) and efflux of transported tryptophane did not occur in the mutant at either 0° or 30°C, although incorporation occurred actively at the latter temperature. The most

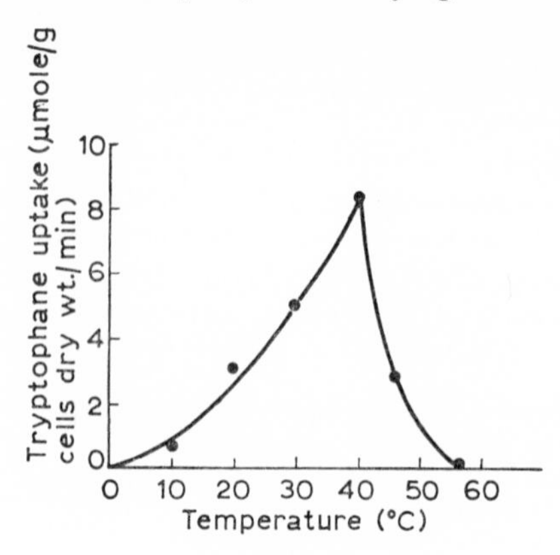

Fig. 8.12. L-tryptophane uptake by *Neurospora crassa* in relation to temperature; note sharp cut off above 40°C. (From Wiley and Matchett)

interesting observations, however, were on the specificity of the process. Steric specificity was absolute and L-leucine and L-phenylalanine showed competitive inhibition, suggesting that there was a common carrier or transport site for these amino acids (Table 8.10). This may be compared with the competitive inhibition of L-tryptophane and L-leucine to L-phenylalanine

Table 8.10. The effect of various amino acids on the uptake of 10^{-5} tryptophane added simultaneously to *Neurospora crassa* in a sucrose-mineral salts medium. (From Wiley and Matchett, 1966)

Competing amino acids	Concentration $\times 10^{-4}$ M	% inhibition initial rate of uptake
L-amino acids		
Leucine	5	90
Methionine	5	82
Ethionine	5	82
Cysteine	5	77
Phenylalanine	5	95
Tyrosine	5	72
Histidine	4	47
Serine	4	53
Isoleucine	5	11
Valine	5	9
Aspartic acid	4	0
Glutamic acid	4	0
Glycine	5	0
Lysine	4	10
Alanine	5	20
D-amino acids		
Tryptophane	5	0
Leucine	5	0

Table 8.11. The effects of various inhibiting amino acids used to investigate the minimal structural requirements for reactivity with the transport site for tryptophane by *Neurospora crassa*. (From Wiley and Matchett, 1966)

Competitor	Effect	Concentration $\times 10^{-4}$ M	Initial rate of uptake of 2×10^{-5} M tryptophane
None		—	10
Indole	Testing indole moiety	4	12
L-leucine	Effect on α-carboxyl	4	0·96
L-isoleucine	of substituting	4	9·6
L-valine	β-carbon	4	9·0
Indole pyruvic acid	Requirement for α-amino	4	12
L-(α) amino butyric acid	group	4	1·96
L-(γ) amino butyric acid		4	9·0
Tryptamine HCl	Requirement for α-carboxyl	4	12
L-glutamic acid	Effect of electrically	4	11
L-lysine	charged side chain	4	10

observed by the de Busks. Substitution in the tryptophane molecule enabled a partial list of requirements for the reactivity of the supposed carrier to be drawn up (Table 8.11). The minimal requirements are related to the side chain of tryptophane and include an α-amino group next to a carboxyl and an unchanged side chain. This high specificity suggests very strongly indeed the existence of a definite chemical carrier, possibly a 'permease', for successful transport of the tryptophane family of amino acids.

In the course of these studies it was also shown that there seemed to be a free pool of both phenylalanine and tryptophane within the hyphae (Fig. 8.13). This had been studied earlier by ZALOKAR (1961) who showed that in *Neurospora* the internally accumulated amino acids could not be freely exchanged with added, exogenous acids. Zalokar suggested that the amino acids entered a pool from which they were withdrawn for protein synthesis and then, if not so utilized, accumulated in an expendable pool.

This observation raises an important issue in connexion with amino acid transport in fungi. The situation in the yeasts can be summarized in the diagram prepared originally from *Candida* (Fig. 8.14). In both *Neurospora* and the yeasts there are clearly two pools of amino acid within the cell, one associated with protein synthesis and of similar composition and one of variable composition not showing this close relationship to protein synthesis. However, in *Neurospora*, entry to the 'expandable pool' appeared to be via the 'internal pool' but in the yeasts entry was independent. The reasons for this are not clear. There may be a real difference, of course, but it should be noted that the cells were in different conditions, the yeasts were in a steady state but not *Neurospora*, and the techniques employed differed greatly. An issue of some importance, however, is that while much is known concerning transport

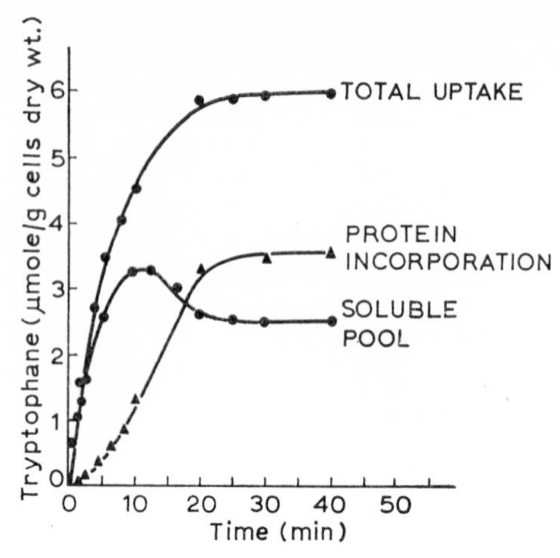

Fig. 8.13. Time-course of L-tryptophane uptake and incorporation by *N. crassa* tryptophaneless-mutant, *td-201* in presence of 20% sucrose at pH 5·8. 'Soluble pool' is the difference between total and trichloracetic acid-insoluble fraction. (From Wiley and Matchett)

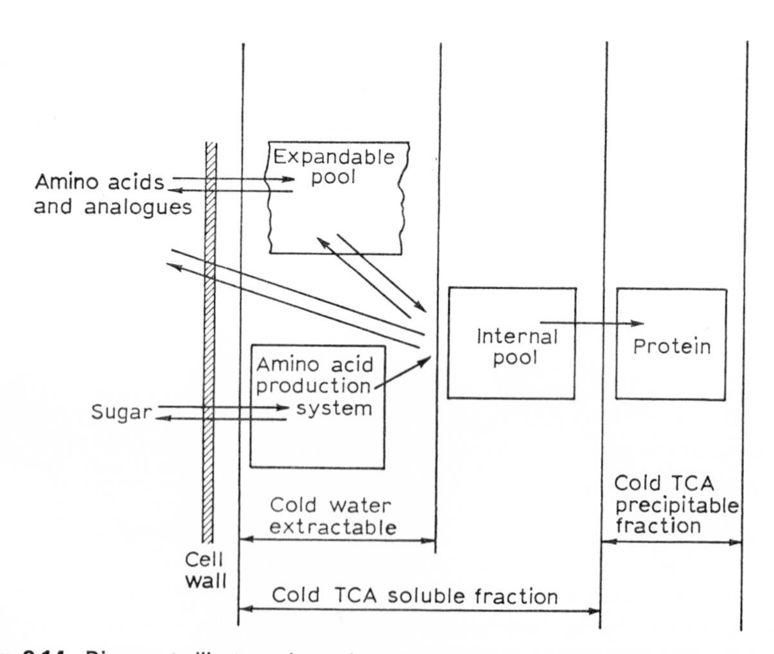

Fig. 8.14. Diagram to illustrate the notions developed by Cowie and co-workers from a study of amino acid and fructose uptake by *Candida utilis*. Note that exogenous amino acids can enter both the 'expandable pool' and the 'internal pool' directly, whereas sugar feeds directly into the 'internal pool'. (From Cowie)

into the 'expandable pool' virtually nothing is known about the mechanism of entry into the 'internal pool'. It is possible that a formulation based on that of HALVORSON and COHEN (1958) may represent the situation more correctly, viz:

$$\begin{array}{ccccccc} & & \text{Expandable} & & \text{Internal} & & \\ & & \text{pool} & & \text{pool} & & \\ \text{AA} & \rightleftharpoons & (\text{AA}-\text{X}) & \rightleftharpoons & \text{AA} & \rightleftharpoons & \text{Protein} \\ & & \upharpoonleft\downharpoonright & & & & \\ & & \text{Internal pool} & & & & \\ & & \text{AA} & & & & \end{array}$$

AA = amino acid (AA–X) = exchangeable fraction of AA

In such a situation the essential transport step is,

$$\text{AA} \rightleftharpoons (\text{AA}-\text{X})$$

and presumably this is common to all the fungi studied.

Another problem concerned with amino acid uptake is the form in which the acids are transported. They are, in general, weak electrolytes and SIMON and BEEVERS (1952) have shown that the undissociated molecule is the typical form in which weak acids and bases are usually transported. On the other hand the pH of the environment will affect the situation; for example, in acid conditions glycine will exist as $CH_2\,(NH_3)^+\,COOH$ and in alkaline conditions as $CH_2\,(NH_2)\,COO^-$, i.e. the changes are on the α-amino and carboxyl groups. Other ionizable groups in amino acids are the imidazole ring, phenolic, hydroxyl, sulphydryl and guanidino groups as in histidine, tyrosine, cysteine and arginine, respectively.

If amino acids are in the ionized form there is the possibility of exchange with other electrolytes and there is some evidence for this. For example, CONWAY and DUGGAN (1958) noted that sodium-rich yeast could exchange Na^+-ions with ornithine, arginine and lysine, all cationic amino acids, although the neutral amino acids citrulline, glutamine, phenylalanine and cysteine could also exchange in that order of decreasing affinity. They suggested that all these amino acids possibly utilized a common cation carrier. Correlated transport between alkali–metal ions and amino acids has been postulated for mammalian cells (cf. CHRISTENSEN, 1961).

Peptides, proteins and amines

Very little is known concerning the transport of larger nitrogenous moieties such as peptides or polypeptides. JONES (1963) noted that DL-leucylisoleucine and glycylglycine were accumulated rapidly but could not be detected within hyphae of *Botrytis* even with high exogenous concentrations and short accumulation periods. However, with glycylglycine an almost complete recovery of the constituent, free glycine, was achieved from the mycelium. He suggests that rapid hydrolysis of peptide bonds occurs but whether this is before or after the transport stage is unknown.

The only experiments reported on protein uptake are those referred to earlier (p. 207) in which pinocytosis was shown not to occur in yeast as studied by the fluorescent antibody technique.

Amine uptake has been observed in *Botrytis* and the rates were generally lower than those for amino acids and appeared to be related to the chain length of the amine (Table 8.12).

Table 8.12. Rates of uptake of amines compared with L-leucine by *Botrytis fabae*. Substrates at 10^{-4} M: rates of uptake μM/g/hr. (From Jones, 1963)

Substrate	Rate of uptake
Methylamine	93
Ethylamine	50
Propylamine	38
Butylamine	20
L-leucine	77

Data such as these can hardly be used as a basis for any firm generalizations.

Nucleotides and nucleic acids

There is a good deal of evidence that fungi can incorporate nucleotides, much of it derived from mutants biochemically deficient for such compounds, e.g. *Penicillium chrysogenum*, *Ophiostoma multiannulatum*, *Ustilago maydis*, etc. (BONNER, 1946; FRIES, 1947; HOLLIDAY, 1961a). However, the only detailed study of uptake in a normal fungus is that of COWIE and WALTON (1956) in *C. utilis*. They were able to show a pattern of 'internal' and 'expandable pools' similar to that already described for amino acids and the uptake showed similar characteristics.

An extraordinary mode of incorporation of high molecular weight RNA, isotopically-labelled with [^{14}C]-uracil, has recently been reported for isolated protoplasts of *Saccharomyces carlsbergensis* (BLOEMERS and KONINGSBERGER, 1967). The RNA was derived from the same species of yeast and its molecular weight was of the order of 560,000 or more. Uptake resembled a damped oscillatory process with a periodicity between peaks of about 7 min at 30°C and 15 min at 20°C. Successive peaks of uptake declined and the total amount of RNA incorporated was about 0·1–1·0% of the total supplied.

MECHANISMS FOR TRANSPORT

The observations discussed in the preceding sections are summarized in Table 8.13 and it can be seen that they include data relevant to most of the possible mechanisms for transport outlined earlier.

Table 8.13. A summary of the characteristics of transport processes in fungi

Substance	H^+ or other exchange	Kinetics	Specificity	Affinity	Affected by pH	Affected by Temperature	Affected by O_2	Affected by Exogenous sugar	Affected by Phosphate	Inhibitors	Carrier	Internal 'pool'
K^+ ions	$+H^+$	[1]M-M	–	K > Rb > Cs > Na > Li	+	–	Below 5 %	–	–	[2]DNP KCN azide	? +	
Mg^{2+} ions (a)	$+H^+$	M-M	–	cf. to Li	+	–		–	–	DNP but depends on anion	? +	
(b)	–		–	Mg > Mn > Ca > Si				+	+		? +	
Phosphate	–	M-M	–	K^+ can affect	+	+	–	+		Azide		+
'Ammonia'	$+H^+$	Diffusion	–	? NH_4 > NO_3 assimilation	+	–	–	–	–	–		+
NO_3 (a)	$+H^+$	Diffusion ?	–	As 'ammonia'	+	–	–	–	–	–	–	+
(b)	$+H^+$		–					+		– ?		
Monosaccharides	–	M-M	Yes	Glucose > others		–	–		–	UO_2^{2+} surface effect	? +	–
Amino acids	$+Na^+$ etc.	M-M	L-forms α-amino		+	+ (–)		+		DNP azide		+

[1] M-M: Michaelis-Menton
[2] DNP: 2,4-dinitrophenol

A number of general points emerge from these experiments. Firstly, because investigators have been concerned to study only a limited number of processes at any one time, comparisons between them and effects due to interactions are difficult to assess. For example, something is known about the interaction of Na^+ or K^+-ions with transport processes of bivalent cations, phosphate and a few amino acids but there is no information concerning the interactions of alkali ions on 'ammonia', nitrate or sugar transport. It would clearly be of great value to monitor simultaneously exchanges of all the principal solutes in at least one set of conditions. Without such information it may be that there is a danger of making a series of special hypotheses where one or a few general ones could suffice, e.g. one for all electrolytes including amino acids, one for small non-electrolytes and one for large molecules.

A second point related to the problem of comparability is that apparently there are different routes through cellular membranes and these may be different from the adsorption sites. The cytoplasmic membrane has to be regarded as a patchwork of impermeable and permeable areas and of binding sites, both reversible and irreversible in their properties. In all these features the various regions differ in their kinds and degrees of specificity. Such a concept, complex as it is, does not adequately convey reality for, in a living hypha, the properties of the cytoplasmic membrane must be in a dynamic equilibrium capable of adapting in response to changed internal metabolic conditions or to external conditions and materials. At present, neither the chemical nor electronmicroscopic knowledge of the cellular membranes enables this complexity to be adequately portrayed and this will continue to handicap studies of transport processes.

Finally, a further point, perhaps related to the last, is that transport processes may not be the same in different circumstances. This is particularly well shown in studies of K^+ transport with and without the presence of exogenous phosphate, or the uptake of sugars and ammonia in conditions when they can or cannot be utilized. It is possible that these differences are real and are due either to different processes occurring in different conditions or to changes in the properties of the cellular membranes with the changed conditions. It is equally possible that in such cases either the basic transport process is the same and the differences are associated with processes immediately succeeding the transport phase, or that the transport process is complex and that variations can arise in later phases. In this connexion the demonstration of different kinds of 'pools' of solutes in a number of cases gives some indication of the complex interactions which can arise from transport through the membrane followed by intracellular transport from one kind of cell compartment to another. Intracellular compartmentalization is clearly an important yet imperfectly understood phenomenon in fungi. It is clear that, at present, a dogmatic assertion of the type 'nitrate is transported by an "x" process' cannot be made. This should be borne in mind in the ensuing discussion where it may seem as if one process were being set uniquely against another.

Passive diffusion does not seem to be a common transport mechanism save, perhaps, for nitrate and 'ammonia'. Even in the latter case the operation of a Donnan-equilibrium system cannot be completely ruled out.

Facilitated diffusion seems a very likely mechanism for the transport of monosaccharides, at least in yeast. CIRILLO (1961a) has discussed a variety of schemes that would bring about facilitated diffusion and a factor common to them all is the notion of a carrier molecule. Such substances have never been isolated, if they exist, and their nature is quite unknown. The most successful theoretical treatment of a sugar transport system in fungi assumes that:

(i) the carrier is a stereospecific substance capable of moving across the membrane in both directions;
(ii) that the carrier combines with the solute to form a solute carrier complex at one side of the membrane and that this dissociates at the other side after transfer;
(iii) the transfer process is probably the rate limiting step (CIRILLO, 1962).

Thus the hypothetical carrier operates in a manner analogous to that of an enzyme and so the system can often be described by kinetics developed to deal with enzyme action. This analogy has been virtually accepted as an homology in the use of the term 'permease' by the workers at the Institut Pasteur for hypothetical proteins concerned with transport through the membranes of bacteria and yeasts. Particularly strong evidence now exists for a high degree of structural specificity of uptake of the tryptophane family of amino acids by *Neurospora.* The transport system's kinetics are comparable with enzyme kinetics and its temperature sensitivity is similar to that expected for an enzyme. However, until a carrier or permease has been isolated little further progress can be achieved.

Evidence for active transport in the strict sense does not exist in many cases in fungi, a notable exception is K^+-ion transport in yeast (ROTHSTEIN and BRUCE, 1958a, b). This is because there are still too little data on influxes and effluxes and it is not always clear that accumulation is against an electrochemical gradient. It is in this kind of study that adequate and accurate measurements of the simultaneous transport of a number of solutes would be particularly valuable. There are a number of cases where exchange processes of some kind or other may occur, e.g. K^+ for Na^+ or H^+, K^+ with amino acids, or associated transport processes, e.g. Na^+ and Cl^--ions in yeast. There does seem, however, to be a *prima facie* case for the active transport of both alkali cations and phosphate. Here respiratory energy is involved in at least some forms of transport. The question arises as to how it is utilized. The two main hypotheses were mentioned earlier (cf. p. 207), that ATP acts as an energy source to drive 'ion pumps' or that redox energy from electron-transport chains is used directly to 'pump' ions. There is no evidence available from fungi that will discriminate unequivocally between these hypotheses.

The proponents of the view that ATP is involved rely especially on the facts that in some experiments phosphate is apparently implicated and in others, 2,4-dinitrophenol acts as an inhibitor of transport. If phosphate is involved in transport then the production of phosphorylated compounds is likely to be involved. Arsenate is said (SUSSMAN and SPIEGELMAN, 1950) to interfere specifically with the action of glyceraldehyde phosphate dehydrogenase responsible for the production of 1,3-diphosphoglyceric acid, a step coupled with ATP production. It is, therefore, significant that arsenate inhibits both phosphate and Mn^{2+}-ion uptake in a similar manner (Table 8.14) and it will be recalled that the supposed Mn^{2+}-ion carrier is identical with that for Mg^{2+} (cf. p. 215). Rothstein concluded that, 'the Mg^{2+} carrier is, or is dependent on, a phosphate compound derived from a phosphorylation reaction in the cell membrane'.

Table 8.14. Comparison of the effects of arsenate on the uptake of phosphate and on the uptake of Mn^{2+} by yeast. (From Rothstein, 1961)

Arsenate	Inhibition	
	Phosphate uptake	Mn^{2+} uptake
(mM)	(%)	(%)
0·1	8	9
0·25	40	42
0·5	60	52
2·0	80	70

50 mg/ml.; Mn^{2+} 1·0 mM; phosphate 1·0 mM; glucose 0·1 M. Yeast used in Mn^{2+} experiment pre-exposed to phosphate and glucose for 30 min, washed and exposed to Mn^{2+}+glucose or Mn^{2+}+glucose+arsenate.

2,4-Dinitrophenol is thought to uncouple phosphorylation from oxidative processes and it generally increases the level of inorganic phosphorus (P_i) in the cell. There is thus a suggestion that alkali cation and amino acid uptake in yeast, phosphate uptake by mycorrhiza and 'ammonia' uptake by *Scopulariopsis* involve phosphorylated intermediates. These have not been isolated and, indeed, such an isolation would be extremely difficult technically. It may be noted, however, that in broken human erythrocyte membranes an enzyme which splits ATP has been detected. It behaves in several respects in the same way as the sodium transport system (POST and ALBRIGHT, 1961). A particulate entity has been discovered in the membrane of crab nerves with specific sites for Na^+ and K^+, each with characteristic affinities and an enzyme that liberates energy from ATP. How such enzyme systems operate is still uncertain and nothing like them has yet been detected in fungi. Moreover, these animal systems are inhibited by *g*-strophanthin and ouabain which do not affect K^+/Na^+ balances in yeast cells (JENNINGS and SOUTAR, in JENNINGS, 1963).

Evidence from DNP inhibition may be misleading and recent, incompletely published data on sulphate transport by yeast shows how this may come about

(SCOTT and SPENCER, 1965). When supplied with $^{35}SO_4^{2-}$ there was an immediate, rapid uptake followed by a temperature sensitive, equally rapid partial loss. Then a more gradual sustained uptake occurred, due it was believed to the utilization of SO_4^{2-}-ions for choline sulphate and amino acids. In a mutant (CMl 38581) which was blocked at its ATP sulphurylase, preventing utilization, it was found that the two initial phases occurred but not the final sustained uptake. This last phase was the one susceptible to inhibition by 2,4-DNP, not the uptake as a whole.

CONWAY (1955) has also claimed that the ATP hypothesis is unlikely since the known relative affinities of yeast for different alkali cations is quite different in order and magnitude from those which would be expected if a phosphate or polyphosphate bound carrier were involved.

CONWAY (1951, 1953, 1955) suggested that an ATP supplying mechanism was unnecessary, for electrolytes at least, and suggested instead a 'redox pump'. The notion is simple. If two redox systems at different potentials are separated by a barrier, electrons can be transferred by an appropriate carrier enzyme from one system to the other. The free energy per equivalent of electrons transferred can be estimated in appropriate electrical terms. This free energy may be lost from the system as heat or it may be utilized in chemical reactions or in osmotic work. Conway suggests that the free energy is used for such osmotic work, namely 'pumping' ions across the barrier. An example of cation transfer will illustrate these concepts.

Suppose that there are two metal redox systems one an electron donor, the other an acceptor. The donor system is reduced and is attached to a cation. It then loses an electron to the acceptor system but simultaneously releases the cation.

The potential E_D of the donor system, relative to the standard electrode potential is:

$$E_D = E_{O_D} - \frac{\mathrm{RT}}{n\mathbf{F}} \log_e \frac{[\mathrm{Red}_D]}{[\mathrm{Ox}_D]}$$

where E_{O_D} = characteristic potential of donor system; n = equivalent of electrons; $\mathbf{F}$ = Faraday constant; $[\mathrm{Red}_D]$ $[\mathrm{Ox}_D]$ = activity of reduced and oxidized states of donor system.

Similarly, for the acceptor system:

$$E_A = E_{O_A} - \frac{\mathrm{RT}}{n\mathbf{F}} \log_e \frac{[\mathrm{Red}_A]}{[\mathrm{Ox}_A]}$$

If an equivalent of electrons is transferred from the donor to the acceptor system then, for each equivalent transferred:

$$\mathbf{F}(E_A - E_D) = \mathbf{F}[E_{O_A} - E_{O_D}] - \mathrm{RT} \log_e \left[\frac{(\mathrm{Red}_D)}{(\mathrm{Ox}_D)} \cdot \frac{(\mathrm{Red}_A)}{(\mathrm{Red}_D)}\right]$$

In other words, the difference is made up of two components, the electrical work (strictly the difference between the standard electron energies) less the 'osmotic work'. It can be shown that if the product of the activities

$$\left[\frac{(\mathrm{Red}_D)}{(\mathrm{Ox}_D)} \cdot \frac{(\mathrm{Red}_A)}{(\mathrm{Ox}_A)}\right]$$

is greater than unity then some of the difference between $\mathbf{F}\,E_{O_A}$ and $\mathbf{F}\,E_{O_D}$ has been used to perform osmotic work.

It can also be seen that the actual values of reductants to oxidants in the two systems may vary yet the ratio remain the same. Further, even if the ratio changes appreciably in different metabolic conditions it should remain relatively constant for any one set of conditions. (A full discussion of these theoretical concepts appears in CONWAY, 1953.)

The sequence of processes envisaged by Conway for cation transport is, therefore, something like this:

(a) union of a reduced carrier enzyme (M_1) with a cation (K^+) to form a complex. (The reducing enzyme system is Ct.)

(i) $\frac{1}{2}\mathrm{Ct\,H_2} + M_1 \rightleftharpoons M_1^- + H^+ + \frac{1}{2}\mathrm{Ct}$ (reduction)

(ii) $M_1^- + K^+ \rightleftharpoons \{M_1K\}$ (complex formation)

(b) transport of carrier cation complex across membrane and release of cation and necessary energy for osmotic work. An electron is then transferred to a second enzyme (M_2) incapable of combining with the reduced cation. The electron may ultimately be transferred through respiratory enzymes (M_R) to oxygen under aerobic conditions

(iii) $\{M_1K\} \rightleftharpoons M_1 + K^+$ (cation release)

(iv) $M_1 + e \rightleftharpoons M_1^- + e \overset{M_2}{\rightleftharpoons} e \overset{M_R}{\rightleftharpoons} e \ldots$ (electron transfer)

(v) $\ldots e + H^+ + \frac{1}{2}O_2 \rightleftharpoons \frac{1}{2}H_2O$

(c) return of carrier enzyme to original side of membrane when the sequence can recur.

Processes (a) (ii) and (b) (iii) presumably take place in the membrane. Fig. 8.15 (based on CONWAY, 1953) provides a diagrammatic representation of how Na^+-ions and K^+-ions could exchange by a 'redox pump' mechanism. The simplicity of this system is, of course, that it depends upon an enzyme involved in the redox system which at the same time acts as a carrier molecule.

This hypothesis has been no more proved, or disproved, than the ATP system previously discussed. There have been few direct tests of the validity of the 'redox pump'. One of the most striking is the effect on the transport of K^+- and H^+-ions by yeast of altering the external redox potential of the medium (CONWAY, 1955). The experiments were done anaerobically at pH 4·5 in the presence of glucose. Redox potentials were modified by adding either Nile blue or phenol indo-2,6-dichlorophenol at a concentration of 1×10^{-4} M, an addition which does not affect anaerobic metabolism appreciably.

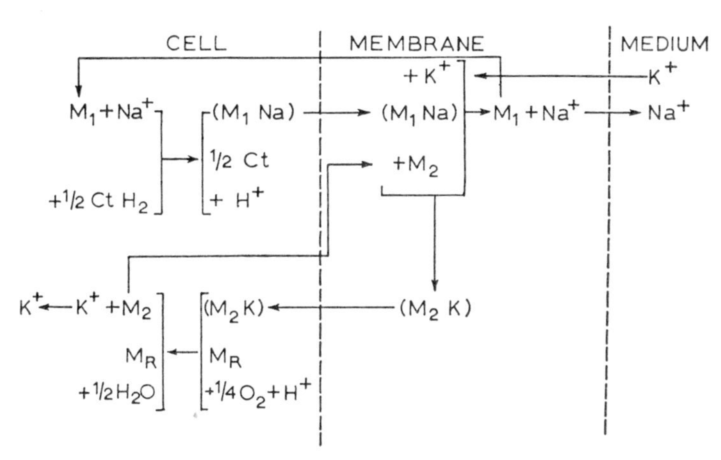

Fig. 8.15. A possible sequence for the exchange of Na^+- and K^+-ions by a redox pump. Symbols as in text. (From Conway)

Untreated cells have a potential of about +180 mV under these conditions and the dyes altered the range from about +100 mV to > +250 mV. At +100 mV, K^+- and H^+-ion transport was effectively stopped and it increased, linearly, over the range +140 to +240 mV, the two ions showing parallel behaviour (Fig. 8.16). He has also shown that there is, on average, a change in potential of 8 mV per unit of pH, over the range pH 4–8, in a yeast suspension under the conditions used in the redox dye experiments. He claims that this indicates that a metal system is involved rather than a metabolic one (i.e. an ATP system), when a change of 30–60 mV per unit of pH would be expected (CONWAY and KERNAN, 1955).

Conway's 'redox pump' hypothesis has much in common with that of Lundegårdh, devised to account for accumulation of salts in higher plant cells. The Lundegårdh hypothesis has been modified over the years but, in essence, it supposes that the cytochrome system transports anions inwards while electrons flow outwards (LUNDEGÅRDH, 1960). Some of the data available for fungi could be interpreted on Lundegårdh's hypothesis but the only case where this has been attempted is in the studies of Harley and his associates on the 'salt respiration' of excised beech mycorrhiza (HARLEY, MCCREADY and GEDDES, 1954; HARLEY *et al.*, 1956; JENNINGS, 1958; HARLEY and JENNINGS, 1958; HARLEY and AP REES, 1959). This phenomenon, an increase in the respiration rate on addition of salts after it had become reduced through storage in distilled water, was thought by Lundegårdh to be associated with the transport of anions. Harley was able to show that excised mycorrhizal roots, if stored for some days in aerated distilled water showed an increased O_2 uptake and CO_2 emission on exposure to a variety of salts at a concentration of 60 mM. The decline in respiration during storage was not due to depletion of carbohydrate reserves. Application of reduced oxygen tensions, cyanide or DNP at appropriate concentrations decreased the uptake of

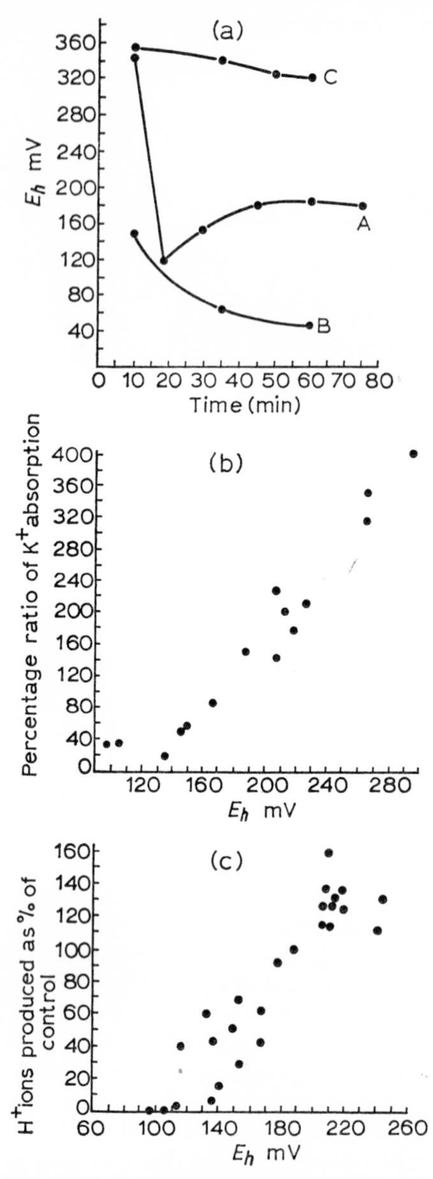

Fig. 8.16. Redox potentials, K^+-uptake and H^+-loss by yeast cells. (a) Redox potential of yeast suspension at pH 4·5 with 5% glucose and no redox dye present (A), with 0·0001M Nile blue, (B) and phenol indo-2,6-dichlorophenol (C); (b), (c) Percentage absorption of K^+-ions and loss of H^+-ions, respectively, at different redox potentials using redox dyes after 90 min fermentation. Values in absence of dyes taken as 100 units. (From Conway)

phosphate while increasing O_2 consumption and CO_2 emission. The respiratory stimulation was greater in stored than in fresh, excised mycorrhizas. These data were interpreted on the hypothesis that the treatments resulted in changes of 'energy-rich' phosphate acceptors in the tissues, in particular a decrease in acceptors during storage. It was also shown that the fungal sheath of mycorrhizas probably contains cytochrome oxidase as well as another electron transport system which is not blocked by inhibitors of cytochrome oxidase. Thus the information available leaves the situation open. Anion transport via a cytochrome system could occur but this is neither the only route nor, necessarily, the preferred route and uptake could well be associated with phosphorylation.

It must be concluded that it is not yet possible to account for the transport processes of any of the common and important electrolytes and non-electrolytes across the cellular membranes of fungi. This is due, in part, to the intrinsic difficulties of such experimentation, in part to the relative paucity of available observations and, in part, to the neglect of such studies in favour of experiments on utilization, assimilation and nutrition in general.

RELEASE OF LARGE MOLECULES

The efflux of substances, both electrolytes and non-electrolytes, can evidently occur but, in addition, many larger molecules are also found in the external medium. Three well known groups of such substances are antibiotics, toxins and extracellular enzymes. Two questions arise, whether such substances are released uniformly by growing fungi and how they are released.

Some of these substances are of considerable molecular size. For example, extracellular celluloses have been isolated with molecular weights of 55,000, 30,000 and 5300 (SELBY and MAITLAND, 1965), although the antibiotics and fungistatic substances are of smaller size, e.g. penicillins, 322; gliotoxin, 329; griseofulvin, 352·8. In some cases the release of these substances is due to autolysis of the older parts, for example, the release of indigo by a mutant of *Schizophyllum commune* or of the anthraquinone pigments of various *Fusarium* spp. However, it is clear that this does not account for all the cases observed and there is a widespread notion that growing hyphal tips secrete substances. The evidence for this view is not great and the mere existence of a diffusing enzyme or antibiotic well ahead of the growing margin of a colony provides little evidence concerning the region from which it has been released.

Some of the best evidence comes from a study of the production of penicillin by *Penicillium chrysogenum* (≡*P. notatum*). WHINFIELD (1947, 1948) was able to show that germinating conidia whose germ tubes averaged 30 μm were capable of producing extracellular penicillin. A concentrated suspension of conidia (C) was prepared and also three dilutions C/10, C/100 and C/1000. They were incubated as 0·5 ml. drops on permeable cellophane and their growth and penicillin production measured. The maximum rate of penicillin production was attained simultaneously at 2–3 days in all the cultures, despite differences in the time of attaining maximum growth rate. In the C and C/10 cultures rapid growth in the first 24 hours was followed by a marked reduction, due perhaps to the accumulation of staling substances. Thus growth is largely due to that of the primary germ tubes. In C/100 and C/1000 cultures the growth rate was slower and more sparse and, after 24 hours, branching commenced. Thus maximum penicillin production coincides with the region where lateral branches may develop whether they do so develop or not. It commences just behind the apex and is maximal further back in the zone 2–3 days old. In the older autolysing parts penicillin was destroyed. This simple yet elegant study shows clearly that the release of penicillin, at least, is not exclusively at hyphal tips. It is unfortunate that this is the only study of its kind.

The other problem is the mechanism of release. It is not clear whether substances are formed in the fungal cell and transported through the membranes, or whether they are actually synthesized at the surface and then released to a greater or lesser extent. In *Polystictus versicolor* there is evidence that a polyphenol oxidase having similar properties to that of the extracellular enzyme occurs within the hyphae (LINDEBERG and HOLM, 1952) and this is also claimed to be true for the β-glucosidase of *Fomes annosus* (NORKRANS, 1957). This suggests the possibility of internal synthesis followed by transport. Unfortunately neither of these enzymes is well characterized and precautions were not taken to separate wall material from cytoplasm. The most intensively studied case is that of invertase production (cf. pp. 201, 205). In intact yeast or *Neurospora* conidia the invertase is principally associated with the membrane (BURGER, BACON and BACON, 1958, 1961; METZENBERG, 1964). The sensitivity of invertase activity to external pH is such that it seems probable that the enzyme is located near the surface of the hyphae of other fungi such as *Myrothecium verrucaria*, ectotrophic mycorrhiza of beech and the fungal component of the lichen *Peltigera polydactyla* (MANDELS, 1953; HARLEY and JENNINGS, 1958; HARLEY and SMITH, 1956). Protoplasts of yeast or *Neurospora* release invertase, further confirmation that *in vivo* it is associated with the outermost region of the cell. The enzymes released contain covalently bound mannan and glucosamine, and hexosamine, respectively, and these are, of course, normal wall components (LAMPEN, 1965). The enzyme in protoplasts has been studied and seems to have the same properties as that obtained externally (LAMPEN, 1965; TREVITHICK and METZENBERG, 1964). However, there is a certain amount of 'molecular sieving' so that the monomer form is released more readily than the complex form (cf. p. 201). In yeast protoplasts, extracellular enzyme production was independent of amino acid supplementation over a 2- to 3-hour period but was temperature dependent and required metabolic energy. Unlike intact cells, the rate of secretion is invariable between pH 4–7 and LAMPEN (1965) has interpreted this as indicating that neither the rate limiting step in synthesis nor the mechanism of release is external to the protoplast. During release the protoplast membrane visibly enlarged, even in the absence of amino acids. Release from protoplasts is quite specific for while invertase was being released intracellular α-glucosidase was retained.

There seems little doubt that this enzyme of molecular weight *c.* 30,000 is synthesized within the cell membrane although once made it can be associated in some way with a region associated with the wall and cytoplasmic membrane. Release can hardly be accounted for by pinocytosis unless this process is very much more specific than seems to be the case (HOLTER, 1965). Its release may well be associated in some way with membrane synthesis, for this would account for its association with wall components in protoplasts and for the observed enlargement of their membranes during release. Since so little is known concerning the process of wall and membrane synthesis in fungi (see p. 46), this suggestion is not very helpful at present. Until the precise sites of synthesis of extracellular enzymes have been located it will prove difficult to develop an adequate hypothesis for their behaviour.

How far the matters discussed in this section apply generally to the release of large molecules is quite uncertain. It seems probable that there is some specific transport system whose nature is at present unknown. The location of the site of release is an equally uncertain matter. Indeed, this uncertainty concerning the sites of transport in the cell wall, or in the mycelium as a whole, for *all* transport processes is one in which only a beginning has been made. Until far more information is available it will hardly be possible to provide a rational account of how a mycelium is physiologically differentiated in relation to transport processes, if indeed, it is so differentiated.

Chapter 9
Translocation and Transpiration

It may seem surprising that these two topics are linked together in the same chapter but it is more than a mere matter of convenience. The mechanism of translocation is not understood but, in certain cases such as translocation of materials to developing basidiocarps, transpiration does seem to be involved more or less directly.

Clear indications have already been given, when dealing with apical growth (cf. pp. 42–46) and basidiocarp development (cf. pp. 134–139) for example, that growth involves translocation of protoplasm and other materials. In this chapter the nature and mechanism of such translocatory processes as have been studied by experimental methods will be discussed.

Translocation in hyphae and strands

It is a remarkable fact that the first study of any significance to this subject was only reported by SCHÜTTE in 1956. He employed a very simple technique of placing one container within another, petri dishes in larger circular pyrex dishes, tins within tins, etc. One of the containers was filled with deficient medium, the other contained all nutrients necessary for growth (Fig. 9.1a). The theory behind the method is that growth on deficient media can only be supported by translocation of essential nutrients transported into that part of the mycelium growing on the complete medium. In this way he studied translocation of glucose, sucrose, nitrogen (derived from NH_4NO_3), phosphate (from KH_2PO_4) and fluorescein in a variety of fungi. These experiments were substantiated by others, involving analysis of the sugar or nitrogen translocated, or the rates of respiration in two halves of a continuous mycelial mat when one was on a nutrient medium and the other on water. The results of all these experiments were remarkably clear-cut and consistent. The fungi studied fell into two groups, one, e.g. *Rhizopus* spp., capable of translocation, and the other, e.g. *Aspergillus* and *Penicillium* spp., incapable (Table 9.1). Sugar

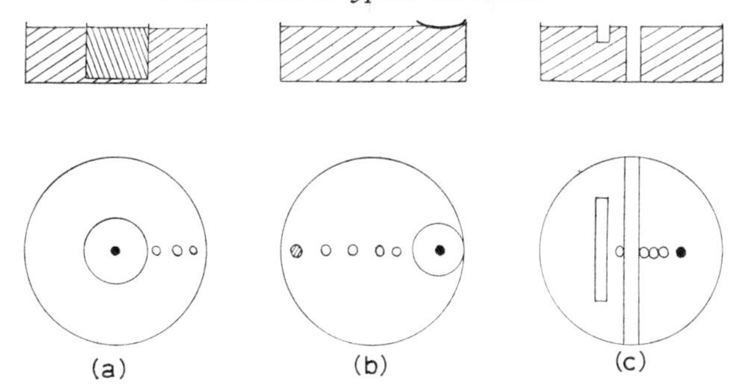

Fig. 9.1. Elevations and plan views of arrangements of containers used to study translocation in fungi. In each case the outer container is shown as a petri plate: ●—the inoculum, ●—alternative position; ○—positions of samples. (a) Schütte's method, the outer container filled with medium lacking a nutrient, e.g. sugar, the inner with complete medium; (b) Monson's and Sudia's method, the watch glass embedded at one side of the medium and to which the radioisotope is added. Inoculation either in the watch glass or diametrically opposite on the medium; (c) Lucas's method (modified by Smith), the central diametrical trough acts as a diffusion barrier, the radioisotope is added to the shallow trough in the medium.

Table 9.1. Observations on translocation in *Rhizopus oryzae* and *Aspergillus niger*; (**a**) growth at 30 °C on deficient medium due to translocation and (**b**) amount of glucose translocated. (From Schütte, 1956)

(**a**) Translocation, when it occurred from old hyphae to young tips at 30 °C

Fungus	Inner container	Outer container	Duration (days)	Observations
A. niger	C	−S	6, 14, 18	No growth, i.e. no translocation
	C	−N	21	
	C	−S	14	
A. niger	−N	C	14	
	+FC	+FC	10	
	C	C		Growth, i.e. can grow over rim of containers
	C	−S	4, 8	Good growth
	C	−N	21	Good growth
R. oryzae	C	−P	14	Growth but over −P is poor
	+FC	−FC	10	Growth, F secreted at tips
	−S	−N	7	No growth, therefore no bidirectional translocation
	−N	−S	7	

C, complete; +S, lacking sugar (glucose or sucrose); −N, lacking nitrogen (as $NH_4NO_3^-$); −P, lacking phosphorus (as $H_2PO_4^-$); ±F, with or without 1% fluorescein.

Table 9.1—*Continued*

(b) Glucose in mycelium of inner and outer containers, the sugar being only present in the medium of the inner container.

Fungus	Mycelium treatment	Glucose (mg/g mycelium) Inner	Outer	
		4·20	4·29	
		7·74	9·23	
		7·49	5·24	Inner < outer
		2·99	6·50	
		6·51	11·80	
		9·10	2·60	
		10·70	3·60	
		5·17	5·02	
R. oryzae	Blotted	7·49	5·24	
		10·40	5·61	
		10·10	5·13	Inner > outer
		5·39	4·29	
		6·85	2·50	
		7·45	2·10	
		8·20	1·25	
		6·51	11·80	
		0	13·50	Inner < outer
	Oven-dried	14·50	7·75	
		14·95	13·40	Inner > outer
		14·50	3·68	
		1·20	10·90	Outer 10% glucose originally*
		0·65	20·10	
		3·27	31·30	
A. niger		0·21	1·97	Outer 5% glucose originally*
		0·13	2·06	
		3·56	2·97	Outer 1% glucose originally*

* Inner container 1% glucose throughout

translocation in *Rhizopus oryzae* resulted in a higher concentration in the growing tips in deficient agar than in the nutrient agar in nearly 40% of the experiments. This suggests that translocation is not along a diffusion gradient. Nitrate nitrogen was translocated in a similar way.

Schütte attempted to investigate bidirectional streaming by inoculating two dishes with *R. oryzae*, the outer lacking sugar and the inner lacking nitrogen. Growth was extremely sparse and no clear-cut results were obtained. With a complete inner medium and a medium lacking sugar but with fluorescein added in the outer container a healthy mycelium developed. However, no fluorescein was translocated backwards to the inner region. He concluded that bidirectional translocation was, therefore, effectively non-existent.

This simple, elegant, pioneering study has been followed by others, using essentially the same technique but with radioactive materials to detect translocation. The results have not always been the same. THROWER and THROWER (1961) repeated Schütte's experiments and report that they found translocating (10 spp.), non-translocating (6 spp.) and indeterminate species (4). It is unfortunate that they neither list the species nor indicate what is meant by an indeterminate translocator. They repeated the experiments supplying [^{14}C]-sucrose to the inner dish and found that translocation could occur in both translocators and non-translocators provided, in the latter case, the outside medium was not sugar-deficient. A more elaborate experiment was carried out to test this in an established mycelium. Here the inner plate was supplied with liquid potato-dextrose+0·1% peptone medium which was replaced after 14–16 days with complete medium containing [^{14}C]-glucose. The outer medium was complete and, in these circumstances, the mycelium grew over both inner and outer plates both in a translocating species, *Rhizopus stolonifer* and in a non-translocating species, *Saprolegnia ferax*. Samples of 1 cm^2 disks were removed from the outer dish at its inner and outer margins and mid-way between. The results confirmed that translocation could occur through an established medium although not so effectively in a non-translocating fungus as in a translocating species (Table 9.2). These results differ somewhat from those observed by LUCAS (1960) who used a slightly different technique to study translocation of ^{32}P-labelled KH_2PO_4. Translocation only occurred if the ^{32}P was present before inoculation when it was taken up and 'carried by the extending hyphae'. Translocation could not be detected through established mycelium either from old to young regions or vice versa. Similar results were obtained with *Absidia glauca* and *Chaetomium* spp. for ^{32}P; with ^{60}Co and ^{137}Cs in connexion with translocation by *Phycomyces blakesleeanus* into developing sporangia (GROSSBARD and STRANKS, 1959) and for similar translocation into conidia of *Aspergillus niger* of ^{32}P (YANAGITA and KOGANÉ, 1963b).

MONSON and SUDIA (1963) carried out more extensive studies of translocation of radioactive isotopes, viz, ^{65}Zn, ^{35}S, ^{89}Sr, ^{32}P, both from young to old hyphae and vice versa in established mycelia of *Rhizoctonia solani*. With the first three radioactive isotopes, translocation occurred from young to old hyphae for at least 5 cm over 5–6 days. There were curious fluctuations, for example more ^{65}Zn was apparently translocated in alternate 24-hour periods, but there is no doubt in all cases that some translocation occurred. More extensive studies were made with ^{32}P; mycelia aged 3, 5, 7 and 9 days were employed. There was little difference between mycelia 3, 5, or 7 days old but the 9-day-old mycelium differed considerably; much greater daily increases occur in uptake but the ^{32}P does not travel as far as in the younger mycelia. This age effect could account for the difference between the results of Lucas and Monson and Sudia. The latter authors also showed clearly that young hyphal tips absorb and translocate more ^{32}P farther than do older hyphae from an 8-day-old mycelium (Fig. 9.2). This is of particular interest since SMITH (1965) has been able to demonstrate simultaneous bidirectional

Table 9.2. Translocation of [^{14}C]-glucose by *Pellicularia filamentosa*, *Rhizopus stolonifer* and *Saprolegnia ferax* using Schütte's technique but sampling from inner (I), outer (O) and middle (M) regions of outer container containing either complete or deficient medium. 14–16 days at 25°C then inner liquid replaced by 5 μCi [^{14}C]-glucose. Results as counts/min/cm^2 above background. (From Thrower and Thrower, 1961)

Time after adding ^{14}C (hr)	*Pellicularia filamentosa*						*Rhizopus stolonifer*						*Saprolegnia ferax*		
	Deficient			Complete			Deficient			Complete			Complete		
	I	M	O	I	M	O	I	M	O	I	M	O	I	M	O
1	0	11		0	1	—	2	0	0	0	0	4	12	2	1
6	9	12		6	1	—	3	0	2	2	0	29	9	1	1
24	23	11	4				9	0	0				40	8	9
54	26	17	17	55	22	—	33	10	6	54	9	7	64	17	2
78	64	23	16	69	28	—				80	16	10	56	16	3
120							56	17	2				50	19	13

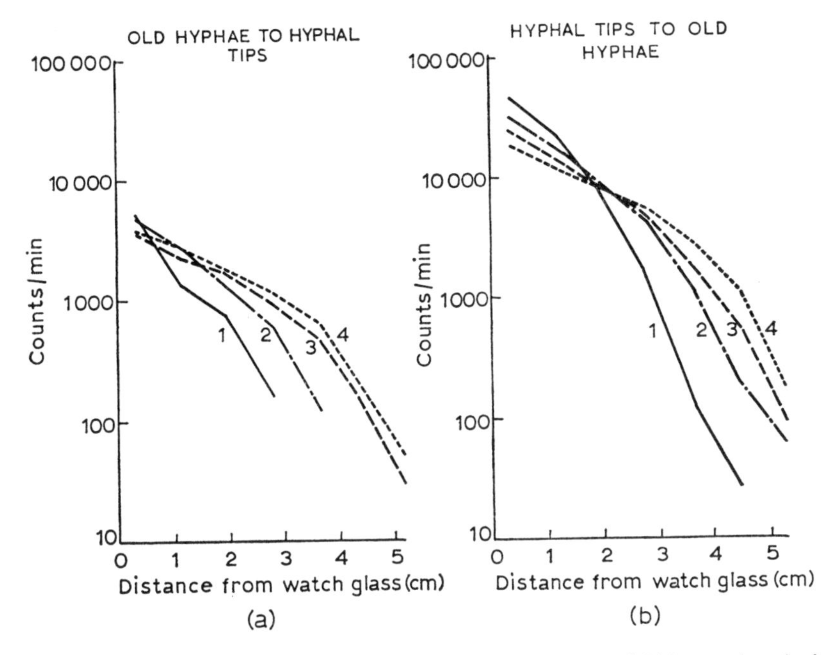

Fig. 9.2. Comparison of radioactivity (counts/min) in hyphae of *Rhizoctonia solani* at different distances from the point of uptake at 4 successive days after application (1–4). The mycelium at the time of application was 8 days old and the radioisotope ^{32}P. (a) translocation from old hyphae to young tips; (b) translocation from young hyphal tips to old hyphae. (From Monson and Sudia)

translocation of ^{32}P in her isolate of *R. solani* (Table 9.3). LITTLEFIELD (1965) extended studies on translocation of ^{32}P in established mycelia of *R. solani* to several strains and also studied the effects of light and temperature. He showed that translocation across a petri dish was greater than lateral transport,

Table 9.3. Bidirectional translocation of ^{32}P by *Rhizoctonia solani* 7-day culture on PDA; 10 μCi ^{32}P, sampled after 24 hr as 5 mm discs (see Fig. 9.1c). (Unpublished data of S. Smith, 1965)

Sample site	Forwards	Backwards
1	425 ± 67	727 ± 92
2	122 ± 25	437 ± 84
3	53 ± 9	291 ± 83
4	34 ± 7	205 ± 62
5	29 ± 10	182 ± 70
Total	685 ± 102	1831 ± 302

that light had no significant effect and that the process was significantly faster between 20° and 30°C than at temperatures below 20°C. He also suggested that the amount of ^{32}P translocated was inversely proportional to the distance travelled from the source. This suggests transport by diffusion.

SMITH (1966, 1967) also demonstrated translocation of both ^{32}P orthophosphate and [^{14}C]-D-glucose in established cultures of *R. repens*, one of the endophytic, mycorrhizal fungi of *Dactylorchis purpurella*, both in pure culture and when the hyphae extended from the mycorrhizic orchid tubers. She provided suggestive evidence that even when [^{14}C]-glucose was supplied to infective hyphae it was translocated as trehalose. Similar evidence suggests that [^{14}C]-sucrose supplied to ectotrophic beech mycorrhiza is converted to trehalose and mannitol and it could well be that sugars are translocated as trehalose in fungal hyphae (LEWIS and HARLEY, 1965a, b, c). Such an hypothesis needs to be investigated carefully but attention may be drawn to the fact that in higher green plants sucrose, another disaccharide, appears to be the form in which carbohydrates are normally translocated in the phloem tissue. The translocation of other nutrients such as ^{32}P, ^{15}N in ammonia or glutamate, ^{45}Ca and other cations by hyphae of ectotrophic mycorrhizal fungi such as *Boletus variegatus* to seedlings of *Pirus* spp. had been shown some years before by Melin and his associates (MELIN and NILSSON, 1950, 1952, 1953, 1954, 1955, 1958; MELIN *et al.*, 1958 and see pp. 344–345).

The nutrient experiments described earlier in connexion with rhizomorph development of *Armillariella mellea* and strand formation in *Psalliota hortensis* or *Merulius lacrymans* (GARRETT, 1953, 1954; BUTLER, 1958) indicate clearly that these act as translocating organs. The only experiment reported on translocation of radioactive materials in such structures is that of WEIGL and ZIEGLER (1960) of [^{32}P]-NaH_2PO_4 and [^{14}C]-glucose by *Merulius lacrymans* (Fig. 9.3). Translocation was demonstrated both by autoradiography and by counts along a strand (Table 9.4). Application of ^{32}P to a mycelial region including a strand provided little or no evidence of uptake by strand hyphae. Thus these organs do seem to be especially specialized for translocation (cf. pp. 83–90). The minimal rate of translocation was estimated at 2 cm/hr for

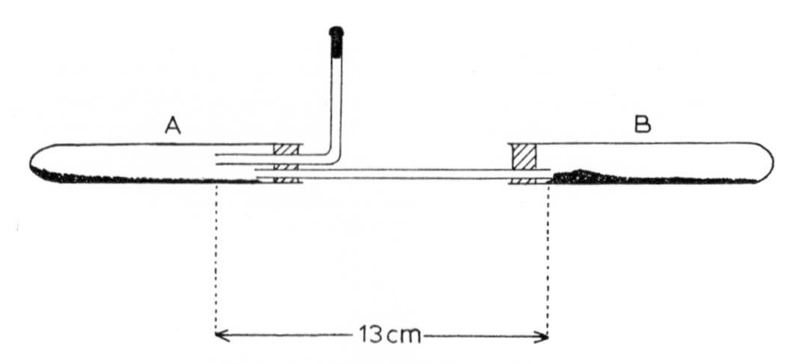

Fig. 9.3. Apparatus to investigate translocation of radioisotopes from mycelium of *Merulius lacrymans* in container A via a 13 cm hyphal strand along a narrow-bore tube to container B. Isotope added by L-tube to container A. (From Weigl and Ziegler)

Table 9.4. Evidence of translocation of ^{32}P through a strand of *Merulius lacrymans* connecting a phosphate-containing with a phosphate-free medium (cf. Fig. 9.3). (From Weigl and Ziegler, 1960)

Experiment	Duration (hr)	Counts/min over control in second medium
1	3	0
2	6	16
3	9	88
4	12	62 *

* Only a weak mycelial development in phosphate-free medium here

both ^{32}P and ^{14}C labelled compounds and the process was inhibited by low temperature (1°C), reduced oxygen tension or an atmosphere of nitrogen. SCHÜTTE (1956) showed that in plate culture, rhizomorphs injected with 1% fluorescein translocated it to their tips at a rate of at least 3 cm/hr at 18°C. No backward translocation was detected by this method. The data for rates in strands and rhizomorphs are remarkably similar. Schütte also studied translocation by subterranean mycelium of fairy rings of several species (*Tricholoma* sp., *Amanita* sp., *Clitocybe* sp. and especially *Marasmius oreades*) by watering the soil with fluorescein solution. Translocation was only really effective from the inside of the ring outwards, i.e. towards and into the basidiocarps, but the rate was said to be comparable with that in the stipe of a basidiocarp, i.e. *c.* 10 cm/hr (see later). This is much faster than through mycelial strands.

Translocation in basidiocarps

Studies have been made of the transport of fluorescent and other dyes and radioactively labelled phosphate in a number of basidiocarps. SCHÜTTE (1956) placed stipes of agaric basidiocarps in 1% fluorescein or Rose Edicol (detected by ultra-violet light) and showed that several had distinct translocating zones (Fig. 9.4). The rate varied with the size and age of the basidiocarp but those of *Armillariella mellea* (pileus diameter, 7·5 cm) achieved a rate of 12 cm/hr at 20°C under 100% R.H. and at 60% R.H. it was 15 cm/hr. Under unsaturated conditions most agarics achieved 6–8 cm/hr. Littlefield has studied these findings further in *Lentinus tigrinus* and *Collybia velutipes* (LITTLEFIELD, WILCOXSON and SUDIA, 1963, 1965; LITTLEFIELD, 1966). In *L. tigrinus* the central translocating zone is composed of elongated (at least 700–800 μm long) coenocytic cells in a loose matrix of frequently branched, interwoven hyphae of smaller diameter. The outer cortex is more of a loose network of 20–30 short cells, smaller to the outside and becoming larger further inwards. Both dyes and ^{32}P were translocated in the central zone which had a higher oxygen consumption than the cortex. Only estimates of rate can be made since translocated materials occurred in the pileus after 30 min and the basidiocarps

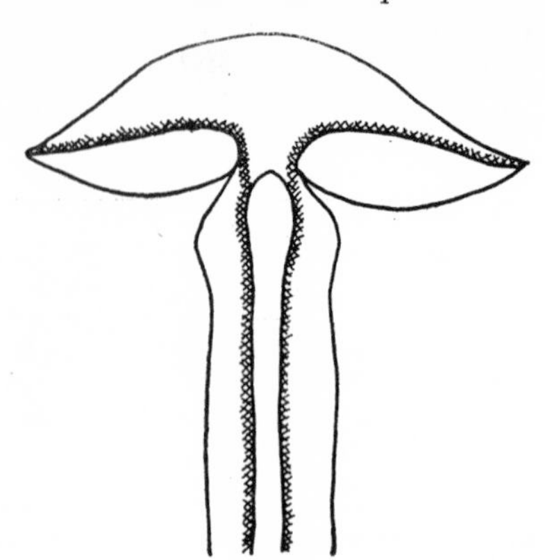

Fig. 9.4. Cross-section of a typical agaric basidiocarp to show region of translocation just inside the hollow stipe. (After Schütte)

ranged in size from 1·5–3·5 cm, giving minimal estimates of 3–7 cm/hr. Translocation ceased in killed stipes and was sensitive to relative humidity (Table 9·5). It is clear that translocation is certainly not impeded in this case by high relative humidity which was ensured by totally enclosing the basidiocarp for some time before the experiment in a polythene bag. Young, unexpanded basidiocarps of many species were shown by Schütte to be capable of translocating throughout their tissues. However, as a basidiocarp expands so its path of translocation becomes restricted to the central translocating zone.

Table 9.5. Translocation of ^{32}P into the pileus via the stipe in *Lentinus tigrinus* at low and high relative humidities and in killed basidiocarps after 30 min. The stipes were divided into four equal segments (1–4) from apex to base. (From Littlefield, Wilcoxson and Sudia, 1965)

Tissue	High R.H. (sample 14)	Low R.H. (sample 27)	Low R.H. dead (sample 10)
	counts/min/mg dry wt. tissue		
Pileus	60	70	0
Stipe 1	900	80	0
2	1430	140	0
3	3170	320	20
(Base) 4	6500	2870	3390

No distinct translocating zone was detected in various species of *Lycoperdon*, *Hydnum* and polypores studied by Schütte and this was confirmed by a careful study of *Polyporus brumalis* by PLUNKETT (1956, 1958). Plunkett used a variety of dyes but was not able to determine whether translocation was strictly intra-hyphal or not. A similar doubt surrounds translocation in agaric basidiocarps. The most striking feature of translocation in *P. brumalis*, however, was its relationship to growth. Rapid growth and development were associated with rapid translocation and both were affected by evaporating conditions. 'Rapid' transpiration, which depended in part on the age of the basidiocarp, ranged from *c.* 6·5–1·5 mg $H_2O/cm^2/hr$ while 'slow' transpiration was from *c.* 0·001–0·0005 mg $H_2O/cm^2/hr$. A change from 'rapid' to 'slow' evaporating conditions resulted in a 75% difference (4·23 ± 0·9 mg: 1·02 ± 0·24 mg) in dry matter increment at the apices of young, developing basidiocarps in 4 days. Schütte had also noticed the effect of relative humidity on the translocation rates of dyes. Vaseline applied to the surface of the basidiocarps, except the gills, reduced transpiration and translocation to about 50 – 33% of the normal rate in *Clitocybe* sp. Even under completely saturated conditions, translocation occurs at appreciable rates in *A. mellea* (Table 9.6); moreover, both transpiration and translocation rates decrease as the basidiocarp develops and ages. There is evidently a clear suggestion that transpiration is associated with translocation; it will, therefore, be convenient to consider transpiration before discussing possible mechanisms for translocation.

Table 9.6. Translocation rates of fluorescein in basidiocarps of *Armillariella mellea* of different sizes and at relative humidities of 60% and 100%. Experiments on basidiocarps stood in water for 3 hr first and at 20°C. (From Schütte, 1956)

Specimen size	Relative humidity 60%	Relative humidity 100%
Large; pileus 7·5 cm across	15·0	12·5
	13·8	11·3
	12·5	10·0
	8·8	10·0
Small, veil unbroken; pileus 2·5–5 cm across	5·0	3·8
	3·8	3·1
	3·8	1·9
	3·8	1·9

Transpiration

With the exception of the larger forms the loss of water from fungi has not received much attention and little is known concerning rates of water loss, or its prevention by fungi.

It is a common experience in plate cultures of fungi to observe droplets of water on the mycelium and these are usually supposed to have been lost from the hyphae either by transpiration or possibly by guttation. Plate 1, Fig. 4 clearly shows droplets of water between the outer lipid cuticle and wall of the sporangiophore of *Phycomyces blakesleeanus*. Here, too, it is uncertain how the water is lost from the hypha. Evidently in this species the well developed cuticle must restrict loss of water from the hypha and this may be the reason why *Phycomyces* does not collapse in conditions of low R.H. Indeed, *Phycomyces* can be grown in an open petri plate in a desiccator with anhydrous $CaCl_2$ present and still forms sporangiophores which grow normally. In these fungi, notably mucoraceous species where droplets of water develop on the sporangiophores, there is usually an extremely active flow of cytoplasm up the sporangial stalk. Insufficient comparative physiological and morphological studies have been made of these fungi.

Very few data concerning loss of water from plate cultures are available; YARWOOD's (1947) are set out in Table 9.7. He compared water loss under the same laboratory conditions (22–26°C and *c.* 48% R.H.) from open control agar plates (mean 0·0095 g/cm²/hr), closed plates (0·0032 g/cm²/hr) and plates covered with mycelium. No diurnal rhythm was detected. In general, cultures in open plates lost less water, and cultures in closed plates lost more water than their appropriate control plates. He suggested that there was also a correlation between the rate of loss of water and the amount of aerial mycelium in closed plates but this is not very high. The lower loss of water from open plates than from closed plates could be due to rapid, initial, drying out so that the occurrence of 'relatively inactive aerial mycelium may reduce water loss by restricting air movement above the agar surface'. Thus the mycelium might behave in a manner akin to that of leaf pubescence in higher plants. It is unfortunate that actual rates are not given in this paper as, without them, it is not possible to determine whether water loss was, in fact, more rapid initially from the open plate cultures.

Table 9.7. Relative water loss by various fungi grown in plate culture (22–26°C; and *c.* 48% R.H. external) with the lids either on (closed) or off (open) compared with evaporation from plain agar. (From Yarwood, 1947)

	Height aerial mycelium mm	Ratios, water loss from mycelial plates: Closed	Open
Rhizopus nigricans	10	1·16	0·80
Botrytis cinerea	8	1·10	0·84
Aspergillus niger	6	1·09	0·93
Sclerotinia fruticola	5	1·02	0·99
Thielaviopsis basicola	3	1·14	0·92
Phomopsis sp.	2	0·93	0·97
Sclerotinia sclerotiorum	2	1·14	0·93
Rhizoctonia solani	1	0·97	0·91
Fusarium solani	1	0·74	0·87

The most carefully compiled data on transpiration rates from a higher fungus are those of PLUNKETT (1956), using *Polyporus brumalis* (Table 9.8). The process was clearly susceptible to the atmospheric R.H. and the stage of development of the basidiocarp but remained unaffected by changes in light intensity. This is evidently a xerophytic fungus, comparable with *Phycomyces*

Table 9.8. Estimated rates of water loss by *Polyporus brumalis* at different relative humidities in relation to rates of flow, light intensity and stage of development of basidiocarp, inception (I) or mature (F). (From Plunkett, 1956)

R.H. (%)	Flow rate (ml./min)	Light intensity (lux)	Rate of H_2O loss (mg/cm²/hr) I	F
0	200		6·76	1·64
75	200	1720	1·22	0·60
96	5		0·0035	0·0013
0	200		6·5	2·09
75	200	430	1·22	0·72
96	5		0·0035	0·0015

in its behaviour to low R.H. Xerophytism may well be a common feature in polypores, as there is evidence of such behaviour from *Polystictus xanthopus* (CORNER, 1932a) and *Ganoderma applanatum* (HOPP, 1938). On the other hand, less detailed studies suggest that the agarics *Collybia velutipes* and *Coprinus lagopus* are much less resistant to desiccation (PLUNKETT, 1956; BORRISS, 1934b) and this may be true for many agarics. Detailed studies of all classes of fungi need to be made before precise and valid generalizations can be made concerning transpiration rates and reactions of fungi to different evaporating conditions.

PLUNKETT (1958) was able to demonstrate a reasonably close correlation in *P. brumalis* between the observed rate of transpiration and the movement of dyes and dry matter into the developing basidiocarp. Not only did trypan blue, methylene blue and neutral red, all water-soluble dyes, ascend stipes of basidiocarps under conditions of high transpiration (R.H.—0%) and fail to ascend under conditions of low transpiration (R.H.—96%) but transfer from one condition to the other was associated with a change in the sense expected, e.g. transfer from low transpiration conditions to high was followed by conspicuous ascent of the dye. However, there were some cases where no dye ascended under high transpiration conditions but these were correlated with abnormal growth. The transpiration stream cannot, therefore, be wholly due to an 'evaporation pull' being transmitted from hyphal surfaces back through the hyphae.

So little is known about transpiration and the movement of water in fungi that little can be gained from its further consideration. By analogy with higher green plants the fact that water-soluble dyes can be translocated does not necessarily provide information about water movement. The possible mechanisms of translocation will, therefore, now be considered.

Mechanisms for translocation

In 1885, DE VRIES first observed cytoplasmic streaming in a fungus, the developing sporangiophore of *Phycomyces nitens*. He considered that it was a process concerned with the transport of nutrients to the growing region. Streaming is still the only hypothesis to account for translocation in fungi. If it is correct, cytoplasmic streaming must be able to account for:

(a) rates of the order of 2–3 cm/hr in hyphae or hyphal strands and of 6–15 cm/hr in stipes of basidiocarps or subterranean mycelium attached to them;
(b) translocation in all types of hyphae, regardless of the presence of different kinds, or absence of septal pores;
(c) the existence of simultaneous bidirectional translocation;
(d) inhibition of translocation by reduction of temperature or anaerobic conditions.

Table 9.9 gives rates observed for cytoplasmic streaming in various fungi. The observations have generally been made by noting the movement of cytoplasmic particles, vacuoles or nuclei along known lengths of hyphae on the assumption that they are carried passively. The rates observed in both aseptate and septate hyphae are of a sufficient or greater order of magnitude than that which is required to bring about translocation. It is surprising that

Table 9.9. Rates of cytoplasmic streaming determined in several fungi with different kinds of septal pore and at different temperatures. Data of Arthur, 1897; Burnett, unpublished; Dowding and Buller, 1940; Jahn, 1934; Schröter, 1905; Ternetz, 1900)

Fungus	Observation	Pore type	Temperature (°C)	Rate (cm/hr)
Phycomyces nitens	CP	—	19	6–12
	CP	—	26–28	19·8
P. blakesleeanus *	CP	—	—	7·2–10·8 up 10·8–16·2 down
Rhizopus nigricans	CP	—	19	6–12
	VAC	—	28	19·8
R. stolonifer	CP	—	26–28	19·8
Pyronema confluens	CP	S	?	1·6
Ascophanus carneus	CP	S	Room ?	10·5
Humaria leucoloma	CP	S	27	90·0
Sordaria fimicola	VAC	S	20	6·0
Gelasinospora tetrasperma	VAC	S	20	3·0
			20	25–40
Polystictus versicolor	CP	D	20	8·6

Observations on cytoplasmic particles (CP) or vacuoles (VAC) in hyphae lacking septa (—), with simple pores (S) or dolipores (D).

* Observations on peripheral downward and central upward streaming in a sporangiophore.

the presence of a simple pore, let alone a dolipore septum, does not appear to reduce the rate of streaming. JAHN (1934) illustrated the eddies he had noted as cytoplasm streamed through the central pore of *Humaria leucoloma* (Fig. 9.5) and ISAAC (1964), using the known dimensions of the dolipore of *Rhizoctonia solani*, has calculated that the pressure drop across each septum with a flow rate of 0·36 cm/hr is at least of the order of 0·5 bar and probably more. The pressure developed by streaming cytoplasm must, therefore, be considerable.

Fig. 9.5. Diagram to illustrate eddies in the streaming cytoplasm in a hypha of *Humaria leucoloma* as it passes through a simple septal pore. (After Jahn)

Bidirectional streaming was first described by ARTHUR (1897) in *Rhizopus nigricans* and it has been confirmed in the hyphae of a number of other mucoraceous fungi (SCHRÖTER, 1905; BULLER, 1933) and in the sporangiophore of *Phycomyces* (POP, 1938). Buller noted that the double flow could be maintained for 1–4 min when the two streams could be seen to reverse simultaneously. Pop has provided the only estimate of the speeds of the two streams, the peripheral, downward stream being some 1·5–2·5 times faster than the main, upward central stream in the sporangiophore. Bidirectional streaming has not been detected in other fungi, notably *Rhizoctonia* spp. in which, it will be recalled, bidirectional translocation of ^{32}P has been detected (see p. 249). It should, however, be appreciated that this is a technically difficult observation to make in narrow hyphae. Reductions of temperature and anaerobiosis are known to reduce or prevent cytoplasmic streaming.

There is thus a good measure of agreement between the observed facts of translocation and cytoplasmic streaming. Even though translocation and cytoplasmic streaming have not been followed simultaneously in .the same hypha, it is a reasonable hypothesis that the latter accounts for the former. This, of course, leads to the problem of the causal mechanism of cytoplasmic streaming.

Mechanisms for cytoplasmic streaming

This is a general phenomenon in living cells which has recently been monographed by KAMIYA (1959) who discusses various mechanisms. In fungi possible theories have been discussed by JAHN (1934), BULLER (1933) and ISAAC (1964).

Jahn has suggested that in hyphae cytoplasm flows according to internal turgor pressure differences. Thus when cultures of *H. leucoloma*, grown on media of different osmotic pressures, were allowed to anastomose, the flow of cytoplasm between them was from that with the higher suction pressure to that with the lower up to a difference of about 4 bar (Table 9.10). Above 7·9 bar difference, the flow was in the other sense. At such osmotic pressures the growth of *H. leucoloma* was notably affected and plasmolysis occurred. These observations resemble earlier ones, that localized application of strong solutions of osmotically active substances, such as KNO_3 and cane sugar, resulted in cytoplasmic flow to such regions (ARTHUR, 1897; TERNETZ, 1900; SCHRÖTER, 1905). It also accounts for the effects of transpiration on rates of translocation of dyes, etc., if they are transported by cytoplasmic streaming. A mechanism for mass-flow of solutes which depends on differences in turgor pressures in different parts of the same system has been suggested by Münsch for translocation in sieve tubes of higher green plants. A similar mechanism could account for Jahn's observations up to pressure differences of the order of 4 bar. However, neither in fungi nor in higher green plants is this hypothesis satisfactorily proved and in fungi it has not been critically tested.

Table 9.10. Directions of flow of cytoplasm in *Humaria leucoloma* after anastomoses between mycelia grown on malt extract media of different osmotic pressures (bar). (From Jahn, 1934)

Osmotic pressure medium A	Direction of flow and number of observations	Osmotic pressure medium B	Differences in osmotic pressures B > A = +
↑	6 →	<0·1	−2·0
	3 →	0·2	−1·8
	3 →	1·3	−0·7
	5 →, ← 5	2·0	0
	← 5	2·7	+0·7
	3 →, ← 9	3·2	+1·2
	← 5	4·1	+2·0
2	← 5	4·7	+2·6
	← 5	5·3	+3·2
	1 →, ← 7	5·9	+3·9
	17 →, ← 9	9·9	+7·9
	4 →, ← 1	13·2	+11·1
↓	2 →, ← 1	>20·3	> +20·3

Buller acknowledged the possible effects of turgor pressure differences but suggested that other possible causes could be vacuolar pressure and pressure exerted by increase in the volume of growing cytoplasm. He noted that there was a close correlation in *R. nigricans* between alterations in the direction of cytoplasmic flow and vacuolar behaviour. Vacuoles in that part of the mycelium from which cytoplasm was flowing increased in size, while those in the receiving region diminished. When the flow reversed the vacuoles also showed a correlated decrease. A very clear-cut example of cytoplasmic pressure being built up due to vacuolation is visible during the development of basidia (hence CORNER's (1948a) term the 'ampoule effect'). Buller calculated that the cytoplasmic stream reached about 10 cm/hr in the developing basidium of *Coprinus sterquilinus* as a consequence of vacuolation. He supposed that in a young, actively growing hypha the pressure of growing cytoplasm was the principal motive force responsible for cytoplasmic streaming but that in older mycelia, vacuolar pressure in the ageing parts was the principal agent responsible for streaming. However, the older, vacuolated parts of a mycelium are not always filled with highly osmotic materials as expected on this hypothesis.

It seems not improbable that vacuolar pressure plays some part in unidirectional flow in mucoraceous fungi but, even in these fungi, some sort of turgor pressure gradient could also operate. However, these mechanisms neither explain the temperature- and oxygen-sensitivity of cytoplasmic streaming nor the observations of bidirectional flow. The former condition suggests the participation of metabolic energy in streaming, the latter either some sort of flow and return mechanism or the occurrence of different routes for translocation.

Two other hypotheses have been considered by Isaac, the sol-gel interface mechanism of KAMIYA (1959) and the 'jet propulsion' mechanism of KAVANAU (1963a, b). Neither of these hypotheses was originally developed to deal with the situation in fungi. In both a difference is assumed to exist between the outer ectoplasm and the inner endoplasm of the cytoplasm. In Kamiya's hypothesis the boundary between these two layers is the site of a shearing force, the ectoplasm being semi-rigid and attached to the cell membrane, the endoplasm semi-fluid. The result of this shearing force is to push the endoplasm forward, the ectoplasm suffering an equal and opposite force and hence a constant rotation of the cytoplasm. In Kavanau's hypothesis there are supposed to be undulations of the endoplasmic reticulum of the endoplasm which pump the enclosed fluid matrix backwards and hence move the reticulum and attached organelles forward.

So far no measurements similar to those made by Kamiya and his colleagues on *Nitella* and *Physarum* in establishing his hypothesis have been made in fungi. However, two kinds of observations have a bearing on Kavanau's hypothesis. Firstly, Girbardt has prepared time-lapse films of growing hyphae of *Polystictus versicolor* in which rhythmical waves can clearly be detected in the cytoplasm. Girbardt believes these to be caused by the sheets of endoplasmic reticulum. Secondly, Isaac has argued that it should be possible to detect the fluid counterflow due to the hypothetical undulations of endoplasmic reticulum in septate hyphae by using an interference microscope. He has detected a density drop across a septum of *Rhizoctonia solani* in which streaming was taking place; the cytoplasm of the upstream side had a dry matter content of 24%, that of the downstream side was only 8·5%. This difference was enhanced by fixation in 4% neutral formalin, which suggests that there is a greater proportion of soluble and extractable material on the downstream side of the septum than on the upstream side. This distribution is not inconsistent with the counterflow hypothesis. Not every hypha showed this phenomenon and, indeed, its manifestation appeared to be related to the nutrient status of the mycelium.

While it is possible that either of these hypotheses could account for cytoplasmic streaming in fungi it is not wholly clear how they could account for all the observed facts of translocation. Both evidently require the participation of energy either to bring about sol-gel interactions or undulations of endoplasmic reticulum. It is not clear how Kavanau's hypothesis can account for bidirectional translocation although presumably some ^{32}P could be bound to membranes or organelles and some could be free in the fluid matrix. Kamiya's hypothesis, although untested, does suggest a situation resembling that observed in bidirectional flow, for example, in *Rhizopus*. On the other hand, there is no reason to suppose that this feature of cytoplasmic streaming is the basis of bidirectional translocation.

Another possible mechanism is transcellular streaming in some form such as suggested for higher green plants by THAINE (1964), or in the more complex hypothesis of CANNY and PHILLIPS (1963). As yet, however, transcellular strands have not been observed in fungi but this is not a strong reason for rejecting the argument since their observation in higher green plants is not easy and is, indeed, still disputed. The rate of transcellular streaming given by Canny is comparable with that in fungi and the hypothesis accounts for bidirectional streaming and dependence of the process on temperature and respiratory energy.

The few facts available concerning translocation in fungi are hardly sufficient on which to make bold generalizations. There does seem to be good reason to suppose, however, that some aspects of translocation could be accounted for by cytoplasmic streaming based on internal turgor pressure gradients. Bidirectional flow, at present only known in a few examples, stresses once again the lack of knowledge concerning the internal compartmentalization of fungal cells referred to in the previous chapter. The conflicting evidence on whether a particular fungus can or cannot translocate and how effectively it can perform this activity would seem to arise from two causes. The first is differences in the actual technique employed; the second is that differences seem to arise from different environmental and nutritional conditions. This latter point would seem to support the notion that turgor pressure gradients, however caused, may determine translocation to a great extent. Nevertheless, it is clear that this mechanism is unlikely to be the sole one involved.

Chapter 10
Carbohydrate Catabolism

The role played by fungi, notably yeast, in the elucidation of the biochemistry of fermentations and respiration hardly needs to be reiterated (see HARDEN, 1923, for an historical review). It is, therefore, surprising that knowledge concerning the utilization of carbohydrates by filamentous fungi has lagged somewhat behind studies with other organisms. There are two reasons for this, both technical. The first is the difficulty experienced by several investigators in producing satisfactory, homogeneous and reproducible samples of mycelial material. Spore suspensions are not necessarily typical of hyphal material and, because of erratic and non-synchronous germination, may not produce results with a good standard of reproducibility. Cultures shaken in liquid media derived from spores or hyphal fragments can often be handled readily but constitute highly heteromorphic entities which may well be physiologically heterogeneous as well. A lightly blended mycelium derived from surface or submerged growths enables some of the heterogeneity to be randomized but raises problems of loss of material from cut ends of hyphae. Nevertheless, this is probably the best type of material to use and, in manometric studies for example, the coefficient of variability of samples can usually be kept to, or within 10%. Such samples have the added advantage that they can be bulked and so a reasonable amount of living material obtained for study.

The second reason for the lag in developing respiratory studies with fungi is their high endogenous respiration. This makes the results of experiments with exogenous substrates difficult to interpret and imposes restrictions upon the interpretation of inhibitor studies. As with nearly all biochemical studies in fungi, published work is heavily biased in favour of common moulds such as *Aspergillus*, *Penicillium*, *Fusarium*, *Neurospora* and mucoraceous fungi. Very little is known about other Phycomycetes, although *Blastocladiella emersonii* is an exception, and not a great deal about the Homobasidiomycetes. Fortunately, the similarity of respiratory processes in all organisms investigated is such that it is improbable that any strikingly novel features will be discovered. There seems little doubt that the Embden–Meyerhof–Parnas (EMP) pathway and the pentose-phosphate sequence, one of the hexose monophosphate (HMP) pathways occur in many fungi. This is true also of the tricarboxylic acid (TCA) cycle and terminal oxidation mechanisms involving cytochromes. Valuable reviews of these topics in relation to fungi have recently been written (COCHRANE, 1958; BLUMENTHAL, 1965; NIEDERPRUEM, 1965; LINDENMEYER, 1965) and excellent accounts of respiration in plants are also available in special texts (JAMES, 1953; BEEVERS, 1961) or general biochemical texts (e.g. MAHLER and CORDES, 1966). It would seem superfluous, therefore, to reiterate the detailed matter described so fully elsewhere. Here an outline will be given of the processes involved and attention directed to topics which are either unresolved or of especial interest in fungi.

PHOSPHORYLATIVE CARBON CATABOLISM

The catabolism of carbohydrates usually involves three main stages which can be related to their energy-yielding properties. In Stage I the carbohydrates are converted, if necessary, to an appropriate form, such as hexose, and phosphorylated. The process is effectively unidirectional, no useful energy is liberated and, indeed, adenosine triphosphate (ATP) is usually required. In Stage II degradation proceeds from a 6-C compound to 3- or 2-C compounds by various routes with the liberation of about one-third of the available free energy. In the final Stage III, the degradation is completed to 1-C compounds, notably CO_2, and intermediate products are available as substrates for anabolic processes. The TCA cycle is intimately involved in this stage. Throughout Stages II and III oxidation processes, usually of several linked enzyme and carrier systems, result in the transfer of electrons and ultimately in the reduction of oxygen as a terminal acceptor. The processes will be considered in this sequence.

Phosphorylation

In Chapters 7 and 8 it was pointed out that while exogenously supplied carbohydrates are utilized by many fungi, they are usually reduced to monosaccharides by extracellular enzymes before being transported into hyphae or cells. The numbers and kinds of extracellular carbohydrases known to be produced by fungi are very great and virtually any carbohydrate can be so hydrolysed by one fungus or another and several by many fungi. For example, *Aspergillus niger* is known to be able to produce α-amylase, α-glucosidase, glucoamylase, 'limit' dextrinase, cellulases, β-glucosidases, laminarinase, β-1:6 glucan hydrolase, invertase, β-1:4 xylanase, pentosanases, endopolymethylgalacturonase, endopolygalacturonase and exopolygalacturonase. Doubtless it produces others. Details of the extracellular enzymes of many fungi are usefully summarized by DAVIES (1963) in general and for cellulolytic and related enzymes in REESE (1963). Thus, while much of the carbohydrate transported into fungal cells is in the form of glucose, free glucose is difficult to demonstrate in hyphae. This is because it is phosphorylated by hexokinase, an intracellular enzyme which must, nevertheless, operate very near to the cell membrane, to glucose-6-phosphate. This enzyme functions in a similar manner with D-fructose and D-mannose where C-3 to C-6 have similar configurations, viz.:

$$\text{(glucose: } CH_2OH\text{, H, O, OH, HO, H)} + ATP \xrightarrow{Mg^{2+}} \text{(} CH_2OPO_3^{2-}\text{, H, O, OH, HO, H)} + ADP$$

The hexokinase of yeast has been prepared in crystalline form, and is known to have this broad specificity although the dissociation constants and velocity constants vary for each sugar, e.g. glucose $1{\cdot}5 \times 10^{-4}$ M, 13,000 min^{-1}; fructose $1{\cdot}5 \times 10^{-1}$ M, 26,000 min^{-1}; mannose 1×10^{-4} M, 6000 min^{-1}. Many yeasts cannot utilize galactose and those that do have to adapt to it; the same enzyme is not, therefore, involved in phosphorylation. In fact there is a specific galactokinase which acts in an analogous manner to hexokinase but is specific for galactose and produces galactose-1-phosphate. This is converted to glucose-1-phosphate and finally to glucose-6-phosphate. Di- and tri-saccharides may be hydrolysed but it will be recalled that maltose and trehalose can be transported intact and can accumulate. They may then be hydrolysed internally, as is trehalose in *Neurospora* ascospores, or conceivably, phosphorylated directly. Certain lactobacilli are capable of phosphorylating maltose direct to glucose-1-phosphate without the participation of ATP but there is no evidence for this in any fungus.

Once hexose-6-phosphate has been formed the next stage in the catobolic process can take place.

Partial degradation

In this stage the phosphorylated C_6-compounds may be broken down by several different metabolic pathways illustrated diagrammatically in Figs. 10.1 and 10.2.

THE EMP PATHWAY

The characteristic reaction is the cleavage of fructose,1,6-diphosphate to give a mixture of triose phosphates and the principal rate-controlling step is said to be the oxidation of glyceraldehyde 3-phosphate to 1,3-diphosphoglyceric acid by triose phosphate dehydrogenase. During this reaction $NADH_2$ is generated, coupling phosphorylation and dehydrogenation. There is a net

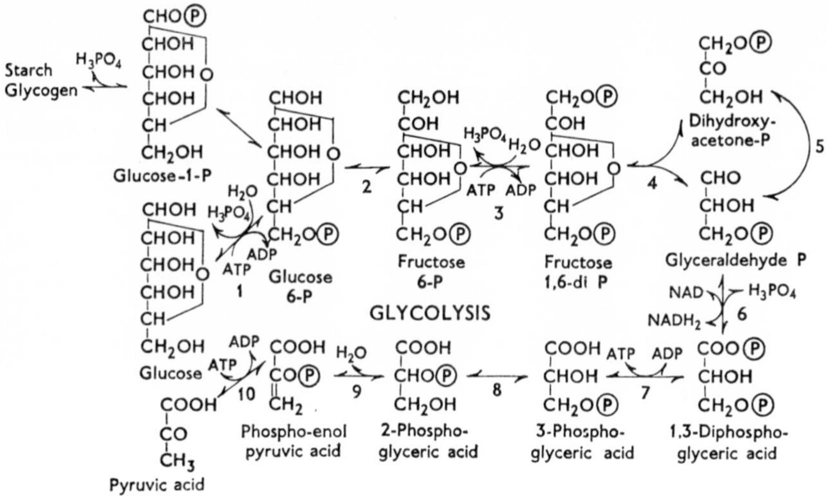

Fig. 10.1. The Embden–Meyerhof–Parnas pathway of glycolysis. Enzymes 1–10 as in Table 10.1; ADP/ATP, adenosine di- and tri-phosphate; NAD/$NADH_2$, nicotinamide-adenine dinucleotide, oxidized and reduced forms; H_3PO_4, inorganic phosphate (P_i) (After Hawker *et al.*)

gain of two moles of ATP for every mole of glucose consumed at the stages

1,3-diphosphoglyceric acid $\longrightarrow$ 3-phosphoglyceric acid

and phosphoenolpyruvic acid $\longrightarrow$ pyruvic acid

The overall balance (including the initial phosphorylation step) can be expressed as

$$C_6H_{12}O_6+2NAD+2P_i+2ATP+2ADP \longrightarrow 2CH_3CO.COOH+2NADH_2+4ATP$$

THE HMP PATHWAYS

A number of alternative pathways exist in which the essential compound that is broken down is glucose-6-phosphate. The best known in fungi is the pentose-phosphate sequence. Here the glucose-6-phosphate is converted via 6-phospho-gluconate to the pentose compound ribulose-5-phosphate and CO_2. Thereafter a number of phosphorylated intermediates with 3-, 4-, 5-, 6- or 7-C atoms may be produced. These are, of course, valuable intermediates in all kinds of synthetic processes but the sequence can accomplish a total break-down to trioses such as glyceraldehyde 3-phosphate and, thereafter, the same sequence as in the EMP pathway may obtain. In other reactions hexoses can be regenerated. Pentoses can, presumably, enter the sequence at the appropriate point. Since the first two steps involve the reduction of NADP, this coenzyme can act as a limiting factor.

Recently, an alternative sequence, the Entner–Doudoroff (ED) sequence has been found to exist in a few fungi. The first stages in the sequence are similar to those of the pentose-phosphate sequence but the 6-phosphogluconate is converted to 2,keto-3,deoxy-6,phosphogluconate, thence to glyceralde-

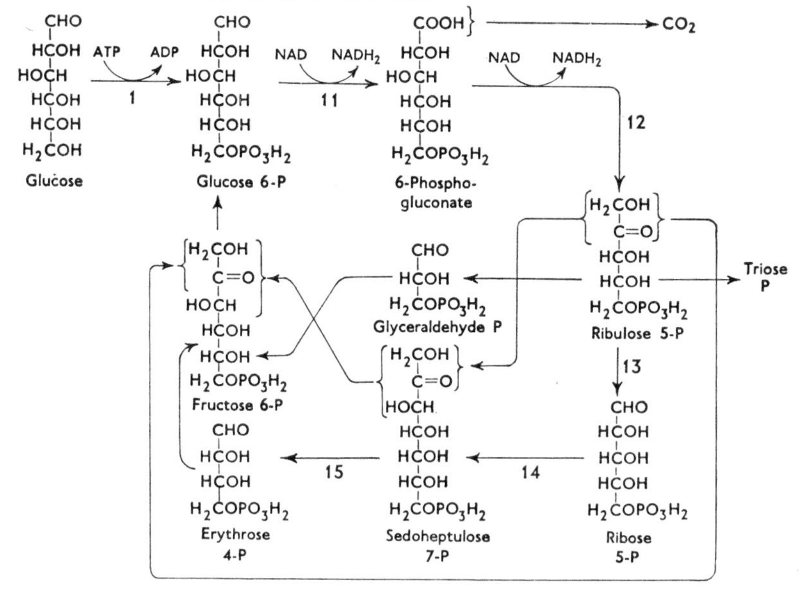

B. THE ENTNER-DOUDOROFF SEQUENCE

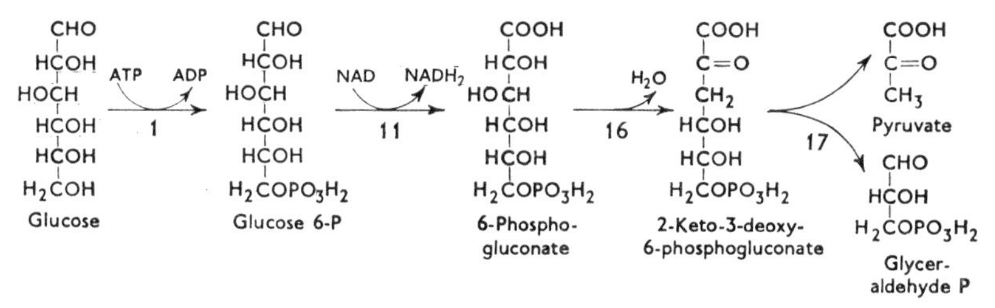

Fig. 10.2. The hexose monophosphate pathways. A. Pentose phosphate sequence; B. Entner–Doudoroff sequence. Enzymes 1, 11–17 as in Table 10.1; ADP/ATP, NAD/$NADH_2$ as in Fig. 10.1. (After Hawker *et. al.*)

hyde-3, phosphate and, ultimately, by the same stages as the later ones of the EMP pathway, to C-3 compounds.

DISTINCTION BETWEEN THE PATHWAYS

Table 10.1 lists the enzymes involved in the three degradative processes described for phosphorylated monosaccharides. It can be seen that they have a number of enzymes in common and, in each case, a number of distinctive enzymes.

The presence of phosphogluconate dehydrogenase or of transketolase or transaldolase is highly suggestive of the occurrence of the pentose-phosphate sequence while 6-phosphogluconate dehydrase and 2-keto-3,deoxy-6,phospho-

Table 10.1. Enzymes in the EMP and HMP pathways including the pentose phosphate (PP) and Entner-Doudoroff (ED) sequence, together with fungi in which they have been found, and some indications of the relative importance of the pathways. The reactions are those shown in Figs. 10.1 and 10.2. (Adapted from Blumenthal, 1965)

(**a**) Enzyme distribution

Pathway EMP	PP	ED	Enzyme	Reaction	Example
+	+	+	Hexokinase	1	*1, 3, 5, 6, 7, 8, 9*
+	+	−	Phosphohexoisomerase	2	*1, 3, 5, 6, 7, 8,*
+	−	−	Phosphofructokinase	3	*1, 3, 9*
+	+	−	Aldolase	4	*1, 3, 4, 5, 6, 7, 8, 9, 9m*
+	+	+	Triosephosphate isomerase	5	*9*
+	+	+	Phosphoglyceraldehyde dehydrogenase	6	*1, 4, 5, 6, 7, 9, 9m*
+	+	+	Phosphoglycerate kinase	7	*4, 7, 9, 9m*
+	+	+	Phosphoglycerate mutase	8	*1, 2, 4, 5, 6, 7, 9*
+	+	+	Enolase	9	*1, 2, 4, 5, 6, 7, 9*
+	+	+	Pyruvic kinase	10	*1, 2, 4, 5, 6, 7, 9*
−	+	+	Glucose-6-P dehydrogenase } 6-Phosphogluconate }	11	*1, 3, 5, 6, 7, 8, 9a*
−	+	+			
−	+	−	6-Phosphogluconic dehydrogenase	12	*1, 3, 5, 6, 8, 9a*
−	+	−	Phosphoriboisomerase	13	
−	+	−	Transketolase	14	*1, 3, 6, 8, 9*
−	+	−	Transaldolase	15	*1, 3, 6, 8, 9*
−	−	+	6-Phosphogluconate dehydrase	16	*10*
−	−	+	2-Keto-3-deoxy-6-phosphogluconate aldolase	17	

1. Aspergillus niger. 2. Caldariomyces fumago. 3. Claviceps purpurea. 4. Fusarium lini. 5. Microsporum canis. 6. Penicillium chrysogenum. 7. Rhizopus MX. *8. Tilletia caries. 9. Ustilago maydis* spores; *9a.* Absent until 12 hr after germination; *9m. U. maydis* mycelium. *10. Aspergillus flavus-oryzeae.*

(**b**) Percentage pathway utilization in different fungi referred to above.

Fungus	EMP	PP	ED
Aspergillus niger	78	—	—
Caldariomyces fumago	—	35	65
Claviceps purpurea	90·96	10–4	—
Fusarium lini	83	17	—
Penicillium chrysogenum	56·70	46–30	—
	42	58	—
	77	23	—
Rhizopus MX	100	—	—
Tilletia caries			
Spores	—	—	100
Mycelium	66	34	—

gluconate aldolase are equally suggestive of the ED sequence; phosphofructokinase is the only enzyme in the EMP pathway that is not involved in either of the HMP pathways. However, while the presence of an enzyme or, indeed its substrates, may be suggestive, it is not clear evidence of the operation of the pathway. Another method commonly employed is that of supplying glucose whose carbon atoms are specifically replaced by radioisotopic ^{14}C and then investigating the distribution of such labelled carbon in the products of dissimilation (Fig. 10.3.)

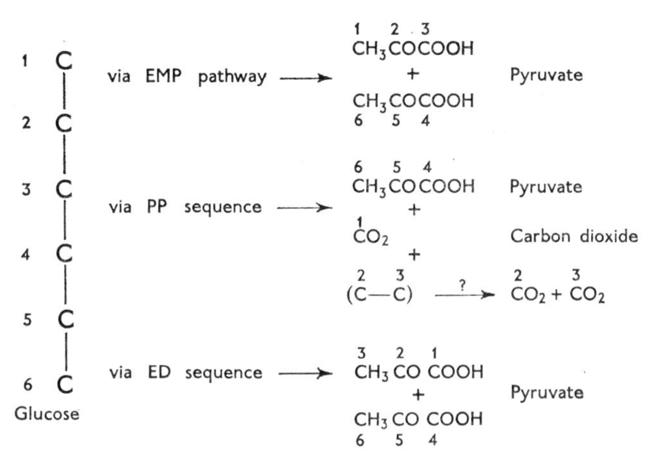

Fig. 10.3. The distribution of C-atoms of glucose in the products of the EMP and HMP pathways

In the EMP pathway glucose is cleaved symmetrically so that C-1, C-2 and C-3 are equivalent to C-4, C-5 and C-6. If CO_2 is released by the subsequent action of the TCA cycle then the first CO_2 to be released is expected to be derived from C-3 and C-4, the next from C-2 and C-5 and finally from C-1 and C-6. In the pentose phosphate sequence, however, the first CO_2 released is expected to be derived from C-1 and if only [1^{14}-C]-glucose is supplied the CO_2 will be radioactively labelled and the other intermediates not. In the ED sequence C-1 and C-4 positions provide the carbon atoms in the pyruvate produced so that the CO_2 produced will be derived from C-1 and C-4, C-2 and C-5, and C-3 and C-6 in that order. Thus, in principle, it should be possible not only to determine whether or not a particular sequence is participating in the breakdown of glucose, for example, but also the extent of such participation. In practice the quantitative evaluation of the relative contributions of different pathways has proved to be fraught with difficulties. For example, any block in the operation of the TCA cycle would reduce the C-1 released as CO_2 from the EMP or ED pathways but would have no effect on the pentose-phosphate sequence. In a similar manner the operation of this latter sequence could confuse the issues because, through its cyclic properties, hexose phosphate can be reformed and this, of course, could be utilized by any degradative process then operating. Such re-formed hexose phosphate would be derived from any of the combinations C-2-3-2-4-5-6, C-2-3-3-4-5-6, or C-6-5-4-4-5-6. Despite these difficulties attempts have been made to study the operation of these processes by examining the radioactive labelling and carbon sequence involved in CO_2 production (e.g. WANG *et al.*, 1958, and Table 10.1b) and the radioactive labelling of intermediates with allowance for recycling in both the pentose phosphate sequence (BLUMENTHAL *et al.*, 1954; DAWES and HOLMS, 1958) and the ED sequence as well (RACHMACHANDRAN and GOTTLIEB, 1963).

OCCURRENCE OF THE PATHWAYS IN FUNGI

By studies of the occurrence of enzymes and intermediates and of radioisotopic distributions in CO_2 or intermediates after supplying [^{14}C]-glucose in variously labelled forms, it has proved possible to infer the presence and contribution of the different pathways in a number of fungi.

The EMP pathway is apparently the most widespread and is apparently responsible for the bulk of the degradation of hexose phosphate material. There seem to be no very unusual situations except perhaps the differences between the glucose catabolism of spores versus mycelium in *Neurospora crassa* and *Tilletia caries* respectively, and the total absence of the EMP pathway in *Caldariomyces fumago*, an imperfect sooty mould found on leaves of various trees. It is perhaps of significance that in both *C. fumago* and spores of *T. caries* the cells were in a 'resting' condition when the EMP pathway was apparently totally inoperative.

The occurrence of a number of enzymes common to all these pathways provides an interlocking device which could act as a control mechanism under certain conditions. For example, RACKER (1965) has pointed out that glucose-phosphate isomerase, triose-phosphate isomerase and aldolase are operating to produce pyruvate in glycolysis but to synthesize glucose-6-phosphate in the pentose phosphate sequence. The availability of substrates for forward or backward reactions governs the events which actually occur, so that there is competition for substrate. However, in some organisms (although this has not been shown for fungi) several intermediates in the pentose-phosphate sequence, such as phosphogluconate and sedoheptulose-7-phosphate, are most effective inhibitors of glucose-phosphate isomerase. Thus one pathway can have a profound effect on another acting here as a negative feed-back mechanism (cf. Chapter 11, pp. 314–317).

It is clear that, at present, wholly satisfactory evidence for all the appropriate enzyme systems and for estimates of the quantitative operation of the various pathways does not exist for any fungus. Moreover, the fungi studied represent the common moulds whereas information concerning the larger Basidiomycetes or lower Phycomycetes, for example, is conspicuously lacking.

Final degradation to carbon dioxide

During the partial processes of degradation described in the previous section some CO_2 may be released but the majority of energy available in hexose has not yet been made available. This is achieved in the final phase in which pyruvate is dissimilated usually aerobically, via the TCA and glyoxylate cycles (Fig. 10.4), with the formation of a variety of di- and tricarboxylic acids which can function as intermediates for syntheses or, in certain conditions, accumulate and remain stored and metabolically inactive. The excellent and critical review by NIEDERPRUEM (1965) has already been mentioned and only a few comments need be made here.

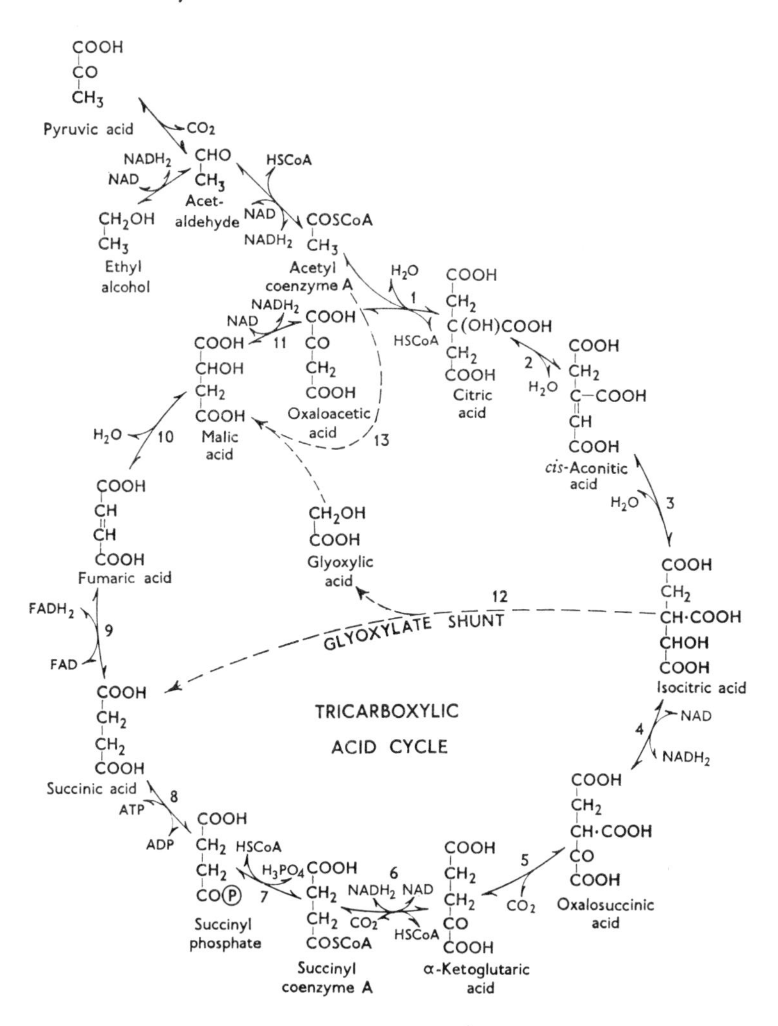

Fig. 10.4. The tricarboxylic acid cycle. Enzymes involved: 1. Citrate condensing enzyme; 2, 3. Aconitase; 4, 5. Isocitric dehydrogenase; 6, 7, 8. Succinyl CoA synthetase; 9. Succinic dehydrogenase; 10. Fumarase; 11. Malic dehydrogenase; 12. Isocitritase; 13. Malate synthetase; ADP/ATP, NAD/$NADH_2$, H_3PO_4 as in Fig. 10.1. (After Hawker *et. al.*)

Great difficulty is experienced when attempts are made to investigate the TCA cycle by the classical techniques of feeding acids as respiratory substrates such as was done by Krebs. This is due to the high endogenous respiration of fungi. Moreover, as mentioned earlier (cf. pp. 202–204), fungal membranes are impermeable to a number of organic acids such as succinate and also to inhibitors of the TCA cycle, such as malonate, iodoacetate and fluoride. Thus little direct evidence is available to indicate the operation of the TCA cycle in the intact mycelium of many fungi. However, there is now a good deal of evidence for the presence of appropriate enzymes, their location in subcellular fractions and the operation of the cycles in cell-free extracts. In many cases, notably *Aspergillus niger*, a number of the TCA cycle enzymes have been isolated and purified to a greater or lesser degree. Enzyme activity has usually been found to be associated with the mitochondrial fraction, as would be expected on grounds of comparative biochemistry, provided that the disruptive procedures were mild and appropriate media were used for isolation. Particulate fractions with TCA cycle enzymes have now been isolated from most of the major groups of fungi.

The point of entry into the TCA cycle of the products of glucose catabolism is the conversion of pyruvate in the presence of NAD and Coenzyme A (CoA) to acetyl CoA and $NADH_2$ with liberation of CO_2. CoA is the carrier for acetyl groups, to which it is linked by a thioester linkage. The combination of acetyl CoA with oxaloacetic acid to form citrate constitutes the first step of the cycle and oxalacetic acid is eventually reformed provided that intermediate acids are not utilized for other processes. In practice this frequently, if not always, occurs, so that the loss in oxalacetic acid has to be made good in order to maintain the cycle. This is achieved in two ways. Firstly, oxaloacetic acid can be provided by the carboxylation of pyruvate to malate followed by its conversion, the enzymes involved being malic enzyme and malic dehydrogenase. Alternatives are the conversion of phospho-enol-pyruvate (PEP) by either PEP carboxykinase or PEP carboxylase to oxaloacetic acid. These reactions are summarized below:

(i) $\text{Pyruvate} + NADPH_2 + CO_2 \rightleftharpoons \text{malate} + \text{NADP}$

$\text{Malate} + \text{NAD} \rightleftharpoons \text{Oxaloacetate} + NADH_2$

(ii) $\text{Phospho-enol-pyruvate} + \text{ADP} + CO_2 \rightleftharpoons \text{Oxaloacetate} + \text{ATP}$

(iii) $\text{Phospho-enol-pyruvate} + CO_2 + H_2O \longrightarrow \text{Oxaloacetate} + P_i$

Little is known concerning the distribution of malic enzyme or of PEP carboxylase or carboxykinase in fungi. There is evidence for their existence in yeast and *A. niger* (CANNATA and STOPPANI, 1963a, b; WORONICK and JOHNSON, 1960). These reactions involve heterotrophic CO_2 fixation and there is evidence that this is more widespread in fungi than is suggested by the enzymes distribution just mentioned. CO_2 fixation occurs in Mucoraceae (FOSTER *et al.*, 1941), *Allomyces arbuscula* (LYNCH and CALVIN, 1952), *Puccinia recondita* (STAPLES and WEINSTEIN, 1959), yeasts and Aspergilli and may follow one of these routes. In other organisms, for example *Neurospora*, CO_2 is incorporated only in the presence of nitrogen sources as well.

A second source of oxaloacetic acid is its production via the glyoxylate shunt due to malate synthetase and isocitritase. These enzymes are now known from a number of fungi. Isocitritase is an adaptive enzyme and is not formed in the presence of glucose or succinate in the medium (KREBS and LOWENSTEIN, 1960). Thus the glyoxylate shunt only operates when there is a shortage of succinate.

TERMINAL ELECTRON-TRANSFER

So far the discussion of this final stage of degradation has been concerned with the production of decreasingly small C-fragments, ultimately CO_2. The energy so released is, however, available in the CoA compounds and can be transferred to other acceptors of acetyl groups in the cells. The electrons removed at the four steps of dehydrogenation in the TCA cycle are, in three cases, transferred to NAD and, in the fourth case, to the active group of succinic dehydrogenase which is a flavoprotein. These electrons can then be transferred through a terminal respiratory chain involving cytochromes and finally to oxygen, so reducing it to water. The overall reaction in many organisms can be summarized as

$$NADH_2 + \tfrac{1}{2}O_2 \longrightarrow NAD + H_2O$$

Cytochromes have been identified by their absorption spectrum at low (−195°C) temperatures and in the presence of $Na_2S_2O_4$ in all groups of fungi (BOULTER and DERBYSHIRE, 1957) and by their susceptibility to inhibition by cyanide and CO, the latter being reversed by exposure to high light intensities. In various fungi different cytochromes have been detected, isolated and purified to various degrees, e.g. *Ustilago sphaerogena* produces up to 1% of its dry weight as cytochrome C under certain conditions (NEILANDS, 1952; GRIMM and ALLEN, 1954) so that it has been isolated fairly readily. The precise sequence in which the cytochromes operate is not yet known for any fungus. It has been claimed that 'the cytochrome system of the fungi is remarkably similar to that of mammalian and avian cells and unlike the systems found in bacteria and green plants' (LINDENMEYER, 1965).

Terminal oxidation could, conceivably, be mediated by other systems. A cyanide-insensitive respiration is known in several fungi, e.g. *Myrothecium verrucaria* (DARBY and GODDARD, 1950), *Ustilago sphaerogena* (GRIMM and ALLEN, 1954), on *Candida albicans* (WARD and NICKERSON, 1958) and mycorrhizal roots of beech (HARLEY and AP REES, 1959). This phenomenon has been ascribed to the operation of a flavoprotein oxidase in other organisms but this view has not received much support (BEEVERS, 1961). A more plausible explanation has been provided for the phenomenon in higher plants by the discovery of cytochrome b_7, which is cyanide resistant. This cytochrome is thought to be operative only under conditions inhibiting the normal cytochromes. There is as yet no evidence for the occurrence of cytochrome b_7 in any fungus. Another common oxidase type in fungi is polyphenol-oxidase. The possibility that this could act as a terminal oxidase in *Polystictus versicolor* and *Neurospora crassa* has been examined by BOULTER and HURST (1960) and rejected on the grounds that a large part of the O_2 uptake was inhibited, reversibly by CO_2 even though polyphenol oxidases were present. This is, however, suggestive rather than conclusive.

OXIDATIVE PHOSPHORYLATION

The transfer of electrons through the terminal processes is accompanied by changes in free energy and these are thought to be utilized in the synthesis of ATP from ADP and P_i. Thus oxidation and phosphorylation are coupled and this can be measured by determining ATP production (as P_i esterified) and oxygen uptake. The results are usually expressed as the P/O ratio (micromoles P_i esterified: microatoms O absorbed), values of which can be predicted for different substrates. Thus terminal respiration can better be described by the equation:

$$NADH_2 + \tfrac{1}{2}O_2 + 3ADP + 3P_i \longrightarrow NAD + 3ATP + 4H_2O$$

Few measurements of P/O ratios have been made in fungi. Particles, presumably mitochondrial, which effect oxidative phosphorylation have been extracted from yeast (VITOLS and LINNANE, 1961), *Aspergillus oryzeae* (IWASA, 1960) and *Allomyces macrogynus* (BONNER and MACHLIS, 1957). In all of these the values of the ratio were lower than those predicted but how far it is due to the consequences of extraction and resuspension in the experimental situation is not known. Evidently in some fungi, and probably in many others where phosphorylation can be uncoupled by 2,4DNP, the oxidative processes are responsible for utilizing the major part of the energy released from the breakdown of hexose.

NON-PHOSPHORYLATIVE CARBON CATABOLISM

A number of fungi possess the ability to oxidize glucose without any prior phosphorylation. This is due to a glucose oxidase, better termed glucose aerodehydrogenase, an enzyme first discovered in *A. niger* (MÜLLER, 1926). In *Penicillium chrysogenum* it is a flavoprotein enzyme with high, specific activity towards β-D-glucose. The first product of the reaction is glucono-δ-lactone and then gluconate. The overall reaction can be written

$$Glucose + O_2 + H_2O \longrightarrow gluconate + H_2O_2$$

The enzyme utilizes molecular oxygen as a hydrogen acceptor (BENTLEY and NEUBERGER, 1949) and was at one time described as an antibiotic 'notatin', although its action derived from the production of H_2O_2 (COULTHARD *et al.*, 1945). Because there is similar specific activity towards mannose and galactose to give mannoic and galactonic acids respectively, it has been supposed that other similar aerodehydrogenases exist (KNOBLOCK and MAYER, 1941). However, the fact that several fungi are known to be capable of producing gluconic acid but apparently lack glucose aerodehydrogenase, e.g. *Mucor racemosus*,

Dematium pullulans (FRANKE and DEFFNER, 1939), suggests that claims for such enzymes should be treated with caution.

The role, if any, of this oxidation in respiration is unclear. Gluconate could be phosphorylated and enter an HMP pathway; however, in Pseudomonads and *Leuconostoc* spp. it is broken down further to oxogluconate and after phosphorylation, to phosphoglycerate. Since 5-oxogluconate has been found in *A. niger* it is possible that a sequence similar to that in the bacteria may occur here. The H_2O_2 produced in the primary reaction could act as an oxidizing agent in the presence of catalase. This enzyme is known from yeast, *Neurospora* spp., *Aspergillus niger* and *A. oryzae*. The latter two also possess glucose aerodehydrogenase activity. The natural role of this enzyme, whose presence is utilized in a commercial production of gluconic acid by *A. niger* (MARTIN, 1963), has yet to be elucidated and related to phosphorylative catabolism of carbon compounds.

ENDOGENOUS RESPIRATION

Attention was drawn at the outset of this chapter to the high rate of endogenous respiration shown by many fungi. There is no reason to suppose that the same substrate is used at all times and in all parts of the mycelium. Indeed, experiments by MANDELS (1963) on spores of various species provided results suggesting that different substrates were utilized at different phases of germination and growth. BLUMENTHAL (1963) studied the inhibitory effects of glucose and acetate separately and together on endogenous respiration in *Penicillium chrysogenum*. It was concluded that these substances inhibited different endogenous substrates although there was no evidence for the utilization of amino acids. In contrast to this, MIZUNUMA (1963) has studied the effects of varying the C/N ratio on the endogenous respiration of *A. sojae*. He claimed that lipid or carbohydrate was the main endogenous substrate with high C/N ratios, being replaced later by nitrogenous compounds with low C/N ratios; however, pool amino acids, protein and nucleic acids were used preferentially to lipid and carbohydrate.

It is likely, therefore, that there is more than one endogenous substrate but in the presence of suitable exogenous materials the utilization of endogenous substances may be highly suppressed, e.g. 70% by exogenous glucose in the case of *Neurospora* conidia (BLUMENTHAL, 1963).

Chapter 11
Accumulated and Synthesized Products and their Metabolism

A bewildering diversity of substances can be found within the mycelia of the fungi. In the previous chapter a wide range of degradation products which arise from the breakdown of carbohydrates has been described. In some circumstances their products accumulate and appear to play no immediate functional role, e.g. organic acids. These can be described as primary accumulated products. A second class of substances, primary metabolites, are of general distributions and are associated with the maintenance and growth of the living cell and its structure. They are derived more or less directly from primary synthetic processes involving the degradation products of carbohydrate catabolism, e.g. amino acids and proteins, fatty acids, polysaccharides. Finally, there is a vast and ever-increasing number of secondary metabolites, derived in various ways from the primary metabolites but differing in that their functions are either highly specific or unknown while their production is often confined to one, or a few, fungal species.

It is not possible to consider all these substances here and, indeed, it is unnecessary for admirable and detailed reviews or books are already available. Here an indication of the pathways will be given, together with reference to more detailed studies, and finally, an outline of the regulations imposed upon cellular processes. The general relationship of these processes to the main catabolic sequence for carbohydrates is set out in Fig. 11.1.

PRIMARY ACCUMULATED PRODUCTS

The best known of these are the organic acids, their immediate derivatives and alcohol. Some of these, notably ethanol and citric acid, are the basis of

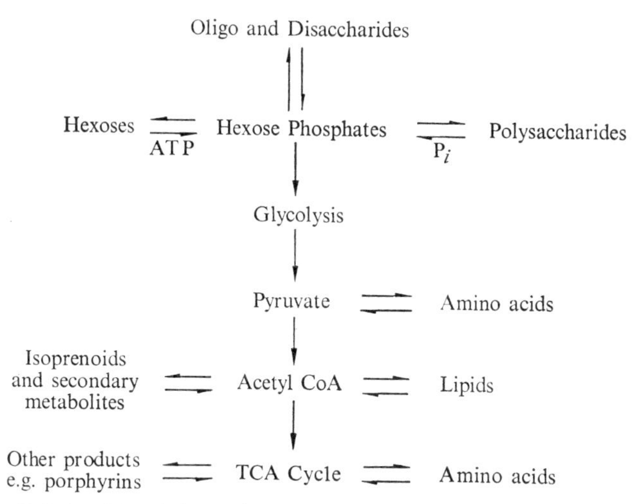

Fig. 11.1. Relationship of the main glycolytic sequence to catabolic and synthetic abilities

industries of world-wide importance (WILKINSON and ROSE, 1963; MARTIN, 1963). Accumulations of metabolic intermediates may arise for a number of reasons.

Environmental conditions may favour the direction of metabolism in a particular way and the internal regulation of cellular processes may have a similar effect. Alternatively, there may be an actual metabolic block due to the loss or malfunction of a particular enzyme or step in a sequence. Whatever the cause, accumulations can only occur if the metabolite is in some way isolated from the metabolic systems which would normally utilize it. This could be due either to internal partitioning into a metabolically inaccessible pool (the physical basis of which is at present unclear) or to the actual loss of material, e.g. ethanol from the cell.

Accumulation of alcohol and related compounds

The accumulation of ethanol as the principal product, by yeast, exemplifies processes determined by an external environmental factor, in this case, anaerobiosis. Thus the overall equation is,

$$C_6H_{12}O_6 \longrightarrow 2CH_3CH_2OH + 2CO_2$$

Fermentation of glucose to produce ethanol is a property which many fungi share with yeast, notably the common mould genera, e.g. *Aspergillus*, *Fusarium*, *Mucor*, but also a wide range of Ascomycetes, Basidiomycetes and Fungi Imperfecti. Ability to produce and accumulate ethanol is highly variable. For example, TAMIYA and MIWA (1928) compared the fermentative ability (as ml./CO_2/mg/dry wt.) of different species of *Aspergillus*. *Aspergillus clavatus* gave a value of 4161 compared with 200–300 for *A. niger*, and 0·10 for *A. nidulans*. Moreover, not all fungi require anaerobic conditions to produce alcohol and in some cases the yield approaches 50% of that expected for complete fermentation of glucose, e.g. *Fusarium oxysporum* 53%, *A. flavus* 48·5% (LOCKWOOD *et al.*, 1938; KLUYVER and PERQUIN, 1933). The production of ethanol, whether aerobically or anaerobically, is also a function of the age of the mycelium as the data of TAMIYA (1928) show clearly for *A. oryzae* (Table 11.1) under anaerobiosis.

Table 11.1. The effect of age of cultures on the ratio of aerobic to anaerobic ethanol production by *Aspergillus oryzae* grown in surface liquid culture. (From Tamiya, 1928)

Age (days)	1	2	3
Anaerobic culture	5·2	7·9	16·6
Aerobic culture	1	1	1

It is now generally held that ethanol is formed as a consequence of the breakdown of pyruvate:

$$\text{(i)} \qquad CH_3CO\ COOH \xrightarrow[\text{Co-carboxylase}]{\text{Pyruvate decarboxylase}} CH_3CHO + CO2$$

$$\text{(ii)} \qquad CH_3CHO + NADH_2 \xrightarrow[\text{Dehydrogenase}]{\text{Alcohol}} CH_3CH_2OH + NAD$$

Evidence for these reactions arises from the detection of the enzymes involved in several fungi and by the effects of Mg, Zn and thiamin deficiency on fermentation. Mg is a constituent of yeast carboxylase and alcohol dehydrogenase is a Zn-protein (VALLEE *et al.*, 1955). Thiamin in the form of the pyrophosphate is co-carboxylase and hence in thiamin deficient cultures pyruvate accumulates, e.g. *Fusarium* spp. (WIRTH and NORD, 1942), while excess thiamin, which inhibits growth, results in increased yields of co-carboxylase and alcohol, e.g. *Rhizopus nigricans* (FOSTER and GOLDMAN, 1948). In this latter case, however, the ethanol was ultimately oxidized and utilized once the exogenous glucose had all been used. Here, therefore, is an example of how the internal regulation of an enzymic system within a fungal cell can be determined by the presence of external metabolites.

In yeast fermentation, which has been intensively studied because of its industrial importance, the products are affected by the pH and the presence of external metabolites in addition to those just mentioned. For example, glycerol production could be enhanced greatly by either operating at an alternative pH (Table 11.2) or by addition of sulphite (NEUBERG and HIRSCH, 1919a, b). Under alkaline conditions the dynamic equilibrium between the triose derivatives, 3-phosphoglyceraldehyde and dihydroxyacetone phosphate, due to isomerase and normally in the latter's favour is shifted further in favour of glycerol-3-phosphate production. This is brought about by transfer of hydrogen from $NADH_2$ and, finally, phosphate removal by phosphatase

Table 11.2. The effect of different pH values on the products of fermentation of 5% glucose by *Saccharomyces cerevisiae*. Ammonium hydroxide was added automatically to maintain the pH value. (Extracted from Neish and Blackwood, 1951)

Data as mM product/100 mM glucose fermented

Product	pH 3	4	5	6	7
Ethanol	171·5	177·0	172·6	160·5	149·5
CO_2	180·8	189·8	187·6	177·0	161·0
2,3-Butanediol	0·75	0·48	0·46	0·53	0·45
Acetoin	—	—	—	—	0·07
Glycerol	6·16	6·60	7·82	16·2	22·2
Acetic acid	0·52	0·69	0·84	4·03	8·68
Butyric acid	0·13	0·32	0·25	0·36	0·25
Formic acid	0·36	0·42	0·63	0·82	0·35
Lactic acid	0·82	0·38	0·47	1·63	1·93
Succinic acid	0·53	0·26	0·32	0·49	0·23
Glucose C assimilated	12·4	16·1	14·0	12·4	—
% Glucose fermented	98·5	97·0	96·5	98·0	98·3
% Fermentation time (hr)	29	14·5	17·5	15·5	35
% C recovered	93·8	98·0	96·3	96·4	92·5

produces glycerol. There is thus a deficiency in the amount of $NADH_2$ available to reduce acetaldehyde which undergoes dismutation into acetate and ethanol. Thus the overall reaction is:

$$2C_6H_{12}O_6 \xrightarrow{+H_2O} 2CH_2(OH)\,CH\,(OH)\,CH_2OH + CH_3COOH + CH_3CH_2OH + 2CO_2$$

the two alternatives sequences involved being:

(a) Dihydroxy acetone-P

$$\begin{array}{c} CH_2OH \\ | \\ CO \\ | \\ CH_2O(P) \\ + \\ NADH_2 \end{array} \rightleftharpoons \begin{array}{c} CH_2OH \\ | \\ CHOH \\ | \\ CH_2O(P) \\ + \\ NAD \end{array} \underset{}{\overset{\text{Phosphatase}}{\rightleftharpoons}} \begin{array}{c} CH_2OH \\ | \\ CHO + P_i \\ | \\ CH_2OH \end{array}$$

(b) Acetaldehyde $2CH_3CHO \xrightarrow{+H_2O} CH_3CH_2OH + CH_3COOH$

The products rarely show these precise stoichiometric relationships. In the presence of sulphite the acetaldehyde formed combines with the sulphite and cannot act as an hydrogen acceptor for $NADH_2$. Once again dihydroxy acetone phosphate replaces acetaldehyde as an acceptor and the same reduction sequence to glycerol ensues. Thus,

$$C_6H_{12}O_6 + SO_3^{2-} \longrightarrow CH_2(OH)\,CH(OH)\,CH_2OH + CH_3CHO.SO_3^{2-} + CO_2$$

This process was employed during the First World War to produce glycerol in Germany. It is not now used.

In some fungi, notably species of *Rhizopus*, ethanol formation is associated with an approximately equimolar yield of lactic acid. GIBBS and GASTEL (1953) supplied radioactive glucose either as [^{14}C-1] glucose or [^{14}C-3,4] glucose to *R. oryzae*. C-1 and C-6 were found in methyl groups of lactate and ethanol, and the carboxyl of lactate and the CO_2 were derived from C-3 and C-4. This suggests that a common pool of triose compounds gives rise to pyruvate which could be followed by dismutation, viz:

$$2CH_3CO\ COOH \xrightarrow{4H} CH_3CHOH\ COOH + CH_3CH_2OH + CO_2$$

It is not clear how this step is brought about nor how the equimolar relationships are maintained. In *Rhizopus* MX far more lactic acid is produced under aerobic conditions, as can be seen in Table 11.3 (WAKSMAN and FOSTER, 1939).

Table 11.3. The aerobic and anaerobic dissimilation of glucose by *Rhizopus* MX grown in surface-liquid culture. (After Waksman and Foster, 1939)

Condition	Age (days)	Glucose consumed (g)	Lactic acid produced (g)	Ethanol produced	Glucose: Lactic: ethanol ratio
Aerobic	15	82·8	61·5	4·3	1:1·49:0·2
Anaerobic	15	53·9	28·5	13·1	1:1·05:0·59
Anaerobic	20	57·3	29·4	15·8	1:0·98:0·93

The reasons for this are not clear. In contrast, several genera and species of water moulds, e.g. *Allomyces arbuscula*, *Blastocladia pringsheimii* (INGRAHAM and EMERSON, 1954; CANTINO, 1949) accumulate lactic acid without the accompaniment of ethanol. Accumulation is not enhanced in aerobic conditions and over 50% of any glucose supplied is so converted. Very recently (GLEASON *et al.*, 1966) an NAD-linked, D-lactic dehydrogenase has been dected in several Leptomitales as well as in *Blastocladiella* (CANTINO and LOVETT, 1960). This suggests that in these fungi lactic acid is produced in exactly the same way as in animal muscle, viz:

$$NADH_2 + CH_3CO\ COOH \xrightleftharpoons{\text{lactic dehydrogenase}} CH_3CHOH\ COOH + NAD$$

In these fungi, therefore, the substance accumulated reflects a particular enzymatic endowment not apparently possessed by the majority of the fungi.

Accumulation of organic acids

The accumulation of lactic acid has already been discussed; in addition several acids of the TCA cycle also accumulate in many fungi, notably citric acid in the 'black moulds' (*A. niger* and allied species). A full list of acids accumulated by various fungi is given by COCHRANE (1958).

CITRIC ACID

Since about 1930 the majority of commercial citric acid produced has been made by mould processes using especially *Aspergillus niger*. It is used in effervescent drinks; as a buffer to ensure setting of gels; and as an emulsifier, stabilizing agent and antioxidant in various foods. The conditions for citric acid accumulation are now well known but the precise pathways involved in its formation are not. It seems probable that the final step is that in the TCA cycle (cf. Fig. 10.3) but it is not clear through which routes the considerable amounts of oxaloacetic acid required are formed. The practical conditions are a high (15–20%) glucose medium and a relatively low pH, *c.* 2·0, which can be as high as 4·5 in the highly aerated conditions frequently employed. The high aeration factor suggests that oxaloacetic acid is probably formed irreversibly by PEP carboxylase since this operates at low CO_2 tensions.

The actual accumulation process is not understood. Best yields occur when growth is somewhat restricted and this may reflect the fact that there is an impairment of primary synthetic processes which utilize TCA cycle intermediates, e.g. for fatty acids or proteins. High-accumulating strains of *A. niger* are slow-growing but low-accumulating strains grow extremely rapidly (GARDNER *et al.*, 1956; ERKAMA *et al.*, 1949). Strain variation indicates that a genetic element is involved but its locus is unknown. RAMAKRISHNAN *et al.* (1955) have noted that isocitrate dehydrogenase activity is low or absent and that this enzyme can be inhibited by addition of citrate or ferrocyanide, an excess of the latter usually being added in commercial operations (MARTIN, 1957). This inhibitor also inhibits growth (MARTIN, 1955). Inhibition of this enzyme would not prevent synthesis of oxaloacetic acid but could account for the inhibition of growth through a reduction in the supply of L-ketoglutaric acid and hence of amino acid production. It is not yet clear whether such a simple explanation is entirely satisfactory and, indeed, it cannot be fully assessed until the pathway of citric acid synthesis is ascertained with certainty.

ACONITIC AND ITACONIC ACIDS

The latter acid is formed from the former, a component of the TCA cycle, by *cis*-aconitic decarboxylase (BENTLEY and THIESSEN, 1957).

$$
\begin{array}{lcl}
\mathrm{COOH} & & \mathrm{CH_2COOH} \\
| & & | \\
\mathrm{CH_2} & \rightleftharpoons & \mathrm{C\text{-}COOH} + \mathrm{CO_2} \\
| & & | \\
\mathrm{C\text{-}COOH} & & \mathrm{CH_2} \\
| & & \\
\mathrm{CH} & & \\
| & & \\
\mathrm{COOH} & &
\end{array}
$$

Originally discovered in *A. itaconicus* (KINOSHITA, 1929), it is now known to be produced by strains of *A. terreus* (LOCKWOOD and REEVES, 1945). This species has been exploited by means of ultra-violet induced mutants and yields some 15% better than the parent strain have been obtained. Here too, therefore, there is a genetic element but accumulation is also dependent upon precise control of certain exogenous factors. The most critical is the low and narrow pH range, 1·6–2·0 in agitated culture and 2·1–2·5 in still cultures (LOCKWOOD and NELSON, 1946). There is also a large requirement for $MgSO_4.7H_2O$ (4·75 g/l.) and it has been suggested that this provides tolerance to acidity (FOSTER, 1949).

FUMARIC ACID AND RELATED ACIDS

Species of *Rhizopus* and other Mucoraceae convert a great deal of glucose supplied into fumaric acid but the property occurs also in *Aspergillus* spp. It is believed to arise from its precursor in the TCA cycle, succinic acid and this also accumulates in a few fungi, notably *Fusarium* spp. (LOCKWOOD *et al.*, 1938) and *Blastocladia pringsheimii* where some 10% of the glucose consumed is so accumulated.

A high glucose concentration and appreciable trace amounts of Zn (Table 11.4) are critical for accumulation by *R. nigricans*, which has been studied extensively (FOSTER and WAKSMAN, 1939a, b). The accumulation process is evidently an equilibrium position, although it may persist for some time. As cultures age there is a tendency for fumarate to disappear and for malate to accumulate. It is not clear how fumaric acid accumulates nor what is its origin. There are two main suggestions concerning the latter. FOSTER (1949) has drawn attention to the common association of accumulated C_2 compounds with fumarate. He suggests, therefore, that the sequence involved is:

2 Alcohol → 2 Acetaldehyde → 2 acetic acid → succinic acid → fumaric acid

Table 11.4. The effect of trace amounts of zinc and different glucose concentrations on the consumption of glucose, fumaric acid production and the efficiency of carbon conversion by *Rhizopus nigricans*. (Foster and Waksman, 1939b)

Glucose concentration (%)	2·5		10	
Zinc (p.p.m.)	0	1·2	0	1·2
Glucose consumed (mg)	2369	4752	4530	9795
Fumaric acid produced (mg)	891	474	1040	2214
Conversion (%)	37·6	10	22·9	22·6

This is supported by the disappearance of C_2 compounds in sugar cultures and formation of C_4 acids and the ability to convert alcohol and acetate to fumarate in organisms which accumulate that substance. On the other hand, there is evidence that isotopically-labelled $^{14}CO_2$ is incorporated into fumarate and this could come either from the reductive carboxylation of pyruvate to oxaloacetic acid or malic acid. In some fungi both routes may occur since fumarate can be produced aerobically or anaerobically.

OXALIC ACID

This is not, of course, a compound of the TCA cycle but it is extremely widespread in the fungi, especially in the basidiocarps of the Agaricales and Polyporales. The necessary conditions are a high glucose concentration, adequate aeration and a relatively high initial pH. In *A. niger* it seems probable that ^{14}C is incorporated via oxaloacetate into oxalate and acetate (BOMSTEIN and JOHNSON, 1952). The enzyme responsible for the conversion is soluble and requires Mn^{2+} for its action, it is oxaloacetate hydrolase (HAYAISHAI *et al.*, 1956). If the pH is low there is a further conversion of the oxalate to formate and CO_2 due to formation of oxalic decarboxylase (SHIMAZONO, 1955). This last reaction accounts for the pH specificity associated with accumulation but it is not clear how the possible eventual fate of the oxaloacetic acid is determined.

SUGAR ACIDS

The principal acids apparently derived by direct oxidation of sugars are gluconic (considerable accumulations of which occur in *Aspergillus* and *Penicillium* spp.) and kojic acid, also formed by Aspergilli, notably *A. oryzae* and *A. flavus*.

Gluconic acid may well arise from the action of glucose aerodehydrogenase via gluconolactone which is quite stable in aequous solution (cf. Chapter 10, pp. 272–273). Accumulation is favoured by a relatively high pH, a limiting amount of phosphate, vigorous aeration and, of course, a high concentration of glucose. Almost complete conversions of glucose can occur in *A. niger* (GASTROCK *et al.*, 1938). It might be supposed that the condition of low exogenous phosphate accounted for the accumulation since phosphorylated gluconate is capable of utilization in an HMP pathway. Such an explanation is, however, too simple for it is almost certain that sufficient endogenous phosphate would be available for the appropriate phosphorylations, in *A. niger* at least.

Kojic acid appears from experiments with [^{14}C-1]-glucose to be derived by a direct conversion of the C-skeleton of glucose although the nature of the reaction is unknown (ARNSTEIN and BENTLEY, 1953a, b, c). However, it can also be synthesized from pentoses, acetate, pyruvate and glycine so that its synthesis could also involve common C_3 intermediates. Very little is known of its origin or of the reason for its accumulation even though yields of 50–60% of the weight of glucose consumed can be obtained (KLUYVER and PERQUIN, 1933). Indeed, these investigators obtained 78·0% of the theoretical yield at pH 1·9 although at pH 2·2 this was reduced to 49·6% of theory. The total turnover of glucose at pH 1·9 is about one half of that at pH 2·2 and the best yield is in highly aerobic, replacement cultures free of nitrogen.

The accumulation process

In all these cases certain features appear to be common. Highest yields occur late in the growth cycle when carbohydrate is present in high concentration, often with a low C/N ratio, where normal growth is inhibited for some reason and aerobic conditions normally obtain. This has led to the concept of overflow and shunt metabolism. The notion is that in the early stages of growth, metabolites are used for synthesizing essential cellular metabolites but, when such syntheses slow down or cease, conversion of carbohydrate fragments can still lead to acid accumulation. These diversions are supposed to arise from 'not only a saturated and overloaded enzyme system, but also on a second enzyme system, normally latent or subdued, whose activity becomes manifest or accentuated through the availability of overflow intermediates' (FOSTER, 1949). This viewpoint suggests that, ultimately, the accumulated products will be utilized and there is some support for this prediction (cf. p. 276). Nevertheless, there is not always evidence that this can, in fact, occur. It also raises the problem of why the 'second enzyme system' is 'normally latent or subdued', i.e. it has a low affinity for the substrate. Such situations do not occur widely in living organisms and it is difficult to see why it should apparently be so highly developed in fungi when compared, for example, with higher green plants. Although acid accumulation can be regarded as a form of reserve material it is the case that polysaccharides and lipids which subserve this function in other organisms also occur, in fungi. The functional problem then is the significance of two kinds of reserve material in fungi not present in other organisms. A parallel may, perhaps, be drawn between acid accumulations in fungi and in higher green plants. RANSON (1965) reviewed the problem in higher plants and pointed out that acids might well accumulate in vacuoles and the significant feature determining accumulation could then be the acid transport system, whatever its nature. He has also pointed out that accumulation does not occur in animal cells which are not usually vacuolated. Fungi, of course, are vacuolated cells and could, therefore, qualify for Ranson's hypothetical system.

This is a particular example of the general hypothesis that materials are compartmentalized within cells and it can only be tested by examining the intracellular localization of the accumulated acids. This has yet to be done.

PRIMARY METABOLITES

It will be convenient to consider primary metabolites approximately in the sequence in which they originate along the route of carbon catabolism (Fig. 11.1). Thus oligo- and polysaccharide synthesis will be considered first, then nitrogen assimilation, amino acid and protein synthesis and finally, fatty acids and lipids.

Carbohydrate synthesis

Many fungi are capable of synthesizing glycogen and a variety of structural polysaccharides (STACEY and BARKER, 1960), as well as several oligosaccharides which appear to accumulate as reserves. The synthesis of chitin and glucans has already been referred to in Chapter 3 and will not be dealt with here.

GLYCOGEN

Most studies have been concerned with yeast glycogen. It occurs distributed uniformly throughout the cell as particles *c.* 200 10^{-10}m in diameter. Chemically it is composed of *c.* 10^3 unit chains, each of 11–13 α1:4-linked glucose residues with α1:6-linkages scattered at random; the exterior chains are, on average, of 8 glucose residues (MANNERS and KHIN MAUNG, 1955; NORTHCOTE, 1953; PEAT, WHELAN and EDWARDS, 1955). Recently EATON (1963) has provided evidence that two components are involved, a heavy component with an average of 9·4 residues external to the 1:6-linkages and a light component with an average of 4·2 such external units. Under aerobic conditions only the component external to 1:6-linkages can be utilized.

The detailed steps in the synthesis of this complex molecule are not known but the general mechanism is now understood.

In 1940 the Cori's (CORI and CORI, 1940) showed that yeast contained a phosphorylase capable of carrying out the synthesis of glycogen from a glucose 1-phosphate; the reverse of a reaction they had demonstrated in animal tissues a few years earlier. The reaction

$$\text{Glucose-1-phosphate} \underset{P_i}{\overset{\text{Phosphorylase}}{\rightleftharpoons}} \text{glycogen}$$

can certainly occur but it now seems that sugar nucleotides are more likely to be the substrate for polysaccharide synthesis. The relevant nucleotide for glycogen synthesis, uridine diphosphate-D-glucose was isolated from yeast by LELOIR (1951), having originally been discovered in yeast extract by MUNCH-PETERSON *et al.* (1953), and an enzyme responsible for synthesis isolated from yeast in 1960 (ALGRANATI and CABIB). The reactions are:

$$\text{UTP} + \text{glucose-1-phosphate} \rightleftharpoons \text{UDP} - \text{glucose} + \text{pyrophosphate}$$

$$x\text{UDP} - \text{glucose} + (\alpha 1{:}4\text{-glucose})_n \rightarrow (\alpha 1{:}4\text{-glucose})_{n+x} + x\text{UDP}$$

This is more likely to be the synthetic pathway than that due to phosphorylase since the equilibrium constant at the optimum pH of 7·5 is about 250 for the nucleotide-mediated reaction compared with 3 for the phosphorylase system. Indeed, at pH 7·5 phosphorylase is more effective in converting glycogen to glucose-1-phosphate. The UDP glucose-glycogen transglucosylase, the enzyme responsible for synthesizing glycogen, has two deficiencies. Firstly, it requires some 'starter' polysaccharide material, to which it can add glucose residues and, secondly, it is incapable of forming α1:6-linkages. In animal and higher plant tissues these linkages are formed by a distinct enzyme now called amylo (1,4 → 1,6) transglucosidase. Evidence for this enzyme has not been obtained in yeast or other fungi. Synthesis of glycogen is as yet far from being completely understood in fungi and, indeed, the presence of glycogen often only rests upon simple biochemical tests such as Bauer's stain or iodine in KI solution (ZALOKAR, 1959b).

OLIGOSACCHARIDES

Many oligosaccharides are known from fungi, including a number with very remarkable configurations, e.g. galactocarolose of *Penicillium charlesii* where galactose units probably in furanose form, are joined by 1:5-linkages probably in the β-configuration (HAWORTH, RAISTRICK and STACEY, 1937). It seems likely, however, that their basic synthesis will follow lines similar to those described for starch, i.e. appropriate sugar nucleotides and trans-glucosylases. For example, LELOIR and CABIB (1953) have shown that there is an enzyme in yeast which can transfer glucose from UDP-glucose to glucose-6-phosphate to yield trehalose-6-phosphate.

It is evident that the exploitation of this field of study along the lines laid down by workers with higher plants, animals and bacteria should rapidly increase understanding of carbohydrate synthesis in fungi.

Nitrogen assimilation

Although nitrogen containing compounds of various types can be transported into fungal hyphae (Chapter 8, pp. 218–222), inorganic compounds must be converted to 'ammonia' before they can be assimilated. Recent work on this topic has largely employed *Neurospora crassa* and other fungi and may conveniently be considered here. An admirable detailed account has been prepared by NICHOLAS (1965) who has contributed much to the subject.

Although the details are not wholly clear it seems probable that nitrate undergoes the following sequence, suggested originally for plants by MEYER and SCHULTZE (1884):

$$\begin{array}{c} \text{nitric oxide} \\ \upharpoonleft\downharpoonright \\ \text{Nitrate} \rightarrow \text{nitrite} \rightarrow \text{hyponitrite} \rightarrow \text{hydroxylamine} \rightarrow \text{'ammonia'} \end{array}$$

Enzymes are known from fungi which will catalyse all these reactions. Nitrate reductase occurs in *Neurospora* as an inducible enzyme and it has also been identified in *Hansenula anomala* (NICHOLAS *et al.*, 1954; SORGER, 1963). It has a flavine adenine dinucleotide (FAD) prosthetic group and contains molybdenum, phosphate and sulphydryl groups. The molybdenum is the essential factor in transporting electrons which are donated preferentially by $NADPH_2$ in *Neurospora* or by either $NADPH_2$ or $NADH_2$ in *Hansenula* (SILVER, 1957). The action of the nitrate reductase is intimately linked with the cytochrome system, in particular cytochrome *c*, and their induction and activities parallel each other (KINSKY and MCELROY, 1958; KINSKY, 1961). Mutants lacking nitrate reductase are known in both *Neurospora* and *Aspergillus* (SILVER and MCELROY, 1954; SORGER, 1963). It proved possible to demonstrate that the FAD/Mo protein complex and the NADP/FAD/Cytochrome *c* were separately determined by non-allelomorphic genes *nit-1* and *nit-2* in *Neurospora*. These two reactions are, therefore, probably functionally distinct.

Other FAD-containing enzymes in *Neurospora* are nitrite reductase which utilizes $NADH_2$ as hydrogen donor and hydroxylamine reductase which uses the same donor (NASON *et al.*, 1954; NICHOLAS *et al.*, 1960; ZUCKER and NASON, 1955). Although the former enzyme is affected by Fe or Cu deficiency and the latter by Mg or Mn deficiency, the important metals implicated in electron transfer are thought to be the copper and manganese, respectively. In addition, hyponitrite reductase and nitric oxide have also been detected in *Neurospora*. Nitric oxide could dimerize to hyponitrite. Since some mutants of *Neurospora* incapable of utilizing nitrite were capable of accumulating hydroxylamine it is possible that both pathways of reduction are possible. SILVER and MCELROY (1954) found that exogenous pyridoxine was necessary for growth in some nitrite mutants. They suggested that free hydroxylamine and pyridoxal phosphate condensed to form the oxime, thence the amine and, by transamination the α-keto acid. In normal cultures the hydroxylamine could be

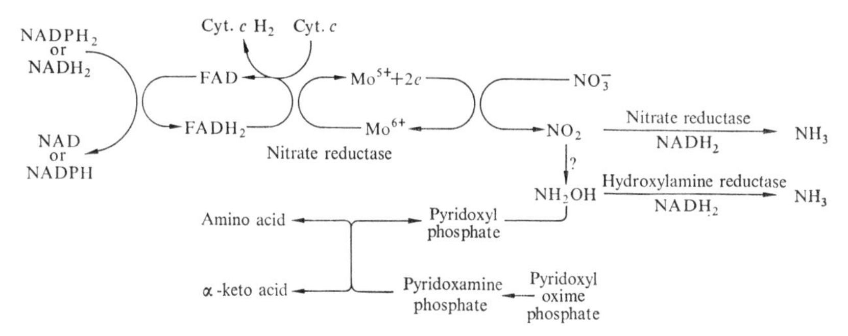

Fig. 11.2. Possible reaction sequences involved in reduction of nitrate to ammonia by fungal cells

either reduced directly or via the oxime with oxaloacetic acid and hence to aspartic acid. There is, however, no strong supporting evidence for this route (NICHOLAS, 1959a, b). The possible reactions are summarized in Fig. 11.2.

Amino acid and protein synthesis

A great many studies are available which have a bearing on the synthesis of amino acids. On the one hand are the results of biochemical studies using mutants of *Neurospora*, *Aspergillus*, yeasts and a few other fungi; on the other are studies in which compounds radioactivity labelled with ^{15}N and ^{14}C have been fed and the amino acids and proteins later isolated and identified. The results of both kinds of studies agree in demonstrating the key role of TCA cycle acids which combine with 'ammonia'. The consequences of such reactions are to provide a number of primary amino acids, glutamic and aspartic acids, alanine, serine, glycine from which other amino acids and metabolites can be derived.

There is also clear evidence (cf. Chapter 8, p. 231) of at least two 'pools' of amino acids within *Neurospora* and *Candida*, at least. One of these closely resembles that of proteins in its composition. Thus the two problems, amino acid synthesis and protein synthesis, will be considered separately here.

AMINO ACID SYNTHESIS

Supply of [^{14}C]-fructose to *Candida* (*Torulopsis*) *utilis* is followed very rapidly by the detection of ^{14}C-labelled amino acids in the yeast (ROBERTS *et al.*, 1955; COWIE and WALTON, 1956). Since fructose is catabolized by the EMP pathway to the TCA cycle it suggests that somewhere on this route the C-fragments become incorporated into amino acids. THORNE (1950) showed that for yeasts, in general, 'ammonia' is a better source of nitrogen than single amino acids, with the exception of aspartic and glutamic acids and the same acids are the best for growth of *A. niger* (STEINBERG, 1942), although closely followed by glycine. He also showed that the amino acids most readily assimilated were those with an α-amino group. It will be recalled that 'ammonia' is rapidly assimilated by *Scopulariopsis brevicaulis* and *Neocosmospora vas-infecta* either in the presence of exogenous glucose or by high-sugar mycelium (cf. pp. 218–220). All these facts suggests a close link with TCA cycle intermediates for, not only does respiration give rise to them, but glutamate, aspartate and glycine can readily be obtained by reactions between 'ammonia' and appropriate intermediates. Studies on biosynthetic processes in germinating spores leave little doubt that the products of carbohydrate catabolism provide skeletal material for amino acids. For example, [^{14}C-2]-acetate was supplied to uredospores of *Uromyces phaseoli* and *Puccinia helianthi* and, shortly after, isotopically labelled glutamate, aspartate and alanine were detected in both and, in addition, lysine in the latter species (STAPLES *et al.*, 1961). The glutamate was labelled within 3 sec and aspartate within 10 sec. Moreover, $^{14}CO_2$ fixed by germinating spores of *A. niger* is found mainly in

Table 11.5. The origins of α-amino acids in *Candida utilis* supplied with ^{15}N-ammonia in exponentially growing cultures. (From Sims and Folkes, 1963)

Primary (from NH_3)	Secondary (from glutamic acid)	Tertiary (from ornithine: apparent secondary from glutamic)	Quaternary (from citrulline: apparent tertiary)
Glutamic acid	Alanine	Citrulline	Arginine
Glutamine	Serine		
	Threonine		
	Aspartic acid		
	Lysine		
	Histidine		
	Proline		
	Glycine		
	Valine		
	Leucine		
	Isoleucine		
	Phenylalanine		
	Tyrosine		
	Methionine		
	Ornithine		

the cell extract supernatant after centrifuging to 105,000 **g**. This fraction contains soluble proteins. The careful studies of SIMS and FOLKES (1963) leave little doubt that in *Candida utilis*, under steady-state conditions in the exponential phase, glutamate is the first formed amino acid. When supplied with $(^{15}NH_4)_2HPO_4$, free glutamic acid and glutamine rapidly became labelled and its ^{15}N-content was equally rapidly reduced when the yeast was transferred to isotope-free medium. This pattern differentiated glutamic acid from all others including alanine, glycine and aspartic acid, all of which could be primary products. They have suggested, therefore, that glutamate is the first formed amino acid and that the others derive their α-amino nitrogen from it and glutamine. By means of an appropriate kinetic analysis, these investigators have been able to suggest how 18 amino acids have been derived (Table 11.5).

The reactions postulated are firstly amination of α-ketoglutaric acid by glutamic dehydrogenase and, thereafter, a variety of transamination reactions.

$$\text{(a)}\quad \begin{array}{c} COOH \\ | \\ C \\ | \\ C \\ | \\ C{=}O \\ | \\ COOH \end{array} + NH_3 + NADPH_2 \xrightleftharpoons[]{\text{Glutamic dehydrogenase}} \begin{array}{c} COOH \\ | \\ C \\ | \\ C \\ | \\ C{-}NH_2 \\ | \\ COOH \end{array} + H_2O + NADP + H^+$$

α-Ketoglutaric acid — L-glutamic acid

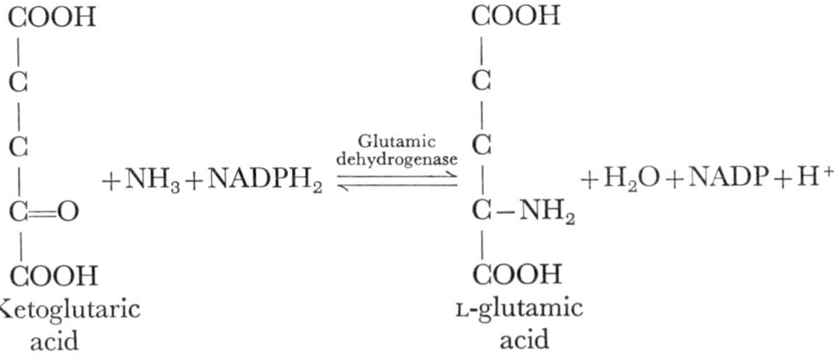

Amination reaction

$$\text{(b)}\quad \text{L-glutamate} + NH_3 + ATP \xrightleftharpoons[]{\text{Glutamine synthetase}} \text{L-glutamine} + ADP + P_i + H_2O$$

Amine formation

$$\text{(c)}\quad \begin{array}{c} COOH \\ | \\ C \\ | \\ C \\ | \\ C{-}NH_2 \\ | \\ COOH \end{array} + \begin{array}{c} C \\ | \\ C{=}O \\ | \\ COOH \end{array} \xrightleftharpoons[]{\text{Transaminase}} \begin{array}{c} COOH \\ | \\ C \\ | \\ C \\ | \\ C{=}O \\ | \\ COOH \end{array} + \begin{array}{c} C \\ | \\ C{-}NH_2 \\ | \\ COOH \end{array}$$

L-glutamic acid — Pyruvic acid — α-Ketoglutaric acid — alanine

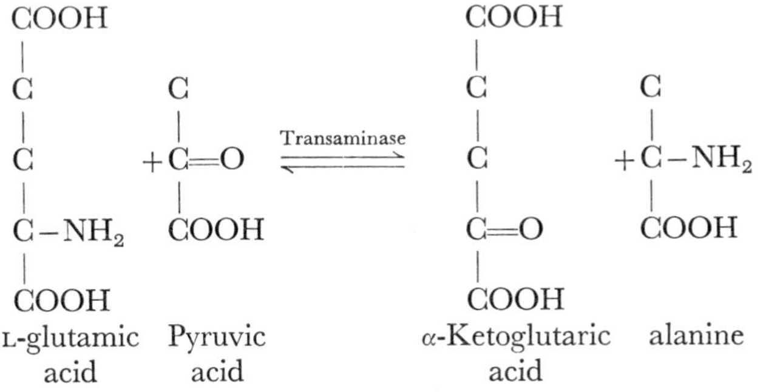

Transamination

Glutamic dehydrogenase is known in several fungi, e.g. 13 transaminations are involved in *Puccinia helianthi* uredospores (SMITH, 1963), glycine-alanine transaminase in *Blastocladiella emersonii* (MCCURDY and CANTINO, 1960). It has also received special study in *Neurospora*, *Ophiostoma* and *Penicillium* where mutants deficient in the enzyme have been obtained. FINCHAM (1956) showed that in *N. crassa*, mutant 32213 will not grow on the keto acid nor with glycine, serine, threonine, lysine, but that it will grow with α-amino acids such as glutamic and aspartic acids or alanine. Ammonia tended to accumulate in the medium if nitrate was supplied exogenously. This kind of information provides clear evidence for amination as a process in normal, wild-type mycelia. He was also able to show that at least four transaminases occurred in both wild-type and mutant strains, viz leucine and phenylalanine transaminate α-keto-β-methyl valerate to isoleucine and α-ketoisovalerate to valine (FINCHAM and BOULTER, 1956). This type of information provided by mutants incapable of carrying out a particular reaction coupled with the 'one-gene, one-enzyme' dictum has proved particularly valuable in resolving many biosynthetic pathways. However, before considering these it will be convenient to consider results obtained which appear to disagree with those of Sims and Folkes. From the distribution and amount of radioactivity in free amino acids in *Candida utilis* after supplying it with [^{14}C]-fructose, American workers concluded that there were a number of primary amino acids from which the others were derived (ROBERTS *et al.*, 1955; COWIE and WALTON, 1956; COWIE, 1962). Thus the concept of amino acid families, groups of acids with intimate biosynthetic relationships, was developed, viz the glutamic, aspartic, alanine and serine families (Table 11.6). This concept is at variance with that of Sims and Folkes since it neither gives primary place to glutamate formation nor requires so great an emphasis on transamination of glutamate. The problem has not been resolved wholly and cannot readily be settled until simultaneous studies are made with ^{14}C- and ^{15}N-labelled compounds in *Candida*. The important point to note is that it is clearly possible to derive the amino acids from the reaction of 'ammonia' with either one, α-ketoglutaric acid, or several acids of the TCA cycle, with or without further transaminations.

Table 11.6. The families of amino acids in *Candida utilis* and *Neurospora crassa* based upon the flow of carbon in synthesis when supplied as $^{14}CO_2$ or $CH_3[^{14}C]$-OONa

A	B	C	D	E
Aspartic acid	Lysine	Glutamic acid	Alanine	Serine
Homoserine		Proline	Valine	Glycine
Threonine		Arginine	Leucine	Cysteine
Methionine				
Isoleucine				

Studies with mutants have been especially valuable in elucidating possible further synthetic pathways. It is not always possible from mutants to be certain that any particular pathway is the pathway normally followed. For example, although amination deficient mutants of *N. crassa* are incapable of growing rapidly (presumably because they lack the major enzymatic pathway via glutamic dehydrogenase) they can grow after a lag period on media lacking glutamate. This is because they possess an $NADH_2$-linked glutamic dehydrogenase which is unaffected by mutation at the *am* locus and, presumably, is not normally functional in wild-type strains (SANVAL and LATA, 1961). It is becoming clear that most organisms possess this kind of replacement system so that an organism is potentially highly buffered against various biochemical lesions.

Figure 11.3 summarizes knowledge gained from biochemical studies and fungal mutants (in the vast majority of cases *Neurospora crassa*) concerning amino acid biosynthesis. The essential technique is to screen the mutant with substances required in the supposed region of the block. In this way the block can be located since substances utilized after it will relieve it, those utilized prior to the block will not. Moreover, the penultimate substance before the block may accumulate and so enable it to be identified, although in several cases such accumulations are dispersed by a variety of shunt mechanisms. For example, in both yeasts and *Neurospora* a block in the synthesis of adenine results in the production of a reddish or purple pigment due to polymerization of accumulated intermediates. A final test is the isolation of the enzyme from the wild type and the demonstration of its absence or altered and reduced activity in the mutant.

Two examples, histidine synthesis and one synthetic pathway for the aromatic ring, will be considered in some detail.

Histidine mutants of *Neurospora* are only detected if grown on a minimal medium (sugar, salts, biotin) supplemented with histidine since the presence of a basic amino acid plus various neutral ones or tyrosine, inhibits their growth. Thus the complete or amino acid-supplemented media, frequently used in the detection of biochemical mutants, select against such mutants. The sequence had also to be investigated in cell-free systems since the three phosphate esters of the imidazole derivatives cannot be transported across the cell wall. Dephosphorylated intermediates which accumulated in the culture media could enter the hyphae but apparently could not be phosphorylated there.

A technique employed was to make all possible pairs of combinations of *his-1* to *his-7*. The object in preparing double mutants was to determine their order of action. If one, *a*, was blocked before another, *b*, which accumulated a distinctive compound then this would not accumulate in the double mutant if the block due to *b* preceded that due to *a* and vice versa. This example is also of interest since *his-3* appears to specify two enzymatic functions, one prior to the 'Bratton Marshall' compound and the penultimate step in synthesis due to histidinol dehydrogenase. Insufficient study has been given to the early stages of histidine synthesis to interpret this situation, and in particular, these enzymes have not yet been isolated. It should also be noted that these *his-3* mutants appear to occupy different positions in comple-

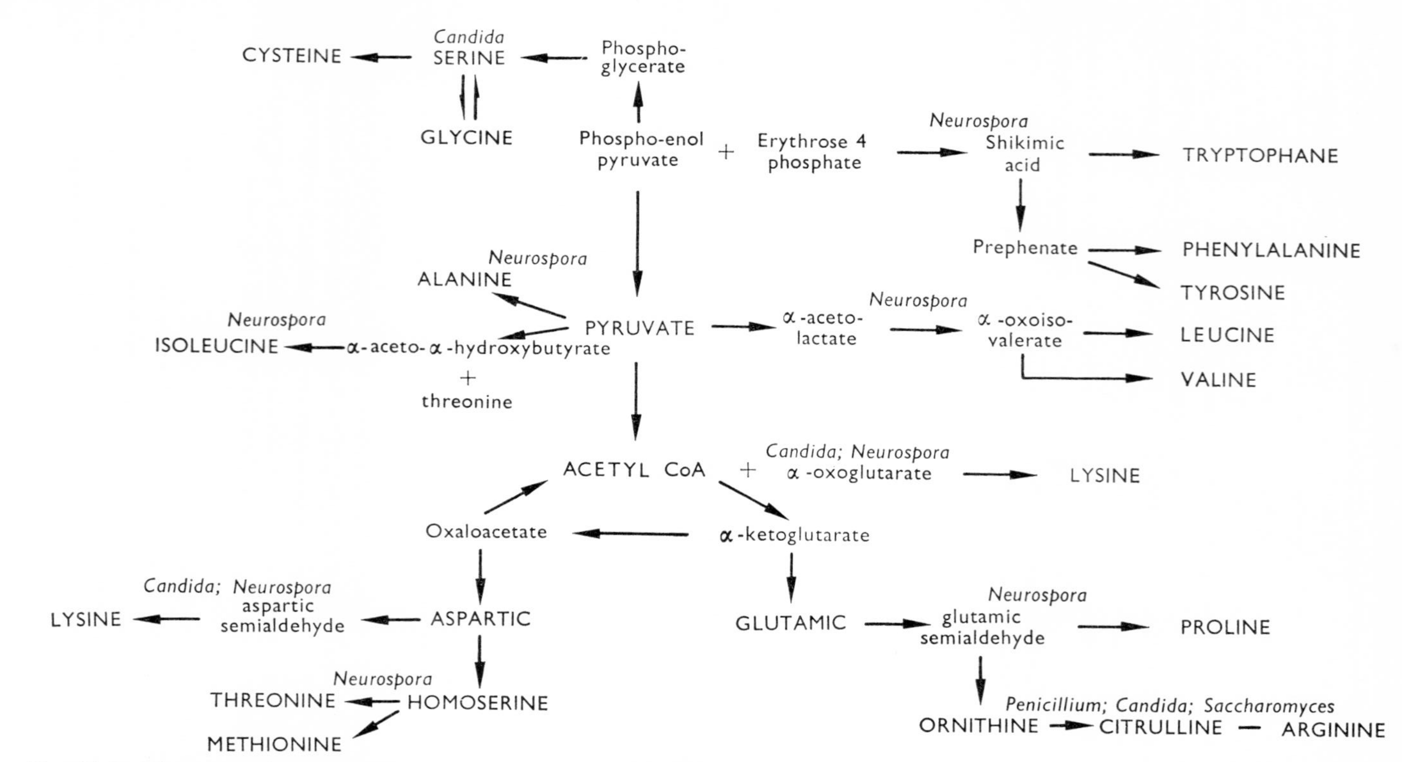

Fig. 11.3. Diagram to illustrate how many amino acids synthesized by fungi are related to pyruvate, acetyl CoA and TCA cycle intermediates

mentation maps, those affecting histidinol dehydrogenase being to the right of the others. Thus it may turn out that *his-3* will really be two closely linked loci. The sequence is summarized in Fig. 11.4.

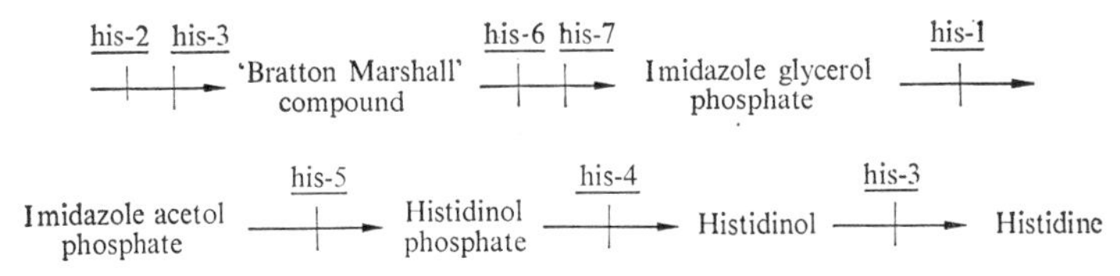

Fig. 11.4. Sequence of synthetic steps in the production of histidine by *Neurospora* to indicate where the 7 histidine-less mutants *his-1* to *his-7* operate. Note that *his-3* apparently regulates enzyme action early in the sequence and at the final step. 'Bratton-Marshall' compounds are precursors which react with the Bratton-Marshall reagent. (Based on Catcheside)

The study of *Neurospora* mutants involved in the early stages of the biosynthesis of aromatic amino acids, provides a good example of the multiple effects of a single metabolic block. The mutant *arom-1* appeared at first to require phenylalanine, tyrosine, tryptophane and *p*-amino-benzoic acid simultaneously. Only later was it found that shikimic acid could replace them since it is a common precursor to a branched system of biosynthesis. The mutant accumulated 5-dehydroshikimic acid, the immediate precursor of shikimic acid, as might be expected and also protocatechuic acid. Moreover it had acquired two new enzymes, dehydroshikimic dehydrase and protocatechuic oxidase, absent from wild-type mycelia. The explanation of these findings is that both these enzymes are inducible. When 5-dehydroshikimic acid accumulates, the dehydrase is induced and protocatechuate is formed. This induces the oxidase. Since 5-dehydroshikimic acid never accumulates in wild-type strains these enzymes are never normally induced. The situation is also of interest since *arom-1*, *3* and *4* are on the same chromosome (Fig. 11.5), *arom1* and *4* being quite close to each other. 5-Dehydroshikimic synthetase and 5-dehydroquinase are both lacking from *arom-2* which will not complement with *arom-1* or *4* although it will do so with *arom-3*; *arom-2* is, therefore, most probably a deletion which includes the loci of *arom-1* and *4*.

These examples show clearly the kinds of difficulty which may arise in the use of mutant strains to identify the routes of biosynthesis of various compounds, in this case the amino acids. Difficulties of a similar nature arise in other cases and reference to them may be made in various text books or in compilations of lists of mutants (e.g. WAGNER and MITCHELL, 1964; ESSER and KUENEN, 1965; BARRATT *et al.*, 1954).

PROTEIN SYNTHESIS

The general outline of protein synthesis which is believed to obtain in all organisms is now widely known. The primary specification resides in the base sequence of the chromosomal DNA. This is copied by the base sequence in

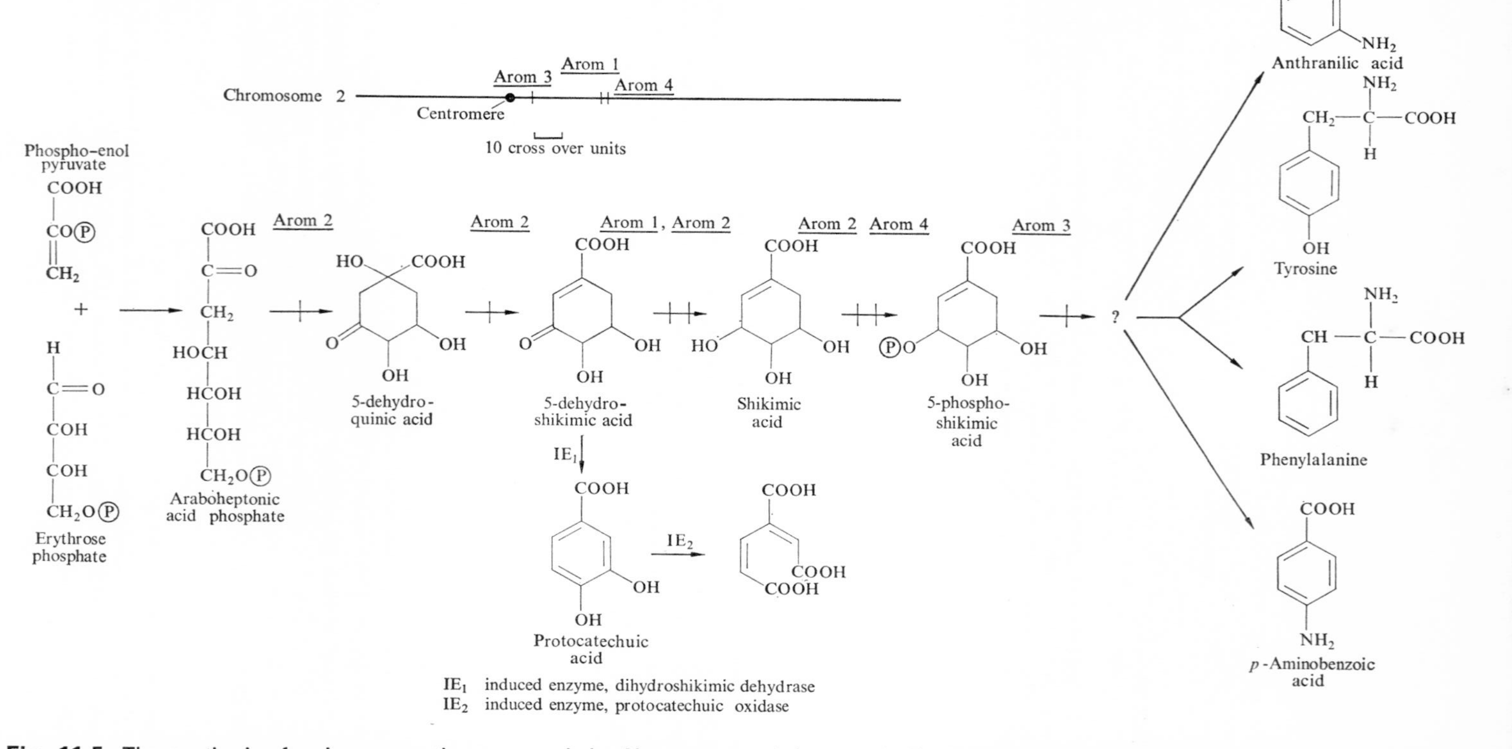

Fig. 11.5. The synthesis of various aromatic compounds by *Neurospora* and the genetic blocks induced by *arom* mutants. Since all these operate before the problematical precursor following 5-phosphoshikimic acid they all seem to have a multiple requirement for anthranilic acid, etc. Since *arom-1* blocks shikimic acid production, 5-dehydroshikimic acid accumulates and so induces the two enzymes IE_1 and IE_2 which result in a new sequence arising. Note linkage relationships of *arom*-1, 3 and 4. (Based on Gros and Fein)

messenger-RNA (m-RNA) and this molecule passes from the nucleus to ribosomes in the cytoplasm. Here it is associated with clusters of ribosomes, the polyribosomes, and the individual transfer-RNAs (s-RNA), a specific one for each amino acid, also become associated with these complex structures. The acetylated amino acids react with s-RNAs to form aminoacyl-s-RNAs, each activated amino acid being attached to one end of an s-RNA. The m-RNA is thought to act as a template whose bases are recognized by the three, unpaired bases at the opposite end from the attached amino acid of each aminoacyl s-RNA molecule. Polymerization of the amino acids then occurs in a manner not fully understood and ultimately a polypeptide strand is released from the ribosome.

Studies with fungi have not so far contributed much to this general scheme, although there seems to be little doubt that this kind of mechanism obtains. Here, therefore, attention will be drawn to those instances which appear to provide evidence for such a scheme in fungi.

One of the results of studies of exogenous amino acid transport by filamentous fungi and yeasts is the discovery of two 'pools' called, in *Candida utilis*, the 'expandable' and the 'internal pools' respectively (cf. COWIE, 1962). The composition of the 'internal pool' is remarkably similar to that of the cellular proteins whereas the other is highly variable and appears to reflect the 'pool' of free amino acids. It will be recalled (Chapter 8, pp. 229–231) that entry to the 'internal pool' was not identical in all fungi but *Blastocladiella*, *Candida*, *Neurospora* and *Saccharomyces* had in common the property of direct entry of some exogenous amino acid into the 'internal pool'. In *Candida* the 'internal pool' is resistant to osmotic shock and can only be rendered more sensitive by exposure to pressures of 2068 bar. After such treatment the 'internal pool' loses some 31% of its ^{14}C-labelled contents compared with 3·8% after 30 min, in distilled water at normal pressures. The 'expandable pool' loses some 80% of its ^{14}C-labelled contents with similar precise treatments (COWIE and WALTON, 1956). ZALOKAR (1965) has suggested that the spatial location of the 'internal pool' may be either the vacuole or the cisternae of the endoplasmic reticulum. No evidence exists to support either of these suggestions. The fact that vacuolar membranes are apparently highly elastic in protoplasts of *N. crassa* suggests that this organelle might be the preferred site of the 'internal pool' (EMERSON and EMERSON, 1958). On the other hand it is perhaps relevant to recollect that MOOR and MÜHLETHALER (1963) detected vesicles covered with particles and aggregations of particles on the cytoplasmic side of the membrane in frozen-etched yeast cells. These particles, of diameter 100 10^{-10}m were tentatively identified as ribosomes. Whatever its location it is clear that there is an internal, cellular compartmentalization which divides those amino acids destined for protein synthesis from the rest. To this 'pool' both exogenous and endogenous sources can contribute at least in yeasts (HALVORSON and COWIE, 1961, and see Fig. 8.13).

Presumably the 'internal pool' is closely related to that stage in protein synthesis where the amino acids are polymerized to form protein since the composition of the pool is so similar to that of cellular proteins (cf. Table 8.8).

There is little doubt that synthesis occurs in ribosomes. ZALOKAR (1960a, b 1961) inoculated *Neurospora* on a top-shaped structure so that the hyphae grew out radially from the apex to the base. This was, in fact, the rotor of an air-driven centrifuge and the contents of the cells could be centrifuged to 50–6000 **g** in 1 min. After stratification the cells recovered satisfactorily. The mycelium-bearing rotor was dipped for a few seconds into [^{3}H]-DL-leucine, washed, centrifuged and autoradiographs made of the hyphae. It was clear, both qualitatively and quantitatively, that maximum incorporation of isotopically-labelled material in the first 10–20 sec was in the ribosomal fraction. Kinetic studies suggested that the assembly of a protein molecule occurred in a few seconds at the most (Table 11.7). Protein synthesis by ribosomes in yeast has also been demonstrated and here the rate of synthesis is more rapid if the ribosomes are associated with fragments of membrane (HAUGE and HALVORSON, 1962). The significance of this is unclear but the association of ribosomal-like particles with membrane-enclosed vesicles in yeast, mentioned earlier, should be recalled. It may be that such units provide a closer association of 'internal pool' amino acids with ribosomes than occurs if the latter are free. In *Neurospora* clear evidence has now been provided for the occurrence of polyribosomes in extracts from germinated spores and mycelium; none could be detected in dormant ascospores or conidia (HENNEY and STORCK, 1964).

Table 11.7. Distribution of leucine and uridine labelled with ^{3}H in different fractions of *Neurospora* hyphae after rapid centrifugation and autoradiography. (From Zalokar, 1960b)

Substrate	Period of application (sec)	Exposure time (days)	Cell fraction (counts/layer/100 μm): Super-natant	Nuclei	Mito-chondria	Ribosomes
[^{3}H]-DL-leucine	10	30	0	26	18	*38*
	20	9	2	18	13	*64*
	60	4	33	53	33	*81*
	120	2	54	50	50	*100*
[^{3}H]-uridine	60	9	0	*146*	0	0
	16 × 60	3	1	*40*	5	*17*
	60 × 60	1	1	*51*	9	*53*

Figures in italic type indicate sites of synthesis referred to in the text.

Using the same ingenious technique [^{3}H]-uridine, a precursor of RNA, was also fed to *Neurospora* by Zalokar. After 1 min, 90% of the isotopically-labelled RNA detected was in the nuclear fraction. Transferred to non-radioactive uridine for 60 min, after a 1 min exposure to ^{3}H-uridine, the isotope was found to be about equally distributed in nuclear and ribosomal material with some in the mitochondrial fraction which lay between these two (Fig. 11.6 and Table 11.7). This is suggestive evidence for the synthesis and movement of, presumably, m-RNA. It bears out the classic observations of CASPERSSON (cf. 1950) that the cytoplasm of yeast cells actively synthesizing protein becomes filled with materials which, from their absorption spectrum, are evidently RNA. That there is some sort of complementarity between DNA and m-RNA was shown by the existence of natural DNA/RNA complexes extracted from *Neurospora* (SCHULMAN and BONNER, 1962). By analogy with other organisms, such as T-2 bacteriophages, this indicates some considerable similarity of nucleotide sequences in the two nucleic acids. Moreover

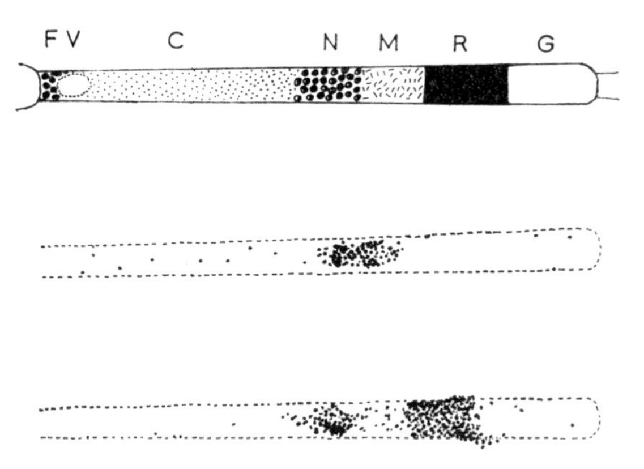

Fig. 11.6. Autoradiographs of a cell of a highly centrifuged *Neurospora* hypha after supplying [^{3}H]-uridine. Note heavy autoradiographs in areas overlying nuclei and ribosomes. Key: F, fat droplets; V, vacuoles; C, cytoplasm; N, nuclei; M, mitochondria; R, ribosomes; G, glycogen particles. (× *c.* 600, after Zalokar)

YCAS and VINCENT (1960) supplied ^{32}P as orthophosphate to yeast, extracted the newly synthesized RNA and compared its base composition with that of DNA and the overall values for RNA. The composition of the latter was determined chiefly by ribosomal and, perhaps, nucleolar RNA. Table 11.8

Table 11.8. Comparison of molar base composition of DNA, overall RNA (chiefly ribosomes) and newly synthesized RNA in *Saccharomyces cerevisiae*. (From Ycas and Vincent, 1960)

Bases	DNA (%)	Newly synthesized RNA (%)	Ribosomal RNA (%)
Adenine	31·5	32	25
Thymine	33·0	—	—
Uracil	—	29	28
Guanine	18·5	19	23
Cytosine	17·0	20	24

shows a clear correspondence in base composition between the newly synthesized, presumably, m-RNA and the DNA.

Finally, there is evidence for the existence of s-RNA and activated amino acids in fungi. HOLLEY *et al.* (1965) have not only isolated but determined the complete nucleotide sequence of the s-RNA responsible for alanine transfer from *Saccharomyces*, although its configuration is still somewhat uncertain. Further, cycloheximide is known to inhibit the incorporation of L-alanine into protein in the Basidiomycete, *Schizophyllum commune*. This compound is believed to inhibit the transfer of activated amino acids to ribosomes in other organisms (NIEDERPRUEM, 1964). Moreover, in germinating spores of *Penicillium atrovenetum* and *A. niger*, puromycin inhibits protein synthesis and the incorporation of L-leucine into protein, respectively (TRIPATHI and GOTTLIEB, in GOTTLIEB, 1966; STAPLES *et al.*, 1962). This substance is said to act by causing the premature release of polypeptides from ribosomes.

Thus, although the full mechanism for protein synthesis is not known from any one fungus, different steps exist in different fungi. This can most readily be explained by supposing that the general scheme for protein synthesis described at the outset applies to the fungi as a whole.

Synthesis of fatty acids and lipids

Lipids are a large class of somewhat heterogeneous compounds. They are basically derived from fatty acids and glycerol, the former may be unsaturated or saturated and of variable chain length. Complex lipids include the phospholipids, with one hydroxyl group of glycerol esterified with phosphoric acid, and the glycolipids where the glycerol is linked to a sugar by a glycosidic bond:

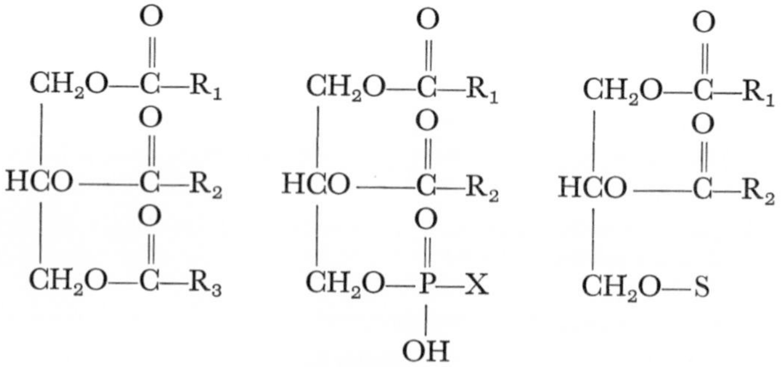

$$\begin{array}{l} CH_2O\text{—}C(=O)\text{—}R_1 \\ | \\ HCO\text{—}C(=O)\text{—}R_2 \\ | \\ CH_2O\text{—}C(=O)\text{—}R_3 \end{array} \qquad \begin{array}{l} CH_2O\text{—}C(=O)\text{—}R_1 \\ | \\ HCO\text{—}C(=O)\text{—}R_2 \\ | \\ CH_2O\text{—}P(=O)(OH)\text{—}X \end{array} \qquad \begin{array}{l} CH_2O\text{—}C(=O)\text{—}R_1 \\ | \\ HCO\text{—}C(=O)\text{—}R_2 \\ | \\ CH_2O\text{—}S \end{array}$$

Simple lipid — Phospholipid — Glycolipid

R_1–R_3 = fatty acids; P = phosphate; S = Sugar
X = various, including choline, ethanolamine, serine, sphingosine, inositol.

Here attention will be concentrated on the derivation of glycerol, fatty acids and simple lipids.

Fat production, readily seen as globules in the hyphae, is a common feature of rapidly growing fungi in high sugar concentrations (cf. Fig. 2.3c, p. 32). This ability was originally utilized in the First World War to provide glycerol from yeast but since then *Endomyces vernalis* and *Geotrichum* (*Oospora*) *lactis* have also been used. Other common mould species were employed for the same purpose in the Second World War in Germany (cf. FOSTER, 1949).

Lipid content can vary from 1–50%, excluding bound lipids, of the dry weight. High sugar content and high C/N ratios in the medium promote lipid synthesis and, under these conditions, almost 90% of the carbohydrate supplied can be converted to fat in *A. niger*, for example (TERROINE and BONNET, 1927). The origin of glycerol during fermentation by yeast, probably from the diversion of dihydroxyacetone phosphate from the EMP pathway, has already been described (pp. 276–278). The origin of fatty acids has also been studied in yeast and a number of pathways have been demonstrated.

In 1957 KLEIN noted that in a cell-free extract from yeast, lipid synthesis was greatly enhanced by provision of CO_2. This was found to be due to the combination of CO_2 with acetyl CoA to form malonyl CoA which was then condensed by a complex of soluble enzymes, the so-called fatty acid synthetase, with acetyl CoA or acyl CoA of a higher fatty acid (DEN and KLEIN, 1961; LYNEN, 1961). Thus:

$$CH_3CO.S-CoA+CO_2+ATP \underset{}{\overset{\text{acetyl carboxylase } Mn^{2+}}{\rightleftharpoons}} COOH.CH_2.CO.S-CoA+ADP+P_i$$

$$COOH.CH_2.COS-CoA + CH_3CO.S-CoA+2NADPH_2 \underset{}{\overset{\text{Fatty acid synthetase}}{\rightleftharpoons}} CH_3CH_2CH_2COOH+2CoA.SH + CO_2+2NADP$$

Lynen has, in fact, described reactions of the second type shown here as being a sequence of seven reactions which can undergo a cycle. At the final stage the butyryl/enzyme complex can condense once more with malonyl CoA and so the chain length of the fatty acid can be increased. Indeed, it is not clear what finally terminates the cycle. In yeast the principal product is palmitic acid ($CH_3(CH_2)_{14}COOH$) plus a little myristic and stearic acid. Confirmation of this sequence has come from the studies of Vagelos with *Escherichia coli* where he has isolated five of the six enzymes required for the seven postulated steps. In yeast, however, separate enzymes have not been isolated and Lynen has described the situation as one of 'six different enzymes arranged round a functional sulphydryl group which firmly binds the intermediates of fatty acid synthesis in close proximity to the active sites of the component enzymes'. The location and organization of such a multimacromolecular complex in the yeast cell is clearly a matter of great interest and it is, perhaps, surprising that it is not associated with any known particulate fraction.

Unsaturated acids are apparently synthesized in yeast by an aerobic particulate system which could be obtained free of RNA although sedimenting with ribosomes at 100,000 **g**. Either $NADPH_2$ or $NADH_2$ could be utilized and the conversion could take place via the CoA derivatives of the acids, e.g.:

$$CH_3(CH_2)_{16}CO.S—CoA + 2NADPH_2 + O_2 \rightleftharpoons CH_3(CH_2)_7CH{=}CH(CH_2)_7CO.S—CoA + NADP + 2H_2O$$

Stearic acid — Oleic acid

The mechanism of this reaction is not yet completely understood. A similar system is responsible for desaturating oleic acid to linoleic ($CH_3(CH_2)_4CH{=}CH—CH_2—CH{=}CH(CH_2)_7COOH$) in *Candida* (MEYER and BLOCH, 1963).

There are, in addition to these well documented cases in yeast, a few other fragmentary items concerning lipid synthesis in fungi but they do not amount to much. Nothing is known of the systems responsible for synthesizing complex lipids but, by analogy with animal and bacterial systems, it seems probable that UTP, and cytidine triphosphate (CTP) are likely to be involved in order to provide carrier molecules for synthesis. For example, in *E. coli*, the complex phospholipid, phosphatidyl choline is formed as shown:

$$\alpha\text{-phosphatidic acid} + \text{CDP-choline} \rightleftharpoons \text{Phosphatidyl choline} + \text{CMP}$$

$$\text{CMP} + 2\text{ATP} \rightleftharpoons \text{CTP} + 2\text{ADP}$$

$$\text{Phosphorylcholine} + \text{CTP} \rightleftharpoons \text{CDP choline} + \text{pyrophosphate}$$

This is an area of study in which mutants have contributed little; two classes are known in *Neurospora*, one requires acetate, the other long chain unsaturated acids such as oleic or linoleic (LEIN and LEIN, 1950). These could correspond to blocks in either of the two systems already described but this is not known.

It will be a matter of some interest to determine how widespread such systems are in other fungi, especially the highly efficient lipid producers. It will also be of great interest to elucidate the ultrastructural characteristics and genetic control exercised over the 'fatty acid synthetase' enzyme complex. Indeed, it will be important to learn how widespread it is since, in one bacterium at least, it appears to exist as separable enzymes. This is, of course, no guide to the situation in fungi. Tryptophane synthetase for example, is a single enzyme in *Neurospora* determined by a complex chromosomal locus whereas, in *E. coli*, two enzymes each determined by a distinct locus carry out the same function.

SECONDARY METABOLITES

It is hardly possible to do more than touch upon this complex subject here. A number of general accounts is available of the whole field and particular areas of it (reviews: BIRKINSHAW, 1965; JENSEN, 1965; CIEGLER, 1965; gibberellins: GROVE, 1963; antibiotics: BU'LOCK, 1961; HOCKENHULL, 1963; alkaloids: VINING and TABER, 1963).

Here attention will be paid to the better known isoprenoids, alkaloids, antibiotics and toxins produced by fungi.

Isoprenoids

These include carotenoids, sterols and gibberellic acid. The carotenoids are the best known fungal pigments and are widely distributed in all the main groups (GOODWIN, 1963; CIEGLER, 1965). The commonest appear to be β-carotene and acidic carotenoids, e.g. neurosporoxanthin. In mixed + and − cultures of the Choanephoraceae (Mucorales) the amount of β-carotene produced is so great that it is employed for commercial production, yields of *c.* 140 mg carotene/100 ml. medium/6-day period having been obtained (HESSELTINE, 1960; CIEGLER, 1965). Fungi have also been employed in the elucidation of carotenoid biosynthesis.

Sterols can be up to 2% of the dry weight of the mycelium of a fungus, most of it usually in the form of ergosterol, and many fungi are capable of carrying out a variety of specific transformations of steroids both exogenously and endogenously (FOSTER, 1949; PETERSON, 1953). However, very little is yet known concerning their biosynthesis.

Gibberellins, although discovered as a product of *Gibberella fujikuroi* (*Fusarium moniliforme*), the cause of a soil-borne disease of rice, are now known to be produced by higher plants and attention has, therefore, been directed recently to them rather than to the fungi.

Isoprenoid biosynthesis is now known to follow a common pathway with mevalonic acid playing a key role.

SYNTHESIS OF MEVALONIC ACID

GROB *et al.* (1951, 1956) showed that radioactively-labelled [^{14}C]-acetate was incorporated into β-carotene by *Mucor hiemalis* and $^{14}CO_2$ is also incorporated by *Phycomyces blakesleeanus* in the presence of leucine (GOODWIN, 1959). [^{14}C]-mevalonate is also incorporated into β-carotene by *Mucor hiemalis*, *Phycomyces blakesleeanus* and *Neurospora crassa* (GROB, 1957; BRAITHWAITE and GOODWIN, 1960; KRZEMENSKI and QUACKENBUSH, 1960). It is suggested, therefore, that mevalonic acid synthesis follows the following scheme (Fig. 11.7). It will be seen that the immediate precursor of mevalonate can be derived directly from acetoacetyl-CoA or from leucine. Since this amino acid is not apparently required by *M. hiemalis* for carotenogenesis, it presumably uses the direct synthetic route. However, there seems little doubt that *P. blakesleeanus* utilizes leucine and, indeed, in a partially purified cell-free system, the only one so far obtained in fungi, acetate and β-hydroxy-β-methylglutarate had to be supplied together with Mn^{2+}, CoA, ATP, $NADH_2$, NADP and $NADPH_2$ (YOKOYAMA *et al.*, 1962). Presumably there is an enzyme which activates hydroxy-methylglutarate in *P. blakesleeanus* but no such enzyme is known from yeast.

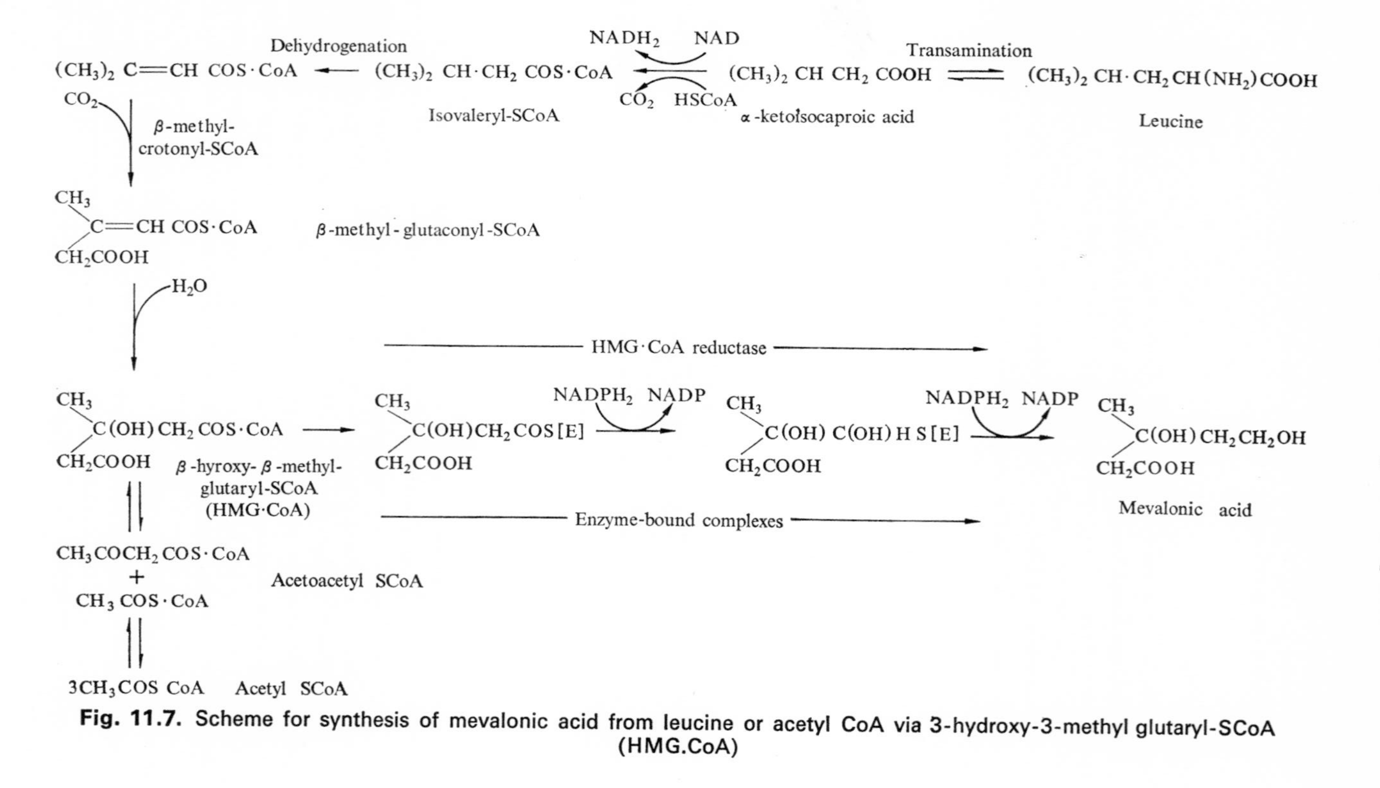

Fig. 11.7. Scheme for synthesis of mevalonic acid from leucine or acetyl CoA via 3-hydroxy-3-methyl glutaryl-SCoA (HMG.CoA)

CAROTENOID SYNTHESIS

After the production of mevalonate the next important product is believed to be isopentyl pyrophosphate (IPP) and the C_2 of mevalonate forms the methyl group of this latter compound:

Mevalonic acid $\quad HOO\overset{1}{C}.\overset{2}{C}H_2.\overset{3}{C}OH(\overset{4}{C}H_3).\overset{5}{C}H_2.\overset{6}{C}H_2OH$

$ATP \longrightarrow ADP \quad \rightleftharpoons$

5-phosphomevalonic acid $\quad HOOC.CH_2.COH(CH_3).CH_2.CH_2Ⓟ$

$ATP \longrightarrow ADP \quad \rightleftharpoons$

5-diphosphomevalonic acid $\quad HOOC.CH_2.COH(CH_3).CH_2.CH_2ⓅⓅ$

$ATP \longrightarrow ADP \quad \rightleftharpoons$

Isopentyl pyrophosphate (IPP) $\quad \overset{2}{C}H_2{=}\overset{3}{C}(\overset{4}{C}H_3).\overset{5}{C}H_2.\overset{6}{C}H_2ⓅⓅ$ $+CO_2+H_3PO_4$

An isomerase enzyme has been isolated from yeast that is sensitive to iodoacetate and causes the production of the isomer of IPP, dimethylallyl pyrophosphate. The subsequent stages of carotenogenesis, up to the basic C_{40} saturated intermediate from which the unsaturated coloured carotenoids are derived, are believed to be these: (basic C-skeleton only)

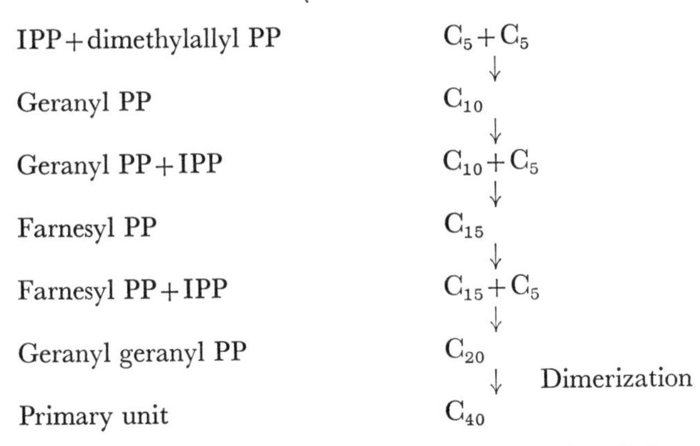

There is evidence to support this sequence in *Phycomyces*. In cell-free extracts, iodoacetate inhibits the incorporation of mevalonate in β-carotene, conceivably by acting on the isomerase (YOKOYAMA *et al.*, 1962). Moreover, as positive evidence, these extracts incorporate both [1-^{14}C]-IPP and [^{14}C]-farnesyl PP into β-carotene (VARMA and CHICHESTER, 1962; YOKOYAMA *et al.*, 1961).

There is some disagreement about the primary C_{40} unit, lycopersene and phytoene being possible contenders. GROB *et al.* (1961) claimed to have isolated a particulate enzyme from *Neurospora* that synthesized lycopersene from mevalonate. Further, in diphenylamine cultures, where colourless precursors tend to accumulate, it was said to accumulate (GROB and BOSCHETTI, 1962).

This has not been substantiated by other workers either with the same fungus or in inhibited *Phycomyces* cultures (DAVIES *et al.*, 1963; GOODWIN, 1958). Colourless mutants of *Neurospora* also accumulated phytoenes not lycopersene (HAXO, 1956). Phytoene is, therefore, the preferred primary unit at present but it will be realized that few fungi have been examined.

From phytoene the coloured unsaturated compounds are derived, probably by one of the sequences:

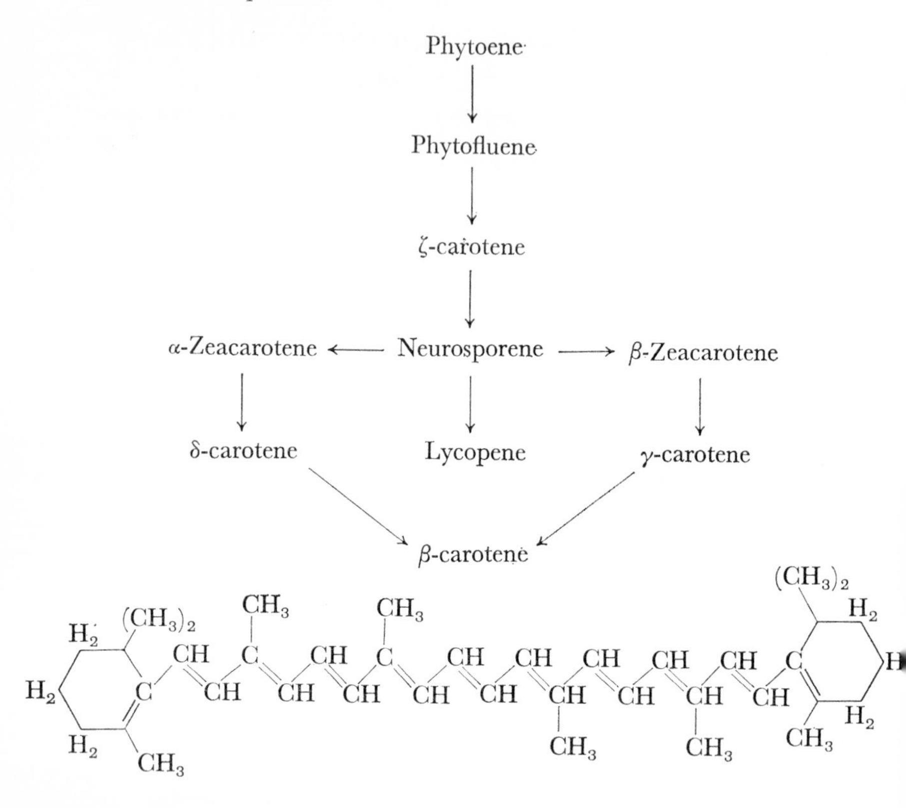

It is not yet clear whether neurosporene can act as a branch point. Diphenylamine-inhibited *Phycomyces* cultures contain β-zeacarotene and in the aquatic Phycomycete *Rhizophlyctis rosea*, lycopene is synthesized in young cultures but, after this has reached its maximum level, γ-carotene is synthesized. In *Rhizophlyctis*, therefore, it seems as though both routes are operative but at different stages of the life cycle (DAVIES, quoted in GOODWIN, 1965). Temperature is also known to affect the balance between γ-carotene and β-carotene (SIMPSON *et al.*, 1964).

Other environmental variables are known to affect carotenogenesis, the most striking in their effects being light and β-ionine.

As little as 60 seconds, or less, exposure to white light of dark-grown cultures of *Neurospora* is sufficient to bring carotene synthesis to a level almost as high as that of continuously illuminated cultures (ZALOKAR, 1955). The effect is presumably on the later stages of carotene synthesis but this is not known with certainty. A similar effect occurs with *Fusarium oxysporum* (CARLILE, 1956) but in other fungi the overall level or quantitative balance of pigments is altered by light, e.g. *Phycomyces* (GARTON *et al.*, 1951) and *Rhodotorula gracilis* (PRAUS, 1952) respectively. Addition of β-ionine enhances the production of β-carotene enormously in mixed + and − cultures of *Blakeslea trispora* (CIEGLER *et al.*, 1959; and Table 11.9), although it is toxic to mycelial growth if added at the outset of the experiment. It might be supposed that it was incorporated into the molecule since a β-ionine group occurs at each end of the carotene molecule. The action of this substance is not known but it has been suggested that it affects the sequence 5-phospho-mevalonic acid to dimethylallyl pyrophosphate, possibly by eliminating a negative feedback control imposed by the pyrophosphate compound (REYES *et al.*, 1964). At present this view is highly speculative.

Table 11.9. The stimulation of β-carotene production in mixed + and − cultures of *Blakeslea trispora* in shaken-flask culture by addition of β-ionine. (From Ciegler *et al.*, 1959)

β-Ionine (mg/100 ml.)	*Mycelium* (mg dry wt./100 ml.)	*Carotene yield* (mg/g dry wt.)	(mg/100 ml.)
0	6·14	1·6	10·1
0·0034	6·49	1·5	10·0
0·094	6·31	1·7	11·0
0·94	5·84	3·4	19·7
9·4	5·45	5·4	29·4
94·0	5·23	7·0	36·8
188·0	4·78	8·0	38·2

Attention has also been paid to the reasons for the enhanced production of β-carotene in mixed + and − cultures of Choanephoraceae. It is claimed that this is due to the production of trisporic acids A, B, C, of which the last is the most abundant (80% of the whole) and has been assigned the formula (CAGLIOTI *et al.*, 1964, 1966):

Trisporic acid $C_{18}H_{26}O_4$

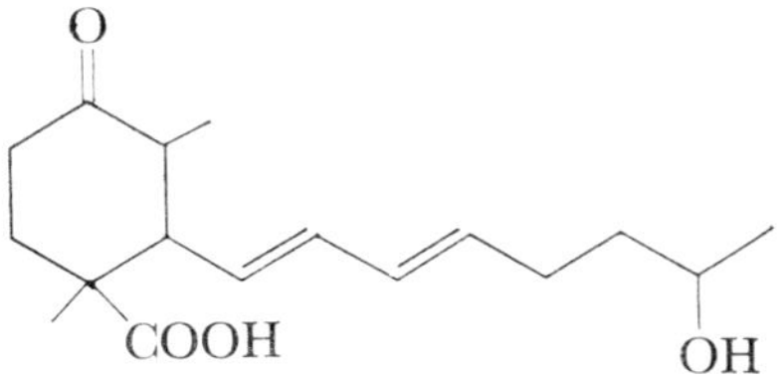

O
COOH
OH

There is, as yet, no account of how these acids operate, by which strains they are produced nor of the extent of the stimulation they can produce on one or the other mating type (see also p. 413).

HEIM (1947) has claimed that carotenoids arise in mitochondrial-like structures although, later, the pigments may become dispersed and associated with fat droplets, e.g. in *Mucor hiemalis* β-carotene is 70–75% in free lipids and 20–25% in bound lipids (HOCKING, in BURNETT, 1965a). These claims for the site of origin are largely based upon histochemical studies although carotene-containing 'gamma-particles' have been detected in zoospores of *Blastocladiella emersonii* (CANTINO and HORENSTEIN, 1956). Attention could usefully be paid, in future, to the location of carotenoid synthesis and its sites of accumulation in the fungal cell.

SYNTHESIS OF STEROLS AND ISOPRENOID DERIVATIVES

Much of the preceding section is relevant to sterol synthesis in fungi. The essential branch point appears to be after farnesyl pyrophosphate, since the precursor for sterols, analogous to phytoene for carotenoids, is the C_{30} compound, squalene. It has been claimed that the cell-free extract from *Phycomyces*, already mentioned, can also synthesize ergosterol but only in the

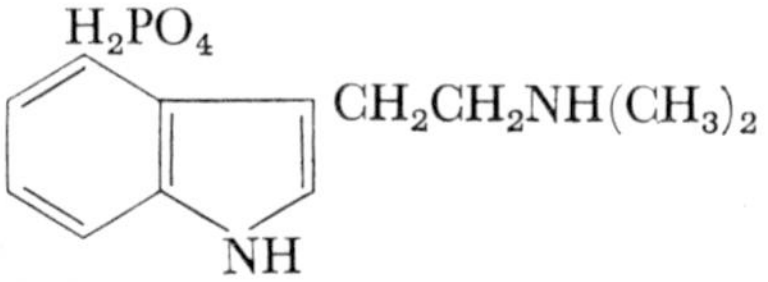

Psilocybin

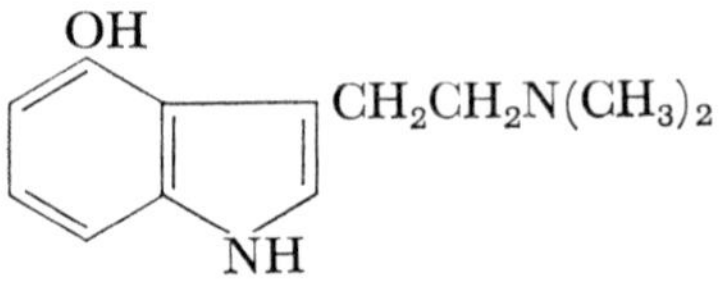

Psilocin

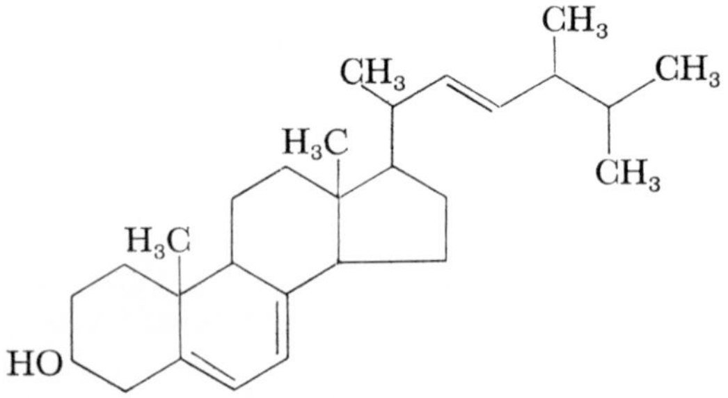

Ergosterol

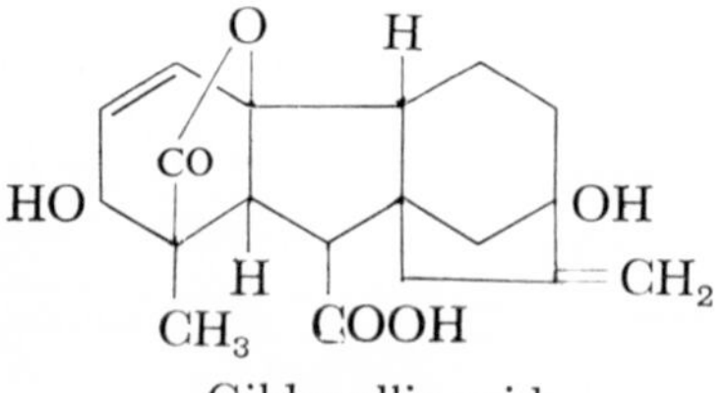

Gibberellic acid

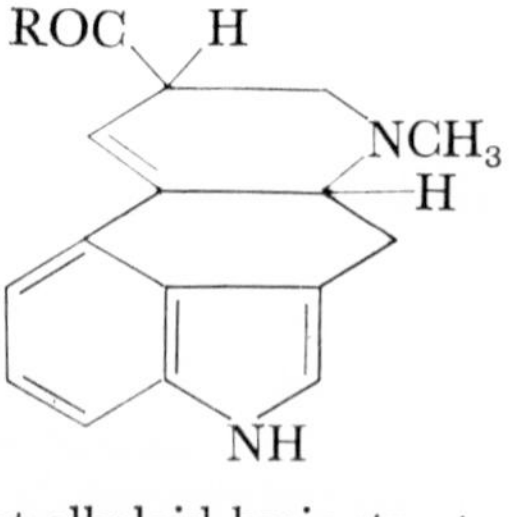

Ergot alkaloid basic structure

presence of NAD. If this is the true point at which regulation operates it could do so through the availability of NAD. There is no confirmation of this distinction from other studies.

Four kinds of compounds are of considerable biological interest. There are the indole derived hallucinogens psilocybin and psilocin found in basidiocarps of the agaric genus *Psilocybe*; the ergot-alkaloids from the sclerotia of the Ascomycete *Claviceps purpurea* which stimulate smooth muscle and affect the sympathetic nervous system; ergosterol, the precursor of vitamin D and the commonest fungal sterol, and the gibberellins which are now known to have growth-regulating properties in higher, green plants. The functions of these classes of compounds in the fungi themselves are quite unknown.

It is a remarkable fact that the *Psilocybe* and *Claviceps* compounds appear to be formed naturally in basidiocarps or sclerotia and this is true of the hallucinogens in culture also.

However, since Abe succeeded in producing ergot-like alkaloids in culture in 1951 it has become apparent that they can be produced in appropriate cultural conditions in mycelia. There is some evidence that *Claviceps* show host specialization in nature but the position is very confused. ABE (1951) classified *C. purpurea* strains by their host genera, e.g. *Agropyron*-, *Elymus*-, *Secale*-types, etc. Their yields in vivo and in vitro of alkaloids varied both qualitatively and quantitatively but there seemed to be a qualitative resemblance, at least, between normal ergot alkaloids and those produced synthetically.

Because ergot alkaloids can be produced in mycelial cultures they have received much attention. Different strains have different requirements but, in general, the highest yield of alkaloids was achieved in still (not agitated) cultures in the late growth phase when the polyols, carbohydrates and lipids were maximal or just declining (TABER and VINING, 1961; VINING and TABER, 1963). These investigators regarded alkaloid production as a shunt mechanism which operated when non-nitrogenous shunt compounds were restricted.

In both classes of compounds an obvious component is tryptophane and supply of radioactively-labelled material is followed by the incorporation of radioisotope in the molecule (GRÖGER *et al.*, 1959 in ergot alkaloids; BRACK *et al.*, 1961, for psilocybin). Nothing else is known concerning the synthesis of *Psilocybe* compounds. In the ergot alkaloids it was suggested that the rest of the molecule could come from an isoprenoid intermediate and this was tested by adding either [1-^{14}C]-acetate or [2-^{14}C]-mevalonate to cultures. Incorporation was very dependent on cultural conditions but it was successfully achieved, especially if the labelled precursor was added just prior to the onset of alkaloid synthesis, e.g. 90% of mevalonate added appeared in ergot alkaloids, $>10\%$ in lipids (TAYLOR and RAMSTAD, 1960). Since the addition of pyridoxal phosphate increased incorporation of tryptophane it seems as though a decarboxylation step is also involved (GRÖGER *et al.*, 1959).

Ergosterol is the most common sterol found in fungi. It may occur in the pure form in a particulate fraction, as in yeast (KLEIN and BOOHER, 1956), as crystalline inclusions, as in *Neurospora* (TSUDA and TATUM, 1961) or as the

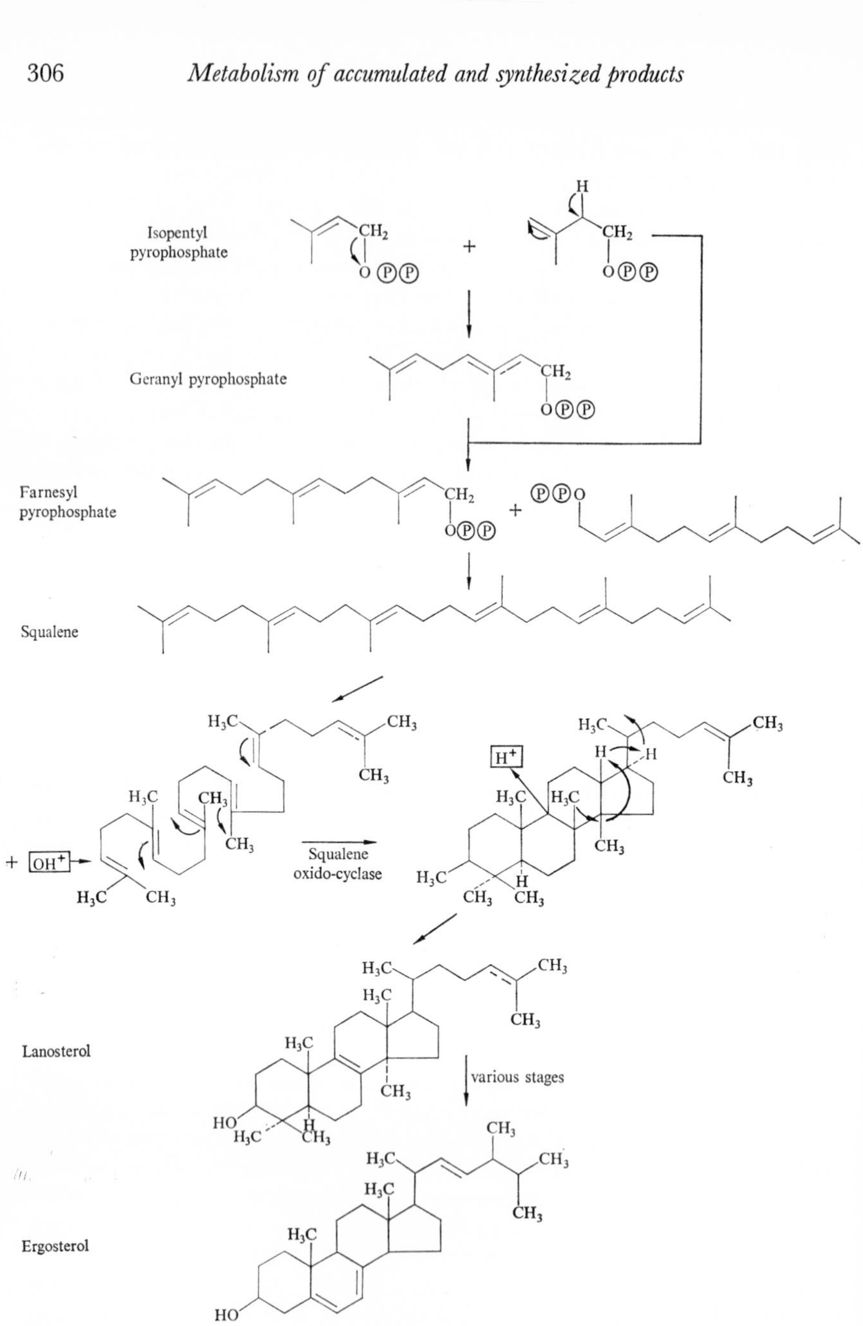

Isopentyl pyrophosphate

Geranyl pyrophosphate

Farnesyl pyrophosphate

Squalene

Squalene oxido-cyclase

Lanosterol

various stages

Ergosterol

Fig. 11.8. The probable derivation of ergosterol, very common in fungi, from isopentyl pyrophosphate via squalene acted on by the cyclizing enzyme squalene-oxidocyclase

palmitate compound, as in *Penicillium chrysogenum* (MIYATAKI *et al.*, 1962). Its precursor is undoubtedly squalene which can be synthesized from mevalonate via farnesyl pyrophosphate by yeast homogenates (SCHWERK and ALEXANDER, 1958; LYNEN *et al.*, 1958; HENNING *et al.*, 1959). This molecule has the ability to cyclize and become folded in a variety of ways. The enzymes responsible have not been adequately characterized but it seems probable that a precursor of ergosterol is lanosterol. This is known in other organisms to be brought about by squalene-oxidocyclase (ARIGONI, 1960). In yeast, the origin of the methyl group on C-24 is apparently from methionine due to transmethylation (PARKS, 1958). The process is illustrated in Fig. 11.8.

The first of the gibberellin compounds was isolated and crystallized from cultures of *Gibberella fujikuroi* by Yabuta and his collaborators in 1938 but not until the work was repeated twenty years later was gibberellic acid isolated (YABUTA and SUMIKI, 1938; CURTIS and CROSS, 1954). In addition to the acid, nine other similar gibberellins are now known but only gibberellins A1, A2, A4, A7 and A9 from fungi.

As with many secondary metabolites the best yield is obtained after a

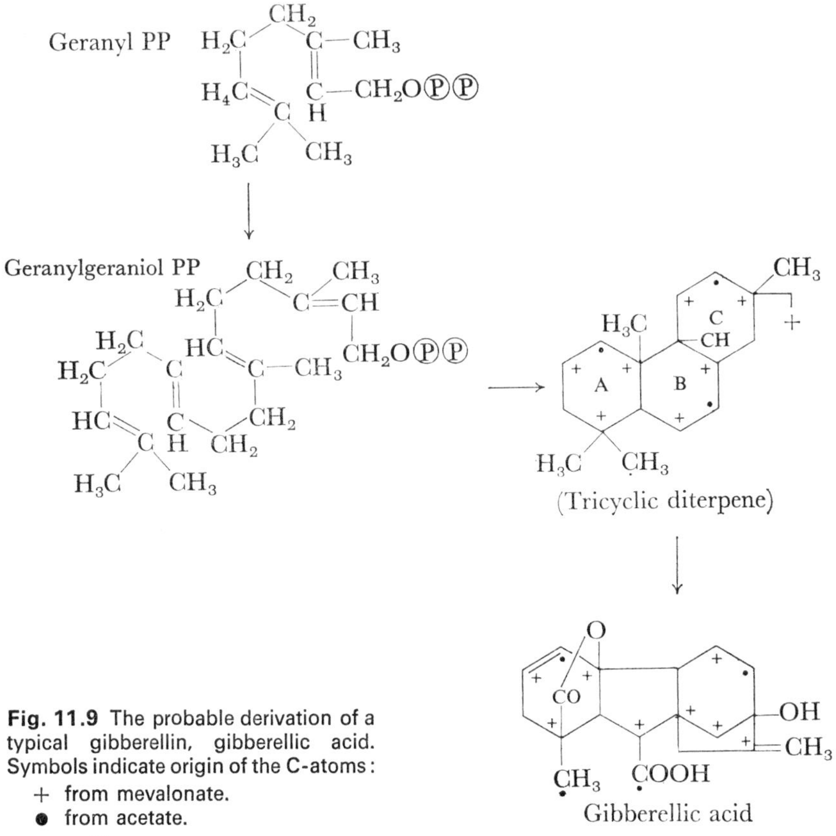

Fig. 11.9 The probable derivation of a typical gibberellin, gibberellic acid. Symbols indicate origin of the C-atoms:

+ from mevalonate.

● from acetate.

primary phase of active growth has been checked; commercially by means of a nutritionally unbalanced medium. The culture is submerged, stirred and CO_2 is introduced; the addition of mevalonic acid is said to improve yield, which may be up to 1 g/l. medium (GROVE, 1963). In 1955 WENKERT suggested that gibberellins could be derived from geranylgeraniol by (a) loss of C-17, (b) contraction of Ring B to a 5-membered ring with extrusion of the C-9 as a carboxyl group and (c) the formation from ring C and its substituents of a bridged ring structure (Fig. 11.9).

BIRCH *et al.* (1959) tested these speculations by feeding cultures of *Gibberella* with $CH_3\,^{14}COOH$ and [2-^{14}C]-mevalonic lactone. If the hypothesis were correct the labelling would be expected to be distributed as shown in Fig. 11.9. The labelling did indeed follow this pattern. Thus, although the enzymatic basis is not known in this case, the technique of supplying labelled precursors and then degrading the resultant molecule and determining the final position of the radioisotopic carbon atoms has provided an outline of the processes involved. (The probable metabolic basis is discussed on p. 311.)

Antibiotics and toxins

Unlike the last group of compounds these do not have any common biosynthetic pathway to bind them together. They are included here because of their biological interest and also to illustrate the diversity of fungal products.

The most celebrated antibiotic, and still one of the most effective, is penicillin, detected originally in *Penicillium notatum* plate cultures. This is effective against gram+ve bacteria. Griseofulvin, produced by a range of *Penicillium* species including *P. griseofulvum* has an antibiotic effect on fungal hyphae which has already been described (Chapter 3, p. 59). Gliotoxin has both bacteriostatic and fungicidal properties. It is produced by *Trichoderma viride*, a soil fungus, and may well play a role in competition between this fungus and other soil organisms. Several fungi are pathogenic to higher plants as a consequence of producing extracellular toxins. Some of these are of relatively low molecular weight, e.g. lycomarasmin, produced by *Fusarium oxysporum*.

PENICILLIN

Although penicillin has now been made in part by chemical synthesis for two decades its original production was entirely from the mould *Penicillium chrysogenum*. The related antibiotic cephalosporin is produced by another genus of Fungi Imperfecti, *Cephalosporium*. By selection of induced mutants yields were raised dramatically from about 200 units/ml. to over 1000 units/ml. (SYLVESTER and COGHILL, 1954). Cultural conditions are typical of those of shunt metabolites, the initial rapid rate of mycelial growth having levelled off, but high phosphate and pH control are essential. This is because penicillin is very sensitive to pH; values over 7·5, especially in the presence of ammonia, resulting in its destruction. The phosphate requirement is not fully understood.

The basic structure of penicillin can be regarded as being, theoretically, composed of D-valine, L-cysteine and a substituted acetic acid viz:

```
H3C   CH3 |                  |      (CH3)2
   \  /   |   'cysteine'     | acyl   |
    C----S                   |        CH          CH2SH
    |     \                  |        |           |
    |      C                 |        |           |
    |     / \                |        |           |
   HC----N   CH.NH           | R      CH(NH2)     CH(NH2)    CH3COOH
    |     \  /               |        |           |
    |      CO                |        |           |
   COOH   |                           COOH        COOH
 'valine' |                           valine      cysteine   acetic acid (R)
 Basic penicillin structure                 constituent molecules
```

R may include aromatic rings
e.g. C_6H_5CHCO—Penicillin G,
or aliphatic groups
e.g. H—6-amino penicillic acid,
or $CO(CH_2)_3CH(NH_2)COOH$—cephalosporin N

The cysteine moiety was shown to be involved by feeding with radioactively-labelled L-[β-^{14}C, ^{13}N, ^{35}S]-cysteine or ^{14}C-labelled serine or glycine (ARNSTEIN and GRANT, 1954). Thus the biosynthesis of cysteine seems to follow the usual route already described for fungi (Fig. 11.3, p. 290). The origin of the SH-group is still obscure. In yeast, sulphate is reduced to sulphite and thence to H_2S which can be incorporated into cysteine. Mutants of *P. chrysogenum* have been used to demonstrate the reduction of sulphate to sulphite (HOCKENHULL, 1948) but further steps have not been identified in this fungus. The inclusion of valine was also demonstrated by the use of ^{14}C- and ^{15}N-labelled acids, although surprisingly L-valine was found to be the form most readily utilized (DEMAIN, 1956). The synthesis of L-valine by this fungus has not been demonstrated but presumably is similar to that already described (Fig. 11.3, p. 290). The amino acids are believed to be linked together to form the dipeptide L-cysteinyl-L-valine and the acyl group is then added. The evidence for this is the discovery of a tripeptide δ-amino-adipyl-cysteinyl-valine from the mycelium of a *Penicillium* known to synthesize Penicillin G. This tripeptide is converted to Penicillin G by acyl exchange provided that a phenylacetic acid derivative is available and there is evidence that ^{14}C-labelled derivatives are so incorporated (GORDON *et al.*, 1953; ARNSTEIN and MORRIS, 1960). That this tripeptide is probably involved in the final synthesis is suggested by the fact that penicillin synthesis can be doubled by supplying exogenous L-aminoadipic acid (SOMERSON *et al.*, 1961). The enzymatic basis of peptide formation is not known but peptide formation appears to be a common property of a wide range of fungi. In *Neurospora* extracts, γ-glutamyl transferase activity has been demonstrated and this could be responsible for synthesizing peptides containing glutamic acid (THOMPSON, *et al.*, 1962). Analogous enzymes could exist for other amino acids and, indeed, a variety of peptides accumulate in *Penicillium* under conditions favourable to penicillin formation (BU'LOCK, 1967).

LYCOMARASMIN

It is convenient to consider this toxin, which induces wilt in tomato plants here since it too is probably a dipeptide. WOOLLEY (1948a, b) has proposed that it is N-(α(α-hydroxypropionic acid)-glycyl asparagine:

HOOC—CH—NH—CH_2—CH COOH
| |
HOOC—CH_2 NH
|
CH_2 $CONH_2$

It is produced by *Fusarium oxysporum* f. *lycopersici* both on infected plants and in culture, although it has been claimed that in nature its production is only one eighth of that required to produce symptoms (DIMOND and WAGGONER, 1953). It forms a chelated compound with iron and if this is applied to cells the complex greatly increases their permeability (GÄUMANN and BACHMANN, 1957). Nothing is known of its synthesis.

GRISEOFULVIN

This antibiotic has a deleterious effect on many fungi and is employed in the treatment of dermatophytes and, in a very limited manner, against plant diseases. How exactly it interferes with hyphal growth is not yet understood (BENT and MOORE, 1966 and cf. p. 59). Its chemical structure is remarkable:

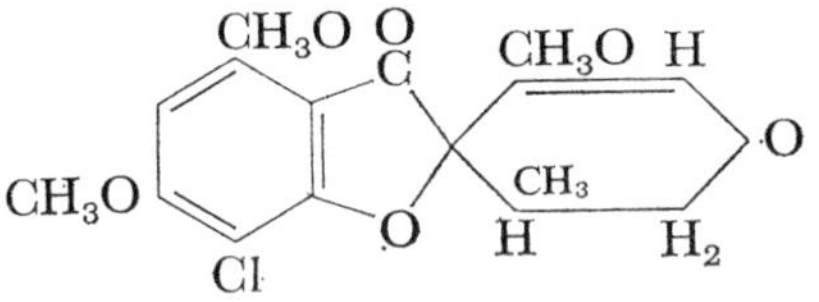

CH_3O O C CH_3O H
O
CH_3O CH_3
O H H_2
Cl

Very little is known about its biosynthesis. It is a typical shunt metabolite. The carbon skeleton can be built up from acetyl residues which join together head to tail but the first aromatic ring precursor known is griseophenone-A. If the fungus (usually in commercial culture a strain of *Penicillium patulum*) is grown on a chloride-free medium then griseophenone-C, the non-chlorinated precursor, accumulates. Addition of chloride to the culture medium is followed by its immediate incorporation into the molecule (RHODES *et al.*, 1961).

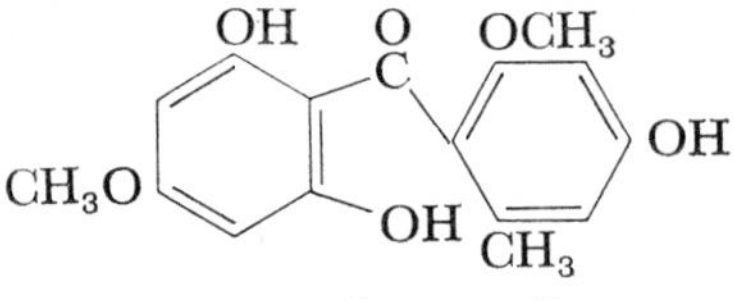

OH O OCH_3
C
OH
CH_3O
OH CH_3

Griseophenone-C

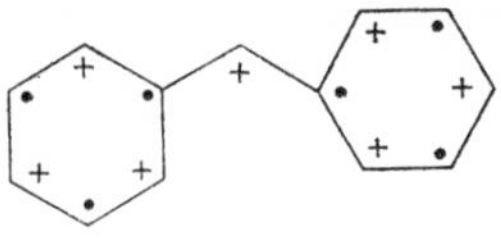

Labelling of griseofulvin molecule
+ = C from COOH of acetate
• = C from CH_3 of acetate

Although the enzymes involved in this synthetic pathway are not known in these fungi there are parallels with the basic processes in other fungi.

For example, an enzyme has been isolated from *Caldariomyces fumago* which catalyses the formation of a carbon-chloride bond utilizing chloride chlorine (SHAW *et al.*, 1959).

The aromatic ring can evidently be synthesized in two ways, one, already described, is the shikimic acid pathway (Fig. 11.5, p. 292) which seems to be that employed for biosynthesis of aromatic amino acids. A second, utilized here, is the head-to-tail association of acetate units followed by their cyclization. The first stage is, presumably, identical with that employed in the biosynthesis of fatty acids, namely, the formation of malonyl CoA by acetyl carboxylase from acetyl CoA and CO_2 followed by the condensation of malonyl CoA and acetyl CoA (cf. p. 297). However, the chain so assembled must be stabilized in some way not involved in synthesis of fatty acids since reduction does not occur. This is followed by a cyclization effect, the nature of which is unknown but which can be distinguished from the chain formation process. Evidence for these two processes has been provided by Bu'lock and his colleagues. For example, 6-methylsalicilic acid is a product of *P. griseofulvum* containing a single aromatic ring. BIRCH *et al.* (1955) showed that [1-^{14}C]-acetate units were incorporated in the ring, head-to-tail but, when [2-^{14}C]-malonate is supplied, it appears that three out of the four 'acetate units' are in fact derived from malonate (BU'LOCK and SMALLEY, 1961). Moreover, if the molecule is degraded and the relative activities of the carbon atoms determined, it is possible to distinguish the 'starter' atom, i.e. the first C-atom at the end of the chain (Fig. 11.10).

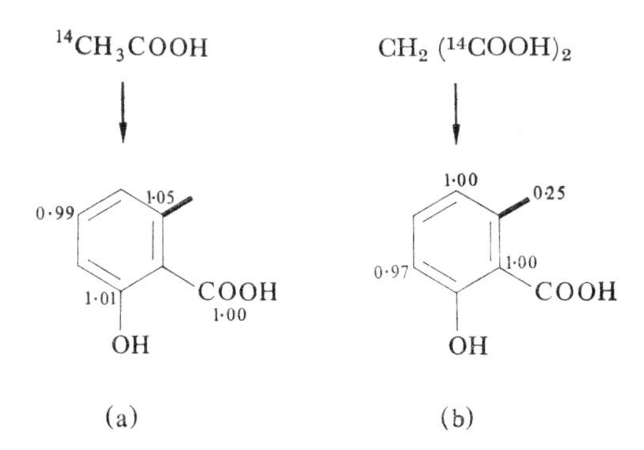

Fig. 11.10. Relative activities in ^{14}C atoms of 6-methylsalicilic acid synthesized from (a) [1-^{14}C]-acetate; (b) [2-^{14}C]-malonate. In each case the remaining carbon atoms carry negligible radioactivity. Note the slightly higher activity of the 'starter group' in (a) and its markedly lower activity in (b). (After Bu'Lock)

ALLPORT and BU'LOCK (1960) also studied the production of some chromones from the Ascomycete *Daldinia concentrica*. Here the aromatic ring is built up in the same way as in 6-methylsalicilic acid but in a number of strains the cyclization process was defective. Here, therefore, although the process of chain assembly was effective it was distinct from the subsequent phase. Other strains were discovered in which the cyclization process was partially defective at a later step so that dimerization or oxidative coupling of phenolic rings was impaired. It is through processes such as these, that complex polycyclic compounds and quinones (such as the anthraquinones—an important group of bright fungal pigments) are formed.

It will be clear that secondary metabolites, however bizarre, can arise in fungi from the same basic processes as are involved in producing fatty acids, amino acids and so on. But they have, in addition, some peculiar or restricted ability which diverts the metabolite from its usual biosynthetic end-point. This is essentially a problem of cellular regulation and this topic will now be considered briefly.

REGULATION OF METABOLISM

Metabolic regulation in fungi is probably not very different from that in other organisms.

Reserve energy

Primary catabolic and anabolic routes will be determined ultimately by the availability of oxidation-reduction energy conserved in 'energy-rich' bonds such as phosphate in ATP or the thiol-ester link of CoA. The role of other nucleoside phosphates and thiol esters is not negligible but, ultimately, their energy is channelled from ATP and CoA. Calculations for bacteria (e.g. LEHNIGER, 1965) suggest that there are insufficient molecules of ATP to maintain biosynthetic activities for more than a very short time so that regeneration of ATP from ADP and P_i is of great importance. Thus oxidative phosphorylation, both of substrates and in the oxidation chain in mitochondria, is of the utmost importance in ATP production. The efficiency of such reactions and the availability of ADP and P_i will be critical. Glycolysis and oxidation processes will be restricted by high levels of phosphorylation, associated with low rates of turnover of ATP; hence biosynthetic processes will also be restricted.

Pacemakers and diversions

In addition to this general control exercised by the availability of a cell's energy reserve, particular controls will operate at various points; such reactions can act as pacemakers. An example is the first step in glycolysis, the phosphorylation of glucose by transfer of phosphate from ATP, mediated by hexokinase. Anything which effects the amount or activity of hexokinase, or the availability of ATP will obviously have a profound effect upon the whole catabolic sequence. Thus the rate of production of ATP in the TCA cycle will effect the rate of reaction of hexokinase. In general this will function in the form of a positive feed-back mechanism. Increased hexokinase activity increases the rate of supply of phosphorylated intermediates to the glycolytic systems and hence an increased rate of ATP production.

Other points in the catabolic sequence of importance are the 'branch' points. An example of such a system was discussed earlier in relation to ethanol production. It will be recalled that the balance between 3-phosphoglyceraldehyde and dihydroxyacetone phosphate could be altered in the latter's favour by alkaline pH resulting in production of alcohol rather than pyruvate. Sometimes the balance may be determined by kinetic considerations. For example, in yeast low substrate concentrations of pyruvate are converted to acetyl CoA because of the high affinity of the condensing enzyme. At high pyruvate concentrations this enzyme is saturated and the pyruvate carboxylase increasingly converts the pyruvate, virtually irreversibly, to acetaldehyde (HOLZER, 1961). This altered sequence can clearly have important secondary consequences for the biosynthetic processes.

Similar kinds of diversions, consequent upon the primary biosynthetic processes must be responsible for secondary metabolism. Such diversions seem to be initiated by a check on growth when its maximum rate has either been achieved or is just declining. In such circumstances there should be a considerable supply of carbon fragments for utilization. The situations equivalent to those used in commercial processes for the production of secondary metabolites can readily be envisaged in natural conditions. Phosphate deficiency is almost universal in all natural situations and considerable changes in pH can arise as a consequence of fungal activity; both these are conditions known to affect secondary metabolism under controlled conditions. The function, if any, of secondary metabolites is obscure and it is not clear why the primary synthesis should be followed by the restricted and unusual activities so frequent in secondary metabolism. Some products are apparently of value to the fungus, for example, antibiotics and toxins. The former can be recovered from sterile soil inoculated with appropriate fungi and shown to be produced in sufficient quantity to affect other organisms, yet it is very doubtful how far such substances are really effective in a natural environment (BRIAN, 1957; BAKER and SNYDER, 1965). BU'LOCK (1961) has argued the intriguing hypothesis that secondary metabolism 'serves to maintain mechanisms essential to cell multiplication in operative order when cell multiplication is no longer possible'. His argument is based upon the observation that secondary metabolites arise from general synthetic processes in odd combinations, although often with unusual additional features, and the belief that substrate stabilization is a general phenomenon. Thus, on his view, secondary metabolism prolongs the activities of fundamental enzyme systems and prevents their loss so that they do not have to be reconstituted whenever the constraint on growth is relaxed. This hypothesis cannot readily be tested but it would be surprising if it were true. If, as seems to be the case for many organisms, enzyme synthesis recommences again in favourable metabolic circumstances then much of the *raison d'être* of Bu'lock's argument is lost. It would seem to be simpler to suppose that secondary metabolism arises from the effects of normal enzyme systems on abnormal accumulations of intermediates although this does not account for the special additional synthetic pathways which occur. Such a situation is analogous perhaps to that described in connexion with the *arom-1* mutant of *Neurospora* (cf. p. 291). It will be recalled that the accumulation, as a result of the genetically-determined block, of dehydroshikimic acid resulted in the induction of dehydroshikimic dehydrogenase and protocatechuic oxidase, two enzymes not normally detectable in *Neurospora*. This viewpoint is essentially that originally put forward by FOSTER (1949) as 'shunt metabolism' and it indicates that secondary metabolism must be regarded as a consequence of the lack of effective regulation of metabolism under abnormal conditions. While this seems to be the most plausible hypothesis yet made, it is unattractive for, in general, organisms seem to have the most remarkable ability to effectively regulate their activities over a wide range of conditions (cf. Chapter 7).

Enzyme regulation

This discussion of secondary metabolism has raised a further issue concerning the regulation of metabolism, namely, the processes which regulate the action and synthesis of enzymes. This topic has received much attention recently as a consequence of studies with bacteria. Some findings with these organisms are applicable to fungi but probably not all. At present, caution should be exercised in transferring, in an uncritical manner, explanations derived from one organism to others, particularly since the structure of bacterial and fungal cells are so dissimilar.

The principle enzyme regulating processes described for bacteria are shown in Fig. 11.11.

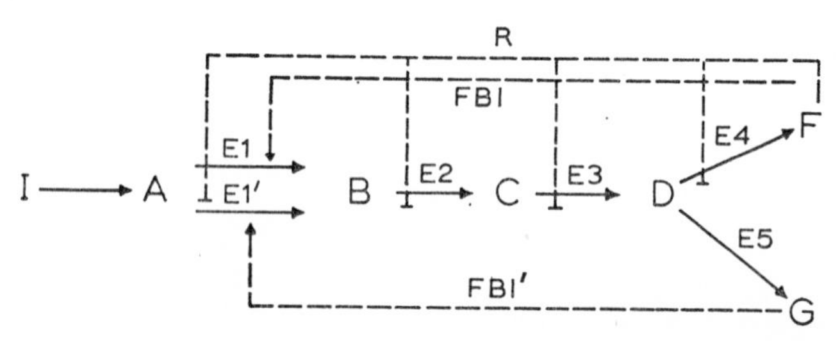

Fig. 11.11. Scheme to illustrate enzyme feedback inhibition and repression, and induction. The pathway bifurcates so that isoenzymes (E1 and E1′) are involved. Key: I, inducer; A–D, successive steps in a synthetic sequence with two alternative end-points, F–G; E1, E1′, isoenzymes for step A–B; E2–E5, enzymes controlling subsequent steps. Feedback inhibition (FBI) shown by arrowed lines, feedback (co-ordinate) repression (R) by barred lines

In feedback inhibition, the end-product inhibits an enzyme responsible for an early step in a multi-stage process. In enzymic repression the end-product prevents the formation of an enzyme or all the enzymes (co-ordinate repression) catalysing earlier steps in the pathway. Both types of control may operate in the same pathway, but, in general, enzyme repression is the most economical since the production of unnecessary enzymes is wholly prevented, rather than inhibition of existing but unnecessary enzymes. The effect of feedback inhibition, however, should be immediate while repression will have a gradual effect until existing enzymes cease to function and fail to be replaced. A superficially different phenomenon, the opposite of repression, is enzyme induction of either one or all the enzymes in a pathway (sequential induction). In certain bacterial systems JACOB and MONOD (e.g. 1961) have provided a unifying explanation for these two phenomena. Enzyme formation is regulated by the presence of a repressor substance in the cell whose action can be blocked by some form of interaction with an inducer, so relieving the repression. There is little evidence from fungi of any of these phenomena except that of enzyme induction.

Enzyme induction is a well known and important phenomenon in many fungi, ranging from the adaptation of yeast in culture from hexose to other sugars, e.g. maltose, to the production of cellulases and pectic and macerating enzymes by many parasitic fungi. The induction of enzymes in these cases is brought about by very small amounts of inducing substance indeed. The studies of Spiegelman and his associates on the induction of 'galactozymase' and maltase in yeast provide a typical example (SPIEGELMAN and HALVORSON, 1953). He was able to show that yeast cells are capable of adapting to the utilization of a new sugar in a medium with or without exogenous nitrogen. The new enzyme, maltase, arose from the free amino acid pool of yeast but, in the absence of exogenous nitrogen, this could involve the breakdown of pre-existing enzymes to free amino acids which were then utilized. The normal necessary conditions for protein synthesis had to be maintained so, for example, anaerobic or nitrogen-starved yeast cells showed reduced ability to respond to induction, or none at all. Induction is a competitive process; attempts to adapt yeast cells simultaneously to galactose and maltose resulted in the galactose-fermenting system predominating despite an initial rapid production of maltase (Fig. 11.12). Further genetic work has provided evidence concerning the induction of galactozymase. It has been shown that galactokinase, transferase and epimerase are induced simultaneously on exposing cells to galactose and that these enzymes are under the control of three closely linked genes, *ga*, *ga*7 and *ga*l0 (DE ROBICHON-SZULMAJSTER, 1958; DOUGLAS and HAWTHORNE, 1964). A situation similar to this is known in *E. coli* where groups of linked genes control enzymes in sequential steps of a pathway, the group as a whole being controlled by an 'operator' gene. This gene's action can be inhibited by an hypothetical repressor substance. No such gene has been detected in yeast. This is the nearest approach to the bacterial situation yet found in fungi.

Feedback inhibition is probably quite widely exhibited but it has not been searched for systematically. Examples occur in yeast and *Neurospora*. For example, in the sequence:

$$\text{Ornithine} \longrightarrow \text{Citrulline} \longrightarrow \text{Arginino-succinic acid} \longrightarrow \text{Arginine}$$

Ornithine transcarbamylase, which couples ornithine and carbamoyl phosphate to form citrulline, is inhibited by arginine (BECHET *et al.*, 1962).

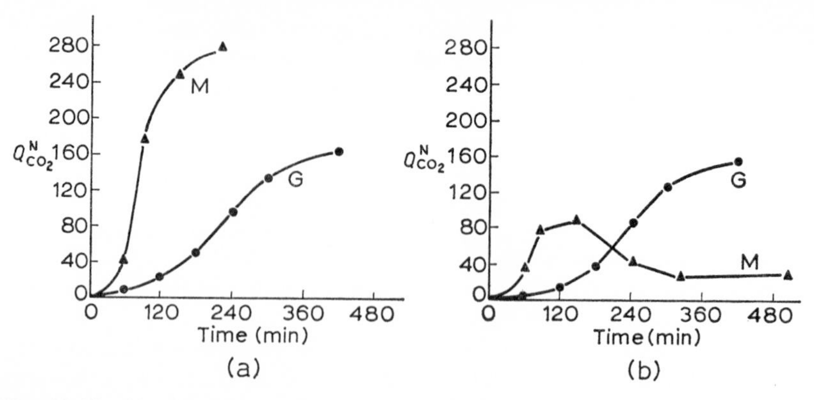

Fig. 11.12. Simultaneous adaptation to galactose and maltose of *Saccharomyces cerevisiae* grown previously in a glucose medium. (a) M, curve of adaptation to maltose alone; G, curve of adaptation to galactose alone. (b) M, G curves of adaptation to maltose and galactose respectively during simultaneous induction. Note only slight reduction of galactozymase activity compared with (a) but considerable inhibition of maltozymase activity. Mixture contained equimolar proportions of sugars in M/15 KH_2PO_4 but lacked exogenous nitrogen. (After Spiegelman)

In a branched pathway, one end-product could inhibit an enzyme early in the common pathway and so reduce the end-product of the other branch of the pathway, regardless of the metabolic requirements for this latter end-product. It is usually found, however, that in such cases the enzyme to be inhibited in the common pathway exists in more than one form, each form being uniquely inhibited by a different end-product. Such a branched pathway occurs in the biosynthesis of threonine and methionine:

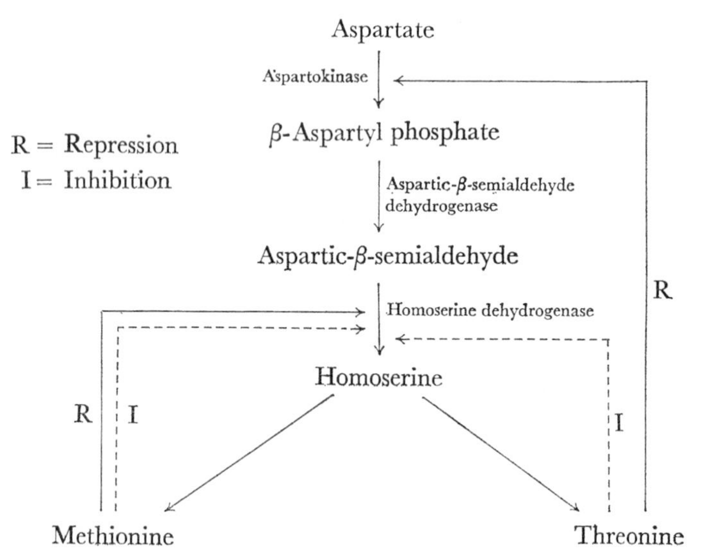

Both methionine and threonine inhibit homoserine dehydrogenase in yeast, the former being most effective (KARASSEVITCH and DE ROBICHON-SZULMAJSTER, 1963). It is not known whether there are two isoenzymes. Repression is also shown; methionine represses the same enzyme but threonine represses aspartokinase (DE ROBICHON-SZULMAJSTER *et al.*, 1965). A combination of both feedback inhibition and repression is not uncommon in bacteria; another example in fungi, although the effects are of an altogether lesser magnitude, occurs with aspartyl carbamoyl transferase in *Neurospora* (DONACHIE, 1964).

It is clear that systems exist in fungi that lead to the formation, loss or inhibition of enzymes and such co-ordinated changes are bound to have a profound effect on metabolism. Certain striking examples, the development of enzyme activity at spore germination (Chapter 6, pp. 182–187) or the alterations in enzyme activities in *Blastocladiella emersonii* in relation to alternative developmental pathways (Chapter 5, pp. 101–102) have already been mentioned. *Blastocladiella emersonii* is one of the few fungi for which well documented data on enzyme levels exists and from which it is possible to try to relate metabolic cause to developmental effect. This type of detailed study is necessary if the study of function is to be related to that of structure. A few other examples of quantitative data exist, e.g. ROTHERY *et al.* (1962) have estimated that about 40% of the oxygen uptake is involved in the TCA cycle; and peptidase and proteinase levels rise during interphase and fall at bud formation in synchronously dividing cultures of yeast (SYLVEN *et al.*, 1959).

Structural aspects

The actual formation of enzymes is not always sufficient to ensure metabolic activity. They must be at the right site and in the right configuration. It will be recalled that the apparent inactivity of trehalase in *Neurospora* ascospores has been attributed to their being in some way separated from the trehalose. The most striking example of this type of situation known in fungi occurs in yeast. Anaerobically grown yeast possess either no mitochondria or structurally simplified ones, as seen in electronmicrographs (POLAKIS *et al.*, 1964) and there is little evidence for enzymes of the TCA cycle or the cytochrome chain (SLONIMSKI, 1953; HIRSCH, 1952). If such cells are homogenized and the contents separated by density centrifugation, particles can be isolated with succinic dehydrogenase activity (SCHATZ, 1963). If anaerobic yeast is restored to aerobic conditions there is a dramatic induction of fumarase, succinic-dehydrogenase, cytochrome *c* reductase and, indeed, the whole respiratory system including the oxidation chain. The mitochondria also develop and there is an inverse relationship between them and the content of respiratory particles. Despite this enormous development of enzyme activity, associated with the organization of mitochondria, no new protein is said to develop (HEBB and SLEBODNIK, 1959). It must be concluded that the enzyme systems were in existence but incorrectly distributed and so ineffective. The ultra-structural repatterning of mitochondria consequent upon aerobic conditions presumably rendered them effective once more. This is a particular example of the general principle that fungal cells, like all living cells, have a definite pattern of structure and compartmentalization that must be achieved if metabolism is to proceed normally. In respect of large cell organelles such as mitochondria, some knowledge is available concerning the distribution of enzymes and other materials. There is evidence that other, smaller structural patterns exist. Glycolytic enzymes occur in the supernatant of fractionated and centrifuged cells of yeast, yet ROTHSTEIN *et al.* (1959) have provided evidence that a structural unit can be isolated (by slowly drying cells after lyophilization) that is capable of more rapid fermentation than soluble enzyme preparations. Lynen's 'fatty acid synthetase' enzyme complex is another example of a structural entity which is not yet understood.

Metabolic studies of fungi will, no doubt, continue to reveal new end-products and pathways. The great task now, however, is the quantitative determination of the parameters which regulate metabolic processes and the analysis of their distribution and configurations in space and time within the living fungal cell.

Chapter 12
Reactions and Interactions

All the metabolic activities of a fungus are, in a sense, a response to chemical and physical environmental factors but, in this chapter, particular responses to certain factors of the environment and reactions with other organisms will be considered.

The environmental factors to be considered are light, gravity, contact and chemicals and the appropriate tropistic and other specific responses which they invoke in fungi; in the second part of the chapter the whole range of interactions between fungi and other organisms will be considered in a general way in relation to both microorganisms and higher plants and animals.

REACTIONS TO ENVIRONMENTAL FACTORS

Photo-effects

Light has profound effects on many fungi, despite the absence of photosynthesis. The consequences of such responses are variable. They may be morphogenetic, such as the initiation of reproductive structures (cf. Chapter 5, pp. 102–105, 142); or on metabolism, e.g. the light-induced uptake of glucose by *Blastocladiella britannica* or light-stimulated polysaccharide and protein synthesis in *B. emersonii* (HORENSTEIN and CANTINO, 1964; GOLDSTEIN and CANTINO, 1962); or on growth, e.g. change in growth rate of stage 4b sporangiophores of *Phycomyces* in response to a change in light intensity (SHROPSHIRE, 1963); or they may provoke an orientated response, e.g. positive phototropism of many mucoraceous sporangiophores or of asci in several Discomycetes (cf. Chapter 6, pp. 154, 158). Some useful compilations of such effects have recently been made (MARSH, TAYLOR and BASSLER, 1959; CARLILE, 1965; PAGE, 1965).

All such responses must depend upon a photoreceptor of some sort, associated with an appropriate metabolic link to transfer the perception of the light stimulus to the site of response.

With the exception of the morphogenetic and metabolic consequences of exposure to light, which have not been fully analysed, all the other responses involve growth of the hyphae, either singly or, as in the response of complex structures, of groups of hyphae. Whether in such cases the response transcends the reaction of individual hyphae or merely reflects the summation of their individual responses is not known. Since growth is confined to hyphal tips the site of response in these cases, at least, can be defined with some precision.

The problem of the photoreceptor has usually been attacked by investigating the spectral sensitivity of the response and then seeking to correlate the action spectrum with the absorption spectrum of compounds known to be present in the hyphae. These are technically difficult measurements to make. SMITH and FRENCH (1963) wrote, 'an action spectrum measured to a precision of 12% at 10 nm wavelength intervals with a half-band width of 5 to 10 nm is considered good work. The precision now ordinarily obtained in absorption spectroscopy is far greater than in action spectroscopy'. Very few action spectra have been determined for fungi with anything approaching this accuracy. A further difficulty arises because the absorption spectrum of a compound in an appropriate solvent in vitro may be very different from that in vivo, where the compound may be complexed with other substances, e.g. the β-carotene-protein complex from spinach leaves (NISHIMURA and TAKAMATSU, 1957).

There is considerable agreement that all photoresponses, phototropic, light-growth effects and some morphogenetic responses (e.g. trophocyst formation by *Pilobolus* (PAGE, 1956)), have a similar action spectrum. A typical example is that determined for phototropic and growth response of individual sporangiophores of *Phycomyces* with peaks at 4850, 4550, 3850 and 2800 10^{-10}m (DELBRÜCK and SHROPSHIRE, 1960, and Fig. 12.1). So long as the

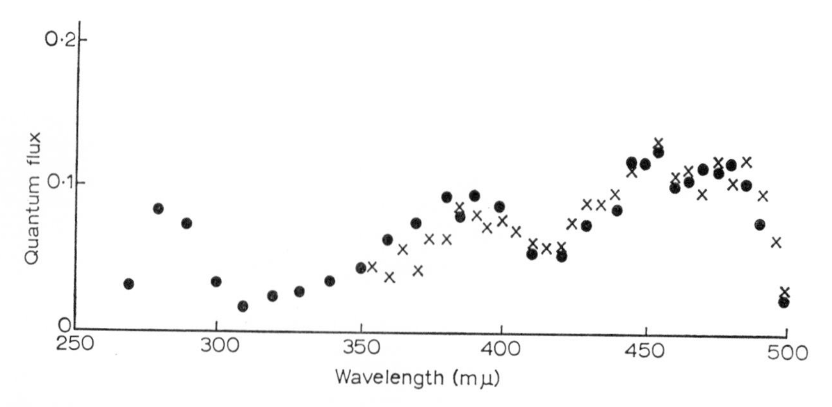

Fig. 12.1. The growth response and tropic response of individual sporangiophores of *Phycomyces*; o = growth response, x = tropic response. (After Delbrück and Shropshire)

action spectrum in the visible range is considered there are two possible contenders for the photoreceptor system, carotenoids and flavins, especially riboflavin. If the peak in the ultra-violet is taken into account, however, this favours the riboflavin hypothesis because this compound has a single peak at *c.* 4500 10^{-10}m and also a high peak at 2650 10^{-10}m. However in an alga, *Platymonas*, a peak at 2750 10^{-10}m has been said to be due to an aromatic amino acid moiety in a protein combined with carotenoids (HALLDAL, 1961). It is obvious that simple comparisons of action and absorption spectra will not resolve this problem.

Attempts have been made to find fungi lacking the proposed photoreceptor or to impair their efficiency by specific inhibitions. Riboflavin appears to be universally present and carotenoids may often turn out to be detectable even where they had not been thought to exist. However, there is evidence that photo-inductive macroconidium-formation in *Fusarium macrosporium* and photo-induction of sexual reproduction in *Pyronema confluens* occurs in the absence of carotenoids (CARLILE, 1956; CARLILE and FRIEND, 1956). An albino mutant of the latter fungus required a photostimulus to develop apothecia but totally lacked colourless polyenes. Attempts have also been made to investigate photoresponses of carotenoid-forming fungi inhibited by diphenylamine (cf. Chapter 11, p. 301). The results are equivocal since this inhibitor reduces rather than eliminates carotenoids, e.g. in diphenylamine-treated *Phycomyces* the treated/normal ratios were (μg): β-carotene 25/586; phytofluene 80/15; ζ-carotene 47/5. Similar experiments have been done using lyxoflavine and mepacrine to inhibit riboflavin. Mepacrine is probably not a specific inhibitor of riboflavin so that results obtained with it are also equivocal (HEMKER and HÜLSMANN, 1960). PAGE (1956) obtained inhibition of trophocyst formation in *Pilobolus* cultures with added lyxoflavin which could be reversed by additional exogenous riboflavin (Fig. 5.5, p. 103). CARLILE (1962) obtained depressed growth of illuminated *Phycomyces* cultures in the presence of mepacrine and lumichrome but this was found by HOCKING (1963) only to be effective in stationary cultures, not in shake cultures. The situation is, therefore, not yet resolved.

Other possible photoreceptors may occur. CARLILE (1960) has suggested a pteridine in relation to the induction effects due to ultra-violet light. PAGE (1965) has described the response curve for the discharge of conidia by *Entomophthora coronata* with maximum response at 4050 10^{-10}m and responses up to 6300 10^{-10}m. A porphyrin-like pigment could be extracted with absorption peaks at 4120, 5060, 5430, 5800 and 6300 10^{-10}m.

Studies have also been made to attempt to locate possible photoreceptors. In the sporangiophores of *Phycomyces* and the germ tubes of *Botrytis* and *Puccinia graminis tritici* there is clear evidence of a lens effect, the light being concentrated at, or near the inner surface of the cell wall farthest from the source (JAFFE, 1960; JAFFE and ETZOLD, 1962; GETTKANDT, 1954). JAFFE's (1960) experiments employed polarized light and led him to postulate the existence of dichroic, longitudinally orientated, photoreceptor units arranged periclinally and about 0·5 μm from the hyphal surface. DELBRÜCK and VARJU

(1961) produced a similar hypothesis. A similar lens effect is shown by the sub-sporangial vesicle of *Pilobolus kleinii*. If the focusing effect is prevented by immersing the sporangiophore in an optically more dense medium (oil of refractive index 1·47) the near side receives a higher intensity of illumination and the resulting curvature is away from the light source (BÜDER, 1918; BULLER, 1934). This behaviour is susceptible of the same explanation concerning the location of the photoreceptors and it may be significant that the sub-sporangial swelling is rich in β-carotene (Fig. 12.2).

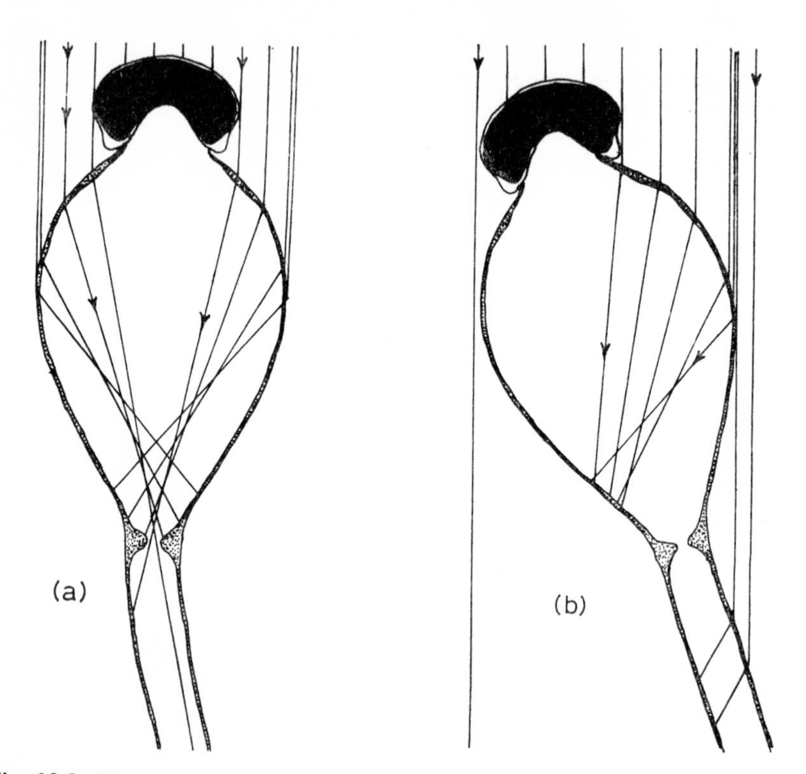

Fig. 12.2. Diagrams to illustrate the path of unilateral light rays which enter the sub-sporangial swelling of *Pilobolus* sp. when (a) vertical and (b) inclined from the vertical. No rays penetrate the sporangium itself. The sub-sporangial swelling, especially the constricted region near the base, is rich in β-carotene. (After Buller)

Since the photoreceptor molecule or molecules are not known, it is not easy to consider subsequent reactions leading to the response. Advocates of riboflavin as photoreceptors have related its action to that of indole-acetic acid but there seems no reason to suppose that this is an effective growth regulator in fungi (GRUEN, 1959b). The only fungus where there is documentation of metabolic events subsequent to photo-induction is *B. emersonii*. The effective wavelengths appear to be between 4000 and 5000 10^{-10}m and the effect is associated with CO_2 incorporation. The metabolic pathways affected were speeded up; they appeared to be:

Ketoglutarate → isocitrate → glyoxylate → glycine → thymine → DNA

and the enzymes and products involved in this sequence all showed the appropriate changes (CANTINO and TURIAN, 1961). Recently, CANTINO (1965) has noted that light also alters the distribution of a haemoprotein resembling a cytochrome. In light-grown individuals it is present in soluble form but in dark-grown individuals it was bound to a pellet-fraction that was brought down at 1000 **g**. Whether, in fact this porphyrin is in any way associated with the photo-perception or response is not yet clear but it is of interest that its absorption spectrum is not dissimilar from that described by Page for *Entomophthora*, mentioned earlier, having peaks at 3825, 5120, 5400 and 6400 10^{-10}m in acetone extract. As yet it has only been obtained in a relatively impure condition.

It may be suggested that further progress in the study of the effects of light on fungi are likely to be rewarding if studies similar to those of Cantino and his colleagues namely, the quantitative metabolic changes subsequent to the light-induced effect are pursued. At present, the search for the photoreceptor seems to be the least attractive starting point.

The effect of gravity

Even less is known of this phenomenon. There is no doubt that various fungal structures, notably basidiocarps and their organs, both perceive and respond to gravity. The very precise adjustments of the gills of agarics is an outstanding example of such a response. In general in the basidiocarps of many fungi the stipe is negatively geotropic and the gills, spines or dissepiments of the pores, apparently positively geotropic. Many such basidiocarps show both photo- and georesponses. In some cases the georesponse will only occur in the light, e.g. *Lentinus lepidus* (TABER, 1966), when it is presumably dependent upon some morphogenetic development that can only be induced by light. PLUNKETT (1961) has made a detailed study of the interaction of the photo- and georesponses of *Polyporus brumalis* which has already been discussed (Chapter 5, pp. 142–143). The fact that in early stages of stipe development photoresponses override georesponses in relation to stipe elongation, suggest that both kinds of responses must ultimately be mediated through the same metabolic processes.

The only recent detailed experimental study of geotropism is that made by DENNISON (1959) of the *Phycomyces* sporangiophore. He either displaced sporangiophores from the vertical or spun them slowly on a centrifuge. He detected three types of responses. Displacement was followed by transient growth curvatures in the opposite direction except when the centrifugation was done in a medium of greater density than the cell contents. In these circumstances it was in the same direction as the force. He attributes these curvatures to 'flexing stresses' on the sporangiophore walls. The third type of response was a long-term one and resulted in a curvature which was always opposite to the applied force, regardless of the density of the external medium. He attributed this to the displacement of geotropic receptors although their nature was unknown.

The present position is that while geotropism is known to be a widespread phenomenon in many fungi and inferred in at least as many, there is in fact great ignorance of how far geotropic stimuli really do affect the behaviour of fungi. So far as any known case of georesponse is concerned there is a total ignorance of the mechanisms involved.

Contact sensitivity

The contact sensitivity of hyphae is at least as widespread a phenomenon in fungi as are georesponses and they are equally little understood. The importance of contact reactions in the formation of multi-hyphal structures has been discussed earlier (Chapter 4, pp. 84–90). These occur in strands (Fig. 12.3c) and rhizomorphs, sclerotia, and such specialized structures as

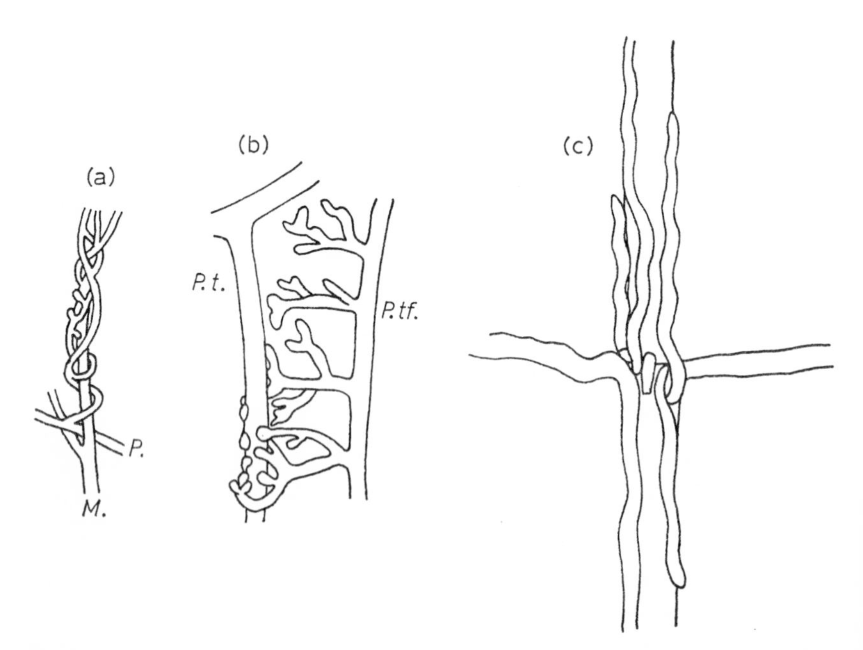

Fig. 12.3. Thigmotropic or contact reactions. (a) Between *Mucor* (*M.*) and *Peziza* (*P.*); (b) between *Peziza tuberosa* (*P.t.*) and *P. trifolii* (*P.tf.*); and (c) *Coniophora cerebella*, backward- and forward-directed fine tendril hyphae on a main, branched hypha. (a, × *c.* 200; b, × *c.* 300, from Reinhardt; c, × 800, from Falck)

appressoria and nematode traps. Thigmotropic responses also occur in interactions between hyphae of the same or different species (Fig. 12.3), whether or not this is followed by a successful fusion, and between conjugant structures at reproduction, e.g. gametangia of mucoraceous fungi. The nature of the actual stimulus which initiates response is not clear. The only evidence comes from the sensitive hyphae of the nematode trapping fungi, which are responsive to friction induced by stroking (COMANDON and DE FONBRUNE, 1938). Presumably this could be a general cause of thigmotropic responses but experiments are needed. The rheotropic responses shown by germ tubes of *Rhizopus nigricans* and *Botrytis cinerea* could be due to a frictional stimulus imparted by flowing medium to the hyphal walls but explanations based on a diffusible chemical stimulus have been preferred (cf. Chemotropism). Alternatively, or subsequent to the initial stimulus, chemical factors may be involved. Aggregation between myxamoebae of the Acrasiales, for example, is believed to be mediated by aggregation antigens (SONNEBORN *et al.*, 1964), possibly of a peptide-polysaccharide nature (WHITFIELD, 1964). In fungi, studies of aggregation have largely been confined to yeasts, either during sexual reproduction or when flocculating during fermentation.

BROCK (1959) studied agglutination between opposite mating types, *A* and *a*, of *Hansenula wingei*. This is due to a reaction at the wall surfaces. Treatment with trypsin affected one conjugant but not the other, treatment with 0·001 M sodium periodate affected the latter strain but not the former. He interpreted this to indicate that the walls carried complementary components, one a protein, the other a polysaccharide, which underwent a reaction analogous to an antigen/antibody reaction. Since neither conjugant would agglutinate with the diploid cell so formed this seems to be a specialized reaction but it could provide a model for general contact reactions. Support for this comes from the work of Eddy and his colleagues (EDDY, 1958a, b; EDDY and RUDIN, 1958a, b, c). They showed that isolated cell walls of flocculating yeasts retain this property and that this property probably resided in the presence of a mannan-protein complex. This was absent from young cells, which do not flocculate, and loss of flocculating ability in older cells or their walls could be correlated with enzyme and chemical treatments which destroyed the mannan or protein moieties. Thus, however, detected, it does seem as if there may well be a chemical basis, perhaps analogous to that between antigen and antibody, to the response to contact stimuli.

Chemotropism

Such considerations lead naturally to the phenomenon of chemotropism, a phenomenon widespread in the younger parts of fungi, i.e. in the hyphal tips. Chemotropic responses may be of some importance in the establishment of propagules on hosts of various types. The chemotropic stimulus may be responsible for initiating germination or for directing the growth of germ tubes to a host. Chemotropic responses are also involved in the attraction of gametangia towards each other and, presumably, in similar responses between vegetative hyphae prior to fusion as well as in respect of zoospore orientation. The former will be considered later in connexion with reproduction (Chapter 15, pp. 410–413); the last two situations have already been discussed (Chapter 4, pp. 70–73 Chapter 6, pp. 178–179). Here the general situation will be considered.

Many early experiments were carried out on the responses of fungal hyphae, usually germ tubes to a variety of chemicals, nutrient and otherwise but much of this work was done without adequate controls. The studies of *Rhizopus nigricans* by GRAVES (1916) showed that if spores were employed, a source of error was the failure to recognize that germ tubes usually grew away from each other because of staling products which they produced and regardless of the chemotropic stimulus to which they appeared to be reacting. STADLER (1952, 1953) has examined this behaviour more fully. His technique was to employ two layers of agar separated by plastic coverslips with holes drilled through the plastic (Fig. 12.4). When spores were inoculated in one agar

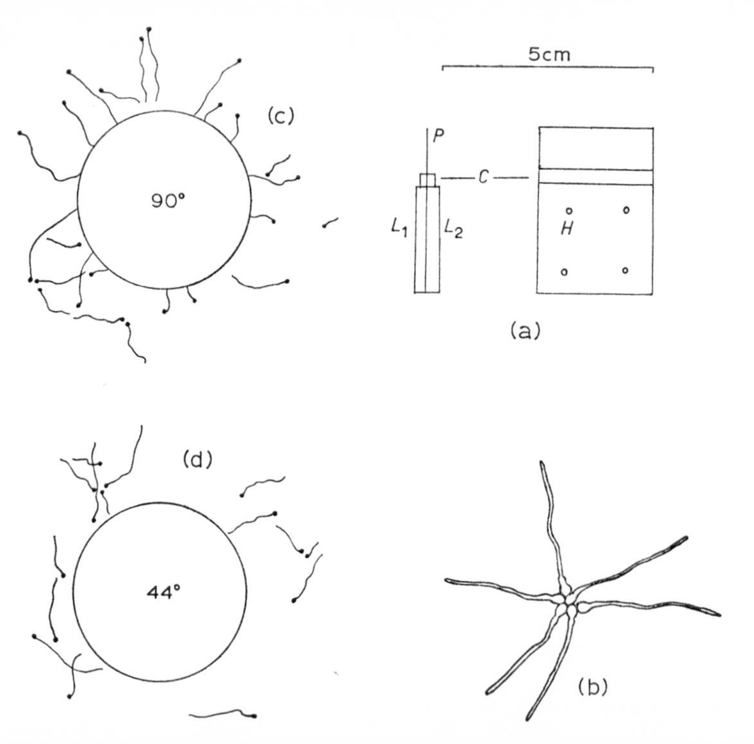

Fig. 12.4. Technique for studying autotropic (chemotropic) responses of germ tubes of *Rhizopus nigricans*. (a) Apparatus—L_1, L_2, layers of medium on opposite sides of perforated plate (*P*) with 1 mm diameter holes (*H*), the whole held together by a clamp (*C*); (b) clump of germinating spores to show orientation away from each other due to staling reaction growth; (c), (d) camera-lucida drawings of a test region around a hole. In (c) spores in synthetic medium in both layers (L_1, L_2); in (d) synthetic medium in both layers but spores only in L_1. Angles indicate mean value of angles between line of tip of hypha and line from spore to centre of hole. (From Stadler)

layer only, a majority were orientated towards the holes but, if both layers had equal concentrations of spores, a majority of germ tubes were orientated at random. With various plant juices added to one layer, orientation of germ

Table 12.1. Orientation of germ tubes of *Rhizopus nigricans* in relation to spore concentrations, staling substances and the presence of vegetable juices (see Fig. 12.4). (From Stadler, 1952)

Test layer	Opposite layer	Mean orientation angle (°)
90 spores/mm^3 synthetic	90 spores/mm^3 synthetic	92 (4)
90 spores/mm^3 synthetic	0 spores—synthetic	46 (4)
90 spores/mm^3 synthetic	0 spores—H_2O agar	44 (4)
60 spores/mm^3 synthetic	0 spores—turnip juice	16·5 (4)
60 spores/mm^3 turnip juice	0 spores—turnip juice	23 (4)

All data differ significantly from 92° reading at 1% level of probability.
Synthetic medium: glucose 1%: $MgSO_4$, 2×10^{-3} M: KH_2PO_4, $1{\cdot}1 \times 10^{-2}$ M: Asparagine, $7{\cdot}6 \times 10^{-3}$ M: Agar 2%
Nos. of replicates in parentheses

tubes was improved but, if the concentration of spores was reduced the degree of orientation fell off (Table 12.1). He accounted for these observations by supposing that germ tubes repelled each other by a diffusible chemotropic stimulus so that they always grew towards uncolonized medium. Plant juices affected the situation by inactivating the diffusible factor, hence the reduced orientation in the low spore concentration condition. He tested the hypothesis of a diffusible material by allowing liquid medium to flow over the surface of agar on which groups of spores were germinating. Instead of showing the normal orientation of all germ-tubes away from each spore clump, the germ tubes grew 'upstream'. Stadler claimed that this was because the diffusible inhibitor had been swept 'downstream' by the flow of the liquid medium and this was borne out by the reduced germination of spores 'downstream' compared with those 'upstream'. The opposite effect was found by Jaffe and his colleagues (JAFFE, 1966; MÜLLER and JAFFE, 1965) in *Botrytis cinerea*. Here germ tubes from condial clumps grew towards each other and in flowing culture-liquid experiments, the germ tubes tended to develop in a downstream direction. These results were attributed to a diffusible growth accelerator. The responses were also found to be highly susceptible to CO_2 concentration and at 0·3–3·0% the reactions between pairs of spores at low concentrations were reversed. No generalizations can be made from these observations but the responses of germ tubes, sometimes called autotropisms, can clearly be positive or negative and this may depend upon variable environmental factors. How far such autotropic responses can be extrapolated to the analysis of chemotropic responses of hyphal tips of a vegetative mycelium is unknown, so that interpretation of such experiments is not possible.

INTERACTIONS WITH OTHER ORGANISMS

The title of this section might be held to cover the whole subject of fungal ecology but consideration will only be given here to the basic kinds of interactions which can occur between fungi and other organisms. Much of the discussion concerning heterotrophic nutrition in Chapter 9 is of relevance to this topic.

Competition

It will be realized that interaction can occur without actual contact of organisms. The production of primary and secondary metabolites by fungi and their transport outside the hyphae, whether by secretion or leakage, are bound to affect the environment. The cause of synergistic or antagonistic interactions may well lie in such reactions. Classic experimental demonstrations of such situations are the inhibition of *Saccharomyces cerevisiae* by *Schizosaccharomyces kefir* through the lower tolerance of the former to alcohol produced by the latter (Fig. 12.5) and the mutual growth of *Mucor*

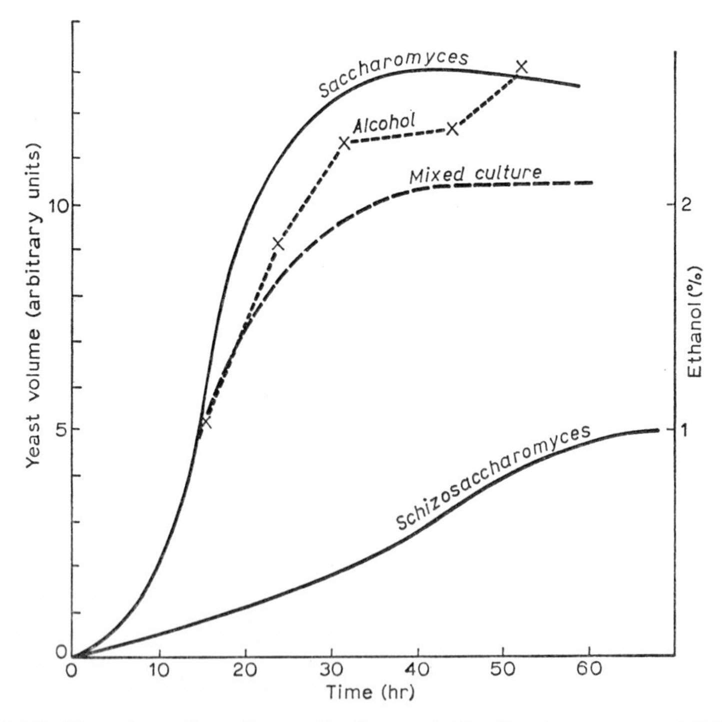

Fig. 12.5. The volume (in arbitrary values) occupied by *Saccharomyces* and *Schizosaccharomyces* in single and mixed culture, and ethanol production. Note that growth of *Saccharomyces* is limited in relation to ethanol production but not that of the latter. (After Gause)

ramannianus and *Rhodotorula rubra*, the former utilizing thiazole produced by the yeast, the latter utilizing pyrimidine produced by the *Mucor*; both require thiamin for normal growth (GAUSE, 1932; MÜLLER, 1941). This kind of association is presumably of importance to all saprophytic fungi which are heterotrophic for vitamins and other growth factors. Another type of interaction, of the same general nature, seems to determine the sequence of fungi in the decomposition of plant residues. Here, because of their more limited catabolic activities and rapid rates of utilization of simple carbon compounds, the first agents of decomposition, or 'sugar fungi', are mostly Phycomycetes; these are followed by Ascomycetes with or without Basidiomycetes which utilize cellulose; the final group includes the majority of lignin-utilizing species and Basidiomycetes predominate (GARRETT, 1951). This statement, not surprisingly, represents an over-simplification of the substrate relationships of fungi but it has proved of value for example in general ecological considerations of soil fungi.

The phenomenon of antagonism, often detected as the exclusion of a particular fungus from a habitat where it might be expected to grow on a basis of the substrate(s) available, is a complex problem. It involves a variety of interactions; competition for nutrients and, in some circumstances, space; parasitic attack by other fungi or predation by other organisms; antibiosis, whether by the production of general toxic materials, e.g. alcohol, or by specific antibiotics, e.g. griseofulvin, or toxins. Mention has already been made of the fact that specific antibiotic production in vitro does not necessarily bear any relation to that in vivo or, indeed, to its subsequent effects. There is no doubt that antibiotics can be extracted from soil, and that antibiotic-producing fungi occur in soil and can be shown to produce their antibiotics there (BRIAN, 1957). The evidence that they antagonize antibiotic-sensitive forms in the soil is less impressive; nevertheless, there is some evidence. Large masses of mycelium are often associated with a reduction in the associated bacterial and fungal floras, e.g. *Marasmius oreades*, fairy-ring mycelium (EVANS, 1967) or *Psalliota hortensis* var. *bisporus* spawn (EAGER in BAKER and SNYDER, 1965). A more subtle interaction is shown in the difference in ability of *Curvularia ramosa* and *Helminthosporium sativum* to colonize straws in unsterilized soil. This appears to be due to the greater sensitivity of the latter to antibiotics rather than to substrate requirements or rates of tissue penetration (BUTLER, 1953a, b; MACER, 1961). There is reasonable evidence that antibiotics are produced and are probably effective in association with organic matter in the soil such as straw and seeds (WRIGHT, 1956a, b). This property has been utilized to protect oat seed from *Fusarium* blight by inoculating it, prior to sowing, with *Chaetomium* spp. of known antibiotic producing ability towards *Fusarium* (TVEIT and WOOD, 1955). However, although a great many soil fungi possess the ability to produce antibiotics those which are most effective are not necessarily the most successful, e.g. *Trichothecium roseum* (PARK, 1956). In other experiments in natural soils, antagonism between *Pythium mamillatum* and other sugar fungi has been shown not to involve specific antibiotics. The inability of *P. mamillatum* to compete with these fungi in colonizing wood chips has been attributed to their production of general 'staling substances', i.e. the complex mixture of primary and secondary metabolites which so many fungi produce (BARTON, 1960). Actual parasitic attack may also be involved, as in the case of *Ophiobolus graminis* and *Didymella exitialis* (SIEGLE, 1961). *Didymella exitialis* can penetrate and kill *O. graminis*, permeating the cells. It may also reduce the pathogenicity of the same fungus by 60% through secreting amino acids and other ninhydrin-positive substances, apparently causing an imbalance of the nitrogen nutrition of *Ophiobolus*. Insufficient cases of these more complex types have been studied to do more than indicate possibilities up till now.

Fungi and higher green plants, especially their roots

Situations of more interest and perhaps indicative of some form of specific fungal response are those where the habitat is more specialized. This can sometimes be related to substrate specificity but there are interesting examples of such situations in the association of fungi with higher green plants. Thus, for instance, there are fungi which predominate in the region immediately around the roots—the rhizosphere; others on the root surface—the rhizoplane and yet others which actually penetrate the roots—the mycorrhizal and root-pathogenic fungi. Analogous situations occur in the aerial parts—those on the leaf surface, the phylloplane, and the pathogenic forms. Since so much is known about soil fungi, attention will mostly be concentrated here on situations relating to the roots of higher plants model examples.

Various classifications have been developed, based either on the biology or spatial location of the fungi under consideration; it is convenient to employ the latter criteria here and to recognize the rhizosphere, rhizoplane and root-inhabiting fungi.

THE RHIZOSPHERE

This is usually a region rich in nutrients due to their loss or excretion from the roots, and sugars, amino acids, vitamins, phosphatides and various aromatic substances are known to be available in this region at some stage or other in the life of a plant. LOCKHEAD (1958) showed that there were many bacteria heterotrophic for vitamins and amino acids in the rhizosphere populations but a similar clear selective effect has not been detected for fungi. One procedural difficulty with fungi is to determine which of the fungi obtained in the rhizosphere samples are those actually growing there and which are present in a dormant condition. This is of course, a general problem in soil ecology but away from the rhizosphere it is less difficult, in so far as soils appear to demonstrate a widespread mycostatic effect (DOBBS and HINSON, 1953), so that most of their fungal inhabitants are, indeed, dormant. In the rhizosphere, by contrast, there is a notable reduction in the general mycostatic effect so that the problem of discriminating between physiologically active and dormant fungi is of greater importance. There is evidence to suggest that exudates of the roots of higher plants exert a selective effect upon the soil fungi. Most of the published examples refer to comparisons between pathogenic versus non-pathogenic forms, or races, of the same fungus but there seems no reason to suppose that this is the only type of discrimination exercised. A classic example was TIMONIN's (1941) discovery that flax plants resistant to *Fusarium oxysporum* f. *lini* produced root exudates which inhibited spore germination of the fungus, and, in which, HCN was detected. However, it seems improbable that the HCN is directly concerned in this selective effect (TRIONE, 1966). A similar situation has been described by BUXTON (1962) in relation to *Fusarium oxysporum* f. *cubense*, the cause of Panama disease, and the banana varieties Gros Michel, a susceptible form and Lacatan, a highly resistant form. The root exudates contained 18 amino acids; 13 were common to both varieties but Gros Michel lacked cysteine or threonine while Lacatan lacked leucine, serine or tyrosine. Gros Michel exudate contained 8 sugars and one and a half times as much carbohydrate as did that from Lacatan; this predominance persisted at all stages of root growth. Materials could be isolated chromato-

Table 12.2. Actions of factors in root exudates of higher green plants which affect fungi. (After Rovira, 1965)

	Effect	Higher plant	Fungus
Mycelium	Growth stimulation	Raddish Lettuce Strawberry	*Rhizoctonia solani* *Rhizoctonia* sp.
	Growth inhibition	Oats	*Byssochlamys fulva*
		Potato *Datura* spp.	*Spongospora subterranea*
		Tomato Turnip Peas	*Colletotrichum atramentarium* *Pythium mamillatum* *Aphanomyces euteiches*
Non-fertile spores	Germination stimulation	*Allium* spp.	*Sclerotium cepivorum*
		Tomato Radish Lettuce	*Fusarium* spp.
		Beans Banana (Gros Michel)	*F. solani* f. *phaseoli* *F. oxysporum* f. *cubense*
	Germination inhibition	Banana (Lacatan) Peas	*F. oxysporum* f. *cubense* *F. oxysporum* f. *pisi*
Zoospore	Attractant	Strawberry Peas Solanaceae Avocado	*Phytophthora fragariae* *P. erythroseptica* *P. parasitica* *P. cinnamomi*
Micro-sclerotia	Germination stimulator	Tomato Wheat	*Verticillium albo-atrum*

graphically from exudates of Lacatan that inhibited conidial germination in vitro, whereas Gros Michel exudate enhanced germination. This close similarity between behaviour in vivo and in vitro suggests strongly that the root exudate operates as a selective factor in relation to the growth of the fungus. The most striking experimental demonstration however, is that due to TIMONIN (1941). He showed that two varieties of Flax, Bison and Novelty had different rhizosphere mycofloras; in the former variety, *Mucor*, *Cladosporium*, *Penicillium* and *Trichoderma* spp. predominated; in the latter variety, *Alternaria*, *Cephalosporium*, *Fusarium*, *Helminthosporium* and *Verticillium* spp. Seedlings of each variety were grown in sterile culture solution for 25 days and this was then allowed to diffuse through a colloidon sac, mimicking a 'root', into the surrounding soil. The differences in mycoflora developed around the artificial 'roots' so produced resembled those described above for the actual varieties. ROVIRA (1965) has summarized much of this work and provided a table of the factors which affect fungi in root exudates (Table 12.2). Such a complex form of substrate-specificity effect may be of general occurrence but a simple pattern of this sort is unlikely. KEITT and BOONE (1956) prepared mutants of *Venturia inaequalis*, the cause of apple scab, which were heterotrophic for various amino acids. These acids were available in apples. Nevertheless these mutants were less virulent than the non-mutant wild strains of this fungus, even although they penetrated the cuticle and became established in the flesh of the apple. Too few experiments of this type have yet been done with fungi for any generalization to be made.

THE RHIZOPLANE

One of the features of the rhizoplane fungi, in particular, is that several of them appear to be confined to this habitat. A very striking group is that of the dark sterile forms which can be found on the root surfaces of most plants if appropriate techniques are employed. Some of these fungi are believed to be Basidiomycetes but there are others, as yet unidentified. These forms appear to have no special nutrient requirements. For example, SINGH (1963) isolated some 25 distinct dark sterile forms from the roots of *Calluna vulgaris* and other ericaceous plants; they grew readily on 2% malt extract agar. Despite this they could not be detected in soil plates nor in the soil of the area from which the host plants came. This kind of situation is probably not uncommon and is certainly true of many pathogenic root fungi and mycorrhizal forms, although not all. The selection exercised here cannot be a simple one of substrate specificity. GARRETT (1950) has coined a phrase 'competitive saprophytic ability' to describe this situation. His notion is that root-inhabiting fungi are less capable of competing, saprophytically, for dead organic matter in the soil than are the soil-inhabiting fungi. This term evidently covers a wide range of physiological attributes and it is not yet clear exactly which of these determines the competitive saprophytic ability of any particular fungus. Garrett suggested that, (a) high growth rate and rapid germinability of spores, (b) good enzyme-producing equipment, (c) production of antibiotic toxins and (d) tolerance of antibiotics produced by other organisms, would characterize high saprophytic ability. This concept has enabled experimental assessments of competitive ability to be made in a number of semi-natural situations. It has drawn attention to the fact, 'that it seems possible that quite small differences in saprophytic ability, acting over a wide span of space and time in the natural environment, may be a decisive factor in saprophytic competition, and may hence determine the particular ecological niche of an individual fungus species' (GARRETT, 1960). Yet, so long as the physiological bases of this complex property of a fungus remain undefined a fundamental analysis of the important differences between fungi in different niches will be prevented. It must be concluded, with regret, that at present there is insufficient knowledge of any one fungus to account for its competitive saprophytic ability.

ROOT-INHABITING FUNGI

Not all fungi which penetrate the host tissue are pathogenic and not all root pathogens are normal inhabitants of the root. The former include the mycorrhizal forms, the latter the relatively unspecialized, facultative parasites such as *Pythium* or, on aerial parts, *Botrytis*.

FACULTATIVE PARASITES

The facultative parasites share a number of common features. They are usually fast-growing fungi with great ability to produce extracellular enzymes that break down the walls of the host plant and lead to rapid autolysis and disintegration. Thus their effects are widespread and devastating and they can utilize their host as readily after its death as before. On the other hand they appear to be restricted to infecting young plants, older plants becoming resistant. The only remarkable property which distinguishes this group of fungi is their ability to penetrate the host cells and the rapid subsequent damage which they cause.

Although such parasites are often regarded as soil fungi, many of them lie dormant in the soil and germinate only when stimulated by root exudates, e.g. *Pythium mammilatum* (BARTON, 1957). In those fungi with motile zoospores, chemotactic attraction to the root surfaces may also be involved, e.g. *Pythium aphanidermatum* (see Chapter 6, pp. 178–179).

Penetration is almost certainly mechanical, the best evidence coming from *Botrytis cinerea*, cause of soft rot of fruits. In a series of classic experiments W. Brown and his colleagues showed that there is no diffusible material that enters and kills the tissue before penetration. When indeed, conidia were placed in drops of water on petals or leaves there was a marked increase in the conductivity of the water which was not found in drops without spores but this phenomenon preceded the discoloration and disintegration of the underlying tissues and coincided with the penetration of the germ tube (BROWN, 1915, 1916, 1922). He also showed that conidia could penetrate films of paraffin wax or alcohol/formalin-treated gelatin of varying hardness and thickness, to different degrees, and regardless of the presence or absence of nutrients on the far side of the membrane (BROWN and HARVEY, 1927). The germ tube becomes attached to the surface of the plant, membrane, or even glass, in the region of its tip and ceases to grow in length. This thigmotropic reaction is followed by penetration by an extremely fine growth which swells out once the cuticle has been passed. It seems probable that the wall-disintegrating and macerating enzymes are only secreted after penetration. These enzymes have proved difficult to disentangle but it now seems clear that they include cellulases, pectinesterases (—pectinmethylesterases—PME), polygalacturonases (PG), polymethylgalacturonases (PMG) and macerating enzymes, whose function is not quite clear but which could break down 'insoluble pectates' (WOOD and GUPTA, 1958). It should be noted that the ability to produce all, or any of these enzymes in culture is no guide to what may be produced in vivo. For example, *Pythium debaryanum* on ordinary potato decoction and most natural media shows a negligible secretion of enzymes but the reverse behaviour is shown by *B. cinerea.* When tested on living potato tissue, however, the damage by *Pythium* is far greater than that due to *Botrytis.* In this connexion the water content of the host is of great importance and the greatest parasitic activity is usually correlated with full turgidity (BROWN, 1965). Under such conditions, of full turgidity *B. cinerea* will vigorously attack potato tissue. The dissolution of the tissue is largely to be attributed to the destruction of middle lamellar material by the enzymes but the death of the cells may be a purely osmotic phenomenon consequent upon the destruction of the wall (TRIBE, 1955).

Speculations about how such facultative parasites are distinguished from other soil saprophytes and why they are only capable of infecting young tissues can now be made. Entry is mechanical and requires the ability to produce an appressorium and infection hypha; not all fungi are capable of this. The hardness of the cuticle seems to be the limiting factor, hence the greater susceptibility of young tissues. Moreover, those tissues which exude most nutrients are likely to be the subject of any chemotropic response by a fungus. SCHROTH and SNYDER (1961) have shown for broad bean (*Vicia faba*) that maximum exudation of ninhydrin-positive substances (mainly amino acids) is from the seed hilum early in germination, and from the zones of elongation of the roots. These are young regions where cuticle is likely to be thin and where, indeed, maximum infection is often found.

OBLIGATE PARASITES

The true root-inhabiting fungi show a very different behaviour. They frequently demonstrate what GARRETT (1960) has termed an ectotrophic growth habit. This has been studied in some detail in *Rhizoctonia solani*, strains of which can occur in the soil, or the rhizoplane, or as pathogenic or mycorrhizal forms (FLENTJE, 1957, 1959). The hyphae are closely appressed to the surface of the root or hypocotyl and grow along the lines of junction between epidermal cells. Short, lateral branches give rise to appressoria from which infection pegs arise (Fig. 12.6). In resistant hosts, it is claimed that the fungus either grows irregularly on the surface or in an aligned manner, but in neither case are appressoria formed. The phenomenon is susceptible of an explanation in terms of thigmotropism, or chemotropism or both. E. BUTLER (1957) has shown that hyphae of a strain of *R. solani* parasitic on other fungi respond to the contact stimulus provided by glass tubing or cotton fibres by coiling round them, although the reaction was not as strong as with the normal parasitic reaction. He suggested that factors other than thigmotropic response were involved. Flentje's work has supported this view in the case of *R. solani*

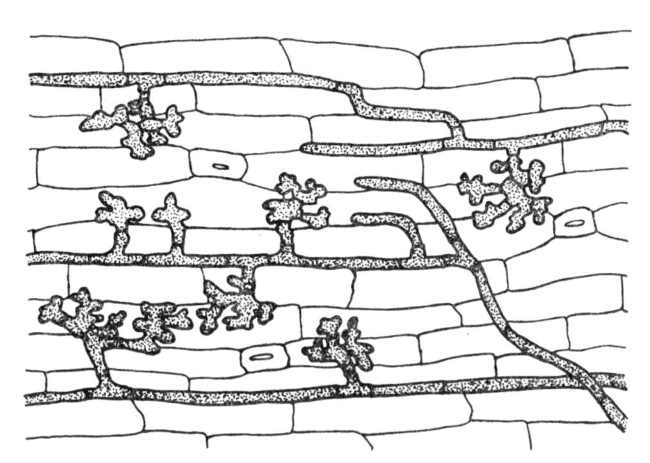

Fig. 12.6. *Rhizoctonia solani* on surface of tomato stem. Note contact-stimulus effect, hyphae growing along lines of cell boundaries, and development of appressoria. (×200; from Flentje)

isolates which infect higher plants. If the roots are enclosed in permeable cellophane sheaths the same patterns are noted as in free roots and it has been argued that this demonstrates the intervention of a diffusible material or materials. The same results were obtained on cellophane covering filter-paper impregnated with root exudates, on agar plates.

The ectotrophic growth habit may not be associated with penetration. For example, in *Fomes lignosus*, the cause of white root disease of rubber, the superficial rhizomorph produced by the fungus can extend some 1·5–4·5 m along the root ahead of the region actually penetrated. Unlike those soil-inhabiting species which are capable of developing an ectotrophic habit, e.g. *R. solani*, the true root-inhabiting species are always to be found attached at some point to a living host, or hosts, which acts as a food base. If isolated from their hosts they are incapable, under normal conditions, of competing with other fungi, although they may grow quite effectively in vitro on simple nutrients, e.g. *Armillariella mellea*. Although this has been denied, it has often been found that in cases where a mycelium appears to be free-living in the soil, it is in fact attached to such fine rootlets that the attachment may be missed, e.g. *Fomes annosus* on *Pinus* (RISHBETH, 1951). Since such fungi can grow saprophytically, free of competition in culture, they cannot be described as truly obligate pathogens but are best described as ecologically-obligate pathogens.

Since these fungi are characterized by their dependence upon an obligate association with living plants, it is essential that they should penetrate and establish an effective physiological connexion. Penetration can presumably, be effected by mechanical means through the cell wall and this is usually confined to root hairs or the younger parts of roots. In some cases, e.g. *Armillariella mellea*, entry is through wounds or cracks in the root surface, although this fungus is capable of forming lateral, aggregated hyphal strands which just penetrate the cortical layers superficially and, thereafter, entry is effected by the sheer mechanical pressure of the lateral branch on the epidermis and periderm. GARRETT (1960) has drawn attention to the fact that penetration is repeated along the length of the region covered by ectotrophic growth so that 'actual infection normally takes place immediately in advance of preceeding infection, in an unbroken wave'. The rate of dissolution of infected cells and regions is altogether slower than that due to the facultative parasites discussed earlier. *A. mellea*, it is true, produces some toxic material which appears to spread beyond the limits of the advancing mycelial front and, in susceptible hosts, further penetration of the cortex and cambium from the initial foci of infection may lead to the relatively rapid poisoning and death of the tree. In many cases, however, the pathogen may live with the host for a long period without any great damage becoming evident, e.g. *Ophiobolus graminis* in roots of perennial grasses (GARRETT, 1956).

Much the same pattern is shown by ecologically and truly obligate pathogens of aerial parts of plants. Many of these infect through natural openings, such as stomata, lenticels and nectaries. In the rusts, downy mildews and *Phytophthora*, the cell wall is penetrated by an infection hypha and then develops an intracellular haustorium which does not apparently, penetrate the cytoplasmic membrane of the host cell (cf. Chapter 4, pp. 77–80 and Fig. 4.13). The consequences of such infection are not necessarily dramatic; there may be some increase in the rate of respiration and metabolic activity which may, sooner or later, have an effect upon photosynthesis. On the other hand the pathogen may persist for an extremely long period, or even become systemic without showing symptoms, e.g. *Epichloë typhina* in *Lolium* spp. (SAMPSON, 1933). Even in the case of those pathogens which ultimately kill their hosts the process is slow; typical examples are the polyporaceous brown and white rots of standing timber. These Basidiomycetes are only found infecting trees and do not occur on dead timber as saprophytes. Some are root pathogens like *F. annosus* but others infect through wounds like broken branches, e.g. *Ganoderma applanatum*, or *Fomes ulmarius* in parklands where trees have been damaged by deer. The xylem is penetrated and slowly utilized. Hyphae secrete extracellular cellulases and glucanases and their progress can be followed by the 'boreholes' which they develop in the walls which are little larger than the penetrating hyphae. The loss due to these boreholes is quite insufficient to account for the amount found to be lost in practice, e.g. a 25% loss in weight is only associated with the equivalent of a 1·5% loss due to borehole formation (COWLING, 1961). There is evidence of the development of sub-microscopic pores within the walls by both brown and white rot fungi (Fig. 12.7) and the latter also show a gradual thinning of the walls from the lumen towards the middle lamella (MEIER, 1955; WATERMAN and HANSBROUGH, 1957), as well as the expected degradation of lignin in the middle lamella. Boreholes are often seen to lack hyphae in microscopical preparations and this could well be due to a progressive loss of autolysed old hyphae and transfer of the material to the actively growing ones. One reason for supposing that this may occur is the low content of nitrogen (0·05%) in the wood attacked compared with the requirements for chitin and protein

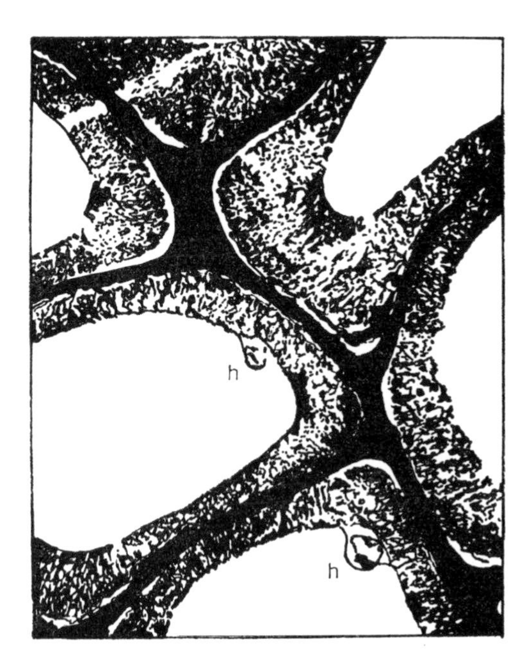

Fig. 12.7. Electronmicrograph-tracing through walls of xylem fibres infected by brown rot fungus. The clear areas in the wall are where carbohydrates have been rotted away. Note hyphae (h) in cell-lumen. (After Meier)

nitrogen by the fungus. It is possible that nitrogenous compounds are translocated from those parts of the mycelium still in living parts of the plant but this has not been demonstrated. Even so, fungal material only accounts for about 2% of the total weight of wood. The process is extremely slow since although some fungi can rapidly utilize the amorphous cellulose, the crystalline components are only broken down very slowly (Fig. 12.8).

It is amongst those pathogens which infect the aerial parts of plants that truly obligate pathogens exist. Some of the downy mildews have been cultivated successfully on callus-culture tissue of appropriate hosts and limited success has been achieved with *Erysiphe cichoracearum* on sunflower tumour tissue but not on normal callus tissue (MOREL, 1944; HEIM and GREIS, 1953).

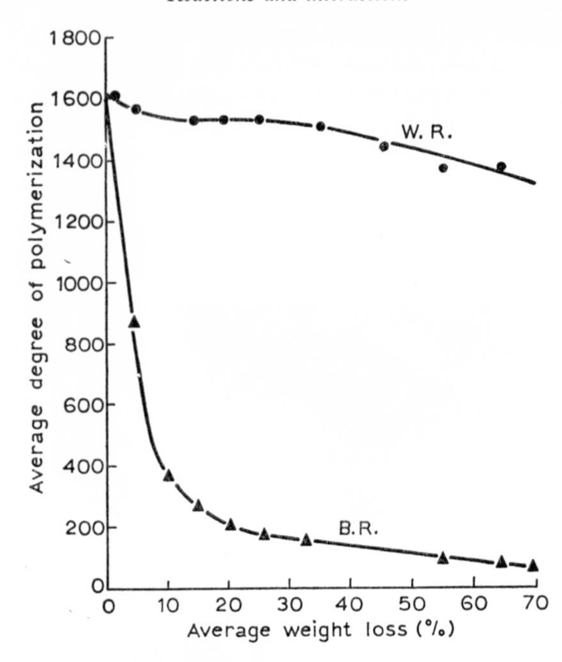

Fig. 12.8. Reduction in average degree of polymerization of holocellulose of wood of Sweet Gum (*Liquidambar styraciflua*), by the white rot *Polystictus versicolor* (W.R.) and the brown rot fungus *Poria monticola* (B.R.). Extrapolation of curve of latter fungus to zero weight loss indicates resistant (presumably crystalline) fraction of holocellulose. (After Cowling)

The most spectacular claims have been made by Cutter (HOTSON and CUTTER, 1951; HOTSON, 1953; CUTTER, 1951, 1959, 1960). He employed a simple medium of mineral salts and sugar supplemented by ascorbic acid and yeast extract; biotin was later reported to be the only growth factor required. Blocks of infected host tissue containing either of the rusts *Gymnosporangium juniperi-virginianae* or *Uromyces aritriphylli* were left in a sterile manner on the agar for several months. A very few gave rise to hyphal growths in the agar, which were isolated, grown on fresh agar and used to re-infect host tissue or, in the case of *Gymnosporangium*, the alternate host, Hawthorn (*Crataegus* sp.) as well. The rarity of these saprophytic hyphae was so great that originally Hotson and Cutter suggested that they had arisen as mutants from the original rusts. This work has been received with suspicion and doubt and there are few published records of attempts to reproduce the work. CONSTABEL (1957) failed to obtain any saprophytic hyphae of *Gymnosporangium* spp. from 1000 gall tissues. In general, it is supposed that obligate pathogens have some special nutritional relationship with their hosts and this is borne out by the wide-spread phenomenon of 'physiological races' shown by many of them.

Physiological races are by no means confined to obligate parasites; for example, they are known in *Phytophthera infestans* (potato blight), *Cladosporium fulvum* (tomato leaf-mould), *Claviceps purpurea* (ergot), *Helminthosporium gramineum* (barley leaf-stripe) and *Fusarium* spp. (wilts, etc.), which includes saprophytic free-living species as well as facultative and ecologically-obligate pathogens. Nevertheless, the delimitation of such races appears to be more restricted in obligate pathogens, although this may reflect the closer study made of them than of other fungi. The nature of these races is discussed elsewhere (cf. Chapter 17, pp. 462–469).

A matter which has not been discussed so far is the nature of the hosts' responses to parasitic attack. These are extremely varied and have been discussed in detail in various pathology texts (GÄUMANN, 1950; WOOD, 1967). Defence against penetration is due either to exudates (cf. pp. 329–330) or to mechanical resistance of cuticle and cell walls (cf. pp. 333–334). Once penetration has been achieved defence reactions may be passive, involving the presence of pre-existing toxic materials, e.g. protocatechuic acid in cells of onions resistant to *Colletotrichum circinans*, or phenols and tannins in the xylem which inhibit wood-destroying fungi. Active defence reactions can involve the development of suberized tissue which effectively seals off the invasion, e.g. lignitubers of *O. graminis*; the immediate death of the cell before the parasite has time to spread further, e.g., *Erysiphe martii* on *Trifolium pratense*; or the development of particular or general mycotoxins. This last reaction has received a more thorough analysis in recent years than the others. MÜLLER (1956) proposed that non-specific, anti-fungal compounds were induced in all cells infected by a fungus. The difference between a resistant and susceptible variety lay in the speed of the response. The compounds were named 'phytoalexins' and one of them, pisatin, has been isolated from pea-pod tissues. In such tissues pisatin appears to meet all the claims made for phytoalexins by Müller (CRUICKSHANK, 1965). It is a relatively simple chromanocoumarone:

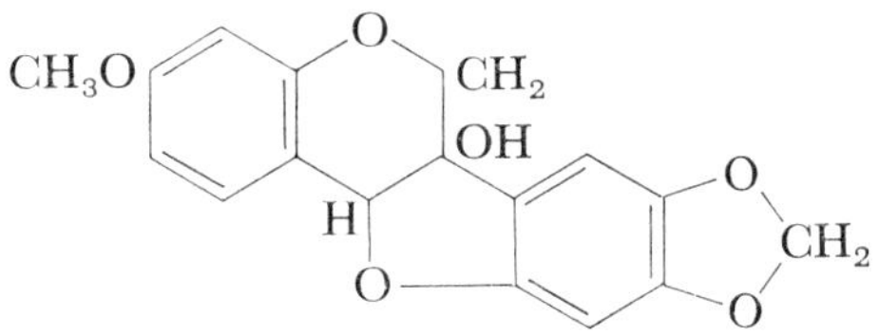

MYCORRHIZA

It will have been noted that two features characterize the interaction of host and pathogen in ecologically-obligatory, pathogenic fungi; these are their ectotrophic habit and the fact that their metabolic processes are more readily in balance with those of their host, at least initially. These features, especially the latter, are shown by the mycorrhizal fungi. It is convenient to recognize ectotrophic and endotrophic mycorrhiza, although intermediate conditions do occur. A recent book discusses mycorrhiza in detail (HARLEY, 1959) and a good historical introduction is provided by RAYNER (1927).

Ectotrophic mycorrhiza develop between woody hosts and basidiomycetous fungi in a wide range of habitats. The ectotrophic growth habit is extremely well developed and the fine roots in the litter layer become invested by a sheath of hyphae which frequently covers the root apex and suppresses growth of the root hairs (Fig. 12.9). In many cases the morphology of the root is changed so that the roots may appear swollen and coralloid as in beech

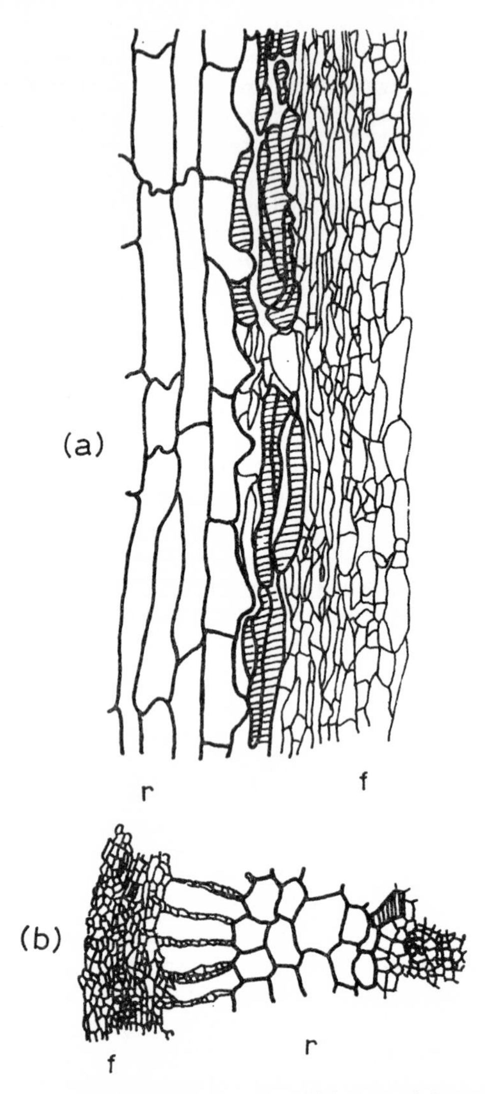

Fig. 12.9. L.S. (a) and T.S. (b) through mycorrhizic root of *Fagus sylvatica* (beech). In (a) the shaded cells are sloughed epidermal cells between ectotrophic fungal tissue (f) and root cells (r); in (b) note Hartig net hyphae penetrating between radial walls of epidermal cells. Shaded cells are tannin-filled cell of endodermis. (× 350, after Clowes)

(*Fagus*), or dichotomously branched at the tips as in pine, or elongated, swollen and turbinate as in the date palm (*Phoenix*). There are, of course, hyphal connexions extending into the soil from the sheath but, in addition, inward growing hyphae penetrate the outer layers of the cortex. The majority of penetrations are intercellular but intracellular penetration can occur although this is rare. Thus the mycelium forms an enclosing thimble-like mass of hyphae from which a three-dimensional network (the Hartig net, from its discoverer) enmeshes the cells within. The stele is not normally penetrated. Host-specificity is variable. *Cenococcum graniforme* is said to form mycorrhiza with 7 genera of conifers and 9 genera of deciduous broad-leaved trees including oak, beech, hickory, lime and willow. By contrast some species of *Boletus* are so discriminating that they can be associated with one particular species of host tree (SINGER, 1962).

Endotrophic mycorrhizha are more variable. It seems probable that most shrubby and herbaceous plants, including lower green plants, can form these associations under appropriate environmental conditions. Especial attention has been focused, historically, on Orchidaceae, Ericaceae and bryophytes but it is clear that other groups show the phenomenon at least as frequently, e.g. Gramineae. The fungi forming these associations include Basidiomycetes, such as *Armillariella mellea*; sterile dark forms, some being *Rhizoctonia* spp.; aseptate forms, believed to belong to the genus *Endogone*, a probable sterile form of this genus, *Rhizophagus*, and, conceivably *Pythium* (HAWKER, 1962; MOSSE, 1963). The ectotrophic growth habit is not always as well developed in endotrophic mycorrhiza.

The mycorrhizal fungi of Ericaceae invest the roots with a more or less closely applied hyphal weft and infect it intracellularly in several places. Within, the hyphae grow from cell to cell developing irregularly distributed but connected masses of hyphae sometimes within, sometimes between, the cortical cells. The stele is rarely penetrated and, as the hyphae enter the cells of the inner cortex, they tend to become irregular and to lose their form. In the Orchidaceae a similar infection pattern obtains although, especially in the case of seedlings, the invasive hyphae may completely parasitize the host. In root tubers there is a clear development into an outer zone, an inner 'digested' zone and a central zone, including the stele, free of the fungus. In the outer zone the fungal hyphae form coils and loops but in the 'digested' zone they collapse to form clumps of unorganized fungal matter, sometimes surrounded by a membrane. BURGEFF (1909, 1932, 1936) has made detailed and precise studies of the mycorrhiza of orchids. He has shown that there are hyphal connexions between the endophytic fungus and the soil. Infection of new roots usually occurs internally from the hyphae of the active, outer layer but the growing points of roots are not normally infected.

The aseptate, presumably phycomycetous, mycorrhizal fungi are now receiving increasing attention (BUTLER, 1939; NICOLSON, 1959, 1960; MOSSE, 1963). They appear to be extremely widespread and even form mycorrhizal associations with shrubs and small trees of the genera *Coprosma*, *Griselina*, *Pittosporum* and the conifer, *Podocarpus* with small lateral nodules on its roots (BAYLISS,

1962). Mycorrhiza of this type also appear to have occurred in fossil club mosses (*Lepidodendron* spp.), of the early Devonian period (KIDSTON and LANG, 1921). They are characterized by several infection points and the development of some coils of intracellular hyphae but, especially, by arbuscules, or vesicles or both these structures. They are often called vesicular-arbuscular mycorrhiza. Arbuscules resemble haustoria; they arise by repeated dichotomous branching until the extremes are so fine that they become invisible and they extrude cytoplasm. Vesicles are terminal or intercalary swellings in hyphae and they may often develop intercellularly. As they age their walls thicken and oil drops accumulate. There is also a remarkable development of external mycelium. This is often markedly dimorphic with large, thick-walled aseptate hyphae and fine, thin-walled hyphae which frequently become regularly septate. The large hyphae can develop large spores or aggregations of up to 10 spores to form a sporocarp 1 mm in diameter (MOSSE, 1953). The external mycelium can spread out up to 1 cm in the soil but sporocarps readily become detached (Fig. 12.10). So far it has not proved possible to grow these fungi in culture so that they represent truly obligate mycorrhizal forms. GODFREY (1957) has seen and illustrated germination of the spores by fine germ hyphae of limited growth in plate culture but, once severed from the spore, growth ceased in these hyphae. Infection from a spore has been achieved in aseptic culture provided that nitrogen was limiting and certain bacteria, or filtrates from them, were present. This latter condition enabled an appressorium and infection to develop, the former was a prerequisite for germination. Subsequently, infection from a root occurred after vesicle development had commenced.

The mycorrhizal fungi offer a fascinating field of interactions between host and fungus. Obviously thigmotropic and, probably, chemotropic factors are involved in the infection stage. Particular attention has also been paid to some of the morphogenetic consequences of the association, especially in ectotrophic mycorrhiza.

CLOWES (1949) showed that the difference in anatomy between mycorrhizal and non-mycorrhizal roots of beech (*Fagus*) lay in a difference in cell shape. In infected roots some cortical cells showed reduced longitudinal extension and increased transverse expansion and, in addition, meristematic activity might be reduced. Differentiation of all the cells, cortical and stelar, was precocious. These observations accounted adequately for the swollen shape of the mycorrhizal roots. In Pinaceae, a different sequence is followed. HATCH (1937) has described an extended period of growth and branching of roots and a delayed development of the endodermis. SLANKIS (1948a, b, 1949, 1951, 1958) has shown that *Boletus variegatus*, or its culture filtrates in the same medium as *Pinus* roots increased the frequency of short-root laterals which often showed dichotomous forking. Similar effects were obtained with synthetic and natural growth regulators including indole acetic acid. The fungus can produce indole acetic acid in culture. He investigated the dichotomously branched laterals anatomically and compared them with true mycorrhizal roots. However, this work is highly suggestive of a possible morphogenetic

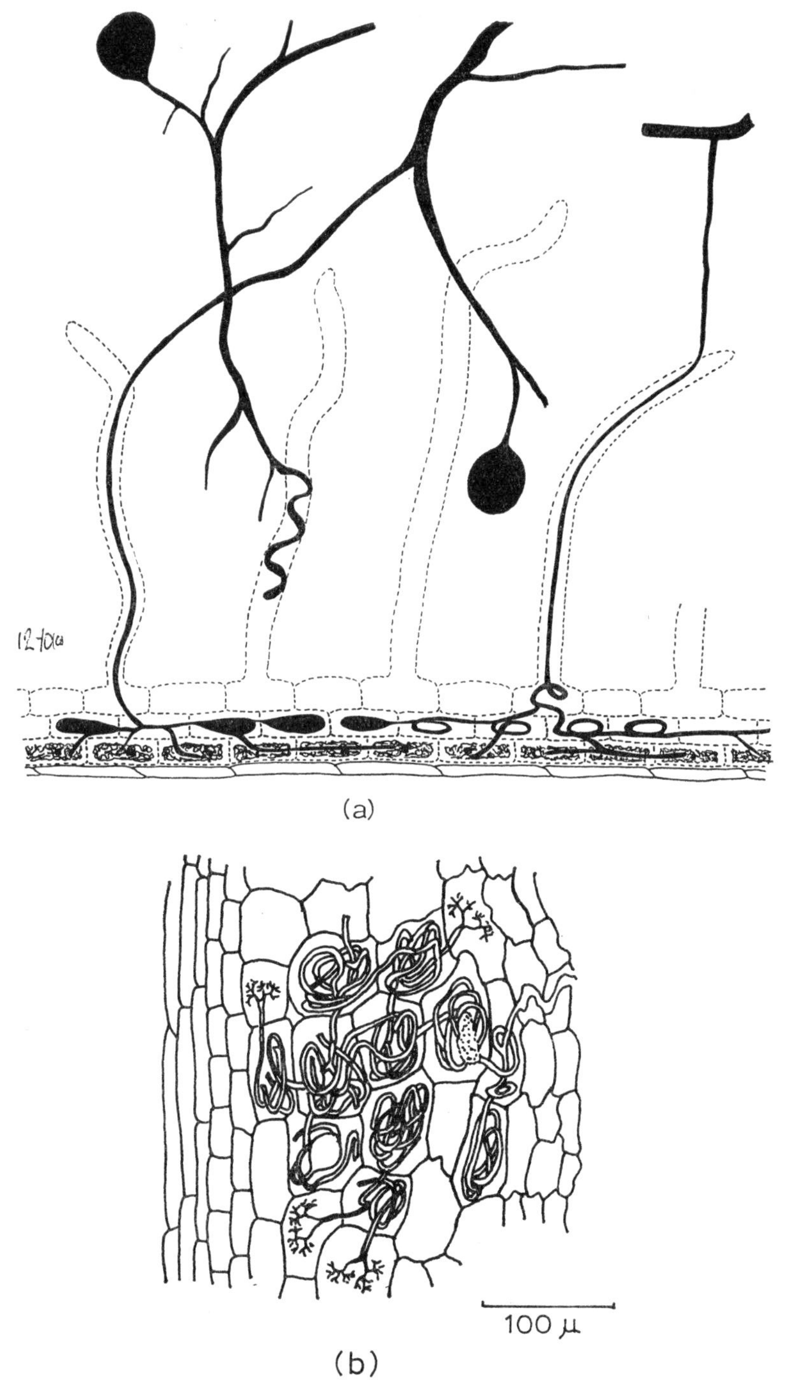

Fig. 12.10. '*Rhizophagus*-type' mycorrhiza. (a) Diagram of infection of root by *Endogone* sp. Sporocarps are borne on external hyphae of two kinds, thick and thin. Internal hyphae are either contorted, vesiculate or branched finely; (b) Coils and branching hyphae in roots of *Tamus*. (a, after Nicolson; b, from Galland)

effect of the fungus. MELIN (1963) reviewed his observations of the effect of the host on the fungus. He pointed out that exudates (M-factors) from living pine roots had a growth-promoting effect on various mycorrhizal fungi and that, in addition, an inhibitory factor was also produced. None of these factors was uniquely specific although they affected different species to different degrees.

Studies have also been made of the nutritional requirements of mycorrhizal fungi (cf. HARLEY, 1959). As a group, fungi of ectotrophic mycorrhiza are not effective in utilizing either lignin or complex polysaccharides, including cellulose, for nutrition in pure culture. They can effectively utilize simple sugars and this goes some way to explain how it is that effective infection is correlated with high sugar content in roots of the host and hence, via photosynthesis, on the light intensity (BJÖRKMAN, 1949). In this respect these fungi resemble obligate parasites (ALLEN, 1954). Other aspects of their nutrition are not remarkable, although many of them, like many other soil- and root-inhibiting fungi, are deficient for thiamin or its constituents. Melin's M–factors, already referred to, are additional to these. Endotrophic mycorrhiza by contrast are not so restricted in their nutritional requirements (excluding *Endogone/Rhizophagus*). They readily utilize polysaccharides, both cellulose and starch and many of them break down lignin. In addition, several strains of *Rhizoctonia*, saprophytic, parasitic or isolated from *Dactylorchis* spp., produced PMG, PME and PG enzymes in vitro and grew successfully on media with pectins as a carbon source. There was no correlation between ability to produce pectin-utilizing enzymes in vitro and the natural mode of life of the strain (HADLEY and PÉROMBELON, 1963; PÉROMBELON and HADLEY, 1965). There is thus a marked contrast in the nutrition of these two great mycorrhizal groups of fungi and, presumably, a third group the *Endogone*-type, represents an even more nutritionally exacting group.

Once infection has been achieved, the question arises as to the interrelationships of the fungus and its host. Firstly, the consequences could be negligible for the host. Secondly, the fungus could obtain nutrients or growth factors from the host under conditions of reduced competition as compared with the external soil environment. Finally, the host could benefit from the fungus by conduction of materials via the hyphae, by increased availability of nutrients either as a consequence of their breakdown by the fungus in the soil or as the result of the increased absorbing area, or from materials synthesized by the fungus. There seems little doubt that some or all of the processes described in the second and third alternatives are involved.

Amongst ectotrophic forms it is clear that each partner obtains materials from the other. HATCH (1937) noted that the effect of infection on pines was to increase the available root absorbing area (in the litter layer) compared with uninfected plants. He noted also that mycorrhiza developed most vigorously where there was a deficiency of some mineral nutrient. Direct studies by MELIN and NILSSON (1950; 1952; 1955; 1957) have shown that seedling *Pinus* infected with their mycorrhizal fungus are capable of absorbing and transporting ^{32}P, ^{15}N (from ammonia) and ^{45}Ca from exogenous sources via the soil hyphae, fungal sheath and so into the root. STONE (1950) has shown the same thing for ^{32}P with mature *Pinus halopensis*. HARLEY (cf. 1959; 1965) and his colleagues have made a detailed study of the uptake of phosphate and other salts by excised beech mycorrhiza. They have shown that the ability of such compound organs to absorb $H^{32}PO_4^{2-}$ is greater than that of uninfected roots and that this phosphate is accumulated in the tissues of the sheath. Much of the success of this work has been due to an ingenious dissection by which it has proved possible to remove the sheath of fungal tissue from the 'core' of root tissue. While, of course, the hyphae of the Hartig net are still retained within the 'core' the amount of fungal material involved is quite negligible compared with that in the sheath or with the amount of host tissue. When external phosphate is low or absent the accumulated phosphate can be remobilized and transported to the root tissues. In intact pine seedlings a similar pattern of behaviour has been demonstrated by MORRISON (1962). Thus ectotrophic mycorrhiza, whether or not the increased absorbing area of the roots is significant, are capable of absorbing and accumulating ions in periods of high exogenous concentration and passing them on to the host plant immediately or, in the case of phosphorus at least, over a prolonged further period. In this case, therefore, the host gains by having an assured and continuing supply of an essential nutrient, phosphorus; one that is most frequently in limiting supply in natural conditions.

Ectotrophic mycorrhiza also gain carbon compounds from the host. This was shown first with pine seedlings by MELIN and NILSSON (1957) who supplied $^{14}CO_2$ to the host plants and recovered organic carbon compounds from the external mycorrhizal mycelium. LEWIS and HARLEY (1965a, b, c) have reported on the consequences of supplying [^{14}C]-sucrose either to host tissue or to the mycorrhizal sheath tissue of excised beech mycorrhizas. When sucrose was fed via host tissue it appeared in the sheath principally as labelled mannitol and trehalose. Similarly, when fed to the sheath, it appeared in the same form but only as radioactively-labelled sucrose, glucose and fructose in the host tissue (Fig. 12.11). Host tissue was shown to be unable to utilize either mannitol or trehalose so that here is a situation where a high proportion of the available carbon compounds in the fungus are not available to the host. Carbon compounds obtained from the host in this sense may pass through a 'one-way door', so long as they are converted to polyol or trehalose by the fungus.

Other mycorrhizal situations have not been investigated to such a great extent but recently S. SMITH (1966, 1967) has been able to show that hyphal strands of *Rhizoctonia* spp., continuous with the endotrophic mycorrhiza of seedlings of *Dactylorchis purpurella*, can translocate ^{14}C- and ^{32}P-labelled compounds from exogenous sources to the seedlings. When the *Rhizoctonia* strains were supplied with filter paper, greater growth was shown by the seedlings than when carbohydrates were withdrawn from either fungus or seedlings (Table 12.3). She was also able to show that [^{14}C]-glucose supplied to hyphae, external to the seedlings, was translocated and incorporated in the seedlings. Uninfected tissue was found to contain sucrose, glucose and fructose; isolated

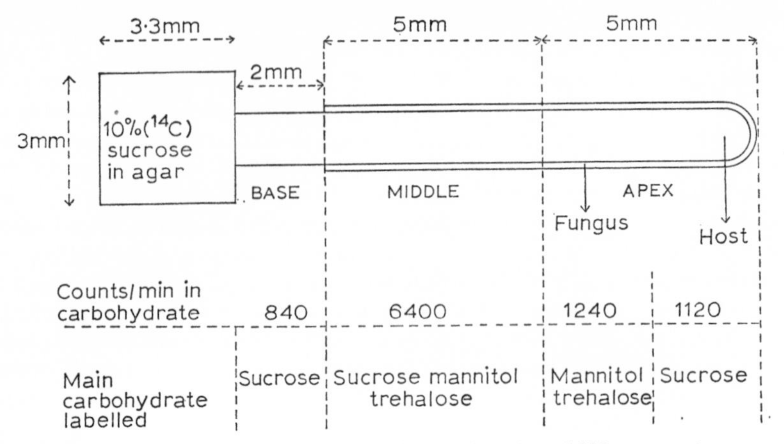

Fig. 12.11. Experimental technique to study translocation of ^{14}C sucrose by root and mycorrhizal tissue in beech. Note absence of mannitol and trehalose from base region from which mycorrhizal sheath has been removed and that they occur in the sheath after separation in apical region. (From Harley)

Table 12.3. Growth of *Dactylorchis purpurella* seedlings after 14 weeks with or without inoculation by *Rhizoctonia repens* in the presence or absence of cellulose. All measurements in μm. Figures in parentheses are replications. (From S. Smith, 1966)

	Uninoculated	Inoculated + Cellulose	Inoculated − Cellulose
Total seedlings	14 (4)	797 (4)	865 (3)
Length	248 ± 9	820 ± 52	625 ± 43
Width	206 ± 10	517 ± 53	444 ± 25
Healthy seedlings	—	226 (4)	324 (3)
Length	—	1170 ± 119	800 ± 72
Width	—	692 ± 44	519 ± 39
'Brown' seedlings	—	571 (4)	532 (3)
Length	—	471 ± 7	450 ± 17
Width	—	343 ± 37	371 ± 15

The experiment was done in 'Schütte' dishes, the inner container with cellulose, if added, otherwise with mineral nutrient medium (Knudson B) in both inner and outer dishes. Orchid seedlings in the outer dish, fungus in the inner.

mycorrhizal mycelium contained glucose and trehalose and, in two of the three tested, mannitol as well (Table 12.4). In experiments with labelled glucose over seven days, sugars were determined in the infected seedlings. Trehalose was the first sugar to become labelled and the proportion so labelled declined as ^{14}C was incorporated into other sugars (Fig. 12.12). Thus there is good circumstantial evidence that a sugar transport system operates in this quite different mycorrhizal situation similar to that which

Table 12.4. Components of the neutral soluble fractions of uninfected orchid tissue, fungal mycelium and mycorrhizal seedlings. (From S. Smith, 1967)

Tissue	Carbohydrate				
	mannitol	trehalose	glucose	sucrose	fructose
Orchid leaves and tubers	−	−	+	+	+
Fungal mycelium					
R. repens	(+)	+	+	−	−
R. solani D_a	−	+	+	−	−
R. solani Rs 10	+	+	+	−	−
Seedlings					
+*R. repens*	−	+	+	+	−
+*R. solani* D_a	−	+	+	+	+
+*R. solani* Rs 10	+	+	+	+	+

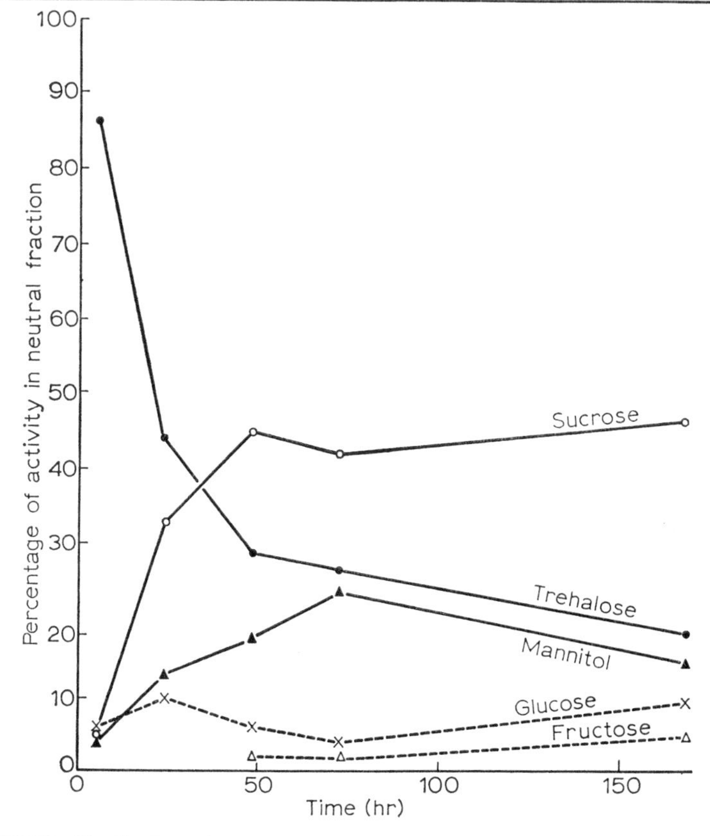

Fig. 12.12. Distribution of percentage activity in a neutral extract from tubers of *Dactylorchis* sp. infected with *Rhizoctonia solani* Rs 10. Note labelling first appears in trehalose although glucose was supplied. (From S. Smith)

occurs in ectotrophic mycorrhiza. However, in these endotrophic mycorrhiza, it may well be that the host normally gains sugar derived exogenously from the breakdown of complex polysaccharides by the fungus and transported inwards. In ectotrophic mycorrhizas, on the other hand, it seems possible that the main flow of carbon is in the opposite direction. It is not yet possible to provide more than an outline of potential activities at this stage.

No such detailed studies of mechanisms have been made with vesicular arbuscular mycorrhiza but in *Rhizophagus* BAYLIS (1959) has been able to show marked differences in total dry weight and mineral composition between mycorrhizal and non-mycorrhizal plants of *Griselina littoralis* in soil deficient in several mineral nutrients, especially phosphorus (Table 12.5). MOSSE (1963) has discussed other circumstantial evidence for the beneficial effects of *Endogone* infections on a wide range of host plants. As with Hatch's original claims for the quite different ectotrophic *Pinus* mycorrhiza, the most striking effects have been under conditions of low nutrient availability. Evidence in support of this has recently been provided by DAFT and NICOLSON (1966) who inoculated tobacco, tomato and maize plants with *Endogone* spp. isolated from cultivated soils near Dundee (Scotland). The plants were grown on in sand culture and watered with complete or phosphate-free nutrient solution, or with solutions containing different levels of phosphate. The dry weight was markedly affected by the level of phosphate, in relation to the degree of infection achieved by the *Endogone* (Table 12.6). Indeed, as with ectotrophic mycorrhiza, fertile soils are inimical to the development and continuance to the mycorrhizal condition. MOSSE (1959), however, noted that structures reminiscent of arbuscules developed from external *Endogone* mycelium in peat particles in potting compost and that disintegration of peat particles was correlated with successful inoculation of the host. This observation, coupled with DOWDING's (1959) claim that *Endogone* spores are common in soils high in organic matter, suggests that mycorrhiza derived from these fungi occur in all soils of low base status, whatever the cause of this condition.

It will be seen that mycorrhizal fungi have much in common with pathogenic root fungi. All forms, like pathogens, must establish entry to the host; some have a well developed ectotrophic growth habit, others a well developed haustorial habit and, in the orchids in particular, the mycorrhizal infection is remarkably reminiscent of and, indeed, at times actually is a pathogenic attack. Recently Gäumann and his colleagues have demonstrated that digestion is associated with a phenolic compound of molecular weight 256 which they term orchinol. It was strongly antifungal, occurred in great quantities in tubers of orchids, e.g. 3 g/kgm *O. militaris* tuber, and was developed in response to fungal infection. Orchinol formation, however, could be induced by a variety of soil fungi both saprophytic and parasitic, although, surprisingly, it was not induced by an isolate of *Rhizoctonia solani* known to be orchinol-sensitive and capable of developing a mycorrhizal association with *Dactylorchis* spp.

It is clear that the interaction of host with parasite or mycorrhiza-forming fungus has many parallels and many complexities. It is equally clear that a

Table 12.5. Chemical compositions in respect of P, N, K, Ca, Na and Mg of *Griselinia littoralis*, *Pinus radiata* and *Myrsine australis* with or without mycorrhiza, the latter grown either in *Griselinia*-soil or sterile soil. (From Bayliss, 1959)

Element	*Griselinia*			*Pinus*			*Myrsine*	
	M	NMa	NMb	M	NMa	NMb	NMa	NMb
Phosphorus								
Expt. 1	0·45		0·31					
Expt. 2	0·56	0·40	0·38	0·45	0·28 +	0·31	0·35	0·34
Nitrogen								
Expt. 1	0·60		1·23					
Expt. 2	1·74	3·53		0·74	1·58 +	1·26	1·23	1·02
Potassium								
Expt. 1	0·90		0·77					
Expt. 2	0·65	0·25 [1]	0·39	0·49	0·29 +	0·49	0·53	0·56
Calcium								
Expt. 1	1·79		1·87					
Expt. 2	1·60	1·46 [1]	1·67	1·36	1·25 [1]	1·60	1·61	1·49
Sodium								
Expt. 1	0·86		0·66					
Expt. 2	0·74	0·51 [1]	0·61	0·28	0·36 [1]	0·36	0·40	0·42
Magnesium								
Expt. 1	0·30		0·25					
Expt. 2	0·23	0·18 [1]	0·17	0·11	0·14 [1]	0·13	0·18	0·19

M = Mycorrhizal ; NMa = Non-mycorrhizal, pre-treated in *Griselinia* soil; NMb = Non-mycorrhizal, pre-treated in sterilized soil
[1] One determination upon the combined dry matter of three plants

Table 12.6. Dry weights, total phosphate contents and infection levels of tomato (var. Eurocross) grown in six bone-meal concentrations and infected with *Endogone* endophyte No. 1. Plants grown for 65 days and supplied with nutrient solution minus the phosphate salt in two series **A** and **B**. (From Daft and Nicolson, 1966)

	Relative phosphate level		Mean dry weight (g)	Total phosphate content (μM/100 mg dried plant material) Roots	Leaves	Mean infection (%)
	0·25	Mycorrhizal	0·393 [1]	6·2	4·7	90·5
		Control	0·088	3·3	3·2	—
	0·5	Mycorrhizal	0·383 [1]	6·8	5·6	82·4
		Control	0·127	2·5	2·9	—
A	1	Mycorrhizal	0·344 [1]	9·3	5·4	77·2
		Control	0·115	3·6	3·3	—
	2	Mycorrhizal	0·483	14·0	3·7	63·9
		Control	0·355	2·6	1·8	—
	2	Mycorrhizal	0·516 [2]	13·6	9·2	73·4
		Control	0·301	3·8	5·0	—
B	4	Mycorrhizal	0·730	6·1	3·3	72·4
		Control	0·707	1·8	3·0	—
	8	Mycorrhizal	0·737	6·2	3·9	61·4
		Control	0·702	2·7	2·6	—

[1] Significantly different from controls at 1%
[2] Significantly different from controls at 5%

fungus may behave differently in different situations. Perhaps the most remarkable recorded example is that of *Armillariella mellea* in Japan which was growing as a pathogen on conifers whilst at the same time associated with the mycorrhizal tubers of the orchid *Gastrodia* (KUSANO, 1911). This similarity extends to a number of features and it is clear that the solution of any of them could come from studies with pathogens or mycorrhizal fungi. The outstanding problems are these:

(a) The nature of the stimulus to the ectotrophic growth habit.
(b) The mechanism of host penetration.
(c) The restriction, or otherwise, of internal morphological development of the host.
(d) The exchange of materials between host and fungus.

In addition to interactions with terrestrial green plants and other fungi as illustrated by their behaviour in the soil, fungi show interactions with algae and with animals. These will now be briefly reviewed.

Fungi and algae

The most remarkable consequence of the interaction between fungi and algae is the 15–20,000 species of lichens! Considerable interest has always been shown in these plants since they were first recognized as dual organisms by WALLROTH in 1825. Recently, as a consequence of increased knowledge of the biochemistry and physiology of fungi and algae, the experimental investigation of lichens has been renewed (cf. QUISPEL, 1959; D. SMITH, 1962, 1963b; AHMADJIAN, 1965).

The majority of lichens are formed between fungi of ascomycetous affinities, mostly Disco- and Pryenomycetes, and green or blue-green algae, notably the genera *Trebouxia*, *Trentepohlia* and *Nostoc*. Attention has been focused on three features: (a) the nature of the alga and fungus involved and the origin of the association, (b) the interactions between the partners and (c) the properties of the dual organism.

Many attempts have been made to induce the synthesis of a lichen thallus. In nature and in culture, it is not difficult to bring about a loose, readily separable association of a wide range of algae and fungi. The common association of pleurococcoid algae and fungi in the green drip-courses on tree trunks or walls is a common example. AHMADJIAN (1962, 1965) has carried out many experiments on *Acarospora fuscata*, a crustaceous lichen found on rocks. He inoculated the partners (having previously isolated them) on a range of nutrient agars including glucose, soil infusion water and mineral salts and exposed them to a range of light intensities. As the nutrient content was reduced, a closer association was formed. In the presence of glucose the two organisms grew quite independently; without glucose, some hyphae encircled algal cells and developed a few haustoria at the lowest light intensities. The number of contacts was increased on agar of mineral salts only while on plain agar association was rapid, especially at low light intensities. Pseudo-parenchymatous tissue developed within thirty days wherever algal cells occurred but no further significant progress was then made for over a year. The low level of association was improved by the addition of biotin and a régime of alternate wetting and drying. Ahmadjian attributed this restricted development to excessive moisture in the agar. He, therefore, transferred 'lichenized' cultures several months old to sterile fragments of the same rock from which the original lichen had been taken. They were alternately wetted and dried and illuminated. 'After several months the synthetic cultures resembled naturally occurring thalli. The algal and medullary layers were defined clearly but, although a brown colour (the natural pigmentation of the lichen) was present in some parts of the thallus surface, a cortex was not present'. He has suggested that further differentiation might have been achieved by imposing an alternating temperature régime as well.

This success represents the outstanding achievement in attempts to synthesize a lichen thallus but there is clear evidence from other failures that low nutrient status, especially of organic materials, and an alternating water régime are important if not essential requirements (THOMAS, 1939; QUISPEL, 1959).

The partners can be separated much more readily and each one subcultured alone. The fungi form extremely slow-growing, erumpent colonies on agar, often brightly coloured. They rarely develop any morphological features resembling those of the lichen thallus from which they have been derived, although there are exceptions, e.g. some *Cladonia*-fungi form stalks and podetia in culture (AHMADJIAN, 1963), and only a few produce any kind of reproductive structure, e.g. *Buellia*-fungus gives pycnidia and pycnidiospores (AHMADJIAN, 1964). It seems probable that most lichen fungi need growth factors; biotin and thiamin, being the commonest deficiencies, as in free-living fungi (QUISPEL, 1943). Although there seems to have been no systematic survey, they are described as utilizing relatively simple polysaccharides only, including a range of mono- and disaccharides.

The algal components can usually be grown separately, they are capable of photosynthesis and, in the case of the blue-green form *Nostoc*, can fix nitrogen. However, unlike many free-living unicellular or simply filamentous algae, they are facultative, if not preferential heterotrophs and some, e.g. *Trebouxia*, grow better in media supplemented with sugar (AHMADJIAN, 1960). They also differ from apparently similar free-living forms by reproducing only by simple division.

Thus the fungal and algal components show some differences from their free-living analogues, most significantly, the very slow growth rate of the fungi, the preferential heterotrophy of the alga and the slow growth rates of both components.

At the establishment of a lichen there is little evidence of any specificity either of attraction of fungal hyphae to algae or of association of particular algae with fungi. This is a subject which needs further exploration and there are a few exceptions to the previous generalization. The most striking is *Solorina saccata*, where both a blue-green and a green alga are involved. SCOTT (1964) described the natural development of this lichen as only proceeding to the juvenile stage with the blue-green alga but to full development with the green alga. The significance of this difference is unknown.

The lichen thallus possesses properties which are not shown by either partner alone. It has a characteristic morphological form in which the algal partner may be distributed regularly (homoiomerous) or confined to a narrow band, usually near the top surface (heteromerous). The algae are encircled by fungi and haustoria are not uncommon. They are said to be of two kinds; intracellular, especially in homiomerous forms, and intra-membranous, not entering the algal protoplast, in heteromerous and some homoiomerous forms. The situation would repay investigation and comparison with that in higher plants infected by haustoria-forming fungi. The thallus can be filamentous, gelatinous, upright or hanging, branched or flat and leaf-like or forming a crust. It is held in place by fungal hyphae forming rhizoids. Reproduction is by soredia, little groups of algal cells and fungal hyphae or smaller, similar structures called isidia. In addition, the fungal partner reproduces by ascospores, or rarely, conidia or both. They do not seem to grow more rapidly than the components do alone and this may range from 1 mm/annum to 2–4·5 cm/annum, as in *Peltigera* spp. For this reason some of the crustose lichens in arctic-alpine regions, where growth is slowest, are believed by some to be 1000–4500 years old (BESCHEL, 1961). They have very low net rates of assimilation and Smith (1962) has suggested that for part of the year, at least, some lichen may live purely saprophytically. Smith (Harley and Smith, 1956; Smith, 1960a, b, c, 1961, 1963a, b; Smith and Drew, 1965) has investigated the relationships of alga and fungus in the leafy lichen *Peltigera polydactyla*. This is an heteromerous form so that the upper algal plus fungal layer, the 'algal' zone, can be separated from the lower, purely fungal layer, the medullary zone. This provides the opportunity of investigating the separate activities of the components as was achieved by Harley and his co-workers for ectotrophic mycorrhiza of beech. Discs of lichen thallus were punched out and used for study. Sucrose uptake was shown to follow on the lines already described for fungi (cf. Chapter 8, p. 275) and greater uptake was shown by the algal than the medullary zone, the ratio ranging from 3·1–6·8:1 (Fig. 12.13). 'Ammonia' was absorbed preferentially over

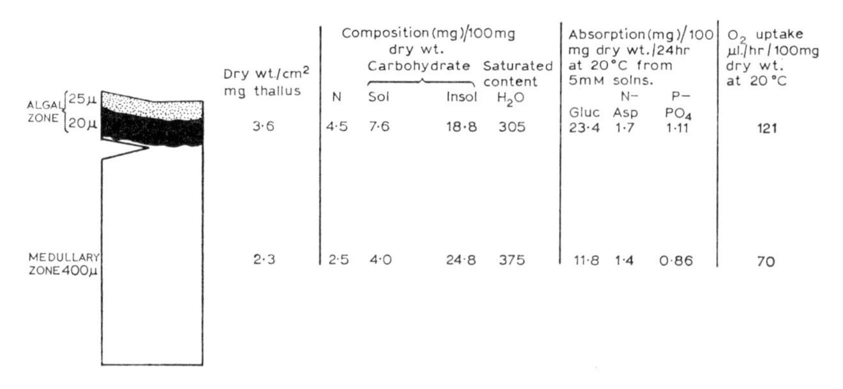

Fig. 12.13. Diagram to illustrate distribution of activities between different parts of thallus of the lichen *Peltigera canina*. The algal zone is divided into an upper cortical layer and a lower algal zone. The line of dissection is illustrated by the indentation. (From D. C. Smith)

nitrate and thalli retained a good deal as ammonia, little being converted to insoluble nitrogen compounds. A similar pattern was shown with asparagine, glutamine and their analogous acids. Here too the insoluble nitrogen content did not greatly increase and the organic molecule, apparently absorbed whole, accumulated, although some was converted to ammonia. Phosphate was also accumulated. In all these cases the algal zone showed a greater accumulation than the medullary zone, the latter only exceeded the former in water content (Fig. 12.13). When discs were supplied with $NaH[^{14}C]O_3$, CO_2 was fixed both in the light and in the dark. Some 40–50% of the photosynthetic carbon accumulated in the medulla within 30 min and this movement continued in darkness for some hours. Smith has suggested that, since most of the carbon fixed by photosynthesis cannot be utilized by the algae and since their storage capacity is limited, the movement to the medullary zone is very largely an overflow effect. When $[^{14}C]$-glucose was supplied to the discs some 50% was converted to mannitol. No free glucose was found in the discs here or in the $NAHCO_3$ experiments; 5% also occurred in the medullary zone as a glycoside. Mannitol could be a storage product or respiratory substrate. This is not always the case, however, for BEDNAR (in AHMADJIAN, 1965) found that $^{14}CO_2$ was mostly converted to labelled sucrose, and also some glucose, fructose, organic acids and amino acids in *Peltigera apthosa*. This species has a different algal partner, a *Coccomyxa* sp., from *P. polydactyla* but how significant this is is not known.

The striking features which appear from these metabolic studies are the high rates of accumulation of all nutrilites by the lichen thallus (a feature borne out by their high accumulation of radioactive isotopes from fall-out compared with higher green plants (GORHAM, 1959)) and their low rates of assimilation and, especially, protein turnover. Indeed, Smith has suggested that this last feature could be a principal cause of their slow rates of growth. The water régime must also effect growth indirectly at least in part, because photosynthesis and light absorption are profoundly affected by the water content of the thallus, being greatly restricted by less than about 60% of maximum (RIED, 1960; ERTL, 1951). In this context the drought-resisting capacity of lichens is very high and when dry they can resist high temperatures up to 50–60°C (LANGE, 1953). This may account for the occurrence of numerous lichens in such regions as the Central Asiatic deserts.

The two other remarkable features of the physiology of lichens are their great sensitivity to atmospheric pollution, perhaps a reflection of their outstanding absorptive powers, and the production of lichen acids. These remarkable substances often have some antibiotic effect and include the depsides and depsidones, derived from 2- or 3-orcinol or β-carboxylic acids joined by ester-linkages, which are virtually restricted to the lichens (SHIBATA, 1948). MOSBACH (1964a, b, c) has shown that their biosynthesis involves the acetyl CoA/malonyl CoA route (cf. Chapter 11, pp. 297, 300) so common in the synthesis of the aromatic ring by fungi. It is not known whether these acids can be produced by either partner growing alone.

The lichens provide perhaps the greatest challenge in the study of inter-relationships between fungi and other organisms. The morphogenetic processes involved are not so very different from some of those shown by free-living forms in the development of multi-hyphal structures. This suggests that lichen fungi may be forms which have lost an internal stimulus, or part of a biosynthetic sequence, this being replaced by the activities of the algal cell. Such fungi might be useful for comparative studies with free-living forms in the same way as biochemical mutants are of value in elucidating processes in wild-type strains.

Fungi and animals

WARM-BLOODED ANIMALS

Increasing numbers of fungi are found to be pathogenic to warm-blooded animals (e.g. SIMONS, 1954; AINSWORTH and AUSTWICK, 1959) and cause dermatophytoses or mycoses of a more or less deep-seated nature. Most of the fungi involved are Fungi Imperfecti. They are frequently dimorphic, having mycelial and yeast-like phases in culture and in vivo respectively, although, frequently, in deep-seated mycoses, pathogenic fungi assume a granular appearance (MARIAT, 1964). Mycotic infections are usually confined and the fungus does not reproduce but this is not so in dermatophytes. For instance, in some species which attack hair, arthrospores are developed in peculiar and characteristic patterns.

Considerable attention has been given to the dimorphic behaviour of some of these fungi, notably *Candida albicans* by Nickerson and his colleagues (e.g. NICKERSON, 1963). Mycelial growth is most readily achieved on a starch or natural medium, such as potato agar but yeast-like growth could be induced by adding glucose or cysteine. Cysteine has been implicated in the provision of SH-groups in cell walls of the yeast form although they are lacking in the walls of filaments. The significance of these groups lies in the conversion of the mannan-protein whose properties have been summarized diagrammatically as:

Covalent bonded; presumably elastic; necessary condition for extension growth.	Mannan-protein S-S	Filament form
	↓ reduction ↑ oxidation	
Undergoes plastic deformation and fibrillar ordering; necessary condition for division.	Mannan-protein-SH	Yeast form

(cf. Chapter 3, pp. 46–47)

Most of the pathogens of warm-blooded animals appear to be facultative and many of them are now being isolated as free-living saprophytes from soil (AINSWORTH, 1965).

INSECTS

The other main group of animals associated with fungi are insects although it seems probable that most animals can suffer some form of facultative parasitic attack in appropriate circumstances. In the insects, however, there are some remarkable examples of interrelationships ranging from casual associations to obligatorily pathogenic or symbiotic ones. Fungi parasitic on insects have been described recently by MADELIN (1963; in STEINHAUS, 1966) and these associations are very similar to those already described in relation to facultative root pathogens. Infection is, of course, somewhat different and is usually a consequence of ingestion or contact with a fungal propagule although, as in the Entomophthoraceae, there are devices for discharging the conidia violently (cf. Chapter 6, pp. 154–155). Entry is via damaged intestines, spiracles or by chitinase activity on the exoskeleton. Once within the insect the parasite spreads through the haemolymph system and usually invades the tissues. Most of these fungi can be cultivated on natural or synthetic media although the chytridiomycete genus *Coelomyces* is an obligate parasite of mosquitoes and has not been cultivated; this is true of some Entomophthoraceae also (MADELIN, 1966).

Obligate parasites which have never been cultured are the ascomycetous family, the Laboulbeniales. These occur as minute exoparasites on insects, mostly beetles of the Carabidiae and Staphylindeae. They have been superbly monographed by THAXTER (1895, 1908, 1924, 1926, 1931). The mycelium varies from a few to several hundred cells, which are short and broad, with thick, often darkly pigmented walls each perforated by a conspicuous pore (Fig. 15.9). In some cases there are distinct male and female plants. The ascospores are usually sticky and can adhere to insects. Perhaps the most extraordinary feature of these fungi is the nature and location of their attachments. Most merely seem to be attached by a relatively blunt, broad basal foot which just penetrates the exoskeleton of the insect. A few species have a penetrating haustorium and one, *Trenomyces histophtorus* a parasite of a chicken louse, produces a copiously branched system of intermittently swollen hyphae and bulbs which becomes embedded in the fat-body of the louse. The problem of how the many hundreds of species, apparently attached only by a foot, are nourished has not been investigated much less solved. RICHARDS and SMITH (1956) do not consider it possible to obtain nourishment from an insect cuticle without its penetration. It may well be, however, that a system analogous to that described for the utilization of cell walls by wood-destroying fungi may operate (cf. pp. 336–338). This must await electron-microscopic and other investigations.

Not only is there the problem of nutrition but there is also the problem of site. Some Laboulbeniales can occur more or less indiscriminately at various places on various hosts but this is exceptional. Most are highly specific to a particular position on a specific host. How this comes about is not clear. THAXTER (1895) suggested that, since the ascospores were sticky, transfer in particular positions could come about during mating. Thus, in *Laboulbenia formicaria*, the male and female plants are found in positions such as could arise from contact during mating. However, in *Herpomyces stylopagae*, on cockroaches, RICHARDS and SMITH (1955) observed ascospores all over the body yet mature fungi were found only on the antennae and palps. Selective mechanisms can be suggested, thigmo- or chemotropism, selective brushing off and so on but, at present, this problem is quite unsolved. Transmission is usually presumed to be from insect to insect but LINDROTH (1948) has claimed that infection can be from the soil.

Equally remarkable but somewhat more fully studied are the associations between yeast-like fungi contained in special organs of various arthropods, so-called mycetomes or mycetocytes, and those between insects which utilize fungi for food in various tropical regions such as the ambrosia beetles or the termites. These situations have been monographed by BUCHNER (1953) and useful reviews provided by KOCH (1960), BROOKS (1963) and BAKER (1963).

Mycetomes are usually derived as pouch-like structures from some part of the gut. Infection may occur from ingestion of the fungus or, in some cases, as a result of the contamination of the egg surface by the fungus. In cases such as the latter it is possible to eliminate the fungus by sterilizing the egg and the insects raised after such treatments usually require to have their diets supplemented. In the case of two beetles (*Lasioderma serricorne* and *Stegobium paniceum*) it proved possible to isolate the fungi concerned from the egg and to replace its function by addition of B-group vitamins and yeast sterols. Alternatively, if sterile eggs were smeared with the fungus, a yeast-like organism, the normal situation was restored. Each beetle had a different yeast partner but the beetles could be successfully reared after cross-infection; the yeasts retained their identity in the unusual hosts (BLEWETT and FRAENKEL, 1944; PANT and FRAENKEL, 1950, 1954). It seems probable that the fungi release the vitamins and sterols to their hosts, although they could also be digested by the host. This latter event seems improbable because in some insects, at least, there is some degree of strain specificity. For example, when free-living yeasts or strains from other insects were fed to *Stegobium paniceum* they were not retained in the special mycetomic cells and, when the larvae pupated, they were expelled (FOECKLER, 1961). No attempts have been made with fungal components of mycetomes to see whether materials from the host insect are incorporated in them. Presumably, fungi obtain their basic carbohydrates and salts, at least, from the host. Perhaps the most remarkable feature of this situation is that the potential mycetome tissues are frequently differentiated in embryos, prior to fungal infection.

This interrelationship seems somewhat one-sided in that the insect exercises selection on the fungus. It is not yet clear whether some fungal inhabitants of mycetomes are obligatorily so or not. This is because few cultures have been made and little identification carried out. Many forms show pleomorphic behaviour analogous to that in mycotic infections so that it is possible that some mycetomal 'yeasts' may turn out to be morphologically modified forms of filamentous soil microorganisms as well.

The last case to be considered is that of the 'ambrosia' fungi usually associated with wood-eating insects such as beetles and termites, or with ants. The significant feature here is the regular transmission and cultivation of such fungi by the insects. As with most associations various degrees of specialization can be seen.

Many common moulds can be isolated from tunnels of wood-destroying beetles whence they have been carried fortuitously by the insects or through passive means such as downwash. In some cases these fungi may be highly pathogenic to the tree, as the cause of Dutch Elm disease, *Ceratostomella ulmi*. This is an important but fortuitous association. A similar situation of some pathogenic importance occurs with wood wasps, notably *Sirex gigas*. *Stereum sanguinolentum*, a basidiomycetous wood-destroying fungus usually confined to dead timber in Europe, is present in special pouches close to the ovipositor of the female so that the egg becomes infected before it is laid in the wood. Thereafter, *Stereum* infects the wood and grows indiscriminately. Although the larva can feed on mycelia in culture it seems as though, in nature, they utilize the decomposed wood (CARTWRIGHT, 1938). Here, therefore, there is transmission but not cultivation. The fungus seems to have no special nutritional requirements, it requires thiamin but can otherwise be grown on a simple salts-sugar medium. On 5% glucose medium it tends to break up and form short segments which resemble those found in the pouches of *Sirex* (PARKIN, 1942).

In the true ambrosia fungi, transmission may be by specialized organs at various positions on the body surface but they are often glandular and filled with oil in female Scolytid beetles. In Playpodid beetles there are fewer specialized organs but even when the spore masses are carried on the body surface they are associated with fatty or oily droplets. While this association with oil may be fortuitous and enable the spores to adhere to the insect, it may be more profound. The majority of ambrosia fungi studied have complex nutritional requirements, they are usually assigned to genera like *Trichosporium, Monilia, Cephalosporium* or yeasts. They do not grow or germinate readily on malt agar but do so if hydrolysed casein or peptone is added. Others have been found to require B-group vitamins and some are unable to utilize inorganic nitrogen and utilize fats as carbohydrate, e.g. *Trichosporium tingens* (MATHIESON-KÄÄRIK, 1953, 1960). Species of *Monilia*, which grew on peptone/ Malt agar, only produced sterile hyphae unless olive oil was poured on, when typical moniloid conidiophores and conidia developed. There seems to be strong evidence here for a clearly nutritional basis, on the part of the fungus to account for its obligate partnership.

Once transmitted to bore holes, the fungus invades the wood surrounding the tunnel for a few millimetres, the cells becoming filled with a mass of hyphae. A stroma usually develops on the wall of the tunnel where spores develop. In some cases these are replaced by a yeast-like phase. It is clear, therefore, that these fungi are capable of utilizing cellulose and perhaps lignin. Nevertheless, it is probable that they obtain lipid material from the beetles, which secrete oil on the surface which is brushed off during their movements in the tunnels. In deserted tunnels other fungi often enter and overgrow the ambrosial forms.

There is little doubt that ambrosia beetles and their larvae browse on the fungi and their spores but there are only a few cases in which the complete life cycle of a beetle has been carried out in vitro on a fungal substrate; many undoubtedly eat wood as well, at least as adults (GADD, 1947).

Many fungi are used as food by insects and this is the situation in the termite mounds, galls of certain midges and ants' nests. Less is known about the associations. In termites there is controversy over the function of the fungus and it is sometimes regarded as a pest which is controlled. At all events certain agaric species of the genus *Termatomyces* are confined to such locations (HEIM, 1948). South American leaf-cutting ants strip leaves from plants, cut them up into small pieces, chew them and leave them to be infected by fungus. This develops small spherical swellings (bromatia) on which the adult ants feed and which also nourish the larvae. It has been claimed that other fungi are weeded out but this is doubtful. When a new colony is started the virgin queen apparently carries a fungal pellet with her as an inoculum for a new culture (BELT, 1874; MÖLLER, 1893; BAYLISS-ELLIOTT, 1915). Nothing seems to be known of the nutrition of these forms.

It will be clear that a vast range of problems lie ahead for those concerned to investigate the interrelationships of fungi and insects. They include problems of chemotropism, thigmotropism, nutrition, change of form and adaptability to a wide range of physical and chemical conditions as well as biological problems of the most general importance to evolutionary thought. It is not surprising, therefore, that a completely new and common group of fungi, the Trichomycetes, has been recognized since 1948; their ecological niche is the hind-gut of arthropods. They are very simple, septate, filamentous forms which reproduce by amoeboid or non-motile spores and, in some cases, develop structures reminiscent of zygospores. A few have been cultured. This group and its recent discovery indicate clearly the vast areas of unknown fungal biology which have not yet been revealed.

The situations just discussed are of immense importance for the biology of the fungi and there is still no clear indication of the ways in which antagonism, parasitism, or mutually balanced growth relationships arise or are sustained, for any one fungus with another organism. In some ways the simplest situations are those in the fungi themselves, where the whole gamut of relationships is displayed within the same groups of organisms (BARNETT, 1963). Throughout, the essential steps in the establishment of any interaction appear to be governed by contact reactions and/or nutritional relationships. This is not surprising for the whole basis of fungal sociology, as BULLER (1931) pointed out so clearly, is hyphal fusion and the fact that fungi are heterotrophic hardly needs to be restated. Contact stimuli and their responses would appear to be a key to the understanding of much in the vegetative and reproductive life of fungi and to their relationships with other organisms. Particular attention might usefully be paid to these processes in Phycomycetes where vegetative fusions are minimal and yet where reproductive fusions and fusions with other organisms can, and do, occur. Some of the data bearing on this latter problem will be discussed in the next section.

SECTION III Recombination

Chapter 13
Nuclear Division

The study of crossing-over in fungi has provided a great many unexpected and new facts in the last thirty years. Unhappily, knowledge of the associated cytological events has lagged far behind and fungal cytology is badly in need of intensive and critical investigations. It is true that the small size of fungal nuclei and chromosomes and the frequent inability of the latter to take the more usual chromosomal stains has been a serious handicap and deterrent to their study. However, studies such as those described in this chapter demonstrate that these hurdles can be overcome.

Broadly speaking there is general agreement, based largely on studies with Ascomycetes, that meiosis in fungi is similar to that in higher organisms, at least from diplotene onwards. Mitosis, however, is variously described and interpreted. For some (OLIVE, 1953) all divisions can be fitted to the procrustean bed of 'normal' mitosis; for others, none are 'normal' and several different, unique and bizarre modes of division can be observed at different stages within the same organism (DOWDING and WEIJER, 1960, 1962; WEIJER, KOOPMANS and WEIJER, 1965; WEIJER and MACDONALD, 1965; WEIJER and WEISBERG, 1966). It will be clear from this account that the former view, at least, is hardly tenable.

SOMATIC NUCLEAR DIVISION

Three well documented cases of somatic nuclear division are illustrated in Plates V and VI; these are *Macrophomina phaseoli*, a Deuteromycete (KNOX-DAVIES, 1966) and the Phycomycetes *Mucor hiemalis* and *Basidiobolus ranarum* (ROBINOW, 1957, 1962, 1963). The first is a more or less typical mitosis, as generally understood; the second is a most unusual type of division which is well documented for a number of fungi; the last is somewhat intermediate in its features.

Macrophomina phaseoli, an important root parasite in tropical countries, produces its asexual spores in a flask-shaped pycnidium. The single nuclei of immature pycnospores are relatively large. Repeated divisions occur, the later ones synchronous, and the mature spores are multinucleate. During these divisions the nuclei become progressively smaller. Figures 1–8 in Plate V are of the earliest divisions in the spore and were stained preparations made by a Giemsa technique similar to that developed originally by ROBINOW (1942) and followed by squashing. Giemsa is not as specific for nucleic acid as the Feulgen stain is said to be but it gives very comparable results.

Interphase nuclei show chromatin threads and a deeply staining nucleolus; as the nucleus becomes smaller in successive divisions it stains more intensely. At full prophase individual chromosomes can be recognized; they are derived by thickening of the original chromatin threads. They shorten; heterochromatic and clear regions, perhaps centromeres, are recognizable and the nucleolus generally becomes detached from the nucleolar chromosome. At metaphase a top-shaped spindle with spindle fibres and terminal centrioles develops. As the spindle develops the nucleolus loses its ability to stain but the chromosomes continue to shorten, thicken, and stain more deeply. The centrioles are usually located at the margin of an area stained more heavily than the surrounding cytoplasm and Knox-Davies suggests that this area is enclosed in a membrane. No distinct metaphase-plate arrangement of the chromosomes has ever been recognized. Early anaphase is short, to judge by its infrequent detection, but is followed by a long anaphase. At this stage individual chromosomes cannot be distinguished, each pair of nuclei is associated with a single, intensely staining nucleolus and, typically, there is a well marked band of interzonal fibres between each chromosomal group. These fibres disappear at telophase, the nuclei re-form and stain less intensely and the nucleolus finally disappears. Presumably a new nucleolus is developed in each nucleus during interphase.

As divisions proceed it becomes more difficult to recognize centrioles and sometimes only a single, minute body can be seen near the metaphase chromosomes. This is especially true of divisions occurring in hyphae, which were studied less fully.

Somatic divisions in *Mucor hiemalis* (Plate V, Figs. 9–14) and the Mucoraceae, contrasts greatly with that in *Macrophomina*. The interphase nucleus of *Mucor* stained with Feulgen (as in Plate V, Figs. 11–14) or Giemsa appears as an object of variable shape with numerous, small stained particles in its outer region which may even assume a reticular appearance. The nuclear material separates as two half-cups or shells which stain more intensely and separate. Sometimes a granular strand stretches for a time between the separating halves but, once clear, two nuclei re-form. The whole process takes 2–4 min as observed under phase contrast in living spores or hyphae.

When stained with iron haematoxylin a very different appearance is seen. The centre of the nucleus is occupied by a deeply staining central body, sometimes called a nucleolus (cf. p. 364), the periphery is unstained. As division proceeds the central body may come to lie at the side of the nucleus, it elongates, becomes constricted in the middle, separates and two new nuclei are re-formed. This kind of 'direct' nuclear division had been described by many earlier investigations of both Mucors and *Saprolegnia* (SMITH, 1923 and Fig. 13.1).

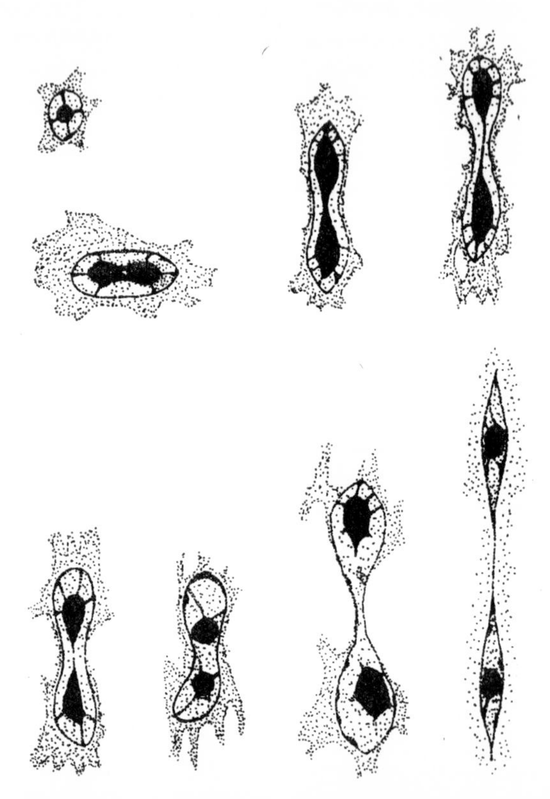

Fig. 13.1. Somatic division in *Saprolegnia* material fixed and then stained with iron haematoxylin. Note intact nuclear membrane and heavily staining central body (× 2025 from F. Smith)

Rather more detail can be made out in the divisions in germinating spores. Here there is a loose arrangement of Feulgen-positive threads and granules which appear to be embedded in a matrix of material derived from the central body.

The third example, *Basidiobolus ranarum* (Plate VI) is somewhat intermediate between these two examples. It has a well defined nuclear membrane and a vacuole-containing nucleolus which stains heavily with iron haematoxylin and is clearly separated from the peripherally located Feulgen-positive chromosomes which tend to cluster at the poles. The nuclear membrane disappears, the outline of the nucleolus softens and some of its material appears to be incorporated into fibres which form part of a spindle. The rest of the nucleolar material develops into caps or end-plates (as identified by staining with iron haematoxylin). The spindle is more rounded-ovoid than top-shaped but can be seen clearly by phase contrast microscopy. A metaphase stage occurs when the closely packed, minute chromosomes form a 'Saturn's ring' on the equator of the spindle. At anaphase the chromosomes separate a little way into the material of the end-plates and the spindle then elongates. Its end-plates round up causing the spindle fibres to diverge in the equatorial region and then they become dissipated. At telophase the end-plates break up into blobs of material and two new nuclei reform. It has been estimated that the whole process takes twenty minutes.

Somatic division in these three fungi is compared in Table 13.1.

Table 13.1. A comparison of the various features exhibited at somatic division by *Macrophomina*, *Mucor* and *Basidiobolus*

Feature	*Macrophomina*	*Mucor*	*Basidiobolus*
Intranuclear division	+	+	+
Individual chromosomes recognizable	+	± ?	+
Centromeres	+	−	−
Nucleolus	+	−	−
Central body	−	+	+
Top-shaped spindle	+	−	−
Spindle-like form	−	−	+
Spindle fibres	+	−	+ ?
Centrioles	+	−	−
End-plates	−	−	+
Metaphase-plate configuration	−	−	+

It can be seen that the principal features which differentiate them are: the degree of distinctness of individual chromosomes, the presence or absence of centrioles and spindle, the behaviour of the chromosomes (e.g. arrangement on the equator or not) and the behaviour of the central body or nucleolus. The most significant features seem to be the presence or absence of a spindle, the behaviour of the central body or nucleolus and, as a corollary to these, the behaviour of the chromosomes during division.

The resolution of individual chromosomes reflects in part the technique employed, in part their actual size. Inappropriate or unsuitable techniques have often led to the failure to recognize chromosomes of the usual type. A recent example is the *Mucor*-type division described by BAKERSPIGEL (1959b) for *Neurospora*. Lighter staining and other differences in technique have now enabled this to be recognized as a more or less normal mitosis with chromosomes distinctly visible (WARD and CIURYSEK, 1961). The difficulty of recognizing individual chromosomes because of their intrinsic features is well shown in *Macrophomina*. As successive nuclear divisions in the pycnospore reduce the size of the nuclei while their intensity of staining increases, the recognition of individual chromosomes becomes increasingly difficult.

Much the same kind of behaviour is shown in *Macrophomina* by centrioles. It is possible that centrioles, if present, will only be detected readily in cells where both the nuclei and the amount of cytoplasm are large. This may rationalize some of the difficulties which are raised by observations such as those of ZIEGLER (1953) and BAKERSPIGEL (1960) in Saprolegniaceae. The former detected and figured centrioles in the developing oogonium, a large cell with a relatively large nucleus, but Bakerspigel could not recognize such structures in somatic divisions in hyphae, where the nuclei are much smaller and more crowded. The analogy with *Macrophomina* is clear. As has been mentioned earlier (see pp. 30–31), centrioles have been detected in the sporangium of *Albugo* and it seems likely, by analogy with green plants, that centrioles may well be found in all situations where divisions give rise to motile spores. That the absence of a visible centriole in no way prevents the development of a normal spindle is attested by the situation in the Angiosperms where centrioles are almost entirely lacking. In the fungi, *Basidiobolus* shows this behaviour, but that part of the central body which forms the end-plates appears to play an analogous role to a centrosome.

Some of the discrepancies mentioned, especially those between appearances at meiosis and somatic division, have led some observers to deny the reality of the *Mucor*-type of division (OLIVE, 1953; WARD and CIURYSEK, 1962). The only important discrepancy at present, however, is that concerning Saprolegniaceae, for, in those groups where a *Mucor*-type division has been described, meiosis has either not yet been studied adequately, e.g. Mucoraceae, or it has been found that the somatic division is not of the *Mucor*-type after all, e.g. *Neurospora*. There is even some uncertainty about the discrepancy in Saprolegniaceae for SANSOME (1963) has cast some doubt on ZIEGLER's (1953) interpretations of division in the oogonium and his work does not include photographic evidence comparable with that available for other fungi.

At present it seems best to accept the reality of the *Mucor*-type division. Attention may be drawn to the similarity between such divisions and those described by LEEDALE (1958a, b; 1959) in the Euglinineae (Flagellata). In both kinds of organisms there is a persistent central body (the 'endosome' of Leedale) which divides separately from the chromosomes. Centromeres and spindle are lacking in both and it may well be that, like the Euglinineae, the nuclear membrane persists through the division in the Mucoraceae. On the other hand there are differences for, in Mucoraceae, individualized chromosomes neither appear clearly nor duplicate themselves longitudinally, the sister chromatids segregating to opposite poles. The important common feature, however, is that there is a structure similar in staining reactions (and, perhaps, in containing RNA) which may well perform some role in the process of nuclear separation.

That a central body or nucleolus can perform a quasi-tractile function is suggested by *Basidiobolus* where, it will be recollected, part of the material goes into the development of spindle fibres and end-plates. In this sense *Basidiobolus* can be regarded as intermediate between the *Mucor*-type and the more normal type of mitosis.

Even in those fungi whose somatic divisions most resemble normal mitosis there are some differences from other organisms. Metaphase is usually a short phase but really well-condensed metaphase plates are probably unusual; an exception is *Rhizidiomyces* (FULLER, 1962). This fungus also shows another feature which may well characterize somatic division in a number of fungi, namely the persistence of the nuclear membrane which finally constricts between the two poles in telophase after daughter chromosomes have separated. A similar suggestion was made earlier for *Macrophomina* and Mucoraceae and much of the older cytological work suggested the occurrence of intranuclear spindles. In this respect *Basidiobolus* is unusual in that its nuclear membrane does seem to disappear. The retention of the nuclear membrane, where it occurs, is a further cause of technical difficulty in the study of somatic divisions, since its presence prevents the adequate spreading of the chromosomes (EVANS, 1959).

MOORE (1964) has studied the division of the nuclear membrane at ascospore formation in *Cordyceps militaris* using electron microscopy. His interpretation supposes that the inner electron-dense layer invaginates first and that the nucleus finally divides by the invagination of the outer electron-dense layer into the space produced between the invaginated and fused inner layers (Fig. 13.2). His observations are not entirely conclusive and it is most

Fig. 13.2. Diagram to illustrate manner in which the nuclear membrane may ultimately divide in fungi. (After Moore)

desirable that this aspect of nuclear division should receive further study in a wide range of fungi.

Certain additional issues are still not clearly resolved in the somatic divisions of fungal nuclei: these are, the situation in the Plasmodiophoromycetes, division in yeast and behaviour in Basidiomycetes.

Nearly all investigators of *Plasmodiophora*, the club-root organism, and its allies described a rather remarkable sequence of events in the early nuclear divisions of the plasmodia of these organisms. This has been termed the 'promitotic' or 'cruciform' division; good accounts are given by HORNE (1930) and WEBB (1935).

The interphase nucleus possesses a large central body, the subsequent behaviour of which is similar to that described for *Mucor*. Linearly differentiated, individual chromosomes become visible, however, and tend to become arranged in one plane and eventually associated end-to-end, when they form a 'Saturn's ring' stage surrounding the now elongated central body at metaphase. The chromosomes are detectably double at this stage, they separate longitudinally as two parallel rings and their approach to the ends of the nuclei coincides with the development of a constriction in the central body (Fig. 13.3). The nuclear membrane remains intact throughout this division,

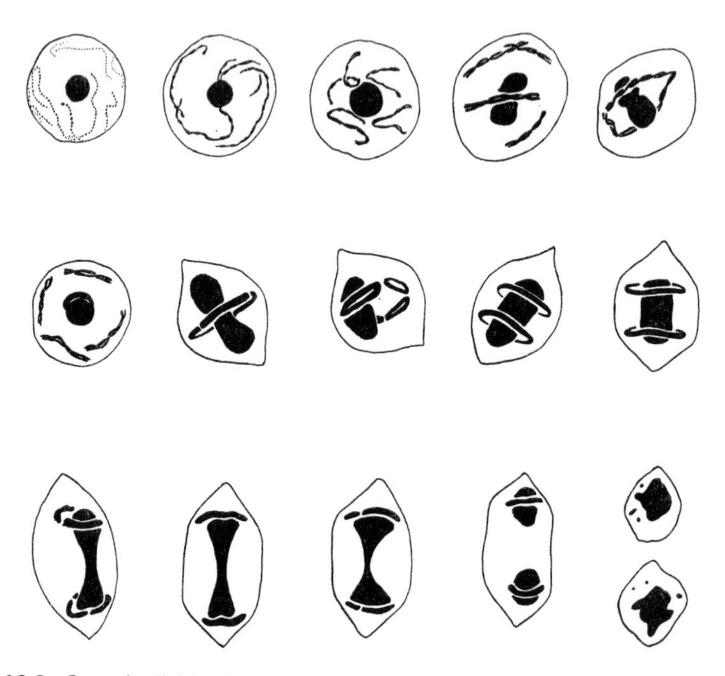

Fig. 13.3. Somatic division in *Spongospora subterranea*, a plasmodiophoromycetous form. Note intact nuclear membrane, deeply staining central body and 'Saturn's ring configurations' of dividing chromosomes. (×3275, from Webb)

constricting and dividing at the very end. The earlier workers, largely employing iron haematoxylin, also figured an intranuclear spindle and spindle fibres. After this division an akaryotic phase frequently occurs in which the conspicuous central body loses its ability to take stain and eventually disappears for a time, reappearing at the onset of what is thought to be meiosis, e.g. in *Spongospora* and *Sorosphaera* (HORNE, 1930; WEBB, 1935). This meiosis is of the usual type and the central body has the appearance and shows the behaviour of a normal nucleus. A recent study, using the Feulgen stain, confirms the existence of a normal meiosis in *Plasmodiophora* but provides no information concerning the 'cruciform' division (HEIM, 1955). At its face value, somatic division in Plasmodiophoromycetes appears to combine some features of *Mucor*-type and normal mitosis with a characteristic alignment and separation of the chromosomes at metaphase and anaphase respectively.

The problem in *Saccharomyces* has been both to identify the nucleus and to describe its division. NAGEL (1946), in a valuable and painstaking comparative study, employed a wide range of stains and fixatives as well as living cells and decided that the nucleus was the Feulgen-positive, small body (her 'parvicorp') lying to the side of a central vacuole: this view was confirmed much later by electron microscopy (MOOR and MÜHLETHALER, 1963). The nucleus is bipartite (Fig. 13.4). The larger part is Feulgen-positive and during division

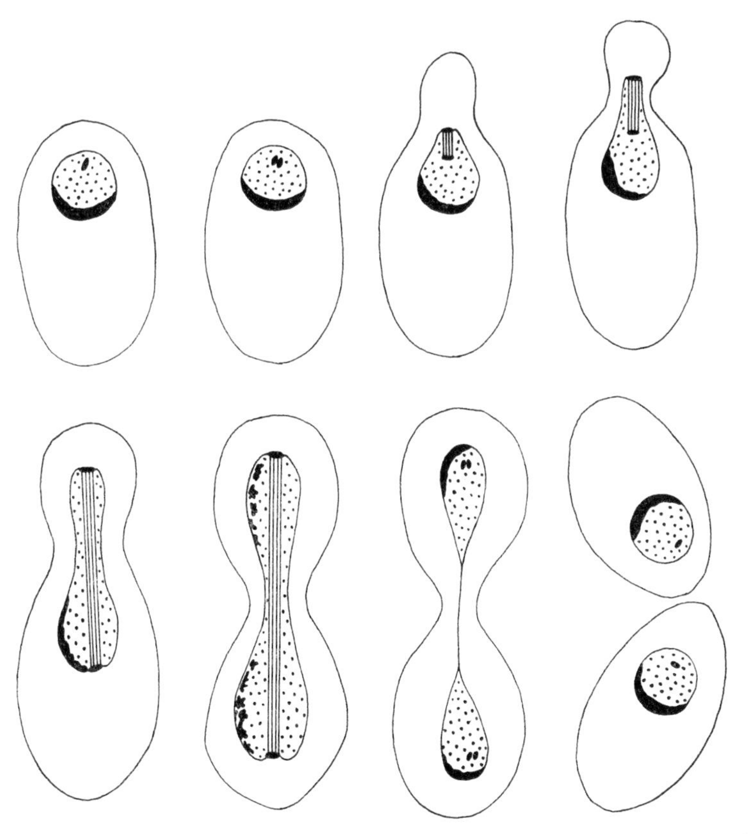

Fig. 13.4. Diagram to illustrate nuclear division in yeast cell. The dots indicate 'chromosomal' material. Note behaviour of basiphilic crescent (shown black), this is Feulgen-negative. (Based on unpublished information of Robinow)

becomes elongated, constricted in the centre and eventually separates into two halves. Individual chromosomes are not usually recognizable. A second, crescentic, smaller part is strongly basiphilic but Feulgen- (and Giemsa-)

negative (Nagel's 'companion body'). This body becomes attenuated, sometimes breaking into small, irregular pieces and eventually is more or less evenly divided between the daughter nuclei. Recently, ROBINOW and MARAK (1966) has provided evidence that there is a single, intranuclear fibre. Opposite the crescentic, basiphilic body within the Feulgen-positive material is a small bead-like structure, which stains readily in 2–3 min with very dilute acid fuchsin (1:20,000). This bead or fibre initial, expands into a long, straight fibre which becomes anchored at each end to the nuclear membrane. This development can also be seen by phase microscopy of yeasts grown in 21% gelatine plus nutrients. By electronmicroscopy the fibre has been resolved into a bundle of cytotubules 150–180 10^{-10}m in diameter, anchored to the nuclear membrane by 'centriolar plaques' (Fig. 13.5). It is tempting to equate this

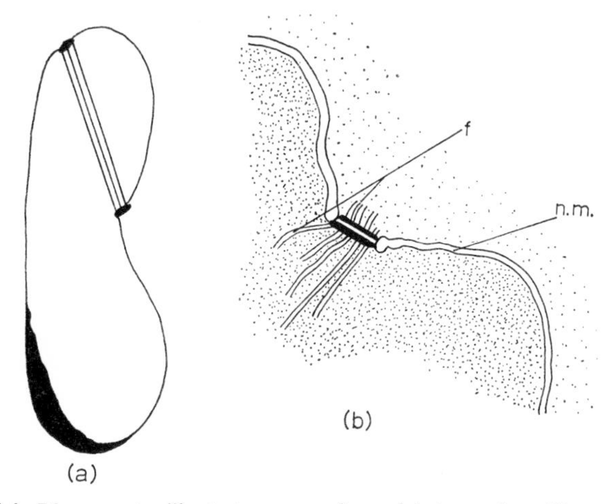

Fig. 13.5. (a) Diagram to illustrate connexion of intranuclear fibres to centrioles attached to nuclear membrane. (b) Tracing of electronmicrograph through nuclear membrane (n.m.) including centriole; (f) intranuclear fibres which extend a little into cytoplasm (×41,500) (b, traced from electronmicrograph of Robinow)

fibre to spindle fibres and, indeed, those of *Basidiobolus* stain in a similar manner with extremely dilute acid fuchsin. It is, perhaps, not unreasonable to apply the terms nucleolus and centriole to the basiphilic crescent and fibre-initial respectively. These discoveries do much to render somatic division in *Saccharomyces* intelligible although many problems, such as the mode of chromosome replication, still remain.

Division in the yeast-like *Blastomyces* is said to be of the *Mucor*-type (BAKERSPIGEL, 1957) so that it may well prove rewarding to look for an intranuclear fibre in fungi which show this type of division. On the other hand, other yeasts, notably *Lipomyces* (ROBINOW, 1961), differ from *Saccharomyces* and *Blastomyces* in possessing distinct, countable chromosomes. They show a remarkable alignment at the end of a normal prophase in which sister pairs form a parallel stack whose long axis marks the plane of the division. This is followed by their coalescence to a linear mass which separates, by a constriction, into two halves connected by a thin strand. This mode of division has something in common both with *Saccharomyces* and *Plasmodiophora*.

The concept of an intranuclear fibre receives support from observations on somatic divisions associated with clamp formation in the dikaryon of the Basidiomycete *Polystictus versicolor* (GIRBARDT, 1961, 1962). The leading nucleus usually enters division earlier than the rear nucleus in the so-called conjugate division.

The nucleus which is to migrate into the branch hypha of the clamp elongates (Fig. 13.6). A thread of material stretches from the central body to

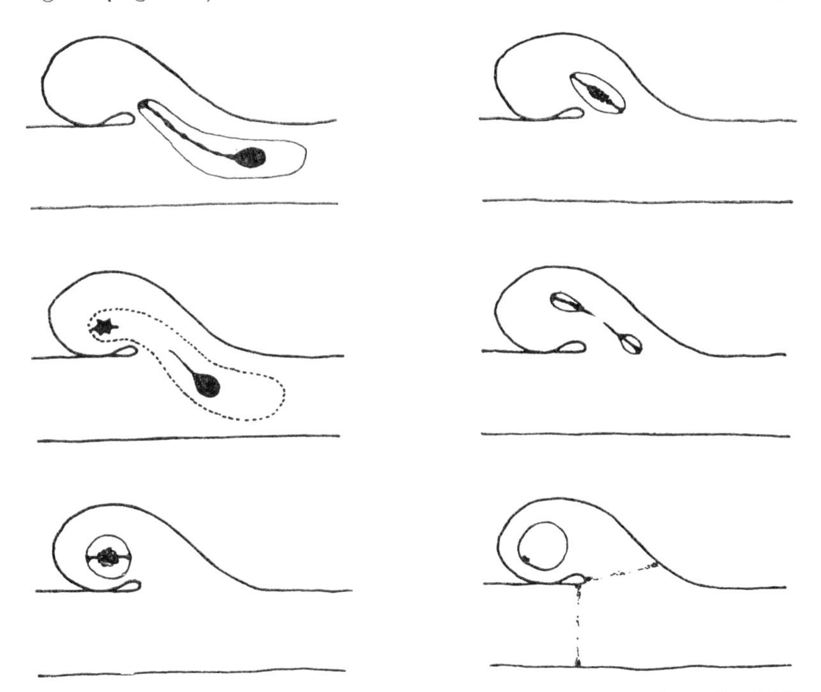

Fig. 13.6. Diagram to illustrate somatic division of one of two conjugately dividing nuclei at clamp connexion formation in *Polystictus*. Note intranuclear fibres with chromosomes associated at division. (After Girbardt)

a basiphilic granule on the membrane at the leading tip of the nucleus. Virtually no chromosomal material is detectable. The thread seems to contract and coalesce towards the forward part of the nucleus which becomes rounded and irregularly fibrillar while granular, Feulgen-positive material becomes visible. This material now collects in a mass at the centre of an intranuclear fibre (Girbardt's 'Zentralstrang'), clearly visible by phase microscopy, which develops along the long axis of the now elongated nucleus. This fibre is attached to the nuclear membrane by strongly basiphilic bodies which Girbardt describes as 'centriole-like'. In the electron microscope this intranuclear fibre, like that of yeast, can be resolved into a number of tubular filaments, *c.* 200 10^{-10}m in diameter, clearly attached to the nuclear membrane (Plate IX, Fig. 1). There is a suggestion that similar structures also extend into the cytoplasm beyond the membrane attachments but this is not certain. The nucleus now elongates, separating the chromosomal material, which has become spread along the fibre, into two terminal masses. The fibre and nuclear membrane constricts and two daughter nuclei are formed. Their contents are not stainable save for a small, Feulgen-positive mass against the nuclear membrane near the former position of attachment of the intranuclear fibre. A basiphilic central body develops in the subsequent interphase. LU (1964) has described the comparable processes in the Gasteromycete, *Cyathus stercoreus*. In his material the nucleus appears to be somewhat larger and distinct, contracted chromosomes can be recognized in the migrating nucleus as well as a centriole at the leading tip. The intranuclear fibre is not, however, shown so clearly. WARD and CIURYSEK (1961) working with an unknown Basidiomycete (recognized by the presence of clamp connexions) have provided evidence of a possible metaphase plate-like configuration but their account is not fully documented. The earlier paper of BAKERSPIGEL (1959a) on *Schizophyllum*, a gilled Basidiomycete, is now seen to be of especial interest. The divisions studied do not appear to be associated with clamp formation. Those which illustrate the earliest divisions in a binucleate, germinating spore are remarkably reminiscent of divisions of the *Mucor*-type but later figures, from multinucleate hyphae, are more reminiscent of the divisions illustrated by Girbardt in *Polystictus*. This, perhaps, adds further weight to the suggestion

Plate V. Somatic division in fungi

Figs. 1–8. Somatic division in pycnospores of *Macrophomina phaseoli.* Sections and squashes fixed acetic-ethanol (1:3), hydrolysed in N.HCl and stained in Giemsa at pH 6·9.
1. Typical interphase nucleus; 2. Full prophase; 3. Prometaphase. The nucleolus has become separated from the nucleolar chromosome. Note the marked heteropycnosis of the chromosomes. 4. Metaphase, note centrioles apparently within a nuclear membrane (or possibly on the periphery of an unstained nucleolus). 5, 6. Further metaphase configurations, note evidence of centrioles and fibres. 7. Late anaphase, note intensely-stained intrazonal fibres and nucleoli. 8. Telophase, note nucleoli apparently left outside of re-forming nuclei. (All ×4000)

Figs. 9–14. Somatic division in *Mucor hiemalis* in germ hyphae. All fixed, hydrolysed in N.HCl and stained either in Giemsa (9–10) or Feulgen (11–14). Compare with Fig. 13.1 (p. 362) which shows similar divisions in *Saprolegnia* stained with iron haematoxylin.
9. Resting and dividing (elongated and constricted) nuclei. Note peripheral 'halo' of chromatin material and lack of evidence of spindle or centrioles. 10. In these nuclei, which have been sheared between slide and coverslip, the peripheral chromatin can be seen to be corpuscular or filamentous. 11–14. Details of division, note dumb-bell like constriction of nucleus and formation of peripheral arcs of chromatin material. (All ×*c.* 3400)

[Figs. 1–8 supplied by P. S. Knox-Davies, from Knox-Davies, 1966. Figs. 9–14 by C. F. Robinow, from Robinow, 1957 and unpublished].

made earlier that on reinvestigation, fungi showing *Mucor*-type division may be found to have an intranuclear fibre, or similar structure.

Another puzzle which cannot yet be resolved is the remarkable behaviour of the chromosomes of the rust fungi of the genus *Puccinia*. In three tantalizingly brief reports MCGINNIS (1953, 1954, 1956) has described end-to-end attached chromosomes apparently forming twisted or spiral chains, or even rings and pair-like arrangements suggestive of somatic pairing or secondary association. These configurations cannot be explained at present and warrant further study (see Fig. 17.9, p. 481).

Certain features shown by *Puccinia* are reminiscent of recent observations on *Neurospora*, *Gelasinospora* and *Aspergillus* concerning which a further, highly controversial interpretation of somatic division has been presented by Weijer and his associates (DOWDING and WEIJER, 1960, 1962; WEIJER, KOOPMANS and WEIJER, 1963, 1965; WEIJER and MACDONALD, 1965; WEIJER and WEISBERG, 1966). Their proposals are similar to those made 30 years earlier by VARITCHAK (1928, 1931), originally for *Ascoidea rubescens* but later for other Ascomycetes. The essence of their interpretation is that the nucleus gives rise to chromosomes connected by a filament which is terminated at one end by a nucleolus and at the other by a centrosome (described as a 'centriole') which initiates a process of longitudinal division. Sometimes the filament curves round to form an incomplete or apparently complete ring. At longitudinal division, therefore, U-, V- and Y-like configurations can be seen as well as partial double crescentic or ring-like configurations. In *Neurospora* further variations are imposed upon this basic pattern by the phases of the life cycle and the state of the cytoplasm. The former affects principally the overall size of the configurations, e.g. the 'juvenile phase' at microconidium formation is far smaller than 'mature phase I' in aerial hyphae; the latter affects their appearance and mechanism of division, e.g. in streaming cytoplasm the configurations are drawn out in the direction of streaming.

Plate IX. Structures associated with nuclear division

Fig. 1. Electronmicrograph of ultra-thin section through lateral hypha of clamp connexion of *Polystictus versicolor.* Note fibrils which form the 'Zentralstrang' across the nucleus. They appear to extend through a complex in the nuclear membrane into the cytoplasm. (×14,600)

Fig. 2. Electron-micrograph of ultra-thin section through pachytene nucleus of *Neottiella* (*Humaria*) *rutilans.* Note nucleolus and bivalents with synaptinemal complexes visible in several cases. Note absence of detail in chromosomal material. (×*c.* 7600)

Fig. 3. Enlarged photograph of synaptinemal complex from central bivalent of Fig. 2. Note regular alternation of thick and thin bands in complex adjacent to chromosomal material, and central common, narrow bands opposite each outer one. (×*c.* 39,200)

[Unpublished photographs supplied by M. Girbardt (1) and M. Westergaard and von Wettstein (2, 3)].

Weijer suggests that the characteristic features of such somatic division arise from the location of the 'centriole' which is in line with the equatorial plate rather than at right angles to it. He writes, 'Fibres extending from the centriole are therefore parallel to one another and consequently the microscopic appearance of the spindle is a thread differing in thickness towards the nucleolus. Therefore, it seems that the function of the spindle fibres is different in each system; in the vegetative mitotic nucleus these fibres are employed to ensure the entity of the nucleus in a fast streaming and multinucleate environment and hence the mechanism of division is passive, i.e. separation is brought about by cytoplasmic streaming. . . . In the specialized reproductive structures, however, the spindle fibres assume their typical function of separating the metaphasic sister chromosomes' (WEIJER *et al.*, 1963).

WESTERGAARD (1965) has criticized Weijer's methods which involve heat shocks that could distort chromosomal configurations. Certainly to dry material on a slide at 60°C as an initial procedure seems to be rather drastic. Interpretation and comparison of the photographic data are difficult because of the absence of scales in some of the published work. It should also be noted that WALKER (1935) did not confirm VARITCHAK's (1928) original observations of a similar process in *A. rubescens*. Such configurations as appeared to be comparable to his were interpreted by Walker as occurring in 'degenerating' nuclei. No wholly satisfactory interpretation of any fungal somatic division can yet be made but there seems to be no compelling reasons, as yet, to accept Weijer's highly unconventional and novel proposals. However, despite these criticisms of Weijer's observations and interpretations there does seem to be an increasing amount of evidence for the occurrence of an end-to-end association of fungal chromosomes followed by their separation as a whole or fragmented unit. Attention has already been drawn to this kind of behaviour in Plasmodiophorales, *Lipomyces* and *Puccinia* species. Recent studies have extended such claims.

LAANE (1967) has presented suggestive but not wholly convincing evidence for a compound chromosome strand in *Penicillium expansum*. Normal strains and *clock* mutants (cf. pp. 67–68) of *N. crassa* were employed by NAMBOODIRI and LOWRY (1967); the mutants because nuclear divisions occur with great frequency in hyphae at the beginning and end of the growth periods. They have described the development of a straight or coiled (three-gyred) filament from interphase nuclei. This despiralizes and duplicates from one end as the nuclear membrane becomes disrupted. Bead-like bodies, later resolvable as seven chromosomes, develop along each filament and are attached by interchromosomal strands. One chromosome, not necessarily the same one on every occasion, becomes displaced laterally in each strand and the strands separate. Interzonal strands are said to be detectable between such separating compound chromosomal strands but the photographic evidence available is not convincing. Wilson and his colleagues (AIST and WILSON, 1965, 1967; WILSON *et al.*, 1966; BRUSHABER *et al.*, 1967) have provided convincing evidence from a wide range of Ascomycetes (especially *Ceratocystis fagacearum* and including *N. crassa*), Fungi Imperfecti and Basidiomycetes (*Lenzites saepiaria*,

Rhizoctonia solani) that as prophase proceeds the chromosomes apparently contract and form two compound filaments or bars each formed by end-to-end association. They claim that the nucleolar chromosomes lie at the end of the chain which is orientated parallel to the long axis of the hyphae. Sister chromosomes are said to separate individually or in groups and a spindle or spindles develop between such units. The evidence for this latter claim is not convincing.

These more recent studies have been carried out by well-tried and unobjectionable methods such as HCl-giemsa or acetic-orcein and are not, therefore, susceptible to the type of criticism made by Westergaard of Weijer's work.

It should be clear that the study of somatic division in fungi is at present in an active, almost explosive, state. These discoveries in fungi will, sooner or later, have to be incorporated into the framework of 'normal' cytological behaviour. The study of somatic division in fungi is thus of great intrinsic interest and is bound to have an important bearing on the phenomenon of mitotic recombination, which seems to be not unusual in fungi (see p. 397). This involves, as a rare event, nuclear fusion and segregation in hyphae and, at present, there is no cytological basis for the phenomenon. A number of investigations have illustrated somatic division proceeding in highly elongated, crowded, migrating nuclei, e.g. *Gelasinospora*, *Neurospora* (DOWDING and WEIJER, 1962), *Cyathus* (LU, 1964a). It is possible that fusion or exchange of genetic material might be promoted in such circumstances.

Another possibility arises from the re-examination of stages just prior to meiosis in *Neottiella* (=*Humaria*) *rutilans* by ROSSEN and WESTERGAARD (1966). They showed that competence to fuse between haploid nuclei in the ascogenous hyphae was associated with the occurrence of precocious chromosomal replication compared with the condition of nuclei in adjacent cells. In electronmicrographs (WESTERGAARD and VON WETTSTEIN, 1966) the annular pores were better developed in the nuclear membranes of such pre-fusion nuclei than in the adjacent nuclei which did not fuse. How such observations can be related to the dikaryotic condition in Basidiomycetes or the heterokaryotic condition in general is not yet clear. Indeed, the control of fusion of nuclei, whether somatic or immediately prior to meiosis, e.g. in the ascus or basidium, is not understood at all at present.

One of the paradoxes of the present situation is the apparent diversity of types of somatic division compared with the remarkable uniformity of meiotic behaviour. This latter will now be reviewed briefly.

MEIOTIC DIVISION

Amongst the best authenticated accounts of meiosis in any organism are those of MCCLINTOCK (1945) and SINGLETON (1953) for the Ascomycete *Neurospora crassa*. A selection of their figures is displayed in Plates VII and VIII. There are now several other descriptions of similar clarity but in no other fungus have the pachytene chromosomes been so beautifully demonstrated or analysed. Ascomycetes have proved particularly valuable material for the application of modern 'squash' and 'smear' techniques since asci can be fairly readily flattened and the divisions well displayed.

A series of conjugate divisions in the hook-like ascogenous hyphae precedes fusion in a swollen sub-terminal cell which develops into a clavate ascus. Synapsis occurs at about this time while the ascus grows more rapidly in length than breadth until it is virtually cylindrical. Some correlation can be drawn between the stage of meiosis and the size and shape of the ascus as shown in Table 13.2. Meiosis follows a normal course save in prophase. In many organisms synapsis, at zytogene, occurs between greatly elongated chromosomes but, in *Neurospora,* this is not so and the chromosomes are considerably contracted when pairing begins. They elongate as synapsis proceeds and, by pachytene, structural differentiation can be clearly recognized in the large, 10×20 μm nucleus. This possesses a prominent nucleolus, some 6 μm in diameter, clearly attached to a chromosome (designated Chromosome 2). After attaining their maximum length and degree of linear differentiation at late pachyene the paired chromosomes condense and contract until they reach their smallest size at metaphase I. Diplotene nuclei are difficult to observe because of a characteristic fuzzy appearance. The seven bivalents are readily recognized at diakinesis as are chiasmata. An intranuclear spindle now develops although centrioles have not been observed at this stage. The bivalents become arranged at the periphery of the equator of the spindle and this is followed by disjunction and spindle elongation. The nucleolus usually becomes detached during anaphase I and is lost although it may persist until anaphase II. Nuclei are reconstituted after telophase. In division II, spindles are formed in tandem, parallel with the long axis of the ascus, the chromosomes having contracted rapidly in prophase. In telophase II the spindles elongate much more than in telophase I. The third division follows rapidly, the spindles being orientated obliquely across the ascus. Centrioles first become clearly visible at interphase II. They stain with acetocarmine and are flexible, equilaterally-triangular plates, each 2·5 μm long and 0·5 μm thick. The spindle fibres are attached to one face of the plate. The enormous increase in size of the centrioles from Metaphase I to interphase II

Table 13.2. The correlation between ascus size and stage of meiosis in *Neurospora crassa.* (From Singleton, 1953)

Ascus length (μm)	Ascus diameter (μm)	Stage of division
30	5	Synapsis
60	9	Chromosome elongation
75–100	10–12	Full pachytene
150–170	15–18	Diplotene to division III
170	10–15	Uninucleate spores

as shown by their increased visibility seems to be associated with their function in spore delimitation which now occurs. A group of fibrils radiates out from each centriole and surrounds a fusiform mass of cytoplasm including both a nucleus and the centriole. A hyaline wall develops around the delimited mass and the nucleus undergoes a final division to produce a binucleate spore. Divisions III and IV are in all respects apparently normal mitoses. After further differentiation of the wall it becomes pigmented and eventually eight dark spores lie freely in the ascus.

WESTERGAARD (1965; ROSSEN and WESTERGAARD, 1966; WESTERGAARD and VON WETTSTEIN, 1966) has laid considerable stress on the special features of meiosis in *Neurospora* and has extended observations to the Discomycete *Neottiella* (*Humaria*) *rutilans*. The characteristic features of meiosis of the '*Neurospora*'-type are that nuclear fusion is followed immediately by meiosis so that leptotene does not occur and the chromosomes pair in a highly contracted condition, replication having occurred before nuclear fusion. Subsequently the chromosomes become less contracted (uncoil) and reach their maximum extension at pachytene. Diplotene onwards is similar to that in other organisms showing the '*Lilium*'-type of meiosis, the predominant pattern 'in haplontic and haplodiplontic species in which caryogamy and meiosis are separated by diploid mitotic divisions' (WESTERGAARD, 1965).

There is some support for this view for in other Ascomycetes, e.g. *Sordaria fimicola* (CARR and OLIVE, 1958), synapsis occurs when the chromosomes are even more condensed than in *Neurospora* and this is also true of *Hypomyces*, *Venturia* and *Cochliobolus* (EL-ANI, 1956; DAY *et al.*, 1956; HRUSHOVETZ, 1956). Another character widely shown by these fungi and shared with *Neurospora* is the fuzziness of their diplotene chromosomes. This does not seem to occur in *Neottiella*.

Although data concerning the time of replication of the chromosomes prior to nuclear fusion is only available for *Neottiella*, Westergaard believes that a number of observations made in support of the hypothesis of brachymeiosis bears upon this problem. Brachymeiosis was first proposed by HARPER (1895) to support his observation on downy mildews (Erysiphales) where a sequence of two nuclear fusions, the first in the ascogenous hyphae, the second in the ascus, was supposed to be followed by a double reduction division in the ascus. Vigorous support was given by H. C. I. Gwynne-Vaughan (née Fraser) and her school. However, since 1945 modern cytological methods applied to those species where brachymeiosis had been claimed, e.g. *Pyronema* (HIRSCH, 1950) have provided no evidence for the double fusion and reduction processes and the hypothesis is largely of historical interest. However, Westergaard suggests that since chromosomal replication occurs in prefusion nuclei in ascogenous hyphae the double number of *chromosomes* recorded by Gwynne-Vaughan might in fact represent the *chromatid numbers* which would, of course, be twice those of the non-replicated nuclei in the ascogenous hyphae.

WESTERGAARD and VON WETTSTEIN (1966) have also been able to show that synaptinemal complexes are developed between condensed chromosomes that are pairing in *Neottiella*. They are similar to those which have been described between paired homologous chromosomes in '*Lilium*'-type meiosis. Thus pairing, whether of condensed or highly elongated chromosomes appears to be associated, ultimately, with the development of these characteristic structures (Plate IX, Figs. 2, 3; Fig. 13.7). They also occur in *Coprinus* (LU, 1967).

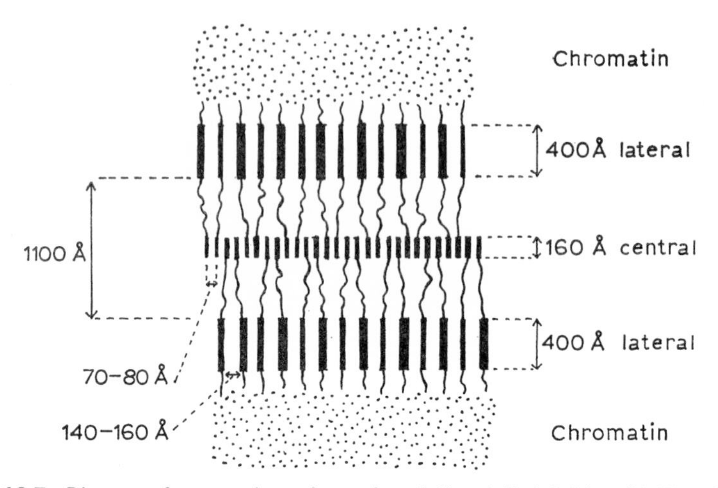

Fig. 13.7. Diagram of a synaptinemal complex of *Neottiella* (cf. Plate IX, Figs. 2 and 3). It is formed between bivalents in the meiotic nucleus. The arrangement of connecting fibrils between the lateral and central components is quite hypothetical. (From Westergaard and von Wettstein)

A number of other variations occur in Ascomycetes but, in all of them, intranuclear spindles occur and the centrioles play an important role in spore delimitation as was first described so accurately by HARPER (1905) for the downy mildews. The extent and consequences of variations in meiosis are usually minor. Aneuploidy, pachytene irregularity, ring chromosomes, stickiness and bridges have been recorded in *Trichometasphaerica turcica* (KNOX-DAVIES and DICKSON, 1960) and the consequences of translocations in *Neurospora* (BARRY, in FINCHAM and DAY, 1965).

The situation at meiosis in yeast is still unclear. Chromosome-like filaments have been seen to arise from the Feulgen-positive body of the nucleus and intranuclear fibres occur as in somatic division (ROBINOW, unpublished). Other details have not yet been worked out. Tetraploid yeasts are known and their genetical behaviour suggests that tetraploid meiosis, similar to that in other organisms, may occur but nothing is known of their cytology. In this connexion *Aspergillus nidulans* is of interest because the cytology of both haploids and diploids has been studied (ELLIOTT, 1960). The chromosomes are exceedingly minute. Early prophase is too crowded for much detail to be seen but the post-pachytene behaviour is unusual. Although fully paired at pachytene the halves of each bivalent become widely separated except at one end, giving a V-shaped appearance, by early diakinesis. As this phase proceeds the halves come together again and the bivalents appear as two parallel rods with no sign of chiasmata. A spindle develops although no centrioles have been seen (orcein was employed as stain) and a metaphase configuration develops. The haploids complete this division and divisions II and III although the small size of the chromosomes precludes detailed observation. In the diploids, however, most asci fail at anaphase I, the chromosomes clump and only a few asci develop. In these anything up to 16 spores may develop whose precise origin is obscure. However, of 567 ascospores isolated, 92 germinated and, of these, 91 were haploid. This suggests that diploid strains develop apogamously and this is borne out by the fact that early stages of division I are identical with those in the haploid and no multivalents, for example, were ever seen. Prior to ascus formation normal crozier-hyphae do not develop and the hyphae giving rise to asci are unusually basiphilic and may be one- to four-nucleate. The situation here is evidently different from that suggested by genetic data from polyploid yeast.

A more striking variation is shown by *Neurospora tetrasperma* (DODGE, 1927) and some other Ascomycetes in respect of spindle orientation. In this fungus the spindles of division II overlap so that spore delimitation results in two nuclei being incorporated in each spore and only four are formed. This has important genetic consequences (see Chapter 15, pp. 418–419).

Investigations of meiosis in other groups of fungi using modern methods are few. This reflects the great difficulty met in attempting to 'squash' basidia, oogonia, zygotes and other, similar structures. Amongst the most recent studies are those of Lu (LU and BRODIE, 1962, 1964; LU, 1964a, b) on species of the bird's-nest fungi of the genus *Cyathus*. In these fungi meiosis takes a normal course but, unlike the Ascomycetes, synapsis occurs at a stage when the chromosomes are fully elongated. Centrioles and spindles are clearly formed at metaphase I. As with most Basidiomycetes, the nuclei migrate into the basidiospores after division II where one or more somatic divisions may then occur. Lu was able to demonstrate multivalent formation and secondary pairing in one species, *C. stercoreus*, and this is suggestive evidence for the occurrence of autopolyploidy, although allopolyploidy cannot be ruled out. Toadstools have been studied cytologically since at least 1884 (STRASBURGER) but much of the work is now thought to be technically unreliable. An exception is probably that of WAKAYAMA (1930) and it is of interest that he figured a great many prophase situations in which the chromosomes appeared to be polarized to form a 'bouquet' stage. It is not yet known how characteristic this is of these fungi.

Cytological aberrations, translocations, fragments and bridges are not uncommon, e.g. *Psalliota hortensis* (EVANS, 1959), *Puccinia kraussiana* (SANSOME, 1959), *Collybia velutipes* (BURNETT, unpublished).

As with the Ascomycetes, some variability is shown in spindle arrangements. JUEL (1898) recognized that division II spindles could be parallel, the stichobasidial type, or crossed, the chiastobasidial type, and thought that these configurations were of taxonomic significance. EVANS (1959) has made an especial study of spindle orientation in the cultivated mushroom, *Psalliota hortensis* var. *bisporus.* In some basidia the spindles of division II are either (a) parallel to each other and the long axis of the basidium, (b) parallel but at right angles to the long axis of the basidium, (c) at right angles to each other and to the long axis of the basidium, or (d) irregular (Fig. 13.8). He

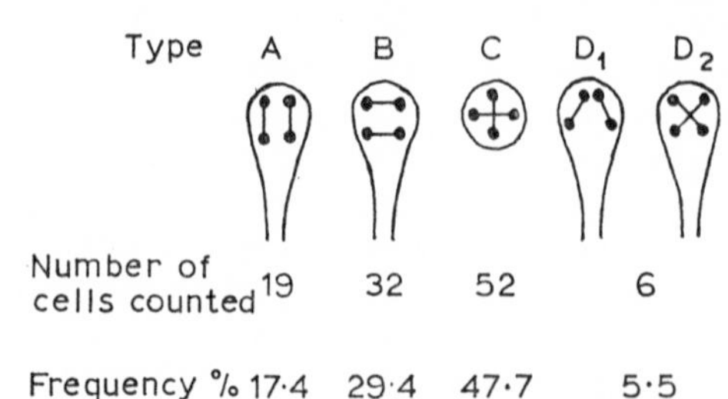

Fig. 13.8. Diagrams to illustrate the frequencies of different spindle arrangements at the second meiotic division of *Psalliota hortensis* var. *bisporus.* (From Evans)

has discussed the possible genetical consequences of these arrangements (see p. 433). It now seems improbable that spindle arrangements are of great significance for classification. A different type of variation, having genetical results of some importance, is the precocious occurrence of division III in the basidium to give an eight-nucleate structure (BURNETT and BOULTER, 1963 and see p. 434). This is not uncommon in the white-spored agarics (KÜHNER, 1945).

The great dearth of information about meiosis in Basidiomycetes is equally true of the Phycomycetes. Virtually nothing is known about meiosis in the Chytridiomycetes save for the brilliant work of WILSON (1952) on *Allomyces* which represents the first detailed study of any Phycomycete. As in other fungi, meiosis is an intranuclear process and involves a spindle with fibres, but centrioles and asters have not often been seen. This is of particular interest since ROBINOW (unpublished) has recently demonstrated an intranuclear spindle diverging from a pair of centrioles during somatic nuclear division in *A. macrogynus*. Another extraordinary feature is the arrest of meiosis in late prophase I until the resting sporangia, in which the process occurs, have germinated. This may be a period of months or even years.

Renewed interest in meiosis in the Oomycetes has come from SANSOME'S (1961, 1963) claim that meiosis takes place in the gametangia so that the vegetative mycelium is, unusually for fungi, diploid. Unfortunately, apart from chromosome number, the details of meiosis are not sufficiently documented for a full discussion. Another recent discovery is the demonstration of an intranuclear spindle in *Albugo bliti* (BOWEN, BERLIN and PEYTON, unpublished) by electronmicroscopy. It will be recalled that the best evidence for a centriole is available from electronmicroscopy in another species of *Albugo* (see p. 33). There thus seems to be an opportunity of correlating behaviour in meiosis with that in mitosis in this group of fungi.

The position in the other groups of Phycomycetes is obscure. Plasmodiophoromycetes are said to have a normal meiosis (e.g. WEBB, 1935) but the situation is obscure and has been discussed earlier in this chapter. An outstanding gap is the inadequate documentation of the Zygomycetes, especially *Mucor* and its allies. Although there are some comparatively recent papers on this group (CUTTER, 1942; SJÖWALL, 1945) in which modern methods were used, the descriptions do not carry conviction. It is unfortunately true that there is not yet adequate cytological or genetical evidence for the occurrence of nuclear fusion or meiosis in this group (see also p. 413). In view of the importance of the *Mucor*-type of somatic division, discussed earlier, a clarification of the meiotic situation is most desirable.

Meiosis in fungi does not appear to pose the same problems, at the moment, as somatic division but there is a woeful lack of knowledge even in the most fully studied group, the Ascomycetes. Now that it is clear that modern methods and microscopes can be used successfully it is to be hoped that there will be a rapid increase of knowledge concerning meiosis as well as somatic division.

Chapter 14

Heteroplasmons, Heterokaryons and the Parasexual Cycle

BLAKESLEE (1906) was the first investigator to report on the germination of zygotes of the Mucoraceae obtained under controlled conditions. He obtained a number of 'neutral' mycelia from the germination of zygotes of *Phycomyces blakesleeanus*. They grew abnormally, tended to produce zygophore-like branches (pseudophores) and were bright yellow as a consequence of the abundant carotene in their mycelia. It was supposed that such mycelia carried nuclei of different mating types, + and −, in their hyphae since such spores could occasionally be isolated from sporangia and this was proved experimentally by BURGEFF (1912, 1914) who had found the same phenomenon in the same species. He removed the tip of a hypha of + mating type and inserted it into the cut end of a hypha of − mating type, so that cytoplasm and nuclei intermingled. The resultant mycelium, derived from this 'myxochimaera', grew in a manner similar to those of Blakeslee's 'neutral' cultures. Burgeff described such mycelia as heterokaryotic and this is the first published example of an heterokaryon, i.e. the co-existence of genetically different nuclei in the same cytoplasm. Heterokaryosis, as a natural phenomenon, is virtually confined to the fungi. Its significance for the biology of the fungi was not appreciated until pathologists realized that it could account for some of the natural variation shown by phytopathogenic fungi (e.g. BRIERLEY, 1929). Particular attention was drawn to this possibility by the study of *Botrytis cinerea* made by HANSEN and SMITH (1932). They showed that if single spores of *B. cinerea* were isolated and cultured, they gave rise to colonies with various attributes but that, in general, they could be grouped into three main types, *a*, *b* and *x*. Types *a* and *b* were different but constant in their cultural and morphological characteristics. Type *x*, however, was inconstant in the sense that on agar slopes it gave rise to sectors of types *a* and *b* in addition to areas sharing its characteristic, irregular morphology; this type of segregation was also shown by conidia isolated from type *x* cultures. They obtained strains of types *a* and *b* that were constant for several sub-cultures and then grew them together. As a consequence of hyphal fusions which they saw and in which they detected nuclei, presumably migrating, they reconstituted a mycelium of type *x* (Fig. 14.1). In 1938, HANSEN reported

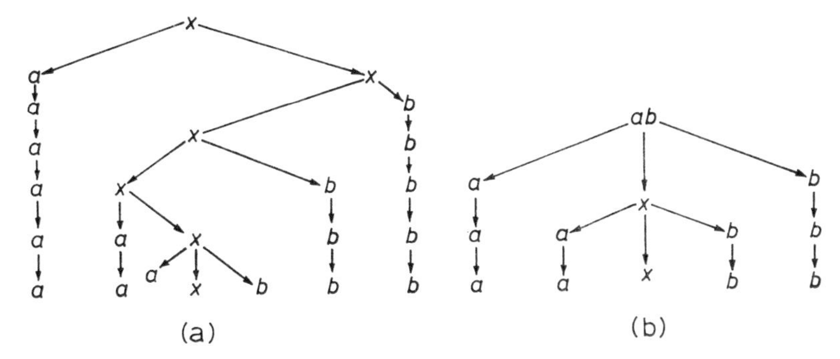

Fig. 14.1. Diagram to illustrate (a) segregation of strains *a* and *b* from strain *x* and (b) production of strain *x* from a mixture of strains *a* and *b* in *Botrytis cinerea.* (From Hansen and Smith)

similar studies on 35 genera of Fungi Imperfecti, in which 32 showed this phenomenon; it, therefore, seemed to be widely distributed. The assumption was made that two distinct types of nuclei can co-exist in *B. cinerea* hyphae and that when spores which are multinucleate are abstricted, differing proportions of these nuclei may be incorporated. Sometimes by chance all nuclei will be the same (types *a* and *b*), but more usually they will be different (type *x*). The products of spore germination will be correspondingly stable or unstable. As JINKS (1959b) pointed out, another explanation is possible, since the origins of the nuclei which entered the spores was not determined; this is the somatic segregation of different cytoplasmic determinants (plasmons) from an heteroplasmon. Such segregations have been demonstrated by Jinks and his co-workers and by others and will be described later. It is unfortunate that there have not been many critical studies of heterokaryons or heteroplasmons in natural populations of fungi; most of the studies from which knowledge of these phenomena is derived have used synthesized mutant combinations.

A heterokaryon or heteroplasmon can arise in one of two ways; either mutation occurs within an originally, genetically homogeneous mycelium and the mutant and original nuclei or plasmons co-exist, or genetically distinct mycelia fuse, the nuclei and cytoplasms intermingle and co-exist in further hyphae. Propagation of this condition can occur through spore formation which involves incorporation of both kinds of nuclei or plasmons. Thus attention may be directed to the frequency with which these heterogeneous systems originate, whether by mutation or hyphal fusion; the number and extent of intermingling of the distinct components; the stability of the mixed system, its maintenance under different conditions and the ways in which it differs (if at all) from its components and, finally, the causes, frequency and consequences of its breakdown or segregation.

Many studies have been made on the assumption that heterokaryosis is the basis of the system investigated. It is convenient, therefore, to consider first the kinds of tests which should be applied to differentiate a heterokaryon from a heteroplasmon and also the properties of the latter. Heterokaryosis will be considered later with the parasexual cycle with which it is associated.

HETEROPLASMONS

In heteroplasmons the basis of distinction resides in the cytoplasm. It may be a difference in the cell organelles, e.g. the mitochondria which differ in their properties in the *poky* strains of *Neurospora* from those in normal strains, or in some undefined but probably particulate entity, e.g. growth rate in *Aspergillus* spp. Cytoplasmic differences can be detected in sexually reproducing organisms by making reciprocal crosses, the progeny often being different, or by detecting non-mendelian segregations in the progeny, e.g. *poky* in *Neurospora* and *dwarf* growth in *Coprinus lagopus*, respectively (MITCHELL and MITCHELL, 1952; QUINTANILHA and BALLE, 1940). In heterokaryons cytoplasmic differences can be detected by having a marker gene in each component nucleus which can be followed in any segregation; any dissociation or reassociation of feature originally associated with either homokaryotic nucleus is evidence of a cytoplasmic difference (Table 14.1). A simple example provided by Jinks involves sexual and non-sexual strains of *Aspergillus nidulans* and the nuclear marker-genes *y* and *Y* for yellow and green conidia. The test is made as follows:

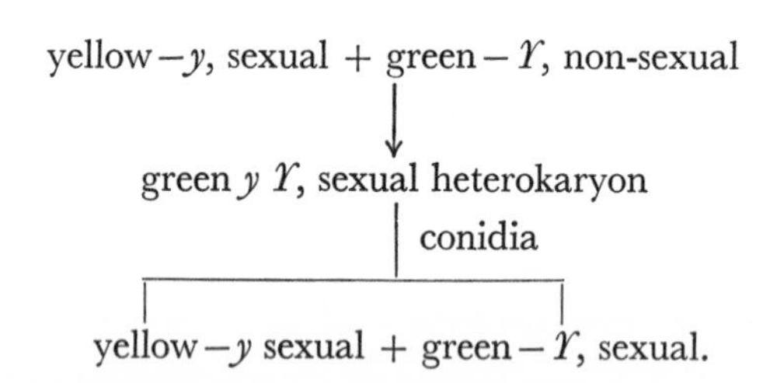

The colonies derived from the conidia are homokaryotic, as demonstrated by their colour, but the reaction of the *Y*-strain has evidently been altered as a consequence of its association with the *y*-strain in the heterokaryon. This alteration must lie in the cytoplasm (JINKS, 1954). The more fully the nucleus is genetically marked, the greater the certainty with which any manifestation of a cytoplasmic factor can be detected and assessed.

Genetically altered, non-nuclear factors may arise spontaneously or can be

Table 14.1. Separation of differences of nuclear and cytoplasmic origin. (From Jinks, 1958)

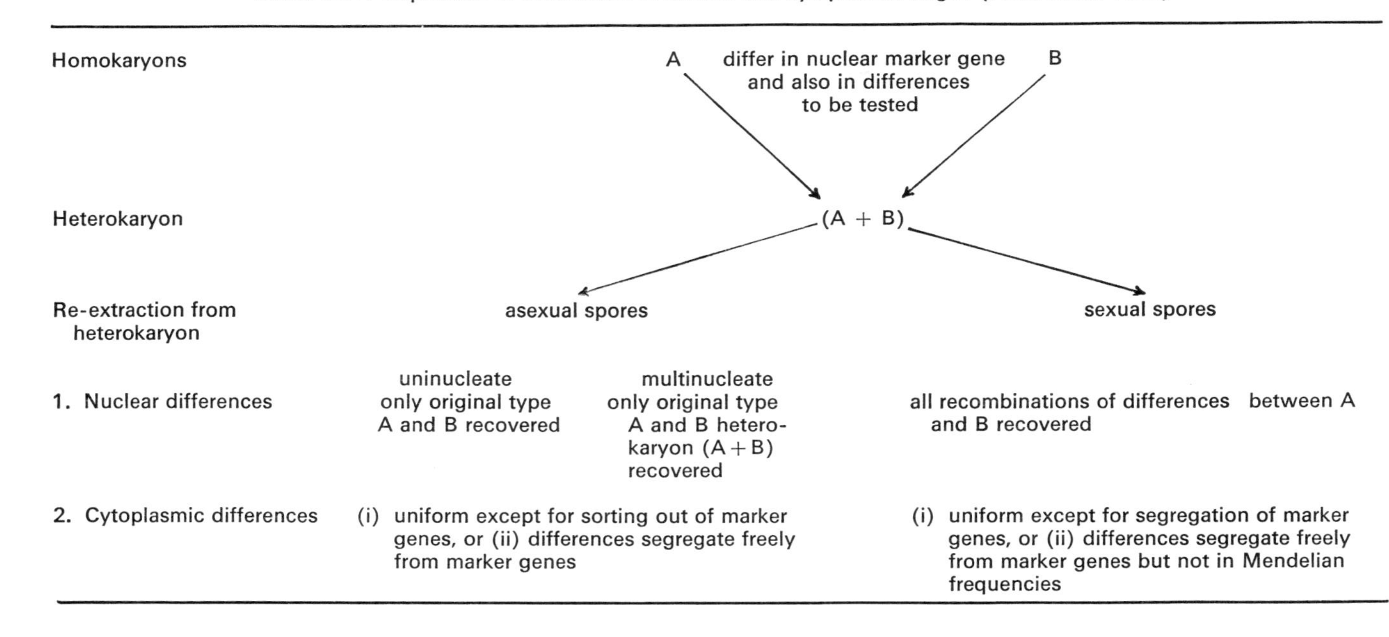

Homokaryons	A	differ in nuclear marker gene and also in differences to be tested	B
Heterokaryon		(A + B)	
Re-extraction from heterokaryon	asexual spores		sexual spores
	uninucleate	multinucleate	
1. Nuclear differences	only original type A and B recovered	only original type A and B hetero-karyon (A+B) recovered	all recombinations of differences between A and B recovered
2. Cytoplasmic differences	(i) uniform except for sorting out of marker genes, or (ii) differences segregate freely from marker genes		(i) uniform except for segregation of marker genes, or (ii) differences segregate freely from marker genes but not in Mendelian frequencies

induced. An excellent example is the *petite* character in yeast. In Baker's yeast a small proportion shows this character spontaneously, but it can readily be induced in all cells by treatment with euflavine (EPHRUSSI, 1953). The character is manifest in plate culture as colonies one half to one third the size of the normal. The individual cells were found to be deficient in a number of enzymes normally associated with mitochondria and the electron acceptor chain (SLONIMSKI, 1953). Appropriate crosses have shown that the nuclear genes are unaltered (although there is, in addition, a similar but distinct condition due to mutation of a nuclear gene). Other cytoplasmically altered conditions have been isolated from old cultures as '*senescent*' in *Podospora anserina* (RIZET, 1957). In this condition the cytoplasmic factors accumulate and eventually become lethal. *Senescent* and a number of other cytoplasmic characters have been shown to be transmitted both by spores and via hyphal fusions. This was elegantly shown by an experiment of MARCOU and SCHECROUN (1959). An hyphal fusion was observed between a dominant colourless (*I*) *non-senescent* strain and a wild type (*i*) *senescent* strain. Fragments of mycelium were cut off at different distances a certain time after the hyphal fusion had taken place; each was sub-divided and cultured to test its colour reaction and whether or not it showed the *senescent* character. The results clearly demonstrated transmission of *senescent* and it was shown that the expression of the character could occur even in very small fragments provided they were taken from near the point of hyphal fusion. If they were taken further away, small sub-cultures could also be isolated lacking *senescent*. This suggested a particulate basis for the character (Fig. 14.2).

There is a good deal of evidence from other fungi that the basis of different cytoplasmic characters is not only particulate when the character is associated with a cell organelle, such as a mitochondrion, but also when it is described as being due to a mutant plasmon. It is now known that mitochondria in other organisms possess DNA in a circular molecular form (NASS, 1966). It is not known if mitochondrial DNA in *Neurospora* exists in this form (LUCK and REICH, 1964). If this proves to be universally true for mitochondria, it would provide a basis for the genetic continuity of cytoplasmic characters associated with this organelle analogous to that known for bacteria and bacteriophages, although the amount of DNA involved is far smaller. Mitochondria in *Neurospora* are believed to arise from pre-existing mitochondria for, when pulse-labelled with choline just prior to a great increase in numbers, all the resultant mitochondria were found to be equally radioactive. Had any mitochondria arisen *de novo* then the radioactivity would not be expected to be equally distributed. The structural or chemical cause for other plasmon characters is unknown but a viral basis is possible.

The stability of plasmon characters is very variable and there is evidence that one character can suppress the expression of another. One of the most striking examples of a change in a cytoplasmic character is that associated with the loss of ability to produce perithecia in *Aspergillus nidulans*. If successive sub-cultures are made by means of hyphal tips or conidia alone this ability of the sub-cultures slowly but steadily declines. However, if at any

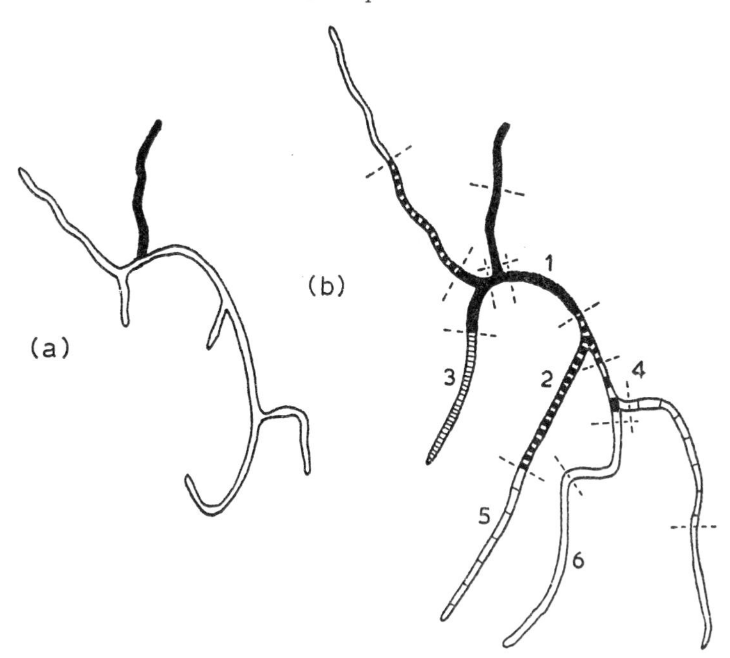

Fig. 14.2. Fusion between a recessive coloured (i) + strain and a coloured (I) − strain of *Podospora anserina*; (a) at the moment of fusion; and (b) later at the time of cutting up the mycelium into six classes of fragments prior to sub-culturing. The results were:

Fragment No.	1	2	3	4	5	6
Growth of sub-culture (cm)	0	<1	4	8	12	N
(3 replicates; N = normal)	0	<1	4	8	12	N
	0	<1	4	8	12	N

(After Marcou and Schecroun)

stage before the total loss of perithecial production, a sub-culture is initiated from an ascospore the full ability to produce perithecia can be restored (Fig. 14.3 and MATHER and JINKS, 1958). This is analogous to the consequences of crossing haploid *petite* and normal yeast cells and isolating wholly normal ascospores from the progeny. It suggests that there is either some form of major re-patterning of the cell at sexual reproduction or a highly selective process operating in the cytoplasm. This process is not, however, confined to the reproductive cells for JINKS (1956a) has been able to show differences in rates of growth, due to cytoplasmic factors, between sub-cultures of *A. nidulans* or *A. glaucus* maintained by ascospores, conidia and hyphal-tip transfers.

There is now little doubt that cytoplasmic variability can be demonstrated in a number of fungi. It is of great importance that heteroplasmons have been shown to be present in two out of four isolations from nature, of *Penicillium* spp. of the Assymetrica group (JINKS, 1959) and probably in monokaryotic progeny derived from wild dikaryons of the agaric *Collybia velutipes* (CROFT and SIMCHEN, 1965). The evidence for the spontaneous origin of cytoplasmic

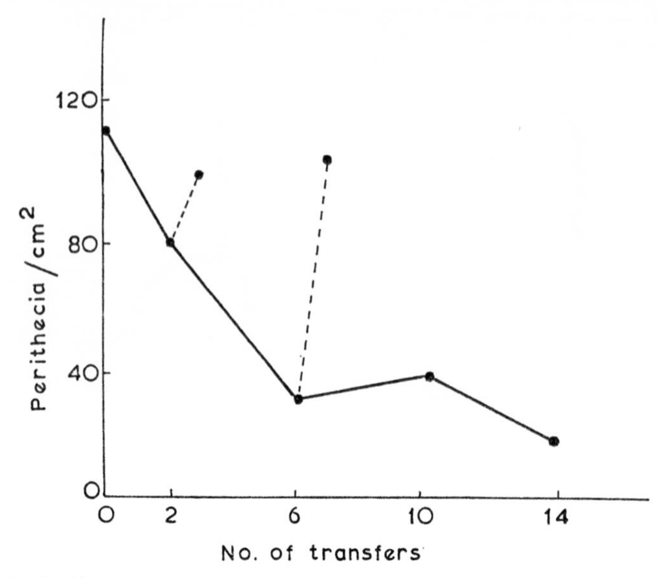

Fig. 14.3. Perithecial production by successive transfers of conidia only in *Aspergillus nidulans*. Dotted lines show effect of an ascospore transfer. (From Mather and Jinks)

variants in culture, their existence in natural isolates and the fact that reciprocal crosses of *Puccinia graminis* produce different progeny (JOHNSON, 1946), suggests that heteroplasmons are likely to occur quite widely in nature. This implies that observations believed to demonstrate heterokaryosis should always be tested by means of Jink's 'heterokaryon test' (described earlier) as a check for the existence of an heteroplasmon.

HETEROKARYONS

There is no doubt that heterokaryons do occur in nature and can be induced and propagated in culture. The original situation described by Burgeff in *Phycomyces* is a special one in so far as the nature of the genetic differences between the nuclei lies in their different mating-type factors. These situations will be discussed in more detail in the next chapter, in connexion with mating systems. BURGEFF (1914) also described the consequences of isolating sporangiospores from normal cultures, growing them on and then repeating the process. In this way he was able to isolate homogeneous forms which differed morphologically from the wild type, e.g. the form *piloboloides*, which developed a distinct sub-sporangial swelling, reminiscent of that in *Pilobolus*. During this selection procedure he obtained forms intermediate in their morphology between the wild type and the mutant forms. He suggested that the wild type was in fact heterokaryotic and that his procedure had isolated homokaryotic mutant forms. The intermediate mycelia represented those with different ratios of the nuclear types, the balance between the nuclei determining the morphology. Since hyphal fusions are rare, or non-existent, in Mucoraceae (except perhaps in *Mortierella*) he could not test his hypothesis by re-synthesizing an heterokaryon from his supposed homokaryons except by using his grafting technique. This he only reported on in relation to plus/minus heterokaryons. He was, however, able to show that his mutant forms retained their identity and segregated in crosses with each other and with the wild type (BURGEFF, 1915, 1928). In 1935, KÖHLER, a pupil of Burgeff, carried out a similar analysis of *Mucor mucedo* and provided evidence that the original strain '40(+)' was heterokaryotic for a number of nuclear types, recognized by the morphology of the homokaryotic mycelia obtained by selection.

These somewhat neglected studies are of great importance because they provide clear circumstantial evidence for the origin of heterokaryotic mycelia in nature through mutation, since hyphal fusion is excluded. It is desirable that this point of view should be tested by attempting to force heterokaryons in mucoraceous fungi between strains with more clearly circumscribed genetic markers. There is one report that this has been achieved with *Rhizopus javanicus* (MINAMI and IKEDA, 1962) but this needs to be repeated and confirmed. It is clear from the work of Burgeff and Köhler that the multinucleate condition of the sporangiospores of *Phycomyces* and *M. mucedo* does not arise from the divisions of one original spore nucleus; otherwise each sporangiospore would give rise to an homokaryotic mycelium. Multinucleate spores may not always arise from incorporation of several nuclei and this may be true of the 'microsporangia' of *Phycomyces*, which have small spores with fewer nuclei. Such a dimorphism could provide a wide range of heterokaryotic and homokaryotic propagules on which natural selection could operate.

In Fungi Imperfecti, Ascomycetes and Basidiomycetes hyphal fusions are usually possible and, while mutation or, possibly, recombination, initiates the genetically distinct nuclei it is as a result of fusions that new heterokaryons arise quickly and readily. The frequency with which hyphal fusions occur has already been discussed (Chapter 4, pp. 70–76); it is subject both to genetic and to environmental control. However, hyphal fusion is only effective when it is followed by nuclear migration, at least into the fusion cell. This phenomenon is a widespread and characteristic feature of these fungi and is worth more detailed investigation.

NUCLEAR MIGRATION

BULLER (1931) first described this phenomenon in *Coprinus* spp., using nuclei of different mating types as his markers. He showed that, in compatible matings, nuclei can apparently migrate from a small mycelium into a large one, cross its diameter and be detected on the opposite side. The presence of both nuclei in newly formed cells could readily be detected in his fungi by the presence of clamp connexions.

Since then the phenomenon has been studied in several other Homobasidiomycetes and in the Ascomycetes *Gelasinospora tetrasperma*, *Neurospora tetrasperma* and *Ascobolus stercorarius* (Table 14.2). The technique has been to

Table 14.2. Relative rates of nuclear migration (NM) and hyphal tip growth (TG). (From Snider, 1965)

Fungus	Temperature (°C)	Rates (mm/hr) NM	Rates (mm/hr) TG	Relative Rate (NM/TG)	References
Basidiomycetes					
Coprinus lagopus	22	1·51	0·15	10·0	Buller, 1931
	28	1·0	0·25	4·0	Swiezynski and Day, 1960
Coprinus macrorhizus	30	3·2	0·15	21·3	Kimura, 1952
Coprinus radiatus	23	0·58	0·09	6·4	Prévost, 1962
Cyathus stercoreus	R.T.	0·37	0·16	2·2	Fulton, 1950
Schizophyllum commune	22	0·5[2]	0·13	3·8	Snider and Raper, 1958
	32	3·0[2]	0·22	13·6	Snider and Raper, 1958
Ascomycetes					
Gelasinospora tetrasperma	Not given	4·0	2·00	2·0	Dowding and Buller, 1940
	Not given	10·5	*c.* 3·00	3·5	Dowding and Bakerspigel, 1954
	30–33	(40·0)	(*c.* 3·00)	(13·3)[3]	Dowding, 1958
Ascobolus stercorarius	22	(20·0)	(*c.* 1·50)	(13·3)	Snider, unpublished

employ nuclear 'markers' recognizable either by a manifestation of the mating reaction, such as clamp formation in Basidiomycetes, or perithecium formation in Ascomycetes, or by a genetically determined biochemical lesion.

A primary problem is the frequency with which nuclei migrate. The only experimental data are those of SNIDER (1965, reported more fully in RAPER, 1966), who used non-allelic, biochemically deficient mutants as markers in *Schizophyllum commune*. A 2m confrontation was made between the two mutant mycelia, which were growing towards each other by laying an agar strip of one culture on the other. At various time intervals the strip was removed and the resident mycelium divided into 2 mm strips, perpendicular to the axis of confrontation. If fusion had been followed by nuclear migration in a 2 mm strip, the fungal cells should contain both nuclear types which could complement each other nutritionally and would thus be able to grow on minimal medium. About 10^5 hyphal tips were opposed in the 2m confrontation, some 400–500/cm of advancing hyphal margin. During the first 24 hr the frequency of migration was about 5/cm of advancing hyphal margin, or 3×10^{-5}–2×10^{-3} per hyphal pair. This is a very low incidence, but it is not known how typical or atypical this may be. This figure also represents a minimal value because in many cases the cells of the heterokaryon contain more than one nucleus, all of which might have migrated.

However, in this experimental design they would be recorded as one migrant nucleus. It is a notable feature of the monokaryons of many Basidiomycetes and it is not uncommon to find in several Ascomycetes and Fungi Imperfecti, e.g. *Aspergillus*, that there are many nuclei near the apical region of a hypha (YEN, 1950; ROBINOW in DUBOS, 1947). In such a condition, a hyphal fusion could be followed by migration of several nuclei, possibly tens of nuclei, through a favourable region for transfer.

Migration proceeds at remarkably rapid rates (Table 14.2) and has a high Q_{10}. It may or may not be affected by light and it does not seem to be affected by an imposed electric field aligned along, or across the direction of migration. SNIDER and RAPER (1958) have measured migration in *S. commune* either after confrontation at mycelial margins (a radiate mycelium) or after plating a mycelial macerate of hyphae containing different nuclear types (a reticulate mycelium). Sampling was made by transferring plugs of mycelium from the points of intersection of a superimposed 1 cm grid. Each plug contained *c.* 10^5 cells and, at first, in radiate mycelia, there were many plugs, discontinuously distributed, lacking migrant nuclei. Raper suggested that this indicated rapid migration of small numbers of nuclei to hyphal tips leaving only daughter nuclei behind occasionally. In reticulate mycelia the advancing front of migrating nuclei assumes a radially symmetrical pattern. Studies of the distribution of patterns of cells with clamp connexions in Basidiomycetes after compatible, radiate confrontations shows a wide pattern of behaviour and this is also found in *Neurospora* and *Gelasinospora*, using perithecial patterns as a guide to migration routes (Fig. 14.4). A particularly interesting pattern is the unilateral one. The determination of which mycelium is to be the donor and which the acceptor may be genetic or environmental. For example, in many cases of confrontations between biochemically normal mycelia and morphological mutants or biochemically deficient mutants, migration is from the former to the latter (RAPER and SAN ANTONIO, 1954). On the other hand, in *Gelasinospora tetrasperma* it was shown that migration took place from a darkened mycelium to an unshaded one (DOWDING and BULLER, 1940).

The mechanism of nuclear migration is not known. It is often assumed that the nuclei are carried passively by cytoplasmic streaming. This could account for their transport to hyphal tips in radiate mycelia but it is difficult to see how this would account for the behaviour observed in reticulate mycelia or in unilateral nuclear migration. Moreover, as Buller originally noted, it is not uncommon to see nuclei stationary while the cytoplasm flows past and the reverse can sometimes appear to be the case. DOWDING and BAKERSPIGEL (1954) have described changes in shape of migrating nuclei and self-propulsion cannot be entirely ruled out, however improbable it may seem hydrodynamically. Any mechanism must also account for the somewhat erratic routes followed during migration. The cytoplasmic flow hypothesis is not unreasonable here for it is a common experience to observe different rates of flow in adjacent hyphae, or even in different parts of the same hypha.

A further problem of great importance to the phenomenon of heterokaryosis is the passage of nuclei through septa. Where there are simple pores,

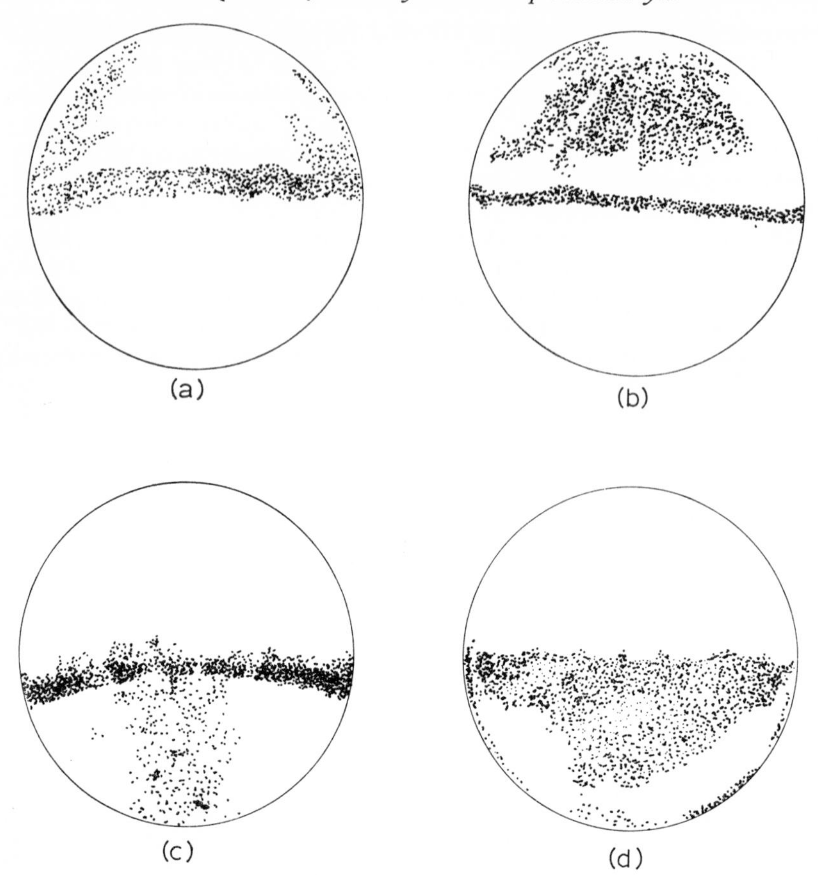

Fig. 14.4. Patterns of perithecia produced after apposition of + and − strains derived from *Neurospora tetrasperma*. (a) Perithecia mostly at junction, some peripheral unilateral nuclear migration; (b) similar to (a) but considerable uniform, unilateral nuclear migration; (c) similar to (a) but highly localized nuclear migration; (d) complete unilateral migration avoiding nuclear fusion at zone of contact. (After Dodge)

nuclei have been observed to pass through in vivo and electronmicrographs of *Neurospora* appear to show this happening (SHATKIN and TATUM, 1959, and Frontispiece). In Basidiomycetes with dolipore septa, however, there is controversy. GIRBARDT (1965) claims that it is not possible for a nucleus to pass through such a complex septal pore. He claims that the electronmicrograph of BRACKER and BUTLER (1964), showing a nucleus constricted in such a pore, is an artefact due to hydrostatic pressures, consequent on fixation, forcing the nucleus into this abnormal position. His criticisms are all the more cogent since GIESEY and DAY (1965) have shown the apparent breakdown of complex septal pores in *Coprinus lagopus* and the migration of nuclei through these secondarily—simple pores (Plate III, Figs. 4 and 5). On the other hand, SANFORD and SKOROPAD (1955) claim to have seen at least three nuclei migrate successively through a septum in *Rhizoctonia solani*, a fungus with a dolipore septum. Clearly, the question of how nuclei migrate through complex septal pores is not yet resolved.

Finally, there is the situation described by BUXTON (1954) in *Fusarium oxysporum* f. *gladioli*, where all the cells except the apical one are uninucleate and yet there is evidence of heterokaryosis. Although the septum possesses a pore in *Fusarium*, Buxton could provide no evidence of nuclear migration between cells. If this is indeed the case, then the formation and maintenance of an heterokaryon would be dependent upon the apical cells being multinucleate and becoming heterokaryotic via anastomoses. As growth proceeds uninucleate cells would be cut off behind the apex since the nuclei apparently do not migrate and so the mycelium would develop as a mosaic ot potentially, genetically-distinct uninucleate cells. Buxton has provided some evidence for

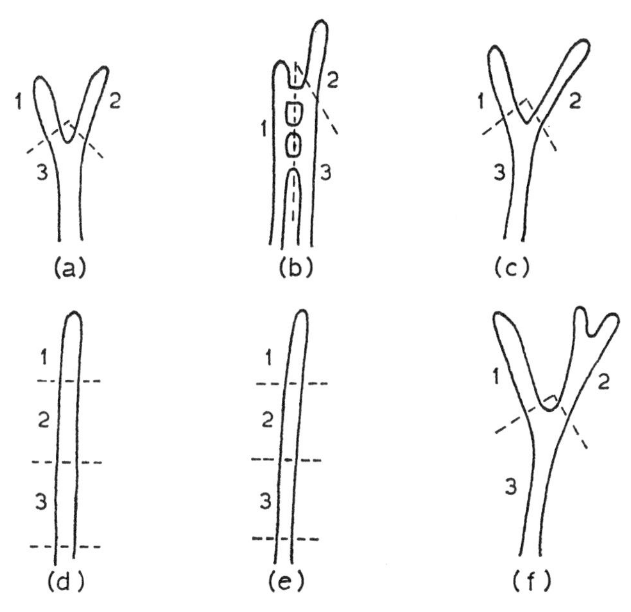

Fig. 14.5. Heterokaryosis in *Fusarium oxysporum*. (a)–(d), Heterokaryons between strains 81A and 90: (e), (f), heterokaryons between strains 81A and 81F. The dotted lines indicate excisions. Strains recovered from each sector are :

	(a)	(b)	(c)	(d)	(e)	(f)
1	81A+90	81A	90	81A	81A+81F	81A
2	81A+	90	81A	81A+90	81A	81A
3	90	no growth	no growth	81A+90	81A+81F	81A+81F

(From Buxton)

this situation but it needs a more thorough examination since this cytological difference between the apical cells and the rest is not uncommon (Fig. 14.5).

NUCLEAR RATIOS

In those fungi where migration is effectively unrestricted a problem arises as to how the ratio of different nuclear types is maintained or altered. This has received some experimental study. A major technical problem is that of determining the kinds and distributions of nuclei in the hyphae. Four techniques have been employed, of which the commonest is to sample the nuclear types in the spores. If these are uninucleate or, if multinucleate, have had a uninucleate origin, and if it be supposed that nuclei are abstricted or enter the spores at random, then the spore types will reflect the nuclear ratios in the cells from which the spores have developed. It should be noted that sporangiophores and conidiophores usually develop from behind the growing margin of a colony (cf. Fig. 4.11, p. 77), so that this technique almost certainly does not sample the nuclear population of growing hyphal tips. If the mycelium is grown in a Ryan tube, so that it is effectively growing forward as a narrow strip, it is possible to sample at different points along the tube and see if changes occur between the older and younger parts. Otherwise, no kind of check on possible differences between parts of different ages has been devised for this technique. A second technique, rarely employed, is to macerate the mycelium and plate out the fragments. In a non-sporulating mycelium, such as the monokaryons of many Basidiomycetes, this is the only general sampling method available, e.g. *Schizophyllum commune* (SNIDER, 1963). A refinement is to isolate individual hyphal tips, this at least ensures that the growing margin is sampled. However, it is technically restricting, the mycelium has to be grown on a medium where hyphae are well spaced out and the isolation procedure is relatively slow and laborious. Thus samples can only be obtained in certain circumstances and the sample size is likely to be small, compared with conidial sampling for example. Once hyphal tips have been obtained, they can be sub-cultured and conidial samplings made of the sub-cultures to assess the kinds of nuclei so isolated. In *Aspergillus nidulans*, a fourth method has been developed which could be used to determine ratios directly. Nuclei have been stained by the use of Acridine Orange, a vital stain that can be detected by its fluorescence in ultra-violet light. Estimates have been made of the relative numbers of diploid and haploid nuclei in a diploid/haploid heterokaryon (CLUTTERBUCK and ROPER, 1966). However, the significance of this technique lies in the possibility of exposing one component of an heterokaryon to the stain before hyphal fusions have occurred. A subsequent exposure of the heterokaryon to ultra-violet light should enable the ratio of fluorescent to non-fluorescent nuclei, i.e. of the two components, to be determined, at least at the hyphal tips.

In many fungi the first technique has to be modified to take account of the possibility of homokaryotic sectors being interspersed amongst the heterokaryotic hyphae and of the fact that many spores are multinucleate. The first point can be dealt with by sampling at random over the mycelial surface and comparing the frequency of homokaryotic and heterokaryotic heads, e.g. *Aspergillus* (RAPER and FENNELL, 1953). Estimates can be made of the frequency of homokaryotic and heterokaryotic heads, and multinucleated spores expected from various mycelial nuclear ratios. These can then be compared with the ratios observed (PROUT *et al.*, 1953).

Thus there are serious drawbacks to all the sampling methods; in the first two cases the apical cells cannot be sampled and assumptions have to be made concerning the entry of nuclei into spores; in the third case, the actual technique is not very easy, but, interpretation is less dependent upon assumptions and the fourth case has only limited applicability.

In *A. nidulans* both conidial sampling and the direct observation technique have been used. The latter method provided evidence for the occurrence of hyphae carrying widely different frequencies of the two nuclear types. PONTECORVO (1946, 1953) has concluded that the frequent sectoring of heterokaryons indicates that there is a wide range and mixture of homo- and heterokaryotic hyphae at the advancing margin of the colony. Conidial sampling indicates that homokaryotic areas are not uncommon in some heterokaryons but in others the ratios are remarkably constant. There is thus some evidence that conidial sampling may reflect the actual condition in apical cells. The early work of PONTECORVO (1946), using colour and biochemical marker genes in *Penicillium* and *Aspergillus*, drew great attention to the importance of the selective advantage of heterokaryons over both homokaryotic components if the former were to be maintained.

Clear evidence for such selective responses have been obtained in *Penicillium cyclopium* where a natural heterokaryon having two types of nuclei, 4A and 4B differing in their control over growth rate, was isolated from a chance contamination of a plate (JINKS, 1952a, b). The heterokaryon was stable on a Czapek agar/apple-pulp medium, the latter component being necessary for stability. The percentage composition was varied and the results showed clearly that the nuclear ratios varied with the composition of the medium (Fig. 14.6a, b; Table 14.3). The nuclear ratios were related to the comparative growth rates of the homokaryons on any medium of given composition and in a heterokaryon the 4A nuclei increased until the rate of growth of the 4A homokaryon equalled that of the heterokaryon, when the latter sectored to give homokaryotic sectors.

It is unfortunate that similar experiments with other fungi have used biochemical-mutant marker genes in the heterokaryons, since it renders difficult comparison with the results obtained from *P. cyclopium* where neither mutant nucleus had an absolute nutritional requirement. In *Neurospora* a range of behaviour is manifest. Some are similar to the *Penicillium* case described, in that growth rates were reduced when one kind of nucleus was so scarce that the amount of nutrilite it synthesized was sub-optimal. Above this level nuclear ratios were stable. Nevertheless, hyphal tip isolation gave rise to sub-cultures which varied greatly in growth rates and nuclear composition as determined by sampling conidia. This was taken to indicate that different nuclear ratios could occur in hyphal tips without affecting the growth of the

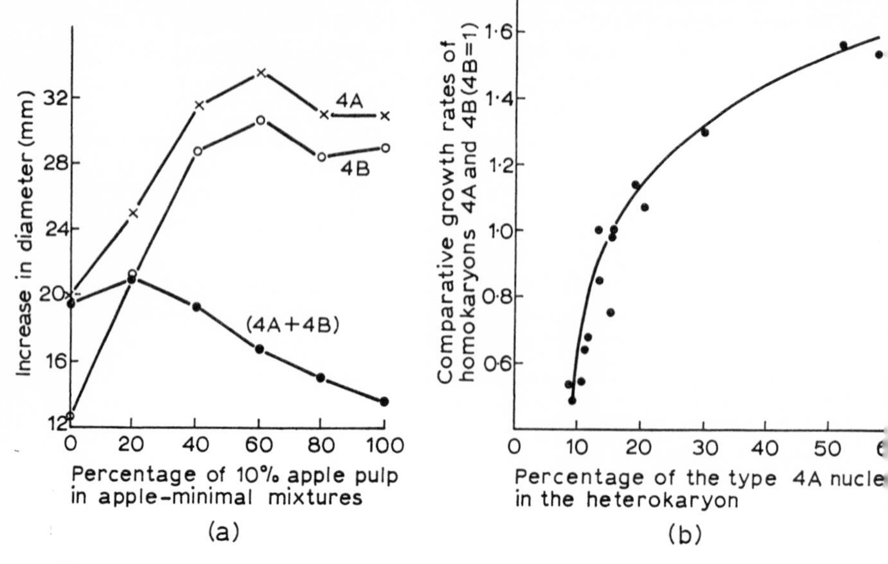

Fig. 14.6. (a) The effect of the 10% apple pulp: minimal medium on growth rates of monokaryons 4A, 4B and a heterokaryon between them over 2–8 days in *Penicillium cyclopium*; (b) correlation between comparative growth rate of homokaryons and estimated percentage of type 4A nuclei in heterokaryotic hyphae on medium as in (a). (After Jinks)

Table 14.3. The effect of various 10 % apple: minimal medium mixtures on the percentage of type 4A nuclei obtained by sampling heterokaryons of *Penicillium cyclopium*, and the comparative growth rates of the two homokaryons 4A and 4B. (From Jinks, 1952b)

Medium 10 % apple	Medium minimal	No. of colonies scored	% of 4A nuclei in heterokaryon	Comparative growth rates of homokaryons 4A and 4B (4B = 1)
100	0	324	8·55	0·47
80	20	381	7·75	0·53
60	40	167	11·11	0·54
40	60	830	12·66	0·67
20	80	831	13·51	1·00
0	100	632	51·81	1·56

mycelium as a whole. The rapid rate of cytoplasmic streaming and, by inference translocation, in *Neurospora* was believed to be the factor which stabilized the response of the mycelium as a whole, despite its nuclear heterogeneity (PITTENGER *et al.*, 1955; ATWOOD and PITTENGER, 1955). The occurrence of nuclear heterogeneity with physiological stability is also shown by *lys/pan* (lysine-less/pantothenic-less) and *lys/paba* (lysine-less/*p*-amino-benzoic-less) heterokaryons of *Neurospora* where there appeared to be an excess of homokaryotic hyphal tips and multinucleated spores compared with a theoretical random distribution (PROUT *et al.*, 1953).

DAVIS (1960a, b) has described a different situation in *pan*/*pan-m* heterokaryons. The difference between these markers is that *pan-m* is less pantothenate-requiring than *pan* and can remove pantothenate from the medium at lower concentrations. Thus in limiting conditions of exogenous pantothenate, *pan-m* would be expected to have an advantage over *pan*. In fact this does not occur. There is rapid growth followed by slow growth or cessation of growth, then a further increase, etc., the sequence proceeding cyclically. The nuclear ratios show a similar cyclical sequence, a majority of *pan-m* nuclei alternating with an increase in *pan* nuclei. Under non-limiting conditions, neither the growth rate nor the nuclear ratios vary, the latter having been determined at intervals in a Ryan tube. Davis accounts for these facts by supposing that *pan* nuclei can utilize pantothenate within the hyphae more efficiently than *pan-m* nuclei, even though the latter are more efficient at uptake. Thus *pan* nuclei-containing hyphae proliferate in all conditions but once the utilization of pantothenate exceeds its rate of supply, determined by *pan-m* nuclei, hyphal growth will be reduced. As *pan-m* nuclei increase, the intramycelial pantothenate concentration increases once more and so growth starts up again.

It is a general property of *Neurospora* hyphae that selection to eliminate one nuclear type is extremely difficult and this may be because of the homogeneous distribution of nuclei throughout the mycelium (PITTENGER and ATWOOD, 1956), although this does not seem to have been proved.

A problem in connexion with the observed changes in nuclear ratios is that of the mechanism which brings it about. There are two contenders, either different nuclei can divide at different rates or, by chance, hyphal tips can acquire different nuclear ratios and those best adapted to the immediate situation will have a selective advantage over tips with other nuclear ratios. A decision between these hypotheses is not, at present, possible. Although PONTECORVO (1946) claimed that nuclei did not divide synchronously in Fungi Imperfecti, subsequent investigations strongly suggest that they do so divide both in homokaryons and heterokaryons (REES and JINKS, 1952; ROBINOW, in DUBOS, 1947, and unpublished). On the other hand, the stable growth of *Neurospora*, despite hyphal tips with variable nuclear ratios, tells against the hypothesis of hyphal selection. It is probably fair to say that, at present, there is a majority belief, not yet proven, that the hyphal selection hypothesis is likely to be the more probable mechanism (e.g. DAVIS, 1966; CLUTTERBUCK and ROPER, 1966).

GENETIC CONTROL OF HETEROKARYOSIS

Hitherto, only selection due to external or internal environmental factors has been considered; genetic factors are also involved.

In *Neurospora* the actual development of an heterokaryon is determined by two pairs of alleles, *C*/*c* and *D*/*d*. Only mycelia carrying like alleles will form an heterokaryon, i.e. *CD* and *CD*, *Cd* and *Cd*, *cD* and *cD* or *cd* and *cd* but not *CD* with *cD*, *Cd* or *cd* and so on. Hyphal fusions are not prevented but the fusion cell is rapidly sealed off. In a short time, bubble-like structures appear and the cell dies and autolyses (GARNJOBST, 1953, 1955; GARNJOBST and WILSON, 1956). Later Wilson extracted cytoplasm from *CD* mycelia and injected it by a micropipette into hyphae of unlike genotype with respect to *C*, *D* or both. The same response was observed and there is evidence that the reaction was due to a protein moiety (WILSON *et al.*, 1961). HOLLOWAY (1955) discovered, independently, a similar system in a different strain of *Neurospora* determined by four pairs of genes at least. The relationship of this system to the *CD* system is not known. These systems are unrelated to mating-type factors which also often operate to prevent mycelial heterokaryon formation although, of course, in the reproductive cells, this restriction does not obtain. This is true of *Phycomyces*, as mentioned earlier and of *Neurospora*, in which such heterocaryons can be forced in highly selective conditions with ratios as disparate as 2000:1 (GROSS, 1952).

'Heterokaryon incompatibility', as it has been called, is also found in *Aspergillus nidulans* (GRINDLE, 1963a) and other species and similar phenomena occur in other fungi, in some cases the reaction being similar to that of the *CD* system, e.g. *Rhizoctonia solani* (FLENTJE and STRETTON, 1964). In some species, attempts to make heterokaryons have failed completely, e.g. *Venturia inaequalis* (KEITT and LANGFORD, 1941), and *Cladosporium fulvum* (FINCHAM and DAY, 1965) and this is a common experience of all workers with certain strains of fungi which normally form heterokaryons. The causes of this behaviour have not usually been ascertained. Discussion of the significance of these systems will be deferred until Chapter 17. Similarly, the genetic control exercised over the dikaryons of Basidiomycetes will be considered as a manifestation of the mating system in Chapter 15.

PITTENGER and BRAWNER (1961) have described a pair of alleles *I*/*i* in *Neurospora* in which the only stable heterokaryons are those with less than 30% *I*-genotype nuclei. If *I*-nuclei exceed 30% then they inevitably continue to increase until a mycelium homokaryotic for *I*-nuclei arises. The mechanism is not understood but a similar kind of situation has also been described by Holloway.

PROPERTIES OF HETEROKARYONS

Stable heterokaryons are those with a selective advantage, usually in rate of growth over their component homokaryons. This may arise in the case of heterokaryons containing nuclei with nutritional defects as a result of complementation, i.e. the deficiency of one nucleus is made good by the ability of the other and vice versa. In general, this effect must operate in the cytoplasm. In some cases only partial complementation occurs. For example, PRASAD (unpublished) has obtained heterokaryons between nuclei deficient for histidine and hypoxanthine respectively, in *Aspergillus niger*. These are incapable of producing conidia although the mycelium grows almost as well as the wild type and is apparently normal. If the nuclei are combined experimentally in a diploid nucleus, a mycelium containing such nuclei grows and produces conidia normally. In this case, therefore, complementation, as a physiological process, can only occur at the intranuclear level, not in the cytoplasm (see Parasexuality). Similar effects may be seen in relation to dominance. Thus *pan-1*$^+$/*pan-1* heterokaryons grow as fast as wild type (*pan-1*$^+$/*pan-1*$^+$) mycelia, but not faster, provided that there are 5% *pan*-1$^+$-carrying nuclei. As the ratio declines from 0·5–0%, the growth rate falls off on minimal medium. Thus, the quantitative expression of dominant genes can be studied (PITTENGER and ATWOOD, 1956).

Many of these interactions are manifested in what DODGE (1942) called 'heterokaryotic-vigour' and this has been compared with heterosis in diploid organisms. It is clear that the phenomena are rather different, the heterokaryotic situation usually being explicable in terms of complementation. Nevertheless, complex situations do arise involving non-allelic genes and their suppressors and little study has yet been made of such complications. An exception is that of sulphonilamide-requiring strains of *Neurospora* and their suppressor strains (EMERSON, 1952).

Finally, attention should be drawn to the fact that recessive genes, including lethals, can arise in heterokaryons and persist, masked, until they segregate in spores, for example. ATWOOD and MUKAI (1955) showed that of 26 spontaneous mutants detected in *Neurospora*, 24 could not be isolated and grown as homokaryons; presumably these were lethals on the complete medium employed and not due to simple metabolic lesions. Adequate screening has not been carried out to determine how widespread this phenomenon is in different fungi.

THE PARASEXUAL CYCLE

Earlier, mention has been made of synthesizing, experimentally, diploid nuclei but, in *Aspergillus nidulans* this was found to occur spontaneously in hyphae and to be followed by recombination and, sometimes, the breakdown of the diploid nuclei to haploid ones. These last two phenomena, somatic recombination and haploidization, are independent processes so far as is known. The 'parasexual cycle' was a term applied by PONTECORVO (1954) to the sequence, heterokaryosis, fusion of genetically dissimilar nuclei, and recombination and segregation of the diploid nuclei so formed, the whole process occurring in hyphae.

PARASEXUALITY IN ASPERGILLUS

Heterokaryosis has already been discussed and so has the distribution of nuclei within an heterokaryon. The way in which nuclear fusion occurs is a mystery. It will be recalled that somatic division usually takes place within the nuclear membrane (cf. Chapter 13, p. 365). It might be thought that in a crowded hyphal apex with 20–30 nuclei dividing synchronously (see Plate IV, Fig. 1, p. 51), it might be possible for overlapping divisions to become intermingled and for a new nuclear membrane to enclose two haploid sets of chromosomes. This is possible and seems indeed to have been envisaged by Weijer and his colleagues in relation to their so-called 'streaming-mitoses'. ROPER (1952) first synthesized heterozygous, diploid nuclei in hyphae of an *A. nidulans* heterokaryon exposed to *d*-camphor vapour. This substance seems to increase the yield of diploids but not necessarily the frequency of nuclear fusions, although the evidence is not easy to interpret (PONTECORVO and ROPER, 1953). Using multinucleated conidia of *A. sojae* and *A. oryzae*, ISHITANI *et al.* (1956) produced diploids by camphor and UV treatment. With the latter, at a level of 1% survival after treatment, the frequency of diploids recovered rose dramatically to >1 in 10^2. The action of either of these treatments on nuclear fusion or selection of diploid nuclei is not understood, but ROPER (1966) favours a selective action by these agents.

The frequency of spontaneous somatic diploids is usually low, in *A. nidulans* about 1 in 10^6–10^7 conidia, in *A. niger* 3·5 in 10^5, in *Penicillium chrysogenum* 2·5 in 10^8 and, in *Coprinus lagopus* 1 in 10^3–10^4 basidiospores (PONTECORVO, 1956; PONTECORVO *et al.*, 1953; PONTECORVO and SERMONTI, 1953; CASSELTON, 1965). In the smut fungus, *Ustilago maydis*, however, diploids have been obtained as a consequence of failure of normal meiosis in as many as 10% of the spores (HOLLIDAY, 1961a). Mitotic recombination is also a rare process, although there are great difficulties in estimating its frequency. The only fungus where a rational suggestion can be made is *A. nidulans* and it is one cross-over per fifty somatic divisions (KÄFER, 1961). In a comparison, at the same genetic locus, of somatic and meiotic crossing-over, PRITCHARD (1955) suggested that the former was as much as 104 times less frequent.

The mechanism also differs (Fig. 14.7). Crossing-over occurs at the equivalent of a four-strand stage in normal mitosis and is almost without exception confined to a single reciprocal exchange in one arm of one chromosome. Centromeres separate mitotically so that the recombinant chromosomes may segregate in one of two ways. Either both recombinant and both non-recombinant daughter chromosomes segregate together, or one recombinant daughter chromosome segregates with one non-recombinant one. In the former case the nuclei, still diploid, are still heterozygous; the alternative segregation pattern results in genes distal to the point of exchange being in the homozygous condition. Thus, if segregation is random and this seems to be the case, only 50% of the nuclei in which a cross-over has occurred will exhibit homozygosity and even then this will only affect the distal region of one chromosome arm in the whole chromosome complement. Thus this rare event is manifested phenotypically with even greater rarity (PONTECORVO *et al.*, 1953).

Some of the consequences of somatic crossing-over which are not detectable become so if the nuclei are haploidized. For example, the products of the first segregation pattern described (Fig. 14.7a) will be separated and could give the genotypes $+++$, *abc*, $+bc$ and $a++$ in haploid nuclei; of these $+bc$ and $a++$ are novel genotypes. Thus haploidization is of importance in revealing hidden segregants. The frequency of this process is, however, low in

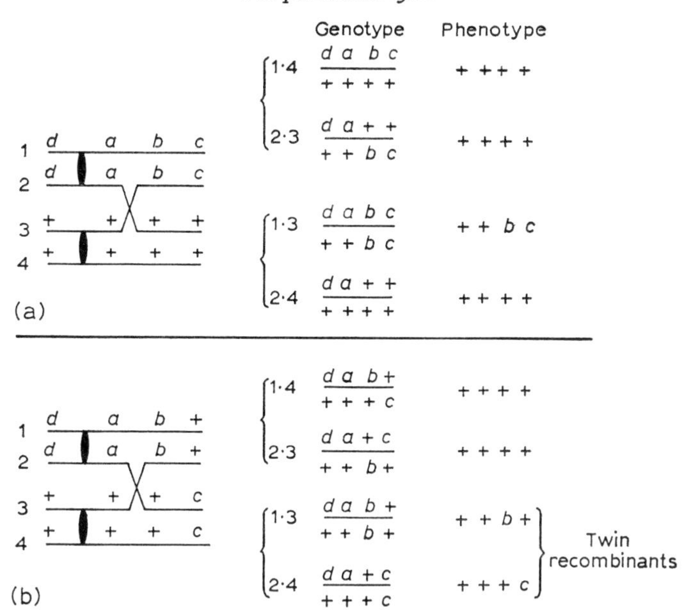

Fig. 14.7. Diagram to illustrate the consequences of mitotic recombination between the 4 chromatids of a pair of homologous chromosomes. *a*/+, *b*/+, *c*/+ are on one side of the centromere, *d*/+ on the other. In (b) the system used to detect 'twin' recombinants is illustrated. These provide evidence of reciprocal recombination. (After Pritchard)

A. nidulans where it has been studied very fully (PONTECORVO *et al.*, 1954; PONTECORVO and KÄFER, 1958; KÄFER, 1961). Käfer estimates its frequency as one in every fifty mitoses. She has also suggested how haploidization occurs. Non-disjunction at division results in two daughter nuclei with $2n+1$ and $2n-1$ chromosomes respectively. The hyperdiploid can revert to the $2n$ condition but the hypodiploid continues to lose chromosomes, probably one at a time, until the haploid condition is restored. In *A. nidulans* with $n = 8$, types carrying appropriate marker genes have been detected with $n = 17(2n+1)$, $15(2n-1$ or $n+7)$, 12, 11, 10, 9 and 8 (haploid). It should be noted that the hypodiploids or aneuploids all showed reduced growth compared with diploid or haploid mycelia; as the haploid condition was approached sectors of increasingly normal growth arose. It is difficult to judge how significant this process would be in conditions other than the rather special ones employed in these experiments.

PARASEXUALITY IN OTHER FUNGI

Claims have been made for the occurrence of the parasexual cycle in a number of other fungi and, also, of claims that it does not occur. It is now quite certain that parasexuality occurs in several Fungi Imperfecti as well as in those Aspergilli, like *A. niger*, which lack a normal sexual cycle. *Verticillium albo-atrum* and *Penicillium chrysogenum* have been adequately studied (HASTIE, 1962; 1964, 1967; PONTECORVO and SERMONTI, 1953, 1954; SERMONTI, 1957) and there is evidence compatible with a parasexual cycle in *Fusarium oxysporum* (BUXTON, 1956, 1962) and *Cephalosporium mycophilum* (TUVESON and COY, 1961). These demonstrations are important for they provide evidence of a further genetic system, in addition to heterokaryosis, which may be available to fungi lacking normal sexual reproduction. Since many of this group are notable for their variability it is, perhaps, reassuring to find that a possible basis of chromosomal recombination may exist to help to account for their behaviour.

Parasexuality has also been detected in Ascomycetes and Basidiomycetes with a normal sexual cycle as well: *Cochliobolus sativus* (TINLINE, 1962), *Ustilago maydis* (HOLLIDAY, 1961a, b), *Coprinus lagopus* and *C. radiatus* (CASSELTON, 1965; SWIEZYNSKI, 1962, 1963; PRUD'HOMME, 1965) and *Schizophyllum commune* (ELLINGBOE and RAPER, 1962; ELLINGBOE, 1963, 1964). There is also suggestive evidence for a parasexual cycle in *Puccinia graminis tritici* (ELLINGBOE, 1961). The significance of two modes of recombination, somatic and meiotic, in the same fungus is problematical. HOLLIDAY (1961a) has shown that in *U. maydis* the diploids are capable of infecting maize plants but the haploid monokaryotic strains are not. Such diploids seem to account for the solopathogenic lines, well known for many years to pathologists. Nevertheless, as has already been pointed out, the origin of these diploids is very different from that in most fungi exhibiting parasexuality. In *U. maydis*, the diploids arise by failure of meiosis so far as is known, not in an heterokaryotic mycelium.

A further anomaly has been recorded in *Schizophyllum* where the somatic recombination processes does not exactly match that described for *Aspergillus*. In particular, it has sometimes seemed to resemble meiosis (CROWE, 1960) and at other times has only affected the mating-type factors and no other adjacent marker genes. These unusual situations have been succinctly reviewed and assessed by ELLINGBOE (1965), who has been responsible for much of the original work on the topic.

In those fungi where a parasexual cycle has not been detected, some are said to be incapable of forming heterokaryons but, amongst those that can, notably *Neurospora*, it has not proved possible to detect the cycle or even diploid nuclei as yet (CASE and GILES, 1962). The reasons for this are obscure.

In this chapter, an attempt has been made to describe the non-sexual mechanisms responsible for generating variability in the fungal mycelium. They are diverse and far from being adequately understood. Moreover, it will be clear that only a small number of fungi have yet been examined. It is, therefore, a rash procedure to generalize on these matters at present, but an attempt will be made to assess their significance after an account has been given of the manifestations and controls of the normal sexual processes in fungi.

Chapter 15
Sexual Reproduction

The fungi are remarkable for the diversity of their sexual processes. Sexual reproduction is widespread in the fungi and involves the essential features of nuclear fusion and meiosis which follow each other closely, usually in that order. Superimposed upon these basic events are a variety of regulating mechanisms and a range of morphological developments, the effects of which are to determine with more, or less precision, the kinds of nuclei that will fuse at fertilization. Such mechanisms may be termed mating systems, but their nomenclature is not at present generally agreed (WHITEHOUSE, 1949a; BURNETT, 1956; ESSER and KUENEN, 1965; RAPER, 1966a). Difficulties arise partly from the wide range of sexual expression in fungi and partly from ignorance of the basis or causes of the regulating mechanisms.

Most recent classifications have been attempts to describe the genetic bases of these phenomena and three classes of mating system can be defined with reasonable clarity. In many fungi there is apparently little or no control over the kinds of nuclei which fuse, so that self-fertilization is not only quite possible but regularly occurs. For example, a single spore of *Rhizopus sexualis* can germinate to a mycelium on which gametangia can develop and, after their fusion, fertilization can occur. This situation was termed homothallism by BLAKESLEE (1904), because he was impressed by the fact that sexual reproduction could be confined to a single thallus without differentiation, of necessity, of male and female organs. In some fungi, however, the same process occurs, save that two kinds of morphologically differentiated gametangia develop on the same thallus, e.g. *Pyronema confluens*. In this account such conditions will be described as homomixis, not homothallism.

A second kind of situation is shown by fungi such as *Mucor mucedo* or *Neurospora crassa* where there are two kinds of conjugant individual which may, or may not, have developed morphologically distinct gametangia. In *Mucor* the individuals and their gametangia are indistinguishable. In

Neurospora each conjugant type develops protoperithecia with trichogynes and conidia which act as fertilizing agents but the two kinds of conjugant are indistinguishable. Blakeslee used the term heterothallism to describe the situation in *Mucor* and it has since been extended to cover situations such as those found in *Neurospora*. WHITEHOUSE (1949a) has distinguished them by the terms 'morphological heterothallism' and '2-allele physiological heterothallism', respectively. An important feature in these situations is that the two kinds of individual are determined by their genetic constitutions which differ as if they were due to a pair of allelomorphs. Such genetic determinants can be called mating-type factors and are variously designated $+/-$, A/a, a/α, in different organisms. It should be emphasized that this is a formal way of describing the inheritance of mating-type factors which may well be complex supergenes (see later, e.g., pp. 420–422 and 432). Such a condition will be called dimixis here to emphasize the nature of the genetic control and the fact that there are normally *only two* conjugant types. Several situations are known in Phycomycetes which look as if they could be described in these terms but for which there is either insufficient information available to be certain, or in which additional modes of expression confuse the pattern, e.g. *Achlya* spp.

A third and final mode of regulation, shown only by some Basidiomycetes, is due to what can be described as a multiple allelomorphic series of mating-type factors. There are many kinds of genetically determined individuals and any two may act as conjugant partners provided that they carry different mating-type factors. This situation of *many different* genetically determined mating types is here called diaphoromixis. The mating-type factors may be at a single locus or at two, unlinked loci so that after meiosis there can be two or four kinds of genetically differentiated basidiospores produced on a single basidium, e.g. *A*1 and *A*2 in a one-locus form; *A*1*B*1, *A*2*B*2, *A*1*B*2, *A*2*B*1 in a two-locus form. These are distinguished as bipolar and tetrapolar diaphoromixis, respectively.

Morphological expression in relation to sexual reproduction also varies. The actual conjugant structures may be free-swimming gametes, gametes with gametangia, gametangia with gametangia or undifferentiated hyphae. Most gametes are indistinguishable or only slightly different from the zoospores which are frequently produced by the same fungus. Gametangia may range from clearly differentiated oogonia, producing one or more eggs, and antheridia, producing male gametes, motile or non-motile, to less well differentiated structures which are similar in shape and size, but recognizably distinct from vegetative hyphae. Thus, there is discernible in the fungi a range from fully differentiated, dimorphic gametangia to undifferentiated hyphae and in certain groups, notably the Ascomycetes, this has sometimes been regarded as a deterioration of sexual reproduction, e.g. by CHURCH (1920). This is clearly not the case since fertilization and meiosis are still involved despite the lack of morphological sexual differentiation. Sexual organs have the effect of imposing a restriction on potentially conjugant regions. If such differentiation is reduced, the restriction of conjugation to particular regions is also lost. Since the region where fertilization, meiosis and subsequent carpophore development takes place need not be the same as that in which the primary act of conjugation has occurred, a phase of nuclear migration is frequently interposed between these events. Thus, in respect of nuclei carrying mating-type factors, a more or less extended phase of heterokaryosis supervenes between conjugation and fertilization, especially in those species with reduced morphological sexual differentiation. The most striking situations are found in Homobasideae where the dikaryotic mycelium, which may persist for several hundreds of years, in fairy-ring fungi for example, is just such a special heterokaryon.

Conjugation cannot occur unless the conjugants come close to each other and this may either be a matter of chance or physiological mechanisms may exist to regulate such encounters. There is a wide range of such physiological co-ordinating processes, but they all show, potentially, three kinds of control: telemorphosis, the induction of potentially conjugant regions at a distance; zygotropism, the directed growth of such regions and, thigmotropism, responses once the regions are in contact leading to delimitation of gametangia, for example. The first two processes operate over greater or lesser distances, ranging from a few millimetres to 10 μm or less; they are mediated by diffusible materials, often called sex hormones. The role of thigmotropism is less obvious and may not always be involved.

In the subsequent account of sexual reproduction in fungi, the order followed will be broadly that of the main taxonomic groups rather than by kinds of mating system. It will become clear, however, that the two kinds of classification are not wholly uncorrelated.

SEXUAL REPRODUCTION IN PHYCOMYCETES

In the Phycomycetes a full range of sexual expression is manifested. Amongst the Chytridiomycetes and many groups of the Oomycetes, motile gametes occur as well as differentiated gametangia, especially in the latter group. In the Zygomycetes gametangial differentiation occurs, but it is greatly reduced and dimorphism is often lost. Sexual reproduction may be either strictly dimictic, as in many Zygomycetes; may show an approach to this condition, e.g. *Achlya* spp., or be either ill defined or apparently homomictic.

Chytridiomycetes

Many fungi of this class produce motile zoospore-like swarmers, which appear to be capable of fusing in pairs under certain circumstances. In some cases differences in size or behaviour occur between the swarmers and this may be reflected in differences between the thalli from which they have come. The situation is usually confused so that in 1950 EMERSON wrote '. . . there is not yet a single well established example in the Chytrids of a differentiation of the sexes clearly attributable to segregation of genetic factors at the time of meiosis in the life-cycle'. There is one example, *Dictyomorpha dioica* (MULLINS, 1961, and in RAPER, 1966b), now known where conjugation occurs between dimorphic gametes derived from different thalli and in which the original mating types can be recovered in the progeny of the zygote. More typical is the situation in the genus *Blastocladiella*. In *B. variabilis* two kinds of thalli develop and morphologically indistinguishable swarmers from the two kinds fuse. However, if a single unfused swarmer is induced to develop parthenogenetically, it can give rise to plants of both morphological types (EMERSON, 1950). This is true also of the fused swarmers. This suggests that differentiation is determined by the environment rather than by a genetic regulating mechanism. *B. stübenii*, by contrast, produces morphologically indistinguishable thalli which can be differentiated by the behaviour of their swarmers. Evidently there are two kinds of thalli in approximately equal numbers and swarmers from the two types are compatible. This situation is not fully resolved, because a single swarmer has not yet been induced to develop parthenogenetically nor has segregation been observed in the zygote (STÜBEN, 1939). Two other species are known, *B. emersonii* and *B. britannica*, which produce colourless and orange-coloured sporangia as in *B. variabilis*. However, the zoospores produced do not seem to conjugate although colourless zoospores from these sporangia do remain in contact with orange-coloured zoospores derived from a third kind of thick-walled resistant sporangium found on these species. Cytoplasmic exchange is believed to occur and temporary cytoplasmic bridges have been seen between the swarmers (CANTINO and HORENSTEIN, 1954), but there is no evidence for sexual reproduction.

In another genus in the family, *Allomyces*, a good deal is known about the physiological regulation of sexuality. In *Allomyces* male and female gametangia develop in juxtaposition, the former being orange-coloured due, as in *Blastocladiella*, to γ-carotene, the latter colourless. The male gametangium may be epigynous or hypogynous, relative to the female gametangium, e.g. *A. macrogynus* and *A. arbuscula* (Fig. 15.1). The gametes differ in size, colour and behaviour; the male gametes being smaller, orange-coloured and altogether more active; female gametes move very sluggishly (EMERSON, 1955). The genus includes polyploid forms and interspecific hybrids, e.g. *A.* × *javanicus*, and this accounts for much of the considerable range in morphological expression of the gametangia. By crossing and selection, EMERSON and WILSON (1954) were able to isolate effectively male and female plants, e.g. male strains have less than one colourless gametangium to every 1000 orange-coloured gametangia. These strains have been used to study sexual differentiation and the course of reproduction.

TURIAN (1961a, b, c) showed that a gradient of RNA differentiated potentially male from female regions in a hypha, female regions having a higher RNA content. The DNA/RNA ratio is 1–(1·5)–2·0:1 in coloured: colourless gametangia. The male gametangium is believed to have a lesion in the Krebs cycle so that isoctritase activity is increased and it employs the glyoxylate shunt, unlike female gametangia. A similar kind of situation has

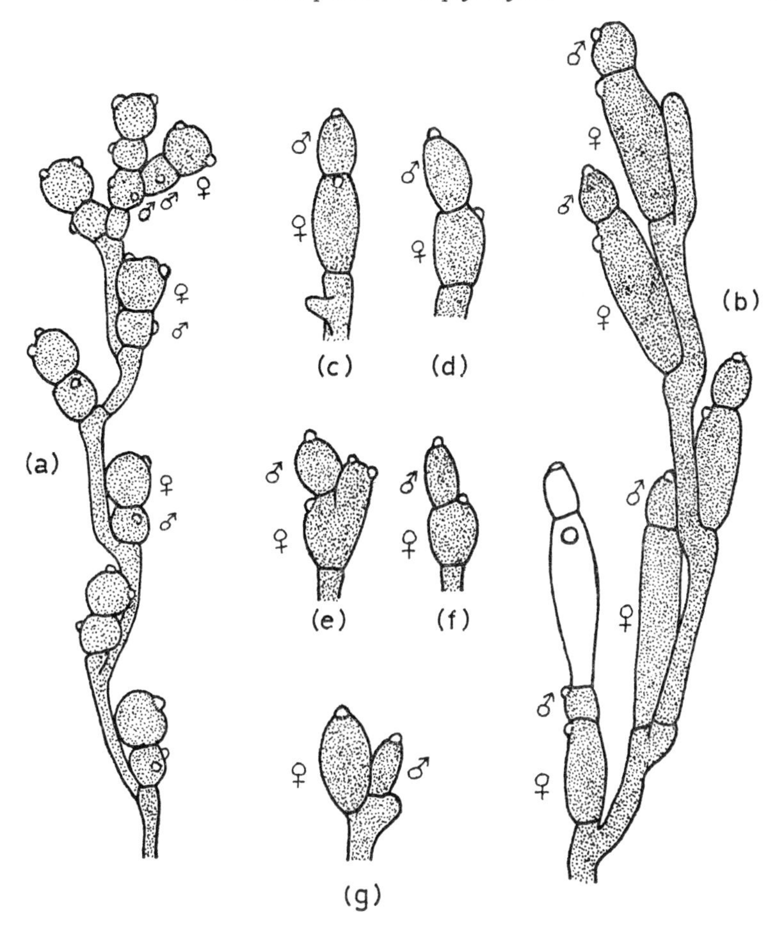

Fig. 15.1. Relative positions of the gametangia in *Allomyces* spp. (a) An Eu-Allomyces-type, *A. arbusculus* with hypogynous male gametangia; (b) a Cystogenes-type, *A. macrogynus* with epigynous male gametangia; (c)–(g) gametangia of *A.* × *javanicus*, now known to be a putative hybrid form, showing various arrangements and positions of gametangia between those of the parents *A. arbusculus* and *A. macrogynus*. Note in all cases exit papillae and clear circular pore in empty gametangia in (b). (a, × 165; b–g, × 200, from Emerson)

been found in *Blastocladiella* where, it will be recalled, the zoospores produced are apparently sexually incompetent (CANTINO, 1966).

Machlis and his collaborators (MACHLIS, 1958a, b, c; 1966) have shown that male gametes are attracted to female gametangia whether on the thallus or detached and embedded in agar. This is true if male gametes are separated by a dialysing membrane from the culture fluid in which female plants have been grown. A diffusible zygotropic hormone was postulated as the cause and termed sirenin. This has now been considerably purified and is known to have only carbon, hydrogen and oxygen present in the molecule. At the moment, the best estimate of the formula and molecular weight is $C_{15}H_{24}O_2$ and 236. Using 236 as the appropriate figure, male gametes can respond to concentrations of 10^{-10} M of sirenin. CARLILE and MACHLIS (1965) have shown that male gametes actively remove and inactivate sirenin and optical observations showed that the behaviour of the gametes was explicable as a topotaxis as in *Pythium* zoospores (cf. Chapter 6, pp. 178–179).

Oomycetes

The most fully documented and remarkable examples of hormonal regulation of the sexual processes occurs in the Saprolegniales, notably in the genus *Achlya*. It was DE BARY (1881) who first suggested that hormonal control was exercised through a specific chemical induction of antheridial hyphae by oogonia and, thereafter, the growth of such hyphae was chemotropically determined. COUCH (1926) and BISHOP (1940) provided additional evidence for *Dictyuchus* and *Sapromyces reinschii* and RAPER (1954) for *Achlya* and other genera and species. The system was, in fact, worked out in *S. reinschii* but fully explored in *A. bisexualis* and *A. ambisexualis*. The thalli of these species grown in isolation bear few or no gametangia but, if brought into proximity with other thalli, gametangia develop. Female plants produce extracellular hormones which induce the formation of antheridia in males; in *Achlya* this was due primarily to Hormone A (Fig. 15.2). The male plants now produce

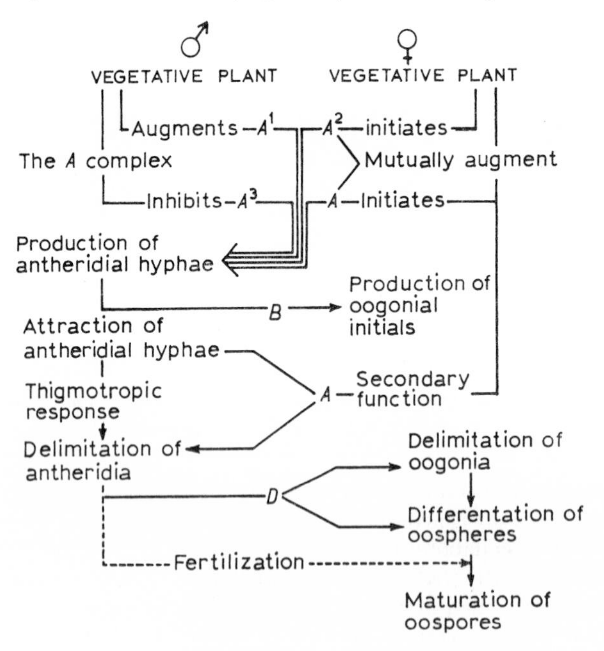

Fig. 15.2. Diagram to illustrate the actions of the sexual hormones in *Achlya* spp. (modified from Raper)

another hormone which induces oogonia (Hormone B) and the antheridial branches are chemotropically attracted to these oogonia. In *Achlya* this attraction was at first thought to be due to a further hormone (C) but BARKSDALE (1960, 1963) has been able to show that this is, in fact, a further action of Hormone A. As with sirenin, the male strain takes up and inactivates Hormone A. Polystyrene particles adsorbed Hormone A and, if these were placed on a permeable cellophane membrane separated from a male strain, not only were antheridial hyphae developed but they responded zygotropically to the particles. Once contact had been achieved, the antheridia were delimited by a cross-wall and the oogonial initials became delimited, perhaps due to a further hormone, D. The delimitation of antheridia was detected by BARKSDALE (1963) in her polystyrene-particle/membrane experiments so that Hormone A may be responsible for this also.

Raper's elegant experimental demonstrations of this system left no doubt of the way in which it operated and are classic in their simplicity and certainty. Unfortunately, little further development has occurred. Attempts to isolate Hormone A (RAPER and HAAGEN SMIT, 1942; BARKSDALE, 1960) have not been successful, although greatly purified preparations have been obtained.

A good deal is known about the biology of the system. RAPER (1936, 1947) had earlier investigated the mating patterns of *A. ambisexualis*. He showed that ten isolates could be arranged in six groups on a basis of interstrain matings. One, E87, behaved as a male to the other nine isolates and three, 78, 80 and 184, always behaved as females to all other isolates. The other six isolates showed an intermediate behaviour. For example, strain 190 in the presence of E87 and 184 could react, in part of its mycelium, as a female to E87 and as a male, in another part of the mycelium, to 184. In *A. bisexualis*, however, as its specific name implies, the behaviour of isolates was more strictly male or female. Barksdale was able to show that the 'stronger' the maleness of an isolate the more efficiently it inactivated Hormone A. Raper had shown earlier that Hormones A, and A^3 are produced by male plants and augment and decrease, respectively, the action of Hormone A. Hormone A^2, secreted by the female, also initiates antheridial hyphae but less effectively than Hormone A. The roles of A^1, A^2 and A^3 in relation to sexual expression are not known. RAPER (1950) has shown that sexually reproducing species showing some sort of approach to a dimictic pattern will also interact with other homomictic species of *Achlya* and, indeed, other genera and species. It is clear, therefore, that regardless of the kind of genetic regulation of the mating system in the Saprolegniales, the physiological regulation of the processes is identical or very similar throughout the group. The analysis of the genetic situation depends upon the ability to germinate zygotes and this is not easy. COUCH (1926), using *Dictyuchus*, showed that males, females and intergrades were produced and this has been confirmed by MULLINS and RAPER (1965) and extended to *A. ambisexualis*. It seems clear that a simple 2-allele system will not account for these findings but, of course, the ploidy of these fungi is still uncertain. It will be recalled (Chapter 1, pp. 4, 8) that SANSOME'S

(1963) interpretation has been provisionally accepted here, that these fungi are diploid, meiosis taking place in the gametangia. If this is so, it may prove simpler, by analogy with other bisexual organisms, to account for the presence of males, females and intergrades. No evidence for a sex-chromosomal mechanism has been obtained.

The difficulty experienced in germinating the zygotes of Saprolegniales has inhibited the study of sexuality in other Oomycetes. One of the most problematical genera, from this point of view, is *Phytophthora*. In this genus and its allies, the mycelium develops oogonia and antheridia, usually in fairly close juxtaposition, and homomixis would seem to be the natural system. However, for many years it has been known that oogonia can be induced or their frequency greatly increased if different strains of the same species or even different strains are grown together. In some cases the sexual response can be ambivalent and this has led to claims of 'relative sexuality', e.g. *Perenospora parasitica* (BRUYN, 1937), *Phytophthora parasitica* (LEONIAN, 1931; KOUYEAS, 1953). Three recent discoveries provide cause for believing that this confused situation may soon be resolved. The first, already referred to, is the probable diploid condition of this fungus. The second, is the discovery by LEAL *et al.* (1964) and others (ELLIOTT *et al.*, 1964, 1966) that sterols promote the development of gametangia. Thus the release of such compounds, produced by Phytophthoras could account for some of the interstrain and interspecific interactions described previously as sexual interactions. Finally, it is now clear that there is a genetic basis for a dimictic system in *P. infestans*. Two mating types, *A*1 and *A*2, have been described and successful sexual fusions are restricted to fusions between these mating types. The surprising thing is that at present the system only appears to operate in Central Mexico and perhaps northern South America. Outside this region only *A*2 forms occur. Thus, in such regions mating is either no longer dimictic or else it has been replaced by homomixis or amixis (≡haploid apomixis) (SMOOT *et al.*, 1958; GALINDO and GALLEGLY, 1960). Heterokaryosis or heteroplasmons are not excluded as possible means of maintaining variability (JINKS and GRINDLE, 1963).

Zygomycetes

The Zygomycetes occupy a special place in the study of sexual reproduction in fungi because it was here that BLAKESLEE (1904) laid the genetical foundation of homo- and heterothallism and he and BURGEFF (1924) initiated the study of the physiology of mating reactions.

The Mucorales were shown by Blakeslee and his colleagues to be either strictly homomictic or dimictic. The gametangia developed were frequently identical but could differ in size or shape in homomictic forms, e.g. *Zygorhynchus heterogamus* (BLAKESLEE, 1913) (Fig. 15.3). Conjugation takes place between gametangia. By assuming that the larger gametangium was the female equivalent and by a series of interstrain and interspecific crosses, BLAKESLEE (1915) attempted to equate the + and − strains of dimictic species with maleness or femaleness. In general, he found a measure of agree-

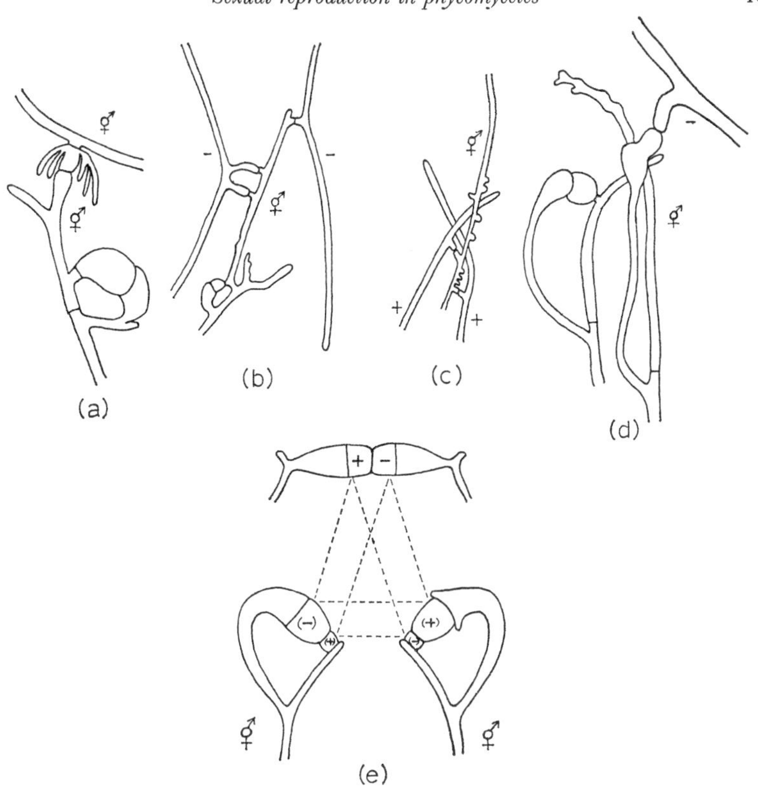

Fig. 15.3. Imperfect sexual reactions in Mucoraceae. (a) Large gametangium of *Zygorhynchus heterogamus* × large gametangium of *Absidia spinosa*, note circinate outgrowths from latter; normal reaction of *Zygorhynchus* below (×165): (b) small progametangia of *Z. heterogamus* × *A. glauca* −; normal reaction of *Zygorhynchus* below (×28): (c) small progametangia of *Zygorhynchus* sp. × *A. glauca* + (×28): (d) large gametangium of *Zygorhynchus* sp. × *A. glauca* −; normal *Zygorhynchus* reaction to left (×165): (e) diagram to illustrate types of reaction found between heterogamous homomictic species and homogamous dimictic species. Note that in one heterogamous species the large gametangium behaves as +, in the other as −. (From Satina and Blakeslee)

ment in his results, namely that + strains usually reacted with the smaller gametangium, but this was by no means always the case, even in the same genus. For example, the larger gametangium of *Z. heterogamus* reacted weakly with + strains, the small gametangium reacted strongly with − strains but, in *Z. moelleri* the large gametangium reacted weakly with − strains and the small gametangium strongly with + strains (SATINA and BLAKESLEE, 1930). On the other hand, in a most extensive series of intra- and interspecific crosses between + and − strains of dimictic species, no strain acted in an ambivalent manner (BLAKESLEE *et al.*, 1927; BLAKESLEE and CARTLEDGE, 1927). There was considerable variation in the strength of the sexual reaction, as measured by the relative numbers of zygotes produced, and some races were classified as neutral, i.e. showing no reaction to either + or – tester strains. Thus the situation resembles that in the Saprolegniales in that interspecific and even intergeneric reactions occur and that reactions vary in their strength, but it differs notably in the total absence of strains showing an ambivalent reaction. The widespread occurrence of 'imperfect sexual reactions' (Fig. 15.3) suggests a common physiological regulation and this aspect has received attention.

Blakeslee illustrated the induction at a distance of gametangia, their tropistic curvature towards each other and the delimitation of the gametangia after contact for *Mucor mucedo* and *M. hiemalis* in his classic paper in 1904. In *M. mucedo* the phenomenon is very clear since the potential gametangia grow out of the agar from margins of the opposed + and – mycelia when they are 2–3 mm apart. They curve towards each other in the air and meet more or less at their tips. In 1924, Burgeff showed that the induction of gametangia could occur through a colloidon membrane separating a + strain in a block of agar from a – strain in agar below. The zygophores, swollen hyphae which would develop into gametangia in normal circumstances, were developed by both strains and grew towards each other so far as was possible. Those above the membrane, for example, curved over and downwards towards the membrane. Membrane experiments with *M. hiemalis* failed but strips some 2 mm apart developed zygophores which grew through the air towards each other. Burgeff postulated that each strain produced a diffusible or volatile substance which induced and directed the growth of zygophores.

Several investigators have attempted to repeat and develop these findings, which have been confirmed for several mucoraceous species. LING YONG (1929) extended the findings by claiming that in *Phycomyces blakesleeanus* a zone of aversion, 1·0–1·5 mm wide developed between colonies of the same mating type (Fig. 15.4). A similar phenomenon has been detected over about 1 mm between zygophores of *M. mucedo* which curve away from each other, if previously induced (BANBURY, 1955). This would not normally occur since they would never be induced in a reaction between mycelia of identical mating types.

The most extensive and successful studies of the regulation of the reaction have been made with *M. mucedo* by Plempel and his associates (PLEMPEL, 1957; 1960a, b; 1963a, b; PLEMPEL and BRAUNITZER, 1958; PLEMPEL and DAVID, 1961). The system he has discovered is set out in Fig. 15.5. Each strain, if grown separately in liquid culture produced a 'progamone' and fungus-free culture medium applied to the opposite strain then induced it to synthesize a 'gamone', i.e. medium, containing + progamone added to – mycelium induced formation of – gamone and vice versa. The gamones induced zygophores in the opposite strain quite specifically, i.e. + gamone was without effect on a + mycelium. GOODAY (1967, unpublished) has isolated what appears to be a single substance from strains of *Mucor mucedo* other than those used by Plempel, which can induce zygophores in both + and – strains of this fungus. The relationship of this substance to Plempel's gamones is not

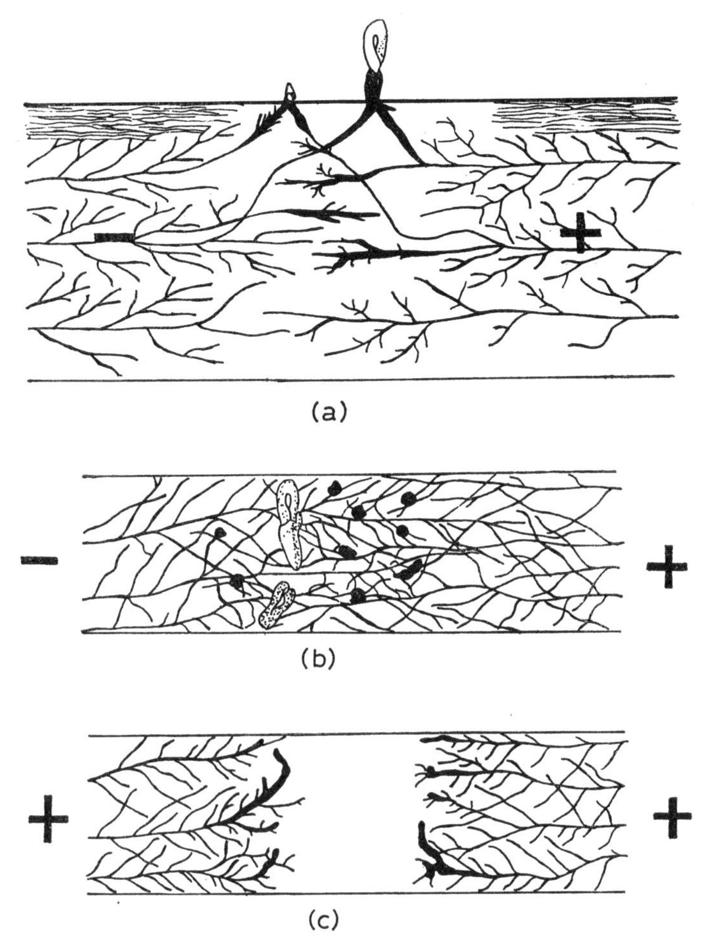

Fig. 15.4. *Phycomyces blakesleeanus.* (a,b) Sexual reactions between + and – strains in elevation (a) and plan (b) view and, (c) between two + (or –) strains in plan view. Note in (a) induction of progametangia at a distance followed by zygotropic responses. In (c) note induction of gametangial-like structures followed by clear aversion reaction. (× *c.* 15, after Burgeff and Ling Yong)

yet clear. Once induced, the zygophores showed zygotropic curvatures towards each other because of two further zygotropic hormones. These, in *M. mucedo*, are believed to be volatile compounds. For example (Fig. 15.6), two agar blocks containing opposite mating types, in which zygophores had been induced, were set up 2·5 mm apart but separated by a mica membrane perforated by a hole 3·5 mm in diameter. After 6–8 hr, the – zygophores had begun to grow towards the hole and, after 14 hr, had contacted the + zygophores. To do this some had had to bend up or down to grow through the

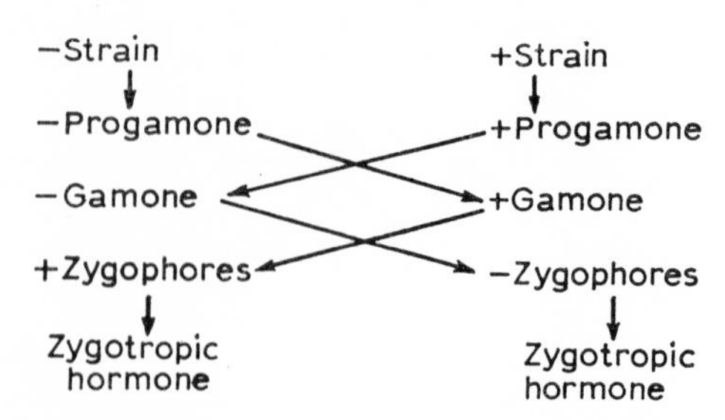

Fig. 15.5. Diagram to illustrate induction of progamones, gamones and zygotropic hormones by interaction of + and − strains of *Mucor mucedo*. (Based on Plempel)

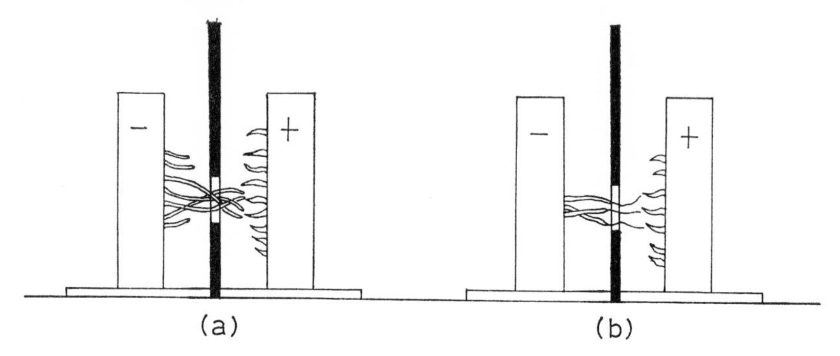

Fig. 15.6. Zygotropic interactions between zygophores of *Mucor mucedo* which have been previously induced by treatment with gamones. (a) Strains separated by mica-membrane with hole. Note that − zygophores opposite hole have grown through and those further away show slight curvature towards hole; (b), the − zygophores have de-differentiated after being opposed by another − culture but this has now been replaced by a + culture. Note elongation reaction of − strain and responses, directly opposite of + hyphae. (From Plempel)

hole. The reaction differed when the blocks were 4 mm apart, those opposite the hole in the membrane elongating more than those at the periphery. If + strain was replaced by a block with a zygophore-carrying, − strain, the reaction did not develop and the elongating zygophores from the original block of − strain de-differentiated into sporangiophores. However, if the block containing the + strain was replaced within 12 hr, attenuated hyphae grew out of the tips of the partially de-differentiated − zygophores towards the + zygophores, as in the normal reaction. When zygophores from colonies of like mating type were opposed, the repulsion effect, described for the same species by Banbury, was not observed.

Considerable progress has been made in isolating the gamones. Chromatographic methods were employed but these did not separate + from − gamones, despite the difference in their biological activities. The most recent estimates of formula and molecular weight are $C_{20}H_{25}O_5$ and 314–374. Thus, like Sirenin, the molecule is of fairly low molecular weight and contains only carbon, hydrogen and oxygen; no aromatic ring appears to be present, but there are one or more hydroxyls, an ester and a C=C bond. Gooday's zygophore-initiating substance appears to have several properties in common with trisporic acid, a substance known to enhance β-carotene production in another mucoraceous fungus, *Phycomyces blakesleeanus* (cf. p. 303). This observation is of particular interest in view of the frequent and massive accumulations of β-carotene in the gametangia of several Mucoraceae. It does not, of course, clarify the supposed correlation between carotene and sexuality in these fungi (BURNETT, 1965).

Three lines of evidence suggest that, just as in the Achlyas, the same physiological control is operative in all mucoraceous fungi. Firstly, as already mentioned, imperfect sexual reactions including tropistic responses occur between different species and genera. For example, Plempel showed that in a confrontation of *M. mucedo* by *P. blakesleeanus*, the former develops aerial zygophores which curve over and grow towards the agar surface, where each is encircled by *Phycomyces* zygophores which have developed below the surface (cf. Fig. 15.4) and grown upwards. Secondly, it has been shown that discs of mucoraceous fungi show enhanced respiration in a Warbung respirometer with compatible mixtures, i.e. + and − strains both in intra- and interspecific combinations (BURNETT, 1953a, b). Finally, Burgeff induced zygophores in a − strain of *M. mucedo* after passing moist air over a + culture of *Rhizopus nigricans* although spores and mycelial fragments were excluded.

This system of physiological control, involving two or four diffusible substances and two volatile ones, is the most complex yet discovered. It seems likely that a thigmotropic response is also involved since gametangia are not delimited until after contact, nor do they develop in + or − zygophores separated from each other by a permeable membrane (BURGEFF, 1924). Moreover, once zygophores of *M. mucedo* have actually made contact for 15–45 min, they can only be separated by tearing them apart; this strongly suggests a surface–contact effect (BURNETT, unpublished).

Once conjugation has taken place the gametangial walls perforate in the middle and the cytoplasms come into contact. The fusion product develops a massive, pigmented wall. In the homomictic *Rhizopus sexualis* this development can be inhibited by volatile substances given off by the mycelium of the same or other mucoraceous species. The significance of this is not known.

Unequivocal evidence that fertilization does actually occur in the zygotes and that segregation of + and − mating types then takes place at meiosis is lacking. The cytology of the zygotes is not easy to study. CUTTER (1942) described four patterns:

(a) In some, e.g. *Mucor genevensis*, all the nuclei fuse in pairs and meiosis takes place before the zygote germinates to give a germ sporangium;

(b) In *R. nigricans* some nuclei fuse, others degenerate and meiosis occurs at the germination of the zygote;

(c) In *Phycomyces blakesleeanus* the nuclei associate in groups of variable numbers, some fuse in pairs just before the zygote germinates. Meiosis occurs as the germ sporangium develops so that the spore nuclei may or may not have undergone fertilization and meiosis;

(d) Some are amictic, like *Syzygites grandis* lacking nuclear fusion altogether and meiosis.

This complex study needs to be repeated and confirmed. Some of the claims have been made before, e.g. the amictic situation in *Syzygites* (KEENE, 1914), and the confused genetic evidence from *Phycomyces* suggests that the products of the germ sporangium may include parental haploid, recombinant haploid and unreduced diploid nuclei (BURGEFF, 1912, 1914, 1915, 1925, 1928). Burgeff provided suggestive evidence also that all the germ-sporangium spores were derived fairly frequently from the meiotic products of a single zygote nucleus. There is no cytological observation which bears on this observation. Until zygotes can be germinated readily, which is certainly not the case at present, and appropriate genetic markers are available it will be difficult to get unequivocal evidence of fertilization and segregation in these fungi.

In the Phycomycetes there is a wide range of sexual expression and regulatory processes. Amongst the Chytridiomycetes the general impression is that homomixis is most common and that cross-fertilization is achieved by chance meetings of motile gametes. However, in *Allomyces* and *Blastocladiella*, for example, a more precise determination of sexual expression occurs in some species and this is further developed in the Saprolegniales. Associated with this increased differentiation of distinct sexual dimorphism, is an increasingly sophisticated system of physiological regulation. Such a system, initiated by a single, diffusible agent, Hormone A, enables a sequence of sexual development to be triggered off whenever the appropriate apposition of potentially male and female thalli occurs. This sophistication, associated with an apparently strict, genetical regulation, reaches its peak in the mucoraceous fungi. In addition to these more clearly defined situations, there is a host of others intermediate in various ways. For example, although gametangial conjugation is the rule in the Oomycetes and Zygomycetes and gamete/gamete or gamete/gametangium conjugation in the Chytridiomycetes described, this is not always the case. Many examples are known from amongst the simplest chytrids of the conjugation of whole thalli at sexual reproduction, e.g. *Rhizophidium* (Fig. 1.2, p. 6). Such conjugation may be between heterogamous or isogamous gametangia and nothing is known of the regulation of such behaviour. Indeed, the study of the genetics of sexuality in the group as a whole is made almost impossible by the great difficulty experienced in almost every species in germinating the zygotes.

SEXUAL REPRODUCTION IN ASCOMYCETES

In the Ascomycetes the mating systems are regulated with much greater precision than in the Phycomycetes. Many of the species are homomictic, but others are dimictic with a well defined genetic control of the mating system. There are, however, a number of groups and species such as the yeasts, *Glomerella* isolates and *Chromocrea spinulosa*, whose mating systems are less clearly defined. In other cases there seems to be a well defined system, the basis of which is not known, e.g. in the Laboulbeniales. A further phenomenon, shown by Ascomycetes such as *Neurospora tetrasperma* for example, is the apparent imposition of inbreeding upon a potentially out-breeding, dimictic form; this will be termed homodimixis.

Dimictic systems

In many of the best known Ascomycetes, e.g. *Neurospora* spp. the site of fertilization is pre-determined, occurring before conjugation, or immediately after, by the development of protoperithecia or protoapothecia. As a consequence, conjugation is either restricted to these sites (a situation exactly analogous to that in *Achlya* or *Mucor*), or, if it is not, then a heterokaryotic phase, in respect of nuclei carrying different mating-type factors, is interpolated between conjugation and fertilization. In *N. crassa* small coiled structures develop all over the mycelium of a single mating type. They are the protoperithecia and from them develop slender, usually branched, hyphae forming trichogynes. Conjugation is restricted to the fusion of a trichogyne with a macro- or microconidium derived from a culture of opposite mating types. Conjugation can be reciprocal, i.e. *A* conidia with an *a* trichogyne and vice versa, but self-fertilization is not possible. Hyphal fusions occur within the mycelium but not between hyphae from mycelia of opposite mating types except under most abnormal circumstances. Such heterokaryons can be achieved using biochemical mutants and minimal medium (GROSS, 1952, and see Chapter 14, p. 396). They are unstable and their growth is most irregular. It is clear that there is an imbalance in the somatic situation, which does not arise in the trichogyne or ascogonial hyphae. The basis of this difference is not known. It could be due to the cytoplasm. Experiments to investigate the inheritance of cytoplasmic characters provide little evidence to suggest that anything but the conidial nucleus is transferred after conjugation (MITCHELL and MITCHELL, 1952). It is also noteworthy that in those cases where conjugation is effected through fusion of an antheridial hypha and a trichogyne, the former is extremely slender, e.g. *Chaetomium* spp. (GREIS, 1941). Thus there is less possibility of transfer and confrontation of cytoplasm at sexual reproduction in these cases than as a consequence of normal hyphal fusion.

The initial restriction of compatible nuclei to a few ascogonial hyphae is by no means the rule in Ascomycetes. In Discomycetes a range of behaviour is exhibited. In *Ascobolus magnificus* conjugation occurs between specialized, short, ovoid branches which develop on the complementary strains as a consequence of a telemorphotic stimulus. After a phase of zygotropic attraction they become intimately coiled around each other (DODGE, 1920). In *Lachnea melaloma* there is but a single, coiled and swollen structure which conjugates with an undifferentiated vegetative hypha and in *Humaria granulata* conjugation is replaced by hyphal fusions although the nuclei then migrate to the sites of the protoapothecia, in which further development occurs (GWYNNE-VAUGHAN and WILLIAMSON, 1930; GWYNNE-VAUGHAN, 1939). In *H. granulata*, therefore, there is no restriction on the presence of both types of compatible nuclei in common cytoplasm but only in the ascogenous hyphae can further development and, ultimately, fertilization occur. In other species of *Ascobolus*, e.g. *A. viridulus*, the ascogonial coil develops without exogenous stimulation, or conjugation and self-fertilization occurs.

A further species, *Ascobolus stercorarius*, has been studied by BISTIS (1956, 1957, 1965; BISTIS and RAPER, 1963) in an attempt to elucidate the sexual regulatory processes. In this fungus the ascogonium is a swollen and coiled branch, the antheridium an extremely slender hypha which may become detached. The ascogonium can develop a short blunt trichogyne and conjugation is between this and either an antheridium or an oidium (Fig. 15.7).

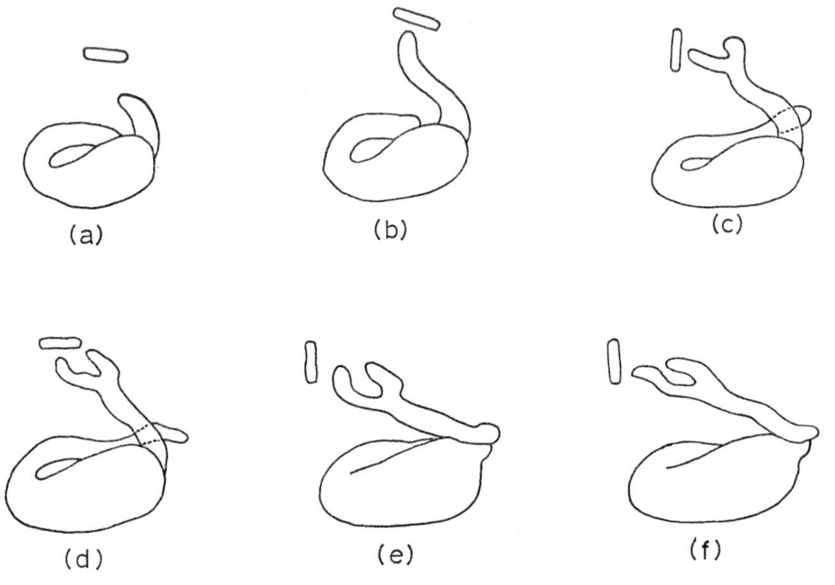

Fig. 15.7. Interactions between conidium of *Ascobolus magnificus* and the trichogyne, details of the ascogonium are not shown. After (b), (c) and (d) the conidium was moved. Note directed growth and branching of trichogyne. (After Bistis)

The fungus is dimictic and conjugation in nature occurs between *A* and *a* strains. If *A* and *a* cultures are in close juxtaposition antheridial hyphae are induced after several hours and ascogonial hyphae and ascogonia after several more hours. Trichogynes develop and show directed growth to antheridia in about half an hour and about seven days after conjugation, the apothecium develops. Since the first phase was telemorphotic, detached hyphae were exposed to culture filtrates of opposite or identical mating type. The only ones which showed development were *A* hyphae in secretions of *a* mycelium. Thus mating type *a* initiates the reaction in *A* mycelium and, after that, *A* mycelium can induce antheridia in *a* mycelium. The zygotropic action of the trichogyne was studied by putting an oidium of compatible mating-type within 100 μm.

The trichogyne grew towards the oidium after a period of induction and made contact. The oidium was immediately moved away. The trichogyne responded to this and subsequent moves by the production of laterals directed towards the odium (Fig. 15.7). It was noted that trichogyne tip growth stopped just prior to contact with the oidium (or if the oidium was removed). Substitution of one oidium by a fresh one also led to cessation of trichogyne growth or, if contact actually occurred, to its irregular growth and the dissolution of its tip with a consequent loss of cytoplasm. Bistis inferred that the normal reactions between trichogyne and oidium (or antheridium) were reciprocal but he did not actually claim that more than one substance was involved. He suggested that oidium substitution followed by trichogyne bursting indicated that normal conjugation involved a specific thigmotropic reaction. Thus in *A. stercorarius* there is a series of reciprocal reactions analogous to those described for *Achlya* or *Mucor* but operative over much smaller distances. Hormonal mechanisms were clearly indicated, the whole sequence being initiated by *a* mycelia.

However, Bistis carried out further substitution experiments. He induced antheridia in *a* and *A* mycelia by appropriate induction treatments and then transferred them to mycelia of the same mating type either purely vegetative, or with ascogonia possessing trichogynes already present. No response occurred in the former case but in the latter conjugation occurred between antheridium and trichogyne, even although the crosses were $A \times A$ and $a \times a$. Ascocarps began to develop but ceased to grow after 24 hr. Thus the mating-type factors probably serve to initiate and maintain sexual reactions and their actions are complementary; in the absence of both factors, processes initiated can only be maintained for a short time. In these experiments the trichogyne/antheridium reaction had been initiated and went on for a time even in a/a or A/A combinations but could not be sustained. In natural conditions, of course, such a sequence is unlikely to arise. Since *a* initiates the reaction, most of the crosses will be $a \times A$ in mixed cultures because of the timing difference. When oidia are transferred by insects, such as mites, from one mating type culture to another, the difference in bulk is such as to ensure that the oidia will only be induced to function as 'males' if they are of appropriate mating type.

This very elegant analysis has not been equalled in any other Ascomycete. There are indications that other species show similar behaviour, in some stages at least. For example, a zygotropic reaction is quite common between trichogynes and conidia in *Neurospora* and *Bombardia lunata*. In the latter the microconidia appear to produce a diffusible chemotropic substance that attracts the trichogyne tip if it comes near enough in its wide-ranging growth to a microconidial mass. The substance is specific to trichogynes of compatible mating type only (ZICKLER, 1937, 1952). It will also be recalled that Esser has demonstrated mutant blocks in the normal sexual reactions of *Podospora anserina* (cf. Chapter 5, pp. 131–133 and Table 5.2); one of these prevented trichogyne attraction of conidia while the other blocked conjugation. Thus this species seems to have a reaction sequence similar to that found in *Neurospora*. In *Saccharomyces cerevisiae*, where conjugation is between morphologically whole cells, LEVI (1956) showed that cells of one compatible type elongated towards the other. This elongation could be induced by removing the cells and replacing them by fresh ones; only those of the same mating type as had responded before showed the elongation reaction. This provides evidence both for an induction effect and an ensuing zygotropic reaction. The nature of the surface contact reaction has been investigated in *Hansenula* and described earlier (Chapter 12, p. 325).

Homodimixis

Attention may now be focused on some of the unusual mating systems. Several species, *N. tetrasperma*, *Gelasinospora tetrasperma* and *Podospora anserina* have 4-spored asci. A single ascospore is capable of producing a mycelium on which further generations of perithecia and ascospores can develop. DODGE (1927) first observed occasional asci with more than 4 spores, usually 5, in *N. tetrasperma*. Two of the 5 ascospores were smaller than the other 3 and, if isolated, were found to be capable of developing protoperithecia but not perithecia or asci. If microconidia were exchanged between the 2 mycelia, conjugation and fertilization occurred; the spores were evidently of different mating type. Normal spores of these species, it seems, are normally heterokaryotic for nuclei of a different mating type. This is now known to be correct for *N. tetrasperma* and the other Pyrenomycetes with 4-spored asci.

There are two points of interest in this condition. The first is that in a species closely related to normal dimictic Neurosporas, a mycelium heterokaryotic for nuclei of different mating type can develop. Here the restriction normally imposed on such heterokaryosis is quite lacking. The second point is the mechanism whereby such homodimictic conditions are brought about and how they have originated. Two mechanisms have been discovered which bring about this condition. In *N. tetrasperma*, the asci are much broader and shorter than those in *N. crassa* and, perhaps because of this, the second and third division spindles overlap. Since the locus of the mating type factors is near the centromere, they virtually always segregate at the first division. These two features ensure that when four spores are delimited, each contains two nuclei carrying different mating type factors (DODGE, 1927; DODGE *et al.*, 1950; SANSOME, 1946). In *P. anserina* quite a different mechanism is involved (RIZET and ENGELMANN, 1949; FRANKE, 1957). The mating type factors segregate at the second meiotic division in 98% of the asci and the spindles of this division lie across the long axis of the ascus. The nuclei separate at the third division and are delimited as four large spores (Fig. 15.8) arranged more or less transversely to the long axis of the ascus.

Crosses have been made between *N. tetrasperma*, *N. crassa* and *N. sitophila* by DODGE (1928). In the perithecia which developed asci, the 8-spored character was dominant but two generations of back-crossing to *N. tetrasperma* restored the 4-spored condition. In 1939 Dodge reported a lethal gene, *E*, in *N. tetrasperma* and studied it further with colleagues (DODGE *et al.*, 1950). Most of the asci aborted but occasionally they developed, in which case they were 8-spored. Finally, PATEMAN (1959) carried out selection experiments on

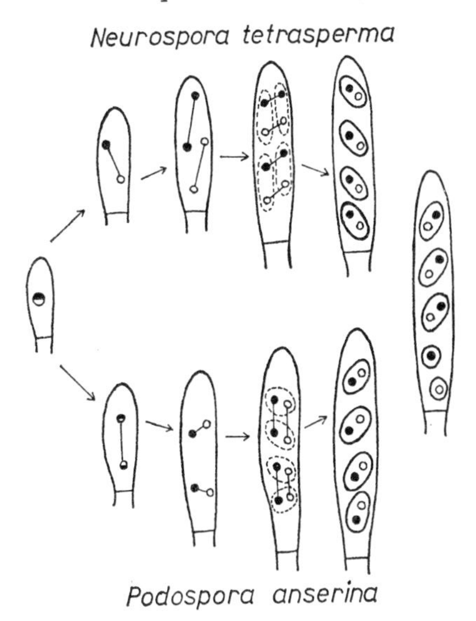

Fig. 15.8. Spindle orientation in the asci of *Neurospora tetrasperma* and *Gelasinospora tetrasperma*, respectively. In the former, mating type factors (indicated by o and ●) segregate at the 1st meiotic division, in the latter segregation is at the 2nd meiotic division. (After Fincham and Day)

N. crassa, selecting in successive generations the largest ascospores from which to breed. In this way he succeeded in rapidly reducing the incidence of 8-spored asci by about the eighth generation, after this the number of 4-spored and aborted asci increased. Between the third and sixth generations of selection, 5 of the strains with 4-spored asci proved to be self-fertile. A large-spored strain crossed with *N. crassa* again gave rise to an 8-spored ascus. Thus something resembling *N. tetrasperma* can be obtained by selection from

Table 15.1. The mean ascospore size of *Neurospora* species compared with *N. crassa* selected for large spore size. (After Pateman, 1962)

Species	No. of ascospores per ascus	Ascospores range (μm)	Mean or range of majority (μm)
N. crassa	8	23–32 × 11–16	27–30 × 14–15
N. crassa large spored strains	8 and 4	23–60 × 14–24	37 × 17
N. tetrasperma	4	29–35 × 14–16	30–31 × 15

N. crassa (Table 15.1), and genes in *N. tetrasperma* are capable of producing the 8-spored condition. This work suggests that the species differ polygenically and thus it may be conjectured that somehow, in the course of time, such differences arose and became genetically fixed in an originally 8-spored species to produce what is now called *N. tetrasperma*.

Anomalous situations

Anomalous situations are known in several Ascomycetes, notably the yeasts. There have been numerous claims concerning the mutability of the a/α mating-type factors of yeasts (LINDEGREN, 1943, 1944; AHMAD, 1953, 1965). This is certainly unusual, for in no other Ascomycete has this situation been recorded, not even after some three decades of intensive study of *Neurospora*, for example. Some of these claims are explicable if ascospores in yeast can be heterokaryotic for mating types although behaving as if of one mating type, i.e. a would have to be dominant to α or vice versa. Supernumary somatic divisions in the ascus could account for the apparently full complement of ascospores and these are known to occur (WINGE and ROBERTS, 1954). Such an ascospore could germinate and the nuclei eventually segregate, the two kinds of nuclei at bud formation apparently giving rise to a clone, mutant for the opposite mating type. In a cross between *S. cerevisiae* and *S. chevalieri*, a gene D segregated and, as a consequence, the ascus had two spores which gave diploid cultures and two of either a or α mating type. D is evidently responsible for diploidizing but, recently, HAWTHORNE (1963) has reported that it also induces mutation of one mating-type factor to the other, about 1–2 times/30 divisions. In this work heterokaryosis seems to have been ruled out. Another complex situation occurs in *Schizosaccharomyces pombe* studied by LEUPOLD (1950, 1958). Three alleles, h^+, h^- and h^{90} were discovered, the first two resembling mating-type factors in normal dimictic forms. Strains carrying h^{90} produced ascospores in about 90% (87%) of these asci and these were all self-fertile. Mutation from one allele to another was not uncommon. Leupold has explained much of the data by supposing that the locus is complex and bipartite, the two components being about one cross-over unit apart. One component confers the h^+ ability on a yeast cell carrying it; the other component is functionless. Thus h^+ cells are, in fact, either h^+h^+ or h^+h°, the h^- cells, are h^-h° and the h^{90} cells are h^+h^-. Thus on crossing an h^+ with an h^- strain the consequence of a cross-over would be to give h^+h^+ and h^-h°, the h^+ and h^- parental types, together with h^+h° and h^+h^-, an apparently h^+ type plus a homomictic h^{90} type. The crosses provided just such evidence and also indicated that there were two kinds of genetically distinct h^+ cell types. An alternative explanation is that there is a specific modifier gene s^{h} which effects only the h^- locus. On this view the cell types are $h^+ = h^+s$ or h^+s^{h}; $h^- = h^-s$; and $h^{90} = h^-s^{\mathrm{h}}$. The data available could support either view. The latter is, perhaps, more probable since suppressors of mating-type factors are known elsewhere in Basidiomycetes (cf. later, p. 432). The situation in the yeasts is clearly not simple and appears to be quite unlike that in any other fungus.

In *Chromocrea spinulosa*, MATHIESON (1952) has described a situation which has some similarity to the directed mutation of mating type factors in yeast by the D gene. The asci are 16-spored, 8 spores are large, 8 small. The latter, l, are self-sterile but the former, l^+, are self-fertile and give rise to stroma containing scattered perithecia with 16-spored asci, as before. If l^+ and l^- ascospores are inoculated together, the number of perithecium-bearing stromata is 4–5 times greater than from an l^+ ascospore alone. In such mixed cultures most of the perithecia arise from $l^+ \times l$ crosses; $l^+ \times l^+$ crosses give no such enhanced response, $l \times l$ crosses were not made. Mathieson thinks that there must be a directed mutation of $l^+ \rightarrow l$ in the hyphae. The l^+ and newly derived l nuclei then fuse to give the scattered perithecia typical of the l^+ ascospores. This remarkable situation was found in a single isolated strain and has not been reinvestigated. It serves to illustrate the unsolved complexities of sexual reproduction in fungi. A more explicable, although unstable, situation occurs in *Glomerella cingulata*. In this Pyrenomycete, self-sterile and self-fertile stocks have been found (WHEELER and MCGAHEN, 1952). Basically, the species seems to be homomictic, mutations having occurred at two loosely linked loci (44·7%), designated A and B. The A locus controls the formation of protoperithecia and conidia, the B locus the distribution of protoperithecia and their fertility (Table 15.2). Since conjugation in this species is restricted to that between trichogynes and conidia, certain crosses, e.g. $A^1B^2 \times A^2B^+$, are highly fertile whereas others, e.g. $A^1B^+ \times A^2B^2$ will be much less fertile or will not occur at all, e.g. $A^1B^2 \times A^2B^2$. Whenever a cross takes place, even if most of the perithecia are heterozygous, some self-fertilized perithecia will also occur. The situation is made more complex by the probable existence of diffusible substances which promote perithecium formation. However, although the addition of filtrates to different strains, or their separation by permeable membranes, led to enhanced perithecial production in some cases, no clear-cut results emerged (MARKERT, 1949). Moreover, the relationship of these phenomena to the genetic controls is not at all clear. The

Table 15.2. The mating system of some isolates of *Glomerella cingulata*. (After Wheeler and McGahen, 1952)

Genotype or cross	Expression or result
A^2B^2	Very slightly self-fertile; produced scattered protoperithecia—a 'female'
A^1B^+	Abundant conidia only—a 'male'
A^1B^2	Abundant conidia only—a 'male'
A^2B^+	More self-fertile than A^2B^2: produces protoperithecia—a 'female'
$A^1B^+ \times A^2B^2$	Intermediate perithecial production
$A^1B^2 \times A^2B^2$	No perithecia
$A^2B^+ \times A^2B^2$	Intermediate perithecial production
$A^1B^+ \times A^2B^+$	Least abundant perithecial production
$A^1B^2 \times A^2B^+$	Highest perithecial production

situation needs to be re-examined. It is an important one, in so far as it represents a situation potentially capable of developing into one resembling that in a typical dimictic Ascomycete such as *N. crassa* or, alternatively, a strictly homomictic form. For example, if A^2 were to become epistatic to B^+ as well as to B^1 and B^2, where it is already effective, then self-fertilization would be suppressed in strains carrying A^2. In a similar manner any mechanism leading to tighter linkage between the A and B loci would tend to perpetuate certain combinations such as the original A^2B^2 and A^1B^+ genotypes. A combination of these two possibilities would result in an apparently 'protoperithecial' and a 'conidial' strain. In such strains the A^2B^2 and A^1B^+ genotypes would appear to function as single alleles, through the hypothetical tight linkage, although in fact they would be complex, allelomorphic supergenes (BURNETT, 1956). This notion has been developed by OLIVE (1958) and, in the homomictic form, *Sordaria fimicola*, he has been able to synthesize an effectively dimictic condition (EL ANI and OLIVE, 1962). Two mutants, *a*-3 and *st*-59, were employed. Both require arginine, that for *a*-3 being only a partial requirement. When grown alone their asci or ascospores abort or are inviable; grown together they can complement each other and, because they are exceedingly closely linked, in 504 asci analysed, only the two parental genotypes segregated. Thus this pseudoallelic situation provides a model of the way in which a dimictic system might arise. Clearly, a case in which the pseudoalleles controlled features such as those shown by A and B loci in *Glomerella* would provide an even more impressive example of how obligatory cross-fertilization can occur.

A further isolated observation has some relevance to the situation. This is the phenomenon termed 'relative heterothallism' by HEMMONS *et al.* (1953). *Aspergillus nidulans* is an homomictic species but it can form heterokaryons through hyphal fusions. In such a case self-fertilization would be expected to result in association of similar nuclei in 50% of the perithecia (assuming that each arises from one binucleate cell) and of dissimilar nuclei in the other 50% of perithecia. In the 27 heterokaryons investigated, however, 16 gave rise to more than 50% perithecia with different genetic markers, 10 of which were exclusively derived from heterozygous fusion nuclei. Thus there is some mechanism in this fungus which either favours association of unlike nuclei, the subsequent growth of heterokaryotic cells, or those with heterozygous diploid nuclei over homokaryotic homozygous cells. The basis of the phenomenon has not been resolved but it is not confined to *Aspergillus*, since it has also been found once in *Sordaria fimicola* (OLIVE, 1954). The phenomenon provides a mechanism which would clearly favour selection for heterozygosity.

Finally, there is the enigmatic position of the Laboulbeniales; nothing is known of the regulation of their sexual activities. In some cases, the mycelium bears antheridia and ascogonia and conjugation is by microconidia. Presumably these are shaken on to the trichogyne. In other cases, each mycelium is either male or female and the two thalli may be developed at different sites on the same or different insect hosts (Fig. 15.9). It is quite unknown whether such dimorphic development is regulated genetically or environmentally, or

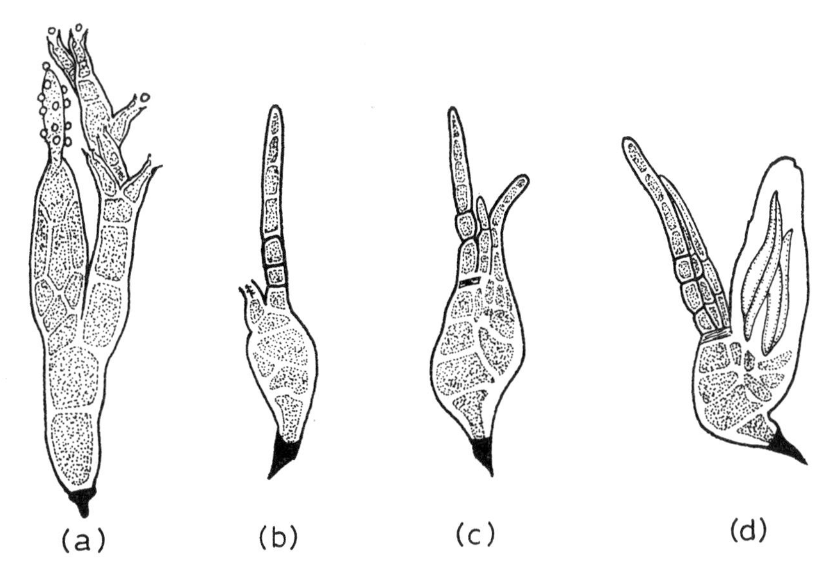

Fig. 15.9. Two species of Laboulbeniales. (a) Homomictic, *Stigmatomyces baeri*, note trichogyne covered in microconidia derived from antheridia on lateral branch; (b), (c), (d) male, female and fertilized female thalli of dimictic *Laboulbenia formicarum* (a, × 580, from Thaxter; b–d, × 410, from Benjamin and Shanor)

how conjugation is effected. Until these fungi are better known and cultivated in vitro it will be difficult to resolve their sexual regulating mechanisms (cf. Chapter 12, p. 356).

The Ascomycetes, on the whole, present simpler modes of sexual reproduction than the Phycomycetes. Dimixis and homomixis are usually clear-cut when they occur. Homodimixis raises, in an acute form, the problem of how fungi regulate heterokaryosis in relation to mating type factors in the different parts of their mycelia. The resolution of such a problem is likely to be of general interest and value to the whole problem of nucleo-cytoplasmic relationships. The problem of the evolution of mating systems is also raised by the situation in fungi such as *Glomerella* and, in particular, the role of recurrent mutation in the origin and stability of mating type factors as exemplified by the situation in yeasts and *Chromocrea*. It is difficult to see how selection has operated to convert an inbreeding species to outbreeding or vice versa but, at least, the potentialities seem to exist in some Ascomycetes.

SEXUAL REPRODUCTION IN BASIDIOMYCETES

More is known about the genetics of sexual reproduction in these fungi than in any others but far less about the ways in which either genetic or physiological regulators operate.

Amongst outbreeding forms, Uredinales are dimictic; the Ustilaginales can be described as di- or diaphoromictic, although the latter description is preferred here and, such other fungi as have been studied, are diaphoromictic. There are still groups in which the regulation of sexual reproduction is not understood, usually because of difficulties experienced in germinating the spores or culturing the mycelium, e.g. several Gasteromycete orders such as Lycoperdales (puff-balls) or Phallales (stink-horns). Many species in all the groups of Basidiomycetes are homomictic but it is possible that more are outbreeding than in Ascomycetes or Phycomycetes (RAPER, 1966a).

Uredinales

Conjugation is usually by means of pycnidiospores and flexuous hyphae and BULLER (1950) has documented these processes in great detail (Fig. 1.10, p. 14). Pycnidiospores only develop in close proximity to flexuous hyphae of compatible mating types which show zygotropic curvatures towards the spores. The essential observation was made by CRAIGIE (1927) who showed that the pycnidiospores are carried passively by insects which suck the 'nectar drops' (exuded by pycnidia), in which the spores are held. Each pycnidium arises from a single uninucleate basidiospore infection and, in any one dimictic species these are of two kinds only. He was also able to show that fusions could occur between monokaryotic hyphae of compatible types; later, BROWN (1932) showed that monokaryotic hyphae could fuse with hyphae already heterokaryotic for compatible mating-type nuclei. After conjugation the hyphae are, of course, heterokaryotic in respect of mating type but, from the development of the aecidiospores onwards the compatible nuclei are associated in pairs. This condition is manifested in both uredospores and teleutospores and probably persists in the intervening mycelial stages, the nuclei dividing synchronously (or conjugately). This especial type of heterokaryon containing paired compatible nuclei and usually showing conjugate divisions is termed a dikaryon. In the Uredinales, therefore, a prolonged post-conjugation phase is exhibited prior to fertilization in the germinating teleutospore. In such fungi it may persist for several months, from the aecidial stage in the spring until the teleutospores are formed in late summer and autumn. In dimictic forms with a perennial dikaryotic mycelium it may last for years, e.g. *Puccinia suaveolans*, and it is of some interest that this phenomenon is also shown by homomictic species.

There is evidence that both heterokaryosis and somatic recombination can also occur during the dikaryotic phase (see Chapter 15, p. 439).

Ustilaginales

Products of a single meiosis, basidiospores, derived from single brand spores (≡teleutospore) are often budded off in a yeast-like manner from a germ tube and may either be released or germinate *in situ* in pairs, e.g. *Tilletia caries*. If separate basidiospores are obtained they can be germinated and grown in culture. When paired, two by two, the monokaryotic cells or hyphae fall into two compatible groups. These are determined genetically, by a pair of allelomorphs (*A*, *a* or *A1*, *A2*). In *Ustilago maydis*, mycelia derived by fusion fall into two groups. They are either irregular in growth, branching frequently and developing uninucleate spores within 12 hours, or regular, little branched and do not produce spores. The latter mycelia are capable of infecting the host plant, maize (*Zea mais*), the former are not. This capability for stability and infectivity of the dikaryon is regulated by at least 13 multiple-allelomorphic genes, *B1–B13*, not linked to the *A* locus. A successful infective dikaryon and hence one capable of ultimately undergoing fertilization, differs in its genotype at both loci, i.e. *A1 Bx*+*A2 By* where $x \neq y$, e.g. *A1 B7*+*A2 B13* (ROWELL, 1954). This type of analysis has not been carried out in such detail in other smut fungi but there is evidence that a similar situation occurs in other species. For example, BAUCH (1932) described the production of mycelia of two types by *U. longissima* after conjugation, 'Suchfaden' and 'Wirrfaden'. The former were normal infective hyphae, the latter consisted of a finely branching mass of tangled hyphae. Conjugation was controlled by a pair of allelomorphs, the subsequent development by at least six allelomorphic genes at another locus. In *Sphacelotheca sorghi* (the cause of covered smut of *Sorghum*) some 66 'sex groups' have been described which can be accounted for, formally, by a two-locus system similar to that already described (VAHEEDUDDIN, 1942). It is, perhaps, a semantic matter whether such a system of regulation is described as dimictic or diaphoromictic. WHITEHOUSE (1951) has written, 'However, just as with *U. longissima*, there is no justification for regarding the form of the hyphae after conjugation as an essential part of the mating system'. He regards the mating system as being that part of the process which is determined by the pair of allelomorphs, e.g. *A1* and *A2* in *U. maydis*; the other locus is concerned with ability to infect. On the other hand, if no infection takes place, fertilization will not succeed and the mating system will not really be regulating sexual reproduction. For such reasons it may be thought preferable to describe the system as diaphoromictic, there being more than two potentially compatible mating types, e.g. *A1 B1* is compatible with *A2 B2*, *A2 B3*, . . . *A2 B13* in *U. maydis*. Very little is known about how the *A* or the *B* factors regulate the development of the mycelium. There is evidence from a number of investigators (e.g. DICKINSON, 1927) that hyphae or germ tubes from isolated basidiospores show zygotropic curvatures towards each other if they are compatible; BAUCH (1932) noted that such germ tubes developed more rapidly in mixtures of compatible spores in liquid culture than in spores of one mating type. Thus there seem to be both telemorphotic and zygotropic processes involved, comparable with those described in Phycomycetes and Ascomycetes. Another point of similarity between Phycomycetes and Ustilaginales comes from the work of THREN (1937, 1940) who studied *U. nuda*. When + mating-type cultures were opposed they intermingled but two − mating-type cultures when opposed developed a narrow zone of aversion *c.* 1·0 mm wide between them. This is reminiscent of the aversion reaction of *Mucor* and *Phycomyces* mentioned earlier, although these develop in both +/+ and −/− confrontations. All these reactions are presumably regulated by the *A*-locus since the *B* factors operate after a successful fusion has been accomplished.

Tremellales

Not many investigations have been made of the remaining groups of the Heterobasideae. In the Jew's ear fungus, *Auricularia auricula-judae* and in *Tremella mesenterica*, tetrapolar diaphoromixis is known to occur although it has not been studied extensively. Other fungi in the order are known to have mating systems but they have not yet been defined with precision.

Homobasideae

In Agaricales, Polyporales and Nidulariales (bird's-nest fungi) either bipolar or tetrapolar diaphoromixis is the rule in outbreeding forms. The numbers of mating-type factors involved may run into tens or even hundreds (cf. Table 15.3); there is some slight reason to suppose that the tetrapolar species may

Table 15.3. Numbers of mating-type factors identified in diaphoromictic fungi and estimated total numbers based on these finds. (From Burnett, 1965b)

BIPOLAR SPECIES

	Monokaryons examined	Alleles identified	Estimated total alleles	Author
Auricularia auricula-judae	10	10	very large	Barnett, 1937
Coprinus comatus	11	9	30	Saunders (unpublished)
	22	11	15	Brunswik, 1924
Fomes roseus	10	9	50	Mounce & Macrae, 1937
Fomes subroseus	20	20	very large	Mounce & Macrae, 1937
Polyporus betulinus	201	28	30	Saunders, 1956, & Burnett (unpublished)
Mycocalia denudata				
Race I	6	4	7	Burnett & Boulter, 1963
Race II	14	7	9	

TETRAPOLAR SPECIES

		A	*B*	*A*	*B*	Author
Coprinus fimetarius	27	27	27	very	large	Brunswik, 1924
Coprinus lagopus	14	14	14	very	large	Hanna, 1925
Coprinus macrorhizus	20	19	19	240	240	Kimura, 1952
Polyporus abietinus	28	23	26	100	200	Fries and Jonasson, 1941
Polyporus obtusus	48	39	39	85	85	Eggertson, 1953
Polystictus versicolor	46	20	20	25	25	Partington, 1959
Schizophyllum commune	114	96	56	339	64	Raper *et al.*, 1958
Crucibulum vulgare	30	3	11	5	15	Fries, 1936
Cyathus striatus	18	4	5	5	5	Fries, 1936, 1940

have larger numbers. The Nidulariales have the smallest numbers of mating type factors and it is noteworthy that their basidiospores are normally dispersed enclosed within a peridium. Thus some inbreeding is enforced physically, since compatible basidiospores will be dispersed together and in close proximity. It is not clear whether these features are correlated or whether they are associated by chance (WHITEHOUSE, 1949b; BURNETT and BOULTER, 1963). Compatible matings are those between monokaryons carrying different mating-type factors in their nuclei, e.g. *A1*+*A2* in a bipolar form such as *Marasmius oreades*, or *A1 B1*+*A2 B2* in a tetrapolar form such as *Coprinus lagopus*. In many fungi a compatible conjugation is followed by nuclear migration and association into compatible pairs which divide conjugately, at least in an apical cell. Patterns of nuclear migration have already been discussed in Chapter 14; the kinds of pattern found in plate cultures of compatible diaphoromictic fungi have received considerable study and are illustrated in Fig. 15.10. In some cases nuclear migration is extensive and

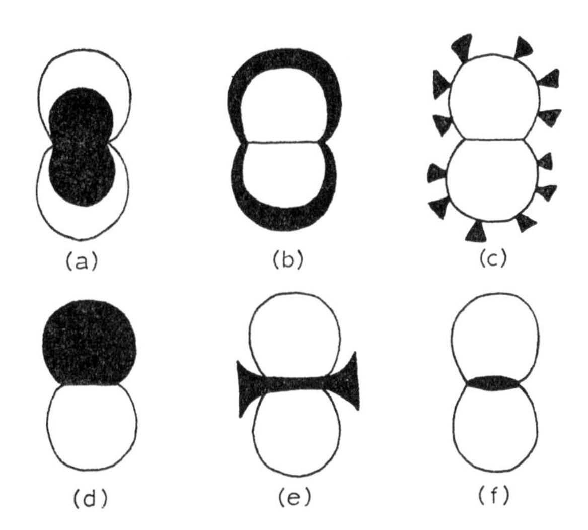

Fig. 15.10. Patterns of dikaryotization in compatible confrontations of diaphoromictic Basidiomycetes. The dark regions represent dikaryotic hyphae. (a)–(c) Reciprocal dikaryotization, in (a) dikaryotic hyphae develop over the surface of the monokaryons; in (b) the nuclei migrate and dikaryotic hyphae only become visible at the growing margins; (c) is similar to (b) but the migration is confined to certain routes and is not general; (d) unilateral migration of nuclei, unilateral patterns analogous to (b) and (c) may also develop; (e) and (f) restricted dikaryotization, in both the dikaryotic hyphae are restricted to the region of contact but in (e) their growth rate is high enough to enable them to fan out at the edges of the zone of contact.

equal, in others unequal or sometimes unilateral. In some cases it is very restricted and dikaryotic hyphae only grow away from the free margins of the zone of migration and dikaryosis (Fig. 15.10e). The dikaryotic condition may be associated with the formation of clamp connexions which involve hook-cell formation, septation and fusion (see Fig. 4.9, p. 73). The mycelium produced may grow at a different rate from the parental monokaryons, either faster or more slowly and may differ from them in its form. An often quoted example of this is the difference between monokaryons and dikaryons of *C. fimetarius* (≡*C. lagopus*) concerning the angle between main and lateral hyphae. This is wide in the monokaryons but very narrow in dikaryons. However, this kind of difference is, at best, species specific and often even strain specific. Another difference, and a much more constant one, between the two types of hypha, is the suppression of conidial or oidial formation in dikaryons but there are exceptions to this, especially in the Thelephoraceae (crust fungi) where conidia are borne on both hyphal types (MAXWELL, 1954).

The presence or absence of clamp connexions is a useful guide to the occurrence of a compatible fusion but they are not essential and various aberrant expressions occur in some non-compatible situations (see later). *Collybia tenacella* is an example of a species with a dikaryotic mycelium and conjugate nuclear division but wholly lacking in clamp connexions (YEN, 1948, 1950).

Dikaryons can persist in some cases for many hundreds of years without conjugation occurring. Some evidence for this has been provided in the bipolar fairy-ring fungus *Marasmius oreades* where it has also been shown that the mating-type factors involved remain stable over the same period (BURNETT and EVANS, 1966). Ultimately, however, fusion of compatible nuclei occurs in a basidial cell and meiosis follows. It is not clear what stimulus leads to the onset of nuclear fusion but it must, presumably, reside in the cytoplasm which, in the basidium, is always intensely basophilic. This suggests that it is rich in RNA but the significance of this is not clear. There is certainly no evidence that in fungi RNA-rich cytoplasm promotes nuclear fusion *per se*. Hyphal cells of the common cultivated mushroom are highly basophilic and RNA-rich but there is no cytological evidence for promotion of nuclear fusion in such a situation (THOMAS *et al.*, 1956).

Although fertilization and meiosis are confined to the basidial cells there is evidence, discussed earlier, for parasexual phenomena in Basidiomycetes. Basidiocarp formation is not necessarily dependent upon the prior formation of a compatible dikaryon. In some fungi monokaryons have never been known to produce basidiocarps on monokaryotic hyphae; in other cases, e.g. *Polyporus betulinus*, they only produce a gill-like or pore-like surface. *Schizophyllum commune* can produce basidiocarps in certain non-compatible situations as a rare event and also in monokaryotic cultures; the process is very dependent upon the genetic background of the strain (RAPER and KRONGELB, 1958). Nuclear fusion and meiosis apparently do not occur in such basidiocarps.

The sequence of development described for a typical diaphoromictic fungus is also followed by many homomictic species, e.g. *Coprinus sterquilinus* (BULLER, 1934). A spore of this fungus, originally uninucleate, germinates and, after some non-synchronized divisions, the nuclei begin to divide synchronously and associate in conjugate pairs. Thereafter, they behave as in a dikaryon and clamp connexions are associated with nuclear division. In the basidium, the conjugate nuclei fuse and undergo two divisions although it is not known whether these are meiotic. In other fungi the dikaryotic condition is not evident, the cells may be multinucleate and lacking clamp connexions but nuclear fusion and meiosis seem to occur in the basidium, e.g. *Psalliota hortensis* var. *bisporus,* the common cultivated mushroom (EVANS, 1959).

Against this wide range of permutations of behaviour and morphological expression in monokaryons and dikaryons, attempts have been made to determine the action of the mating-type factors in diaphoromictic forms. Nothing is known about the action of these factors in bipolar species but more success has attended the study of tetrapolar ones.

If a diploid nucleus has a constitution, in respect of mating-type factors, of *A1 A2 B1 B2*, then four sorts of basidiospore can be produced, the *A* and *B* loci not being linked. These are:

A1 B1 : *A2 B2* : *A1 B2* : *A2 B1*

The consequences of intercrossing them are shown in Table 15.4 and Figs. 15.11 and 15.12. A full discussion is presented by RAPER (1966a) in his important book on the genetics of sexuality in higher fungi. He has introduced a useful notation which will be employed here, viz:

Homokaryon (monokaryon)	$Ax\ Bx$	e.g. *A1 B1*
Common –*AB* heterokaryon	$A{=}\ B{=}$	e.g. *A1 B1*+*A1 B1*
Common –*A* heterokaryon	$A{=}\ B{\neq}$	e.g. *A1 B1*+*A1 B2*
Common –*B* heterokaryon	$A{\neq}\ B{\neq}$	e.g. *A1 B1*+*A2 B1*
Dikaryon	$A{\neq}\ B{\neq}$	e.g. *A1 B1*+*A2 B2*

Table 15.4. Consequences of crossing basidiospores of various genotypes in a tetrapolar, diaphoromictic fungus. (See also Figs. 15.11 and 15.12)

	A1B1	*A2B2*	*A1B2*	*A2B1*	*A3B3*
A1B1	—	C	(F)	(B)	C
A2B2		—	(B)	(F)	C
A1B2			—	C	C
A2B1				—	C
A3B3					—

C = clamps and dikaryon; (F) = 'flat' mycelium; (B) = barrage mycelium.
Flat and barrage are names given to these heterokaryons in *Schizophyllum*.

It became clear from studies with *Cyathus stercoreus* (a bird's-nest fungus), *S. commune* and *C. lagopus* that when the $A{\neq}$ condition obtained either clamp connexions or false clamp connexions formed and, similarly, the $B{\neq}$ condition was always associated with nuclear migration (FULTON, 1950; PAPAZIAN, 1950; SWIEZYNSKI and DAY, 1960). It was concluded, therefore, that the *A*-locus controls clamp connexion formation, the *B*-locus, nuclear migration. RAPER (1966a) has drawn attention to the fact that some developments are regulated by interactions of the *A* and *B* factors also and, by comparisons of

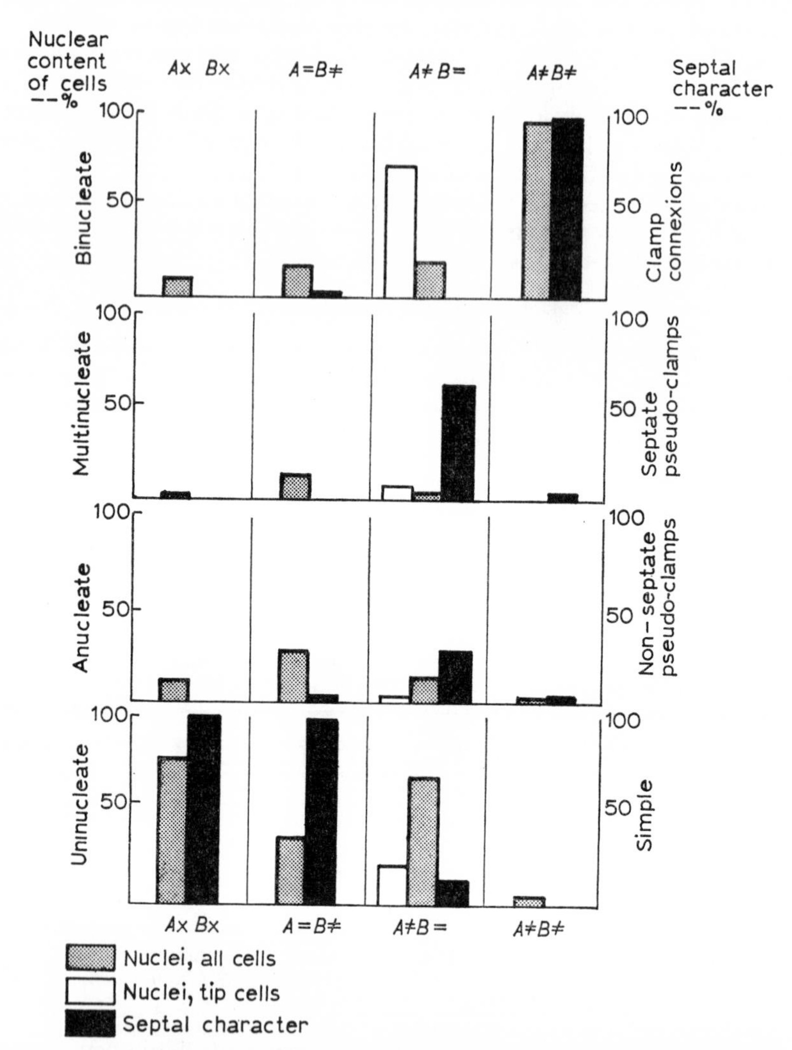

Fig. 15.11. Distribution of nuclei and clamp, or clamp-like condition of mycelia of *Schizophyllum commune*, homokaryotic or heterokaryotic for one or both mating-type loci. Minimally 200 cells were scored for each mycelial type. Tip cells are only plotted separately in $A \neq B =$ heterokaryon since here there is a marked difference from the mycelial average. (After Raper)

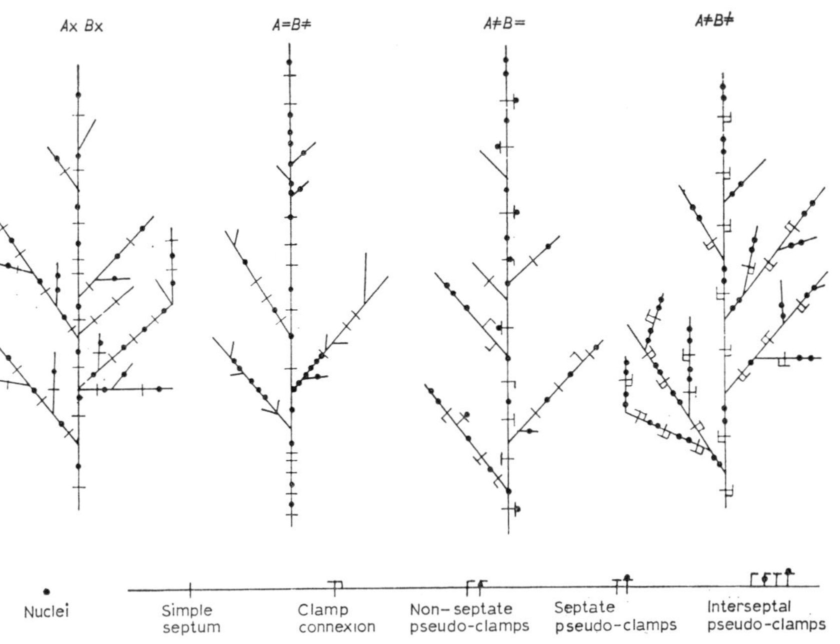

Fig. 15.12. Diagrams to illustrate nuclear conditions and septal developments in the same homokaryons and heterokaryons of *S. commune* as in Fig. 15.11. (From Raper)

normal $A \neq B \neq$ mycelium with $A = B \neq$ and $A \neq B =$ heterokaryons, he has assigned regulating roles to the factors for each event, viz:

(a)	Nuclear migration	$B \neq$
(b)	Nuclear association	$A \neq$
(c)	Conjugate division	$A \neq$ or nuclear proximity
(d)	Hook-cell formation	$A \neq$
(e)	Hook-cell septation	$A \neq$
(f)	Hook-cell fusion	$B \neq ? + A \neq$

He has further developed an hypothesis of how this regulation is exercised but it has not yet been tested. In essence, his hypothesis is based upon the supposition that (a) *A* and *B* factors are master, or switching, regulating genes; (b) single *A* and *B* factors control by inhibiting dikaryosis and (c) association of different *A* or *B* factors in the same cytoplasm results in release of the inhibition on dikaryosis imposed by the presence of single *A* or *B* factors. Thus, in a monokaryon the *A* factor elaborates a repressor which 'switches-off' the genes responsible for the dikaryotic functions (b)–(e), outlined above, and the *B* factor likewise represses the genes responsible for functions (a) and (f). In a heterokaryon or dikaryon, however (with $A \neq$ or $B \neq$ or $A \neq B \neq$), interactions between the gene products of the different *A* and *B* factors, leads to the inhibition of the repressor effects; thus the dikaryotic behaviour is free to develop. This hypothetical system must be remarkably versatile, for the products of any one factor must be able to interact specifically with those of all other factors at the same locus.

The situation is even more complex than first appears for the *A* and *B* factors in *S. commune* are themselves complex loci, there being two closely linked sub-units, e.g. $A\alpha$ and $A\beta$, at each locus. This is true of a number of other tetrapolar species (see TAKEMARU, 1961; FINCHAM and DAY, 1965) and may even occur in all of them. Extensive tests have not yet been made on a range of bipolar forms but there is no suggestion to date of a complex locus, in the only one examined fully for recombination, i.e. *P. betulinus* (BURNETT and SAUNDERS, unpublished, and in RAPER, 1966a). Estimates have been made of the number of sub-units in *S. commune*; these are $A\alpha$—9, $A\beta$—50, $B\alpha$—7 and $B\beta$—7 (RAPER, 1966; KOLTIN *et al.*, 1967) although the *B* factors are not yet resolved satisfactorily. It will be clear, therefore, that it is not strictly true to describe mating-type factors as being multiple allelomorphic series. In a fungus such as *S. commune*, the mating-type factors are each made up of two integrated series of multiple alleles and it is the further integration of the actions of these that results in the regulation of mating. Thus, on Raper's hypothesis the $A\alpha$, $A\beta$ sub-units must act together to co-repress and their products must interact with those of another pair of sub-units in a di- or heterokaryon; a similar situation obtaining at the same time with the $B\alpha$ $B\beta$ sub-units. Since the primary products of any one sub-unit are wholly unknown at present (although presumably protein), it is clear that the full elucidation of regulation in diaphoromictic systems will be a considerable problem.

Mutants of the mating-type factors, or specific suppressors of them are known which cause them not to operate, i.e. self-fertility becomes the rule. Presumably, in these circumstances the supposed inhibitory action of single factors is in some way repressed and the dikaryotic behaviour pattern develops. Incidentally, although the spontaneous or induced inception of new fully functional mating-type factors has been sought, they have not yet been found. New factors can arise by crossing-over between sub-units. For example, in *S. commune* *A41* and *A42* are designated $\alpha 1$–$\beta 1$ and $\alpha 2$–$\beta 5$, respectively, in respect of their sub-unit alleles. (The full designations are written *A41* $\alpha 1$–$\beta 1$ and *A42* $\alpha 3$–$\beta 5$.) These should give cross-over types $\alpha 1$–$\beta 5$ and $\alpha 3$–$\beta 1$ and these have, indeed, been detected and found to be already present in wild populations designated as *A62* and *A31*, respectively. Anomalies in the frequency of crossing over between the α and β loci occur but this may be regulated by genes other than mating-type factors.

Homo-diaphoromixis

In addition to the basic out-breeding systems there is a variety of superimposed modifications which result in a greater or lesser degree of inbreeding. These homo-diaphoromictic systems have not been very fully investigated.

A common feature of Agaricales is the occurrence of species with 2-spored basidia. They are often extremely similar to 4-spored species and were described as varieties of them by taxonomists. Careful studies of them have been made by SASS (1929), SMITH (1934) and LANGE (1952). Lange, especially, in his studies with *Coprinus* spp., showed that crosses were not possible between 2- and 4-spored, apparently very similar, species. When large numbers of basidiospores are isolated from 2-spored species a small percentage gives rise to monokaryotic mycelia, the majority being typically dikaryotic. Mated in all combinations these monokaryons often show compatibility patterns suggesting bipolar or tetrapolar behaviour. For example, 6% of all basidiospores of *Galera tenera* f. *bispora* are monokaryons and these fall into two compatibility groups; the 'related' 4-spored form, *G. tenera* is a bipolar species. This situation could arise if, after meiosis in the 2-spored form, there was some mechanism which tended to bring nuclei of compatible genotypes into the same basidiospore. Thus most basidiospores would be, for example, *A1*+*A2* but a few (the 6% found) either *A1*+*A1* or *A2*+*A2*. The association cannot be random since this would give 50% *A1*+*A2*: 25% *A1*+*A1*: 25% *A2*+*A2*, which is not found. This bias in favour of the association of genetically different nuclei is not dissimilar to the situation already described in the homomictic Ascomycete, *Aspergillus nidulans* (p. 422). In the cultivated mushroom, a 2-spored form, a suggestion has been made by EVANS (1959) that differences in the manner of alignment of spindles at the second meiotic division could account for 60–80% of heterokaryotic spores for a gene near the centromere (cf. Fig. 13.8, p. 378 and Fig. 15.13). The frequency of

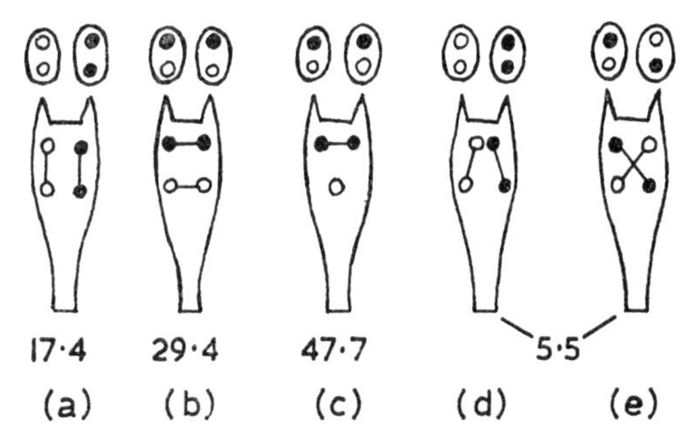

Fig. 15.13. Diagram to illustrate the expected distribution of a pair of alleles, indicated by ○ and ●, into basidiospores of *Psalliota hortensis* var. *bisporus* with the observed frequencies of spindle arrangements. Note *c*. 20% (17·5+2·75) are likely to be homokaryotic for the alleles. (After Evans)

different spindle alignments was measured in a small sample and it was assumed, reasonably enough, that the nuclei nearest the sterigmata would migrate first. He showed that 20–30% of basidiospore cultures do not produce basidiocarps and noted that this is not dissimilar from the expected figures of 20–40% homokaryotic spores expected from his spindle-alignment mechanism. The mating system of *P. hortensis* var. *bisporus* is, however, not known and the coincidence to which Evans has drawn attention, although suggestive, could be quite spurious.

A different system has been described in *Myocalia denudata* which might also apply to other fungi (KENNEDY and BURNETT, 1956; BURNETT and BOULTER, 1963). A number of fungi are known in which a further somatic division follows meiosis in the basidium to give an 8-nucleate condition (KÜHNER, 1945). Some fungi with 4-sterigmata give rise to basidiospores homo- and heterokaryotic in respect of mating-type factors. In *M. denudata* both these factors operate since some strains carry a gene *Pd* which induces a precocious division of the meiotic nuclear products in the basidium, others carry its allele *pd*, which causes this division to occur in the basidiospores. Strains carrying *Pd* are homo- and heterokaryotic in respect of mating type in their basidiospores but *pd* strains are always homokaryotic and give rise to normal monokaryotic mycelia (Fig. 15.14). In this species, therefore, while bipolar

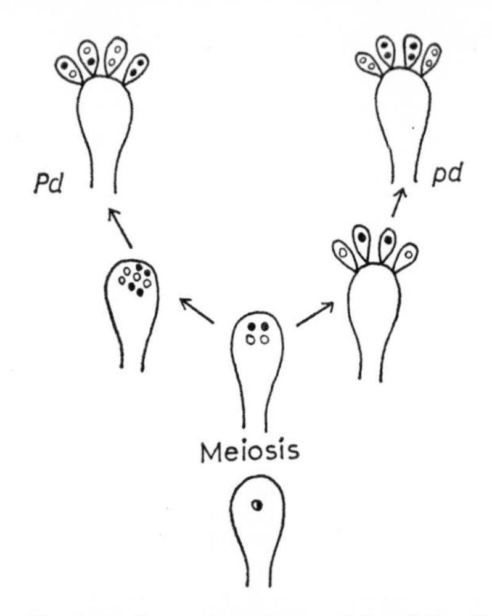

Fig. 15.14. Diagram to illustrate the action of the alleles *Pd*/*pd* in *Mycocalia denudata*. Those strains carrying *Pd* have 8 nucleate basidia which give, on average, 50% spores homokaryotic: 50% heterokaryotic in respect of any alleles indicated by ○ and ●. (From Burnett and Boulter)

diaphoromixis may be the usual system, this can be modified, facultatively, to homo-diaphoromixis if the unlinked gene *Pd* is present in the genotype of the dikaryon. It is also possible in these fungi to produce propagules, heterokaryotic for mating type, as a result of the abortion of developing basidia; these aberrant basidia are commonly found in the peridiola of *Mycocalia* spp. (PALMER, 1957, and *in litt.*). Both processes may contribute to the inbreeding imposed on *M. denudata* in nature.

In the Basidiomycetes, therefore, a full range of obligate outbreeding systems is arrayed. It is a striking fact that the largest and most differentiated of the fungi appear to possess the most complex genetical, regulatory control over reproduction. Moreover, these fungi are unique in the living world in their possession of an extended post-conjugation, pre-fertilization phase. In no other organisms are the potential gametic nuclei maintained in such close proximity, synchronized in function and behaviour for so long before fusion occurs.

The mating systems of the fungi are full of remarkable and challenging situations, many of which are not readily explicable in physico-biochemical, biological or evolutionary terms. These latter considerations will be discussed in the following chapters.

Chapter 16

The Occurrence and Significance of Recombination Systems in Fungi

Various systems which bring about recombination of plasmons, nuclei or genes, have been described in Chapters 14 and 15. It may be asked how far these systems, frequently studied in the laboratory, actually occur in nature and, if they occur, how effective are they in promoting recombination. In this chapter an attempt will be made to answer these questions.

THE OCCURRENCE OF FUNGAL RECOMBINATION SYSTEMS IN NATURE

Heteroplasmons

The best evidence for the occurrence of these in nature comes from demonstrations of the role of cytoplasmic factors both in freshly isolated organisms which have not been cultured for any great length of time and from obligate pathogens which cannot so be cultured.

Jinks, as mentioned in Chapter 14, has been able to demonstrate the existence of heteroplasmons in newly isolated strains of *Penicillium* spp. exhibiting the 'dual phenomenon', i.e. the production of predominantly mycelial or sporing strains. In the case of *Aspergillus glaucus* also, characters such as growth rate and senescence, including the condition 'vegetative death' occurred in strains isolated from the wild.

Other good evidence for the natural occurrence of cytoplasmic factors is the phenomenon of invasive spread. Such a condition is exhibited by the *fluffy* variant of *Coprinus macrorhizus* which infects adjacent mycelia when isolated (DICKSON, 1935; PAPAZIAN, 1958).

Maternal inheritance, although not absolutely conclusive, is highly suggestive and there are clear examples of this in obligate parasites where artificial selection is excluded. For example, JOHNSON and NEWTON (1940) studied the pathogenicity of F_1 aecidiospores, derived from crossing Races 7 and 11 of *Puccinia graminis* var. *avenae*, on different varieties of oats. On some the behaviour was such as might be expected from dominance interactions and independent of which parent had possessed the pathogenic expression. On Joannette and Sevenothree, however, the infection type which developed resembled that of the receptor parent on the variety. Since the donor parent, the pycnidiospore, transmits at most only a little cytoplasm whereas that from the protoaecidial receptive mycelium is enormous in comparison, the reciprocal differences were attributed to the cytoplasmic component (Fig. 16.1). Support for such an interpretation comes from LAMB's (1935) cytological observations that the nucleus alone appears to enter a flexuous hypha from a pycnidiospore in *P. phragmitis* and also from pathogenicity patterns in crosses between races of *P. graminis tritici*. In these crosses, reciprocal differences detected in the F_1 persisted to the F_2 and F_3 generations on some hosts (NEWTON and JOHNSON, 1932; JOHNSON *et al.*, 1934).

Although the number of cases in which a heteroplasmic condition has been detected in a genuinely natural situation is small this is, in part, because it has not really been looked for. There is no reason to suppose that the spontaneous cytoplasmic conditions which have been detected in fungi in culture do not occur in an analogous manner in nature. For instance, the cytoplasmic condition *petite colonie* almost certainly occurs in wild species of yeasts since natural isolation always produces a proportion of such small colony forms. Their basis has not, however, been examined save in domesticated strains and species.

Heterokaryons

There is much less doubt about the occurrence of heterokaryosis in nature since it is clear that even if some of the examples of the 'dual phenomenon' are due to heteroplasmons—at least as many are due to heterokaryons, since this condition has been recorded in almost every Fungus Imperfectus that has been examined. It seems that heterokaryosis is probably widespread although perhaps too often inferred uncritically in the past (CATEN and JINKS, 1966). The general occurrence of hyphal fusions and nuclear migration also provides appropriate circumstantial evidence. Nevertheless, the number of occasions in which nuclear segregation from heterokaryosis has actually been detected in a natural situation is rare. This is to be expected if selection favours heterokaryosis and thus its non-detection by no means indicates its non-existence.

Parasexuality

No unequivocal demonstration has yet been made of the parasexual cycle in nature. The fact that several pathogenic fungi have been shown to be capable

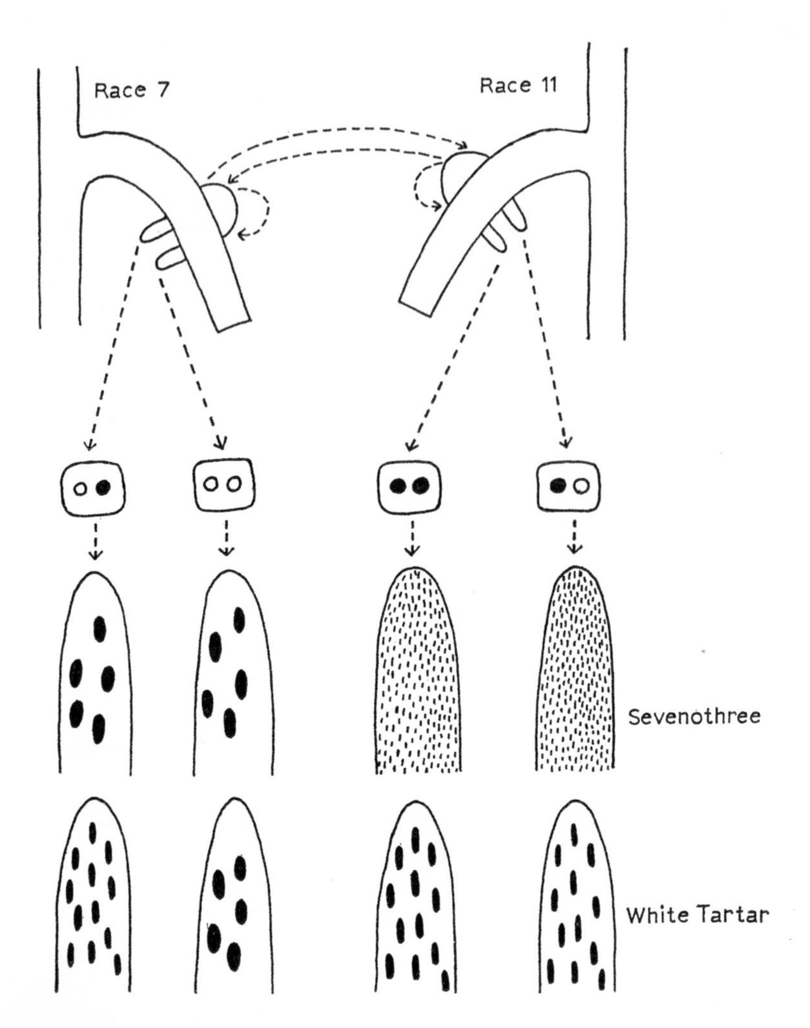

Fig. 16.1. The consequences of self-fertilization and reciprocal crosses of Races 7 and 11 of *Puccinia graminis avenae* on the oat varieties Sevenothree and White Tartar. Note that in White Tartar the type 2 infection of Race 11 is dominant in the cross 7 × 11 but that in Sevenothree the behaviour is determined by the infection pattern of the receptive race, i.e. 7 × 11 gives a type 4 infection and 11 × 7 gives a type 1 infection. (Small dots represent type 1 infection, large dots type 4 and intermediate dots type 2.) (See Fig. 17.6 for illustrations of infection types.) (Based on Johnson and Newton)

of exhibiting the full cycle in quasi-natural conditions is suggestive of its actual occurrence. A good example is BUXTON's study of the pea-wilt and banana-wilt Fusaria (1956, 1962). In these he induced biochemical mutant markers and combined them with natural markers for colour and pathogenicity. A typical example is shown below:

Parental homokaryons	Race 1:	*arg*$^+$	*meth*$^-$	*red*	*intol*
	Race 2:	*arg*$^-$	*meth*$^+$	*white*	*tol*
			↓		
Heterokaryon phenotype	(Race 1:	*arg*$^+$	*meth*$^+$	*red*	*intol*)
			↓		
Diploid phenotype	Race 1, 2:	*arg*$^+$	*meth*$^+$	*red*	*intol*
			↓		
Some segregant phenotypes	*Race 1, 3*:	*arg*$^+$	*meth*$^+$	*white*	*intol*
	Race 1, 2, 3:	*arg*$^-$	*meth*$^-$	*red*	*intol*
	Race 1, 2:	*arg*$^+$	*meth*$^+$	*white*	*intol*
	Race 0:	*arg*$^-$	*meth*$^-$	*white*	*tol*

Biochemical markers: *arg*$^-$ = arginine-less; *meth*$^-$ = methionine-less.
Natural markers: *red/white* = colour; *tol/intol* = tolerance to Actinomycete attack.
Race types: 1, 2; 1, 2; 1, 3; 1, 2, 3; 0.

Remarkably similar results have been obtained with races of *Puccinia* spp. (WATSON, 1957; VAKILI and CALDWELL, 1957). In *P. recondita tritici*, for example, infection with races 2 and 122 resulted in sixteen known races and seventeen completely new races of this rust fungus being isolated. It will be recalled that ELLINGBOE (1961) has provided evidence for somatic recombination in *P. graminis tritici* under experimental conditions. Nevertheless, it is a striking fact that no evidence exists for spontaneous mitotic recombination in any heterozygous race of rusts propagated by successive uredospore transfers, i.e. under the best potential conditions for such recombination. There is also no evidence for spontaneous recombination in natural perennial dikaryons of Basidiomycetes but it must be admitted that a search has rarely been made for such segregation. Parasexuality as a natural phenomenon may, therefore, be rare outside Fungi Imperfecti.

Sexual reproduction

There is no doubt of the occurrence of these systems in nature. The morphological evidence of zygotes, asco- and basidiocarps is overwhelming even if, strictly, only circumstantial. More significant is the lack of such structures in many situations. For example, the recent studies of WEBSTER (1964) have shown that common soil Fungi Imperfecti do have perfect stages but, in the soil, they must persist for long periods without apparent recourse to sexual reproduction, e.g. *Trichoderma viride*. Another common mould is *Monilia sitophila*, better known now as *Neurospora sitophila* from its reproductive structures, but most usually found in nature, or in bakeries, in the domesticated *Monilia* condition.

THE EFFECTIVENESS OF RECOMBINATION SYSTEMS

Precise measurements cannot be made of the effectiveness of non-sexual recombination systems with the exception of the parasexual cycle. Here, as has already been mentioned, the best estimate for one fungus only (*A. nidulans*) is that crossing-over is 500 times less frequent than in normal meiosis. Empirical studies can, of course, be made. In sexually reproducing systems, not only can potential outbreeding be calculated, but investigations can be made of their actual efficacy in nature. The potential outbreeding efficiency of different systems is set out in Table 16.1 and Fig. 16.2.

Table 16.1. Potential self- and cross-fertility of different mating systems in fungi

Mating system	Self-fertility %	Cross-fertility %
Dimixis	50	50
Bipolar diaphoromixis	50	100
Tetrapolar diaphoromixis	25	100

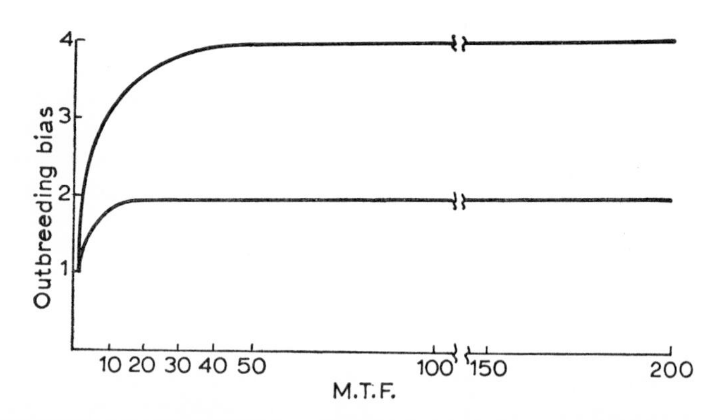

Fig. 16.2. Diagram to illustrate the relationship between the numbers of mating-type factors (M.T.F.), the number of loci involved (upper curve, 2; lower curve, 1) and the outbreeding bias. Outbreeding bias is Potential number of non-sister matings : Potential number of sister matings. (From Burnett)

It will be seen that self-fertilization is never avoided in a random breeding population in the sense that sib-mating is always possible between some of the progeny of a single meiosis. In this respect the fungi compare unfavourably with some of the breeding systems exhibited by higher plants where self-fertilization can be absolutely excluded. In diaphoromictic forms, however, a

Table 16.2. The distribution of +, − and neutral strains amongst 1098 races of mucoraceous fungi. (After Blakeslee *et al.*, 1927)

Organism	+ Mating type	− Mating type	
Absidia blakesleeana	19	18	3
caerulea	4	13	5
cylindrospora	1	1	0
dubia	4	20	0
glauca	4	6	0
ramosa	2	1	0
repens	8	1	0
sp. (whorled)	14	18	2
Blakeslea trispora	1	1	0
Chaetocladium brefeldii	?	?	0
Choanephora sp. A	1	1	0
cucurbitarum	5	28	0
Circinella spinosa	34	14	7
umbellata	1	1	0
Cunninghamella sp. A	22	29	2
bertholletiae	12	69	8
echinulata	10	8	0
elegans	25	16	1
Helicostylum piriforme	6	3	0
Mucor griseo-cyanus	3	6	0
hiemalis	1	1	0
dispersus	1	1	0
mucedo	14	11	14
sp. N	1	1	0
sp. III	3	2	0
sp. IV	6	13	0
sp. V	3	2	0
sp. VI	1	1	0
sp. VII	1	5	0
sp. VIII	1	1	0
Parasitella simplex	1	3	0
Phycomyces blakesleeanus	11	1	3
Rhizopus nigricans	89	62	85
Syncephalastrum racemosum	37	39	4
Other races of above species	7	9	2
Races of species not illustrated	40	64	99
Totals	**393**	**470**	**235**

very high degree of outbreeding can be obtained, depending upon the number of mating-type factors involved. WHITEHOUSE (1949b) has shown that for x unlinked loci and a mating-type factors at each locus (assumed equal) then, with random mating, the mean frequency of compatible pairs y, is given by

$$y = \left(\frac{a-1}{a}\right)^x$$

Such an increase in outbreeding does not reduce the incidence of inbreeding between segregants of a meiotic division.

In diaphoromictic systems outbreeding is far more efficient than in dimictic systems, even though self-fertility is not reduced. On the other hand, the probability of sexual reproduction occurring at all is much greater with diaphoromixis than in dimixis. In the former, almost any two spores or mycelia are likely to be of different mating types and, therefore, compatible but the chances are much less in a dimictic system.

Actual investigations of the effectiveness of recombination systems are few, but there are some well authenticated cases. Amongst dimictic fungi the most extensive studies are those of BLAKESLEE *et al.* (1927), who investigated the numbers of strains of different mating types (Table 16.2). They showed that there was a large number of neutral strains (*c.* 20%) that do not presumably participate in sexual reproduction and that extremely unequal numbers of mating types were isolated in some species, e.g. *Absidia dubia*, 4+:20−; *Choanephora cucurbitarium*, 5+:28−. Moreover, general experience of isolating mucoraceous fungi from substrates such as soil, does not suggest that strains of both mating types of a species are often found in close proximity. Zygotes can be detected but their inability to germinate is well known (BLAKESLEE, 1906). Thus even if the mean potential outbreeding efficiency is 50%, in fact, the actual incidence of sexual reproduction seems to be very low. Surprisingly, perhaps, the Mucorales appear to be no less and, indeed, possibly rather more variable than other groups where sexual reproduction would seem to occur more readily.

In diaphoromictic species there are data for a few tetrapolar species and one bipolar species (Table 16.3). The striking factor in common to all these

Table 16.3. Potential outbreeding efficiency of populations of tetrapolar and bipolar fungi. (Data from Roshal, 1950; Miles *et al.*, 1966; Eggertson, 1953; Partington 1959; Burnett, 1965b)

Organism	Sample size (basidiocarps)	Area or distance between populations	Potential outbreeding bias [1]	% Potential outbreeding efficiency
Schizophyllum commune	?12	15·8 ha	3·96	98·5
S. commune	15	11·7 ha	3·79⁺	96·8
Polyporus obtusus	24	?	3·89	96·9
Polystictus versicolor	37	*c.* 6·5 km	3·68	91·6
P. versicolor	12	50 cm	3·24⁺	80·5
Polyporus betulinus	100·5	Br. Isles	3·86	96·4
P. betulinus	33	3·8 ha	3·61⁺	78·3
P. betulinus	29	0·76 ha	2·78⁺	76·4

[1] Outbreeding bias calculated on assumptions of equal numbers, equally common, of all mating types and random mating; only last assumption for ⁺ populations.

examples is the high degree of outbreeding achieved, which is always more than 75% and, in the tetrapolar forms, a good deal higher (in samples of comparable size). Two points of interest arise from these studies. The first has been mentioned already in a different context; it is the means whereby potential conjugants are brought together or separated. The second is the numbers of mating-type factors in relation to outbreeding efficiency in diaphoromictic fungi.

BONNER (1958) has pointed out the close association, in plants, of spore formation and recombination. This situation obtains in most sexually reproducing fungi. The association enables the immediate products of recombination to be effectively distributed and exposed to selection in new environments. It also, of course, enables sister spores to be separated and for non-sister spores to be brought together by chance at some distance from the parental spore-producing structure. For example, the vast distances traversed by uredospores of rust fungi can result in recombination, by one means or another, many hundreds of miles away from the original parent mycelium (cf. Table 6.2, p. 169). It is of some interest in this context that somatic recombination in the hop-wilt organism *Verticillium albo-atrum* appears to occur at nuclear divisions in the phialide which gives rise to the conidia (HASTIE, 1967), i.e. recombination and dispersal are here associated in the parasexual condition.

The situation referred to in diaphoromictic fungi is of a different nature; it seems to be an example of inefficiency rather than efficient biological behaviour. It can be seen that the differences in potential outbreeding efficiency (Fig. 16.2) are not great between populations having 20–30 mating-type factors or 200. There is a gain but it is quite marginal. MATHER (1942) interpreted the mating systems of fungi as a device to promote outbreeding but this alone can hardly account for the great proliferation of mating-type factors in diaphoromictic fungi. In Table 16.3 the only population to show a serious impairment of outbreeding efficiency was the stump population of *Polystictus versicolor* which consisted of twelve basidiocarps carrying various combinations of six *A* and six *B* factors in their nuclei. This is a very small population indeed and its percentage potential outbreeding efficiency is still about 80%. Why then have fungi such as *Schizophyllum commune*, for example, evolved several hundreds of mating-type factors and is this related wholly to the efficacy of their recombination systems? It can be argued that these fungi occur in isolated populations and that different mating-type factors have originated in each so that in any one population the numbers are quite small and it is only when a world sample is studied that the large numbers are detected. It would be surprising if this were so since the efficacy of basidiospore transport seems likely to be considerable (cf. Chapter 6, pp. 166–172), thus tending to reduce the significance of isolated populations. Moreover, in the Nidulariales, where dispersal is probably not so efficient, since the unit is the peridiolum containing several hundreds of basidiospores, the number of mating-type factors seems to be relatively small, *c.* 10–20. It does not appear that the large numbers of mating-type factors found in some diaphoromictic fungi are necessarily maintained as a consequence of selection for outbreeding efficiency alone.

THE SIGNIFICANCE OF RECOMBINATION SYSTEMS

The significance of recombination systems can be considered from the point of view of a particular system, in isolation, or from the point of view of an organism in which different systems may be at least, potentially, capable of operating.

Single systems

It is not easy to assess the significance of heteroplasmons in fungi. In the example cited earlier of *P. graminis avenae*, it became clear that in certain host environments the possession of a particular cytoplasm by a parasitic fungus resulted in a different host-parasite reaction. It is also clear that the possession of infective cytoplasmic factors of a disabling nature can lead to the spread of deleterious conditions, via hyphal fusions in a population. This situation is known in the cultivated mushroom, where the condition known as 'watery stipe' or 'die back' is transmitted by hyphal fusion. In this case the cytoplasmic determinant appears, in fact, to be a virus and particles have been isolated with infective properties (HOLLING *et al.*, 1963). There are other cases of induced changes in fungi, which do not appear to be nuclear in origin and at least one of these is a change in pathogenicity. BUXTON (1957a, b; 1958) has shown that uninucleate conidia of race 1 of *Fusarium oxysporum* f. *pisi* can be grown in root exudates of the resistant pea variety Alaska for 14 days, after which they are now pathogenic to this variety. The condition can be maintained during at least seven passages through the formerly resistant host (Fig. 16.3). Buxton suggested that the change represented some form of

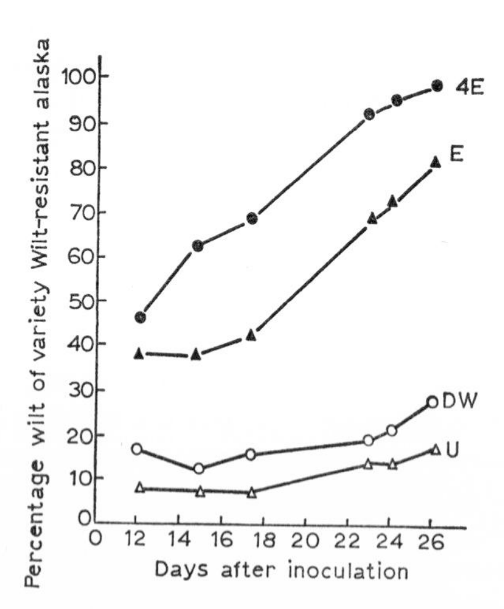

Fig. 16.3. Wilting due to infection by *Fusarium oxysporum* f. *pisi*, Race 1 after 14 days retention in root exudate from the wilt-resistant Pea (*Pisum sativum*) *cv* Alaska, or in water. Key: 4E, Exudate 4 × normal strength; E, exudate normal strength; DW, retained in distilled water; U, uninoculated plants as control. (From Buxton)

persistent enzymatic adaptation. Similar examples are known in other fungi, but none are so well documented (BUXTON, 1960). If this type of change is indeed due to a permanent, induced change in the cytoplasm, then it indicates that such changes could be of considerable significance in nature.

Heterokaryosis is likely to be of some importance in nature for it represents an ideal system ensuring rapid and successive adaptations of a fungus to a continually changing environment. This is a situation that most fungi must constantly be in through their own activities, both catabolic and secretory (JINKS, 1952b). Moreover, experimental studies with heterokaryons have shown that it is exceedingly difficult to eradicate entirely any particular nuclear type. Such nuclei, which may be of little or no adaptive significance in one situation, may be of enormous significance in another. It is for this reason that more experiments on selection with heterokaryons are desirable so that both the possibilities for, and rates of change of, nuclear ratios can be fully assessed. That selection can be efficacious and that new properties can be shown by heterokaryons, compared with those of their parental strains, has been shown in semi-natural situations. For example, in *P. graminis tritici* NELSON (1956) produced a race virulent to a wheat, Vernal Emmer, hitherto only resistant to one of the parental strains of the heterokaryon. BUXTON (1954, 1956) also showed that in the pea-wilt fungus, avirulent parental strains could be combined to produce heterokaryons which were quite as virulent as naturally occurring pathogenic strains.

The effectiveness of heterokaryosis will be controlled by the rate of mutation and the frequency of effective hyphal fusions. Very little is known about the frequency of either of these events in natural conditions and not much about the latter in culture. It will be recalled that Snider's estimates of the frequency of effective nuclear migration in *Schizophyllum*, the only fungus for which such data exist, was very low (Chapter 14, pp. 388–389). Mutation frequencies measured in genetical studies are of the same order of magnitude, per conidium, as in most other organisms. There is some evidence that some of the variability of fungi in culture is due to the accumulation of mutant nuclei. The best authenticated case is the study made of this phenomenon in *Fusarium* by MILLER (1945, 1946a, b). He showed that if single spore cultures are grown, especially in highly nutritious media, they soon develop morphologically abnormal areas, so-called 'patch mutants'. In these areas sporodochia and macroconidia were often developed, as well as any of, flattened patches, fluffy aerial mycelium, plectenchymatous masses, or sclerotia. By contrast the strains isolated from the soil had upright mycelia and produced microconidia almost exclusively. If such strains were maintained in sterile soil tubes, the 'patch mutants' were not detected. Miller rejected heterokaryosis as a possible explanation of the phenomenon since his original strains gave homogenous cultures when started from single, uninucleate microconidia. He suggested that the 'patch mutants' were, indeed, nuclear mutations which in the cultural conditions overgrew the original natural strains. This kind of situation has been encountered in many fungi and has often been lumped under the term 'saltation'. While some cases are undoubtedly segregations from heteroplasmons, heterokaryons or via parasexual phenomena, many cases are similar to Miller's and must be regarded as due to the cultural selection of spontaneous mutants. This phenomenon stresses not only the potential for producing genetically distinct nuclei and hence the production of heterokaryons by fungi but also the restriction and important selective effect due to the natural environment on such mutant nuclei.

Parasexuality is determined by the initial formation of heterokaryons. In Fungi Imperfecti, recombination will be restricted by the numbers of genetically distinct nuclei in a heterokaryon and this will be determined largely by the mutation rate. Thus the number of permutations and recombinations of nuclear types will be relatively low compared with the numbers of genetically different nuclei that can arise via recombination in sexually reproducing forms. Thus parasexuality, which ensures a more rapid production of recombinant nuclei, is likely to be selected for in Fungi Imperfecti and it might well be that its frequency would be higher in them than in sexually reproducing species. There is little evidence for or against this view. It is noteworthy that the spontaneous frequency of diploid nuclei in heterokaryons of *Aspergillus niger* is about 350 times greater than in the sexually reproducing *A. nidulans*. Even if somatic recombination has the same frequency in both species, recombination in *A. niger* will be more frequent than in *A. nidulans* because of greater numbers of diploid nuclei but, in fact, its mitotic recombination index is 100× greater (Lhoas, 1967). This comparison is, unfortunately, based on very limited data but it suggests that parasexuality may well be of considerably greater significance in Fungi Imperfecti than in sexually reproducing species.

Combinations of systems

Hitherto recombination systems have been considered as isolated events and no attention has been paid to those fungi in which the systems are not rigid. A brief consideration will now be given to these latter situations and those where more than one recombination system might be operative.

It has already been pointed out that heterokaryosis provides a means whereby immediate adaptation can be assured. Sexual reproduction, however, provides a means of generating immense variability over a long period. Thus a fungus which combines heterokaryosis with sexual reproduction provides both for the present and the future. It might be expected, therefore, that many homomictic fungi may well turn out to be heterokaryotic. Whether or not preferential fusion occurs at fertilization between genetically dissimilar nuclei, there is always a 50% chance that the diploid fusion nucleus will be heterozygous if the mycelium is heterokaryotic. In this respect *A. nidulans* may well be typical of a great many homomictic Ascomycetes and it would be of interest to compare the frequency of heterozygosity of such a fungus with that of a similar dimictic form. Attention has already been given to the suggestion that as between sexually reproducing forms, whatever their mating system, and asexually reproducing forms, parasexuality could be more frequent in the latter. On the other hand there are many fungi which persist for long periods in an apparently asexual condition, either because of environmentally induced effects, or because of the lack of a compatible strain. An example of this latter situation is *Phytophthora infestans* outside the Central Mexican–northern South American area, where only *A2* mating-type strains occur. This fungus is highly variable in a number of ways. How this is brought about is at present unknown but, clearly, heteroplasmons, heterokaryons, parasexuality or mutation could be involved separately or in combination. Fungi which live in these two states, a prolonged asexual phase and a distinct sexual phase, may in fact, so diverge permanently. The experiments of Jinks have shown how the continued propagation of Aspergilli by asexual conidia can result in a temporary, or even permanent, loss of the ability to develop perithecia (cf. Chapter 14, pp. 384–386). This, it will be recalled was a cytoplasmically-based phenomenon. In such fungi, therefore, the prolongation of an asexual phase beyond a certain period could result in it becoming, in effect, a Fungus Imperfectus. It may be of significance in this context that strains of *Rhizoctonia solani* include those capable of forming a *Corticium*-like basidiocarp, some of which are homomictic, others heteromictic, and other strains which are quite incapable of forming a basidiocarp alone or in combination with other strains (FLENTJE and STRETTON, 1964). In the sterile mycelia, so common on root surfaces, phenomena such as those described for *Aspergillus* spp. concerning loss of sexual reproduction might usefully be sought and analysed. It is desirable, therefore, to investigate how frequently a sexual phase is necessary, simply to maintain sexuality, in fungi in their natural habitats.

The balance of in- and outbreeding

In many fungi a balance between in- and outbreeding arises from defective mating systems. For example, homodimictic or homo-diaphoromictic forms regularly produce a proportion of monokaryotic or homokaryotic segregants which arise as a consequence of the imperfections of the cellular segregation mechanisms, e.g. the 6% monokaryotic basidiospores of *Galera tenera* f. *bispora*. The opposite conditions obtain when normally outbreeding Basidiomycetes produce more, or less than the usual numbers of basidiospores (cf. BURNETT and BOULTER, 1963). In fungi such as *Glomerella* or *Mycocalia denudata* the element of superimposed inbreeding is further randomized. In *Glomerella* the chance association of the genotypes results in more, or less, inbreeding (see Chapter 15, pp. 421–422); in *Mycocalia*, where inbreeding is partly controlled by the alleles *Pd*/*pd*, it could be subject to selection. It will be recalled that Pateman has in fact achieved a change in the mating system in a dimictic species, *Neurospora*, by artificial selection for large ascospores (Chapter 15, pp. 418–420).

In fungi with stable mating systems outbreeding is provided by sexual reproduction, whereas the rapid and effective multiplication of asexual conidia provides a means of exploiting a particularly favourable environment by fungi of a constant and adapted genotype. If the mycelium was heterokaryotic the asexual spores may represent adapted genotypes pre-selected in the mycelial phase as already described.

Thus in the fungi the mating systems, as in other organisms, are subject to selection and can adapt, slowly or rapidly, depending on their nature, to the conditions in which the organism lives (DARLINGTON, 1958).

Diploidy

The most surprising attribute of fungal mating systems is that, the Oomycetes apart, there is little evidence for the persistence of a diploid phase. Mating systems are based essentially on the haploid phase. This may be related to a point already touched upon, the close association of spore production and recombination in time and space. It may be argued that in the Oomycetes, where a diploid phase predominates, dispersal and recombination are separated, the former being very much the perquisite of the asexual zoospores. Sexual reproduction, on the other hand, results in a perennating structure, the oospore, and selection operates to bring about its germination only in appropriate conditions. In other fungi, because selection does not operate on the zygote but on its products, it is desirable that these should be as variable as possible; hence the occurrence of meiosis prior to spore production. An inevitable consequence is the loss, or non-achievement, of diploidy. In addition, it seems probable that the diploid state is not as stable in fungi as in other organisms. The reasons for this are obscure but presumably reflect some sort of nuclear/cytoplasmic interaction. HARDER (1927) showed many years ago that the cytoplasm of *Pholiota mutabilis* was altered in a dikaryon from its monokaryotic condition. A penultimate cell of a dikaryotic hypha was isolated by micrurgery at a time when the nuclei were dividing, so that it was uninucleate. When isolated, such cells continued to grow and produced for some time afterwards, clamp, or false clamp connexions and, in general, showed more resemblance to a typical dikaryotic hypha than to a monokaryotic one. This suggests that the nucleus exerts an effect on the cytoplasm which persists. Such an effect may be exerted by a diploid nucleus but, in an environment of predominantly haploid nuclei the diploid nucleus/cytoplasm balance may not readily be achieved. In Basidiomycetes it is possible to compare monokaryons, heterokaryons, dikaryons and diploid nuclei of the same genotype, in the last two cases in the same cytoplasm. Such comparisons may enable the reasons to be ascertained for the non-establishment of diploid nuclei in fungi as a regular feature. It is of interest, therefore, that CASSELTON and LEWIS (1967b) have shown that a recessive mutant gene is only partially recessive in a heterokaryon, but fully recessive in either a dikaryon or diploid strain of *Coprinus lagopus*. In this and a number of other respects the diploid and dikaryon resembled each other closely yet, in diploid/haploid dikaryons the diploid nucleus is less stable than in a monokaryon and, in a diploid/diploid dikaryon, either one or both nuclei haploidize (CASSELTON and LEWIS, 1967a). Thus the two conditions are, in some respects at least, physiologically comparable so that the dikaryotic condition possesses some of the attributes of diploidy, while not suffering the same instability in relation to the cytoplasm, as does a diploid nucleus.

SECTION IV Speciation and Evolution

Chapter 17

Speciation

Speciation is a neglected topic in fungi despite the fact that cultural studies provide an opportunity of investigating such processes experimentally. Fungal taxonomists are, of course, still concerned to determine what properly constitutes a fungal species and their task is made more difficult by the problem of distinguishing between a population and an individual. This arises from the widespread occurrence of asexual propagation by spores and of hyphal fusion. In all other organisms, even rhizomatous or bulbous higher green plants, it is tolerably easy to define, for operational purposes, a population and an individual. For example, a single bracken spore may give rise to an extensive area of bracken, each shoot being attached, initially at least, to a common rhizome. These shoots compete with each other and can be regarded for many ecological purposes as individuals, the whole plant being a population. If shoots become detached they compete with each other. A single fungal spore can give rise to a similar array of aerial hyphae all connected together. However, intra-mycelial connexions cause the mycelium as a whole to act as if it were a unit. Initially, therefore, the mycelium is effectively an individual. However, if part is detached, or if spores derived from the mycelium germinate close to the original mycelium, they do not compete but, through hyphal fusion, co-operate (cf. Chapter 4 and Fig. 4.10, p. 75). What now is the individual? An extreme case of this probably occurs in certain wood-destroying polypores such as *Polystictus versicolor* (BURNETT and PARTINGTON, 1957). Here bracket-like basidiocarps develop in close proximity on dead stumps. They carry the same mating-type factors in different combinations (Fig. 17.1). In plate culture, mycelia carrying such factor combinations can anastomose freely, monokaryons with monokaryons or dikaryons, dikaryons

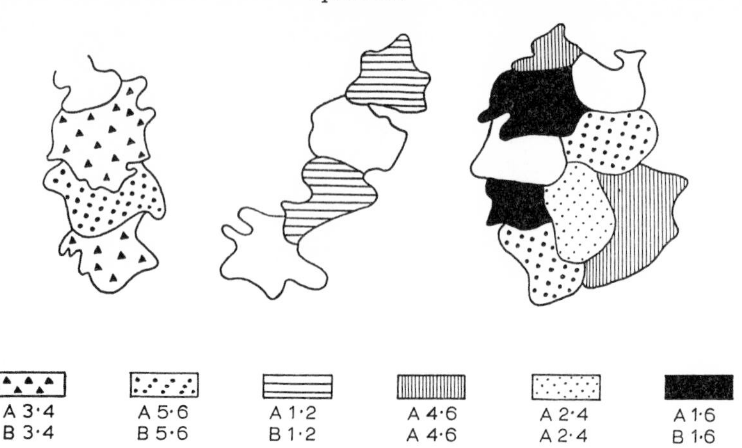

Fig. 17.1. Diagram to illustrate the mating type constitution of three groups of basidiocarps of *Polystictus versicolor*, 30 cm apart on the same stump. The constitution of the unfilled basidiocarps was not determined. (After Burnett and Partington)

with dikaryons, whether genetically alike or different. This suggests that the mycelium, in its natural environment, exists as a single physiological and ecological unit although genetically a mosaic. Since each dikaryon component can give rise to basidiocarp it is convenient, for population studies, to consider each as a distinct individual but this clearly would not apply to physiological studies where the whole mycelium is the unit. So far this kind of situation in the wild has not received much attention but it is fundamental to an understanding of the nature of fungi as Buller explained so clearly nearly four decades ago.

For the purposes of this chapter it will be necessary to regard a fungal individual as a single isolation; in the case of Basidiomycetes, either a dikaryon or a monokaryon, in other fungi a hetero- or homokaryon. This is a necessary simplification at the present time.

ISOLATING MECHANISMS

Speciation processes originate in the action of isolating mechanisms. Although not well understood in fungi these have been recorded over many years by Canadian workers in Basidiomycetes and, more recently, have been recognized in Ascomycetes and Fungi Imperfecti. There is no agreed terminology to describe these situations but here the term sterility barrier will be employed; these may be partial or complete, facultative or obligatory. In essence they operate by interfering either with hyphal fusion or with the events after such fusions have occurred but prior to fertilization. Thus, the recognition of two kinds of isolating mechanism commonly recognized in other organisms as pre-zygotic and post-zygotic mechanisms, is not quite adequate for fungi. In fungi, pre-zygotic mechanisms seem to be the most important and operate in either the pre-conjugation or the post-conjugation phase but before fertilization. It is the location of this latter class of mechanisms that is such a striking feature of fungal isolating mechanisms. Mating-type factors have effects similar to sterility barriers in some fungi and this has led to some confusion in the past. This is understandable since both kinds of reaction, whether mediating conjugation or intersterility, are expressed through the same organ, the hyphal tip.

It seems probable that hyphal fusions, except in Phycomycetes, can never be wholly suppressed whether between mycelia of the same species or of different species, but their frequency can be greatly reduced. CABRAL (1951) made extensive studies of hyphal fusions between different genera, species and strains of Polyporaceae and found that in plate culture on water agar, or agar low in nutrients, some fusions could always be found. Nevertheless, in some cases, the frequency of fusions may be so low that they cannot be detected. This is the case in *Mycocalia denudata* where isolates from northern England fell into two intersterile groups. When opposed or intermixed in plate culture, either as monokaryons or dikaryons, they simply grew past each other, showing neither aversion nor post-fusion reactions (BURNETT and BOULTER, 1963). In many fungi such reactions divide isolations into intersterile groups. In many Basidiomycetes 'barrages' develop, that is, areas between the mycelia in which there may be a zone not penetrated by hyphae but flanked by tangled and piled up hyphae—the 'barrage', e.g. *Marasmius elongatipes*, *Polystictus hirsutus* (ARNOLD, 1935; BOSE, 1934). It is of interest that barrage-like features develop in *P. versicolor*, a fungus taxonomically quite close to *P. hirsutus*. In *P. versicolor*, a tetrapolar fungus, these features are related to the interaction of $B=$ factors (BURNETT, unpublished). This is a good example of how the same phenotypic expression can be determined by quite different genotypic determinants; in *P. hirsutus*, 'barrages' arise quite independently of the action of the mating system, which is bipolar diaphoromictic. Aversion phenomena are shown by fungi other than Basidiomycetes and examples which have been investigated are *Diaporthe perniciosa* and *Podospora anserina* (CAYLEY, 1923, 1931; RIZET, 1952). In *Diaporthe*, different single-spore isolations show different degrees of aversion, as estimated by the zone not penetrated by hyphae. The phenomenon was apparently controlled basically by a pair of alleles, distinct from the mating-type factors, so that 'averting' and 'non-averting' strains were genetically determined. The regulation of the degree of expression was not fully analysed. By contrast, the 'barrage' phenomena in *Podospora* may be due to a cytoplasmic factor. Between S and s strains in plate culture, the hyphae become disorganized in a zone 1–2 mm wide, they fail to fuse, branch and form a thick, deeply pigmented network. When S and s strains are crossed, some of their progeny behave as if of S phenotype but, if such strains come into contact with an S carrying mycelium, they are converted to an s phenotype. These s^S strains are converted as a consequence of infection by a cytoplasmic factor which enters via hyphal fusions (Fig. 17.2). This case is unique.

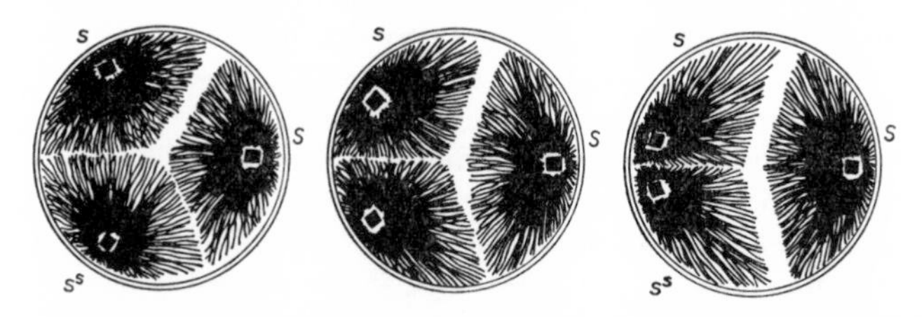

Fig. 17.2. Barrage effects in *Podospora* due to cytoplasmic factors. In each petri plate an s^S strain was confronted simultaneously with an *s* and an *S* strain but at different distances apart with respect to the s^S and *s* strains. Note that the closer s^S and *s*, the longer the barrage between s^S and *S* because more of the s^S strain is converted to *s*-type by contact with an *s* strain. (From Ephrussi, after Rizet)

In some cases hyphal fusion is not prevented but reduced and this may or may not be accompanied by adverse reactions in the fusion cells. Further investigations of 'barrage' phenomena in *P. anserina*, when strains of different geographical origin are opposed, illustrates such reactions (RIZET and ESSER, 1953; ESSER, 1956, 1958, 1959a, b). These 'barrage' phenomena are determined by four unlinked allelic pairs *a*/*a*l, *b*/*b*l, *c*/*c*l, *v*/*v*l so that there are sixteen potential genotypes. However, the homokaryons *a*l*b* or *c*l*v* are inviable. Ascospores produced at meiosis germinate but the mycelia grow slowly, their nuclei degenerate and the whole organism dies in a few days. Even in heterokaryons such as *ab*+*a*l*b*l or *cv*+*c*l*v*l, there is an interaction between *a*l and *b* or *c*l and *v*, although they are in different nuclei and, as a result, the numbers of *a*l or *c*l carrying nuclei decline and homokaryons develop carrying only *ab* or *cv* nuclei respectively. A further property of these gene pairs is the production of 'barrages', zones of unpigmented cells where mycelia come into contact, which may, or may not, develop lines of perithecia at both, at one, or at neither side. The last two conditions arise because of the suppression of conjugation between microconidia and trichogynes, even though the trichogynes show zygotropic curvatures and may make contact with the microconidia. The reactions are set out below.

Mating Type	*A*	*a*	*A*	*a*	*A*	*a*
Microconidial parent	*ab*	*a*l*b*l	*cv*	*c*l*v*l	*ab* *c*l*v*l	*a*l*b*l *cv*
Ascogonial parent	*ab*	*a*l*b*l	*cv*	*c*l*v*l	*ab* *c*l*v*l	*a*l*b*l *cv*
Perithecia (·) or barrage (‖)	⋮‖		⋮‖		‖	

In the first two cases the reaction is unilateral, *ab* ♀ × *a*1*b*1 ♂ and *cv* ♀ × *c*1*v*1 ♂, these being the only successful crosses; the reciprocal inhibition arises from the simultaneous action of both unilateral reactions. The crosses between all viable genotypes are set out in Table 17.1.

Table 17.1. The consequences of crossing the 9 viable genotypes of *Podospora anserina* in respect of genes *a*, *a*1; *b*, *b*1 and *c*, *c*1; *v*, *v*1, which control barrage formation and heterogenic compatibility. (Based on Esser, 1956, 1959a)

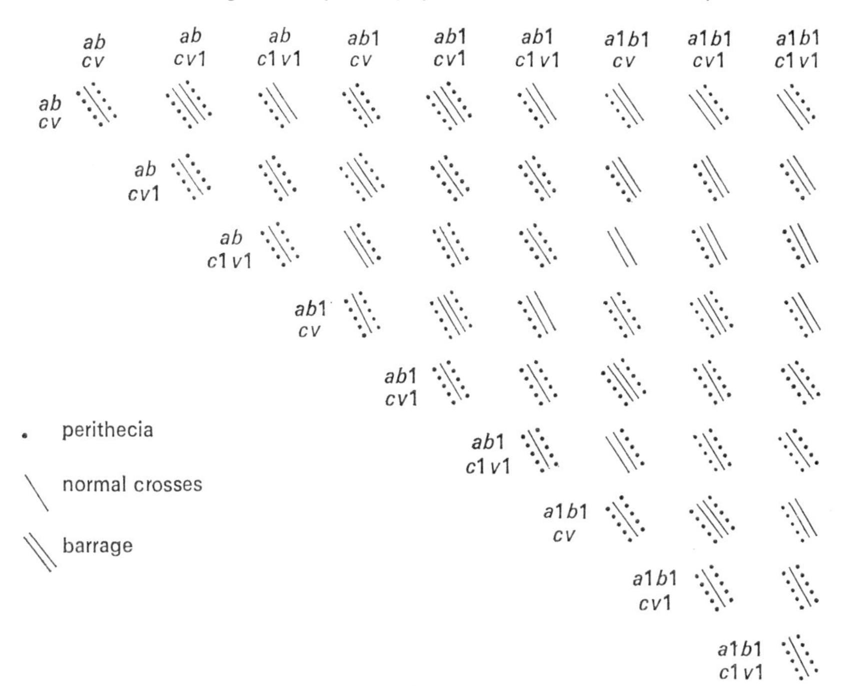

This phenomenon has been termed 'heterogenic compatibility' by Esser and the two kinds of reaction as 'semi-compatibility' and 'reciprocal incompatibility'. As he has pointed out, they result in inbreeding being favoured. The physiological basis of the phenomenon is not fully understood. No diffusible extra-mycelial agents are apparently involved and in mixed cultures of semi-incompatible strains there is a modification of protein specificity as tested serologically in agar-gel diffusion plates.

BERNET (1963; BERNET *et al.*, 1960), studying different strains, has found three gene pairs, *C*/*c*, *D*/*d*, *E*/*e*, which operate in a similar manner, i.e. *Ce* × *cE* and *Cd* × *cD* result in unilateral crosses and barrage development. The *D* gene, however, is temperature dependent and is only expressed at 20°C. The relationship between the series described by Esser and Bernet and their colleagues is unknown. There are thus minimally 4 or 7 loci involved in producing partial or total intersterility between different strains of *P. anserina*, the basis of the isolating mechanism being the presence of unlike alleles in potential conjugant partners. Such behaviour contrasts with that in mating systems, where unlike alleles usually promote crossing.

No other case has been analysed specifically but there is evidence of genetic control in the Aspergilli. Natural isolations of *A. nidulans* and other *Aspergillus* spp. were examined to see whether they would form heterokaryons; the test employed colour mutants so that mixed heterokaryotic heads could be detected (GRINDLE, 1963a, b; JINKS and GRINDLE, 1963; JONES, 1965; CATEN and JINKS, 1966; JINKS *et al.*, 1966). The isolations fell into a number of groups, about 30 for *A. nidulans* in the British Isles, between which heterokaryons were either not formed or only formed very rarely. For example, within groups 2 and 26, paired isolations produced 4–6% heterokaryotic heads but between these groups the incidence was as little as 0·10%. Strains within a group often had the same spore colour and similar pigmentation in their hyphae and the medium. They resembled each other, in plate culture, in their general surface features, growth rate, ability to conidiate and in numbers of conidia per unit area as well as in the frequency of their production of perithecia. These similarities are not always shown, even in *A. nidulans* and, in members of the *A. glaucus* complex, morphological features are hardly ever correlated with heterokaryon-forming ability. While strains of the same group may occur in proximity to each other, this is by no means the rule.

It has been estimated that at least five loci must be common to any two strains if they are to form a successful heterokaryon, although this is a very tentative estimate. This is a figure not dissimilar to that which determines intersterility in *P. anserina* but it must be admitted that this coincidence is quite likely to be fortuitous. Crosses have been carried out between strains which show 'heterokaryon incompatibility', as Jinks has called the phenomenon. The vigour of the progeny of such crosses is considerably lower than that of their parents (Table 17.2). Thus there must be a considerable measure of isolation between such strains in nature, despite their overall similarities, which causes them all to be designated as members of the same taxonomic species.

This kind of behaviour is known but not so fully investigated in many other fungi. In the Phycomycetes the barrier to intercrossing may lie either in the stages before conjugation or in the cytoplasm after conjugation. The complex, extra-mycelial physiological processes described for Saprolegniales or Mucorales (cf. Chapter 15, pp. 406–413) provide several points at which the mating reaction can be impeded and, even if these are overcome, 'imperfect' sexual reactions rarely result in viable offspring or even in zygote formation (but see later).

In some cases the physiological barrier is clearly expressed. FLENTJE and STRETTON (1964) have shown that when cultures of pathogenic *Rhizoctonia solani* (= *Thanatephorus cucumeris*), derived from single spores of different strains, were confronted, anastomoses were either few or, if they occurred, were followed by death of the fusion cell after about 72 hours (Plate IV, Fig. 2, p. 51). Between strains from the same host fusions were usually

Table 17.2. Heterokaryon compatibility of progeny listed against parent strains and growth of parental and progeny strains in *Aspergillus nidulans*. (After Jinks *et al.*, 1966)

Cross-type	Same isolate	Different isolates, same compatibility group		Different isolates of different compatibility group		
		Same locality	Different localities			
Heterokaryon compatibility						
Sample size	40	40	40	59	44	50
Compatible 2 parents	40	40	32	0	0	0
Compatible 1 parent	0	0	8	2	2	4
Compatible 0 parent	0	0	0	57	42	46
Growth rate (mm/day)						
Parental mean	6·0	6·6	6·0	5·3	6·3	4·5
range	5·9–6·1	6·1–7·2	5·7–6·3	5·3–5·3	5·8–6·7	4·2–4·8
Progeny sample	60	98	98	60	82	99
mean	5·93	6·76	6·57	4·30	5·28	4·04
range	5·2–6·3	6·0–7·2	5·5–8·1	3·1–5·9	3·7–6·5	3·1–4·9
Significance of differences amongst progeny (P)	>0·2	0·0·2–0·05	<0–0·001	<0·001	<0·001	<0·001

Table 17.3. Restrictions on mating ability between groups within recognized taxonomic species in relation to differences in mating systems, morphology and ecology

Organism	Differences in mating system	Differences in morphology	Differences in ecology	Inter-sterility	Author
Mycocalia denudata and *M. castanae*	+	+	+	complete	Burnett and Boulter, 1963
Corticium coronilla	+	+	?	complete	Biggs, 1937
Gloeocystidium tenue	+	+	?	complete	Boidin, 1950
Peniphora spp.	–	+	+	complete or partial	McKeen, 1952
Fomes ignarius	–	+	+	complete	Verall, 1937
Auricularia auricula-judae	–	+	+	partial	Barnett, 1937
Polyporus abietinus and forms	–	+	±	partial or none	Macrae, 1941; Raested, 1941
Gloeocystidium tenue	–	+	?	complete	Boidin, 1950
Coprinus micaceus	–	+	?	partial	Kühner, Romagnesi and Yen, 1947
Fomes pinicola	–	–	+	partial	Mounce, 1929
Fomes pinicola	–	–	–	complete	Mounce and Macrae, 1938
Polyporus betulinus	–	–	–	complete	Burnett, unpublished
Mycocalia denudata	–	–	–	complete	Burnett and Boulter, 1963
Coprinus macrorhizus f. *microsporus*	–	?	?	complete	Kimura, 1952
Coprinus callinus	–	–	?	complete	Lange, 1952
Coprinus subimpatiens	–	–	?	complete	Lange, 1952

normal, although in a small percentage of cases the lethal reaction developed. This reaction is reminiscent of that which controls heterokaryon formation in *Neurospora* (see Chapter 14, p. 395). In *Rhizoctonia*, however, this is largely an interstrain reaction, a situation not demonstrated in *Neurospora*.

There is, unfortunately, in Basidiomycetes no evidence concerning the ultimate fate of rare, interstrain dikaryons and very little evidence bearing on the failure to establish dikaryons. Differences between the strains may be undetectable or may affect their morphology, ecology or mating systems, amongst other features, or combinations of these (Table 17.3). These examples provide a range of behaviour between strains which could readily be described as separate species to those which, although fully isolated, cannot be distinguished save by a breeding test. These latter, therefore, represent 'sibling' species such as have been described in other organisms (DOBZHANSKY, 1937).

In cases of partial intersterility, a wide variety of causes probably operates. In *Auricularia*, for example, fusions were common between strains from deciduous hosts or coniferous hosts but between such strains either very few hyphal fusions occurred or when they did no dikaryotic hyphae developed. There were slight but constant morphological differences between strains from broad-leaved or coniferous hosts although all strains had a bipolar diaphoromictic mating system. This situation contrasts with that in *Fomes pinicola*, a wood-destroying polypore common in N. America and Europe but rare in Britain. In culture there is very considerable variation in dikaryotic or monokaryotic isolations but all are recognizably of the same type. In paired cultures of either type, aversion reactions often develop. They may be merely slight inhibitions of aerial hyphae, the development of tangled hyphal growth (which can become pigmented and almost sclerotial-like) or an actual zone of aversion traversed by only a few hyphae. These variations of morphology and aversion pattern are quite unrelated to the formation of dikaryons. Three groups of isolations can be recognized, A, B and C. A and B are confined to N. America and between them dikaryons entirely fail to form. Adjacent basidiocarps on the same host tree may belong to these two groups. Group C lies outside N. America and readily forms dikaryons with Group A strains; yet with Group B strains hyphal fusions and dikaryons are only formed very rarely (Table 17.4 and Fig. 17.3). All the strains have the same bipolar mating system (MOUNCE, 1929; MOUNCE and MACRAE, 1938).

The '*Stereum pini*' and '*Polyporus abietinus*' complexes represent interesting contrasts with that of *F. pinicola*. They are the causes of important rots of conifers in N. America and Europe. Two species, *Stereum pini* and *Peniophora duplex*, were recognized. Collections of these were mated in all combinations. It then appeared that *S. pini* from N. America were intersterile with European *S. pini* or N. American *P. duplex*, although the latter two showed some interfertility. In some of the European *S. pini*/*P. duplex* confrontations, the dikaryotic mycelia were confined to the line of apposition and, in other cases, dikaryotizations were unilateral. Sub-cultures derived from such hyphae often lost their clamp connexions after a few transfers but, unfortunately, the constitution of such derived apparent monokaryons has not been recorded.

Table 17.4. The results of pairings of monosporous mycelia of *Fomes pinicola* from various hosts and localities. (From Mounce and Macrae, 1938)

North American origin Group A (rows 283–5564); bracketed pairs: 1015A/1015B, 1152A/1152B, 1155A/1155B, 1264/1265

	283	285A	285B	694	831A	1002	1015A	1015B	1105	1120	1150	1151	1152A	1152B	1153B	1155A	1155B	1162	1169B	1169C
283		+	+	+	+															
285A	+		+	+	+	+	+			+	+	+	+	+	+	+	+	+	−	
285B	+	+		+	+															
694	+	+	+		+	+				+	+	+	+	+	+	+	+	+	+	
831A	+	+	+	+																
1002		+		+						±	+		+	+	+	+	+	+	+	
1015A		+						∓												
1015B							∓													
1105																				
1120		+		+		±					+			+	+		+	+	+	
1150		+		+		+				+				+	+		+	+	+	
1151		+		+										+			+		+	
1152A		+		+		+								±				+		
1152B		+		+		+				+	+	+	±		+		+		+	
1153B		+		+		+				+	+			+		+	+	+	+	
1155A		+		+		+									+		±	+	+	
1155B		+		+		+				+	+	+		+	+	±				
1162		+		+		+				+	+		+		+	+			+	
1169B		−		+		+				+	+	+		+	+	+		+		+
1169C																			+	
1259		+					+													
1264		+					+													
1265		+					+													
2250		+			±		+		+							+			+	
2380		+					+									+				
3248		+					+									+	+			
5563		+					+									+				
5564		+					+									+				

	1259	1264	1265	2250	2380	3248	5563	5564	5770	5778	6612	6616	6617	6618	6624	6925
283									+	+						
285A	+	+	+	+	+	+	+	+	+	+	+	+	+	±	+	+
285B									+	+						
694									+	+						
831A				±					+	+						
1002																
1015A	+	+	+	+	+	+	+	+			+	+		+	+	+
1015B																
1105				+												
1120										+						
1150										+						
1151										+						
1152A																
1152B																
1153B										+						
1155A				+	+	+	+	+			±	+		+	+	+
1155B						+										
1162																
1169B				+												
1169C																
1259		+	+													
1264	+		±	+	+	+	+	+			+	+	+	+	+	+
1265	+	±														
2250		+														
2380		+														
3248		+														
5563		+														
5564		+														

	562C	1025	1258	1268	2251	2379	3249	6613	6923
283	−								
285A	−	−	−	−	−	−	−	−	−
285B	−								
694	−	−		−	−				
831A	−								
1002	−	−	−	−	−				
1015A	−	−	−	−	−	−		−	−
1015B									
1105									
1120	−	−		−					
1150	−	−		−	−				
1151	−								
1152A	−	−		−					
1152B	−			−					
1153B	−	−		−					
1155A	−	−		−	−	−		∓	−
1155B	−			−					
1162	−	−		−					
1169B	−	−		−	−				
1169C									
1259	−	−	−		−				
1264	−	−	−	∓	∓	−		−	−
1265	−		−		−				
2250	−	−	−	−	∓				
2380	−	−							
3248	−	−							
5563	−	−							
5564	−	−							

	586	928	1005	1006	1008	1339	6895
283	+	±					
285A	+	+	+	+	+	+	+
285B	+	+					
694	+	+	+		+		
831A	+	+		+			
1002	+	+	+	+	+	+	
1015A	±	+	+	±	+	+	+
1015B							
1105							
1120	+	+			+		
1150	+	+			+		
1151	+	+					
1152A	+	+			+		
1152B	+	+			+		
1153B	+	+	+		+		
1155A	+	+			+		+
1155B		+			+		
1162	+	+			+		
1169B	+	+			+	+	
1169C							
1259		+					
1264	+	±	+	+	+	+	+
1265		±					
2250	+	+	+	+	+	+	
2380	+	+				+	
3248	±	−				+	
5563	+	+				±	
5564	+	+				+	

	5770	+	+	+	+	+																									+						
	5778	+	+	+	+	+					+	+	+			+														+							
	6612		+					+									±						+														
	6616		+					+									+						+														
	6617		+																				+														
	6618		±					+									+						+														
	6624		+					+									+						+														
	6925		+					+									+						+														
North American origin Group B	562C	−	−	−	−	−	−	−			−	−	−	−	−	−	−	−	−	−		−	−	−	−	−	−	−	−	−	−	−	−	−	−	±	−
	1025		−		−		−	−			−	−		−		−	−		−	−		−	−		−	−	−	−	−			−	−		−	∓	−
	1256		−				−	−														−	−	−	−												
	1268		−		−		−	−			−	−		−	−	−	−	−	−	−			∓		−						−						
	2251		−		−		−	−				−					−			−		−	∓	−	∓												
	2379		−					−									−						−														
	3249		−																																		
	6613		−					−									∓						−														
	6923		−					−									−						−														
Foreign origin Group C	586	+	+	+	+	+	+	±			+	+	+	+	+	+	+		+	+			+		+	+	±	+	+	+	+	+	+		+	+	+
	926	±	+	+	+	+	+	+			+	+	+	+	+	+	+	+	+	+		+	±	±	+	+	−	+	+	+	+	+	+		+	+	+
	1005		+		+		+	+								+							+		+												
	1006		+			+	+	∓															+		+												
	1008		+		+		+	+			+	+		+	+	+	+	+	+	+			+		+						+						
	1339		+				+	+												+			+		+	+	+	±	+			+	+		+	+	+
	6895		+					+									+						+														

	5770	−								
	5778	−			−					
	6612	−	−							
	6616	−	−							
	6617	−								
	6618	−	−							
	6624	±	∓							
	6925	−	−							
North American origin Group B	562C		+	+	+	+	+	+	+	+
	1025	+		±	+	+	+		∓	∓
	1256	+	±		+	+				
	1268	+	+	+		+				
	2251	+	+	+	+					
	2379	+	+							
	3249	+								
	6613	+	∓							
	6923	+	∓							
Foreign origin Group C	586	∓	−	∓	∓	±	+		±	−
	926	−	∓	±	∓	−	±		±	∓
	1005	−	±	−	±	+				
	1006	−	−	−	−	−				
	1008	−	∓	−	±	±				
	1339	∓	−	∓	∓	±	±		±	−
	6895	∓	±							

	5770	+	+					
	5778	+	+			+		
	6612	+	+				+	
	6616	+	+				+	
	6617							
	6618	+	+				+	
	6624	+	+				+	
	6925	+	+				+	
North American origin Group B	562C	∓	−	−	−	−	∓	∓
	1025	−	∓	±	−	∓	−	±
	1256	∓	±	−	−	−	∓	
	1268	∓	∓	±	−	±	∓	
	2251	±	−	+	−	±	±	
	2379	+	±				±	
	3249							
	6613	±	±				±	
	6923	−	∓				−	
Foreign origin Group C	586		+	+	+	+	+	+
	926	+		+	+	+	+	+
	1005	+	+		+	+	+	
	1006	+	+	+		+	+	
	1008	+	+	+	+		+	
	1339	+	+	+	+	+		+
	6895	+	+				+	

Fig. 17.3. The distribution of the three breeding groups of *Fomes pinicola*. In the crossing diagram ——— indicates full fertility; –·–·–·–·–, partial fertility and no line, complete sterility. (From Mounce and Macrae)

This phenomenon is superficially similar to the behaviour of the $ab + a1b1$ heterokaryons of *P. anserina* and further investigations of the phenomenon are desirable. The morphology of the basidiocarps of *S. pini* and *P. duplex* is very different and European *S. pini* cannot be readily separated from the N. American material. In plate culture, however, the two *S. pini* isolates are clearly separable whereas European *S. pini* and *P. duplex* are barely separable. All the species are tetrapolar diaphoromictic in mating system (WERESUB and GIBSON, 1960).

Similar morphological differences occur between *P. abietinus* and *Irpex fusco-violaceus*. Two morphologically distinct kinds of *P. abietinus* have been recognized, a typical form with the characteristic shape and pores of a polypore, the poroid type; and a form having the bracket-like shape but with the hymenium distorted to form lamellae-like structures, the lamellate form. In addition, a form with a hymenial surface like teeth is often associated with this fungus and believed to be a form of it; this is *I. fusco-violaceus*. In Europe the irpicoid form is confined to the genus *Pinus* and has usually been held to be a distinct species. In N. America none of the morphological forms shows host specialization. The results of attempted dikaryotizations are shown in Fig. 17.4. There are two completely isolated poroid groups in N. America, A and B, both of which are partially fertile with strains from European poroid forms, C, between which there is complete fertility. However, the irpicoid forms differ in mating system, being bipolar; all strains are completely interfertile regardless of their geographical origin. The irpicoid forms are quite isolated from all other supposedly related forms. The most surprising group is that of the lamellate strains, which is virtually wholly cut off from all other strains despite the occurrence of morphologically-intermediate forms. Macrae claimed that three pairings between lamellate and European poroid forms occurred. This is surprising in view of the difference in their mating systems

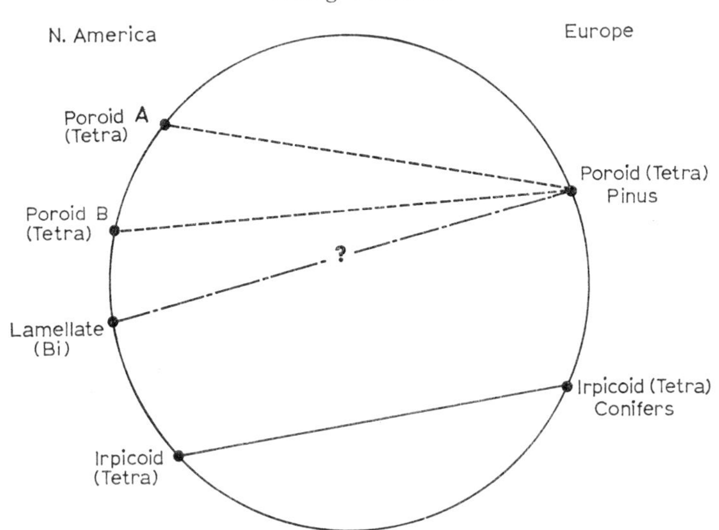

Fig. 17.4. Crossing diagram of the breeding groups believed to be associated with morphologically different forms of *Polyporus abietinus*. ———, fully interfertile; - - - - -, partially interfertile; –·–·–·–, rarely interfertile; no line, complete sterility. (From Burnett, based on Macrae and Raestad)

and further work has not apparently substantiated this claim (MACRAE, 1941, 1967; WERESUB in BURNETT, 1965b). This kind of pattern of interfertility and sterility as determined by dikaryon formation seems not to be uncommon in the Homobasideae but, until further analysis has been carried out, its significance cannot be adequately assessed.

At present these cases are remarkably reminiscent of the situation already described in *Aspergillus* spp. so that throughout the fungi perhaps, failure to form an heterokaryon may prove to be a critical stage in the isolating mechanism. Not all cases are susceptible of analysis, however, in *Mycocalia denudata*, for example, hyphal fusions fail to occur between the morphologically, physiologically and ecologically indistinguishable intersterile groups, although a careful search was made for them (BURNETT and BOULTER, 1963).

Another kind of isolation is imposed by host specificity. This is shown by some of the examples cited in Table 17.3, such as *Peniophora* or *Irpex*. In the former, three species were recognized, *P. heterocystidea*, *P. populnea* and *P. mutata*, but the last was shown to be composed of two groups, III and IV, incapable of forming dikaryons. One of those groups (Gp. IV) and *P. populnea* are confined to *Populus* spp. as host trees; the others, *P. heterocystidea* and *P. mutata* (Gp. III) are found on a variety of broad-leaved trees. Thus host-specialization may or may not be associated with morphological differences but can be associated with sterility barriers (MCKEAN, 1952).

The most remarkable examples of host specialization, however, are the 'physiological races' which have been detected and studied in a number of pathogenic fungi, notably the rusts.

Physiological races

ERIKSSON showed as early as 1894 that, despite the very great morphological similarities between rusts of the same species on different hosts, dikaryotic strains from one host might not be capable of infecting another. He studied *Puccinia graminis*, the cause of stem rust of wheat, barley, oats, rye and over a hundred wild and cultivated grasses. Some six varieties are now recognized, based largely on host specialization (Table 17.5). Thus, among cultivated

Table 17.5. The host plants and uredospore sizes of the 6 varieties of *Puccinia graminis*. (From Stakman and Harrar, 1957)

Variety	Host plant	Uredospore size (μm)
1. *tritici*	Wheat (*Triticum*), barley (*Hordeum*), grasses	32 × 20
2. *secalis*	Rye (*Secale*), barley (*Hordeum*), grasses 1	27 × 17
3. *avenae*	Oats (*Avena*), wild grasses different from 1 and 2	28 × 20
4. *phleipratensis*	*Phleum*, wild grasses	24 × 17
5. *agrostidis*	*Agrostis* spp.	22 × 16
6. *poae*	*Poa pratensis* and related spp.	19 × 16

cereals, *P. graminis tritici* infects wheat and *P. graminis avenae*, barley or oats; they also infect different species of wild grasses. On the other hand, over 14,000 different kinds of cultivated wheat are known and not all strains of *P. graminis tritici* are capable of infecting them equally. Indeed, by using

Table 17.6. Infection types produced on leaves of differential varieties of *Triticum* spp. by physiological races of *Puccinia graminis tritici.* (From Stakman and Harrar, 1957)

Infection type	Description	Reaction
0	Immune	No pustules developed but sometimes small flecks of dead host tissue (;)
0;	Practically immune	
1	Very resistant	Pustules extremely small, surrounded by dead areas
2	Moderately resistant	Pustules small-medium; usually in green islands of tissue surrounded by band of chlorotic/dead tissue
3	Moderately susceptible	Pustules medium, usually separate; no dead areas ± chlorotic areas, especially in unfavourable conditions
4	Very susceptible	Pustules large, often united; no dead tissue ± leaf yellowing especially in unfavourable conditions
5	Heterogenous	Pustule size variable, sometimes includes 0–4 and intermediates on same plant

different varieties of wheat and assessing the degree of infection by the size and shape of the lesions, if any, produced on the leaves, it is possible to subdivide *P. graminis tritici* into over 200 recognizable races or biotypes (Fig. 17.5

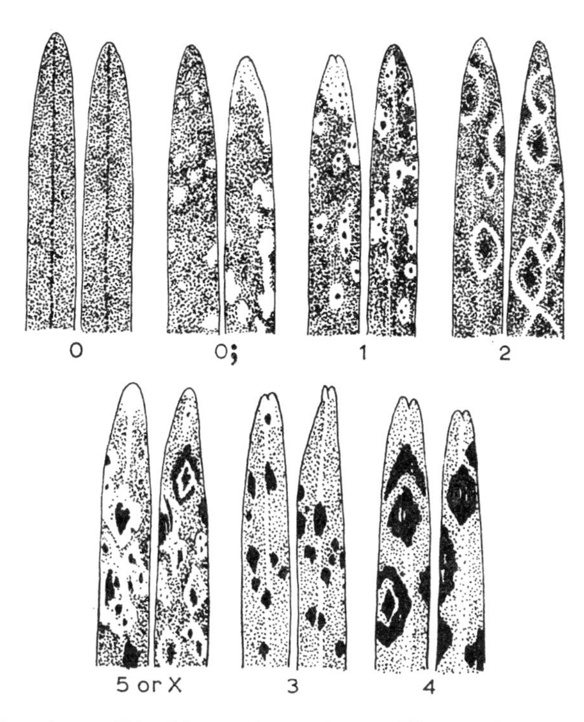

Fig. 17.5. Pustules and blemishes on leaves due to different infection types of wheat stem rust, *Puccinia graminis tritici* on differential varieties of wheat. Key: 0—entirely immune; 0;—practically immune; 1—extremely resistant; 2—moderately resistant; 5—heterogenous, showing variable response; 3—moderately susceptible; 4—completely susceptible. (After Stakman and Harrar)

and Table 17.6). By increasing the numbers of so-called 'differential hosts', i.e. varieties showing different infection responses, the races can be further subdivided. The separations are, therefore, subjective, since they depend in part upon the numbers and kinds of differential hosts employed and the natural variability of the fungus. This work, initiated and developed by Stakman and his colleagues since 1917, has been shown to apply to other rusts, to smut fungi, *Phytophthora*, *Helminthosporium*, *Fusarium* and *Rhizoctonia*, to name some of the most important pathogens.

There is no doubt that these races can persist and that from time to time new races can arise. Moreover, the relative frequencies of different races change both with changes in the varieties of host plants grown and quite independently. A good example is Race 7 of *P. graminis avenae*. This was first identified in Canada in 1928 and found for the first time in the U.S.A. in

Maine in 1933 where it infected barberry, the alternative host. (A similar sudden spread of Race 8 is shown in Fig. 17.6.) By 1950, instead of being an interesting curiosity, it had become of major importance as a pathogen. In other cases even more rapid dissemination has occurred. Race 15B of *P.*

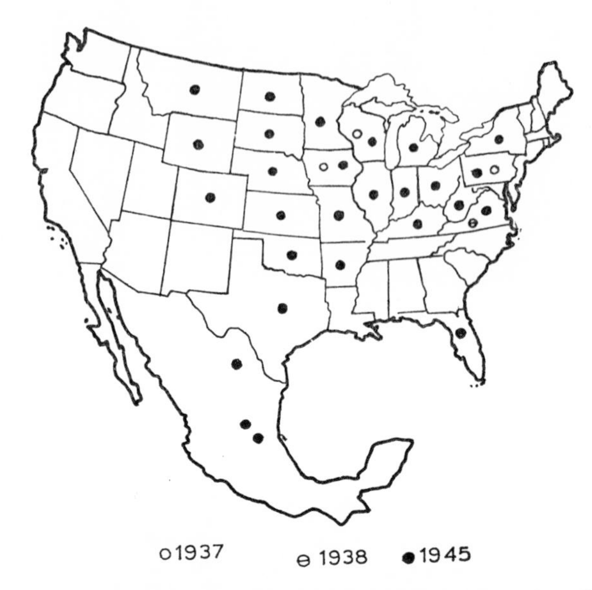

Fig. 17.6. Spread of *Puccinia graminis avenae* Race 8 in the U.S.A. Note dramatic spread between 1938 and 1945. (From Stakman and Harrar)

graminis tritici was detected sporadically between 1939 and 1950 in the U.S.A. In 1950, however, it spread from Texas in the spring northwards to Dakota, Minnesota and Canada. Here it multiplied rapidly on the bread and durum wheats hitherto resistant to the prevailing races of wheat rust. Multiplication was aided by a late harvest and good environmental conditions for rust attack through the late autumn. In the autumn and winter, therefore, the winds carried it back southwards to southern U.S.A. and Mexico. Thus, within a year, it had become established over about $7{\cdot}77 \times 10^6$ km^2, mainly as a consequence of rather unusual climatic conditions coupled with the susceptibility of wheats not previously exposed to this race. The consequences of this fortuitous spread were seen in the enormous losses sustained by susceptible varieties in 1953 and 1954, which were epidemic years. In some cases, it is the change in the pathogen which is wholly responsible; for example, Race 34 of *P. graminis tritici* arose in Australia and attacked the hitherto immune and widely grown variety Eureka (MACINDOE, 1945).

A good deal of evidence has been amassed to indicate that races of rust are usually highly heterozygous (cf. JOHNSON, 1946) although homozygous races, which do not segregate on selfing, are known, e.g. in *Melampsora lini* in Australia (WATERHOUSE and WATSON, 1944). In experimental crosses it is usually found that intra-varietal crosses are highly successful but inter-varietal crosses, although possible, are usually very infertile, e.g. *P. graminis tritici* × *secalis* (STAKMAN *et al.*, 1930). The degree of fertility/sterility varies, *tritici* × *secalis* crosses are easier to make than *tritici* × *avenae* or *poae*, for example. Whenever such crosses are found to be reasonably interfertile, it seems that there is a common host; in the cases just mentioned these are *Hordeum vulgare* and *Poa* spp. respectively. This stresses the importance of the host specialization as a character correlated with isolation, it also raises the possibility of the breakdown of isolation. A thorough study of this situation has been made with the brown rust of brome grass, *P. dispersa* on *Bromus* spp. (BEAN *et al.*, 1954). Five physiological races were recognized, one was restricted to *Bromus* sect. *Stenobromus*, another to sect. *Serrafalcus* and the remaining three were distinguished on various differential hosts. The races remained stable for over 14 years in conditions designed to prevent contamination. In earlier work it had been shown that *B. arduennensis* was a host common to a number of physiological races of brown rust. Accordingly, Races 2a and 5 were cultivated on this host for 164 sub-cultures over a period of 8 years. They were periodically assessed on the differential hosts and throughout this period showed no alteration in their responses as compared with stock cultures maintained on the usual host Bromes. Thus, in this period, selection had neither been effective in leading to their adaptation to this common host, nor in altering their pathogenicity to their normal hosts or to any of the differential host plants.

In other studies, the hybrids and progeny of intervarietal crosses of *P. graminis* were tested on the parental hosts. It has been found that in the great majority of cases the hosts are attacked less vigorously than by the parent varieties of fungus. Exceptions are known, thus *P. graminis secalis* × *tritici* gave rise to a hybrid capable of attacking wheat, rye and barley quite heavily (LEVINE and COTTER, 1931).

It was pointed out earlier that the recognition of physiological races was arbitrary, in the sense that the number recognized depended upon the classes of pathogenicity reaction and numbers of differential hosts used. Attempts have been made to assess the genetic nature of pathogenicity by investigating the genetics of host resistance. The best known example is FLOR's studies with the flax rust *Melampsora lini* (summary, 1956). He developed a gene-for-gene theory: that for every gene capable of mutating to give resistance in the host, there is a gene in the pathogen capable of mutating to a virulent condition which will overcome the host resistance. For example, Races 22 and 24 were tested on the flax varieties Ottowa and Bombay; they were then crossed and the tests repeated with both F_1 and F_2 progeny (Table 17.7). It can be seen that in the F_1, resistance is dominant to susceptibility and in the F_2 there is the segregation expected if two genes are involved. Thus most resistance and fungal virulence parallel each other in the cross. This model has been investigated theoretically by PERSON (1959) in a most penetrating paper. He

Table 17.7. The inheritance of virulence to flax varieties of two races of flax rust, *Melampsora lini*, and of resistance to the same races by the same flax varieties. (From Flor, 1956)

+, − Susceptible, or resistant respectively

Flax host	*Flax rust races, generations, phenotypes and frequencies*						
	P_1		F_1	F_2			
	Race 22	Race 24					
	$p_1p_2^+$	$p_1^+p_2$	$p_1^+p_2^+$	$p_1^+p_2^+$	$p_1p_2^+$	$p_1^+p_2$	p_1p_2
Ottawa (R_1)	+	−	−	−	+	−	+
Bombay (R_2)	−	+	−	−	−	+	+
Frequencies				78 :	27 :	23 :	5

Rust race	*Flax varieties, generations, phenotypes and frequencies*						
	P_1		F_1	F_2			
	Ottawa	Bombay					
	R_1r_2	r_1R_2	R_1R_2	R_1R_2	R_1r_2	r_1R_2	r_1r_2
22(p_1)	+	−	−	−	+	−	+
24(p_2)	−	+	−	−	−	+	+
Frequencies				110 :	32	43 :	9

started with the assumption that there were five genes each in host and parasite and listed the expected interrelationships which would arise in the situation where each host had either one dominant resistant gene, or two, and so on. Since n varieties, each either susceptible or resistant, distinguishes 2^n races, five genes will distinguish 32 races but if, by chance, only four testers are employed, then only 16 races will be recognized (2^4). Moreover, if one variety carries two of the resistant genes, only 24 races will be recognized even though five testers are used, e.g. R_1, R_2, R_3, R_4 and R_1R_5. The implications of this situation are that there might be two or more genes for resistance in the tester varieties. Thus Flor's approach may not give the precision of analysis which it had been hoped to achieve since exactly similar arguments as used for host resistance apply to pathogen virulence. This attempt to analyse the genetic basis of pathogen virulence in different races only gives an indication of the numbers of loci and allelomorphs involved. Nevertheless, this type of analysis is of some value and resistant genes in the host have been used to characterize physiological races in several pathogenic fungi, e.g. *Phytophthora infestans* (BLACK, 1952, 1957 and Tables 17.8 and 17.9), *Cladosporium fulvum*—tomato leaf mould (DAY, 1956).

The important conclusions to be drawn from these studies of pathogens is the fact that there is a close relationship between genotypes of host and parasite. PERSON (1966) has described a polymorphic situation in which the genotypes fluctuate according to the selection pressure. At times there will be a balanced position but the introduction of a new gene for virulence in the pathogen, or for resistance in the host, will result in a change in gene frequencies which may be quite rapid. Apparently constant physiological races

Table 17.8. Interrelationships of 16 strains of potato-blight fungus *Phytophthora infestans*, and major genes (R_1–R_4) which control resistance to the disease in potato (*Solanum tuberosum*) in various combinations. (From Black, 1952)

Genotypes	Strains of *Phytophthora infestans*															
	A	B^1	*H*		*D*	*G*	*E*	B^2		*C*	*I*			*F*		
	0	(1)	(2)	(3)	(4)	(1, 2)	(1, 3)	(1, 4)	(2, 3)	(2, 4)	(3, 4)	(1, 2, 3)	(1, 2, 4)	(1, 3, 4)	(2, 3, 4)	(1 2, 3, 4)
r	**+**	**+**	**+**	+	**+**	**+**	**+**	**+**	+	**+**	**+**	+	+	**+**	+	+
R_1	**−**	**+**	**−**	−	**−**	**+**	**+**	**+**	−	**−**	**−**	+	+	**+**	−	+
R^2	**−**	**−**	**+**	−	**−**	**+**	**−**	**−**	+	**+**	**−**	+	+	**−**	+	+
R^3	**−**	**−**	**−**	+	**−**	**−**	**+**	**−**	+	**−**	**+**	+	−	**+**	+	+
R_4	**−**	**−**	**−**	−	**+**	**−**	**−**	**+**	−	**+**	**+**	−	+	**+**	+	+
R_1R_2	**−**	**−**	**−**	−	**−**	**+**	**−**	**−**	−	**−**	**−**	+	+	**−**	−	+
R_1R_3	**−**	**−**	**−**	−	**−**	**−**	**+**	**−**	−	**−**	**−**	+	−	**+**	−	+
R_1R_4	**−**	**−**	**−**	−	**−**	**−**	**−**	**+**	−	**−**	**−**	−	+	**+**	−	+
R_2R_3	**−**	**−**	**−**	−	**−**	**−**	**−**	**−**	+	**−**	**−**	+	−	**−**	+	+
R_2R_4	**−**	**−**	**−**	−	**−**	**−**	**−**	**−**	−	**+**	**−**	−	+	**−**	+	+
R_3R_4	**−**	**−**	**−**	−	**−**	**−**	**−**	**−**	−	**−**	**+**	−	−	**+**	+	+
$R_1R_2R_3$	−	−	−	−	−	−	−	−	−	−	−	+	−	−	−	+
$R_1R_2R_4$	−	−	−	−	−	−	−	−	−	−	−	−	+	−	−	+
$R_1R_3R_4$	**−**	**−**	**−**	−	**−**	**−**	**−**	**−**	−	**−**	**−**	−	−	**+**	−	+
$R_2R_3R_4$	−	−	−	−	−	−	−	−	−	−	−	−	−	−	+	+
$R_1R_2R_3R_4$	**−**	**−**	**−**	−	**−**	**−**	**−**	**−**	−	**−**	**−**	−	−	**−**	−	+

− = Resistant, **+** = Susceptible — Observed
− = Resistant, + = Susceptible — Theoretical

Table 17.9. The distribution of races of *Phytophthora infestans* in different countries, genotypes of races are given in Table 17.8. (From Black, 1957)

Country	Physiological race															
	0	1	2	3	4	1 2	1 3	1 4	2 3	2 4	3 4	1 2 3	1 2 4	1 3 4	2 3 4	1 2 3 4
Scotland	+	+		+	+			+		+	+			+		
England	+	+		+	+			+		+	+					
Ireland	+	+	+		+	+		+					+			
Holland	+	+	+		+			+		+			+		+	+
Belgium		+						+								
Germany	+	+			+	+	+	+						+		
Denmark					+											
Sweden		+			+			+								
Switzerland					+											
Portugal					+											
Cyprus					+											
Jordan	+				+											
Kenya	+		+		+	+				+	+					
Tanganyika (now part of Tanzania)			+				+			+	+					
S. Rhodesia (now Rhodesia)	+				+			+			+			+		
S. Africa		+		+		+										
Canada	+	+			+	+		+				+	+			+
U.S.A.	+	+	+	+	+	+	+	+		+	+		+			
Colombia	+				+											
Peru	+			+	+											

are maintained by selection for particular genetic combinations. (Incidentally, there is suggestive evidence for the existence of polymorphism in other fungi. Situations such as the purple and brown forms of *Laccaria laccata*, or the supposed species pair *Amanita fulva* and *A. vaginata*, which occur in different proportions in different localities, come to mind.) The significance of heterozygosity in races of rust fungi is now much clearer. New gene combinations can be generated through recombination (somatic or sexual) from heterozygotes. Those races which are detectable in the offspring will give evidence of the potentialities of races in this respect but, in the natural populations, they will be subject to selection and so only a few races will appear to dominate at any one time. The constancy of such races is, in a sense, spurious in that while they are maintained by clonal reproduction, via uredospores for example, they can also be constantly regenerated by recombination between other races and, if at a selective advantage, can dominate. The study of physiological races provides an insight, in fact, into the basic population changes upon which isolation is imposed.

The origins of isolation

In the previous pages examples have been given of the ways in which isolating mechanisms may operate in different fungi. This raises the question of how they arise.

Sympatric and allopatric speciation is recognized in other organisms and distinguishes speciation within populations living in the same area from that in those inhabiting different areas. However, even in sympatric speciation, isolation is usually supposed to originate through some form of ecological (habitat, seasonal or temporal), morphological or physiological barrier. In some fungi it is extremely difficult to see how this arises. The last six entries in Table 17.3 make this point clearly and a notable example is *Mycocalia denudata* strains 144 and 144N. This fungus is related to the larger and better known bird's-nest fungus and its basidiocarp consists of a heap of peridiola covered by an evanescent peridium. Basidiospore isolations from a group of peridiola gave rise to monokaryotic progeny which, when mated in all combinations, fell into two intrafertile, intersterile strains (Table 17.10). It became clear that what had been assumed to be the peridiola of a single basidiocarp, in fact represented two confluent basidiocarps. The growth rate and colour of the mycelia, the colour and size range of the peridiola and the ranges of basidiospore size of the two intrasterile strains were compared. No constant differences could be found between them and they clearly shared the same ecological niche! No hyphal fusions could be detected or provoked between strains 144 and 144N.

Table 17.10. Results of crossing strain 144 and 144N of *Mycocalia denudata* with each other. (From Burnett and Boulter, 1963)

Strain and isolations	144/3	144/12	144N/1	144N/2
144/3	—	+	—	—
144/12	+	—	—	—
144N/1	—	—	—	+
144N/2	—	—	+	—

+ = dikaryon with clamp connections

Situations such as these are extremely unusual and appear to contradict GAUSE's (1934) view that two or more forms with similar ecological requirements cannot co-exist indefinitely in the same environment. In those cases where species seem to co-habit in the same environment it is thought that they will be found to differ in the microhabitats they occupy, be subject to different controlling factors or to be reproductively isolated as a consequence of post-zygotic mechanisms (HARPER *et al.*, 1961; STEBBINS, 1966). There is no

Table 17.11. The six groups of the

Group number	Basidiocarp characters of *C. coronilla* Groups I–IV Spore characters	Basidia	Character of hyphae and subiculum	General appearance of specimen
I	1·8–2 × 3–4·5 μm ellipsoid not curved	Variable, most usually 3–4 × 12–14 μm	Subiculum with some incrustation, hyphae sometimes distinct 3 μm	Delicate, coarsely powdery edge indeterminate, with a general tendency to grandinioid granulations
II	1·8–2 × 3·5–5 μm ellipsoid curved	Very variable, 3–4 × 13–20 μm	Subiculum with conspicuous incrustation. Hyphae 2–5 μm	Rather thick, edge indeterminate, all specimens with distinct grandinioid granulations
IIIA	1·8–2 × 3–3·5 μm ellipsoid flattened on one side	3–4 × 12–14 μm	Subiculum hyphae regular and distinct 5–6 μm. No incrustation	Pellicular with a rhizoidal fibrillose edge
IIIB	2–2·5 × 4–4·5 μm ellipsoid flattened on one side	4–5 × 13–15 μm	Subiculum hyphae regular and distinct 5–6 μm. No incrustation	Pellicular. Edge indeterminate
IIIC	2·5–3 × 5–5·5 μm ellipsoid curved	4·5–6 × 18–20 μm	Subiculum hyphae regular and distinct 5–6 μm. No incrustation	Pellicular. Edge rhizoidal. Masses of pale brown bulbils found in association
IV	2–3 × 3·5–4·5 μm ellipsoid-subglobose flattened on one side	5·5–6 × 12–13 μm	Subiculum hyphae of wide diameter and often irregular 5–6 μm	Very delicate with no grandinioid granulations. Edge indeterminate

evidence that in these fungi any of these factors operate and their isolation cannot, at present, be explained.

In other cases, however, the more usual attributes of sympatric speciation apply. The most striking, perhaps, are those associated with host specificity, i.e. the occupation of different ecological niches. Such situations are comparable with those described as ecotype-differentiation in other organisms. They may, as in the varieties of *P. graminis*, the strains of *R. solani* or the forms of *Auricularia*, be associated with some morphological differences but this is

Corticium coronilla group. (From Biggs, 1937)

Cultural characters of *C. coronilla* Groups I–IV

Hyphae	Heterothallism	Fructification	Growth and general appearance
Usually agglutinated and indistinct	Bipolar	None	Growth rapid submerged, no microscopic distinguishing characters
Often agglutinated and indistinct up to 5 or 6 μm	Homothallic	In both monospore and polyspore cultures	Growth rapid and always some aerial mycelium. Several specimens produce bulbil-like cells
Distinct 5–7 μm	Tetrapolar	None	Growth mediumly rapid with conspicuous aerial rhizoidal strands. Bulbils and oidia produced
Distinct 5–7μm	Tetrapolar	None	Growth rapid and lacking rhizoidal strands. Aerial mycelium cottony. Bulbils and oidia produced
Distinct 5–7 μm	?	None	Growth medium rapid with conspicuous aerial rhizoidal strands. Bulbils and oidia produced
Distinct 5–7 μm	Heterothallic	None	Growth very slow and almost entirely submerged. Conspicuous separable spherical cells produced apically or in chains

not always the case as in *Peniophora mutata* Groups III and IV or European and N. American *Irpex fusco-violaceus*.

In some cases it is not possible to determine the nature of the differences. This is true of *Aspergillus nidulans* where the habitat is so broadly defined that it is not possible to know whether different microhabitats are occupied by each different groups of strains. However, in such cases, allopatric speciation may be invoked, the differences between groups of strains having arisen as a consequence of isolation by distance. This is not as striking a phenomenon in the fungi as it is in some other organisms. For example, Raper's world-wide sample of *Schizophyllum commune* basidiocarps provided no evidence for any form of intersterility, whereas in *Polyporus betulinus*, strains isolated in the British Isles are intercompatible yet intersterile with strains from Norway or Sweden (RAPER *et al.*, 1958; BURNETT, unpublished). Despite the differences in host specialization between N. American and European forms, *Irpex* strains seem to be fully compatible yet N. America includes two completely intersterile groups of *P. abietinus*, both of which show partial fertility with European strains. Allopatric speciation does not seem, therefore, to be a well marked mode in the fungi. This may be related to the widespread dissemination of fungal propagules of all types (cf. Chapter 6).

All the data available, small though it is, supports the notion that isolation is associated with the accumulation of different genes which reduce the compatibility of hyphae, gametes or gametangia prior to or, perhaps more often, after fusion or conjugation. The evidence in support of this view seems to be unambiguous in the rusts and Aspergilli. The case of *Podospora* draws attention to the possibility of such isolation arising quite abruptly as the consequence of one or a few mutations. The widespread distribution of 'aversion' and 'barrage' phenomena throughout Ascomycetes, Basidiomycetes and Fungi Imperfecti at least suggests that selection might well lead to the incorporation of such phenomena in the isolating mechanism of some species. The genetic bases of such phenomena might well repay investigation.

Aversion phenomena could lead to rather abrupt isolation and much the same could occur if one mating system were to be replaced by another. This seems to have taken place amongst the diaphoromictic species. It is not unusual to find that closely similar or sibling species possess different mating systems. *Mycocalia denudata*, for example, is bipolar but *M. castanea* is homomictic and crosses have not, so far, been successful between them. A more striking case is that of *Corticium coronilla* studied by BIGGS (1937). From fourteen isolations, she was able to recognize six groups on a basis of their morphology, cultural characters and pairing reactions (Fig. 17.7 and Table 17.11). Two groups were bipolar, one was homomictic and the third tetrapolar but divided into two intersterile subgroups. A further strain appeared to belong to this group but was intersterile with cultures of the other two subgroups. Observations such as these suggest that in diaphoromictic fungi the homomictic, bipolar and tetrapolar conditions may not be as distinct as they appear to be. It will be recalled that mutants are known in both *Coprinus* and *Schizophyllum* which convert them into self-fertile forms or increase the proportion of fertile matings between sister basidiospores. In anomalous dimictic systems, *Glomerella* also provides an example of a fungus in which certain crosses are more probable than others and so evolution to an obligatory cross-fertile or self-fertile condition could take place (see Chapter 15, pp. 421–422).

A further possible mechanism which could lead to abrupt isolation is polyploidy. Although this seems to occur in fungi (see later) it is not known if it has operated in this way.

There is not much evidence from fungi concerning post-zygotic mechanisms

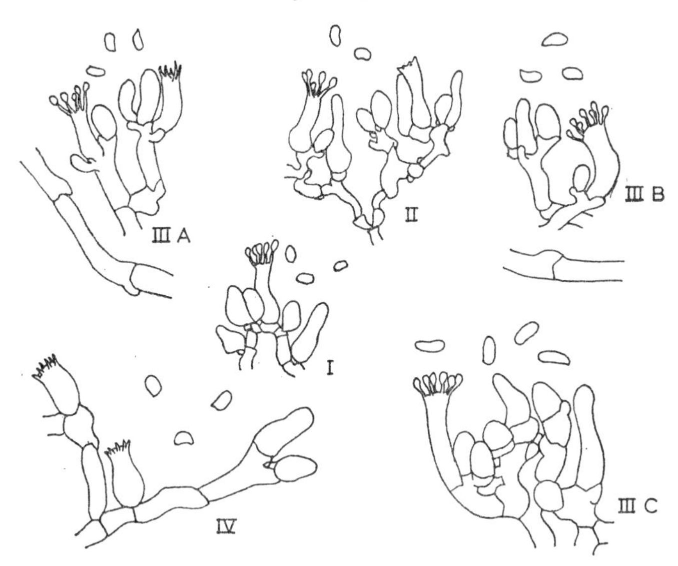

Fig. 17.7. Basidia and basidiospores of intersterile groups of the *Corticium coronilla* complex of species. Note that more than 4 spores are usually produced by each basidium and that the morphological differences are very slight. Numbers refer to groups in Table 17.11. (× 560, from Biggs)

which lead to isolation. This is because of the relative paucity of hybridization in the fungi as compared with green plants. Moreover the interpolation of a prolonged heterokaryotic or dikaryotic phase prior to fertilization enables intracellular isolating mechanisms to operate at this stage. However, there are some clear-cut examples of such post-zygotic mechanisms and these will now be described.

HYBRIDIZATION

Hybrids have been made between many fungi but the number of cases in which they have proved viable or from which progeny have been obtained is low. The groups most studied are the Oomycetes, Mucorales, yeasts, rusts and smut fungi, together with a few Ascomycetes, such as *Neurospora* and *Cochliobolus*. It will be noticed that, with the exceptions of the Ustilaginales, all these are dimictic fungi and even in the smuts conjugation is determined primarily by a pair of allelomorphs. It is a remarkable fact that no unequivocal case of hybridization has been detected between different species of bipolar or tetrapolar diaphoromictic fungi. Claims have been made, such as that of YEN (1948) who believed he had obtained a hybrid basidiocarp between *Agrocybe praecox* and *A. semiorbicularis*. The supporting evidence for this claim is lacking, however, and, for instance, the possibility that it was an induced, monokaryotic basidiocarp was not tested. The reasons for the apparent lack of hybrids in diaphoromictic forms are not known but it may be surmised that it is related to the intracellular controls which regulate the growth and behaviour of dikaryons. These are clearly complex and it is clear from the work on the *Stereum pini* complex, for example, that the establishment of a dikaryon between evidently closely related forms may not be achieved readily.

Phycomycetes

Many crosses have been attempted between different species and genera of the Saprolegniales and in some cases the mating reaction appears to have been completed. However, the great difficulty experienced in germinating the zygotes (oospores) of normal crosses also applies to supposed hybrids. The best authenticated examples are *Thraustotheca clavata* ♂ × *Achlya ambisexualis* ♀, and *T. clavata* ♂ × *A. flagellata* ♀ where oospores were fully developed (RAPER, 1950; SALVIN, 1942). In most other crosses the reactions were 'imperfect', i.e. either gametangia or gametangial initials only were formed (COUCH, 1926; RAPER, 1939, 1950). The same situation is known from the Mucorales but here the cross *Phycomyces blakesleeanus* f. *gracilis-piloboloides* − × *P. nitens* + was not only achieved but the zygote germinated and produced some sporangia similar to those of *P. nitens* and three with apparently normal spores that could not be germinated (BURGEFF, 1925). This is good circumstantial evidence if it be supposed that nuclear fusion and meiosis really does occur even in the normal species but it will be recalled that the position is very obscure (cf. Chapters, 15, p. 413, and 16, p. 442). SAITO and NAGANISHI (1915) obtained zygotes with various *Mucor* spp. but BLAKESLEE (1920) claimed that they were probably strains of the same species, a view accepted by the two main monographers of the group (NAUMOV, 1936; ZYCHA, 1935). Zygotes also formed between the homomictic *Rhizopus sexualis* and *R. nigricans* + and − strains but they too could not be germinated (CALLEN, 1940). As with the Saprolegniales, in many cases gametangia, gametangial initials or azygospores developed in the numerous 'imperfect' sexual reactions referred to earlier.

The possibility of interspecific hybridization has often been raised in the genus *Phytophthora* and there are unequivocal reports of stimulation of oogonium production in mixed cultures of strains and species (e.g. SAVAGE and CLAYTON, 1962). These claims should be viewed with caution, however, in the light of recent discoveries concerning the promotion of gametangium production by sterols, mentioned earlier (see Chapter 15, p. 408). Because of the difficulty of germinating oospores it is not clear whether these 'hybrid' oospores are truly hybrid or are apomictic in origin and due to stimulation by sterols.

The best documented case of hybridity in the fungi is, however, found in the Phycomycetes in *Allomyces*. Three 'species' *A. arbuscula*, *A. javanicus* var. *macrogynus* (*A. macrogynus*) and *A. javanicus* var. *javanicus* were known. The last, in many respects was intermediate between the other two, notably in its irregular distribution of oogonia and antheridia in gametophytes. They are strictly epigynous in *A. arbuscula* and *hypogynous* in *A. macrogynus*. Hybrids were synthesized and two main types A and B were produced, including several that resembled the natural *A. javanicus*. Chromosome counts supported the view that *A. javanicus* was indeed a putative hybrid between the other two species (Table 17.12 and Fig. 17.8). The different types of artificial hybrids

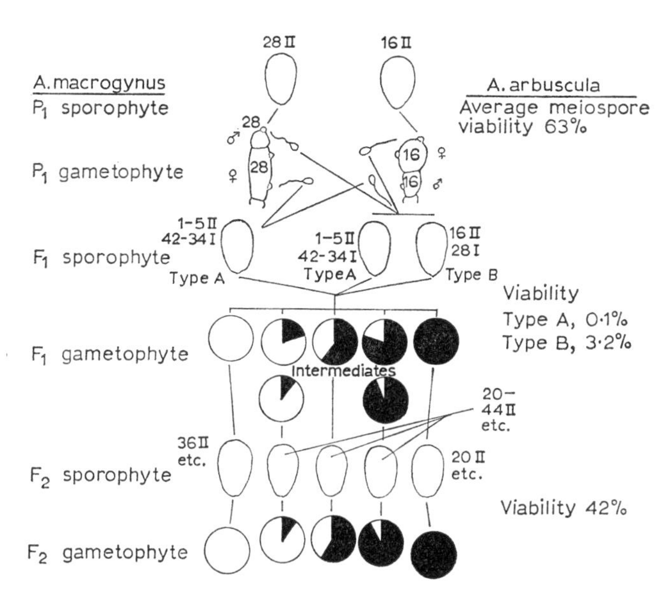

Fig. 17.8. The products of the reciprocal cross *A. macrogynus* × *A. arbuscula*. In the sporophyte generations, I and II refer to univalents and bivalents at meiosis, respectively. In the F_1 and F_2 gametophyte generations the degree of shading indicates the degree of hypogyny of male gametangia

were thought to be due to normal hybridization in the case of the A-type but, with a doubling of the chromosome set of *A. arbuscula* ($2 \times 16 + 28 = 60$), at fusion of the gametes in the B-type. In subsequent generations there was a reduction in chromosome numbers and increased viability.

Ascomycetes

The best and earliest undoubted cases of fungal hybrids were obtained by DODGE (1928, 1931) with species of *Neurospora* and this has been extended with a fuller genetic analysis by FINCHAM (1951). Dodge described the *N. sitophila* × *N. tetrasperma* cross in detail and Fincham that of *N. sitophila* × *N. crassa*. In the former the perithecia of the F_1 were fertile, although only a few were formed, there was much ascus abortion and not all the ascospores matured. The *sitophila* characters appeared to be dominant. In F_2 and F_3 fertility improved but traces of the *tetrasperma* ancestry still persisted. In the second cross no perithecia were produced for about 6 weeks, compared with 10–14 days for a normal cross, and then only a very few perithecia which had neither a full complement of asci nor more than 4 mature spores per ascus, 1–2 being more usual. Further selfing has not been reported in this cross but

Table 17.12. Chromosome counts (haploid numbers) in *Allomyces arbuscula*, *A. macrogynus* and the putative and synthesized hybrid between them, *A.* ×*javanicus*. (After Emerson and Wilson, 1954)

Fungus	Chromosome counts (numbers of isolates tested in parentheses)							
A. arbuscula	8(2)	16(14)	24?(1)	22–26(1)	32(1)			
A. macrogynus	14(1)	28(3)	50^{+}(1)					
A. ×*javanicus*—natural	13(1)	14(1)	14/15(1)	16/17(1)	19(1)	21(2)		
A. ×*javanicus*—synthetic	20(1)	22(1)	27(2)	32(1)	36(1)	42+(2)	42+(1)	44(1)

it is clear that both show very considerable hybrid inviability and this is extended to the F_2 generation.

A similar pattern has been described by Nelson and his colleagues, who have studied the interactions of *Helminthosporium* pathogenic to various grasses, many having a reproductive phase, *Cochliobolus* (reviewed Nelson, 1963, 1964). Sixteen species were studied, 10 being imperfect forms only, and all possible combinations of cross were made. Only 13 of the possible 120 crosses gave viable ascospores although 14 others developed immature asci. Ten of the 13 viable crosses were between species with similar conidial morphology; the other 3 were between species differing in their conidial morphology but 12/14 of the crosses with immature asci were of this type. In later work he showed that interspecific inviability could be reduced by crossing the recombinant progeny derived from intra- or interspecific crosses. In the earlier work the relative frequency of interspecific crosses was only 0·39% but in the later work this was increased to 1·03% in some cases. Nevertheless in many cases the F_1s or recombinants obtained were not very viable when crossed with their parents or other species or strains (Table 17.13).

Neurospora backcrosses and *Cochliobolus* data show that fertility can be restored to F_1s and recombinants in a few generations, of the order of 2–7. This suggests that hybrid inviability is due to several genes or is possibly polygenic. Thus in Pyrenomycetes the phenomenon of hybrid inviability is similar to that in many other organisms and unless there is strong selective pressure in favour of fertility it is unlikely that this will be achieved. So hybrid inviability and weak F_2 and subsequent progeny probably constitute an effective post-zygotic isolating mechanism for these fungi in nature.

The other group of Ascomycetes which has been studied is the yeasts. Within the genus *Saccharomyces* crosses can usually be achieved although the percentage germination of ascospores is often low (WINGE, 1941). On the other hand, germination in intraspecific crosses also is often low and the problem here is to know what the limits of a species really are. In the genus *Hansenula*, for example, where the species limits seem to be more clearly defined, it has been claimed that no intraspecific hybrids have been produced despite numerous trials (WICKERHAM and BURTON, 1956). Nevertheless, crosses between recognizably distinct genera are not impossible, e.g. *S. lactis* and *Zygosaccharomyces ashbyi* or *S. cerevisiae* and *Z. priorianus*. These genera differ in the balance of their life cycles and the critical requirement is to obtain a sufficient number of haploid cells of *Saccharomyces*. Perhaps the best way to interpret the data for yeasts is to suppose that they include a great many races which are not greatly dissimilar and between which various degrees of sterility have developed. There is no doubt that, in nature, different kinds of yeasts occupy specific microhabitats, for example, those of the mycetomes of insects (cf. Chapter 12, pp. 356–357). Speciation in yeasts could, therefore, be allopatric. In such circumstances, sterility barriers would not necessarily arise, since the species in nature live in distinct habitats and never come into contact. Attempts to cross them in vitro are therefore, in a sense, artificial situations and the lack of inviability is biologically meaningless.

Table 17.13. Comparison of the consequences of crossing strains of *Cochliobolus* spp. and an *Helminthosporium* sp. directly with intercrosses of F_1 or other hybrid derivatives. (From Nelson, 1964)

Interspecific cross	Conidial morphology of species	Extent of compatibility	No. of different crosses		Attempted multiple cross	Extent of compatibility	No. of different crosses	
			Attempt.	Succ.[1]			Attempt.	Succ.[1]
C. cynodontis × *C. sativus*	Distinct	Incompatible	782	0	(*C. cyn.* × *C. cyn.*) × (*C. sat.* × *C. sat.*)	Asci	800	3
C. cynodontis × *C. heterostrophus*	Distinct	Incompatible	817	0	([*C. carb.* × *C. carb.*] × *C. hetero.*) × ([*C. hetero.* × *C. hetero.*] × [*C. cyn.* × *C. cyn.*])	Asci	640	2
C. cynodontis × *C. spiciferus*	Distinct	Incompatible	610	0	([*C. cyn.* × *C. cyn.*] × *C. cyn.*) × (*C. spic.* × *C. spic.*)	Asci	580	4
C. heterostrophus × *C. spiciferus*	Distinct	Incompatible	430	0	(*C. hetero.* × *C. hetero.*) × (*C. spic.* × *C. spic.*)	Imm. asci	720	3
C. victoriae × *C. sativus*	Distinct	Imm. asci	724	3	(*C. vic.* × *C. vic.*) × *C. sat.*	Spores	417	2
C. carbonum × *C. sativus*	Distinct	Imm. asci	923	3	(*C. carb.* × *C. carb.*) × *C. sat.*	Spores	600	1
C. heterostrophus × *C. sativus*	Distinct	Imm. asci	1368	2	(*C. sat.* × *C. sat.*) × *C. hetero.*	Imm. spores	400	2
C. miyabeanus × *C. sativus*	Distinct	Imm. asci	747	1	([*C. sat.* × *C. sat.*] × *C. sat.*) × *C. miy.*	Imm. spores	835	3
C. carbonum × *C. cynodontis*	Distinct	Asci	913	5	([*C. carb.* × *C. vic.*] × *C. carb.*) × (*C. cyn.* × *C. cyn.*)	Spores	728	4
C. miyabeanus × *H. sacchari*[2]	Similar	Imm. asci	412	2	([*C. hetero.* × *C. miy.*] × *C. miy.*) × *H. sacc.*	Spores	376	3
C. heterostrophus × *H. sacchari*[2]	Similar	Asci	338	2	(*C. hetero.* × *C. hetero.*) × ([*C. hetero.* × *C. hetero.*] × *H. sacc.*)	Spores	580	6

[1] Succ. = successful with reference to extent of compatibility.

[2] All species of *Cochliobolus* used herein produce *Helminthosporium* imperfect stages; the sexual stage of *Helminthosporium sacchari* is not known.

Basidiomycetes

Reference has already been made to the reduced viability of hybrids between host-specific varieties of rusts (p. 465) and this is in great contrast to the situation in the smuts where interspecific crosses are fairly readily achieved. However, although the crosses are possible, there is a good deal of hybrid inviability. In some cases the promycelium produced by the zygote lyses, e.g. *Ustilago hordei* × *U. bullata* (FISCHER, 1951), in others no spores develop, e.g. *U. avenae* × *U. kolleri* (HOLTON, 1932) or they are of low viability, e.g. *Sorosporium syntherismae* × *Sphacelotheca panici-miliacei* (MARTIN, 1943). In some cases, however, the range of pathogenicity may be increased and this may persist into the F_2 and F_3 generations (Table 17.14). Thus, apart from the fact that crosses of smuts are made more readily than are those between varieties of rusts there is very little difference. The F_1 usually show hybrid inviability and subsequent generations are usually less pathogenic than the parental types. In both groups of fungi, however, there are rare occasions when F_1 or later generations show either a wider range of pathogenicity or even greater virulence to one or both the normal hosts or even to other varieties.

Table 17.14. Alterations in infectivity of hybrid progeny of the cross *Ustilago avenae* × *U. kolleri* on oat plants of different initial susceptibility. (From Holton, 1941)

Oat varieties	Parents		Percentage infection of fungus	
	U. avenae	*U. kolleri*	F_2	F_3 [1]
Gothland	100·0	0·0	22·0	92·0
Monarch	0·0	99·0	21·0	45·0

[1] F_2 and F_3 selected from Gothland

Broadly speaking, hybrid inviability is similar in its manifestations in fungi to that in other organisms. It represents a well developed post-zygotic isolation mechanism but this can break down. The two outstanding examples are the yeasts and the probable natural hybrid of *Allomyces*, *A.* ×*javanicus*. These represent two quite different kinds of phenomenon. The yeasts probably represent a group in which speciation has been allopatric and selection for sterility barriers has not been necessary. In *Allomyces*, however, the natural hybrid appears to be a consequence of hybridization between strains of different ploidy. This does not seem to be a common situation in fungi but it may reflect the paucity of cytological information about them.

POLYPLOIDY

Study of the chromosome numbers reported for fungi provides no very clear evidence for polyploid series in fungi save in a few cases.

WAKAYAMA (1930) was the first worker to suggest that polyploidy might occur in fungi since certain numbers appeared to be multiples of the other, e.g. *Hypholoma fasciculare* $n = 2$, *H. appendiculatum* $n = 6$; *Tricholoma melaleucum* $n = 4$, *T. rutilans* $n = 2$. These counts do not seem to be very accurate, however, and by the use of modern smear techniques, British material of *H. fasciculare* can be seen to have at least 7 chromosomes (BURNETT, unpublished). Three recent reports, using modern methods, are suggestive of polyploidy (Table 17.15).

Table 17.15. Chromosome counts and cytological features which suggest polyploidy in fungi. (Data from McGinnis, 1953, 1954, 1956; Sansome, 1959; Emerson and Wilson, 1954; Lu, 1964b)

Species	Haploid Chromosome No.	Multivalents	Secondary association	Possible ploidy
Puccinia malvacearum	4			Diploid
asteris	4			Diploid
xanthii	4			Diploid
arenariae	4			Diploid
coronata	3			Diploid
graminis	6		+	Allopolyploid
helianthi	6			Allopolyploid
sorghi	6			Allopolyploid
carthami	6			Allopolyploid
kraussiana	20–21–22	–		Highly polyploid
Allomyces arbuscula	8, 16, 24, 32	–	–	Diploid— ?Allopolyploid
A. macrogynus	14, 28, 50+	–	–	(as above)
Cyathus stercoreus	12	+IIs & IVs	+	Autopolyploid

The first, already referred to, is *Allomyces* and here the evidence from chromosome counts has been supplemented by experimental crosses between *A. arbuscula* and *A. macrogynus*. The fact that both species appear to comprise a polyploid series is remarkable. There was little evidence for multivalent associations in the parental strains and, in the F_1 hybrid, pairing is variable and incomplete. This suggests, but does not prove, that allopolyploidy is likely to be the basis of the phenomenon.

In contrast, the other two examples, *Puccinia* and *Cyathus stercoreus*, appear to suggest the occurrence of autopolyploidy. Studies of somatic chromosome numbers of some species showed that some were apparently multiples of others. Moreover, in certain species, such as *P. graminis*, there was clear evidence at mitosis of secondary association due to residual terminal attractions (Fig. 17.9). In earlier stages of division an end-to-end like arrangement was seen and this too is interpreted as reflecting residual homology. In the one published illustration of meiosis, in *P. kraussiana*, the chromosome number seems to be 20–22 and this could fit the possibility of a high polyploid number. However, no evidence of multivalents is available so that if *P. kraussiana* is a

Fig. 17.9. Chromosomes in (a) ungerminated basidiospore, and (b) at early metaphase of somatic division in germinated basidiospore of *Puccinia graminis tritici*. Note in (a) apparent end-to-end association and in (b) residual terminal attraction. (After McGinnis)

polyploid, it is not evidently autoploid. The remaining example is the best documented example of a probable autopolyploid in the fungi. Quadrivalents were found frequently as well as secondary associations and since there was no genetical or cytological evidence for structural hybridity, autopolyploidy was postulated as the cause. How this has arisen is not known. Since there is clear evidence that diploid nuclei can arise somatically in *Aspergillus* and other fungi showing the parasexual cycle, the same process could presumably occur in *Puccinia* or *Cyathus*. However, ELLIOTT (1960) studied tetraploid meiosis in *A. nidulans* and noted that it followed a normal course up to first meiotic metaphase when it inexplicably failed. Eight bivalents ($n = 8$) were formed and there was no clear evidence that the dividing nucleus was not a formal diploid one. Why a $4n$ fusion nucleus was not detected has not been explained. Not only is a $4n$ nucleus evidently selected against but a $2n$ nucleus in cytoplasm of a diploid strain does not seem to behave normally at meiosis. Thus if *Cyathus* is indeed an autopolyploid these adverse conditions must have been overcome.

SPECIATION PROCESSES IN FUNGI

Speciation in fungi both resembles and differs from that in other organisms.

From the study of crosses between races of rust fungi there is clear evidence of the constant formation and selection of new genotypes. Moreover, in heterokaryons of other forms it is clear that, either by chance or by special mechanisms, genetically different nuclei can fuse and recombine. On the other hand, there is clear evidence from all groups of the fungi, including the Fungi Imperfecti, that sterility barriers or heterokaryon compatibility barriers exist, that they are effective and are widespread. In those cases where it has been possible to study them, because they are partial not complete barriers, it appears that the basis is either a number of genes or, possibly, polygenic. In those cases where aversion phenomena have a simple basis isolation may be determined by as little as a pair of alleles. Heterogenic differences appear to be more important in determining sterility barriers than homogenic differences, such as often control sexual reproduction. The consequences of such isolation are, of course, a restriction on recombination. It would seem reasonable to suppose that the hybrid inviability shown to be associated with sterility or heterokaryon compatibility barriers in Phycomycetes and Ascomycetes develops after the initial phase of isolation. The origins of such isolating mechanisms are not always clear. In some cases the isolated groups appear to show no morphological, physiological or ecological differences yet isolation is apparently complete. Whether these are exceptions to Gause's 'rule' that similar species cannot occupy identical habitats indefinitely is not clear. In other cases, although speciation is sympatric, the fungi occupy different microhabitats. This is clearly shown in the varieties of rust fungi and doubtless could be found in saprophytic species if sufficient knowledge were available concerning their precise ecological niches. Such sympatric speciation leads to the establishment of situations similar to those described as ecotypic or ecospecific in other organisms. A surprising feature of the higher fungi, in the small sample available, is the number of cryptic or sibling species. Whether this is a reflection of the relatively simple morphological development of fungi or whether it reflects a more profound cause is not known.

Isolation by polyploidy may have taken place but generalizations on the meagre data available are hardly possible. The breakdown of isolation mechanisms can occur. In *Allomyces* there is clear evidence of hybridization between different polyploids. In other fungi post-zygotic barriers appear to operate fairly effectively in most groups to prevent the breakdown of groups separated by isolation.

The most outstanding exception is the yeasts. It is suggested that the apparent breakdown of isolation and the ease of hybridization is a spurious phenomenon. In nature yeasts are widely dispersed in a variety of micro-habitats and may well have undergone allopatric speciation. In general, however, this mode of speciation in fungi does not seem to be well marked and strains isolated by many thousands of miles retain their morphological similarities and compatibility. Nevertheless, in some fungi, there is evidence of incipient speciation as, for example, host specialization of strains in one geographical locality and its absence in another, e.g. *Irpex*.

The genecology of the fungi is a topic of intense interest not only to mycologists but also to pathologists and evolutionists. The almost total neglect of this great group of organisms by evolutionists is striking and until speciation in fungi has been further studied and assessed it will not be possible to discuss comparative evolution in a balanced manner. In the light of the limited data it is clear that the nature of the biological organization of the fungal individual is such that the primary process in speciation, i.e. genetical isolation, is of a quite different nature in fungi from that in other organisms. The critical stage can already be seen to be at, or just after, the process of fusion or conjugation and before fertilization occurs. In most other organisms isolation is effective either in preventing conjugation ever occurring or in eliminating the effects of an inappropriate fertilization. This biological difference between fungi and other organisms is of considerable significance.

Chapter 18
Phylogenetic and General Considerations

The phylogeny and supposed phylogenetic trends in the fungi have been a fertile field for the imagination in the past (e.g., BESSEY, 1950). By their nature, many of these speculations have been quite incapable of proof or disproof. For example, the similarities between the reproductive parts of Ascomycetes (trichogynes and non-motile, unicellular, uninucleate microconidia) and red algae have led to speculations on the algal ancestry of those fungi and much the same kind of speculation has associated Oomycetes with siphonalean algal types (ATKINSON, 1914; DE BARY, 1884). Other speculations have been based upon the detailed comparative morphology of fungi and the existence of forms intermediate between one taxonomic group and another has received special attention. Such speculations have led to the development of ideas concerning series of fungi which are believed to represent a phylogenetic trend. For example, Gäumann based his arrangement of the Ascomycetes on the supposition that they were derived from the Zygomycetes. He noted that in *Endogone* and other Endogonaceae the gametangia arise as multinucleate swellings which become laterally opposed and that a single pair of nuclei move to the centre of the fusion cell and fuse, the others degenerating. The fusion cell was regarded as an outgrowth of a gametangium, e.g. of *Phycomyces*, and he finally noted that the zygote germinates by a germ sporangium rather than a germ tube. He compared these features with *Dipodascus* and pointed out that this differed from *Endogone* only by forming a sporangium directly after conjugation of coenocytic gametangia, in which only two nuclei undergo fertilization. Subsequently, the reduction of the ascospores in the ascus to eight and the development of the reproductive structures were supposed to have taken place. This type of study can hardly be proved; it must remain a matter of opinion and its plausibility cannot be subjected to scientific testing.

Table 18.1. Proposed evolutionary series connecting Lactartio-Russulas with Astero-gastraceous series. (After Heim, 1952)

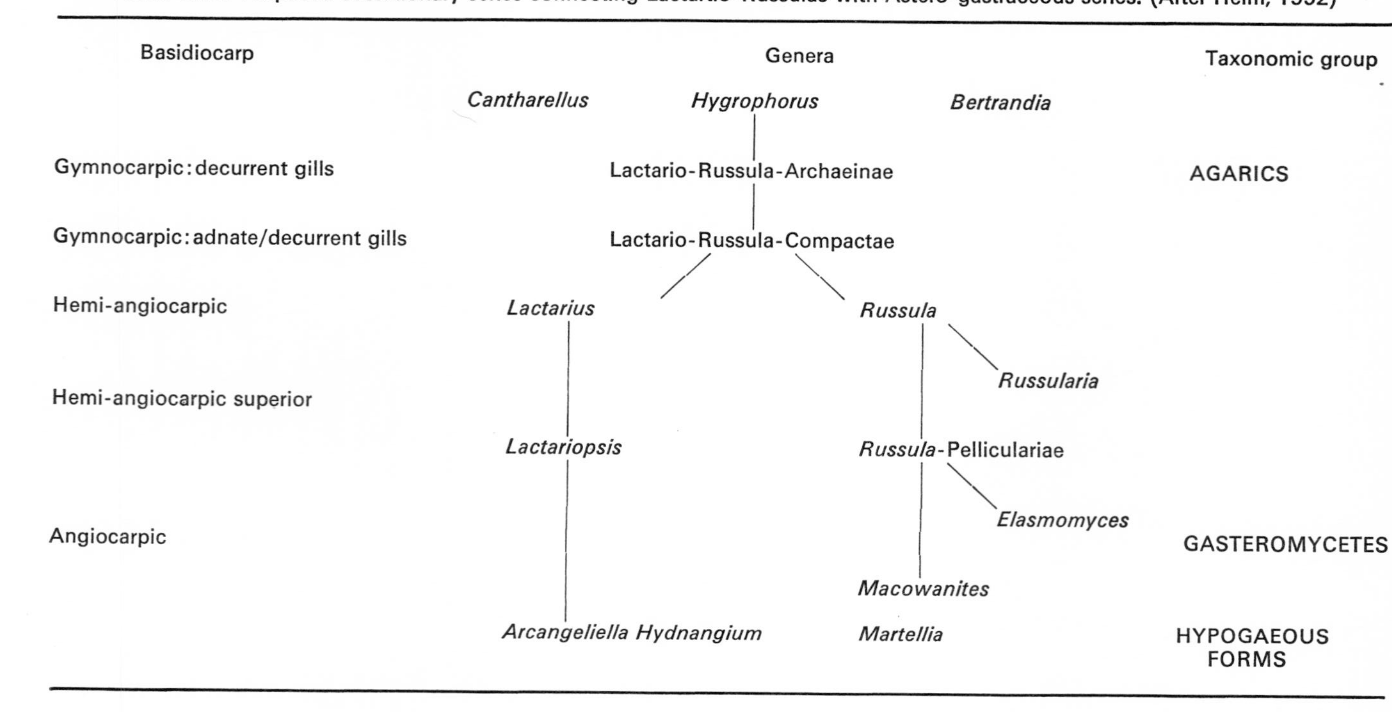

Naturally, some kinds of speculations are more plausible than others. The discovery of forms which, morphologically, appear to be intermediate between agarics and polypores and between agarics and Gasteromycetes suggests strongly that all those fungi share a common ancestry. The precise delineation of a specific sequence such as that due to Heim for the evolution of the Asterogastraceae from the Lactario-Russulaceae (Table 18.1) is far more a matter of opinion (HEIM, 1952). It is typical of such speculations that further, more recent studies of these fungi have led SINGER and SMITH (1960) to claim an almost exactly opposite sequence to that of Le Gal and Heim. It is clear that no unequivocal experiment can, as yet, be devised to test such claims which, however stimulating, must continue to rest upon the plausibility, coherency and integration of comparative structural data viewed in the light of untestable assumptions.

A distinction may be drawn between these broad phylogenetic speculations, such as the derivation of the clamp connexion from the ascogenous hyphae of Ascomycetes, and those based on the detailed studies of much smaller groups. An example of the latter is the study by BENJAMIN (1959) concerning relationships within the Mucorales (Fig. 18.1). Benjamin recognized three levels of advancement of morphological characters:

A. *Primitive.* 1. Vegetative and fruiting hyphae generalized, initially non-septate; simple imperforate septa formed to delimit reproductive structures or laid down adventitiously in ageing hyphae.
   2. Sporangia relatively large multispored, columellate.
   3. Progametangia opposed, highly differentiated morphologically; zygospores usually formed on aerial hyphae; zygosporangium thick-walled, rough, often highly pigmented.

B. *Intermediate.* 1. Vegetative and fruiting hyphae essentially as in A; perforate septa sometimes formed in fruiting structures.
   2. Sporangia containing relatively small numbers of spores; columellae absent or rudimentary.
   3. Progametangia opposed, more or less differentiated; zygospores often as in A, sometimes formed in or at the surface of the substrate; zygosporangium sometimes thin-walled, hyaline, nearly smooth.

C. *Advanced.* 1. Vegetative and fruiting hyphae septate from the beginning; septa highly modified.
   2. Sporangiola containing only one or two spores.
   3. Sexual hyphae relatively undifferentiated; zygospores usually formed in or near the surface of the substrate; zygosporangium thin-walled, hyaline, nearly smooth.

He points out that this scheme can only give an indication of the range of development since the rate of evolution of all morphological characters is not necessarily the same in all lineages. He cites as an example of this the fact that the vegetative and reproductive features of *Cunninghamella* and its allies are relatively primitive (A) but this group possesses sporangiola with one or two spores and therefore is advanced (C). This kind of speculation shows very

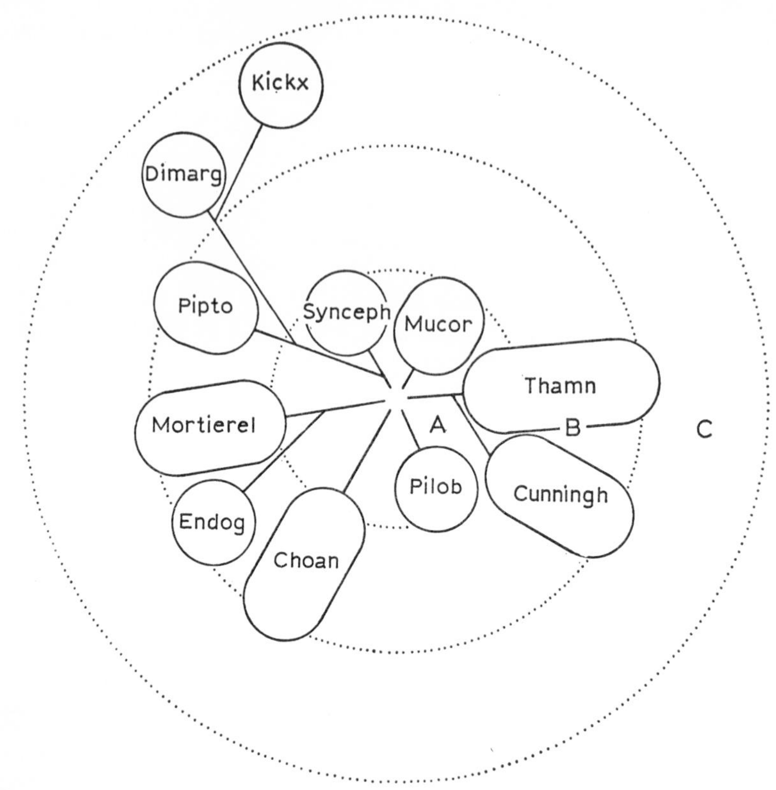

Fig. 18.1. Diagram to illustrate possible phylogenetic relationships of families in Mucorales. A, B and C represent increasingly more complex levels of morphological advancement. Key: Choan., Choanephoraceae; Cunningh., Cunninghamellaceae; Dimarg., Dimargaritaceae; Endog., Endogonaceae; Kickx., Kickxellaceae; Mortierel., Mortierellaceae; Mucor., Mucoraceae; Pilob., Pilobolaceae; Pipto., Piptocephalidaceae; Synceph., Syncephalastraceae; Thamn., Thamnidiaceae. (From Benjamin)

clearly the strengths and limitations of the method. The strength lies in the observations which lead to the grouping together of particular species. These observations can be taken even further using newer methods. For example, electron-microscopic studies of mucoraceous spores has revealed that in those often termed 'conidia', there is frequently a double-layered wall. This adds strength to the view that such 'conidia' were in fact derived from the reduction of sporangia to a single-spored sporangium enclosed in a wall (HAWKER, 1966). On the other hand, does the discovery, by the same method, that conidia of *Botrytis*, which are abstricted quite differently, lead to the same conclusion? In this case there is not an appropriate context into which such a speculation can neatly fit. This stresses the weakness of phylogeny based on comparative morphology alone; it leaves out function. For example, *Pilobolus* is often placed in a distinct group with *Pilaira* on morphological grounds and their sporangiophores are certainly most striking. Yet it will be recalled that Burgeff was able to isolate a form, *piloboloides*, from *Phycomyces* which was remarkably reminiscent of *Pilobolus*. Clearly the specialized form of sporangiophore in *Pilobolus* has a functional significance just as the structure of *Phycomyces* sporangiophores is best considered as a structure functionally adapted to xerophytic conditions. Consideration of function also enables the possibility of parallel evolution to be considered. This possibility may be detected by purely morphological studies, for example the phylogenetic lines connecting Discomycetes with Tuberales delineated by Malençon (Fig. 18.2). The

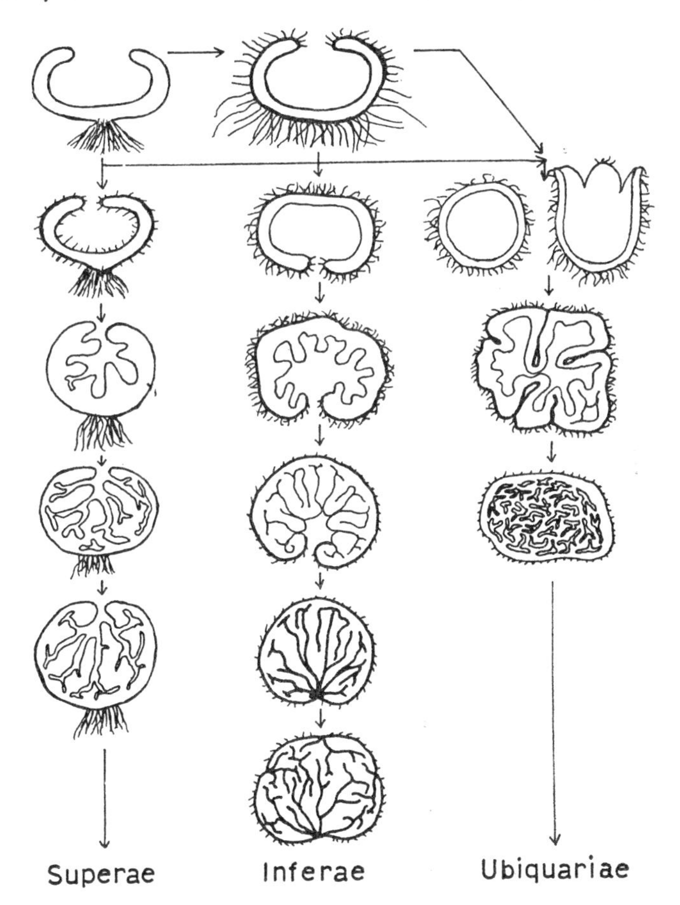

Fig. 18.2. Three proposed phylogenetic lines in the evolution of hypogaeous ascocarps, Superae, Inferae and Ubiquariae. In the Superae the representative species are (from top to bottom):

Galactinia, Genea hispidula, G. sphaerica, Pseudobalsamea, Pachyphlaeus; In the Inferae (*Galactinia*), *Lachnea, Hydnocystis, Stephensia, Tuber excavatum, T. nitidum, T. aestivum*; in the Ubiquariae (*Galactinia*) (*Lachnea*), *Septularia, Geopora, Balsamia.* (From Heim after Malençon)

great advantage of considering function, however, is that it provides a rationalization of structure while at the same time giving an indication of the kinds of selective agencies which are likely to have been operative in the development of a particular feature. Indeed, the study of function is a guide and interpreter to the ecology of organisms and this should play a key role in all attempts to account for evolution.

It must be confessed that it is not always easy to apply this principle. A notable example is the clamp connexion, for which no functional significance has ever been found. It does not exist because conjugately dividing nuclei cannot divide side-by-side in narrow hyphae, for sometimes they do. Moreover, clamp connexions are not necessarily present at every septum nor, indeed, in some cases are they present at all. The discovery of the dolipore septum by electron microscopy makes the whole situation even more baffling. It will be recalled, also, that the dolipore septum probably breaks down to allow migrating nuclei to pass through it. In this situation the simple morphological comparison of the ascogenous hyphae and clamp connexions with the implied phylogenetic descent seems to be more revealing (Fig. 18.3). But, even here, the question must be asked what is the functional significance of

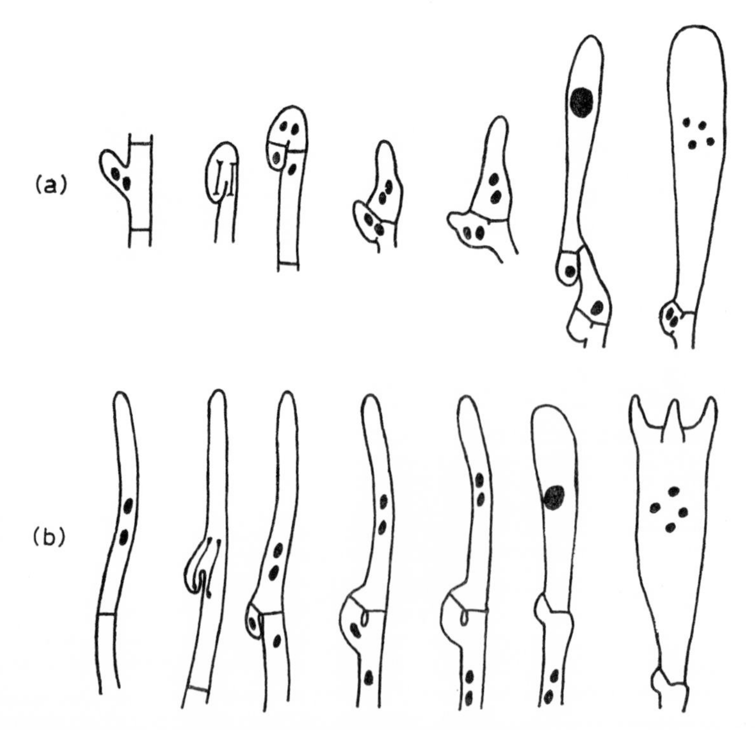

Fig. 18.3. Diagrammatic comparison of development of (a) ascogenous hyphae of a form such as *Pyronema* with the formation of (b) clamp connexions by a Basidiomycete. After Gäumann)

the ascogenous hypha and how does its modification to form a clamp connexion subserve the same or a new function? To these questions there is no clear answer.

This functional approach has provided new information, however, for it has led to a study of the physiology of fungi and hence provides an opportunity for the development of comparative physiological and biochemical studies. The most striking results of such studies are seen in the Phycomycetes. These may be very simple, for example, the comparison of the cell wall chemistry of the Saprolegniales with that of certain green algae with which they could be related (PARKER *et al.*, 1963). This comparison revealed considerable differences, in fact. A much more sophisticated treatment has been attempted by CANTINO (1950, 1955, 1966). He has correlated the scattered and often fragmentary data concerning the nutritional and synthetic capacities of saprophytic water moulds and prepared a scheme to account for their physiological evolution (Fig. 18.4) or, as he has described it, their 'evolutionary flow sheet'. This particular group of fungi is especially amenable to this type of treatment because nutritional factors play an important role in their development and thus nutrition is of great functional significance. He envisages the sequence to be one of successive loss of various functions. He supposes that the primitive ancestors were capable of high synthetic capacity. They are supposed to have been capable of using sulphate S and oxidized and reduced forms of N, and a wide range of carbohydrates and of synthesizing their vitamin requirements. Many Chytridiales are capable of doing all these things but amongst them are some that have lost these capacities. While *Rhizophlyctis rosea* utilizes various carbohydrates (but not sucrose), NO_3^- and 'ammonia', *Chytridium* spp. with a similar nutrition has totally lost the ability to synthesize thiamin. *Chytriomyces* can neither utilize nitrate nitrogen nor synthesize thiamine. Thus there is a tendency to lose the capacity to utilize inorganic nitrogen as the sole nitrogen source and to fail to synthesize vitamins. The Blastocladiales have nearly all lost the capacity to use SO_4^{2-} as sole sulphur source, to utilize NO_3^- and to synthesize vitamins. Since the Monoblepharidales have not been cultured, Cantino has suggested that they are even more demanding in their nutritional requirements and have suffered extensive losses in their synthetic capacities. A parallel sequence leads to the Saprolegniales, Leptomitales and Peronosporales. The Saprolegniales, as a whole, have retained the ability to synthesize vitamins and to utilize 'ammonia' but have lost the capacity to utilize SO_4^{2-}. Some genera are also incapable of using nitrate nitrogen, e.g. *Achlya*, *Saprolegnia*. The Leptomitales have lost the capacity to use either NO_3^- or 'ammonia' although they can still utilize SO_4^{2-}. Two species of *Leptomitus* need exogenous thiamin. Finally, in the Peronosporales virtually the whole group has lost the capacity to synthesize vitamins and several cannot utilize NO_3^-. In some *Pythium* spp. neither NO_3^- nor 'ammonia' can be utilized. The group as a whole, however, can still utilize SO_4^{2-} as its sole source of sulphur. It should be realized that this scheme is often based upon a single study of one organism or, in the case of Monoblepharidales, upon pure conjecture. Nevertheless it is important

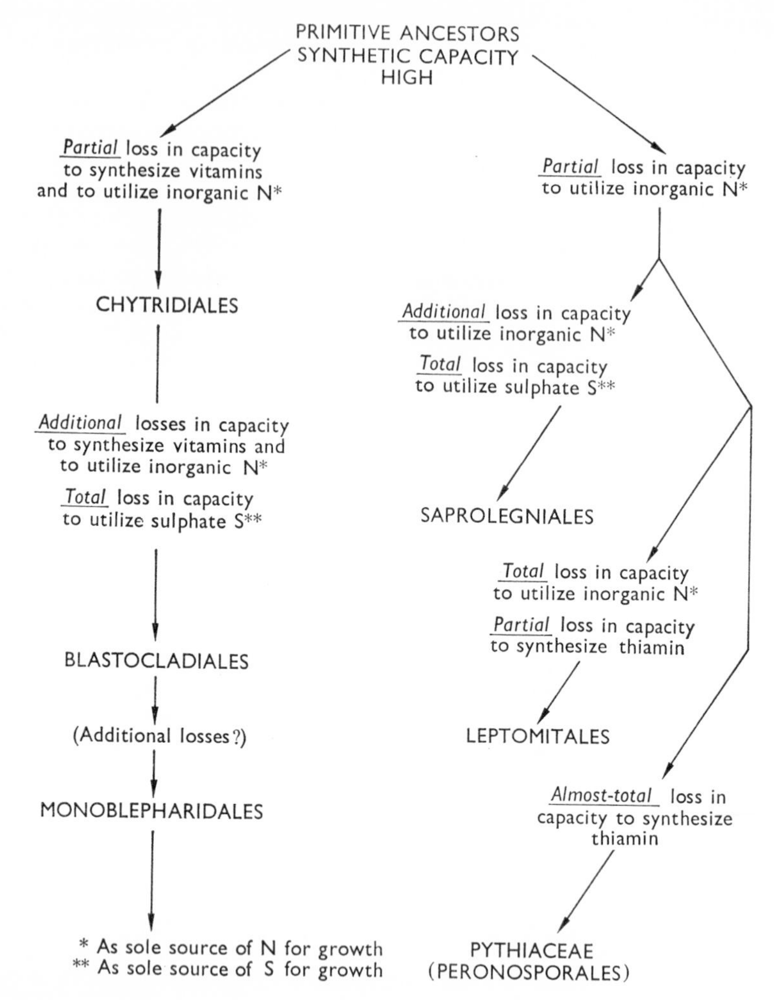

Fig. 18.4. Possible biochemical phylogenetic lines in aquatic Phycomycetes. (From Cantino)

in providing a new way of studying phylogenetic problems and it has led to new suggestions. For example, if the nutritional requirements are taken as indicators it suggests that Leptomitales and Peronosporales have more in common than either has with the Saprolegniales, whereas morphological considerations might have suggested that the Peronosporales were the more isolated group.

Cantino's stimulating hypothesis is contrary to another, more general one concerning the evolution of biochemical syntheses put forward by HOROWITZ (1945). Here the assumption is made that the original entities were dependent upon existing organic molecules and their precursors in the environment. As the essential molecules were depleted, the growth of the organism would become increasingly limited. At this stage any mutant capable of utilizing a precursor molecule and capable of transforming it intracellularly to the essential metabolite would have high selective value. In this way the original organisms would be replaced by mutants and they, in their turn, by others possessing increasingly greater synthetic capacity. Thus biochemical evolution would proceed from complete heterotrophy to increasing autotrophy for various metabolites. This scheme is the reverse of that of Cantino. If weight be given to the morphological complexity of the water moulds as well as their physiology, it would seem reasonable to accept Cantino's scheme. Thus these speculations have a bearing upon even more speculative hypotheses of general biochemical evolution.

It seems doubtful whether Cantino's approach will be of such value in other fungi where the nutritional requirements are not so variable or significant, functionally. Here, however, the synthetic capacities of the species may prove to be a guide to their relationships and the extraordinary range of unusual secondary metabolites may provide useful clues to the development of chemical phylogenies of fungi.

A further technique for obtaining phylogenetic information has the more limited objective of studying species interrelationships. A beginning has been made on this topic. For example, the species of *Neurospora* are well known from genetic studies and it is possible to make comparisons between them (PATEMAN, 1962). Attention has already been drawn to the consequences for selection of ascospore size on the transformation of 8-spored asci to 4-spored ones. An equally possible initial mechanism could have been a change in ascus shape which would have led to overlapping spindles and the possibility of two nuclei being included in a single ascospore. The comparison of the distribution of certain genes along the chromosomes carrying the mating-type factor in *N. crassa* and *N. sitophila* has revealed differences there also. One possible explanation is that the chromosomes differ in the two species by an inversion but there is no clear information on this matter. The difficulty about *Neurospora* species as subjects for phylogenetic speculation is that very little is known about their ecology. A more promising study has been made by LANGE (1952) in the genus *Coprinus*. These agarics are better known ecologically and have been the subject of numerous investigations into their mating systems and, in some cases, their nutritional requirements. Lange studied their habitats and geographical variation, their cultural variation and their mating systems and he attempted crosses. This revealed that amongst forms similar in morphology and in cultural features there were wide differences in variability and mating systems. Insufficient information was available concerning their nutrition or biosynthetic activities to enable generalizations to be made. It will be clear from the information given in Chapter 17 that there are a number of groups in the fungi in which an approach combining morphological, physiological, ecological and cytogenetical studies would provide scientifically testable information concerning their immediate phylogeny. It is true that this would only be a very small and recent part of their phylogenetic history but it would at least be within the realm of scientific method. This is not to denigrate these broad sweeps of educated and informed imagination which have led to the delineation of the major, yet hypothetical, phylogenetic lines. These are valuable as a means of ordering and rationalizing information. Rather, the plea is, that flights of fancy should be left aside for a little and the hard facts of fungal relationships, on however small a scale, be sought by all available means.

This book has, of necessity, followed a progression from considerations of form, through those concerning function and finally to the mechanisms responsible for generating and regulating both form and function. An attempt has been made to show that so far as form is concerned the hyphal apex is the centre of activity and that its activities need to be more fully investigated. The form factors which determine the further changes in size and shape of a mycelium seem to be permutations and combinations of a few superficially simple hyphal processes—branching, fusion and aggregation. Yet it will be clear that no satisfactory hypothesis exists to account for any one of these processes and they have not received the study they deserve. The marvellous structures which these basic processes have generated has led to attention being focused on the final product instead of the origins.

In the study of function much the same situation obtains. The biochemical complexity of fungi springs from the same common processes of carbohydrate catabolism as in most other organisms. In a few fungi, notably *Blastocladiella*, it is possible to obtain a glimpse of the ways in which metabolism and morphology are causally related in a particular process, sporangium formation. Yet the general regulating and co-ordinating processes of growth and translocation have been neglected and their mechanisms can only be speculated upon. Much remains to be done in the study of fungal associations and this is essentially a physiologico-ecological problem. It is not possible to provide a convincing description of the origins of the parasitic habit and the critical point in this mode of life, i.e. entry, is still not understood in full. However, the increasing attention being paid to this topic suggests that some success will be achieved in the foreseeable future.

Much is now known concerning the genetics of a few fungi and of the possible mechanisms of recombination. The significance for their genetics of the unusual and novel cytological features shown by some fungi is not known. Before knowledge of recombination processes can usefully be applied to fungi generally, attempts will have to be made to study these processes as they occur in natural surroundings for, at present, knowledge is based almost entirely on laboratory studies. It is equally clear that studies both laboratory and in the field on fungal recombination and speciation will have much to contribute to general biological knowledge and the broad evolutionary picture.

It may seem that undue stress has been laid throughout this book on the lack of information concerning the fungi and their way of life. This is deliberate. So too has been the playing down of the astonishing diversity of form and structure. These features are so obvious and so captivating, if living fungi are studied, that it has not been felt necessary to elaborate upon them. Moreover, there is clear evidence that in fungi there has been some neglect of the 'wood' for the 'trees'! The study of the form, diversity and biology of individual fungi has drawn attention away from those common properties which make them uniquely different from the animal and plant kingdoms.

The contemplation of a single spore germinating to form a colony, perhaps anastomosing *en route* with a similar germ mycelium, provides a major challenge to the general mycologist. To him, this clear-cut, apparently simple, sequence of events provides intense aesthetic satisfaction and provokes unlimited enquiry and investigation into the hidden complexities which underly them. Once the general features and processes of fungal form and function have been established, it will be possible to attempt to answer the ultimate phylogenetic enquiry concerning fungi—how do fungi exist and what are the mainsprings of their evolution?

Appendix

Classification of Fungi referred to in the Text

Synonyms according to Ainsworth (1961)

FUNGI

PHYCOMYCETES

PLASMODIOPHOROMYCETES

PLASMODIOPHORALES

Plasmodiophora
Sorosphaera
Spongospora subterranea

HYPHOCHYTRIDIOMYCETES

HYPHOCHYTRIALES

Rhizidiomyces apophysatus

CHYTRIDIOMYCETES

CHYTRIDIALES

Dictyomorpha dioica
Olpidium
Rhizophydium
Rhizophlyctis rosea
Synchytrium

BLASTOCLADIALES

Allomyces arbuscula; ×javanicus; macrogynus
Blastocladiella britannica; emersonii; pringsheimii; stübenii; variabilis
Coelomyces(=Coelomomyces)

MONOBLEPHARIDALES

Gonapodya

OÖMYCETES

SAPROLEGNIALES

Achyla ambisexualis; bisexualis; flagellata
Aphanomyces eutiches
Dictyuchus

Saprolegnia ferax
Thraustochytrium
Thraustotheca clavata

LEPTOMITALES

Apodachyla brachynema
Leptomitus lacteus
Sapromyces reinschii

LAGENIDIALES

—

PERONOSPORALES

Albugo blitii; candida
Perenospora calotheca; parasitica; tabacina
Phytophthora cinnamomi; fragariae; erythroseptica; infestans; parasitica
Pythium aphanidermatum; butleri; debaryanum; gracile; mamillatum
Sclerospora philippinensis

ZYGOMYCETES

MUCORALES

Absidia dubia; glauca
Blakeslea trispora
Choanephora cucurbitarium
Coemansia
Cunninghamella
Dispira
Endogone
Kickxella
Mortierella
Mucor genevensis; hiemalis; mucedo; racemosus; ramannianus; rouxii
Phycomyces blakesleeanus, f. *gracilis-piloboloides*
Pilaira anomala
Pilobolus crystallinus; klienii; longipes; sphaerosporus
Piptocephalis
Rhizophagus
Rhizopus arrhizus; oryzae; sexualis; stolonifer
Syncephalastrum
Syzygites grandis
Thamnidum elegans
Zygorhynchus heterogamus; moelleri

ENTOMOPHTHORALES

Basidiobolus ranarum
Conidiobolus villosus
Empusa (= *Entomophthora*)
Entomopthora coronata
Stylopage grandis

ZOOPAGALES

—

TRICHOMYCETES

ECCRINALES

—

ASCOMYCETES

Series HEMIASCOMYCETIDAE

ENDOMYCETALES

Ascoidea rubescens
Byssochlamys fulva
Candida (*Torulopsis*)—see MONILIALES
Dipodascus
Endomycopsis
Eremascus
Hansenula anomala; wingei
Kloeckera—see MONILIALES
Lipomyces
Rhodotorula—see MONILIALES
Saccharomyces cerevisiae; chevalieri
Saccharomycodes
Schizosaccharomyces kefir; pombe
Zygosaccharomyces ashbyi

TAPHRINALES

Ascomyces (= *Taphrina*)

Series EUASCOMYCETIDAE

PLECTOMYCETES

EUROTIALES

Aspergillus glaucus; itaconicus; nidulans—see MONILIALES

MICROASCALES

—

ONYGENALES

—

PYRENOMYCETES

ERYSYPHALES

Erysiphe cichoracearum; graminis; martii

MELIOLALES

—

CHAETOMIALES

Chaetomium aureum; globosum

CALVICIPITALES

—

SPHAERIALES

Bombardia lunata
Ceratocystis fagacearum
Ceratostomella ulmi
Cochliobolus sativus—see MONILIALES
Daldinia concentrica
Diaporthe perniciosa
Diatrype stigma
Didymella exitialis
Gelasinospora tetrasperma
Glomerella cingulata
Hypoxylon
Linospora gleditsiae
Neurospora crassa; sitophila; tetrasperma
Ophiobolus graminis
Ophiostoma (=*Ceratocystis*) *multiannulatum*
Pleurage taenoides
Podospora (=*Sordaria*) *anserina; brumalis*
Sordaria fimicola; macrospora
Trichometasphaeria turcica
Venturia inaequalis

DIAPORTHALES

—

HYPOCREALES

Chromocrea (=*Creopus*) *spinulosa*
Claviceps purpurea (*Sphacelea*)
Cordyceps militaris
Epichloë typhina
Gibberella fujikuroi
Hypomyces
Melanospora zamiae
Nectria cinnabarina; stenospora
Neocosmospora vasinfecta

CORYNELIALES

—

CORONOPHORALES

—

DISCOMYCETES

OSTROPHALES

HELOTIALES

Ciboria

Geoglossum
Helotium
Leotia
Mitrula pusilla
Neobulgaria (= *Ombrophila*) *pura*
Rhytisma acerinum
Sclerotinia fruticola; gladioli; sclerotiorum
Trichoglossum

PEZIZALES

Ascobolus immersus; magnificus; stercorarius; viridulus
Ascophanus carneus
Geopyxis cacabus
Helvella elastica
Humaria granulata; leucoloma (see *Neottiella*)
Lachnea (= *Patella*) *melaloma*
Morchella
Neottiella [= *Patella* (*Humaria*)] *rutilans*
Peziza aurantia
Pyronema confluens
Sarcoscypha protracta

TUBERALES

—

LABOULBENIOMYCETES

LABOULBENIALES

Herpomyces stylopagae
Trenomyces histophtorus

Series LOCULOASCOMYCETIDAE

MYRIANGIALES

Trichosporum tingens

MICROTHYRIALES

—

HYSTERIALES

—

PLEOSPORALES

—

DOTHIDEALES

—

FUNGI IMPERFECTI

DEUTEROMYCETES

SPHAEROPSIDALES

Macrophomina phaseoli
Phoma pirina

Phomopsis
Zythia fragariae

MELANCONIALES
Colletotrichum atramentarium; circinans

MONILIALES
Alternaria solani; tenuis
Anguillospora longissima
Arthrobotrys cladodes
Articulospora tetracladia
Aspergillus flavus; niger; oryzeae; sojae; terreus; wentii—see EUROTIALES
Botrytis alii; athophila; cinerea; fabae
Caldariomyces fumago
Cephalosporium mycophilum
Cladosporium fulvum
Clavariopsis aquatica
Curvularia ramosa
Dactylella aquatica; bembicoides; ellipsospora
Deightoniella torulosa
Flagellospora curvula; penicilloides
Fusarium culmorum; moniliforme; niveum; oxysporum f. *cubense*, f. *gladioli*, f. *lini*, f. *lycopersii*, f. *phaseoli*, f. *pisi*, f. *solani*
Geotrichum (*Oospora*) *candidum; lactis*
Graphium
Heliscus lugdunensis
Helminthosporium geniculatum; gramineum; sativum—see *cochliobolus*
Isaria
Kloeckera brevis—see ENDOMYCETALES
Lemonniera aquatica; brachycladia
Memnoniella echinata; subsimplex
Microsporum
Monilia cephalosporium
Myrothecium verrucaria
Nigrospora sphaerica
Penicillium charlesii; chrysogenum; claviforme; clavigerum; cyclopium; expansum griseofulvum; isariiforme; italicum; notatum; patulum; terrestre; vermiculatum
Phymatotrichum omnivorum
Pullularia (=*Aureobasidium*) *pullans*
Rhodotorula gracilis; rubra; aurantiaca—see ENDOMYCETALES
Scopulariopsis brevicaulis
Sporobolomyces
Stysanus stemonites
Tetrachaetum elegans
Tetracladium marchalianum; setigerum

Thielaviopsis basicola
Tricellula aquatica
Trichoderma viride
Trichophyton
Tricladium gracile; splendens
Tricothecium roseum
Varicosporium elodeae
Verticillium albo-atrum; dahliae
Zygophiala

MYCELIA STERILIA
Cenococcum graniforme
Ozonium
Rhizoctonia repens; solani—see THELEPHORALES
Sclerotium cepivorum

BASIDIOMYCETES

HETEROBASIDIOMYCETIDAE

TREMELLALES
Auricularia auricularia-judae
Helicobasidium purpureum
Hyaloria
Phleogena
Pseudotremellodendron
Sebacina
Tremella mesenterica
Tremellodendron

UREDINALES
Cronartium ribicola
Gymnosporangium juniperi-virginianae; nidus-avis
Melampsora lini
Puccinia dispersa (=*recondita*); *graminis; graminis* f. *agrostidis, avenae, phleipratensis poae, secalis, tritici; helianthi; krausiana; phragmitis; suaveolans*
Uromyces phaseoli

USTILAGINALES
Sorosporium syntherismae
Sphacelotheca panici-miliacei; sorghi
Tilletia caries; tritici
Ustilago avenae; bullata; hordei; kolleri; longissima; maydis; nuda; sphaerogena; violacea

Series HOMOBASIOMYCETIDAE

EXOBASIDIALES
—

HYMENOMYCETES

CLAVARIALES

—

THELEPHORALES

Corticium coronilla
Peniophora duplex; heterocystidea; mutata; populnea
Stereum sanguinolentum; umbrinum; veriforme
Thanetophorus cucumeris (=*Pellicularia filamentosa*)—see MYCELIA STERILIA

POLYPORALES

Aporpium
Fomes annosus; fomentarius; lignosus; rimosus; roseus; subroseus; texanus
Ganoderma applanatum (=*Elfvingia applanata*)
Lenzites sepiarum; trabea
Merulius lacrymans
Polyporus abietinus; agariceus; betulinus; brumalis; mylittae; pargamenus; squamosus
Polystictus hirsutus; microcyclus; versicolor; xanthopus
Poria ambiguum; vaillanti

BOLETALES

Boletus colosseus; variegatus

AGARICALES

Agaricus campestris; bisporus
Agrocybe praecox; semiorbicularis
Amanita fulva; vaginata
Armillariella (=*Armillaria*) *mellea*
Cantharellus infundibuliformis
Clitocybe
Collybia apalosarca; fusipes; radicatus; velutipes
Coprinus atramentarius; bisporus; disseminatus; fimetarius; lagopus; macrorhizus; radiatus; stercorarius; stercoreus; sterquilinus
Cortinarius
Craterellus
Galeola (=*Galerina*) *septentriola*
Galera (=*Galerina*) *tenera* f. *bispora*
Hypholoma appendiculatum; fasiculare
Laccaria laccata
Lactarius torminosus
Lentinus lepidus; regium; tigrinus; tuber
Lepiota
Marasmius androsaceus; elongatipes; oreades
Mycenella
Mycena vitilis
Pholiota mutabilis
Pleurotus corticatus; ostreatus

Pluteus admirabilis
Psalliota (=*Agaricus*) *hortensis* (*Agaricus bisporus*)
Psathyrella candolleana
Psilocybe
Pterygellus
Rimbachia
Schizophyllum commune
Termitomyces
Tricholoma melaleucum; nudum
Trogia
Volvariella

HYDNALES
Asterodon
Auriscalpium
Hydnum
Irpex fusco-violaceus
Pseudohydnum

GASTEROMYCETES
HYMENOGASTRALES
Elasmomyces (=*Secotium*)
Hymenogaster luteus
Hysterangium
Podaxis
Protogaster
Secotium

LYCOPERDALES
Calvatia gigantea
Lycoperdon

SCLERODERMATALES
Battarraea
Calostoma
Tulostoma brumale

PHALLALES
Aseroë
Clathrus
Dictyophora
Mutinus
Phallus

NIDULARIALES
Crucibulum
Cyathus stercoreus; striatus
Mycocalia castanea; denudata

LICHENS

Acaraspora fuscata
Buellia
Cladonia
Peltigera apthosa; polydactyla
Solorina saccata

References

ABE, M. (1951). *A. r. Takeda res. lab.*, **10**, 152. *(305)*
ADAIR, E. J. and VISHNIAC, H. (1958). *Science*, N.S., **127**, 147–148. *(193)*
AGAR, H. D. and DOUGLAS, H. C. (1955). *J. Bact.*, **70**, 427–434. *(19)*
AHMAD, M. (1953). *Ann. Bot.* (Lond.), N.S., **17**, 329–342. *(420)*
—— (1965). In *Incompatibility in fungi*, 13–22, ed. ESSER, K. and RAPER, J. R. Springer, Berlin. *(420)*
AHMADJIAN, V. (1960). *Bryologist*, **63**, 250–254. *(352)*
—— (1962). *Amer. J. Bot.*, **49**, 277–283. *(351)*
—— (1963). *Sci. Amer.*, **208**, 122–132. *(352)*
—— (1964). *Bryologist*, **67**, 87–98. *(352)*
—— (1965). *A. r. Microbiology*, **19**, 1–20. *(351, 354)*
AINSWORTH, G. C. (1955). *J. gen. Microbiol.*, **12**, 352–355. *(355)*
—— (1961). *Ainsworth and Bisby's Dictionary of the Fungi.* 5th edition. Commonwealth Mycological Institute, London. *(4, 124, 494)*
—— and AUSTWICK, P. K. C. (1959). *Fungal diseases of animals.* C.A.B., Farnham Royal. *(354)*
AIST, J. R. and WILSON, C. L. (1965). *Proc. Arkansas Acad. Sci.*, **19**, 32–36. *(372)*
—— —— (1967). *Amer. J. Bot.*, **54**, 99–104. *(372)*
ALASOADURA, S. O. (1963). *Ann. Bot.* (Lond.), N.S., **27**, 125–145. *(154)*
ALEXOPOULOS, C. J. (1962). *Introductory Mycology.* 2nd edition. Wiley, New York and London. *(1, 4, 6, 8, 9, 11, 14, 15)*
ALGRANATI, I. D. and CABIB, E. (1960). *Biochim. biophys. Acta*, **43**, 141–143. *(283)*
ALLEN, P. J. (1954). *A. r. Pl. Physiol.*, **3**, 225–248. *(344)*
—— (1955). *Phytopathology*, **45**, 259–266. *(183)*
—— (1957). *Pl. Physiol.*, **32**, 385–389. *(188)*
—— (1965). *A.r. Phytopathology*, **3**, 313–342. *(181)*
ALLPORT, D. C. and BU'LOCK, J. D. (1960). *J. Chem. Soc.*, 654–662. *(311)*
ARIGONI, D. (1960). *Biochem. Soc. Symp.*, **19**, 32–45. *(307)*
ARLETT, C. F. (1957). *Nature* (Lond.), **179**, 1250–1251. *(69)*
ARNOLD, J. D. (1935). *Mycologia*, **27**, 388–417. *(451)*
ARNSTEIN, H. R. V. and BENTLEY, R. (1953a). *Biochem. J.*, **54**, 493–508. *(281)*
—— —— (1953b). *Biochem. J.*, **54**, 508–516. *(281)*
—— —— (1953c). *Biochem. J.*, **54**, 517–522. *(281)*
—— and GRANT, P. T. (1954). *Biochem. J.*, **57**, 353–359. *(309)*
—— and MORRIS, D. (1960). *Biochem. J.*, **76**, 357–361. *(309)*
ARONSON, J. M. and MACHLIS, L. (1959). *Amer. J. Bot.*, **46**, 292–299. *(20)*
—— and PRESTON, R. D. (1960a). *Proc. R. Soc.*, **B152**, 346–352. *(19, 54)*
—— —— (1960b). *J. biophys. biochem. Cytol.*, **8**, 247–256. *(21, 50, 54, 56)*
ARTHUR, J. C. (1897). *Ann. Bot.* (Lond.), **11**, 491–507 *(256, 257, 258)*
ASTBURY, W. T. and PRESTON, R. D. (1940). *Proc. R. Soc.*, **B129**, 54–76. *(56)*
ASTHANA, R. P. and HAWKER, L. E. (1936). *Ann. Bot.* (Lond.), **50**, 325–344. *(129)*
ATKINSON, G. F. (1914). *Ann. Missouri Bot. Gard.*, **2**, 315–376. *(483)*
ATWOOD, K. C. and MUKAI, F. (1955). *Genetics*, **40**, 438–443. *(397)*
—— and PITTENGER, T. H. (1955). *Amer. J. Bot.*, **42**, 496–500. *(394)*
AVERS, C. J. (1967). *Proc. Nat. Acad. Sci. U.S.*, **58**, 620–627. *(29)*

AVERS, C. J. RANCOURT, M. W. and LIN, F. H. (1965). *Proc. Nat. Acad. Sci. U.S.*, **54**, 527–535. *(29)*
ÄYRÄPÄÄ, T. (1950). *Physiol. Plantarum*, **3**, 402–429 *(218)*

BAKER, J. M. (1963). In *Symbiotic Associations*, 232–265, ed. NUTMAN, P. S. and MOSSE, B., Cambridge University Press. *(173, 356)*
BAKER, K. F. and SNYDER, W. C. (1965). *Ecology of soil-borne plant pathogens*. Murray, London. *(181, 313, 329)*
BAKERSPIGEL, A. (1957). *Canad. J. Microbiology*, **3**, 923–926. *(370)*
—— (1959a). *Canad. J. Bot.*, **37**, 835–842. *(368)*
—— (1959b). *Amer. J. Bot.*, **46**, 180–190. *(363)*
—— (1960). *Amer. J. Bot.*, **47**, 94–100. *(30, 364)*
BANBURY, G. H. (1952). *J. exp. Bot.*, **3**, 86–94. *(59)*
—— (1955). *J. exp. Bot.*, **6**, 235–244. *(410)*
—— (1959). In *Encyclopedia of Plant Physiology*, **17/1**, 530–578, ed. RUHLAND, W. *(141)*
BARKSDALE, A. W. (1960). *Amer. J. Bot.*, **47**, 14–23. *(407)*
—— (1963). *Mycologia*, **55**, 627–632. *(407)*
BARNETT, H. L. (1937). *Mycologia*, **29**, 626–649. *(426, 456)*
—— (1963). *A. r. Microbiology*, **17**, 1–14. *(359)*
—— and LILLY, V. G. (1950). *Phytopathology*, **40**, 80–89. *(103)*
—— —— (1955). *Mycologia*, **47**, 26–29. *(104)*
BARRATT, R. W. and GARNJOBST, L. (1949). *Genetics*, **34**, 351–369. *(64)*
BARTNICKI-GARCIA, S. (1966). *J. gen. Microbiol.*, **42**, 57–69. *(21, 22)*
—— and LIPPMAN, E. (1966). *J. gen. Microbiol.*, **42**, 411–416. *(23)*
—— and NICKERSON, W. J. (1962). *Biochim. biophys. Acta*, **58**, 102–119. *(18, 21, 22)*
BARTON, R. (1960). In *The ecology of soil fungi*, 160–167, ed. PARKINSON, D. and WAID, J. S. University of Liverpool Press. *(329)*
—— (1957). *Nature* (Lond.), **180**, 613. *(333)*
BATEMAN, A. J. (1947). *Heredity*, **1**, 303–336. *(173)*
BAUCH, R. (1932). *Phytopath. Z.*, **5**, 315–321. *(425)*
BAYLIS, G. T. S. (1959). *New Phytol.*, **58**, 274–280. *(348, 349)*
—— (1962). *Aust. J. Sci.*, **25**, 195–200. *(342)*
BAYLIS-ELLIOTT, J. S. (1915). *Trans. Brit. myc. Soc.*, **5**, 138–142. *(359)*
BEAN, J., BRIAN, P. W. and BROOKS, F. T. (1954). *Ann. Bot.* (Lond.), N.S., **18**, 129–142. *(465)*
BECHET, J., WIAME, J. M. and DE DEKEN-GRENSON, M. (1962). *Arch. Int. Physiol. Biochim.*, **70**, 564–565. *(315)*
BEEVERS, H. (1961). *Respiratory metabolism in plants*. Row-Peterson, Evanston, Illinois. *(262, 271)*
BELT, T. (1874). *A Naturalist in Nicaragua*. John Murray, London. *(359)*
BENJAMIN, R. K. (1958). *El Aliso*, **4**, 149–169. *(93)*
—— (1959). *El Aliso*, **4**, 321–433. *(26, 93, 485)*
BENT, K. J. and MOORE, R. H. (1966). In *Biochemical studies of antimicrobial drugs*, 82–110, ed. NEWTON, B. A. and REYNOLDS, P. E. Cambridge University Press. *(310)*
BENTLEY, R. and NEUBERGER, A. (1949). *Biochem. J.*, **45**, 584–590. *(272)*
—— and THIESSEN, C. P. (1957). *J. biol. Chem.*, **226**, 689–702. *(279)*
BERLIN, J. D. and BOWEN, C. C. (1964a). *Amer. J. Bot.*, **51**, 445–452. *(28, 78)*
—— —— (1964b). *Amer. J. Bot.*, **51**, 650–652. *(33, 37)*
BERNET, J. (1963). *C. r. Acad. Sci.* (Paris), **256**, 771–773. *(453)*
——, ESSER, K., MARCOU, D. and SCHECROUN, J. (1960). *C. r. Acad. Sci.* (Paris), **250**, 2053–2055. *(453)*
BESCHEL, R. E. (1961). In *Geology of the Arctic*, 1044–1062, ed. RAASCH, G. A., University of Toronto Press. *(353)*
BESSEY, E. A. (1950). *Morphology and Taxonomy of Fungi*. Constable, London. *(4, 483)*

BEVAN, E. A. and KEMP, R. F. O. (1958). *Nature* (Lond.), **181**, 1145–1146. *(141)*
BHATNAGAR, G. M. and KRISHNAN, P. S. (1960a). *Arch. Mikrobiol.*, **36**, 131–138. *(186)*
—— —— (1960b). *Arch. Mikrobiol.*, **37**, 211–214. *(186)*
BIGGS, R. (1937). *Mycologia*, **29**, 686–706. *(456, 471, 472)*
BIRCH, A. J., MASSY-WESTROPP, R. A. and MOYE, C. J. (1955). *Aust. J. Chem.*, **8**, 539–544. *(311)*
——, RICKARDS, R. W., SMITH, H., HARRIS, A. and WHALLEY, W. B. (1959). *Tetrahedron*, **7**, 241–251. *(308)*
BIRKINSHAW, J. H. (1965). In *The Fungi, an advanced treatise*, **I**, 179–228, ed. AINSWORTH, G. C. and SUSSMAN, A. S. Academic Press, New York. *(198, 298)*
BISHOP, H. (1940). *Mycologia*, **32**, 505–529. *(406)*
BISTIS, G. N. (1956). *Amer. J. Bot.*, **43**, 389–394. *(416)*
—— (1957). *Amer. J. Bot.*, **44**, 436–444. *(416)*
—— 1965. In *Incompatibility in fungi*, 23–30, ed. ESSER, K. and RAPER, J. R. Springer, Berlin. *(416)*
—— and RAPER, J. R. (1963). *Amer. J. Bot.*, **50**, 880–891. *(416)*
BJÖRKMAN, E. (1949). *Svensk bot. Tidskr.*, **43**, 223–262. *(344)*
BLACK, W. (1952). *Proc. R. Soc. Edinburgh*, **B65**, 36–51. *(466, 467)*
—— (1957). *Scot. Pl. Breeding Sta. Rpt.*, 43–49. *(433, 466)*
BLAKESLEE, A. F. (1904). *Proc. Amer. Acad. Arts Sci.*, **40**, 205–319. *(401, 408, 410)*
—— (1906). *Ann. mycol.*, **4**, 1–28. *(380, 442)*
—— (1913). *Myc. Centralbl.*, **2**, 241–247. *(408)*
—— (1915). *Biol. Bull.*, **29**, 87–102. *(408)*
—— (1920). *Science*, N.S., **51**, 375–382. *(474)*
—— and CARTLEDGE, J. L. (1927). *Bot. Gaz.*, **84**, 51–58. *(409)*
——, ——, WELCH, D. S. and BERGNER, A. D. (1927). *Bot. Gaz.*, **84**, 27–50. *(409, 441, 442)*
BLEWETT, M. and FRAENKEL, G. (1944). *Proc. R. Soc.*, **B132**, 212–221. *(357)*
BLOEMERS, H. P. J. and KONINGSBERGER, V. V. (1967). *Nature* (Lond.), **214**, 487–488. *(232)*
BLONDEL, B. and TURIAN, G. (1960). *J. biophys. biochem. Cytol.*, **7**, 127–134. *(36)*
BLUMENTHAL, H. J. (1963). *Ann. N.Y. Acad. Sci.*, **102**, 688–706. *(273)*
—— (1965). In *The Fungi, an advanced treatise*, **I**, 229–268, ed. AINSWORTH, G. C. and SUSSMAN, A. S. Academic Press, New York. *(262, 266)*
——, LEWIS, K. F. and WEINHOUSE, S. (1954). *J. Amer. Chem. Soc.*, **76**, 6093–6097. *(268)*
——, HOROWITZ, S. T., HEMERLINE, A. and ROSEMAN, S. (1955). *Bact. Proc.*, 82–85. *(47)*
BOIDIN, J. (1950). *Bull. soc. mycol. France*, **66**, 204–219. *(456)*
BOMSTEIN, R. A. and JOHNSON, M. J. (1952). *J. biol. Chem.*, **198**, 143–153. *(281)*
BOND, T. E. T. (1952). *Trans. Brit. mycol. Soc.*, **35**, 190–194. *(128, 168)*
BONNER, D. (1946). *Amer. J. Bot.*, **33**, 788–791. *(232)*
BONNER, J. T. (1958). *Amer. Nat.*, **92**, 193–200. *(443)*
——, KANE, K. K. and LEVEY, R. H. (1956). *Mycologia*, **48**, 13–19. *(126, 134)*
BONNER, B. A. and MACHLIS, L. (1957). *Pl. Physiol.*, **32**, 291–301. *(272)*
BORRISS, H. (1934a). *Planta*, **22**, 28–69. *(139)*
—— (1934b). *Planta*, **22**, 644–684. *(142, 255)*
BOSE, S. R. (1934). *Cellule*, **42**, 249–266. *(451)*
BOULTER, D. and DERBYSHIRE, E. (1957). *J. exp. Bot.*, **8**, 313–318. *(271)*
—— and HURST, H. M. (1960). In *The ecology of soil fungi*, 277–285, ed. PARKINSON, D. and WAID, J. S. Liverpool University Press. *(271)*
BOULTON, A. A. (1965). *Expl. Cell. Res.*, **37**, 343. *(203)*
—— and EDDY, A. A. (1962). *Biochem. J.*, **82**, 16P. *(203)*
BOURCHIER, R. J. (1957). *Mycologia*, **49**, 20–28. *(74)*
BRACK, A., HOFMANN, A., KALBERER, F., KOBEL, H. and RUTSCHMANN, J. (1961). *Arch. Pharm.* (Berlin), **294/66**, 230. *(305)*

BRACKER, C. E. (1966). In *The Fungus Spore*, 39–58, ed. MADELIN, M. F. Butterworth, London. *(92)*
—— (1967). *A. r. Phytopathology*, **5**, 343–374. *(111)*
—— and BUTLER, E. E. (1963). *Mycologia*, **55**, 35–38. *(19, 24)*
—— —— (1964). *J. cell. Biol.*, **21**, 152–157. *(24, 390)*
BRAITHWAITE, G. D. and GOODWIN, T. W. (1960). *Biochem. J.*, **76**, 5–10. *(299)*
BRANDT, W. H. (1953). *Mycologia*, **45**, 194–208. *(67)*
BREFELD, O. (1877). *Bot. unters. ü. Schimmelpilze*, **3**, 1–226. *(156)*
BRIAN, P. W. (1949). *Ann. Bot.* (*Lond.*), N.S., **13**, 59–74. *(156)*
—— (1957). In *Microbial ecology*, 168–188, ed. WILLIAMS, R. E. O. and SPICER, C. C. Cambridge University Press. *(329)*
——, CURTIS, P. J. and HEMMING, H. G. (1946). *Trans. Brit. mycol. Soc.*, **29**, 173–187. *(59)*
——, GROVE, F. and MCMILLAN, J. (1960). *Fortschr. Chem. org. Naturstoffe*, **18**, 350. *(59)*
BRIERLEY, W. B. (1929). *Proc. Int. Congr. Pl. Sci.*, **2**, 1629–1654. *(380)*
BRIGGS, G. E. (1967). *Movement of Water in plants.* Blackwell Scientific, Oxford. *(206)*
——, HOPE, A. B. and ROBERTSON, R. N. (1961). *Electrolytes and Plant Cells.* Blackwell Scientific, Oxford. *(204, 206)*
BROCK, T. D. (1959). *J. gen. Microbiol.*, **26**, 487–497. *(325)*
BRODIE, H. J. (1951). *Canad. J. Bot.*, **29**, 224–234. *(151)*
BROOKS, M. A. (1963). In *Symbiotic Associations*, 200–231, ed. NUTMAN, P. S. and MOSSE, B. Cambridge University Press. *(356)*
BROWN, A. M. (1932). *Nature* (Lond.), **130**, 777. *(424)*
BROWN, D. H. (1955). *Biochim. biophys. Acta*, **16**, 429–431. *(47)*
BROWN, W. (1915). *Ann. Bot.* (Lond.), **29**, 313–348. *(333)*
—— (1916). *Ann. Bot.* (Lond.), **30**, 399–406. *(333)*
—— (1922). *Ann. Bot.* (Lond.), **36**, 101–119. *(333)*
—— (1965). *A. r. Phytopathology*, **3**, 1–18. *(334)*
—— and HARVEY, C. C. (1927). *Ann. Bot.* (Lond.), **41**, 643–662. *(333)*
BRUNSWIK, H. (1924). *Bot. abh., K. Goebel.*, **5**, 1–152. *(33, 426)*
BRUSHABER, J. A., WILSON, C. L. and AIST, J. R. (1967). *Phytopathology*, **57**, 43–46. *(372)*
BRÜYN, H. DE (1937). *Genetica*, **19**, 553–558. *(408)*
BUCHNER, P. (1953). *Endosymbiose der Tiere mit pflanzlichen Mikroorganismen.* Birkhauser, Basle. *(173, 356)*
BUDD, K. and HARLEY, J. L. (1962a). *New Phytol.*, **61**, 138–149. *(218)*
—— —— (1962b). *New Phytol.*, **61**, 244–255. *(218, 220)*
——, SUSSMAN, A. S. and EILERS, F. I. (1966). *J. Bact.*, **91**, 551–561. *(181)*
BÜDER, J. (1918). *Ber. deut. bot. Ges.*, **36**, 104–105. *(322)*
BU'LOCK, J. D. (1961). *Adv. appl. Microbiol.*, **3**, 293–342. *(299, 313)*
—— (1967). *Essays in biosynthesis and microbial development*, Wiley, New York and London. *(310)*
—— and SMALLEY, H. M. (1961). *Proc. Chem. Soc.*, 209–211. *(311)*
—— and SMITH, H. G. (1961). *Experientia*, **17**, 553. *(110)*
BULLER, A. H. R. (1909). *Researches on Fungi*, I. Longmans Green, London. *(145, 156, 164, 165, 166)*
—— (1922). *Researches on Fungi*, II. Longmans Green, London. *(145)*
—— (1924). *Researches on Fungi*, III. Longmans Green, London. *(145)*
—— (1931). *Researches on Fungi*, IV. Longmans Green, London. *(70, 74, 145, 359, 387)*
—— (1933). *Researches on Fungi*, V. Longmans Green, London. *(24, 29, 33, 64, 70, 71, 73, 112, 145, 153, 163, 172, 257)*
—— (1934). *Researches on Fungi*, VI. Longmans Green, London. *(113, 128, 145, 158, 322, 428)*
—— (1950). *Researches on Fungi*, VII. University of Toronto Press, Toronto, Canada. *(424)*

BULLER, A. H. R. and HANNA, W. F. In BULLER (1924), 224–230. *(166)*
BURGE, R. E. and HOLWILL, M. E. J. (1965). In *Function and structure in micro-organisms*, 250–269, ed. POLLOCK, M. R. and RICHMOND, M. H. Cambridge University Press. *(177)*
BURGEFF, H. (1909). *Die Wurzelpilze der Orchideen-ihre Kultur und ihre Leben in der Pflanze*. G. Fischer, Jena. *(341)*
—— (1912). *Ber. deut. bot. Ges.*, **30**, 679–685. *(380, 414)*
—— (1914). *Flora*, N.F., **107**, 259–316. *(380, 386, 414)*
—— (1915). *Flora*, N.F., **108**, 353–488. *(3, 38, 59, 387, 414)*
—— (1924). *Bot. abh., K. Goebel*, **4**, 1–135. *(408, 410, 413)*
—— (1925). *Flora*, **118/119**, 40–46. *(414, 474)*
—— (1928). *Z. indukt. Abstamm.-u. Vererbungsl.*, **48**, 26–94. *(387, 414)*
—— (1932). *Saprophytism und Symbiosis*. G. Fischer, Jena. *(341)*
—— (1936). *Samenkeimung der Orchideen*. G. Fischer, Jena. *(341)*
BURGER, M., BACON, E. E. and BACON, J. S. D. (1958). *Nature* (Lond.), **182**, 1508. *(205, 242)*
—— —— —— (1961). *Biochem. J.*, **78**, 504–511. *(205, 242)*
——, HEJMOVÁ, L. and KLEINZELLER, A. (1959). *Biochem. J.*, **71**, 233–24. *(222)*
BURGES, A. (1960). In *The ecology of soil fungi*, 185–191, ed. PARKINSON, D. and WAID, J. S. Liverpool University Press, Liverpool. *(197)*
BURKHOLDER, P. R. and SINNOTT, E. W. (1945). *Amer. J. Bot.*, **32**, 424–431. *(60)*
BURNETT, J. H. (1953a). *New Phytol.*, **52**, 58–64. *(413)*
—— (1953b). *New Phytol.*, **52**, 86–88. *(413)*
—— (1956). *New Phytol.*, **55**, 50–90. *(401, 422)*
—— (1965a). In *Chemistry and biochemistry of plant pigments*, 381–403, ed. GOODWIN, T. W. Academic Press, London. *(304)*
—— (1965b). In *Incompatibility in fungi*, 98–113, ed. ESSER, K. and RAPER, J. R. Springer, Berlin. *(426, 442, 461)*
—— and BOULTER, M. E. (1963). *New Phytol.*, **62**, 217–236. *(378, 426, 427, 434, 447, 451, 456, 461, 467)*
—— and EVANS, E. J. (1966). *Nature* (Lond.), **210**, 1368–1369. *(428)*
—— and PARTINGTON, M. (1957). *Proc. R. phys. Soc. Edinburgh*, **26**, 61–68. *(449)*
BUSTON, H. W., JABBAR, A. and ETHERIDGE, D. E. (1953). *J. gen. Microbiol.*, **8**, 302–306. *(131)*
—— and KHAN, A. H. (1956). *J. gen. Microbiol.*, **14**, 655–660. *(131)*
—— and RICKARD, B. (1956). *J. gen. Microbiol.*, **15**, 194–197. *(131)*
BUTLER, E. E. (1957). *Mycologia*, **49**, 354–373. *(335)*
BUTLER, E. J. (1907). *Mem. Dept. Agric. India, Bot. Ser.* 1, **5**. *(92)*
—— (1939). *Trans. Brit. mycol. Soc.*, **22**, 274–301. *(341)*
BUTLER, F. C. (1953a). *Ann. appl. Biol.*, **40**, 284–297. *(329)*
—— (1953b). *Ann. appl. Biol.*, **40**, 298–304. *(329)*
BUTLER, G. M. (1957). *Ann. Bot.* (Lond.), N.S., **21**, 523–537. *(83, 85, 86)*
—— (1958). *Ann. Bot.* (Lond.), N.S., **22**, 219–236. *(83, 250)*
—— (1961). *Ann. Bot.* (Lond.), N.S., **25**, 341–352. *(65)*
BUXTON, E. W. (1954). *J. gen. microbiol.*, **10**, 71–84. *(391, 445)*
—— (1956). *J. gen. microbiol.*, **15**, 133–139. *(400, 439, 445)*
—— (1962). *Ann. appl. Biol.*, **50**, 269–282. *(331, 400, 439)*
—— (1957a). *Trans. Brit. mycol. Soc.*, **40**, 145–154. *(444)*
—— (1957b). *Trans. Brit. mycol. Soc.*, **40**, 305–317. *(444)*
—— (1958). *Nature* (Lond.), **181**, 1222–1224. *(444)*
—— (1960). In *Plant Pathology, an advanced treatise*, **II**, 359–407, ed. HORSFALL, J. G. and DIMOND, A. E. Academic Press, New York. *(445)*

CABIB, E. and LELOIR, L. F. (1958). *J. biol. Chem.*, **231**, 259–275. *(284)*
CABRAL, R. V. DE G. (1951). *Bol. Soc. Brot.*, **15**, 291–362. *(451)*

CAGLIOTI, L., CAINELLI, G., CAMERINO, B., MONDELLI, R., PRIETO, A., QUILICO, A., SALVATORI, T. and SELVA, A. (1964). *Chimica Ind.*, **46**, 961–966. (*303*)
—— —— —— —— —— —— —— —— (1966). *Tetrahedron*, Suppl. 7, 175–187. (*303*)
CALLEN, O. E. (1940). *Ann. Bot.* (Lond.), N.S., **4**, 791–818. (*474*)
CANNATA, J. J. B. and STOPPANI, O. A. M. (1963a). *J. biol. Chem.*, **238**, 1196–1207. (*270*)
—— —— (1963b). *J. biol. Chem.*, **238**, 1208–1212. (*270*)
CANNY, M. J. and PHILLIPS, O. M. (1963). *Ann. Bot.* (Lond.), N. S., **27**, 397–402. (*260*)
CANTINO, E. C. (1949). *Amer. J. Bot.*, **36**, 95–112. (*278*)
—— (1950). *Quart. Rev. Biol.*, **25**, 269–277. (*489*)
—— (1955). *Quart. Rev. Biol.*, **30**, 138–149. (*489*)
—— (1965). *Arch. Mikrobiol.*, **51**, 42–59. (*102, 323*)
—— (1966). In *The Fungi, an advanced treatise*, **II**, 283–337, ed. AINSWORTH, G. C. and SUSSMAN, A. S. Academic Press, New York. (*100, 405, 489*)
—— and HORENSTEIN, E. A. (1954). *Amer. Naturalist*, **88**, 142–154. (*404*)
—— —— (1956). *Mycologia*, **48**, 443–446. (*304*)
—— and LOVETT, J. S. (1960). *Physiol. Plantarum*, **13**, 450–458. (*278*)
—— —— (1964). *Adv. Morphogenesis*, **3**, 33–93. (*100*)
—— —— LEAK, L. V. and LYTHGOE, J. (1963). *J. gen. Microbiol.*, **31**, 393–404. (*36*)
—— and TURIAN, G. (1961). *Arch. Mikrobiol.*, **38**, 272–282. (*323*)
CARLILE, M. J. (1956). *J. gen. Microbiol.*, **14**, 643–654. (*303, 321*)
—— (1962). *J. gen. microbiol.*, **28**, 161–167. (*321*)
—— (1960). In *Progress in Photobiology*, 566, ed. CHRISTENSEN, B. G. and BUCHMANN, B. Elsevier, Amsterdam. (*321*)
—— (1965). *A. r. Pl. Physiol.*, **16**, 175–202. (*102, 104, 319*)
—— DICKENS, J. S. W., MORDUE, E. M. and SCHIPPER, M. A. (1962a). *Trans. Brit. mycol. Soc.*, **45**, 457–461. (*97, 104*)
—— —— and SCHIPPER, M. A. (1962b). *Trans. Brit. mycol. Soc.*, **45**, 462–464. (*97, 104*)
—— and FRIEND, J. S. (1956). *Nature* (Lond.), **178**, 369. (*321*)
—— LEWIS, B. G., MORDUE, E. M. and NORTHOVER, J. (1961). *Trans. Brit. mycol. Soc.*, **44**, 129–133. (*97, 104*)
—— and MACHLIS, L. (1965). *Amer. J. Bot.*, **52**, 478–483. (*406*)
—— and SELLIN, M. A. (1963). *Trans. Brit. mycol. Soc.*, **46**, 15–18. (*181, 185*)
CARNOY, J. B. (1870). *Bull. Soc. bot. Belg.*, **9**, 157–321. (*18*)
CARR, A. J. H. and OLIVE, L. S. (1958). *Amer. J. Bot.*, **45**, 142–150. (*132, 375*)
CARTWRIGHT, K. ST. G. (1938). *Ann. appl. Biol.*, **25**, 430–432. (*358*)
CASE, M. E. and GILES, N. H. (1962). *Neurospora Newsletter*, **2**, 6–7. (*400*)
CASPERSSON, T. O. (1950). *Cell growth and cell function*. Norton, New York. (*295*)
CASSAIGNE, Y. (1931). *Rev. gén. Bot.*, **43**, 140–167. (*29*)
CASSELTON, L. A. (1965). *Genet. Res.*, **6**, 190–208. (*398, 400*)
—— and LEWIS, D. (1967a). *Genet. Res.*, **8**, 61–72. (*448*)
—— —— (1967b). *Genet. Res.*, **9**, 63–71. (*448*)
CASTLE, E. S. (1937). *J. cell. comp. Physiol.*, **7**, 477–489. (*51, 55, 56*)
—— (1938). *J. cell. comp. Physiol.*, **11**, 345–358. (*57*)
—— (1942). *Amer. J. Bot.*, **29**, 664–672. (*49, 51, 56*)
—— (1953). *Quart. rev. Biol.*, **28**, 364–372. (*54, 56*)
CATEN, C. E. and JINKS, J. L. (1966). *Trans. Brit. mycol. Soc.*, **49**, 81–93. (*437, 454*)
CAYLEY, D. M. (1923). *J. Genet.*, **13**, 353–370. (*451*)
—— (1931). *J. Genet.*, **24**, 1–63. (*451*)
CHADEFAUD, M. (1942). *Rev. Mycol.*, N.S., **7**, 57–88. (*159*)
CHAPMAN, J. A. and VUJIČIĆ, R. (1965). *J. gen., Microbiol.*, **41**, 275–282. (*28*)
CHEVAUGEON, J. (1959a). *C. r. Acad. Sci.* (Paris), **248**, 1381–1384. (*67*)

CHEVAUGEON, J. (1959b). *C. r. Acad. Sci.* (Paris), **248**, 1841–1844. *(67)*
—— (1959c). *C. r. Acad. Sci.* (Paris), **249**, 1703–1705. *(67)*
CHRISTENBERRY, G. A. (1938). *J. Elisha Mitchell Sci. Soc.*, **54**, 297–310. *(103)*
CHRISTENSEN, H. N. (1961). In *Membrane transport and metabolism*, 465–449, ed. KLEINZELLER, A. and KOTYK, A. Academic Press, Prague and London. *(231)*
CHURCH, A. H. (1920). *Elementary notes on the morphology of fungi.* Oxford University Press. *(402)*
CIEGLER, A. (1965). *Adv. appl. Microbiol.*, **7**, 1–34. *(298, 299)*
—— ARNOLD, M. and ANDERSON, R. F. (1959). *Appl. Microbiol.*, **7**, 94–97, 98–103. *(303)*
CIRILLO, V. P. (1961a). *A. r. Microbiology*, **15**, 197–218. *(225, 235)*
—— (1961b). In *Membrane transport and metabolism*, 343–351, ed. KLEINZELLER, A. and KOTYK, A. Academic Press, Prague and London. *(223, 224)*
—— (1962). *J. Bact.*, **84**, 485–491. *(223, 224, 235)*
CLOWES, F. A. L. (1949). *The morphology and anatomy of the roots associated with ectotrophic mycorrhiza.* D.Phil. Thesis, University of Oxford. *(342)*
CLUTTERBUCK, A. J. and ROPER, J. (1966). *Genet. Res.*, **7**, 185–194. *(392, 395)*
COCHRANE, V. W. (1958). *Physiology of Fungi.* Wiley, New York. *(181, 198, 262, 279)*
—— (1960). In *Plant Pathology, an advanced treatise*, **II**, 169–202, ed. HORSFALL J. G. and DIMOND, A. E. Academic Press, New York. *(181)*
—— and TULL, D. L. W. (1958). *Phytopathology*, **48**, 623. *(224)*
COHEN, G. and RICKENBERG, H. V. (1956). *Ann. Inst. Pasteur*, **91**, 693–720. *(226)*
—— and MONOD, J. (1957). *Bact. Rev.*, **21**, 169–194. *(227)*
COMMANDON, J. and DE FONBRUNE, P. (1938). *C. r. Soc. Biol.* (Paris), **129**, 619–625. *(325)*
CONSTABEL, F. (1957). *Biol. Zbl.*, **76**, 385–413. *(338)*
CONWAY, E. J. (1951). *Science*, N.S., **113**, 270–271. *(237)*
—— (1953). *Int. rev. Cytol.*, **2**, 419–445. *(237, 238)*
—— (1955). *Int. rev. Cytol.*, **4**, 377–396. *(237, 238)*
—— and BEARY, M. E. (1958). *Biochem. J.*, **69**, 275–280. *(215)*
—— and DOWNEY, M. (1950). *Biochem. J.*, **47**, 347–355. *(204, 211)*
—— and DUGGAN, F. (1956). *Nature* (Lond.), **136**, 1043. *(215)*
—— —— (1958). *Biochem. J.*, **69**, 265–274. *(211, 215, 231)*
—— and KERNAN, R. P. (1955). *Biochem. J.*, **61**, 32–36. *(239)*
—— RYAN, H. and CARTON, E. (1954). *Biochem. J.*, **58**, 158–167. *(212)*
CORI, G. T. and CORI, C. F. (1940). *J. biol. Chem.*, **135**, 733–756. *(283)*
CORNER, E. J. H. (1929a). *Trans. Brit. mycol. Soc.*, **14**, 263–275. *(105, 109)*
—— (1929b). *Trans. Brit. mycol. Soc.*, **14**, 275–291. *(105, 109)*
—— (1930a). *Trans. Brit. mycol. Soc.*, **15**, 107–120. *(105, 109)*
—— (1930b). *Trans. Brit. mycol. Soc.*, **15**, 121–134. *(105, 109)*
—— (1931). *Trans. Brit. mycol. Soc.*, **15**, 322–350. *(105, 109)*
—— (1932a). *Ann. Bot.* (Lond.), **46**, 71–111. *(105, 121, 255)*
—— (1932b). *Trans. Brit. mycol. Soc.*, **17**, 51–81. *(105, 121)*
—— (1934). *Trans. Brit. mycol. Soc.*, **19**, 39–88. *(115, 116, 117, 119, 139)*
—— (1948a). *New Phytol.*, **47**, 22–51. *(128, 156, 259)*
—— (1948b). *Trans. Brit. mycol. Soc.*, **31**, 234–245. *(105, 119, 121)*
—— (1950). 'A monograph of *Clavaria* and allied genera', *Ann. Bot. Mem*, **1**. *(105, 121)*
—— (1953). *Phytomorphology*, **3**, 152–167. *(24, 105, 121, 125)*
—— (1961). *Trans. Brit. mycol. Soc.*, **44**, 233–238. *(105, 121)*
—— (1966). *A monograph of Cantharelloid Fungi.* Oxford University Press. *(105, 116, 119, 121)*
COUCH, J. N. (1926). *Ann. Bot.* (Lond.), **40**, 848–881. *(406, 407, 474)*
—— (1938). *Science*, N.S., **88**, 476. *(34)*

COUCH, J. N. (1941). *Amer. J. Bot.*, **28**, 704–713. (*34, 176*)
COULTHARD, C. E., MICHAELIS, R., SHORT, W. F., SYKES, G., SKRIMSHIRE, G. B., STANDFAST, A. F. B., BIRKINSHAW, J. H. and RAISTRICK, H. (1945). *Nature* (Lond.), **150**, 634–635. (*272*)
COWIE, D. B. (1962). In *Amino-acid pools*, 633, ed. HOLDEN, J. T. Elsevier, Amsterdam. (*225, 285, 293*)
—— and MCCLURE, F. T. (1959). *Biochim. biophys. Acta*, **31**, 236–245. (*226*)
—— and WALTON, B. P. (1956). *Biochim. biophys. Acta*, **21**, 211–226. (*232, 286, 288, 293*)
COWLING, E. B. (1961). *U.S.D.A. Forest Service, Tech. Bull.*, **1258**, 1–74. (*336*)
CRAIGIE, J. H. (1927). *Nature* (Lond.), **120**, 765–767. (*424*)
CROFT, J. H. and SIMCHEN, G. (1965). *Amer. Naturalist*, **99**, 453–464. (*385*)
CROWE, L. K. (1960). *Heredity*, **15**, 397–405. (*400*)
CRUICKSHANK, I. A. M. (1965). In *Ecology of soil-borne plant pathogens*, 325–334, ed. BAKER, K. F. and SNYDER, W. C. Murray, London. (*339*)
CUNNINGHAM, D. G. (1946). *New Zealand J. Sci. Techn.*, **28**, 238–251. (*123*)
CURTIS, P. J. and CROSS, B. E. (1954). *Chem. Ind.*, 1066. (*307*)
CUTTER, V. M. (1942). *Bull. Torrey bot. Club*, **69**, 592–616. (*379, 413*)
—— (1951). *Trans. N.Y. Acad. Sci.*, **14**, 103–108. (*338*)
—— (1959). *Mycologia*, **51**, 248–295. (*338*)
—— (1960). *Mycologia*, **52**, 726–742. (*338*)

DAFT, M. J. and NICOLSON, T. H. (1966). *New Phytol.*, **65**, 343–350. (*349, 350*)
DAINTY, J. (1964). In *Cell function and membrane transport*, 41–52, HOFFMAN, J. F. Prentice-Hall Inc., New Jersey. (*206, 207*)
DANIELLI, J. F. (1954). *Symp. Soc. Exp. Biol.*, **8**, 502–516. (*203*)
DARBY, R. T. and GODDARD, D. R. (1950). *Physiol. Plantarum*, **3**, 435–446. (*271*)
DARLINGTON, C. D. (1958). *The evolution of genetic systems.* 2nd edn. Oliver & Boyd, Edinburgh. (*448*)
DAVIDSON, E., BLUMENTHAL, H. J. and ROSEMAN, S. (1957). *J. biol. Chem.*, **226**, 125–133. (*47*)
DAVIES, B. H., JONES, D. and GOODWIN, T. W. (1963). *Biochem. J.*, **87**, 326–329. (*302*)
DAVIES, R. (1963). In *The biochemistry of industrial microorganisms*, 68–150, ed. RAINBOW, C. and ROSE, A. H. Academic Press, London. (*362*)
DAVIES, R. R. (1959). *Nature* (Lond.), **183**, 1695. (*148*)
DAVIS, R. H. (1960a). *Amer. J. Bot.*, **47**, 351–357. (*395*)
—— (1960b). *Amer. J. Bot.*, **47**, 648–654. (*395*)
—— (1966). In *The Fungi, an advanced treatise*, **II**, 567–588, ed. AINSWORTH, G. C. and SUSSMAN, A. S. Academic Press, New York. (*395*)
DAVSON, H. and DANIELLI, J. F. (1952). *The Permeability of natural membranes.* Cambridge University Press. (*206*)
DAWES, E. A. and HOLMS, W. H. (1958). *Biochim. biophys Acta*, **29**, 82–91. (*268*)
DAY, P. R. (1956). *Tomato Genet. Co-oper. Rep.*, **6**, 13–14. (*466*)
—— BOONE, D. M. and KEITT, G. W. (1956). *Amer. J. Bot.*, **43**, 835–838. (*375*)
DE BARY, A. (1866). *Morphologie und Physiologie der Pilze, Flechten und Myxomyceten.* Engelmann, Leipzig. (*19*)
—— (1881). *Abh. Senckenberg. Naturforsch. ges.*, **12**, 225–370. (*406*)
—— (1884). *Vergleichende Morphologie und Biologie der Pilze, Mycetozoen und Bacterien.* Engelmann, Leipzig. (*483*)
—— (1887). *Comparative morphology and biology of the Fungi, Mycetozoa and Bacteria*, Oxford. (*19, 126*)
DE BUSK, B. G. and DE BUSK, A. G. (1965). *Biochim. biophys. Acta*, **104**, 139–150. (*225, 227*)
DE ROBICHON-SZULMAJSTER, H. (1958). *Science*, N.S., **127**, 28–29. (*315*)
—— SURDIN, Y., KARASSEVITCH, Y. and CORRIVAUX, D. (1965). In *Mécanismes de régulation des activités cellulaires chez les microorganismes*, 255–269. C.N.R.S., Paris. (*317*)

DE TERRA, N. and TATUM, E. L. (1961). *Science*, N.S., **134**, 1066–1068. *(47, 48)*
—— —— (1963). *Amer. J. Bot.*, **50**, 669–677. *(47, 48)*
DE VRIES, H. (1885). *Bot. Z.*, **43**, 1–6; 17–26. *(255)*
DELBRÜCK, M. and SHROPSHIRE, W. (1960). *Pl. Physiol.*, **35**, 194–204. *(320)*
—— and VARJU, D. (1961). *J. gen. physiol.*, **44**, 1177–1188. *(321)*
DEMAIN, A. L. (1956). *Arch. biochem. biophys.*, **64**, 74–79. *(309)*
DEN, H. and KLEIN, H. P. (1961). *Biochim. biophys. Acta*, **49**, 429–430. *(297)*
DENNISON, D. S. (1959). *J. gen. Physiol.*, **45**, 23–38. *(323)*
DICKINSON, S. (1927). *Proc. R. Soc.*, **B101**, 126–136. *(425)*
—— (1949). *Ann. Bot.* (Lond.), N.S., **13**, 219–236. *(80)*
DICKSON, H. (1935). *Ann. Bot.* (Lond.), **49**, 181–204. *(437)*
DIMOND, A. E. and WAGGONER, P. E. (1953). *Phytopathology*, **43**, 319–321. *(310)*
DOBBS, G. C. (1939). *Nature* (Lond.), **143**, 286. *(148)*
—— and HINSON, W. H. (1953). *Nature* (Lond.), **172**, 197–199. *(181, 330)*
DOBZHANSKY, T. (1937). *Genetics and the origin of species*. 1st edn. Columbia University Press. *(457)*
DODGE, B. O. (1912). *Bull. Torrey bot. Club*, **39**, 139–197. *(33)*
—— (1920). *Mycologia*, **12**, 115–134. *(415)*
—— (1927). *J. agr. Res.*, **35**, 289–305. *(377, 418)*
—— (1928). *J. agr. Res.*, **36**, 1–14. *(418, 475)*
—— (1931). *Bull. Torrey bot. Club*, **58**, 517–520. *(475)*
—— (1942). *Bull. Torrey bot. Club*, **69**, 75–91. *(397)*
—— SINGLETON, J. R. and ROLNICK, A. (1950). *Proc. Amer. Phil. Soc.*, **94**, 38–52. *(418)*
DOMNAS, A. and CANTINO, E. C. (1965a). *Biochim. biophys. Acta*, **97**, 300–309. *(102)*
—— —— (1965b). *Phytochemistry*, **4**, 273–284. *(102)*
DONACHIE, W. D. (1964). *Biochim. biophys. Acta*, **82**, 284–292. *(317)*
DOUGLAS, H. C. and HAWTHORNE, D. C. (1964). *Genetics*, **49**, 837–844. *(315)*
DOWDING, E. S. (1958). *Canad. J. Microbiology*, **4**, 295–301. *(308)*
—— (1959). *Trans. Brit. mycol. Soc.*, **42**, 449–457. *(348)*
—— and BAKERSPIGEL, A. (1954). *Canad. J. Microbiology*, **1**, 68–78. *(389)*
—— and BULLER, A. H. R. (1940). *Mycologia*, **32**, 471–488. *(256, 389)*
—— and WEIJER, J. (1960). *Nature* (Lond.), **188**, 338–339. *(360, 371)*
—— —— (1962). *Genetica*, **32**, 339–351. *(360, 371, 373)*
DRIVER, C. H. and WHEELER, H. E. (1955). *Mycologia*, **47**, 311–316. *(132)*
DUBOS, R. (1947). *The Bacterial Cell*. Harvard University Press. *(389, 395)*

EATON, N. R. (1963). *Ann. N.Y. Acad. Sci.*, **102**, 678–687. *(283)*
EDDY, A. A. (1958a). *J. Inst. Brewing*, **64**, 143–151. *(325)*
—— (1958b). *J. Inst. Brewing*, **64**, 368. *(325)*
—— and WILLIAMSON, D. H. (1957). *Nature* (Lond.), **179**, 1252–1253. *(209)*
—— and RUDIN, A. D. (1958a). *J. Inst. Brewing*, **64**, 139–142. *(325)*
—— —— (1958b). *Proc. R. Soc.*, **B148**, 419–432. *(325)*
—— —— (1958c). *J. Inst. Brewing*, **64**, 19–21. *(325)*
EGGERTSON, E. (1953). *Canad. J. Bot.*, **31**, 710–719. *(426, 442)*
EHRLICH, H. G. and EHRLICH, M. A. (1963a). *Amer. J. Bot.*, **50**, 123–130. *(78)*
—— —— (1963b). *Phytopathology*, **53**, 1378–1380. *(78)*
—— —— (1966). *Canad. J. Bot.*, **44**, 1495–1503. *(28)*
EKUNDAYO, J. A. (1966). *J. gen. Microbiol.*, **42**, 283–291. *(182, 208)*
—— and CARLILE, M. J. (1964). *J. gen. Microbiol.*, **35**, 261–269. *(182, 208)*
EL-ANI, A. S. (1956). *Amer. J. Bot.*, **43**, 769–778. *(375)*
—— and OLIVE, E. S. (1962). *Proc. Nat. Acad. Sci. U.S.*, **48**, 17–19. *(422)*
ELLINGBOE, A. H. (1961). *Phytopathology*, **51**, 13–15. *(400, 439)*
—— (1963). *Proc. Nat. Acad. Sci. U.S.*, **49**, 286–292. *(400)*
—— (1964). *Genetics*, **49**, 247–251. *(400)*
—— (1965). In *Incompatibility in fungi*, 36–47, ed. ESSER, K. and RAPER, J. R. Springer, Berlin. *(400)*

ELLINGBOE, A. H. and RAPER, J. R. (1962). *Genetics*, **47**, 85–98. (*400*)
ELLIOTT, C. G. (1960). *Genet. Res.*, **1**, 462–476. (*376, 481*)
——, HENDRIE, M. R. and KNIGHTS, B. A. (1966). *J. gen. Microbiol.*, **42**, 425–435. (*408*)
—— —— —— and PARKER, W. (1964). *Nature* (Lond.), **203**, 427–428. (*408*)
EMERSON, R. (1950). *A. r. Microbiology*, **4**, 169–200. (*403, 404*)
—— (1955). In *Aspects of synthesis and order in growth*, 171–208, ed. RUDNICK, D. Princeton University Press. (*404*)
—— and WESTON, W. H. (1967). *Amer. J. Bot.*, **54**, 702–19. (*197*)
—— and WILSON, C. M. (1954). *Mycologia*, **46**, 393–434. (*404, 476, 480*)
EMERSON, S. (1952). In *Heterosis*, 199–217, ed. GOWEN, J. W. Iowa State College Press, Ames. (*397*)
—— (1950). *J. Bact.*, **60**, 221–223. (*60*)
—— and EMERSON, M. R. (1958). *Proc. Nat. Acad. Sci. U.S.*, **44**, 668–671. (*293*)
ENGEL, H. and SCHNEIDER, J. C. (1963). *Ber. deut. bot. Ges.*, **75**, 397–400. (*154*)
EPHRUSSI, B. (1953). *Nucleo-cytoplasmic relations in micro-organisms.* Oxford University Press. (*384*)
—— and SLONIMSKI, P. P. (1955). *Nature* (Lond.), **176**, 1207–1208. (*29*)
ERIKSSON, J. (1894). *Ber. deut. bot. Ges.*, **12**, 292–331. (*462*)
ERKAMA, J., HÄGERSTRAND, B. and JUNKKONEN, S. (1949). *Acta chem. Scand.*, **3**, 862–866. (*279*)
ERTL, L. (1951). *Planta*, **39**, 245–270. (*354*)
ESSER, K. (1956). *Z. indukt. Abstamm.-u Vererbungsl.*, **87**, 595–624. (*132, 452, 453*)
—— (1958). *Proc. X Int. Cong. Genetics*, **2**, 76–77. (*452*)
—— (1959a). *Z. Vererbungsl.*, **90**, 29–52. (*452, 453*)
—— (1959b). *Z. Vererbungsl.*, **90**, 445–446. (*452*)
—— and STRAUB, J. (1958). *Z. Vererbungsl.*, **89**, 729–746. (*131, 132*)
—— and KUENEN, R. (1965). *Genetik der Pilze.* Springer, Berlin. (*291, 401*)
EVANS, E. J. (1967). *A study of a fairy ring fungus, Marasmius oreades.* Ph.D. Thesis, University of Newcastle upon Tyne. (329)
EVANS, H. J. (1959). *Chromosoma*, **10**, 115–135. (*365, 377, 429, 433*)

FALCK, R. (1927). *Ber. deut. bot. Ges.*, **45**, 262–281. (*166*)
—— (1912). *Hausschwammforsch.*, **6**, 1–405. (*83*)
FAULL, J. H. (1905). *Proc. Boston soc. Nat. Hist.*, **32**, 77–114. (*111*)
FAYOD, V. (1889). *Ann. Sci. nat.: Boton. Biolog. vegetale*, 7–9, 179–411. (*117*)
FINCHAM, J. R. S. (1951). *J. genet.*, **50**, 220–221. (*475*)
—— (1956). *J. biol. chem.*, **182**, 61–73. (*288*)
—— and BOULTER, A. B. (1956). *Biochem. J.*, **62**, 72–77. (*288*)
—— and DAY, P. R. (1965). *Fungal Genetics*, 2nd edition. Blackwell Scientific Publications, Oxford. (*7, 12, 13, 376, 396, 432*)
FINDLAY, W. P. K. (1951). *Trans. Brit. mycol. Soc.*, **34**, 146. (*17*)
FISCHER, E. (1933). *Die Natürliche Pflanzenfamilien*, **7a**, ed. ENGLER, A. and PRANTL, H., ed. 2. W. Engelmann, Leipzig. (*119*)
FISCHER, G. W. (1951). *Phytopathology*, **41**, 839–853. (*479*)
—— and WERNER, G. (1958). *Z. physiol. Chem.*, **310**, 65–91. (*178*)
FLENTJE, N. T. (1957). *Trans. Brit. mycol. Soc.*, **40**, 322–336. (*334*)
—— (1959). In *Plant pathology, problems and progress 1908–1958*, 76–87, ed. HOLTON, C. S. University of Wisconsin Press, Madison. (*334*)
—— and STRETTON, H. M. (1964). *Aust. J. biol. sci.*, **17**, 686–704. (*396, 447, 454*)
FLOR, H. H. (1956). *Adv. genet.*, **8**, 29–54. (*466*)
FOECKLER, F. (1961). *Z. morph. ökol. Tiere*, **50**, 119. (*357*)
FORSYTH, F. R. (1955). *Canad. J. Bot.*, **33**, 363–373. (*183*)
FOSTER, J. W. (1949). *Chemical activities of Fungi.* Academic Press, New York. (*194, 197, 198, 280, 282, 292, 299, 314*)
—— CARSON, F. S., ANTHONY, D. S., DAVIS, J. B., JEFFERSON, W. B. and LONG, M. V. (1949). *Proc. Nat. Acad. Sci. U.S.*, **35**, 663–672. (*270*)

FOSTER, J. W., CARSON, S. F., RUBEN, S. and KAMEN, M. D. (1941). *Proc. Nat. Acad. U.S.*, **27**, 590–596. (*270*)
—— and GOLDMAN, A. (1948). In FOSTER, J. W. (1949). (*276*)
—— and WAKSMAN, S. A. (1939a). *J. Amer. chem. Soc.*, **61**, 127–135. (*280*)
—— (1939b). *J. Bact.*, **37**, 599–617. (*280*)
FRANKE, G. (1957). *Z. Vererbungsl.*, **93**, 109–117. (*418*)
FRANKE, W. and DEFFNER, M. (1939). *Ann. chem. Liebigs.*, **541**, 117–150. (*273*)
FRENCH, R. C. (1962). *Bot. gaz.*, **124**, 121–128. (*183*)
—— MASSEY, L. M. and WEINTRAUB, R. L. (1957). *Pl. Physiol.*, **32**, 389–393. (*183*)
FREY, A. (1927). *Rev. gén. Bot.*, **39**, 277–305. (*18*)
FREY, R. (1950). *Ber. Schweiz. bot. gesell.*, **60**, 119–230. (*21*)
FREY-WYSSLING, A. (1953). *Submicroscopic morphology of protoplasm.* Elsevier, Amsterdam. (*55*)
FRIES, N. (1936). *Bot. Notiser*, 567–574. (*426*)
—— (1940). *Symb. bot. Upsaliensis*, **4**, 5–39. (*426*)
—— (1961). In *Encyclopedia of Plant Physiology*, **14**, 332–400, ed. RUHLAND, W. Springer, Berlin. (*192*)
—— (1965). In *The Fungi, an advanced Treatise*, **I**, 491–523, ed. AINSWORTH, G. C. and SUSSMAN, A. S. Academic Press, New York. (*192*)
—— and JONASSON, L. (1941). *Svensk bot. Tidskr.*, **35**, 177–193. (*426*)
FRIIS, J. and OTTOLENGHI, P. (1959a). *C. r. lab. Carlsberg, Sér. Physiol.*, **31**, 259–271. (*205*)
—— —— (1959b). *C. r. lab. Carlsberg, Sér. Physiol.*, **31**, 272–281. (*205*)
FULLER, M. S. (1962). *Amer. J. Bot.*, **49**, 64–71. (*365*)
—— (1966). In *The Fungus Spore*, 67–84, ed. MADELIN, M. F. Butterworth, London. (*130*)
—— and BARSHAD, I. (1960). *Amer. J. Bot.*, **47**, 105–109. (*21*)
FULTON, I. W. (1950). *Proc. Nat. Acad. Sci. U.S.*, **36**, 306–312. (*429*)

GADD, C. H. (1947). *Ann. appl. Biol.*, **34**, 197–206. (*358*)
GALINDO, J. and GALLEGLY, M. E. (1960). *Phytopathology*, **50**, 123–128. (*408*)
GARCIA MENDOZA, C. and VILLANEUVA, J. R. (1965). *Proc. 2nd Meeting Fed. Eur. Biochem. Soc.*, Vienna, 122. (*203*)
GARDNER, J. F., JAMES, L. V. and RUBBO, S. D. (1956). *J. gen. Microbiol.*, **14**, 228–237. (*279*)
GARNJOBST, L. (1953). *Amer. J. Bot.*, **40**, 607–614. (*396*)
—— (1955). *Amer. J. Bot.*, **42**, 444–448. (*396*)
—— and WILSON, J. F. (1956). *Proc. Nat. Acad. Sci. U.S.*, **42**, 613–618. (*396*)
GARRETT, S. D. (1946). *Trans. Brit. mycol. Soc.*, **29**, 114–127. (*83*)
—— (1950). *Biol. Revs.*, **25**, 220–254. (*332*)
—— (1951). *New Phytol.*, **50**, 149–166. (*329*)
—— (1953). *Ann. Bot.* (Lond.), N.S., **17**, 63–79. (*83, 87, 89, 250*)
—— (1954). *Trans. Brit. mycol. Soc.*, **37**, 51–57. (*250*)
—— (1956) (reprinted 1960). *Biology of root infecting fungi.* Cambridge University Press. (*333, 334, 336*)
—— (1957). *Canad. J. Microbiology*, **3**, 135–149. (*90*)
—— (1960). *Ann. Bot.* (Lond.), N.S., **24**, 275–285. (*90, 189, 336*)
GARTON, G. A., GOODWIN, T. W. and LIJINSKY, W. (1951). *Biochem. J.*, **48**, 154–163. (*303*)
GASTROCK, E. A., PORGES, N., WELLS, P. A. and MOYER, A. J. (1938). *Ind. eng. chem.*, **30**, 782–789. (*281*)
GÄUMANN, E. A. (1964). *Die Pilze.* 5th Edition, Birkhauser, Basle and Stuttgart. (*4*)
—— (1950). *Principles of Plant Infection.* Crosby Lockwood, London. (*339*)
—— and BACHMANN, E. (1957). *Phytopath. Z.*, **29**, 265–276. (*310*)
GAUSE, G. F. (1932). *J. exp. Biol.*, **9**, 389–402. (*328*)

GAUSE, G. F. (1934). *The struggle for existence.* Williams & Wilkins, Baltimore. (*469*)
GEIGER, R. (1965). *The Climate near the ground.* Harvard University Press, Cambridge, Mass. (*167*)
GERHARDT, P. and JUDGE, J. A. (1964). *J. Bact.*, **87**, 945–951. (*201*)
GETTKANDT, G. (1954). *Wiss. z. Univ. Halle-Wittenberg, Math.-Nat.*, **3**, 691–709. (*321*)
GIBBS, M. and GASTEL, R. (1953). *Arch. biochem. biophys.*, **43**, 33–38. (*278*)
GIBSON, A. and GRIFFIN, D. M. (1958). *Aust. J. Biol. Sci.*, **11**, 548–556. (*69*)
GIESEY, R. M. and DAY, P. R. (1965). *Amer. J. Bot.*, **52**, 287–293. (*19, 26, 50, 390*)
GIRBARDT, M. (1955). *Flora*, **142**, 540–563. (*28, 33*)
—— (1957a). *Planta*, **50**, 47–49. (*28, 33, 49*)
—— (1958). *Arch. Mikrobiol.*, **28**, 255–269. (*19, 24, 26*)
—— (1960). *Flora*, **150**, 427–440. (*29*)
—— (1961). *Exptl. Cell. Res.*, **23**, 181–194. (*369*)
—— (1962). *Planta*, **58**, 1–21. (*369*)
—— (1965). In *Incompatibility in fungi*, 71, ed. ESSER, K. and RAPER, J. R. Springer, Berlin. (*390*)
GLASER, L. and BROWN, D. H. (1957a). *Biochim. biophys. Acta*, **23**, 449–450. (*47*)
—— —— (1957b). *J. biol. Chem.*, **228**, 729–742. (*47*)
GLEASON, F. H., NOLAN, R. A., WILSON, A. C. and EMERSON, R. (1966). *Science*, N.S., 152, 1272–1273. (*278*)
GODDARD, D. R. (1935). *J. gen. Physiol.*, **19**, 45–60. (*182*)
GODFREY, R. M. (1957). *Trans. Brit. mycol. Soc.*, **40**, 203–210. (*342*)
GOLDSTEIN, S. and CANTINO, E. C. (1962). *J. gen. Microbiol.*, **28**, 689–699. (*101, 319*)
GOODMAN, J. and ROTHSTEIN, A. (1957). *J. gen. Physiol.*, **40**, 915–923. (*216*)
GOODWIN, T. W. (1963). In *Biochemistry of industrial microorganisms*, 151–205, ed. RAINBOW, C. and ROSE, H. A. Academic Press, London. (*299*)
—— (1958). *Biochem. J.*, **70**, 612–617. (*302*)
—— (1959). In *Biosynthesis of terpenes and sterols*, 279–291, ed. WOLSTENHOLME, G. E. W. and O'CONNOR, M. Churchill, London. (*299*)
—— (1965). In *Chemistry and Biochemistry of Plant Pigments*, 143–173, ed. GOODWIN, T. W. Academic Press, London. (*302*)
GOOS, R. D. and SUMMERS, D. F. (1964). *Mycologia*, **56**, 701–703. (*41*)
GORDON, M., PARR, S. C., VIRGONA, A. and NUMEROF, P. (1953). *Science*, N.S., **118**, 43. (*309*)
GORHAM, E. (1959). *Canad. J. Bot.*, **37**, 327–329. (*354*)
GOTTLIEB, D. (1950). *Bot. Rev.*, **16**, 229–257. (*181*)
—— (1966). In *The Fungus Spore*, 217–233, ed. MADELIN, M. F. Butterworth, London. (*296*)
GRAINGER, J. (1962). *Trans. Brit. mycol. Soc.*, **45**, 147–155. (*17*)
GRAVES, A. H. (1916). *Bot. Gaz.*, **62**, 337–369. (*326*)
GRAY, J. and HANCOCK, G. J. (1955). *J. exp. biol.*, **32**, 802–814. (*177*)
GRAY, W. D. and BUSHNELL, W. R. (1955). *Mycologia*, **47**, 646–663. (*196*)
GREEN, E. (1927). *Ann. Bot.* (Lond.), **41**, 419–435. (*92*)
GREGORY, P. H. (1945). *Trans. Brit. mycol. Soc.*, **28**, 26–72. (*170*)
—— (1949). *Trans. Brit. mycol. Soc.*, **32**, 11–15. (*149*)
—— (1957). *Nature* (Lond.), **180**, 330. (*166*)
—— (1961). *The Microbiology of the Atmosphere.* Leonard Hill, London. (*145, 161, 167, 169, 170, 172*)
—— GUTHRIE, E. J. and BUNCE, M. E. (1959). *J. gen. Microbiol.*, **29**, 328–354. (*151*)
GREHN, J. (1932). *Jahrb. wiss. Bot.*, **76**, 93–165. (*56*)
GREIS, H. (1941). *Jahrb. wiss. Bot.*, **90**, 233–254. (*415*)
GRIMM, P. W. and ALLEN, P. J. (1954). *Pl. Physiol.*, **29**, 369–377. (*271*)
GRINDLE, M. (1963a). *Heredity*, **18**, 191–204. (*396, 454*)
—— (1963b). *Heredity*, **18**, 397–405. (*454*)
GROB, E. C. (1957). *Chimia*, **11**, 338–339. (*299*)

GROB, E. C., BEIN, M. and SCHOPPER, W. H. (1951). *Bull. soc. chim. Biol.*, **33**, 1236–1239. *(299)*
—— and BÜTLER, R. (1956). *Helv. chim. Acta*, **39**, 1975–1980. *(299)*
—— KIRSCHNER, K. and LYNEN, F. (1961). *Chimia*, **15**, 308–309. *(301)*
—— and BOSCHETTI, A. (1962). *Chimia*, **16**, 15–16. *(301)*
GRÖGER, D., WENDT, H. J., MOTHES, K. and WEYGAND, F. (1959). *Z. Naturf.*, **14b**, 355–358. *(305)*
GROSS, S. R. (1952). *Amer. J. Bot.*, **39**, 574–577. *(396, 415)*
GROSSBARD, E. and STRANKS, D. R. (1959). *Nature* (Lond.), **183**, 310–314. *(247)*
GROVE, J. F. (1963). In *Biochemistry of industrial microorganisms*, 320–340, ed. RAINBOW, C. and ROSE, A. H. Academic Press, London and New York. *(299, 308)*
GRUEN, H. E. (1959a). *Pl. Physiol.*, **34**, 158–168. *(53)*
—— (1959b). *A. r. Pl. Physiol.*, **10**, 405–440. *(59, 322)*
—— (1963). *Pl. Physiol.*, **38**, 652–666. *(139, 140)*
GUILLIERMOND, A. (1920). *C. r. acad. Sci.* (Paris), **170**, 1329–1331. *(29)*
—— (1941). *The cytoplasm of the plant cell.* Chronica Botanica, Waltham, Mass. *(28)*
GWYNNE-VAUGHAN, H. C. I. (1939). *Ann. Bot.* (Lond.), N.S., **1**, 99–105. *(415)*
—— and WILLIAMSON, H. S. (1930). *Ann. Bot.* (Lond.), **44**, 127–149. *(415)*

HADLEY, G. and PÉROMBELON, M. (1963). *Nature* (Lond.), **200**, 1337. *(344)*
HAENSELER, C. M. (1921). *Amer. J. Bot.*, **8**, 147–163. *(192)*
HAGIMOTO, H. (1963a). *Bot. Mag.* (Tokyo), **76**, 256–263. *(139)*
—— (1963b). *Bot. Mag.* (Tokyo), **76**, 363–365. *(139)*
—— and KONISHI, M. (1959). *Bot. Mag.* (Tokyo), **72**, 359–366. *(139)*
—— —— (1960). *Bot. Mag.* (Tokyo), **73**, 283–287. *(139)*
HALLDAL, P. (1961). *Physiol. Plantarum*, **14**, 133–139. *(321)*
HALVORSON, H. O. and COHEN, G. N. (1958). *Ann. Inst. Pasteur*, **95**, 73–87. *(225, 231)*
—— —— (1961). In *Membrane transport and metabolism*, 479–487, ed. KLEINZELLER, A. and KOTYK, A. Academic Press, Prague and London. *(225, 293)*
—— FRY, W. and SCHWEMM, D. (1955). *J. gen. Physiol.*, **38**, 549–573. *(225)*
HAMADA, M. (1940a). *Jap. J. Bot.*, **10**, 151–212. *(189)*
—— (1940b). *Jap. J. Bot.*, **10**, 387–464. *(189)*
HANNA, W. F. (1925). *Ann. Bot.* (Lond.), **39**, 431–457. *(426)*
HANSEN, H. N. (1938). *Mycologia*, **30**, 442–455. *(381)*
—— and SMITH, R. E. (1932). *Phytopathology*, **22**, 953–964. *(380)*
HARDEN, A. (1923). *Alcoholic Fermentation.* Longmans, Green, London. *(261)*
HARDER, R. B. (1927). *Z. Bot.*, **19**, 337–407. *(448)*
HARLEY, J. L. (1959). *The biology of mycorrhiza.* Leonard Hill, London. *(214, 339, 344, 345)*
—— (1965). In *Ecology of soil-borne plant pathogens*, 218–229, ed. BAKER, K. F. and SNYDER, W. C. Murray, London. *(345)*
—— and REES, AP. (1959). *New Phytol.*, **58**, 364–386. *(239, 271)*
—— and BRIERLEY, J. K. (1954). *New Phytol.*, **53**, 240–252. *(216)*
—— and JENNINGS, D. H. (1958). *Proc. R. Soc.*, **B148**, 403–418. *(239, 242)*
—— and LOUGHMAN, B. C. (1963). *New Phytol.*, **62**, 350–359. *(216)*
—— and MCCREADY, C. C. (1950). *New Phytol.*, **49**, 388–397. *(216)*
—— —— and BRIERLEY, J. K. (1963). *New Phytol.*, **52**, 124–132. *(216)*
—— —— —— (1958). *New Phytol.*, **57**, 353–362. *(216)*
—— —— —— and JENNINGS, D. H. (1956). *New Phytol.*, **55**, 1–28. *(216, 239)*
—— —— and GEDDES, J. A. (1954). *New Phytol.*, **53**, 429–444. *(216, 239)*
—— and SMITH, D. C. *Ann. Bot.* (Lond.), N.S., **20**, 513–543. *(242, 353)*
—— and WILSON, J. M. (1959). *New Phytol.*, **58**, 281–298. *(213)*
HARPER, J. L., CLATWORTHY, J. N., MCNAUGHTON, I. H. and SAGAR, G. R. (1961). *Evolution*, **15**, 209–227. *(469)*

HARPER, R. A. (1895). *Ber. deut. bot. Ges.*, **13**, 475–481. *(375)*
—— (1899). *Ann. Bot.* (Lond.), **13**, 467–525. *(92)*
—— (1905). *Carnegie Inst. Washington, Publ.*, **37**, 1–104. *(111, 376)*
HARTIG, R. (1874). *Wichtige Krankheiten der Waldbäume.* Springer, J., Berlin. *(83)*
HASTIE, A. C. (1962). *J. gen. Microbiol.*, **27**, 373–382. *(400)*
—— (1964). *Genet. Res.*, **5**, 305–315. *(400)*
—— (1967). *Nature* (Lond.), **214**, 249–252. *(400, 443)*
HATCH, A. B. (1937). *Black Rock For. Bull.*, **6**, 1–168. *(342, 344)*
HAUGE, J. G. and HALVORSON, H. O. (1962). *Biochim. biophys. Acta*, **61**, 101–107. *(294)*
HAWKER, L. E. (1939). *Ann. Bot.* (Lond.), N.S., **3**, 455–468. *(129)*
—— (1947). *Ann. Bot.* (Lond.), N.S., **11**, 245–259. *(129)*
—— (1948). *Ann. Bot.* (Lond.), N.S., **12**, 77–79. *(129)*
—— (1950). *Physiology of Fungi.* University of London Press, London. *(68, 141, 181)*
—— (1957). *The Physiology of Reproduction in Fungi.* Cambridge University Press. *(100, 128)*
—— (1962). *Trans. Brit. mycol. Soc.*, **45**, 190–199. *(341)*
—— (1963). *Nature* (Lond.), **197**, 618. *(28)*
—— (1965). *Biol. Revs.*, **40**, 52–92. *(29, 78, 93)*
—— (1966). In *The Fungus Spore*, 151–161, ed. MADELIN, M. F. Butterworth, London. *(180, 486)*
—— and ABBOTT, P. MCV. (1963a). *J. gen. Microbiol.*, **30**, 401–408. *(28, 29)*
—— —— (1963b). *J. gen. Microbiol.*, **31**, 491–494. *(18)*
—— —— (1963c). *J. gen. Microbiol.*, **32**, 295–298. *(180)*
—— and CHAUDHURI, S. D. (1946). *Ann. Bot.* (Lond.), N.S., **10**, 185–194. *(129)*
HAWORTH, R. D., RAISTRICK, H. and STACEY, M. (1937). *Biochem. J.*, **31**, 640–644. *(284)*
HAWTHORNE, D. C. (1963). *Genetics*, **48**, 1727–1729. *(420)*
HAXO, F. (1956). *Fortschr. chem. org. Naturst.*, **12**, 169–197. *(302)*
HAYAISHAI, O., SHIMAZONO, H., KATAGIRI, M. and SAITO, Y. (1956). *J. Amer. chem. Soc.*, **78**, 5126–5127. *(281)*
HAYMAN, D. S. (1964). *Canad. J. Bot.*, **42**, 13–21. *(111)*
HEBB, C. R. and SLEBODNIK, J. (1958). *Exp. Cell. Res.*, **14**, 286–294. *(317)*
HEIM, J. M. and GREIS, G. A. (1953). *Phytopathology*, **43**, 343–344. *(337)*
HEIM, P. (1947). *Rev. mycol.* (Paris), **12**, 104. *(304)*
—— (1955). *Rev. mycol.*, **22**, 131–157. *(366)*
HEIM, R. (1931). *C. r. acad. Sci.* (Paris), **192**, 291. *(123)*
—— (1948). *Rev. Sc.*, **80**, 69–86. *(358)*
—— (1952). *Les champignons de l'Europe.* N. Boubée et Cie, Paris. *(484, 485)*
HEIN, I. (1930). *Amer. J. Bot.*, **17**, 197–211. *(134)*
HEMKER, H. C. and HÜLSMANN, W. C. (1960). *Biochim. biophys. acta*, **44**, 175–177. *(321)*
HEMMONS, L. M., PONTECORVO, G. C. and BUFTON, A. W. J. (1953). *Adv. Genet.*, **5**, 194. *(422)*
HENDERSON-SMITH, J. (1923). *Ann. Bot.* (Lond.), **37**, 341–343. *(38)*
—— (1924). *New Phytol.*, **23**, 65–78. *(42)*
HENNEY, H. R. and STORCK, R. (1964). *Proc. Nat. Acad. Sci. U.S.*, **51**, 1050–1055. *(183, 294)*
HENNING, U., MOSTEIN, E. M. and LYNEN, F. (1959). *Arch. biochim. biophys.*, **83**, 259–267. *(307)*
HEREDIA, C., DE LA FUENTE, G. and SOLS, A. (1963). *Atti d. VII giorn. Biochim. Latines. Margherita Ligure* (Genova), 86. *(204)*
HESSELTINE, C. W. (1960). *U.S. Dept. Agric., Tech. Bull.*, **1245** *(299)*
—— WHITEHILL, A. R., PIDACKS, C., TEN HAGEN, N., BOHONOS, N., HUTCHINGS, B. L. and WILLIAMS, J. H. (1953). *Mycologia*, **45**, 7–19. *(193)*

HEVESEY, G. and NIELSEN, N. (1941). *Acta physiol. Scand.*, **2**, 347–354. (*209*)
HICKMAN, C. J. (1958). *Trans. Brit. mycol. Soc.*, **41**, 1–13. (*189*)
HILL, E. P. and SUSSMAN, A. S. (1964). *J. Bact.*, **88**, 1556–1566. (*183, 184*)
HIRSCH, H. M. (1954). *Physiol. Plantarum*, **7**, 72–97. (*131*)
—— (1950). *Mycologia*, **42**, 301–305. (*375*)
—— (1952). *Biochim. biophys. Acta*, **9**, 674–686. (*317*)
HOCKENHULL, D. J. D. (1948). *Biochem. J.*, **43**, 498–504. (*309*)
—— (1963). In *Biochemistry of industrial microorganisms*, 227–299, ed. RAINBOW, C. and ROSE, A. H. Academic Press, London and New York. (*299*)
HOCKING, D. (1963). *β-carotene and sexuality in fungi.* Ph.D. Thesis, University of Durham (King's College). (*321*)
HOHL, H. R. (1965). *J. Bact.*, **90**, 755–765. (*27*)
HOLLEY, R. W., ADGAR, J., EVERETT, G. A., MADISON, J. T., MARQUISEE, M., MERRILL, S. H., PENSWICK, J. R. and ZAMIR, A. (1965). *Science*, **147**, 1462–1465. (*296*)
HOLLIDAY, R. (1961a). *Genet. Res.*, **2**, 204–230. (*189, 232, 398, 400*)
—— (1961b). *Genet. Res.*, **2**, 472–486. (*189, 400*)
HOLLINGS, M., GANDY, D. G. and LAST, F. T. (1963). *Endeavour*, **22**, 112–117. (*444*)
HOLLOWAY, B. W. (1955). *Genetics*, **40**, 117–129. (*396*)
HOLTER, H. (1965). In *Structure and function in microorganisms*, 89–114, eds POLLOCK, M. R. and RICHMOND, M. H. Cambridge University Press, Cambridge. (*243*)
—— and OTTOLENGHI, P. (1960). *C. r. trav. lab. Carlsberg*, **31**, 409–422. (*207*)
HOLTON, C. S. (1941). *J. agr. res.*, **62**, 229–240. (*479*)
HOLWILL, M. E. J. (1964). *J. Protozool.*, **11** (Suppl.), 123. (*176*)
HOLZER, H. (1961). *Cold Spring Harbor Symp. Quant. Biol.*, **26**, 227–288. (*313*)
HOPP, H. (1938). *Phytopathology*, **28**, 356–360. (*255*)
HORENSTEIN, E. A. and CANTINO, E. C. (1964). *J. gen. Microbiol.*, **37**, 59–65. (*319*)
HORNE, A. S. (1930). *Ann. Bot.* (Lond.), **44**, 199–230. (*365, 366*)
HOROWITZ, N. H. (1945). *Proc. Nat. Acad. Sci. U.S.*, **31**, 153–157. (*490*)
HORR, W. H. (1936). *Pl. Physiol.*, **11**, 81–99. (*190*)
HORSFALL, J. G. and DIMOND, A. E. (1960). *Plant Pathology*, **1–3**. Academic Press, New York and London. (*310*)
HOTSON, H. H. (1953). *Phytopathology*, **43**, 360–363. (*338*)
—— and CUTTER, Y. M. (1951). *Proc. Nat. Acad. Sci. U.S.*, **37**, 400–403. (*338*)
HOUWINK, A. L. and KREGER, D. R. (1953). *Antonie von Leeuwenhoek*, **19**, 1–24. (*19, 21, 54*)
HRUSHOVETZ, S. B. (1956). *Canad. J. Bot.*, **34**, 641–651. (*375*)
HUGHES, S. J. (1953). *Canad. J. Bot.*, **31**, 577–659. (*95*)
HYDE, J. M. and WALKINSHAW, C. H. (1966). *J. Bact.*, **92**, 1218–1227. (*27*)

INGOLD, C. T. (1939). *Spore discharge in land plants.* Oxford University Press. (*157, 158*)
—— (1946). *Trans. Brit. mycol. Soc.*, **29**, 108–113. (*128, 168*)
—— (1953). *Dispersal in Fungi.* Oxford University Press. (*145, 159*)
—— (1956). *Nature* (Lond.), **177**, 1242–1243. (*152*)
—— (1959). *Trans. Brit. mycol. Soc.*, **42**, 475–478. (*128*)
—— (1960). In *Plant Pathology*, **3**, 137–168, ed. HORSFALL, J. G. and DIMOND, A. E. Academic Press, New York and London. (*163*)
—— (1961). *New Phytol.*, **60**, 143–149. (*164*)
—— (1965). *Spore Liberation.* Oxford University Press. (*145, 149, 156, 161, 164*)
—— (1966). *Mycologia*, **58**, 43–56. (*173*)
—— and COX, V. J. (1955). *Ann. Bot.* (Lond.), N.S., **19**, 201–209. (*163*)
—— and HADLAND, S. A. (1959). *New Phytol.*, **58**, 46–57. (*163*)
—— and ZOBERI, M. H. (1963). *Trans. Brit. mycol. Soc.*, **46**, 115–134. (*146*)
NGRAH AM, J. L. and EMERSON, R. (1954). *Amer. J. Bot.*, **41**, 146–152. (*191, 278*)
ISAAC, I. (1946). *Ann. appl. Biol.*, **33**, 28–34. (*83*)

ISAAC, P. K. (1964). *Canad. J. Bot.*, **42**, 787–792. *(257)*
ISHITANI, C., IKEDA, V. and SAKAGUCHI, K. (1956). *J. gen. appl. Microbiol.* (Tokyo), **3**, 93–101. *(398)*
IWASA, K. (1960). *J. Biochem.* (Tokyo), **47**, 445–453, 484–491. *(272)*

JACOB, F. and MONOD, J. (1961). *Cold Spring Harbor Symp. Quant. Biol.*, **26**, 193–212. *(315)*
JAFFE, L. (1966). *Pl. Physiol.*, **41**, 303–306. *(327)*
—— (1960). *J. gen. Physiol.*, **43**, 897–911. *(321)*
—— and ETZOLD, H. (1962). *J. cell. Biol.*, **13**, 13–31. *(321)*
JAHN, A. (1934). *Z. bot.*, **27**, 193–250. *(256, 257, 258)*
JAMES, W. O. (1953). *Plant Respiration*, Oxford University Press. *(262)*
JARVIS, F. G. and JOHNSON, M. J. (1947). *J. Amer. chem. Soc.*, **69**, 3010–3017. *(190)*
JENNINGS, D. H. (1958). *New Phytol.*, **57**, 254–255. *(239)*
—— (1963). *The Absorption of solutes by plant cells.* Oliver & Boyd, Edinburgh and London. *(206, 215, 236)*
—— (1964a). *New Phytol.*, **63**, 181–193. *(216)*
—— (1964b). *New Phytol.*, **63**, 348–357. *(217, 218)*
—— HOOPER, D. C. and ROTHSTEIN, A. (1958). *J. gen. Physiol.*, **41**, 1019–1026. *(215)*
JENSEN, S. L. (1965). *A. r. Microbiology*, **19**, 163–182. *(298)*
JINKS, J. L. (1952a). *Heredity*, **6**, 77–87. *(393)*
—— (1952b). *Proc. R. Soc.*, **B140**, 83–99. *(393, 394, 445)*
—— (1954). *Nature* (Lond.), **174**, 409. *(69, 382)*
—— (1956). *C. r. Lab. Carlsberg Sér. Physiol.*, **26**, 183–203. *(69, 385)*
—— (1959a). *J. gen. Microbiol.*, **20**, 223–236. *(69, 385)*
—— (1959b). *Heredity*, **15**, 525–528. *(381)*
—— CATEN, C. E., SIMCHEN, G. and CROFT, J. H. (1966). *Heredity*, **21**, 227–239. *(454, 455)*
—— and GRINDLE, M. (1963). *Heredity*, **18**, 245–264. *(408, 454)*
JOHNSON, T. (1946). *Cold Spring Harbor Symp. Quant. Biol.*, **11**, 85–93. *(386, 464)*
—— and NEWTON, M. (1940). *Canad. J. Res.*, C, **18**, 599–611. *(437)*
—— —— and BROWN, A. M. (1934). *Sci. Agric.*, **14**, 360–373. *(437)*
JONES, D. A. (1965). *Heredity*, **20**, 49–56. *(454)*
JONES, O. T. G. (1963). *J. exp. Bot.*, **42**, 399–411. *(225, 231, 232)*
—— and WATSON, W. A. (1962). *Nature* (Lond.), **194**, 947. *(225)*
JUEL, H. O. (1898). *Jahrb. wiss. Bot.*, **32**, 361–388. *(377)*

KÄFER, E. (1961). *Genetics*, **46**, 1581–1609. *(398, 399)*
KAMIYA, N. (1959). 'Protoplasmic streaming', *Handbuch der Protoplasmaforschung*, **8**, Pt. 3a, ed. HEILBRUNN, L. V. and WEBER, F. Springer, Vienna. *(257, 259)*
KARASSEVITCH, V. and DE ROBICHON-SZULMAJSTER, H. (1963). *Biochim. biophys. Acta*, **73**, 414–426. *(317)*
KAVANAU, J. L. (1963a). *Develop. Biol.*, **7**, 22–37. *(259)*
—— (1963b). *J. Theoret. Biol.*, **4**, 124–141. *(259)*
KEENE, M. L. (1914). *Ann. Bot.* (Lond.), **28**, 455–470. *(414)*
KEITT, G. U. and BOONE, D. M. (1956). *Brookhaven Symp. Biol.*, **9**, 209–217. *(332)*
—— and LANGFORD, M. H. (1941). *Amer. J. Bot.*, **28**, 805–820. *(396)*
KEMPNER, E. S. and COWIE, D. B. (1960). *Biochim. biophys. Acta*, **42**, 401–408. *(226)*
KEMPTON, F. E. (1919). *Bot. Gaz.*, **68**, 233–261. *(99)*
KENNEDY, M. E. and BURNETT, J. H. (1956). *Nature* (Lond.), **177**, 882–883. *(434)*
KIDSTON, R. and LANG, W. H. (1921). *Trans. R. soc. Edinburgh*, **52**, 855–902. *(342)*
KIMURA, K. (1952). *Biol. J. Okayama Univ.*, **1**, 72–83. *(426, 456)*

KINOSHITA, K. (1929). *J. chem. soc. Japan*, **50**, 583–593. *(280)*
KINSKY, S. C. (1961). *J. bact.*, **82**, 898–904. *(284)*
—— and MCELROY, W. D. (1958). *Arch. biochem. biophys.*, **73**, 466–483. *(284)*
KIRCHEIMER, F. (1933). *Planta*, **19**, 574–606. *(59)*
KLAUS, H. (1943). *Phytopath. Ztschr.*, **13**, 126–195. *(207)*
KLEBS, G. (1898). *Z. wiss. Bot.*, **32**, 1–70. *(100)*
—— (1899). *Z. wiss. Bot.*, **33**, 513–597. *(100)*
—— (1900). *Z. wiss. Bot.*, **35**, 80–203. *(100)*
KLEIN, D. T. (1948). *Bot. Gaz.*, **110**, 139–147. *(102)*
KLEIN, H. P. (1957). *J. Bact.*, **73**, 530–537. *(297)*
—— and BOOHER, Z. K. (1956). *Biochim. biophys Acta*, **20**, 387–388. *(305)*
KLEINZELLER, A., MÁLEK, J., PRAUS, R. and SKODA, J. (1952). *Chem. listy*, **46**, 470. *(225)*
KLUVYER, A. J. and PERQUIN, L. H. C. (1933). *Biochem. Z.*, **266**, 68–81. *(276, 281)*
KNOBLOCK, H. and MAYER, H. (1941). *Biochem. Z.*, **307**, 285–292. *(272)*
KNOX-DAVIES, P. S. (1966). *Amer. J. Bot.*, **53**, 220–224. *(360, 370)*
—— and DIXON, J. G. (1960). *Amer. J. Bot.*, **47**, 328–339. *(376)*
KOCH, A. (1960). *A. r. Microbiology*, **14**, 121–140. *(356)*
KOCH, W. J. (1956). *Amer. J. Bot.*, **43**, 811–819. *(36)*
—— (1959). *9th Int. Cong. Bot.* (Montreal), **2**, 196 (abstr.). *(176)*
—— (1961). In *Recent Advances in Botany*, 335–339. University of Toronto Press, Toronto. *(176)*
KÖHLER, E. (1929). *Planta*, **8**, 140–153. *(71)*
—— (1930). *Planta*, **10**, 495–522. *(71)*
KÖHLER, F. (1935). *Z. indukt. Abstamm.-u. Vererbungsl.*, **70**, 1–54. *(387)*
KOLE, A. P. (1957). *Tijdschr. Plantenziekten*, **63**, 361–364. *(36)*
—— (1965). In *The Fungi*, **I**, 77–93, ed. AINSWORTH, G. C. and SUSSMAN, A. S. Academic Press, New York and London. *(177)*
—— and GIELINK, A. J. (1962). *Proc. Kon. Ned. Akad. Wetens. Amsterdam*, **C65**, 117–121. *(36)*
—— and HORSTRA, K. (1959). *Proc. Kon. Ned. Akad. Wetens. Amsterdam*, **C62**, 404–408. *(36)*
KOLTIN, Y., RAPER, J. R. and SIMCHEN, G. (1967). *Proc. Nat. Acad. Sci. U.S.*, **57**, 55–62. *(432)*
KONISHI, M. and HAGIMOTO, H. (1962). *Plant Physiol* (suppl.), **37**, ix–x. *(139)*
KOUYEAS, V. (1953). *Ann. Inst. Phytopath., Benski*, **7**, 40–53. *(408)*
KREBS, H. A. and LOWENSTEIN, J. M. (1960). In *Metabolic Pathways*, **1**, 129–203, ed. GREENBERG, D. M. Academic Press, New York. *(271)*
KREGER, D. R. (1954). *Biochim. biophys. Acta*, **13**, 1–9. *(21, 22)*
KRZEMINSKI, L. F. and QUACKENBUSH, F. W. (1960). *Arch. biochem. biophys.*, **88**, 64–67. *(299)*
KÜHNER, R. (1945). *Bull. soc. Linn., Lyon*, **14**, 160–164. *(378, 434)*
——, ROMAGNESI, H. and YEN, H. C. (1947). *Bull. soc. mycol. France*, **63**, 169–186. *(456)*
KUSANO, S. (1911). *J. Agr.* (Tokyo), **4**, 1–66. *(189, 350)*

LAANE, H. M. (1967). *Canad. J. genet. cytol.*, **9**, 342–351. *(372)*
LAMB, I. M. (1935). *Ann. Bot.* (Lond.), **49**, 403–438. *(437)*
LAMPEN, J. O. (1965). In *Function and Structure in microorganisms*, 115–133, ed. POLLOCK, M. R. and RICHMOND, M. H. Cambridge University Press. *(242, 205)*
—— (1966). In *Biochemical studies of antimicrobial drugs*, 111–130, ed. NEWTON, B. A. and REYNOLDS, P. E. Cambridge University Press. *(202, 203)*
——, ARNOW, P. M., BOROWSKA, Z. and LASKIN, A. I. (1962). *J. Bact.*, **84**, 1152–1160. *(200)*
LANGE, M. (1952). *Dansk. bot. Arkiv*, **14**, 1–164. *(433, 456, 491)*

LANGE, O. L. (1953). *Flora* (Jena), **140**, 39–97. *(354)*
LANGERON, M. and VANBREUSEGHEM, R. (1952). *Précis de Mycologie*. Masson et Cie, Paris. *(95)*
—— —— (1965). *Outline of Mycology*, translated WILKINSON, J. Pitman & Sons, London. *(95)*
LARGE, E. C. (1961). *Trans. Brit. mycol. Soc.*, **44**, 1–23. *(62)*
LEAL, J., FRIEND, A. J. and HOLLIDAY, P. (1964). *Nature* (Lond.), **203**, 545–546. *(408)*
LEEDALE, G. (1958a). *Nature* (Lond.), **181**, 502–503. *(364)*
—— (1958b). *Arch. Mikrobiol.*, **32**, 32–64. *(364)*
—— (1959). *Cytologia*, **24**, 213–219. *(364)*
LEHNIGER, A. L. (1965). *Bioenergetics*. Benjamin, New York. *(312)*
LEIN, J. and LEIN, P. S. (1950). *J. Bact.*, **60**, 185–190. *(298)*
LELOIR, L. F. (1951). *Arch. biochem. biophys.*, **33**, 186–190. *(283)*
—— and CABIB, E. (1953). *J. Amer. Chem. Soc.*, **75**, 5445–5446. *(284)*
—— and CARDINI, C. E. (1953). *Biochim. biophys. Acta*, **12**, 15–22. *(47)*
LEONARD, T. J. and DICK, S. (1968). *Proc. Nat. Acad. Sci. U.S.*, **59**, 745–751. *(141)*
LEONIAN, L. H. (1931). *Phytopathology*, **21**, 941–955. *(408)*
LESTER, G. and HECHTER, O. (1958). *Bact. Proc.*, **58**, 109. *(213)*
LEUPOLD, U. (1950). *C. r. Lab. Carlsberg, Sér. physiol.*, **24**, 381–480. *(420)*
—— (1958). *Cold Spring Harbor Symp. Quant. Biol.*, **23**, 161–170. *(420)*
LEVI, J. D. (1956). *Nature* (Lond.), **177**, 753–754. *(418)*
LEVINE, M. M. and COTTER, R. U. (1931). *Phytopathology*, **21**, 107. *(405)*
LEWIS, D. H. and HARLEY, J. L. (1965a). *New Phytol.*, **64**, 224–237. *(250, 345)*
—— —— (1965b). *New Phytol.*, **64**, 238–255. *(250, 345)*
—— —— (1965c). *New Phytol.*, **64**, 256–269. *(250, 345)*
LHOAS, P. (1967). *Genet. Res.*, **10**, 45–51. *(446)*
LILLEY, V. G. and BARNETT, H. L. (1953). *West Va. Univ. Agr. Exp. Sta. Bull.*, **362 T**, 1–58. *(190)*
—— —— (1951). *Physiology of the Fungi*. McGraw Hill, New York. *(191)*
LIN, C. K. (1945). *Amer. J. Bot.*, **32**, 296–298. *(181)*
LINDEBERG, G. (1944). *Symb. Bot. Upsaliensis*, **8**, 1–183. *(191)*
—— and HOLM, G. (1952). *Physiol. Plantarum*, **5**, 100–114. *(242)*
LINDEGREN, C. C. and LINDEGREN, G. (1943). *Genetics*, **28**, 81. *(420)*
—— —— (1944). *Ann. Mo. Bot. gard.*, **31**, 203–216. *(420)*
LINDENMEYER, A. (1965). In *The Fungi, an advanced treatise*, **I**, 301–347, ed. AINSWORTH, G. C. and SUSSMAN, A. S. Academic Press, New York. *(262, 271)*
LINDROTH, C. H. (1948). *Svensk bot. Tidskr.*, **42**, 34–41. *(356)*
LING YONG, M. (1929). *Étude des phénomènes de la sexualité chez les Mucorinées*, Thèse, Paris. *(410)*
LINGAPPA, B. T. and SUSSMAN, A. S. (1959). *Pl. Physiol.*, **34**, 466–472. *(184)*
LITTLEFIELD, L. J. (1965a). *Phytopathology*, **55**, 536–542. *(249)*
—— (1966). *Physiol. Plantarum*, **19**, 264–270. *(251)*
——, WILCOXSON, R. D. and SUDIA, T. W. (1963). *Phytopathology*, **53**, 881 (abstr.). *(251)*
—— —— —— (1965). *Amer. J. Bot.*, **52**, 599–605. *(123, 251, 252)*
LOCHHEAD, A. G. (1958). *Bact. Rev.*, **22**, 145–153. *(330)*
LOCKWOOD, L. B. and NELSON, G. E. N. (1946). *Arch. biochem.*, **10**, 365–374. *(280)*
—— and REEVES, M. D. (1945). *Arch. biochem.*, **6**, 455–469. *(280)*
——, STUBBS, J. J. and SENSEMAN, C. E. (1938). *Zentr. Bakt. Parasitenk. Abt. II*, **98**, 167–171. *(276, 280)*
LOHWAG, H. (1941). *Anatomie der Asco- und Basidiomyceten* (*Encyclopedia of Plant Anatomy*, **6**, Part 8, ed. LINSBAUER, K., TISCHLER, G. and PASCHER, A.) Gebrüder-Borntraeger, Berlin. *(123)*
LOVETT, J. S. (1963). *J. Bact.*, **85**, 1235–1246. *(36)*
LOVETT, J. S. and CANTINO, E. C. (1960a). *Amer. J. Bot.*, **47**, 499–505. *(47, 101)*

LOWRY, R. J. and SUSSMAN, A. S. (1956). *Arch. biochem. biophys.*, **62**, 113–124. (*200*)
—— —— (1958). *Amer. J. Bot.*, **45**, 397–403. (*181*)
—— —— and VON BOVENTER, B. (1957). *Mycologia*, **49**, 609–622. (*200*)
LU, B. C. (1964a). *Chromosoma*, **15**, 170–184. (*377*)
—— (1964b). *Amer. J. Bot.*, **51**, 343–347. (*377*)
—— and BRODIE, H. J. (1962). *Nature* (Lond.), **194**, 606. (*377*)
—— —— (1964). *Canad. J. Bot.*, **42**, 307–310. (*377*)
—— (1967). *J. Cell Sci.*, **2**, 529–536. (*376*)
LUCAS, R. L. (1960). *Nature* (Lond.), **188**, 763–764. (*247*)
LUCK, D. J. and REICH, E. (1964). *Proc. Nat. Acad. Sci. U.S.*, **52**, 931–938. (*384*)
LUNDEGARDH, H. (1960). *Nature* (Lond.), **185**, 72. (*239*)
LUTTRELL, E. S. (1951). *Univ. Mo. Studies*, **24**, No. 3. (*110*)
—— (1955). *Mycologia*, **47**, 511–532. (*110*)
LYNCH, V. H. and CALVIN, M. (1952). *J. Bact.*, **63**, 525–531. (*270*)
LYNEN, F. (1961). *Fed. Proc.*, **20**, 941–951. (*297*)
——, AGRANOFF, B. W., EGGERER, H., HENNING, U. and MÖSELEIN, E. M. (1959). *Angew. Chem.*, **71**, 657–663. (*307*)
LYTHGOE, J. N. (1961). *Trans. Brit. mycol. Soc.*, **44**, 199–312. (*104*)
—— (1962). *Trans. Brit. mycol. Soc.*, **45**, 161–168. (*104*)

MACDONALD, J. A. and CARTTER (1961). *Trans. Brit. mycol. Soc.*, **44**, 72–78. (*83*)
MACER, R. C. F. (1961). *Ann. appl. Biol.*, **49**, 152–164. (*329*)
MACHLIS, L. (1953). *Amer. J. Bot.*, **40**, 189–195. (*191*)
—— (1958a). *Physiol. Plantarum*, **11**, 181–192. (*405*)
—— (1958b). *Nature* (Lond.), **181**, 1790–1791. (*405*)
—— (1958c). *Physiol. Plantarum*, **11**, 845–854. (*405*)
—— (1966). In *The Fungi, an advanced treatise*, **II**, 415–433, ed. AINSWORTH, G. C. and SUSSMAN, A. S. Academic Press, New York. (*405*)
MACINDOE, S. L. (1945). *Agric. gaz. N.S.W.*, **56**, 530–531. (*464*)
MACMILLAN, A. (1956a). *J. exp. bot.*, **7**, 113–126. (*218*)
—— (1956b). *Physiol. Plantarum*, **9**, 470–481. (*218, 221*)
MACRAE, R. (1941). *Genetical and sexual studies in some higher Basidiomycetes.* Ph.D. Thesis, University of Toronto. (*456, 461*)
—— (1967). *Canad. J. Bot.*, **45**, 1371–1398. (*461*)
MADELIN, M. F. (1956a). *Ann. Bot.* (Lond.), N.S., **20**, 307–330. (*134, 141*)
—— (1956b). *Ann. Bot.* (Lond.), N.S., **20**, 467–480. (*134*)
—— (1960). *Trans. Brit. mycol. Soc.*, **43**, 105–110. (*77, 134*)
—— (1966). *A. r. Entomology*, **11**, 423–448. (*356*)
MAGNUS, W. (1906). *Archiv. J. Biontologie*, **1**, 85–161. (*141*)
MAHLER, H. R. and CORDES, E. H. (1966). *Biological chemistry.* Harper & Co., New York. (*262*)
MANDELS, G. R. (1951). *Amer. J. Bot.*, **38**, 213–221. (*225*)
—— (1953). *Exp. Cell. Res.*, **5**, 48–55. (*205*)
—— (1955). *Amer. J. Bot.*, **42**, 921–929. (*192*)
MANNERS, G. J., and KHIN MAUNG (1955). *J. chem. Soc.*, 867–870. (*283*)
MANOCHA, M. S. and COLVIN, J. R. (1967). *J. Bact.*, **94**, 202–212. (*202*)
MANTON, I., CLARKE, B. and GREENWOOD, A. D. (1951). *J. exp. Bot.*, **2**, 321–331. (*34, 36, 177, 178*)
—— —— —— (1952). *J. exp. Bot.*, **3**, 204–215. (*34, 36, 37*)
MARCHANT, R. (1966a). *Ann. Bot.* (Lond.), N.S., **30**, 441–445. (*180*)
—— (1966b). *Ann. Bot.* (Lond.), N.S., **30**, 821–830. (*201*)
—— BANBURY, G. H. and PEAT, A. (1967). *New Phytol.*, **66**, 623–629. (*27*)
—— and WHITE, M. (1966). *J. gen. Microbiol.*, **42**, 237–244. (*180, 182, 208*)
MARCOU, D. and SCHECROUN, J. (1959). *C. r. Acad. Sci.* (Paris), **248**, 280–283. (*69, 384*)

MARIAT, F. (1964). In *Microbial behaviour 'in vivo' and 'in vitro'*, 85–111, ed. SMITH, H. and TAYLOR, J. Cambridge University Press. *(355)*
MARKERT, C. L. (1949). *Amer. Naturalist*, **83**, 227–231. *(133, 421)*
MARSH, P. B., TAYLOR, E. E. and BASSLER, L. M. (1959). *Pl. Disease Reptr., Suppl.*, **261**. *(142, 319)*
MARTIN, S. M. (1963). In *Biochemistry of industrial microorganisms*, 415–452, ed. RAINBOW, C. and ROSE, A. H. Academic Press, London and New York. *(273, 275)*
—— (1955). *Canad. J. Microbiol.*, **1**, 6–11. *(279)*
—— (1957). *Industr. Eng. Chem.*, **49**, 1231. *(279)*
MARTIN, W. J. (1943). *Phytopathology*, **33**, 569–585. *(479)*
MATHER, K. (1942). *Nature* (Lond.), **149**, 54–56. *(443)*
—— and JINKS, J. L. (1958). *Nature* (Lond.), **182**, 1188–1190. *(385)*
MATHIESON, J. M. (1952). *Ann. Bot.* (Lond.), N.S., **16**, 449–466. *(421)*
—— and CATCHESIDE, D. G. (1955). *J. gen. Microbiol.*, **13**, 72–83. *(225)*
MATHIESON-KÄÄRIK, A. (1953). *Medd. Skogforskn Inst., Stockh.*, **43**, 1. *(358)*
—— (1960). *Symb. bot. Upsaliensis*, **16**, 1–168. *(358)*
MAXWELL, M. B. (1954). *Canad. J. Bot.*, **32**, 259–280. *(428)*
MAY, J. W. (1962). *Exp. Cell Res.*, **27**, 170–172. *(41)*
MCCLINTOCK, B. M. (1945). *Amer. J. Bot.*, **32**, 671–678. *(373)*
MCCLURE, W. A., PARK, D. and ROBINSON, P. M. (1968). *J. gen. Microbiol.*, **50**, 177–182. *(33)*
MCCUBBIN, W. A. (1944). *Phytopathology*, **34**, 230–4. *(166)*
MCCURDY, H. D. and CANTINO, E. C. (1960). *Pl. Physiol.*, **35**, 463–476. *(288)*
MCGINNIS, R. C. (1953). *Canad. J. Bot.*, **31**, 522–526. *(371, 480)*
—— (1954). *Canad. J. Bot.*, **32**, 213–214. *(371, 480)*
—— (1956). *J. Hered.*, **47**, 254–259. *(371, 480)*
MCKEAN, C. V. (1952). *Canad. J. Bot.*, **30**, 764–787. *(456, 461)*
MCKEEN, W. E. (1962). *Canad. J. Microbiology*, **8**, 897–904. *(177)*
MEHTA, K. C. (1940). *Sci. Monogr. Counc. Agric. Res., India*, **14**, 1–224. *(170)*
—— (1952). *Sci. Monogr. Counc. Agric. Res., India*, **18**, 1–368. *(170)*
MEIER, H. (1955). *Holzals Roh-Werkstoff*, **13**, 323–338. *(337)*
MELIN, E. (1963). In *Symbiotic Associations*, 125–145, ed. NUTMAN, P. S. and MOSSE, B. Cambridge University Press. *(344)*
—— and NILSSON, H. (1950). *Physiol. Plantarum*, **3**, 88–92. *(250, 344)*
—— —— (1952). *Svensk bot. Tidskr.*, **46**, 281–285. *(250)*
—— —— (1953). *Nature* (Lond.), **171**, 134. *(250)*
—— —— (1954). *Svensk bot. Tidskr.*, **48**, 555–558. *(250)*
—— —— (1955). *Svensk bot. Tidskr.*, **49**, 119–121. *(250, 344)*
—— —— (1957). *Svensk bot. Tidskr.*, **51**, 166–186. *(344)*
—— —— (1958). *Bot. Notiser*, **111**, 251–256. *(250)*
—— —— and HACSKAYLO, E. (1958). *Bot. Gaz.*, **119**, 243–246. *(250)*
MEREDITH, D. S. (1961). *Ann. Bot.* (Lond.), N.S., **25**, 271–278. *(151)*
—— (1962). *Ann. Bot.* (Lond.), N.S., **26**, 233–241. *(151)*
—— (1963). *Trans. Brit. mycol. Soc.*, **46**, 201–207. *(151)*
METCALFE, G. and CHAYEN, S. (1954). *Nature* (Lond.), **174**, 841. *(191)*
METZENBERG, R. L. (1963). *Arch. biochem. biophys.*, **100**, 503–509. *(205)*
—— (1964). *Biochim. biophys. Acta*, **89**, 291–302. *(201, 242)*
MEYER, F. and BLOCH, K. (1963). *J. biol. Chem.*, **238**, 2654–2659. *(298)*
MEYER, V. and SCHÜLTZE, E. (1884). *Ber. deut. chem. Ges.*, **17**, 1554–1558. *(284)*
MIDDLEBROOK, B. and PRESTON, R. D. (1952a). *Biochim. biophys. Acta*, **9**, 32–48. *(39, 49, 54, 57)*
—— —— (1952b). *Biochim. biophys. Acta*, **9**, 115–126. *(57)*
MILES, P. G., TAKEMARU, T. and KIMURA, K. (1966). *Bot. Mag.* (Tokyo), **79**, 693–705. *(441)*
MILLER, J. J. (1945). *Canad. J. Res.*, **C**, **23**, 16–43. *(445)*

MILLER, J. J. (1946a). *Canad. J. Res.*, **C**, **24**, 188–212. (*445*)
—— (1946b). *Canad. J. Res.*, **C**, **24**, 213–223. (*445*)
MINAMI, Z. and IKEDA, Y. (1962). *J. gen. appl. Microbiol.* (Tokyo), **8**, 92–98. (*387*)
MITCHELL, M. B. and MITCHELL, H. K. (1952). *Proc. Nat. Acad. Sci. U.S.*, **38**, 442–449. (*382, 415*)
MITCHELL, R. and SABER, N. (1966). *J. gen. Microbiol.*, **42**, 39–42. (*23*)
MITCHISON, J. M. (1957). *Exp. Cell Res.*, **13**, 244–262. (*41*)
MIYAZAKI, T., AMANO, Y. and NIINOBE, S. (1962). *Tokyo Vakka Daigaku, Kenkyu Nempo*, **12**, 54. (*307*)
MIZUNUMA, T. (1963). *Agr. biol. Chem.* (Tokyo), **27**, 88–98. (*273*)
MÖLLER, A. (1893). *Bot. Mit. Trop.*, **5**, 1–127. (*359*)
MONSON, A. M. and SUDIA, T. W. (1963). *Bot. Gaz.*, **184**, 440–443. (*247*)
MOOR, H. and MÜHLETHALER, K. (1963). *J. Cell. Biol.*, **17**, 609–628. (*20, 28, 29, 30, 50, 293, 367*)
MOORE, E. J. (1965). *Amer. J. Bot.*, **52**, 389–395. (*109, 123*)
MOORE, R. T. (1963). *Mycologia*, **55**, 633–642. (*28*)
—— (1964). *Z. Zellforsch.*, **63**, 921–937. (*365*)
—— and MCALEAR, J. H. (1961). *Mycologia*, **53**, 194–200. (*26*)
—— —— (1963a). *J. cell. Biol.*, **16**, 131–141. (*27, 28, 50*)
—— —— (1963b). *J. Ultrastruct. Res.*, **8**, 144–153. (*29*)
MOREL, G. (1944). *C. r. acad. Sci.* (Paris), **218**, 50. (*337*)
MORRÉ, D. J. and MOLLENHAUER, H. M. (1964). *J. cell. Biol.*, **23**, 295–325. (*27*)
MORRISON, T. M. (1962). *New Phytol.*, **61**, 10–20. (*345*)
MORTON, A. G. and MACMILLAN, A. (1954). *J. exp. Bot.*, **5**, 232–252. (*218, 221*)
MOSBACH, K. (1964a). *Acta chem. Scand.*, **18**, 329–334. (*354*)
—— (1964b). *Biochem. biophys. res. Commun.*, **17**, 363–367. (*354*)
—— (1964c). *Acta chem. Scand.*, **18**, 2013. (*354*)
MOSES, V. (1955). *J. gen. Microbiol.*, **13**, 235–251. (*189*)
MOSSE, B. (1953). *Nature* (Lond.), **171**, 974. (*342*)
—— (1959). *Trans. Brit. mycol. Soc.*, **42**, 439–448. (*348*)
—— (1963). In *Symbiotic Associations*, 146–170, ed. NUTMAN, P. S. and MOSSE, B. Cambridge University Press. (*341, 348*)
MOUNCE, I. (1929). *Bull. Dept. Agric. Canada*, **11**, 1–77. (*456, 457*)
—— and MACRAE, R. (1937). *Canad. J. Res.*, **C**, **15**, 154–161. (*426*)
—— —— (1938). *Canad. J. Res.*, **C**, **16**, 364–376. (*426, 456, 457, 458, 459*)
MÜLLER, D. (1926). *Chem. Ztg.*, **50**, 101. (*272*)
—— (1954). *Friesia*, **5**, 65–74. (*156*)
—— and JAFFE, L. (1965). *Biophys. J.*, **5**, 317–335. (*327*)
MÜLLER, E. and BIEDERMANN, W. (1952). *Phytopath. Z.*, **19**, 343–350. (*200*)
MÜLLER, F. (1911). *Jahrb. wiss. Bot.*, **49**, 421–521. (*179*)
MÜLLER, F. W. (1941). *Ber. Schweiz. bot. gesell.*, **51**, 165–256. (*328*)
MÜLLER, K. O. (1956). *Phytopath. Z.*, **27**, 237–254. (*339*)
MULLINS, L. J. (1942). *Biol. Bull. Woods Hole*, **83**, 326–333. (*216*)
MULLINS, J. T. (1961). *Amer. J. Bot.*, **48**, 377–387. (*404*)
—— and RAPER, J. R. (1965). *Science*, N.S., **150**, 1174–1175. (*407*)
MUNCH-PETERSON, A., KALCKAR, H. M., CUTOLO, E. and SMITH, E. E. B. (1953) *Nature* (Lond.), **172**, 1036. (*283*)

NABEL, K. (1939). *Arch. Mikrobiol.*, **10**, 515–541. (*21*)
NAGEL, L. (1946). *Ann. Miss. bot. gard.*, **33**, 249–289. (*30, 367*)
NAMBOODIRI, A. N. and LOWRY, R. J. (1967). *Amer. J. Bot.*, **54**, 735–748. (*372*)
NASON, A., ABRAHAM, R. G. and AUERBACH, B. C. (1954). *Biochim. biophys. Acta*, **15**, 159–161. (*285*)
NASS, M. M. K. (1966). *Proc. Nat. Acad. Sci. U.S.*, **56**, 1213–1222. (*384*)
NAUMOV, N. A. (1936). 'Clés des Mucorinées', *Encyclopédie Mycologique*, **9**. Lechevalier, Paris. (*474*)

NEILANDS, J. B. (1952). *J. biol. Chem.*, **197**, 701–708. (*271*)
NEISH, A. C. and BLACKWOOD, A. C. (1951). *Canad. J. Technology*, **29**, 123. (*277*)
NELSON, R. R. (1956). *Phytopathology*, **46**, 538–540. (*445*)
—— (1963). *A. r. Microbiology*, **17**, 31–48. (*477*)
—— (1964). *Evolution*, **18**, 700–704. (*477, 478*)
NEUBERG, C. and HIRSCH, J. (1919a). *Biochem. Z.*, **96**, 175. (*276*)
—— —— (1919b). *Biochem. Z.*, **98**, 141. (*276*)
NEWTON, M. and JOHNSON, T. (1932). *Can. Dept. Agric. Bull.*, **160**. (*437*)
NICHOLAS, D. J. D. (1959a). *Symp. Soc. Exp. Biol.*, **13**, 1–23. (*285*)
—— (1959b). *4th Int. Cong. Biochem., Vienna 1958*, XIII, 307–331. (*285*)
—— (1965). In *The Fungi, an advanced treatise*, **I**, 349–376, ed. AINSWORTH, G. C. and SUSSMAN, A. S. Academic Press, New York. (*206, 284*)
—— NASON, A. and MCELROY, W. D. (1954). *J. biol. Chem.*, **207**, 341–351. (*284*)
—— MEDINA, A. and JONES, D. T. G. (1960). *Biochim. biophys. Acta*, **37**, 468–476. (*285*)
NICKERSON, W. J. (1963). *Bact. Rev.*, **27**, 305–324. (*355*)
—— FALCONE, G. and KESSLER, G. (1961). In *Macromolecular complexes*, 205–208, ed. EDDS, M. V. JR. Ronald Press, New York. (*21, 47*)
NICOLSON, T. H. (1959). *Trans. Brit. mycol. Soc.*, **42**, 421–438. (*341*)
—— (1960). *Trans. Brit. mycol. Soc.*, **43**, 132–145. (*341*)
NIEDERPRUEM, D. J. (1964). *J. Bact.*, **88**, 210–215. (*296*)
—— (1965). In *The Fungi, an advanced treatise*, **I**, 269–300, ed. AINSWORTH, G. C. and SUSSMAN, A. S. Academic Press, New York. (*262, 269*)
NISHIMURA, M. and TAKAMATSU, K. (1957). *Nature* (Lond.), **180**, 699. (*320*)
NORKRANS, B. (1957). *Physiol. Plantarum*, **10**, 198–214. (*242*)
NORTHCOTE, D. H. (1953). *Biochem. J.*, **53**, 348–352. (*283*)
—— and HORNE, R. W. (1952). *Biochem. J.*, **51**, 232–236. (*22, 205*)

OLIVE, L. S. (1953). *Bot. Rev.*, **19**, 439–586. (*360, 364*)
—— (1954). *Bull. Torrey bot. Club*, **81**, 95–97. (*422*)
—— (1958). *Amer. Naturalist*, **92**, 233–251. (*422*)
—— (1964). *Science*, N.S., **146**, 542–543. (*157*)
OORT, A. J. P. and ROELOFSEN, P. A. (1932). *Proc. Akad. sci. Amsterdam*, **35**, 898–908. (*18*)
OTTOLENGHI, P. and LILLEHOJ, E. B. (1966). In *Symposium über Hefe-Protoplasten*, ed. MÜLLER, R. Akademie Verlag, Berlin. (*204*)
OWENS, R. G. and MILLER, L. P. (1957). *Phytopathology*, **47**, 531. (*200*)

PAGE, R. M. (1952). *Amer. J. Bot.*, **39**, 731–738. (*102*)
—— (1956). *Mycologia*, **48**, 206–224. (*102, 320, 321*)
—— (1959). *Amer. J. Bot.*, **46**, 579–585. (*102*)
—— (1965). In *The Fungi, an advanced treatise*, **I**, 559–574, ed. AINSWORTH, G. C. and SUSSMAN, A. S. Academic Press, New York. (*319, 321*)
PALMER, J. T. (1957). *Naturalist*, 1–4. (*434*)
PANT, N. C. and FRAENKEL, G. (1950). *Science*, N.S., **112**, 498. (*357*)
—— —— (1954). *Biol. Bull. Woods Hole*, **107**, 420–432. (*357*)
PAPAZIAN, H. (1950). *Bot. Gaz.*, **112**, 143–163. (*429*)
—— (1958). *Adv. Genet.*, **9**, 41–69. (*437*)
PARK, D. (1956). *6th Int. Cong. Soil. Sci. Soc.*, 3, 23–28. (*329*)
—— (1961). *Trans. Brit. mycol. Soc.*, **44**, 377–390. (*72*)
—— (1963). *Trans. Brit. mycol. Soc.*, **46**, 541–548. (*72, 96*)
—— and ROBINSON, P. M. (1966a). In *Trends in Plant Morphogenesis*, 27–44, ed. CUTTER, E. G. Longmans, London. (*72*)
—— —— (1966b). *Ann. Bot.* (Lond.), N.S., **30**, 425–439. (*204, 208*)
PARKER, B. C., PRESTON, R. D. and FOGG, G. E. (1963). *Proc. R. Soc.*, **B158**, 435–445. (*21, 22, 489*)

PARKER-RHODES, A. F. (1954). *New Phytol.*, **53**, 145–154. *(62)*
PARKIN, E. A. (1942). *Ann. appl. biol.*, **29**, 268–274. *(358)*
PARKS, L. W. (1958). *J. Amer. chem. Soc.*, **80**, 2023–2024. *(307)*
PARTINGTON, M. (1959). *Mating Systems of Fungi with special reference to Polystictus versicolor.* Ph.D. Thesis, University of St. Andrews. *(426, 442)*
PASSOW, H. A. and ROTHSTEIN, A. (1960). *J. gen. Physiol.*, **43**, 621–633. *(212)*
—— —— and LOEWENSTEIN, B. (1959). *J. gen. Physiol.*, **42**, 97–107. *(212)*
PATEMAN, A. J. (1959). *Heredity*, **13**, 1–21. *(418, 491)*
—— (1962). In *The evolution of living organisms*, 203–212, ed. LEEPER, G. W. Cambridge University Press. *(419)*
PEAT, A. and BANBURY, G. H. (1967). *New Phytol.*, **66**, 475–484. *(27)*
PEAT, S., WHELAN, W. J. and EDWARDS (1955). *J. chem. Soc.*, 355–359. *(283)*
PERLMAN, D. (1951). *Amer. J. Bot.*, **38**, 652–658. *(187, 193)*
PEROMBÉLON, M. and HADLEY, G. (1965). *New Phytol.*, **64**, 144–151. *(344)*
PERSON, C. (1959). *Canad. J. Bot.*, **37**, 1101–1130. *(465)*
—— (1966). *Nature* (Lond.), **212**, 266–267. *(466)*
PETERSON, D. H. (1963). In *Biochemistry of industrial microorganisms*, 537–606, ed. RAINBOW, C. and ROSE, A. A. Academic Press, London and New York. *(299)*
PEYTON, G. A. and BOWEN, C. C. (1963). *Amer. J. Bot.*, **50**, 787–797. *(27, 28)*
PFEFFER, W. (1884). *Unt. Bot. Inst. Tübingen*, **1**, 363–481. *(178)*
PHAFF, H. J. (1963). *A. r. Microbiology*, **17**, 15–28. *(22, 46)*
PINCKARD, J. A. (1942). *Phytopathology*, **32**, 505–511. *(151)*
PITTENDRIGH, C. S., BRUCE, V. G., ROSENSWEIG, N. S. and RUBIN, M. L. (1959). *Nature* (Lond.), **184**, 169–170. *(67)*
PITTENGER, T. H. (1956). *Proc. Nat. Acad. Sci. U.S.* **42**, 747–752. *(69)*
—— and ATWOOD, K. C. (1956). *Genetics*, **41**, 227–241. *(395, 397)*
—— and BRAWNER, T. G. (1961). *Genetics*, **46**, 1645–1663. *(396)*
—— KIMBALL, A. W. and ATWOOD, K. C. (1955). *Amer. J. Bot.*, **42**, 954–958. *(394)*
PLEMPEL, M. (1957). *Arch. Mikrobiol.*, **26**, 151–174. *(410)*
—— (1960a). *Planta*, **55**, 254–258. *(410)*
—— (1960b). *Naturwissenschaften*, **47**, 472–473. *(410)*
—— (1963a). *Planta*, **59**, 492–508. *(410)*
—— (1963b). *Naturwissenschaften*, **50**, 226. *(410)*
—— and BRAUNITZER, G. (1958). *Z. Naturforsch.*, **13b**, 302–305. *(410)*
—— and DAVID, W. (1961). *Planta*, **56**, 438–446. *(410)*
PLUNKETT, B. E. (1951). *Some aspects of the physiology of fruit-body production in the Hymenomycetes.* Ph.D. Thesis, University of London. *(135)*
—— (1958). *Ann. Bot.* (Lond.), N.S., **17**, 193–218. *(135)*
—— (1956). *Ann. Bot.* (Lond.), N.S., **20**, 563–586. *(142, 252, 254, 255)*
—— (1958). *Ann. Bot.* (Lond.), N.S., **22**, 237–250. *(135, 252, 255)*
—— (1961). *Ann. Bot.* (Lond.), N.S., **25**, 206–233. *(142, 323)*
—— (1966). *Ann. Bot.* (Lond.), N.S., **30**, 133–151. *(64)*
POLAKIS, G. S., BARTLEY, W. and MEEK, G. A. (1964). *Biochem. J.*, **90**, 369. *(317)*
PONTECORVO, G. (1946). *Cold Spring Harbor Symp. Quant. Biol.*, **11**, 193–201. *(393, 395)*
—— (1953). *Adv. Genet.*, **5**, 141–238. *(393)*
—— (1954). *Caryologia* (suppl.), **6**, 192–200. *(397)*
—— (1956). *A. r. Microbiology*, **10**, 393–400. *(398)*
—— and GEMMELL, A. R. (1944). *Nature* (Lond.), **154**, 532–539. *(67)*
—— and KÄFER, E. (1958). *Adv. Genet.*, **9**, 71–104. *(399)*
—— and ROPER, J. A. (1953). *Adv. Genet.*, **5**, 218–233. *(398)*
—— —— and FORBES, E. (1953). *J. gen. Microbiol.*, **8**, 198–210. *(398)*
—— and SERMONTI, G. (1953). *Nature* (Lond.), **172**, 126. *(398, 400)*
—— —— (1954). *J. gen. Microbiol.*, **11**, 94–104. *(400)*
—— TARR-GLOOR, S. and FORBES, E. (1954). *J. genet.*, **52**, 226–237. *(399)*
POP, L. J. J. (1938). *Proc. Akcad. Sci. Amsterdam*, **41**, 661–672. *(257)*

POST, R. L. and ALBRIGHT, C. D. (1961). In *Membrane transport and metabolism*, 219–227, ed. KLEINZELLER, A. and KOTYSK, A. Academic Press, Prague and London. (*236*)
POTGEITER, H. J. and ALEXANDER, M. (1965). *Canad. J. Microbiology*, **11**, 122–125. (*23*)
PRAUS, R. (1952). *Chem. Listy.*, **51**, 1559; 1939. (*303*)
PRESTON, R. D. (1952). *The molecular architecture of plant cell walls*. Chapman & Hall, London. (*55, 56*)
PRÉVOST, G. (1962). *Etude génétique d'un Basidiomycète: Coprinus radiatus, Fr. ex Bolt.* Thèse, Université de Paris. (*388*)
PRINCE, A. E. (1943). *Farlowia*, **1**, 79–93. (*143*)
PRINGSHEIM, N. (1858). *Jahrb. wiss. Bot.*, **1**, 189–192. (*159*)
PRITCHARD, R. H. (1955). *Heredity*, **9**, 343–371. (*398*)
PROUT, T., HUEBSCHMAN, C., LEVENE, H. and RYAN, F. J. (1953). *Genetics*, **38**, 518–529. (*393, 395*)
PRUD'HOMME, N. (1965). In *Incompatibility in fungi*, 48–52, ed. ESSER, K. and RAPER, J. R. Springer, Berlin. (*400*)
PRUSSO, D. C. and WELLS, K. (1967). *Mycologia*, **59**, 337–348. (*27*)

QUINTANILHA, A. and BALLE, S. (1940). *Bull. soc. Brot.*, **14**, 17–46. (*382*)
QUISPEL, A. (1943). *Rec. trav. bot. néerl.*, **40**, 413–541. (*352*)
—— (1959). In *Encyclopedia of Plant Physiology*, **11**, 577–604, ed. RUHLAND, W. Springer, Berlin. (*351*)

RACHMACHANDRAN, S. and GOTTLIEB, D. (1963). *Biochim. biophys. Acta*, **69**, 74–84. (*268*)
RACKER, E. (1965). *Mechanisms in Bioenergetics*. Academic Press, New York. (*268*)
RAESTAD, R. (1941). *Nyt mag. Naturvid.*, **81**, 207–231. (*456*)
RAISTRICK, H., BIRKINSHAW, J. H., CHARLES, J. H. V., CLUTTERBUCK, P. W., COYNE, F. P., HETHERINGTON, A. C., LILLY, C. H., RINTOUL, M. L., RINTOUL, W., ROBINSON, R., STOYLE, J. A. R., THOM, C. and YOUNG, W. (1931). *Phil. Trans. R. Soc.*, **B220**, 1–367. (*194, 195, 198*)
RAMAKRISHNAN, C. V., STEEL, R. and LENTZ, C. P. (1955). *Arch. biochem. biophys.*, **55**, 270–273. (*279*)
RAMSBOTTOM, J. (1953). *Mushrooms and Toadstools*. Collins, London. (*41, 149*)
RANSON, S. L. (1965). In *Plant biochemistry*, 493–525, ed. BONNER, J. and VARNER, J. E. Academic Press, New York. (*282*)
RAPER, J. R. (1936). *J. Elisha Mitchell Sci. Soc.*, **52**, 274–289. (*407*)
—— (1939). *Science*, N.S., **89**, 321–322. (*474*)
—— (1947). *Amer. J. Bot.*, **34**, 31a (abstract). (*407*)
—— (1950). *Bot. Gaz.*, **112**, 1–24. (*407, 414*)
—— (1952). *Bot. Rev.*, **18**, 447–545. (*71*)
—— (1954). In *Sex in Microorganisms*, 42–81, ed. WEINRICH, D. H. A.A.A.S., Washington. (*3, 6, 406*)
—— (1966a). *Genetics of sexuality in higher fungi*. Ronald Press, New York. (*388, 401, 429, 432*)
—— (1966b). In *The Fungi, an advanced treatise*, **II**, 473–511, ed. AINSWORTH, G. C. and SUSSMAN, A. S. Academic Press, New York. (*401, 404, 424*)
—— and KRONGELB, G. S. (1958). *Mycologia*, **50**, 707–740. (*428*)
—— —— and BAXTER, M. G. (1958). *Amer. Naturalist*, **92**, 221–232. (*426, 472*)
—— and HAAGEN SMIT, A. J. (1942). *J. biol. Chem.*, **143**, 311–320. (*407*)
—— and SAN ANTONIO, J. P. (1954). *Amer. J. Bot.*, **41**, 69–86. (*389*)
RAPER, K. B. and FENNELL, D. I. (1953). *J. Elisha Mitchell Sci. Soc.*, **69**, 1–29. (*393*)
RAYNER, M. C. (1927). *Mycorrhiza*. New Phytologist Reprint **15**. Cambridge University Press. (*339*)
REES, H. and JINKS, J. L. (1952). *Proc. R. Soc.*, **B140**, 100–106. (*395*)

REESE, E. T. (ed.) (1963). *Advances in enzymic hydrolysis of cellulose and related materials.* Pergamon, Oxford. *(263)*
REID, D. A. (1957). *Kew Bull. 1956,* 535–540. *(125)*
REIJNDERS, A. F. M. (1963). *Les problèmes du développement des Carpophores des Agaricales et de quelques groupes voisines.* W. Junk, The Hague. *(113, 118)*
REINHARDT, M. O. (1892). *Jahrb. wiss. Bot.*, **23**, 479–566. *(38, 71)*
REISSIG, J. L. (1956). *J. biol. Chem.*, **219**, 753–767. *(47)*
REUSSER, F., GORINI, P. A. J. and SPENCER, J. F. T. (1960). *Canad. J. Microbiology*, **6**, 17–20. *(225)*
REYES, P., CHICHESTER, C. O. and NAKAYAMA, T. O. M. (1964). *Biochim. biophys. Acta*, **90**, 578–592. *(303)*
RHODES, A., BOOTHROYD, B., MCGONAGLE, M. and SOMERFIELD, G. A. (1961). *Biochem. J.*, **81**, 28–37. *(310)*
RICHARDS, A. G. and SMITH, M. N. (1955). *Biol. Bull. Woods Hole*, **108**, 206–218. *(356)*
—— —— (1956). *Ann. Entomol. Soc. Amer.*, **49**, 85–93. *(356)*
RIED, A. (1960). *Flora*, **149**, 345–385. *(354)*
RISHBETH, J. (1951). *Ann. Bot.* (Lond.), N.S., **15**, 1–21. *(335)*
RIZET, G. (1952). *Rév. cytol. biol. végétales*, **13**, 51–92. *(451)*
—— and ESSER, K. (1953). *C. r. Acad. Sci.* (Paris), **237**, 760–761. *(452)*
—— (1957). *C. r. Acad. Sci.* (Paris), **244**, 663–665. *(384)*
—— and ENGELMANN, C. (1949). *Rév. cytol. biol. végétales*, **11**, 201–304. *(418)*
ROBBINS, W. J. (1937). *Amer. J. Bot.*, **24**, 243–250. *(191)*
—— and KAVANAGH, F. (1938). *Bull. Torrey bot. Club*, **65**, 453–461. *(193)*
—— —— (1942). *Bot. Rev.*, **8**, 411–471. *(192)*
—— and MA, R. (1944). *Science*, N.S., **100**, 85–86. *(192)*
ROBERTS, R. B., ABELSON, P. H., COWIE, D. B., BOLTON, E. T. and BRITTEN, R. J. (1955). *Carnegie Inst. Washington Publ.*, **607**, 1–535. *(286, 288)*
ROBERTSON, N. F. (1958). *Ann. Bot.* (Lond.), N.S., **22**, 159–173. *(49, 204)*
—— (1959). *J. Linn. Soc.* (Bot.), **56**, 207–211. *(49, 96, 206)*
—— (1965). *Trans. Brit. mycol. Soc.*, **48**, 1–8. *(49)*
ROBERTSON, J. J. and HALVORSON, H. O. (1957). *J. Bact.*, **73**, 186–198. *(275)*
ROBINOW, C. F. (1942). *Proc. R. Soc.*, **B130**, 299–324. *(361)*
—— (1957). *Canad. J. Microbiology*, **3**, 771–789; 791–798. *(360, 370)*
—— (1961). *J. biophys. biochem. Cytol.*, **9**, 879–892. *(368)*
—— (1962). *Arch. Mikrobiol.*, **42**, 369–377. *(360)*
—— (1963). *J. cell. Biol.*, **17**, 123–152. *(29, 360)*
—— and MARAK, J. (1966). *J. cell Biol.*, **29**, 129–151. *(368)*
ROBINSON, P. M. and PARK, D. (1965). *Trans. Brit. mycol. Soc.*, **48**, 561–571. *(72)*
ROELOFSEN, P. A. (1950). *Biochim. biophys. Acta*, **6**, 340–356. *(18, 53)*
—— (1951). *Biochim. biophys. Acta*, **6**, 357–373. *(18)*
—— (1959). 'The Plant Cell Wall', *Encyclopedia of Plant Anatomy*, **3**, Part 4. Gebrüder Borntraeger, Berlin-Nikolassee. *(51, 54, 56)*
ROGERS, C. H. and WATKINS, G. M. (1938). *Amer. J. Bot.*, **25**, 244–246. *(83)*
ROMAGNESI, H. (1944). *Rev. Mycol.*, N.S., **9** (suppl.), 4–21. *(125)*
ROPER, J. A. (1952). *Experientia*, **8**, 14–15. *(398)*
—— (1966). In *The Fungi, an advanced treatise*, **II**, 589–617, ed. AINSWORTH, G. C. and SUSSMAN, A. S. Academic Press, New York. *(398)*
ROSHAL, J. Y. (1950). *Incompatibility factors in a population of Schizophyllum commune.* Ph.D. Thesis, University of Chicago. *(441)*
ROSSEN, J. M. and WESTERGAARD, M. (1966). *C. r. lab. Carlsberg*, **35**, 233–260. *(373, 375)*
ROTHERY, W. G., BOWN, A. W. and BOULTER, D. (1962). *New Phytol.*, **61**, 41–43. *(317)*
ROTHSCHILD, LORD (1956). *Fertilization.* Methuen, London. *(179)*
ROTHSTEIN, A. (1954). *Symp. soc. exp. Biol.*, **8**, 165–201. *(223)*

ROTHSTEIN, A. (1955). In *Electrolytes in biological systems*, 65–100, ed. SHANES, A. M. Amer. Physiological Society, Washington. (*209, 217*)
—— (1960). In *Regulation of the inorganic ions of cells*, 53–68, ed. WOLSTENHOLME, K. C. Churchill, London. (*217*)
—— (1961). In *Membrane transport and metabolism*, 270–284, ed. KLEINZELLER, A. and KOTYK, A. Academic Press, Prague and London. (*236*)
—— (1963). *J. gen. Physiol.*, **46**, 1075–1085. (*218*)
—— (1964). In *The cellular functions of membrane transport*, 23–29, ed. HOFFMAN, J. Prentice-Hall, New Jersey. (*211*)
—— (1965). In *The Fungi, an advanced treatise*, **I**, 429–455, ed. AINSWORTH, G. C. and SUSSMAN, A. S. Academic Press, New York. (*206*)
—— and BRUCE, M. (1958a). *J. cell. comp. Physiol.*, **51**, 145–160. (*210, 235*)
—— —— (1958b). *J. cell. comp. Physiol.*, **51**, 439–455. (*210, 211, 235*)
—— and ENNS, L. H. (1946). *J. cell. comp. Physiol.*, **28**, 231–252. (*209*)
—— FRENKEL, A. and LARRABEE, C. (1948). *J. cell. comp. Physiol.*, **32**, 261–274. (*223*)
—— and HAYES, A. D. (1956). *Arch. Biochem.*, **63**, 87–99. (*200, 213, 223*)
—— —— JENNINGS, D. H. and HOOPER, D. C. (1958). *J. gen. Physiol.*, **41**, 585–594. (*215*)
—— JENNINGS, D. H., DEMIS, C. and BRUCE, M. (1959). *Biochem. J.*, **71**, 99–106. (*318*)
—— and MEIER, R. (1951). *J. cell. comp. Physiol.*, **38**, 245–270. (*213, 223*)
—— —— and HURWITZ, L. (1951). *J. cell. comp. Physiol.*, **37**, 57–82. (*223*)
ROVIRA, A. D. (1965). In *Ecology of soil-borne plant pathogens*, 170–184, ed. BAKER, K. F. and SNYDER, W. C. Murray, London. (*331, 332*)
ROYLE, D. J. and HICKMAN, C. J. (1964a). *Canad. J. Bot.*, **10**, 151–162. (*178*)
—— —— (1964b). *Canad. J. Bot.*, **10**, 201–219. (*178*)
ROWELL, J. B. (1954). *Phytopathology*, **44**, 504. (*425*)
RUSSELL, E. W. (1950). *Soil conditions and Plant Growth*, 8th edn. Longmans Green, London. (*196*)

SAITO, K. and NAGANISHI, H. (1915). *Bot. Mag.* (Tokyo), **29**, 149–154. (*474*)
SALTON, M. R. J. (1960). *Microbial Cell Walls*. Wiley, New York and London. (*19*)
—— (1964). *The Bacterial Cell Wall*. Elsevier, Amsterdam and New York. (*20*)
SALVIN, S. B. (1942). *Amer. J. Bot.*, **29**, 674–676. (*474*)
SAMPSON, K. (1933). *Trans. Brit. mycol. Soc.*, **18**, 30–47. (*336*)
SANFORD, G. B. and SKOROPAD, W. P. (1955). *Canad. J. Microbiology*, **1**, 412–415. (*391*)
SANSOME, E. R. (1946). *Nature* (Lond.), **157**, 484. (*418*)
—— (1959). *Nature* (Lond.), **184**, 1820. (*337, 480*)
—— (1961). *Nature* (Lond.), **191**, 827–828. (*378*)
—— (1963). *Trans. Brit. mycol. Soc.*, **46**, 63–72. (*4, 364, 378, 408*)
SANVAL, B. D. and LATA, M. (1961). *Nature* (Lond.), **190**, 286–287. (*289*)
SASS, J. E. (1929). *Amer. J. Bot.*, **16**, 663–701. (*433*)
SASSEN, M. M. A. (1962). *Proc. Kon. Akad. Wet. Amsterdam*, **C65**, 447–452. (*73*)
—— (1964). *Rep. 3rd European Congr. Electron Microscopy*, Prague. (*73*)
SATINA, S. and BLAKESLEE, F. A. (1930). *Bot. Gaz.*, **90**, 299–311. (*409*)
SAUNDERS, M. M. (1956). *The distribution of fungal mating type factors with special reference to Polyporus betulinus*. M.Sc. Thesis, University of Liverpool. (*426*)
SAVAGE, E. J. and CLAYTON, C. W. (1962). *Phytopathology*, **52**, 1220. (*474*)
SCHATZ, G. (1963). *Biochem. biophys. res. commun.*, **12**, 448–451. (*317*)
SCHMIDLE, A. (1951). *Arch. Mikrobiol.*, **16**, 80–100. (*163*)
SCHMITZ, J. (1842). *Linnaea*, **16**, 141–215. (*139*)
SCHNATHORST, W. C. (1965). *A. r. Phytopathology*, **3**, 343–366. (*182*)
SCHRÖTER, A. (1905). *Flora*, **95**, 1–30. (*256, 257, 258*)

SCHOPFER, W. H. (1943). *Plants and Vitamins.* Chronica Botanica, Waltham, Mass. *(192)*
SCHROTH, M. N. and SNYDER, W. C. (1961). *Phytopathology* (Abstr.), **52**, 751. *(334)*
SCHULMAN, H. M. and BONNER, D. M. (1962). *Proc. Nat. Acad. Sci. U.S.*, **48**, 53–63. *(295)*
SCHÜTTE, K. H. (1956). *New Phytol.*, **55**, 164–182. *(46, 244, 245, 251, 253)*
SCHWERK, E. and ALEXANDER, G. J. (1958). *Arch. biochem. biophys.*, **76**, 65–74. *(307)*
SCOTT, G. D. (1964). *Advan. Sci.*, 244–248. *(352)*
SCOTT, J. M. and SPENCER, B. (1965). *Biochem. J.*, **96**, 78P. *(237)*
SELBY, K. and MAITLAND, C. C. (1965). *Biochem. J.*, **94**, 578–583. *(241)*
SERMONTI, G. (1957). *Genetics*, **42**, 433–443. *(400)*
SHARP, L. W. (1934). *Introduction to Cytology*, McGraw-Hill, New York. *(30)*
SHATKIN, A. J. (1959). *Trans. N.Y. Acad. Sci.*, **21**, 446–453. *(48)*
—— and TATUM, E. L. (1959). *J. biophys. biochem. Cytol.*, **6**, 423–426. *(390)*
SHAW, P. D., BECKWITH, J. R. and HAGER, L. P. (1959). *J. biol. Chem.*, **234**, 2560–2564. *(311)*
SHIBATA, S. (1958). In *Encyclopedia of plant physiology*, **10**, 560–623, ed. RUHLAND, W. Springer, Berlin. *(354)*
SHIMAZONO, H. (1955). *J. biochem.* (Tokyo), **42**, 321–340. *(281)*
SHROPSHIRE, W. (1963). *Physiol. Rev.*, **43**, 38–67. *(319)*
SIEGLE, H. (1961). *Phytopath. Z.*, **42**, 305–348. *(330)*
SILVER, W. S. (1957). *J. Bact.*, **73**, 241–246. *(284)*
—— and MCELROY, W. D. (1954). *Arch. biochem. biophys.*, **51**, 379–394. *(284, 285)*
SIMON, E. W. and BEEVERS, H. (1952). *New Phytol.*, **51**, 163–197. *(251)*
SIMONS, R. D. G. P. (ed.) (1954). *Medical mycology.* *(334)*
SIMPSON, K. L., NAKAYAMA, T. O. M. and CHICHESTER, C. O. (1964). *Biochem. J.*, **92**, 508. *(302)*
SIMS, A. P. and FOLKES, B. F. (1964). *Proc. R. Soc.*, **B159**, 479–502. *(286, 287)*
SINGER, R. (1962). *The Agaricales.* Cramer, Weinheim. *(341)*
—— and SMITH, A. H. (1960). *Mem. Torrey bot. Cl.*, **21**, 1–112. *(485)*
SINGH, K. (1963). *Fungi associated with the roots and rhizosphere of Ericaceae.* Ph.D. Thesis, University of Durham (King's College). *(332)*
SINGLETON, J. R. (1953). *Amer. J. Bot.*, **40**, 124–144. *(373, 374)*
SJÖWALL, M. (1945). *Studien über Sexualität, Vererbung und Zytologie bei einigen diözischen Mucoraceen.* Doctoral Thesis, Lund. *(379)*
SLANKIS, V. (1948a). *Physiol. Plantarum*, **1**, 278, 289. *(342)*
—— (1948b). *Physiol. Plantarum*, **1**, 390–400. *(342)*
—— (1949). *Svensk. bot. Tidskr.*, **43**, 603–607. *(342)*
—— (1951). *Symb. bot. Upsaliensis*, **11**, 1–63. *(342)*
—— (1958). In *The Physiology of Forest Trees*, 427–443, ed. THIMANN, K. V. Ronald Press, New York. *(342)*
SLEIGH, M. A. (1962). *The Biology of cilia and flagella.* Pergamon, Oxford. *(37, 176)*
SLONIMSKI, P. P. (1953). *La formation des enzymes respiratoires chez la Levure.* Masson et Cie., Paris. *(317, 384)*
SMITH, A. H. (1934). *Mycologia*, **26**, 305–311. *(433)*
SMITH, D. C. (1960a). *Ann. Bot.* (Lond.), N.S., **24**, 52–62. *(353)*
—— (1960b). *Ann. Bot.* (Lond.), N.S., **24**, 172–185. *(353)*
—— (1960c). *Ann. Bot.* (Lond.), N.S., **24**, 186–199. *(353)*
—— (1961). *Lichenologist*, **1**, 209–226. *(353)*
—— (1962). *Biol. Rev.*, **37**, 537–570. *(351, 353)*
—— (1963a). *New Phytol.*, **62**, 205–216. *(353)*
—— (1963b). In *Symbiotic Associations*, 31–50, ed. NUTMAN, P. S. and MOSSE, B. Cambridge University Press. *(353)*
—— and DREW, E. A. (1965). *New Phytol.*, **64**, 195–200. *(353)*
SMITH, F. E. V. (1923). *Ann. Bot.* (Lond.), **37**, 63–73. *(362)*
SMITH, J. E. (1963). *J. gen. Microbiol.*, **30**, 35–41. *(288)*
SMITH, J. H. and FRENCH, C. S. (1963). *A. r. Pl. Physiol.*, **14**, 181–224. *(320)*

SMITH, S. E. (1966). *New Phytol.*, **65**, 488–499. *(250, 345, 346)*
—— (1967). *New Phytol.*, **66**, 371–378. *(250, 345, 347)*
SMOOT, J. J., GOUGH, F. J., LAMEY, H. A., EICHENMULLER, J. J. and GALLEGLY, M. E. (1958). *Phytopathology*, **48**, 165–171. *(408)*
SNIDER, P. J. (1963). *Genetics*, **48**, 47–54. *(392)*
—— (1965). In *Incompatibility in fungi*, 52–68, ed. ESSER, K. and RAPER, J. R. Springer, Berlin. *(388)*
—— and RAPER, J. R. (1958). *Amer. J. Bot.*, **45**, 538–546. *(389)*
SOMERS, E. (1963). *Ann. appl. Biol.*, **51**, 425–437. *(200)*
SOMERSON, N. L., DEMAIN, A. L. and NUNHEIMER, T. D. (1961). *Arch. biochem. biophys.*, **93**, 238–241. *(309)*
SONNEBORN, D. R., SUSSMAN, M. and LEVINE, L. (1964). *J. Bact.*, **87**, 1321–1329. *(325)*
SORGER, G. J. (1963). *Biochem. biophys. res. comm.*, **12**, 395–401. *(284)*
SPARROW, F. K. (1960). *Aquatic Phycomycetes*, 2nd edition. University of Michigan Press. *(37)*
SPIEGELMAN, S. and HALVORSON, H. O. (1953). In *Adaptation in microorganisms* 98–131, ed. DAVIES, R. and GALE, F. E. Cambridge University Press. *(315)*
STACEY, M. and BARKER, S. A. (1960). *Polysaccharides of microorganisms.* Oxford University Press. *(282)*
STADLER, D. R. (1959). *Nature* (Lond.), **184**, 171. *(67)*
—— (1952). *J. cell. comp. Physiol.*, **39**, 449–474. *(326, 327)*
—— (1953). *Biol. Bull. Woods Hole*, **104**, 100–108. *(326)*
STAKMAN, E. C., LEVINE, M. N. and COTTER, R. U. (1930). *Sci. Agric.*, **10**, 707–720. *(465)*
—— and HAMILTON, L. M. (1939). *Pl. Dis. Reptr.*, Suppl., **117**, 69–83. *(169)*
—— and HARRAR, J. G. (1957). *Principles of Plant Pathology.* Ronald Press, New York. *(462)*
STAPLES, R. C. and WEINSTEIN, L. H. (1959). *Contr. Boyce Thompson Inst. Pl. Res.*, **20**, 71–82. *(270)*
—— BURCHFIELD, H. P. and BAKER, J. (1961). *Contr. Boyce Thompson Inst. Pl. Res.*, **21**, 345–362. *(286)*
—— SYAMANANDA, R., KAO, V. and BLOCK, R. J. (1962). *Contr. Boyce Thompson Inst. Pl. Res.*, **21**, 345–362. *(296)*
—— and WYNN, W. K. (1965). *Bot. Rev.*, **31**, 537–564. *(185)*
STEBBINS, G. L. (1966). *Processes of organic evolution.* Prentice-Hall, New Jersey. *(469)*
STEINBERG, R. A. (1939). *J. agr. Res.*, **59**, 749–763. *(192)*
—— (1942). *J. agr. Res.*, **64**, 455–475. *(286)*
STEINHAUS, E. A. (1963). *Insect Pathology.* Academic Press, New York. *(355)*
STEPANOV, K. M. (1935). *Bull. Pl. Protection* (U.S.S.R.), [2], Phytopath., **8**, 1–68. *(168)*
STONE, E. L. (1950). *Proc. soil. sci. soc. Amer.* (1949), **14**, 340–345. *(345)*
STRASBURGER, E. (1884). *Das botanisches Praktikum.* Gustav Fischer, Jena. *(377)*
STRIGINI, P. and MORPUGO, G. (1961). *Nature* (Lond.), **190**, 557. *(193)*
STRUŅK, C. (1964). *Proc. 3rd Reg. Conf. (Eur.), Electron Microscopy* (Prague), 143. *(203)*
STÜBEN, H. (1939). *Planta*, **30**, 353–383. *(404)*
SURDIN, Y., SLY, W., SIRE, J., BORDES, A. M. and ROBICHON-SZULMAJSTER, H. DE. (1965). *Biochim. biophys. Acta*, **107**, 546–566. *(225, 227)*
SUSSMAN, A. S. (1953). *Amer. J. Bot.*, **40**, 401–404. *(182)*
—— (1954). *J. gen. Physiol.*, **38**, 59–77. *(184, 200)*
—— (1961). *Quart. Rev. Biol.*, **36**, 109–116. *(184, 201)*
—— (1965). In *Encyclopedia of Plant Physiology*, **15**, 933–1025, ed. LANG, A. Springer, Berlin. *(145, 181)*
—— (1966). In *The Fungus Spore*, 235–256, ed. MADELIN, M. F. Butterworth, London. *(180, 182)*

SUSSMAN, A. S. DISTLER, J. R. and KRAKOW, J. S. (1956). *Pl. Physiol.*, **31**, 126–135. *(184)*
—— HOLTON, R. W. and VON BÖVENTER-HEIDENHAIN, B. (1958). *Arch. Mikrobiol.*, **29**, 38–50. *(200)*
—— and LOWRY, R. J. (1955). *J. Bact.*, **70**, 675–685. *(260)*
—— LOWRY, R. J. and DURKEE, T. (1964). *Amer. J. Bot.*, **51**, 243–252. *(67)*
—— —— and TYRRELL, E. (1959). *Mycologia*, **51**, 237–247. *(182)*
—— and SPIEGELMAN, S. (1950). *Arch. biochem.*, **29**, 85–100. *(236)*
—— VON BÖVENTER-HEIDENHAIN, B. and LOWRY, R. J. (1957). *Pl. Physiol.*, **32**, 586–590. *(200)*
SUTCLIFFE, J. F. (1962). *Mineral salts absorption in plants.* Pergamon, Oxford. *(206)*
SUTTON, O. G. (1932). *Proc. R. Soc.*, **A135**, 143–165. *(170)*
SVILHA, G., SCHLENCK, F. and DAINKO, J. L. (1961). *J. Bact.*, **82**, 808–814. *(204)*
SWIEZYNSKI, K. M. (1962). *Acta Soc. bot. Polon.*, **31**, 169–184. *(400)*
—— (1963). *Genet. Poloniae*, **4**, 21–36. *(400)*
—— and DAY, P. R. (1960). *Genet. Res.*, **1**, 129–139. *(429)*
SYLVÉN, B., TOBIAS, C. A., MALMGREN, H., OTTOSON, R. and THORELL, B. (1959). *Exp. Cell. Res.*, **16**, 75–87. *(317)*
SYLVESTER, J. C. and COGHILL, R. D. (1954). In *Industrial Fermentations*, 229, **II**, ed. UNDERKOFLER, L. A. and HICKEY, R. J. Chemical Publishing Co., New York. *(308)*

TABER, W. A. (1966). In *The Fungi, an advanced treatise*, **II**, 387–412, ed. AINSWORTH, G. C. and SUSSMAN, A. S. Academic Press, New York. *(323)*
—— and VINING, L. C. (1961). *Bact. Proc.*, 66. *(305)*
TAKEMARU, T. (1961). *Biol. J. Okayama Univ.*, **7**, 133–211. *(432)*
TALLEY, P. J. and BLANK, L. M. (1941). *Pl. Physiol.*, **16**, 1–19. *(192)*
TAMAKI, T. (1959). *Shikoku Acta Medica*, **15**, 252–254; 254–257. *(19)*
TAMIYA, H. (1928). *Acta Phytochem. Japan*, **4**, 77–218. *(276)*
—— (1942). *Adv. Enzymology*, **2**, 183–238. *(77)*
—— and MIWA, Y. (1928). *Z. Bot.*, **21**, 417–432. *(275)*
TANAKA, H. (1963). *See* PHAFF, H. J. (1963). *(23)*
TATUM, E. L., BARRETT, R. W. and CUTTER, V. M. (1949). *Science*, N.S., **109**, 509–511. *(47, 64)*
TAYLOR, E. H. and RAMSTAD, E. (1960). *Nature* (Lond.), **188**, 494–495. *(305)*
TERNETZ, C. (1900). *Jahb. wiss. Bot.*, **35**, 273–312. *(256, 258)*
TERROINE, E. F. and BONNET, R. (1927). *Bull. soc. chim. Biol.*, **9**, 588–596. *(297)*
—— —— (1930). *Bull. soc. chim. Biol.*, **12**, 10–19. *(193)*
—— and WURMSER, R. (1921). *C. r. Acad. Sci.* (Paris), **173**, 482–483. *(193)*
—— —— (1922a). *C. r. Acad. Sci.* (Paris), **174**, 1435–1437. *(193, 194)*
—— —— (1922b). *C. r. Acad. Sci.* (Paris), **175**, 228–230. *(193)*
—— —— (1922c). *Bull. soc. chim. Biol.*, **4**, 518–567. *(193)*
THAINE, R. (1964). *J. exp. Bot.*, **13**, 152–160. *(260)*
THAXTER, R. (1896). *Mem. Amer. Acad. Arts. Sci.*, **12**, 195–429. *(356)*
—— (1908). *Mem. Amer. Acad. Arts Sci.*, **13**, 219–460. *(356)*
—— (1924). *Mem. Amer. Acad. Arts Sci.*, **14**, 309–426. *(356)*
—— (1926). *Mem. Amer. Acad. Arts Sci.*, **15**, 427–500. *(356)*
—— (1931). *Mem. Amer. Acad. Arts Sci.*, **16**, 1–435. *(356)*
THIMANN, K. V. and GRUEN, H. E. (1960). *Beih. Z. Schweitz. Forst.*, **30**, 237–263. *(49, 51)*
THOM, C. and RAPER, K. B. (1945). *A manual of the Aspergilli.* Williams & Wilkins, Baltimore. *(18)*
THOMAS, E. A. (1939). *Beitr. Kryptogamenfl., Schweiz.*, **9**. *(351)*
THOMAS, P. T., EVANS, H. J. and HUGHES, D. T. (1956). *Nature* (Lond.), **178**, 949–951. *(428)*

THOMPSON, J. F., MORRIS, C. J., ARNOLD, W. N. and TURNER, D. H. (1962). In *Amino Acid Pools*, 54–56, ed. HOLDEN, J. W. Elsevier, Amsterdam. *(309)*
THORNE, G. C. (1950). *Wallerstein lab. comm.*, **13**, 519. *(286)*
THREN, R. (1937). *Z. Bot.*, **31**, 337–391. *(425)*
—— (1940). *Z. Bot.*, **36**, 449–498. *(425)*
THROWER, S. L. and THROWER, L. B. (1961). *Nature* (Lond.), **190**, 823–824. *(247, 248)*

TIMONIN, M. I. (1941). *Soil Sci.*, **52**, 395–413. *(330, 332)*
TINLINE, R. E. (1962). *Canad. J. Bot.*, **40**, 425–437. *(400)*
TOSTESON, D. C. and HOFFMAN, J. F. (1960). *J. gen. Physiol.*, **44**, 169–194. *(205)*
TOWNSEND, B. B. (1954). *Trans. Brit. mycol. Soc.*, **37**, 222–223. *(83, 87)*
—— (1957). *Ann. Bot.* (Lond.), N.S., **21**, 153–166. *(83)*
—— and WILLETS, H. J. (1954). *Trans. Brit. mycol. Soc.*, **37**, 213–221. *(82)*
TREVITHICK, J. R. and METZENBERG, R. L. (1964). *Biochem. biophys. res. Comm.*, **16**, 319–325. *(205, 242)*
—— —— (1966a). *J. Bact.*, **92**, 1010–1015. *(201)*
—— —— (1966b). *J. Bact.*, **92**, 1016–1020. *(201)*
TRIBE, H. T. (1955). *Ann. Bot.* (Lond.), N.S., **19**, 351–368. *(334)*
TRIONE, E. J. (1966). *Phytopathology*, **50**, 482–486. *(331)*
TRIPATHI and GOTTLIEB (1956). In GOTTLIEB, 1966. *(296)*
TSUDA, S. and TATUM, E. L. (1961). *J. biophys. biochem. Cytol.*, **11**, 171–177. *(305)*
TUBAKI, K. (1958). *J. Hattori Bot. Lab.*, **31**, 142–244. *(95)*
TURIAN, G. (1956). *Protoplasma*, **47**, 135–138. *(36)*
—— (1958). *Rev. Cytol.* (Paris), **19**, 241–272. *(36)*
—— (1961a). *Nature* (Lond.), **190**, 825. *(404)*
—— (1961b). *Pathol. Microbiol.*, **24**, 819–839. *(404)*
—— (1961c). *Nucleus* (Calcutta), **4**, 151–156. *(404)*
—— (1962). *Protoplasma*, **54**, 323–372. *(180)*
—— and CANTINO, E. C. (1960). *Cytologia*, **25**, 101–107. *(30)*
TUVESON, R. W. and COY, D. O. (1961). *Mycologia*, **53**, 244–253. *(400)*
TVEIT, M. and WOOD, R. K. S. (1955). *Ann. appl. Biol.*, **43**, 538–552. *(329)*

UEBELMESSER, E. R. (1954). *Arch. Mikrobiol.*, **20**, 1–33. *(163)*

VAHEEDUDDIN, S. (1942). *Minnesota Agr. Exp. Sta., Tech. Bull.*, **154**, 1–54. *(425)*
VAKILI, N. G. and CALDWELL, R. M. (1957). *Phytopathology*, **47**, 536. *(439)*
VALLEE, B. L. and HOCH, F. L. (1955). *Proc. Nat. Acad. Sci. U.S.A.*, **41**, 327–338. *(276)*
VAN BAMBEKE, C. (1892). *Bull. Acad. R. de Belgique*, 3 ser., **23**, 472–490. *(123)*
—— (1894). *Mém. Acad. Roy. Belg.*, **52**, 1–30. *(123)*
VARITCHAK, B. (1928). *C. r. Acad. Sci.* (Paris), **186**, 96–98. *(371, 372)*
—— (1931). *Le botaniste*, **23**, 1–182. *(371)*
VARMA, T. N. R. and CHICHESTER, C. O. (1962). *Arch. biochem. biophys.*, **96**, 265–269. *(301)*
VERRALL, A. F. (1937). *Minnesota Agr. Exp. Sta., Tech. Bull.*, **117**, 1–41. *(456)*
VILLANEUVA, J. R. (1966). In *The Fungi, an advanced treatise*, **II**, 3–62, ed. AINSWORTH, G. S. and SUSSMAN, A. S. Academic Press, New York. *(202)*
VITOLS, E., NORTH, R. J. and LINNANE, A. W. (1961). *J. biophys. biochem. Cytol.*, **9**, 689–699. *(19)*
—— and LINNANE, A. W. (1961). *J. biophys. biochem. Cytol.*, **9**, 701–710. *(272)*
VLK, W. (1939). *Arch. Protist.*, **92**, 157–160. *(34)*
VOELZ, H. and NIEDERPRUEM, D. J. (1964). *J. Bact.*, **88**, 1497–1502. *(180)*
VON WETTSTEIN, R. (1921). *Sitzber. Akad. wiss. Wien, math-naturw. Kl.*, Abt 1, **130**, 3–20. *(21)*
VUILLEMIN, P. (1910). *Bull. soc. sci.* (Nancy), **11**, 129–172. *(93)*
—— (1911). *Bull. soc. sci.* (Nancy), **12**, 151–175. *(93)*

VUILLEMIN, P. (1912). *Les champignons. Essai de classification.* Doin et fils, Paris. (*93, 123*)
VUJIČIĆ, R. and PARK, D. (1964). *Trans. Brit. mycol. Soc.*, **47**, 455–458. (*189*)
VINING, L. C. and TABER, W. A. (1963). In *Biochemistry of industrial microorganisms*, 341–378, ed. RAINBOW, C. and ROSE, A. H. Academic Press, London and New York. (*299, 305*)

WAGNER, R. P. and MITCHELL, H. K. (1964). *Genetics and Metabolism.* Wiley, New York. (*291*)
WAHRLICH, W. (1893). *Scripta bot. Hort. Univ. Imp. Petropolit.*, **4**, 101–155. (*24*)
WAKAYAMA, K. (1930). *Cytologia* (Tokyo), **1**, 369–388. (*479*)
WAKSMAN, S. A. and FOSTER, J. W. (1939). *J. agr. Res.*, **57**, 873–899. (*278*)
WALKER, J. C. (1957). *Plant Pathology.* McGraw-Hill, New York. (*182*)
WALKER, L. B. (1935). *Mycologia*, **27**, 102–127. (*372*)
—— and ANDERSON, E. N. (1925). *Mycologia*, **17**, 154–159. (*154*)
WALKEY, D. G. A. and HARVEY, R. (1966). *New Phytol.*, **65**, 59–74. (*164*)
WALLROTH, F. W. (1825). *Naturgesichte der Flechten* I. Frankfurt. (*351*)
WALSH, J. H. and HARLEY, J. L. (1962). *New Phytol.*, **61**, 299–313. (*225*)
WANG, C. H., STERN, I., GILMOUR, C. M., KLUNGSOYR, S., REED, D. J., BIALY, L. J., CHRISTENSEN, B. E. and CHELDELIN, V. H. (1958). *J. Bact.*, **76**, 207–216. (*268*)
WARCUP, J. H. (1959). *Trans. Brit. mycol. Soc.*, **42**, 45–52. (*197*)
WARD, E. W. B. and CIURYSEK, K. W. (1961). *Canad. J. Bot.*, **39**, 1497–1503. (*364, 370*)
—— —— (1962). *Amer. J. Bot.*, **49**, 393–399. (*364*)
WARD, J. M. and NICKERSON, W. J. (1958). *J. gen. Physiol.*, **41**, 703–724. (*271*)
WATERHOUSE, W. L. and WATSON, I. A. (1944). *J. R. Soc. N.S.W.*, **77**, 138–144. (*465*)
WATERMAN, A. M. and HANSBOROUGH, J. R. (1957). *Forest Products J.*, **7**, 77–84. (*337*)
WATSON, I. A. (1957). *Phytopathology*, **47**, 507–509. (*439*)
WEBB, P. C. R. (1935). *Ann. Bot.* (Lond.), **49**, 41–52. (*365, 366, 379*)
WEBSTER, J. (1952). *New Phytol.*, **51**, 229–235. (*154*)
—— (1959). *Ann. Bot.* (Lond.), N.S., **23**, 595–611. (*173, 175*)
—— (1964). *Trans. Brit. mycol. Soc.*, **47**, 75–96. (*439*)
WEIGL, J. and ZIEGLER, H. (1960). *Arch. Mikrobiol.*, **37**. 124–133. (*250, 251*)
WEIJER, J., KOOPMANS, A. and WEIJER, D. L. (1963). *Trans. N.Y. Acad. Sci.*, **25**, 846–854. (*371, 372*)
—— —— —— (1965). *Canad. J. genet. cytol.*, **7**, 140–163. (*360, 371*)
—— and MACDONALD, B. R. (1965). *Canad. J. genet. cytol.*, **7**, 519–522. (*360*)
—— and WEISBERG, S. H. (1966). *Canad. J. genet. cytol.*, **8**, 361–374. (*360, 371*)
WEINHOLD, A. R. (1955). *Tech. Rept. Office of Naval Research, ONR Contract No. N90 nr.* 82400. 1–104. (*166*)
WEISS, B. (1965). *J. gen. Microbiol.*, **39**, 85–94. (*180*)
WELLS, K. (1965). *Mycologia*, **57**, 236–261. (*128*)
WENKERT, E. (1955). *Chem. Ind.*, 282. (*308*)
WERESUB, L. K. and GIBSON, S. (1960). *Canad. J. Bot.*, **38**, 833–867. (*460*)
WESSELS, J. G. H. (1965). *Wentia*, **13**, 1–113. (*21, 22, 136, 138*)
WESTERGAARD, M. (1965). *C. r. lab. Carlsberg*, **34**, 359–405. (*372, 375*)
—— and HIRSCH, H. M. (1954). *Proc. Symp. Colston Res. Soc.*, **7**, 171–183. (*131*)
—— and VON WETTSTEIN, D. (1966). *C. r. lab. Carlsberg*, **35**, 261–286. (*373, 375*)
WHEELER, H. E. and MCGAHEN, J. W. (1952). *Amer. J. Bot.*, **39**, 110–119. (*421*)
WHINFIELD, B. (1947). *Ann. Bot.* (Lond.), N.S., **11**, 35–39. (*241*)
—— (1948). *Ann. Bot.* (Lond.), N.S., **12**, 111–120. (*241*)
WHITFIELD, F. E. (1964). *Exp. Cell. Res.*, **36**, 62–72. (*325*)
WHITEHOUSE, H. L. K. (1949a). *Biol. Rev.*, **24**, 411–447. (*401, 402*)
—— (1949b). *New Phytol.*, **48**, 212–244. (*427, 441*)

WHITEHOUSE, H. L. K. (1951). *Trans. Brit. mycol. Soc.*, **34**, 340–355. *(425)*
WICKERHAM, L. J. and BURTON, K. A. (1956). *J. Bact.*, **71**, 290–295. *(477)*
WILEY, W. R. and MATCHETT, W. H. (1966). *J. Bact.*, **92**, 1698–1705. *(225, 227, 228, 229)*
WILKINSON, J. F. and ROSE, A. H. (1963). In *Biochemistry of industrial microorganisms*, 379–414, ed. RAINBOW, C. and ROSE, A. H. Academic Press, London and New York. *(275)*
WILLIAMS, P. G. and LEDINGHAM, G. (1964). *Canad. J. Bot.*, **42**, 497–505. *(180)*
WILSENACH, R. and KESSEL, M. (1965a). *Nature* (Lond.), **207**, 545–546. *(24)*
—— —— (1965b). *J. gen. Microbiol.*, **40**, 401–404. *(26)*
WILSON, C. L., BRUSHABER, J. A. and AIST, J. R. (1966). *Proc. Arkansas Acad. Sci.*, **20**, 17–21. *(372)*
WILSON, C. M. (1952). *Bull. Torrey bot. Club*, **79**, 139–159. *(378)*
WILSON, J. F., GARNJOBST, L. and TATUM, E. L. (1961). *Amer. J. Bot.*, **48**, 299–305. *(396)*
WINGE, O. (1942). *Sci. genet.*, **2**, 171–189. *(477)*
—— and ROBERTS, C. (1954). *C. r. Lab. Carlsberg, Sér. Physiol.*, **25**, 285–329. *(420)*
WIRTH, J. C. and NORD, F. T. (1942). *Arch. Biochem.*, **1**, 143–163. *(276)*
WOLF, F. A. and WOLF, F. T. (1947). *The Fungi*, **II**. Wiley, New York. *(158)*
WOOD, R. K. S. (1967). *Physiological Plant Pathology*. Blackwell Scientific Publications. *(339)*
—— and GUPTA, S. C. (1958). *Ann. Bot.* (Lond.), N.S., **22**, 309–319. *(334)*
WOOD-BAKER, A. (1955). *Trans. Brit. mycol. Soc.*, **38**, 291–297. *(182)*
WOOLLEY, D. W. (1948a). *J. biol. Chem.*, **176**, 1291–1298. *(310)*
—— (1948b). *J. biol. Chem.*, **176**, 1299–1308. *(310)*
WORONICK, C. L. and JOHNSON, M. J. (1960). *J. biol. Chem.*, **235**, 9–15. *(270)*
WORONIN, M. (1888). *Mém. Acad. Imp. Sci.* (Pétersbourg), **7**. Sér., 36(b), 1–49. *(147)*
WRIGHT, J. M. (1956a). *Ann. appl. Biol.*, **44**, 461–466. *(329)*
—— (1956b). *Ann. appl. Biol.*, **44**, 561–566. *(329)*

YABUTA, T. and SUMUKI, Y. (1938). *J. agr. chem. soc.* (Japan), **14**, 1526. *(307)*
YANAGITA, T. (1957). *Arch. Mikrobiol.*, **26**, 329–344. *(182, 185, 186)*
—— and KOGANÉ, F. (1962). *J. gen. appl. Microbiol.*, **8**, 201–213. *(77)*
—— —— (1963a). *J. gen. appl. Microbiology*, **9**, 179–188. *(60)*
—— —— (1963b). *J. gen. appl. Microbiology*, **9**, 313–330. *(247)*
YARWOOD, C. E. (1947). *Amer. J. Bot.*, **34**, 514–520. *(254)*
—— and HAZEN, W. E. (1942). *Science*, N.S., **96**, 316–317. *(166)*
YCAS, M. and VINCENT, V. S. (1960). *Proc. Nat. Acad. Sci. U.S.*, **46**, 804–811. *(295)*
YEN, H. C. (1948). *C. r. Acad. Sci.* (Paris), **226**, 1214. *(428, 473)*
—— (1950). *Ann. Univ. Lyon*, Sér. 3, **6**, 5–158. *(389, 428)*
YOKOYAMA, H., NAKAYAMA, T. O. M. and CHICHESTER, C. O. (1962). *J. biol. Chem.*, **237**, 681–686. *(299, 301)*
—— YAMAMOTO, H., NAKAYAMA, T. O. M., SIMPSON, K. and CHICHESTER, C. O. (1961). *Nature* (Lond.), **191**, 1299. *(301)*

ZALEVSKI, A. (1883). *Flora*, **66**, 268–270. *(147)*
ZALOKAR, M. (1954). *Arch. Biochem. Biophys.*, **50**, 71–80. *(77)*
—— (1959a). *Amer. J. Bot.*, **46**, 555–559. *(76)*
—— (1959b). *Amer. J. Bot.*, **46**, 602–609. *(30, 42, 44, 45, 283)*
—— (1960a). *Exp. Cell. Res.*, **19**, 114–132. *(294)*
—— (1960b). *Exp. Cell. Res.*, **19**, 559–576. *(46, 294)*
—— (1961). *Biochim. biophys. Acta*, **46**, 423–432. *(229)*
—— (1965). In *The Fungi, an advanced treatise*, **I**, 377–426, ed. AINSWORTH, G. C. and SUSSMAN, S. A. Academic Press, New York. *(293)*
—— (1955). *Arch. biochem. biophys.*, **56**, 318–325. *(303)*
ZELLER, S. M. (1916). *Ann. Mo. bot. gard.*, **3**, 439–514. *(122)*

ZICKLER, H. (1937). *Ber. deut. bot. Gesell.*, **55**, 114–119. *(417)*
—— (1952). *Arch. Protist.*, **98**, 1–70. *(417)*
ZIEGLER, A. W. (1953). *Amer. J. Bot.*, **40**, 60–66. *(364)*
ZOBERI, M. H. (1961). *Ann. Bot.* (Lond.), N.S., **25**, 53–64. *(146, 148)*
ZUCKER, M. and NASON, A. (1955). In *Methods of enzymology*, **2**, 406–411, ed. COLOVICK, S. P. and KAPLAN, N. O. Academic Press, New York. *(285)*
ZYCHA, H. (1935). *Mucorineae, Kryptogamenflora M. Brandenburg*, **6a**. Gebrüder Borntraeger, Leipzig. *(474)*

Index